Elementary Electronic Structure

Elementary Electronic Structure

Walter A. Harrison
Department of Applied Physics
Stanford University

World Scientific
Singapore • New Jersey • London • Hong Kong

Published by

World Scientific Publishing Co. Pte. Ltd.

P O Box 128, Farrer Road, Singapore 912805

USA office: Suite 1B, 1060 Main Street, River Edge, NJ 07661

UK office: 57 Shelton Street, Covent Garden, London WC2H 9HE

British Library Cataloguing-in-Publication Data
A catalogue record for this book is available from the British Library.

ELEMENTARY ELECTRONIC STRUCTURE

ISBN 981-02-3895-9
ISBN 981-02-3896-7 (pbk)

Printed in Singapore by Uto-Print

To my wife, Lucky,
and to our wonderful extended family

Preface

This text presents an account of *analytic* electronic structure, to be distinguished from *computational* electronic structure. Both are based upon a one-electron approximation, local-density theory, and the determination of the quantum-mechanical electronic states. They both seek to predict the properties of the resulting solids, or molecules on the basis of these states. In the computational mode, the minimum number of approximations are used, and numerical solutions are sought. Here we seek instead to focus on the most important aspects of the solution, making what approximations are necessary in order to proceed analytically and obtain formulae for the properties. This reducing of the problem to basics is almost always less accurate that the computational solution, but has the advantage that it displays the dependence of any property on the parameters of the system. It gives us an understanding of the property in a sense that a numerical solution, or a direct measurement, cannot.

There is a long tradition of such approximate analyses of solids, but often based upon a model of the system, designed to describe a particular property. Usually a few experimental parameters needed adjustment for each specific material. Thus spring constants for the interaction between atoms could be adjusted to fit the elastic constants. Then the vibration spectrum for the solid could be calculated and the vibrational contribution to the heat capacity could be obtained. In more chemically oriented studies, the measured elastic constants of LiCl and KCl might be interpolated to obtain the constants for NaCl since sodium lies between lithium and potassium in the periodic table.

The distinctive feature of the approach described here is that it is an approximate theory of the basic electronic structure, and consequently a single approximate theory of *all* of the properties of the solids. One would not know in advance whether such an approach could be accurate enough to be useful, or how much information would need to be supplied in order to estimate properties. It is turned out in fact that once the solids were understood, often on the basis of the results of major computational efforts, it was possible to estimate all of the needed parameters without further reference to the calculations or to experiment. Over a period of forty years of my own efforts, those of coworkers, and colleagues around the world, have provided the insights needed to make this possible, surprisingly often by recognizing that two very different viewpoints are simultaneously valid. Then the combination provided the needed parameters. Perhaps the clearest case of this, discussed in Chapter 1, is the recognition that the electronic states of a chain of lithium atoms can be described as a linear combination or valence s-states on the atoms (a tight-binding state), or as freely propagating electrons along a line. The consequence was that the parameters essential for the tight-binding description, the coupling between neighboring s-states, must be given in terms of the spacing d and the free-electron mass m by $-(\pi^2/8)\hbar^2/md^2$. In terms of such couplings, all other properties of the chain can be estimated. These forty years of effort have also allowed such approaches to be extended over virtually all types of solids: metals, semimetals, semiconductors, insulators, transition-metal and f-shell-metal systems. All are discussed here, as well as some molecular systems.

This book is organized much as*Electronic Structure and the Properties of Solids* , which was published by W. H. Freeman (San Francisco, 1980) and reprinted by Dover (New York, 1989) but the text is entirely new. Subsequent work has made the predictions much more accurate and general, and at the same time greatly simplified the theory. For example, by introducing more stable coupling parameters it became unnecessary to define different sets of covalent and polar energies for dielectric than for bonding properties. By introducing an algebraic overlap repulsion it became possible to write analytically the total energies, as well as dielectric properties, of semiconductors in terms of these covalent and polar energies. It was also found possible to analytically evaluate the effective interaction between atoms in simple metals in the form $Z^2e^2\cosh^2 r_c e^{-\kappa r}/r$, in terms of the empty-core pseudopotential radius r_c , greatly simplifying the estimates of many properties. Improved parameters and approximations for transition-metal systems made their study much more direct and accurate, and the theory was extended to f-shell-metal systems, including the formation of correlated electronic states in the rare earths.

In spite of the diversity of systems and materials, the approach is systematic and coherent, combining the tight-binding (or atomic) picture with the pseudopotential (or free-electron) picture. This provides parameters as we illustrated above, and conceptual bases for estimating the various properties of all of these systems.

Chapter 1 introduces almost all of the concepts and parameters needed to understand the entire range of solids and illustrates their application. The remainder of the text is like an appendix, filling in the details and expanding the treatment, with extensive tables of parameters and properties. It can be used as a text, but the coverage is so large that only a selected portion could be used in one course. For that purpose there are a limited number of problems at the end of chapters, but it is very easy to generate more by asking for estimates different properties, or different materials, than treated in the text. The list of references is extensive, but focused on the particular publications which bear directly on the analytic approach used here, and not on the more general theory of solids.

I was pleased that World Scientific would publish this text a price considerably below that of other publishers, and with this goal in mind I provided camera-ready copy. Not all readers may consider the result adequate, but I believe it can be suffered for the purposes of economy, particularly for its use as a text.

Walter A. Harrison
January, 1999
Stanford, California

Contents

Elementary Electronic Structure

CHAPTER 1

The Basic Approach

The level of quantum theory needed in this book is very elementary and can be stated in a few pages. In this chapter we do that, and introduce tight-binding theory, and most of the parameters and almost all of the concepts needed in the rest of the book. The remaining chapters explore the consequences of this formulation for a wide variety of systems and properties. Many of these results are also summarized in this first chapter. We begin for completeness with the elementary, but fairly complete, summary of quantum mechanics. It will be unnecessary for most readers, but contains for reference what we shall need later.

1. Quantum Mechanics

A. Waves and Observables

The basic assertion of quantum mechanics is that an electron (as well as everything else) may be regarded as a wave as well as a particle so that it is represented by a *wavefunction*, $\psi(\boldsymbol{r},t)$. If the electron is assumed to move freely within a large volume Ω , it is represented by a wavefunction which can be a plane wave as a function of position $\boldsymbol{r}$ and time t ,

$$\psi(\mathbf{r},t) = \Omega^{-1/2} e^{i(\mathbf{k}\cdot\mathbf{r} - \omega t)} \ . \tag{1-1}$$

This is exactly as a light wave in free space is represented by an electric field which has the same dependence on position and time as the real part of Eq. (1-1), and a magnetic field with the same position and time dependence as the imaginary part. The group velocity of the wave, $\partial\omega/\partial k$, is identified with the particle velocity $\partial E/\partial p$, with p the momentum and $E = p^2/(2m)$ the kinetic energy. This is accomplished by asserting that E is proportional to ω and p is proportional to k, with the same constant of proportionality, $\hbar$, Planck's constant divided by 2π, its universal experimental value of $\hbar = 6.568\text{x}10^{-16}$ eV-sec. applies to light, and all other waves and particles. Thus

$$p = \hbar k \ ,$$
$$E = \hbar\omega \ . \tag{1-2}$$

More convenient combinations of constants for electrons are

$$\frac{\hbar^2}{m} = 7.62 \ \text{eV-Å}^2,$$
$$e^2 = 14.4 \ \text{eV-Å}, \tag{1-3}$$

with m the electron mass and e the magnitude of the electron charge. These will give us values in convenient units for most estimated quantities. For a free electron

$$E = p^2/2m = \hbar^2 k^2/2m. \tag{1-4}$$

We shall make use of this form in Section 5-3-C for establishing the parameters of the theory which we introduce in the next section.

The wavefunction $\psi(\mathbf{r},t)$ is an *amplitude*, just as the electric field in a light wave represents an amplitude. Then just as the intensity, or energy density, of light is proportional to the square of the electric field plus the square of the magnetic field, the probability density for an electron is proportional to $\psi(\mathbf{r},t)^*\psi(\mathbf{r},t)$, (equal to the sum of the squares of the real and imaginary parts) with the asterisk signifying the complex conjugate. For the wavefunction in Eq. (1-1) the electron is equally likely at any point in the volume; correspondingly its probability density is proportional to $\psi^*(\mathbf{r})\psi(\mathbf{r})$. We have chosen the leading factor in Eq. (1-1) such that the probability density is *equal* to $\psi^*(\mathbf{r})\psi(\mathbf{r})$ since total probability $\int \psi^*(\mathbf{r})\psi(\mathbf{r})\, d^3r$ comes out equal to one; this is the statement that we have chosen a *normalized* wavefunction.

Because so many such integrals occur in quantum theory, it is convenient to introduce special notation, called Dirac notation, for them. If we number the states by the index i (which could be the wavenumber) then $|i\rangle$ is used to represent $\psi_i(\mathbf{r})$ and $\langle i|j\rangle$ is defined to be

$$\langle i|j\rangle \equiv \int \psi^*_i(\mathbf{r})\psi_j(\mathbf{r})d^3r, \qquad (1\text{-}5)$$

with the integration over all $\mathbf{r}$ where the wavefunctions are nonzero. In this sense, $\langle i|$ represents $\psi^*_i(\mathbf{r})$ and whenever the two types appear together, an integration is to be performed. Our normalization condition was thus $\langle i|i\rangle = 1$.

An electron can have any wavefunction allowed by the boundary conditions we assume (such as the condition that the electron not lie outside a given volume so that the wavefunction must be zero at, and beyond, the surface of that volume). If we wanted to know the average of the position of an electron (the average of a very large number of measurements of the position, each of which would find it at a particular point, for an electron represented by $\psi_i(\mathbf{r})$ at the time of measurement), it would clearly be given by

$$\langle \mathbf{r}\rangle = \int \psi^*_i(\mathbf{r})\mathbf{r}\psi_i(\mathbf{r})d^3r \equiv \langle i|\mathbf{r}|i\rangle \qquad (1\text{-}6)$$

because $\psi^*\psi$ is the probability density. This mathematical relation is stated in words by saying that $\mathbf{r}$ is an operator representing the position of the particle and the average of a large number of measurements is the expectation value of that operator, $\langle i|\mathbf{r}|i\rangle$, with respect to the state $|i\rangle$. Similarly the operator representing momentum is $(\hbar/i)\nabla$, since operating on the wavefunction of Eq. (1-1) with this gradient form gives a factor of the momentum $\hbar\mathbf{k}$ times the wavefunction, just as the operator $\mathbf{r}$ on $\psi_i(\mathbf{r})$ gave the position times the wavefunction. The average value of the momentum of an electron in state $|i\rangle$ is $\langle \mathbf{p}\rangle = \langle i|(\hbar/i)\nabla|i\rangle$. Thus also the operator for the kinetic energy is $\mathbf{p}\cdot\mathbf{p}/2m = -\hbar^2\nabla^2/(2m)$ and if an electron moves in a potential $V(\mathbf{r})$ the operator representing its energy is

$$H = -\frac{\hbar^2 \nabla^2}{2m} + V(\mathbf{r}) \ . \tag{1-7}$$

This energy operator H is called the *Hamiltonian* and is the most important operator in quantum mechanics. In particular, noting that just as $(\hbar/i)\nabla$ operating on the wavefunction of Eq. (1-1) gives the momentum times the wavefunction, $(-\hbar/i)\partial/\partial t$ extracts $\hbar\omega$, equal to the energy. Combining this with Eq. (1-7) for the energy operator leads to the *Schroedinger Equation*,

$$-\frac{\hbar}{i}\frac{\partial\psi(\mathbf{r},t)}{\partial t} = -\frac{\hbar^2\nabla^2}{2m}\psi(\mathbf{r},t) + V(\mathbf{r})\,\psi(\mathbf{r},t) \ , \tag{1-8}$$

which tells us exactly how the wavefunction evolves in time. It is the counterpart of Maxwell's Equations for electromagnetic fields.

For our analysis here, of even more importance is the use of Eq. (1-7) to obtain the energy of the electrons, and then the energy of an entire system. From that energy we can deduce the properties of the system. Just as Eq. (1-6) gives the average of many measurements of the position, $<i/H/i>$ gives the average of many measurements of the energy for an electron in the state $/i>$. Remarkably, this statement of the average of measurements in terms of the operators and wavefunctions which we have given provides the entire physical content for all of the quantum mechanics which we will need. Everything else follows mathematically.

It is a mathematical fact that given some Hamiltonian operator, and boundary conditions, we can obtain a set of functions $\psi_i(\mathbf{r})$ such that

$$-\frac{\hbar^2\nabla^2}{2m}\psi_i(\mathbf{r},t) + V(\mathbf{r})\,\psi_i(\mathbf{r},t) = \varepsilon_i\,\psi_i(\mathbf{r}) \ . \tag{1-9}$$

The functions $\psi_i(\mathbf{r})$ are called *energy eigenfunctions* and the ε_i are their *eigenvalues*. The eigenfunctions can be taken to be *orthogonal* to each other, $<i/j> = \delta_{ij}$, and form a complete set for expansion of functions which satisfy the same boundary conditions. Obviously the average energy for such a state is $<i/H/i> = \varepsilon_i$. However, we may also ask for the root-mean-square deviation of the measurements from that value, $<(H-\varepsilon_i)^2> = <i/H^2 - 2\varepsilon_i H + \varepsilon_i^2/i> = 0$. Thus *every* measurement of the energy will give exactly ε_i .

Finally, we may expand any arbitrary electronic state in these eigenfunctions, $\psi(\mathbf{r}) = \Sigma_i u_i \psi_i(\mathbf{r})$, and it is easy to show that every measurement of energy will yield exactly one of the eigenvalues, ε_i . [For example, evaluate the expectation value of $\Pi_m(H - \varepsilon_m)^2$, which will be zero for every term in the sum over i , but would be nonzero for any measurement which differed from all eigenvalues ε_m.] In fact each ε_i is found with a probability $u_i^*u_i$. This is the familiar quantization of the energy of electronic states.

We have discussed the behavior of a single particle, moving in a potential $V(\mathbf{r})$ which is a function of its position $\mathbf{r}$. A serious problem arises as soon as we talk about more than one particle. For two particles at positions $\mathbf{r}_1$ and $\mathbf{r}_2$, the wavefunction becomes $\psi(\mathbf{r}_1, \mathbf{r}_2)$ and the problem is in six dimensions. For three or more interacting particles it becomes intractable. The approximation which allows one to proceed is that each electron moves in some average effective potential arising from all the others so that each electron can be treated alone, the *one-electron approximation.* It has turned out to be a very effective solution, and we do that here. If we add the condition that the full wavefunction change sign if any two electrons are interchanged (which leads also to the Pauli exclusion principle), and seek the potential which will give the most accurate energies, this is called the *Hartree-Fock* method. This most accurate potential is obtained from a variational calculation, such as we shall describe in another context at the beginning of Section 1-3. The more usual contemporary method for solid-state problems is called *Local-Density Theory.* For our purposes, we simply proceed to describe the states of individual electrons with our own methods.

This provides the quantum-mechanical basis for almost everything we do in this text. We will return at the end of Section 1-5-B to one of the most useful techniques in quantum mechanics, perturbation theory. For the most part we will work directly with these expansions of general states in terms of eigenstates. Our goal will be to find these energy eigenstates and their energies for many systems and we will understand the properties in terms of them. We begin with a discussion of the electron states in atoms.

B. Atomic States

The electron is deflected by electric fields, and in particular in an atom its motion is bent into an orbit around the nucleus. The wavefunction must of course come out even around such an orbit; it is single-valued. If we think of an orbit of some radius r , then $k \cdot 2\pi r$ must be an integral multiple of 2π , or kr is equal to some integer. This corresponds to the familiar quantization of the angular momentum; $p\times r = \hbar kr$ must be an integral number of $\hbar$'s.

The electron states in an atom, corresponding to a spherically symmetric effective potential, can be constructed to be proportional to the spherical harmonics, $Y_l^m(\theta,\phi)$, the spherical generalization of the circular orbit above. The only dependence upon ϕ (the azimuthal angle around the z-axis) is a factor $e^{im\phi}$ so the same analysis gives the z-component of the angular momentum L_z as $\hbar m$ with m running from $-l$ to l . This corresponds to $2l + 1$ states of the same energy, but with differently oriented total angular momentum. One may evaluate the total angular-momentum squared to find eigenvalues of $L^2 = L_x^2 + L_y^2 + L_z^2$ to be $l(l + 1)\hbar^2$.

The states which will be of principal interest here are for l equal to zero, called *s-states*, or one, called *p-states*. In Section 1-7 we return to states with l equal to two, called *d-states*, and l equal to three, called *f-states*. We may think of the *s* as referring to *spherical* since $Y_0^0 = 1/\sqrt{4\pi}$ is spherically symmetric, and such a state is represented by a circle, as in Fig. 1-1. [The *s* and *p* actually refers to characteristics of the corresponding spectral line.] There are three p-states, of the same energy, with $m = -1, 0,$ and 1 . The $m = 0$ state has angular dependence z/r and no angular momentum around the z-axis and is denoted as the state $|p_z\rangle$. In condensed-matter physics it is most convenient to use the two orthogonal states $(Y_1^1 + Y_1^{-1})/\sqrt{2}$ and $(Y_1^1 - Y_1^{-1})/\sqrt{2}$, rather than the $Y_1^{\pm 1}$ themselves. The first of these has angular dependence x/r, vanishing angular momentum around the x-axis and is called $|p_x\rangle$. The second of these has angular dependence y/r, vanishing angular momentum around the y-axis and is called $|p_y\rangle$

These states are represented by figures "eight", with a plus indicating the direction in which the wavefunction is positive, and are illustrated in Fig. 1-1. We shall see that a general p-state can be divided into these three components, like a vector, so that we may think of the *p* as standing for *polar* . We shall always take the atomic wavefunctions to be real. Each of these states is orthogonal, in the mathematical sense, to the other three and is normalized such that $\langle j|i\rangle = \int \psi^*_i(r)\psi_j(r)\, d^3r$ equals zero if the two states are different and one if they are the same.

The lowest-energy state in every atom is an s-state, called the 1s -state, and the s-states of higher energy are successively numbered $2s, 3s, \ldots ns$. In hydrogen the energies of these are

$$\varepsilon_{ns}(H) = -e^4 m/(2\hbar^2 n^2) , \tag{1-10}$$

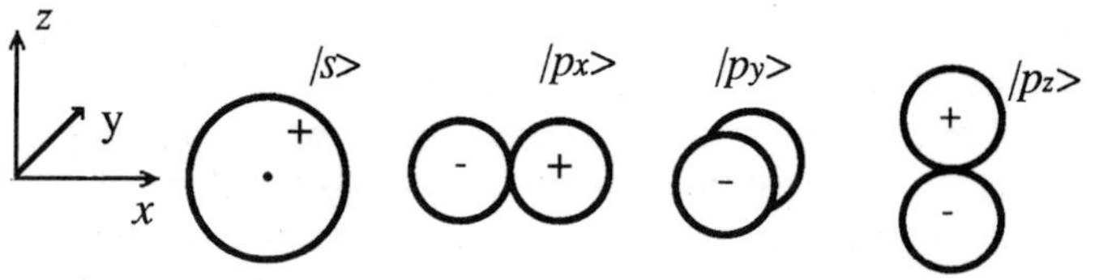

Fig. 1-1. The symbols which are used to represent the s- and p-states of the atom.

equal to -13.6 eV for n=1, the 1s-state. Note that this is the combination of constants, e^2 and $\hbar^2/m$ given in Eq. (1-3), which has the units of energy; the factor $^1/_2$ and the n^2 are appropriate to hydrogen. In the hydrogen atom the lowest-energy p-state has an energy equal to that of the 2s-state, and is called *2p*, with higher-energy p-states called *3p, 4p*, etc., so the same formula, Eq. (1-10) above, applies to p-states. Similarly in hydrogen the

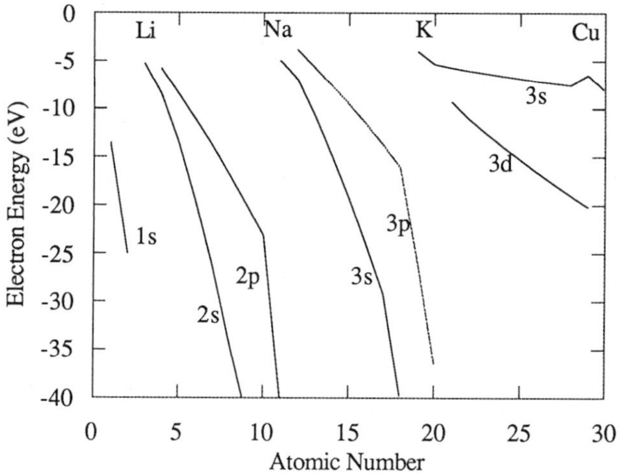

Fig. 1-2. A plot of the atomic term values showing the similarity of the 2s, 2p and the 3s, 3p periods. The alkali metals beginning each period are indicated. After potassium the 3d-level drops below the 4s-level and the transition series begins. It ends with copper , beginning the 4s, 4p series, similar to the other sp-series.

lowest-energy d-state has the same energy as the hydrogen 3s-state and the d-states are numbered *3d, 4d,* etc. There are $2l+1 = 5$ orientations of the d-states (rather than the one for s-states and three for p-states).

The energies of the atomic states for the lighter elements are plotted in Fig. 1-2. The energies of the s- and p-states are also tabulated in Table 1-1, and again in the *Solid-State Table* presented for all of the elements at the end of this volume. We may think of them as the energy required to remove an electron from the corresponding state in the neutral atom since the Hartree-Fock term values which we have used *are* the Hartree-Fock estimates of that removal energy in the approximation that the removal does not modify the remaining electronic wavefunctions. In the neutral atom electrons occupy the states of lowest energy, filling first the 1s-state, then the 2s and 2p-states, then the 3s and 3p-states as we increase the atomic number. We note the strong similarity of the energies in the 2s-2p row and in the 3s-3p row, the similarity which is responsible for the periodic dependence of properties on atomic number.

In fact the systematics of these values is easy to remember. We note first the energy of the 2s-state in lithium, -5.34 eV. The energy of the 2s-state in hydrogen, according to Eq. (1-10), is -3.4 eV. The difference is that lithium has two additional protons in the nucleus and two 1s-electrons making up its "core". The additional electrostatic potential from this combination lowers the 2s-state energy from -3.4 to -5.34 eV. We may nonetheless take this -3.4 eV as a crude estimate of the s-state energy of the lithium, sodium, potassium, rubidium and cesium, column, noting that the correction is only large at the top of the periodic table, or the Solid-State Table at the end of this volume. We then note that the s-state energy is approximately proportional to the column number Z, and that the estimate $\varepsilon_s = -3.4\,Z$ is quite accurate for the row Mg, Al, Si, P, S, Cl, Ar, and the rows below, but underestimates the values in the row Be, B, C, N, O, F, Ne. Finally we note that the p-state energy in each element is approximately half of the s-state energy for that element and we can remember approximate values for all of the elements listed in Table 1-1. These simple systematics are presumably understandable in terms of the atomic pseudopotentials described in Section 1-7 of this chapter, but we have only done that for the universal factor two between ε_s and ε_p. (See Problem 13-3.) In Fig. 1-2 we also note that starting just beyond potassium we begin filling the 3d-state, rather than the 3p-state, proceeding across this *transition* series until all five d-states are doubly

Table 1-1 Hartree-Fock term values for valence levels[a]. The first entry is ε_s, the second is ε_p (values in parentheses are highest core level). The third entry is the sp^3-hybrid energy ε_h. The fourth is U, the intraatomic Coulomb repulsion[b]. All are in eV.

I	II	III	IV	V	VI	VII	VIII	IA	IIA
							He	Li	
							-24.98	-5.34	
								-	
								-	
								8.17	
	Be	B	C	N	O	F	Ne	Na	
	-8.42	-13.46	-19.38	-26.22	-34.02	-42.79	-52.53	-4.96	
	-5.81*	-8.43	-11.07	-13.84	-16.77	-19.87	-23.14	(-41.31)	
	-6.46	-9.69	-13.15	-16.94	-21.08	-25.60	-30.49		
	10.25	10.26	11.76	13.15	14.47	15.75	15.00	6.17	
	Mg	Al	Si	P	S	Cl	Ar	K	Ca
	-6.89	-10.71	-14.79	-19.22	-24.02	-29.20	-34.76	-4.01	-5.32
	-3.79*	-5.71	-7.59	-9.54	-11.60	-13.78	-16.08	(-25.97)	(-36.5)
	-4.57	-6.96	-9.39	-11.96	-14.71	-17.64	-20.75		
	7.28	6.63	7.64	8.57	9.45	10.30	11.12	5.56	6.40
Cu	Zn	Ga	Ge	As	Se	Br	Kr	Rb	Sr
-6.49	-7.96	-11.55	-15.16	-18.92	-22.86	-27.01	-31.37	-3.75	-4.86
-3.31*	-3.98*	-5.67	-7.33	-8.98	-10.68	-12.44	-14.26	(-22.04)	(-29.88)
-4.11	-4.98	-7.14	-9.29	-11.47	-13.73	-16.08	-18.54		
7.07	7.83	6.61	7.51	8.31	9.07	9.78	10.48	5.02	5.71
Ag	Cd	In	Sn	Sb	Te	I	Xe	Cs	Ba
-5.99	-7.21	-10.14	-13.04	-16.03	-19.12	-22.34	-25.70	-3.37	-4.29
-3.29*	-3.89*	-5.37	-6.76	-8.14	-9.54	-10.97	-12.44	(-18.60)	(24.60)
-3.97	-4.72	-6.56	-8.33	-10.11	-11.94	-13.81	-15.76		
6.34	6.95	6.00	6.73	7.39	8.00	8.58	9.13	5.05	5.70
Au	Hg	Tl	Pb	Bi	Po	At	Rn	Fr	Ra
-6.01	-7.10	-9.83	-12.49	-15.19	-17.97	-20.83	-23.78	-3.21	-4.05
-3.31*	-3.83*	-5.24	-6.53	-7.79	-9.05	-10.34	-11.65	(-17.10)	(-22.31)
-3.99	-4.65	-6.39	-8.02	-9.64	-11.28	-12.96	-14.68		
6.75	7.33	6.30	7.03	7.68	8.28	8.85	9.39	4.93	5.54

* Values extrapolated from surrounding values.

(a) J. B. Mann (1967)

(b) W. A. Harrison (1985a)

occupied at copper. This of course repeats in the series ending with silver and that ending with gold. Similarly, in a series beginning with lanthanum, and one beginning with actinium, we successively fill the 4f- and 5f-atomic states.

We noted that these atomic-state energies may be thought of as the energy required to remove an electron from that state in the neutral free atom. It is interesting to see how accurately that is true. The *ionization potentials* of the elements , listed for example in the Chemical Rubber Company Handbook, Weast (1975), p. E68, are experimental values for these energies obtained in various ways. A set are compared in Table 1-2 with the highest-occupied Hartree-Fock energies from Table 1-1. They are not exact, and there are corrections for the ionization energies even within Hartree-Fock theory, but we take the comparison as a general confirmation of the one-electron theory upon which all of our analysis of molecules and solids, as well as atoms, is based.

We note that if we were, for example, to remove a second 3p-electron from silicon, it would take more energy since the repulsion between the first and second electrons is not present once the first is removed. This shows up as a larger *second ionization potential* , listed by Weast (1975) as 16.34 eV for silicon. This indicates a repulsion, called U, of 16.34 - 8.15 = 8.19 eV , of the order of e^2 divided by the "radius" of the atom. In a similar way, the electron affinity of an atom, the energy gained in adding an additional electron to a neutral atom is less than the value we have listed by approximately U . We have compiled a systematic set of such values (Harrison (1985a)) which are listed as the fourth entry in Table 1-1. These values are only occasionally needed since the atoms tend to remain approximately neutral in solids. We shall later introduce U -values for the d-states in transition-elements; the most useful value is the shift in a d-state energy if an s-electron is transferred to the d-shell.

Table 1-2. Comparison of the experimental ionization potentials for free atoms, Weast (1975), with term values ε_s and ε_p from Table 1-1. (in eV)

Atom	$\lvert\varepsilon_s\rvert$	$\lvert\varepsilon_p\rvert$	Ionization Potential
Na	4.96		5.14
Mg	6.89		7.64
Al	10.71	5.71	5.98
Si		7.59	8.15
P		9.54	10.48
S		11.60	10.36
Cl		13.78	13.01
A		16.08	15.75

2. Tight-Binding Theory and the Simplest Molecule

The most important orbitals in any system tend to be the highest-energy states occupied by electrons and the lowest-energy empty states. For atoms these are the states listed in Table 1-1, called *valence states*. The particular case of lithium, and the elements below it, is especially simple since only the s-shell is partly occupied. A molecule of two lithium atoms is the simplest case, and we treat it first.

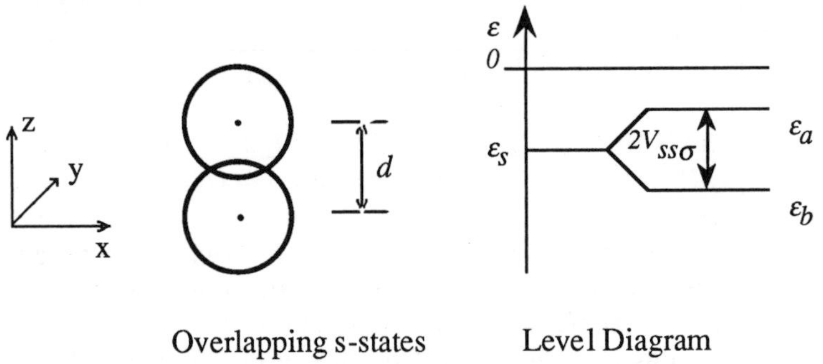

Fig. 1-3. The formation of molecular orbitals in lithium, with $d = 2.67$ Å, as linear combinations of the s-states on the two lithium atoms. The atomic s-levels are split into a bonding state at energy ε_b and an antibonding state at energy ε_a as shown in the energy-level diagram on the right.

A. Molecular Orbitals

If two lithium atoms are brought close enough together so that the atomic states overlap each other appreciably, as illustrated in Fig. 1-3, new *molecular orbitals* arise, which can be approximately described as linear combinations of the atomic states:

$$|MO\rangle = \psi_{MO}(\mathbf{r}) = u_1\,|s1\rangle + u_2\,|s2\rangle\,. \qquad (1\text{-}11)$$

This approximation of the states of a multiatom system by linear combinations of atomic states is called the *Linear Combination of Atomic Orbitals* method. When the various parameters used in the method are obtained from some independent source as we shall do , rather than from actual atomic orbitals, it is called the *Tight-Binding Method*. Our purpose in using the tight-binding method is to simplify the theory sufficiently to make all estimates by hand. It is sometimes used also in heavy computational studies simply to reduce the needed computing time (e. g., Goedecker and Teter (1995) and references therein). In Eq. (1-11) we have used the Dirac notation $|MO\rangle$ for the molecular orbital and for the first and second s-states.

In analogy with Eq. (1-6), if the electron had the wavefunction $|MO\rangle$ the average of energy measurements would yield $\langle MO|H|MO\rangle/\langle MO|MO\rangle$, the appropriate way to evaluate energy for this approximate state. We have included the denominator which would be unity if the wavefunction were normalized so that it will not matter whether we pick the coefficients in Eq. (1-11) so that it is. We have for the energy

$$\varepsilon_{MO} = \frac{\langle MO|H|MO\rangle}{\langle MO|MO\rangle} = \frac{u_1^2\varepsilon_1 + u_2^2\varepsilon_2 + 2u_1u_2\langle s1|H|s2\rangle}{u_1^2 + u_2^2 + 2u_1u_2\langle s1|s2\rangle} \quad . \qquad (1\text{-}12)$$

Here we have noted that $\langle s1|H|s2\rangle = \langle s2|H|s1\rangle$ (writing out the integral, two partial integrations in the kinetic energy term show that they are equal) and have written $\langle s1|H|s1\rangle = \varepsilon_1$ and $\langle s2|H|s2\rangle = \varepsilon_2$.

Completing this calculation in detail for lithium will clarify a number of issues and suggest simplifications which can be used in the remainder of the book. For this simple case, for which the Hamiltonian has reflection symmetry through a plane bisecting the internuclear axis, we can construct states which are even or odd with respect to that plane. This means that $u_2 = \pm u_1$. We also note that $\varepsilon_1 = \varepsilon_2 = \varepsilon_s$ and have $\varepsilon_{MO} = (\varepsilon_s \pm \langle s1|H|s2\rangle)/(1 \pm \langle 1|2\rangle)$. With a little algebra this becomes

$$\varepsilon_{MO} = \varepsilon_s \pm \frac{\langle s1|H|s2\rangle - \langle s1|s2\rangle\varepsilon_s}{1 - \langle s1|s2\rangle^2} - \frac{\langle s1|H|s2\rangle - \langle s1|s2\rangle\varepsilon_s}{1 - \langle s1|s2\rangle^2}\langle s1|s2\rangle \,. \qquad (1\text{-}13)$$

The second term splits the two levels and we write $(\langle s1|H|s2\rangle - \langle s1|s2\rangle\varepsilon_s)/(1 - \langle s1|s2\rangle^2\ )$ as $V_{ss\sigma}$, which will turn out to be negative. We regard it as a "coupling" between two s-states (hence the *ss* -subscript) and the third subscript is added to specify the angular momentum (σ like *s* to indicate zero angular momentum) around the internuclear axis. In this case it is redundant but later it will not be. In terms of the $V_{ss\sigma}$ so defined,

$$\varepsilon_{MO} = \varepsilon_s \pm V_{ss\sigma} - \langle s1|s2\rangle\, V_{ss\sigma}. \qquad (1\text{-}14)$$

Thus the second term splits the two states, which had the same energy initially, into a lower-energy state which is called the bonding state and a higher-energy state which is called the antibonding state, indicated to the right in Fig. 1-3. This is the origin of bonding between the two lithium atoms. The valence electrons, one at ε_s in each atom, go into the bonding state, each having an energy lower by $V_{ss\sigma}$. The third term is a shift in the average energy which we shall see is always positive. It will prevent the two atoms from falling into each other under the bonding interaction. Before discussing Eq. (1-14) as the basic electronic structure of Li_2, we shall continue the molecular-orbital theory a little further in the more traditional form.

We may understand better the parameter $<s1/H/s2>$ by writing the Hamiltonian as a kinetic energy plus the potential v_1 due to the first atom and v_2 due to the second. However, the kinetic energy plus v_2 operating on $/s2>$ gives the energy ε_2 times $/s2>$ so we have $<s1/H/s2> = \varepsilon_2<s1/s2> + <s1/v_1/s2>$. It may be preferable to write this in symmetric form by averaging $<s1/H/s2>$ and $<s2/H/s1>$ which are equal in the case we consider.

$$<s1/H/s2> = \frac{\varepsilon_1 + \varepsilon_2}{2} <s1/s2> + <s1/ \frac{v_1 + v_2}{2} /s2> . \tag{1-15}$$

B. Extended-Hückel Theory

We will work directly with $V_{ss\sigma}$, but we may draw insight from the *Extended-Hückel Theory* of Roald Hoffmann (1963), in which the second term in Eq. (1-15) was dropped. We may then note the qualitative result that the matrix element $<s1/H/s2>$ will always be of opposite sign to $<s1/s2>$ since the atomic energies ε_1 and ε_2 are less than zero. We shall see that this is true also for our $V_{ss\sigma}$, which guarantees that the final term in Eq. (1-14) is indeed positive as we indicated and serves as a repulsion between atoms. Finally, without the final term in Eq. (1-15), the dependence of $<s1/s2>$ upon distance d between atoms is the same as that for the matrix element itself, a feature we shall use in evaluating the final term in Eq. (1-14).

Having dropped the final term in Eq. (1-15), it is possible to evaluate $<s1/s2>$ directly from known free-atom wavefunctions and, using tables such as Table 1-1 for the term values, one has all the parameters which enter the calculation of the total energy. Hoffmann, however, found that the results for organic compounds were not accurate without introduction of a scale factor K such that

$$<s1|H|s2> \approx K\frac{\varepsilon_1 + \varepsilon_2}{2} <s1|s2> . \ (EHT) \qquad (1\text{-}16)$$

A value $K = 1.75$ is commonly used for carbon-row atoms.

It turns out that a value nearer one is appropriate for the heavier elements. We saw this by applying Eq. (1-16) to the covalent energy V_2 which we shall use for tetrahedral semiconductors in Section 2-2. Using the value for diamond, and $K = 1.75$, yields an overlap between the corresponding hybrids of $<h1|h2> = 0.45$. In fact, simple estimates for sp^3-hybrids (e. g., Van Schilfgaarde and Sher (1987)) based upon known atomic wavefunctions give values close to this for diamond and other semiconductors. Then using this value for the overlap and the universal values for covalent energies ($3.22\hbar^2/md^2$) and hybrid energies From Table 1-1 leads to $K = 1.75, 1.06, 0.99,$ and 0.83 for diamond, silicon, germanium, and tin, respectively.

Extended Hückel Theory has comparable accuracy to the methods we shall use here. The difference is that we shall have simple formulas for the couplings such as $V_{ss\sigma}$ which we use, and shall include the nonorthogonalities only through a repulsion $V_0(d)$ which arises from the final term in Eq. (1-14). Then we shall find analytic methods for obtaining the state energies and other properties of the system. Such analytic methods are the essential feature of the approach we shall use for all classes of materials. Extended Hückel Theory has the difficulty that one obtains only numerical results, which are less accurate than more modern density-functional methods. One obtains no more insight than from other numerical methods, so that all one has gained over these more accurate methods is some saving in computing time.

C. Electronic Structure of Li_2

Finally, we may substitute Eq. (1-15) (with $\varepsilon_1 = \varepsilon_2 = \varepsilon_s$) back into the definition of $V_{ss\sigma}$ which we gave just after Eq. (1-13) to obtain

$$V_{ss\sigma} = \frac{<s1|\frac{v_1 + v_2}{2}|s2>}{1 - <s1|s2>^2} . \qquad (1\text{-}17)$$

It is clear that $V_{ss\sigma}$ is negative if $<s1|s2>$ is positive, as was $<s1|H|s2>$, because the potentials are negative and if the atomic wavefunctions are of the same sign where they overlap most, the integral in the numerator will be negative.

The form for the Li_2 electronic structure, Eq. (1-14), corresponds to expanding the molecular orbital in *orthogonal* atomic-like states ($\langle s1|s2\rangle = 0$), with a separate repulsion $V_0(d) = -2\langle s1|s2\rangle V_{ss\sigma}$ absorbing the effects of the nonorthogonality of real overlapping atomic states for the two electrons. The fact that our atomic-like states can be taken to be orthogonal follows from the fact that Eq. (1-14), without the final term, is obtained from $|MO\rangle = u_1|1\rangle + u_2|2\rangle$ based upon orthogonal states of energy ε_1 and ε_2 equal to ε_s and coupled by $V_{ss\sigma}$. A feature of the orthogonality which will be very important in our analysis is that the *sum* all of the energies of coupled orthogonal states, whether occupied or not, is not changed by the coupling, as illustrated to the right in Fig. 1-3.

We have determined the electronic structure of this Li_2 molecule. We regard our molecular-orbital energies as approximations to the Hartree-Fock values, and therefore the removal energy for electrons occupying the corresponding states. In the ground state the two valence electrons occupy the bonding orbital, one with spin-up and one with spin-down.

D. The Total Energy

We next turn to the total energy of the system, which in Hartree-Fock theory *contains* the sum of energies for the occupied states. However, that sum includes the electron-electron interaction twice (once in the energy for the first and once in the energy for the second of two interacting electrons) and one contribution should be subtracted. Also the sum of eigenvalues does *not* contain the Coulomb interaction energy between different nuclei (or different ion cores if the core electrons are included along with the nuclei) so this should be added. If, on the other hand, the total charge density is well approximated by a superposition of neutral atoms, the change in the sum of the two corrections would correspond to a short-range radial interaction (see Section 2-2-F) which could be absorbed in the overlap interaction $V_0(d)$. Similarly, in obtaining Eq. (1-13) we wrote $\langle s1|H|s1\rangle = \varepsilon_1$ and replaced it by an atomic term value. This would in fact be appropriate if the Hamiltonian contained the kinetic energy and only the potential v_1 , but there are also shifts $\langle s1|v_2|s1\rangle$ and we also absorb these in the overlap interaction. Thus *the change in energy of a system of atoms, when the atoms are rearranged, is given approximately by the change in the sum of the energies of the occupied orbitals, plus any change in the sum of these overlap repulsions* in which all the corrections are absorbed.

This overlap interaction becomes a somewhat complicated term, and is probably the weakest link in our theory. However, it will be useful to approximate it further for illustrative purposes. The principal contribution is from the nonorthogonality, in this case - $2V_{ss\sigma}\langle s1|s2\rangle$. We shall see in the following section that $V_{ss\sigma}$ may be expected to vary approximately as $1/d^2$ near the observed spacing. If we again think of the Hückel coupling parameters of Eq. (1-16) as corresponding to our coupling parameters such as $V_{ss\sigma}$, we conclude from Eq. (1-16) that the overlaps $<s1/s2>$ also vary as $1/d^2$ so that this term in V_0 varies as $1/d^4$. This is indeed a very weak justification for the $1/d^4$ form which we use. At some points we use a more general form for $V_0(d)$ and can do even better by retaining the orthogonality explicitly as in Section 20-1-B. (See also Menon and Subbaswamy (1994).) We shall discuss this overlap interaction much more completely for covalent solids in Section 2-2-E with further justifications for $1/d^4$, and again for ionic solids in Section 9-3, and see its counterpart for simple metals, transition, and f-shell metals in later chapters.

Returning to Li_2, the sum of the two energies of the two bonding electrons, relative to the two isolated atoms ($V_{ss\sigma} = 0$) is simply $2V_{ss\sigma}$, and adding a repulsion $V_0(d) = C/d^4$, we may obtain C if we know the observed equilibrium spacing d. We write $\partial(2V` + C/d^4)/\partial d = 0$ again taking $V_{ss\sigma}$ proportional to $1/d^2$. We obtain $-4V_{ss\sigma}/d - 4C/d^5 = 0$ at the equilibrium spacing, which means C/d^4 equals $-V_{ss\sigma}$ at the equilibrium spacing. This leads to the prediction of the binding energy of the Li_2 molecule as $V_{ss\sigma}$, as discussed in Problem 1-1.

From this representation of the electronic structure we can not only obtain the cohesion of the molecule ($-V_{ss\sigma}$) but the dominant optical absorption energy equal to $-2V_{ss\sigma}$ and could even apply a field and obtain the polarizability of the molecule by calculating the shift in the charge density due to the field.

We have used the case of Li_2 to understand the tight-binding theory of electronic structure, to see how it can be represented in terms of couplings $V_{ss\sigma}$ between the two atomic-like states which may be thought of as orthogonal to each other if an overlap repulsion $V_0(d)$ is introduced. We have also seen how this simple representation of the electronic structure allows prediction of almost all properties of the molecule. We move in the next section to a multiatom system, again consisting of lithium atoms, and after that to molecules in which atomic p-states also play a role.

3. The Simplest Many-Atom System, a Metal

We next imagine a long chain of N lithium atoms, separated by a distance d as illustrated in Fig. 1-4. We now seek N composite, or tight-binding, states of the form $/\psi> = \Sigma_{j=1,N}\, u_j\, /j>$. It is a simplification to eliminate the ends of the chain by bending it in a circle so that the state $/1>$ is coupled to the state $/N>$, as shown in Fig. 1-4. It will also be convenient to allow the expansion coefficients u_j to be complex. Then we may extend Eq. (1-12) to this chain as

$$\varepsilon = \frac{<\psi|H|\psi>}{<\psi|\psi>} = \frac{\varepsilon_s \Sigma_j u_j{}^* u_j + V_{ss\sigma} \Sigma_j (u_j{}^* u_{j+1} + u_j{}^* u_{j-1})}{\Sigma_j u_j{}^* u_j} . \tag{1-18}$$

We have dropped the nonorthogonality terms in the denominator, as we did for the molecule, so we now think of the atomic-like states as orthogonal, and have replaced the couplings by $V_{ss\sigma}$ for nearest neighbors and zero for more distant neighbors.

We wish to choose the coefficients to obtain the most accurate energies, and this is accomplished with a *variational calculation* . We seek in particular the lowest-energy state. The approximate state $/\psi>$ could be written as a linear combination of the *real* eigenstates, and each term in $<\psi/H/\psi>$ will contribute at an energy at least as high as the lowest state so Eq. (1-18) must be higher than the energy we seek. The closest we can get is to minimize with respect to each u_i . This gives a set of N simultaneous equations in the u_j . Solving these will give the lowest-energy solution, plus the next-lowest state orthogonal to the lowest, etc., which is the optimum choice of states for this tight-binding approximation. A systematic way of minimizing would to minimize both with respect to the real and with respect to the imaginary parts of $/\psi>$ but the same result is obtained by taking a derivative with respect to $u_j{}^*$ as if it were independent of u_j . The result, which will be the basis of many calculations in this book, is

$$\varepsilon_s u_j + V_{ss\sigma}(u_{j+1} + u_{j-1}) = \varepsilon u_j . \tag{1-19}$$

The solution of such simultaneous equations in general is accomplished by constructing a matrix H_{ij} with diagonal elements H_{jj} equal to ε_s in the case of Eq. (1-18). The off-diagonal matrix elements are $V_{ss\sigma}$ for

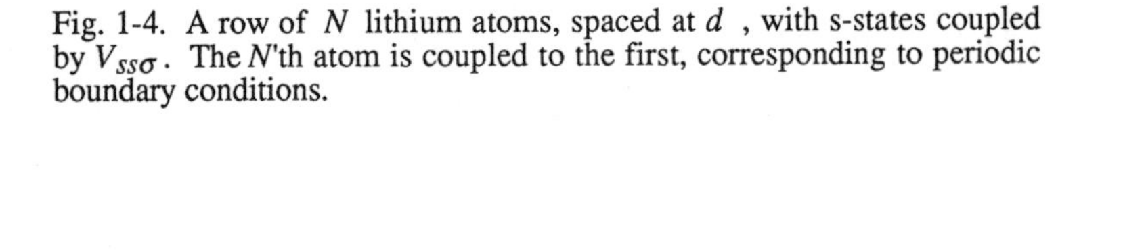

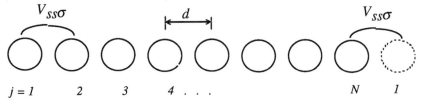

Fig. 1-4. A row of N lithium atoms, spaced at d , with s-states coupled by $V_{ss\sigma}$. The N'th atom is coupled to the first, corresponding to periodic boundary conditions.

elements for which the column number and the row number differ by one (regarding the index $j = N$ also to differ by one from $j = 1$). This matrix is called the *Hamiltonian matrix* and these couplings such as $V_{ss\sigma}$ will be referred to as *matrix elements* in subsequent discussion. The simultaneous solution of such equations is also called the *diagonalization* of the Hamiltonian matrix. In the present very simple case we may write the solutions directly, as we could for the Li_2 molecule.

A. Energy Bands

We can immediately confirm by substitution that there are N normalized solutions of the form of a *Bloch sum* , $u_j = e^{ikdj}/\sqrt{N}$, where the *wavenumber* k is chosen as $2\pi n/(Nd)$ with $-N/2 \leq n < N/2$. More general Bloch sums will be introduced again in Eq. (5-4). This range of wavenumbers is called the *Brillouin Zone* . The condition that n be an integer makes the coefficients come out even around the chain so that the end equations, $j = 1$ and $j = N$ are also satisfied. Solutions with n an integer outside the Brillouin Zone reproduce solutions with exactly the same u_j as some other n (with an integral number of N 's subtracted or added) within the range so that there are only N distinct solutions, as we

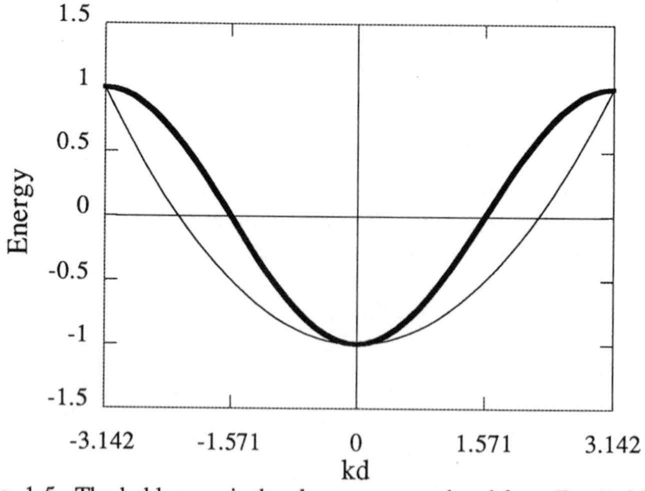

Fig. 1-5. The bold curve is the electron energy band from Eq. (1-20) for a row of lithium atoms. It is measured from ε_s and is in units of $|2V_{ss\sigma}|$. It has been chosen to fit the free-electron parabola, the light line, at $k = 0$ and at the Brillouin Zone edges $k = \pm\pi/d$ by taking $V_{ss\sigma} = -\pi^2\hbar^2/(8md^2)$.

know there must be. The energies of these N states are given by

$$\varepsilon_k = \varepsilon_s + 2V_{ss\sigma} \cos kd \qquad (1\text{-}20)$$

which is plotted in Fig. 1-5. The states correspond to the discrete values of $k = 2\pi n/(Nd)$. For many atoms the states are quite closely spaced and form an almost continuous distribution of energies, called an *energy band*. Note that these solutions of Eq. (1-19) are exact even if there are a small number of atoms; it is only the restriction on k which is changed.

B. Matching the Free-Electron Bands

In most real solids the spacing of atoms is such that the electrons also behave very much as the free electrons which we discussed at the beginning of this chapter, with $\varepsilon_k = \varepsilon_0 + \hbar^2k^2/2m$. If this is to be true in this case, and if the two are to agree at the points $k = 0$ and $k = \pi/d$, then $V_{ss\sigma}$ would need to satisfy $-4V_{ss\sigma} = \hbar^2(\pi/d)^2/2m$ or $V_{ss\sigma} = -(\pi^2/8)\,\hbar^2/(md^2)$. We have also drawn such a free-electron parabola in Fig. 1-5 and chosen $V_{ss\sigma}$ such that these do match. This is an extremely important point since in Chapter 6 we will construct these tight-binding bands from both s- and p-states in semiconductors and choose all of the coupling matrix elements involving these states such that they fit the real, rather free-electron-like, bands of real semiconductors. This, along with the free-atom term values of Table 1-1, will give us the parameters needed for calculating all of the properties of the semiconductors.

One might ask why we select these particular points, $k = 0$ and $k = \pi/d$ to match. The reason is that when we also construct bands from the p-states, which correspond to free-electron states beyond the wavenumbers $\pm\pi/d$, they will have the effect of pushing the tight-binding band downward toward the free-electron band between these fitting points, but will not affect the bands at those end points.

One might also ask why the spacing would turn out in solids to be such that the bands were free-electron like. One intuitive explanation is that as atoms are brought together and the bands begin to broaden from the starting atomic states, the lower-energy states are occupied, lowering the energy and further pulling the system together, until the energy bands reach a free-electron width and then they can get no broader. Further contraction does not lower the energy further and the overlap repulsion prevents further contraction.

The formula, $V_{ss\sigma} = -(\pi^2/8)\,\hbar^2/(md^2)$, and the other similar formulae we shall derive are only valid near the equilibrium spacing, but at that spacing where the bands are free-electron-like, they must apply. Whenever one treats the bands more completely, one finds deviations from this $1/d^2$ dependence. A recent such study, with references, was described by Grosso and Piermarocchi (1995). They incorporated also second-neighbor couplings and found dependences of the nearest-neighbor couplings, $1/d^n$, with n between 2.7 and 5 . This does not at all mean that such higher values are appropriate to the theory with nearest-neighbor coupling only. Since the bands indeed are quite free-electron-like, this must follow also from the second-neighbor theory, but then through cancellation between the more complete set of couplings. Further, where our simple forms give reasonable energy bands, we may expect them to describe the interatomic couplings in molecules also, since those occur at similar spacings. This transferability of the couplings is approximate, but we shall see that it is accurate enough to be very useful.

A further point may be made concerning the free-electron-like behavior. This is behavior as if the atomic potentials were so weak that only the kinetic-energy was important for these electrons. Of course the potentials are not weak; they are so strong that they produce core states, 1s-states in lithium. However, the states which we describe in Eq. (1-20) and Fig. 1-5 are based upon 2s-states which are orthogonal to the 1s-states and the effect of this orthogonality is to replace the strong potentials v_j by weak pseudopotentials. We shall discuss these pseudopotentials from another point of view in Section 1-6 and employ them extensively in Chapters 13 and 14.

C. Cohesion

Returning to our chain of lithium atoms, there is a single valence electron provided per atom and there are as many states in the band as there are atoms. Since each state can be occupied by an electron of spin up and one of spin down, there are just enough electrons to fill half of the band, to the *Fermi wavenumber* which in this case is $\pm\pi/(2d)$. Empty states are available immediately above the *Fermi energy* , which makes this a metal. Under the application of an electric field the electrons will accelerate into these empty states and current will flow.

We might estimate the energy gained in the formation of this chain, the cohesive energy of the lithium chain. Initially each electron was in the state of energy ε_s but now they occupy the lower half of the band shown in Fig. 1-5. It is preferable to use the free-electron parabola for this estimate since if we had included the effect of the p-states, the bands would have more closely resembled that parabola. The average energy relative to the band minimum is then $(1/k_F)\int_{0,k_F}(\hbar^2k^2/2m)dk = \hbar^2k_F{}^2/(6m) = (\pi^2/24)$ $\hbar^2/(md^2)$, but the band minimum is below the s-state energy ε_s by $2V_{ss\sigma} =$

$-(\pi^2/4)\hbar^2/(md^2)$ so the energy relative to ε_s is $-(5\pi^2/24)\,\hbar^2/(md^2)$. The upward shift of the levels due to nonorthogonality, approximated by the overlap repulsion, cancels half of this leaving a cohesion of $(5\pi^2/48)\,\hbar^2/(md^2)$ per atom. We shall see that the corresponding estimates for the cohesion of the three-dimensional metallic structure are good to something like 20% for the elements below the lithium row in the periodic table, and a factor of two too large for the lithium row.

It is useful at this point to tabulate the cohesive energy of the elements, most of which are metals. In Table 1-3 we give Kittel's table, from data supplied by Professor Leo Brewer, (Kittel, 1976, p. 74). The arrangement of elements is not the standard one, but this form is useful for classifying solids as we shall indicate in Section 1-4-E. Values for the transition metals and f-shell metals, which would appear to the left and the right in this arrangement, will be given when those systems are discussed in Chapters 15 and 16.

In the three-dimensional structure, with partially filled bands appropriate to a metal, the free-electron representation of the bands will prove more appropriate than the tight-binding representation, as it was for the chains here. We proceed that way in Chapter 13 and then in Section 13-7-C the effect of the metallic atoms in distorting the bands will be represented by a pseudopotential, allowing estimates of the range of properties of metals.

Table 1-3. The cohesive energies of the elements. The energy required to separate the element into neutral atoms at 0° K at atmospheric pressure, in eV per atom. From Kittel (1976).

I	II	III	IV	V	VI	VII	VIII	IX	X
						H	He	Li 1.63	
	Be 3.32	B 5.77	C 7.37	N 4.92	O 2.60	F 0.84	Ne 0.02	Na 1.113	
	Mg 1.51	Al 3.39	Si 4.63	P 3.43	S 2.85	Cl 1.40	Ar 0.08	K 0.934	Ca 1.84
Cu 3.49	Zn 1.35	Ga 2.81	Ge 3.85	As 2.96	Se 2.25	Br 1.22	Kr 0.116	Rb 0.852	Sr 1.72
Ag 2.95	Cd 1.16	In 2.52	Sn 3.14	Sb 2.75	Te 2.23	I 1.11	Xe 0.16	Cs 0.804	Ba 1.90
Au 3.81	Hg 0.67	Tl 1.88	Pb 2.03	Bi 2.18	Po 1.50	At	Rn 0.202	Fr	Ra 1.66

4. sp-Bonded Systems and Hybrid States

A. Needed Parameters

We return again to a diatomic molecule, but now a nitrogen molecule for which the atoms contain electrons in p- as well as s-states. The term values for these are obtained from Table 1-1 as $\varepsilon_s = -26.22\ eV$ and $\varepsilon_p = -13.84\ eV$. The distance between the nuclei is taken from experiment as $1.09\ \text{Å}$ and we let it lie along the z-axis in our coordinate system. Because of the cylindrical symmetry around this axis, the molecular orbitals can be taken to have angular momentum around that axis of 0, $\hbar$, $2\hbar$, ..., which we designate by σ, π, δ, in analogy with the numbering of atomic states as s, p, d,... A molecular orbital with a particular angular momentum around the axis will be made up of only atomic orbitals of the same angular momentum around that axis, and only atomic orbitals of the same moment around that axis are coupled. (Otherwise the integral which defines the matrix element will have a net factor in the integrand of $e^{im\phi}$ and the integral will vanish.)

We have already written the coupling between two s-states, which have zero angular momentum around *all* axes, as $V_{ss\sigma}$ using this notation. A p-state which is of the form z/r also has zero angular momentum around this z-axis. The coupling between two p-states, oriented in this same direction, as in Fig. 1-6, is called $V_{pp\sigma}$. In addition, the coupling between an s-state and such a p-state is written $V_{sp\sigma}$ as shown. p-states which have angular momentum of $\pm\hbar$ around this axis are of the form $(x/r \pm iy/r)/\sqrt{2}$ and two such "+" states, or two such "-" states have coupling called $V_{pp\pi}$. So also do the sums and differences of these, x/r and y/r, as indicated in Fig. 1-6. Other combinations have no coupling; that is, vanishing matrix elements.

It is easy to understand the signs of the various matrix elements, as it was for $V_{ss\sigma}$. In the defining integrals, such as Eq. (1-17), $(v_1 + v_2)/2$ is negative and we select the sign of ψ according to the diagrams in Fig. 1-6. Thus for $V_{ss\sigma}$ with both wavefunctions positive, we found $V_{ss\sigma} < 0$. Similarly for our definition of $V_{pp\sigma}$, with p-states oriented as in Fig. 1-6, the lobe of one wavefunction is negative where it overlaps the other so the corresponding $V_{pp\sigma}$ is positive. $V_{sp\sigma}$ as defined in Fig. 1-6 is also seen to be positive. Note that if the s-state were moved to the right of the p-state the coupling would be negative, and we would write it $-V_{sp\sigma}$. The final coupling, $V_{pp\pi}$, will be negative, and a matrix element for a π-

oriented p-state (e. g., p_y) and an s-state, or σ-oriented (p_z) state will vanish.

For p-states oriented other than purely σ or π we may decompose the state into a σ-oriented component and a π-oriented component and obtain the matrix elements as combinations of σ- and π-matrix elements, as illustrated at the bottom in Fig. 1-6.

We shall obtain expressions for each of these matrix elements in Chapter 5 when we treat the energy bands of semiconductors. The method is essentially the same as we used to obtain $V_{ss\sigma} = -(\pi^2/8)\,\hbar^2/(md^2)$ in the preceding section, but we shall use the bands of silicon and germanium, because of their importance, to obtain the coefficients. The values we obtain there are

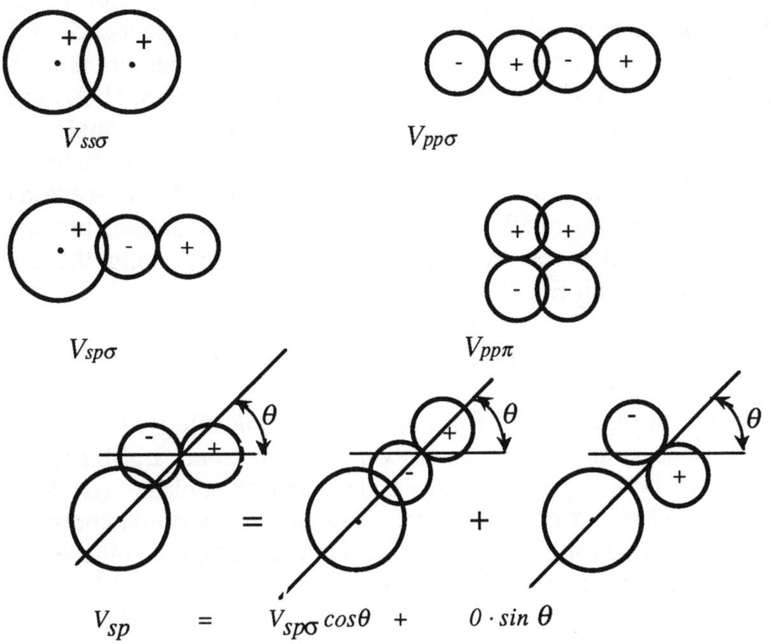

Fig. 1-6. The four types of interatomic matrix elements between s- and p-states. When p-states are not oriented exactly as in the upper four diagrams they can be decomposed as vectors, as illustrated below. The final diagram, by symmetry, must be zero.

$$V_{ss\sigma} = -1.32\ \hbar^2/md^2 \qquad V_{pp\sigma} = 2.22\ \hbar^2/md^2$$

$$V_{sp\sigma} = 1.42\ \hbar^2/md^2 \qquad V_{pp\pi} = -0.63\ \hbar^2/md^2 \tag{1-21}$$

Note that the coefficient for $V_{ss\sigma}$ is not far from the value $-\pi^2/8 = -1.23$ which we obtained for the chain of s-states and the free-electron condition. We shall use these *universal* (independent of element) matrix elements throughout the book, and we use them now for the nitrogen molecule.

The same fitting of free-electron bands for three-dimensional simple-cubic structures leads to $V_{ss\sigma} = -(\pi^2/8)\ \hbar^2/md^2$, as did the chain (Froyen and Harrison (1979)). It also gives matrix elements such as $V_{pp\sigma} = (3\pi^2/8)\ \hbar^2/md^2$ and also a relation $\varepsilon_p - \varepsilon_s = \pi^2\ \hbar^2/md^2$. It is remarkable that this gives a direct prediction of the spacing d of the atoms in the solid in terms of the term values of Table 1-1. With the dependence of d as the fourth-root of the coordination (number of nearest neighbors), which we shall derive in Eq. (2-58), this may be combined with Eq. (1-21) to give *all* of the parameters which enter our theory in terms of that single Table 1-1. However, the predictions of d from this formula are very crude, 5.4 Å for simple-cubic lithium in comparison to 3.0 Å for real body-centered-cubic lithium. They are much better for silicon, for example, but not really accurate enough to be useful and we proceed by using the *observed* spacing, along with Eq. (1-21), in our calculations.

B. The Nitrogen Molecule

Again, nitrogen atoms form a diatomic molecule with a spacing d of 1.09 Å. The s- and p-orbitals on two nitrogen atoms are illustrated to the left in Fig. 1-7. Again, we have taken our internuclear axis along the z-direction and the π-oriented state, $|p_x\rangle$, on one atom is coupled only to the $|p_x\rangle$ state on the other. This is exactly analogous to the bonding and antibonding states on the Li_2 molecule,

$$\varepsilon_\pi = \varepsilon_p \pm V_{pp\pi} \tag{1-22}$$

for both the $|p_x\rangle$ and the $|p_y\rangle$ states. Substituting the values from Table 1-1 and from Eq. (1-21), using $d = 1.09$ Å, we obtain the values designated by π and π^* in Table 1-4.

The σ-oriented states are more complicated. They also will be either even or odd with respect to reflection around the midpoint of the molecule, as were the bonding and antibonding states $(|s_1\rangle \pm |s_2\rangle)/\sqrt{2}$, but they will contain both s-states and σ-oriented p-states. [We shall see that including both gives a lower-energy bonding state, and therefore an improved estimate according to the variational argument given after Eq. (1-18).] The even combination will contain the even combination of p-states, $(|p_{z1}\rangle - |p_{z2}\rangle)/\sqrt{2}$, if we define both p_z-states to have their positive lobes upward. This combination has low energy, $\varepsilon_p - V_{pp\sigma}$, indicated in the middle of the energy-level diagram of Fig. 1-7 by the second diagonal line downward from the energy ε_p . It will also contain the even combination of s-states at the energy below ε_s , the energy $\varepsilon_s + V_{ss\sigma}$, also shown. Thus we may proceed, in analogy with Eq. (1-4), to write the state $|MO\rangle$ = $u_3\,(|s_1\rangle + |s_2\rangle)/\sqrt{2} + u_4\,(|p_{z1}\rangle - |p_{z2}\rangle)/\sqrt{2}$, and to write the energy in analogy with Eq. (1-18) using the matrix elements from Eq. (1-21). Minimizing the results with respect to u_3 and u_4 will be exactly analogous to the minimizing the energy in the case of the polar bond for which the two coupled states have different energy, which we discuss in the next section. We shall find in Eq. (1-27) that levels at $\bar{\varepsilon} \pm V_3$, coupled by

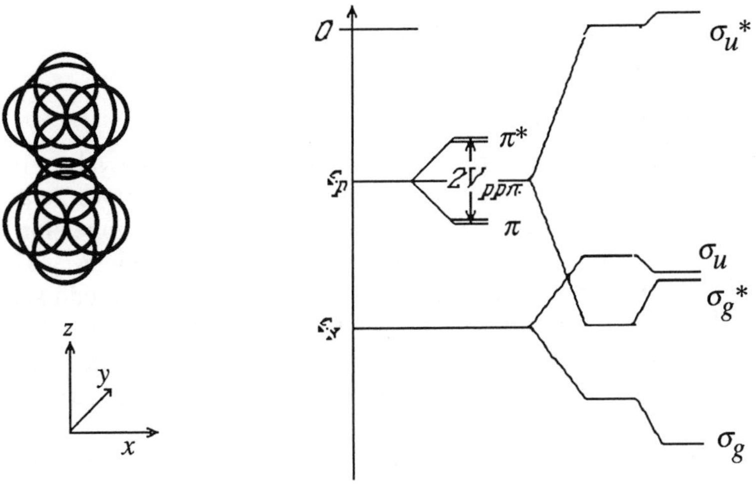

Fig. 1-7. Calculation of the states of N_2. The four p-states which are oriented perpendicular to the nuclear axis form π-bonding and antibonding states at $\varepsilon_p \pm V_{pp\pi}$. Even and odd combinations are formed from the remaining two p-states, and the two s-states, and the bonding and antibonding combinations of the even and of the odd states are formed.

Table 1-4. Energy levels in the N_2 molecule from tight-binding theory (Eqs. (1-22) and (1-23) in g- and u- forms), from the full calculation by Ransil (1960), and as obtained approximately using hybrids.

	Tight-Binding	Full Calculation	With Hybrids
σ_u*	4.2	33.3	0.4
π*	-9.8	8.2	(-9.8)
π	-17.9	-16.7	(-17.9)
σ_u	-21.5	-15.1	-20.0
σ_g*	-21.7	-20.3	-20.0
σ_g	-41.1	-38.6	-40.5

V_2 , give eigenvalues $\bar{\varepsilon} \pm \sqrt{V_2^2 + V_3^2}$. Applying that result to this case we obtain

$$\varepsilon_{(even)} = \frac{\varepsilon_s + V_{ss\sigma} + \varepsilon_p - V_{pp\sigma}}{2} \pm \sqrt{\left(\frac{\varepsilon_s + V_{ss\sigma} - \varepsilon_p + V_{pp\sigma}}{2}\right)^2 + V_{sp\sigma}^2}. \tag{1-23}$$

Substituting the parameters from Section 1-4-A leads to the values listed in Table 1-4. They are indicated as σ_g and σ_g*. The subscript *g* represents the German word *gerade* , meaning *even* with respect to inversion of the wavefunction through the midpoint between the two nuclei. The star indicates the antibonding state, the plus in Eq. (1-23). The odd states are of just the same form, but with the sign in front of each $V_{ss\sigma}$ and $V_{pp\sigma}$ changed. Their values are listed as σ_u and σ_u* in Table 1-4. The *u* represents *ungerade* meaning *odd* . All of these levels are illustrated to the right in Fig. 1-7.

There are a total of 10 valence electrons in the molecule; nitrogen is in column V in the periodic table so there are five electrons from each. In the lowest-energy state of the molecule they fill, with one electron of spin-up and one of spin-down, the lowest-lying levels, σ_g, σ_u, σ_{g*}, and both ($|p_x\rangle$ and $|p_y\rangle$) bonding π-states. We can easily estimate the binding energy of N_2 by subtracting the sum of the energies of the occupied states in the molecule (-240.2 eV) from that for the free atoms (-187.9 eV). Subtracting half the difference to account for the repulsive interaction, as discussed for Li_2, gives 26.2 eV. This is more than twice the experimental cohesive energy of 9.8 eV, as was our estimate for Li_2 and other carbon-row systems. Systems from the lower rows are in better agreement.

Of more interest is a comparison of the entire set of values with those from a detailed quantum-mechanical calculation. Such values are listed in the second column of Table 1-4. In spite of the fact that we have reduced these to trivial calculations, the solution of quadratic equations, with universal coupling parameters coming from semiconductor bands, the tight-binding results for the occupied states are in rather good accord with the more complete theory requiring extensive computer calculation for the states which are occupied. The same formulae apply directly to the other diatomic molecules, e. g., Be_2 C_2, O_2, F_2, and those from higher rows; it is just the spacings d and the term values which are different, and the total number of electrons present is different for each. In Problem 1-2, such calculations are carried out for O_2 in particular.

The level of agreement even for diatomic molecules, which is not as good as for solids, is nevertheless sufficient that it will often be informative - and certainly very easy - to make quick estimates for any system of interest. These not only give the right general magnitudes for energy levels and cohesions, but display the correct trends among most systems. These are the strengths of the methods being given in this book.

The empty states are not well given for N_2 , as seen in Table 1-4, and that is often the case also for solids. However, they do not directly enter the ground-state properties and will be of less concern to us. The difficulty with these states comes mostly from our neglect of excited atomic states (3s- 3p-states, etc.) which are in the same energy range as these antibonding molecular orbitals, but also partly from the effects of nonorthogonality, (Eq. (1-12)).

C. sp-Hybrids

We now make a further simplification of the calculations which will be very important in our studies of solids, though it is of little consequence for N_2 . Application to N_2 will give us an idea of its validity, as the comparison of the energy levels with those of full theory gave us an idea of the validity of tight-binding theory itself. We introduce the familiar concept of *hybrid states* . These are linear combinations of the states on a

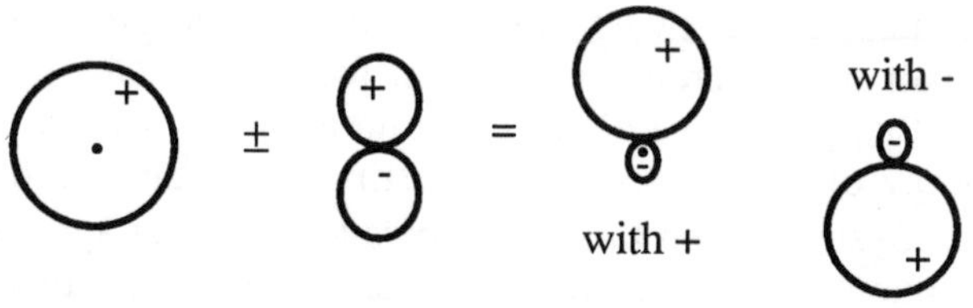

Fig. 1-8. An s-state and a p-state on one atom are combined as in Eq. (1-24) to obtain one (with the plus) which leans in the positive z-direction and one (with a minus) which leans in the negative z-direction.

single atom, here a combination of an s-state and a p-state, illustrated in Fig. 1-8 and given algebraically by

$$|h\rangle = (|s\rangle \pm |p_z\rangle)/\sqrt{2}\,. \qquad (1\text{-}24)$$

Note that it is normalized, and hybrids are always chosen to be orthogonal [$(\langle s| + \langle p_z|)(|s\rangle - |p_z\rangle)/2 = (1 - 1)/2 = 0$] ,but is a combination of states of different energy so it does not satisfy our condition for an eigenstate; it does not satisfy the variational condition (discussed after Eq. (1-18)) with respect to energy. However, expanding our molecular states in terms of these two orthogonal hybrids is entirely equivalent to expanding it in terms of the $|s\rangle$ and $|p_z\rangle$ themselves. We could have constructed our four σ-states by expanding in the two hybrids on each atom and we would have obtained precisely the energies which were given in Eq. (1-23) and Table 1-4. The purpose of the hybrid is to allow us to make a simplifying approximation.

Note that for the hybrid obtained from Eq. (1-24) with the plus sign, the p-state and the s-state wavefunctions add above where both are positive but cancel below, where the p-state is negative. This is illustrated in Fig. 1-8 where a new symbol is drawn for the hybrid; the lobe for the hybrid is drawn large on the side where the two states add and small on the side where the two cancel. We could say that the hybrid with the plus "leans" upward and that with the minus leans downward. A more careful plot of such a hybrid will be made at the end of this section, where it will be seen that the center-of-gravity of the plot lies at a point well into the large circle in the hybrids shown in Fig. 1-8.

That would suggest that the hybrid on each atom which leans toward its neighbor is much more strongly coupled with the neighboring states than that which leans away and would suggest the approximation of including only the coupling between two inward leaning hybrids, and neglecting all other couplings; this is the *Bond-Orbital Approximation.* We shall discuss the full meaning of such an approximation for tetrahedral semiconductors in Section 2-3-A. We may obtain the coupling between the two hybrids by adding up all of the terms: it includes $(1/\sqrt{2})^2$ times $V_{ss\sigma}$ and, noting that the p-states in each case are oriented in the opposite direction from that indicated in Fig. 1-6 for the definition of $V_{sp\sigma}$ and $V_{pp\sigma}$, we have $-(1/\sqrt{2})^2$ times $V_{pp\sigma}$ and $-2(1/\sqrt{2})^2$ times $V_{sp\sigma}$. In fact this makes every term negative and our intuitive guess was correct that the coupling is very strong for these inward-pointed hybrids. The magnitude of the coupling between hybrids is called the *covalent energy* and is given by

$$V_2\ (sp) = \frac{-V_{ss\sigma} + 2V_{sp\sigma} + V_{pp\sigma}}{2} = 3.19\ \frac{\hbar^2}{md^2}\ , \qquad (1\text{-}25)$$

where in the last step we have substituted from Eq. (1-21). We have designated this coupling by "sp" since in solids we will not use equal mixtures of s- and p-states and with a different mixture, a value of the coefficient different from 3.19 will be found.

The coupling between a hybrid pointing outward and one on the other atom pointing inward is readily evaluated (Problem 1-3) and is indeed much smaller, as is that between two outward pointing hybrids. Thus this

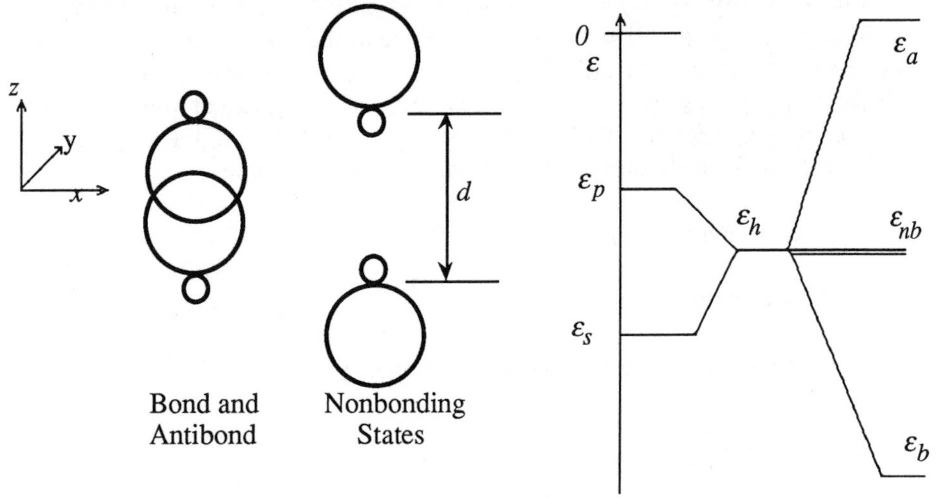

Fig. 1-9. An approximate calculation of the σ-states in N_2. We form hybrid combinations $(|s\rangle \pm |p_z\rangle)/\sqrt{2}$ of energy $\varepsilon_h = (\varepsilon_s + \varepsilon_p)/2$. The first two point inward and form a bonding state at energy ε_b and an antibonding state at energy ε_a. The coupling involving the outward-pointing states is neglected and they are called nonbonding, with energy $\varepsilon_{nb} = \varepsilon_h$.

Bond-Orbital Approximation of including only the covalent energy of Eq. (1-25) seems very promising. We include no coupling involving the two outward-pointing hybrids so they remain as two *nonbonding* states at the energy $\varepsilon_h = (\varepsilon_s + \varepsilon_p)/2$ and the two inward-pointing hybrids form an antibonding and a bonding state at $\varepsilon_h \pm V_2(sp)$, as illustrated in Fig. 1-9. The corresponding values are also compared with the full tight-binding calculation in Table 1-4. They are indeed very close, particularly on the scale of the agreement with the full calculation so that this Bond-Orbital Approximation, which will prove so useful in semiconductors, seems very well justified.

This simplification in the case of the simple N_2 molecule is of little importance, but the use of hybrids becomes more important in the case of a polar molecule, such as carbon monoxide, for which the states on the two atoms have different energies. Then the states are no longer even or odd with respect to reflection and we cannot reduce the problem to the solution of a quadratic equation which was given by Eq. (1-23). The minimization of the energy with respect to the four coefficients u_i for the two s-states and the two σ-oriented p-states yields four simultaneous equations, and the solution of a fourth-order algebraic equation. However, if we form inward and outward pointing hybrids as for nitrogen, that on the carbon will have energy $\varepsilon_h(C) = (\varepsilon_s + \varepsilon_p)/2 = -15.22\ eV$ and that on the oxygen will have energy $\varepsilon_h(O) = -25.39\ eV$. We take these as the energies of the nonbonding (outward-pointing) states and construct bonding and antibonding states as linear combinations of the inward point states, $u_1|h(C)\rangle + u_2|h(O)\rangle$. The energy becomes $\varepsilon = (u_1^2\varepsilon_h(C) + 2V_2u_1u_2 + u_2^2\varepsilon_h(O))/(u_1^2 + u_2^2)$ and minimization conditions become

$$\varepsilon_h(C)u_1 + V_2u_2 = \varepsilon u_1 ,$$
$$V_2u_1 + \varepsilon_h(O)u_2 = \varepsilon u_2 . \quad (1\text{-}26)$$

We may solve these together to obtain the energy

$$\varepsilon = \frac{\varepsilon_h(C) + \varepsilon_h(O)}{2} \pm \sqrt{V_2^2 + V_3^2} , \quad (1\text{-}27)$$

where we have defined a *polar energy*,

$$V_3 = \frac{\varepsilon_h(C) - \varepsilon_h(O)}{2} \quad (1\text{-}28)$$

and V_2 is the same covalent energy as that which entered N_2. The π-states are similarly obtained from solution of a quadratic equation, with the hybrid energies replaced by p-state energies and V_2 replaced by $V_{pp\pi}$. We have the same number of electrons as in N_2 and can similarly estimate the binding energy of CO. We have solved a fourth-order problem with an approximate second-order solution.

We are going to make sufficient use of hybrids, that it may be desirable to see more closely what they look like. For Fig. 1-10 we have constructed

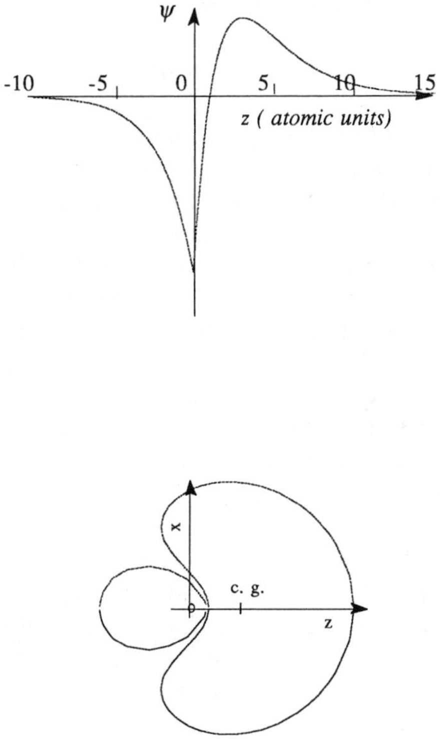

Fig. 1-10. Above is a plot of the wavefunction for an sp-hybrid based upon hydrogen 2s- and 2p-states along the axis of the p-state. Below is a constant-probability-density curve for the same hybrid in a plane containing that axis. The center of gravity of the hybrid state is indicated on the z-axis, three Bohr radii from the nucleus.

sp-hybrids based upon the 2s- and 2p-states of hydrogen. The plot above gives the wavefunction along the axis of the hybrid, with the origin at the nuclear position. Below are shown positive and negative constant-wavefunction contours, or constant probability-density contours. The nuclear position is shown with a small circle. The "c. g." indicates the center of gravity of the entire three-dimensional charge density, which is indeed displaced from the nuclear position. These are what are represented schematically by the figures to the right in Fig. 1-8. If Fig. 1-10 were recalculated for a higher-density contour, the two closed contours would be smaller and they would look more similar to Fig. 1-8, but the two segments would be separated more from each other.

D. Semiconductors

The hybrids provide only a minor simplification even for these polar diatomic molecules, but for a semiconductor they are the essence of understanding the electronic structure and estimating the properties. We shall see that in the crystal structure of the semiconductors, such as silicon and gallium arsenide, each atom has four neighbors. Then we may make four orthogonal sp^3 hybrids (so-called because their probability density is one quarter s-like and three quarters p-like if they are chosen to be orthogonal), oriented in the direction of the four neighbors. If we include only the very large coupling between two hybrids directed into the same bond, we may form independent bonds between each set of neighboring atoms. The energy is obtained just as it was for nitrogen. In the case of a polar bond, as in gallium arsenide, the hybrids formed on the gallium have different energy from those formed on the arsenic but it is a simple task to obtain the bonding states and the energies in order to estimate cohesion, energies required to distort the lattice, and by applying fields we may estimate also the dielectric properties. We shall find that, particularly for the cohesion, the neglected couplings can be important, but they may be included afterward as small corrections to the independent-bond picture.

An important feature of this approach is that formulae are obtained for the various properties which clearly display the trends from material to material. It is not just that we have avoided a major computational task, but also that we obtain a clear intuitive understanding of the results. The next several chapters carry out this analysis of the tetrahedral semiconductors. The hybrids which we introduce for the tetrahedral semiconductors will enable us to treat the bonding, elastic, and dielectric properties of the semiconductors in essentially the same way we treated the corresponding properties of diatomic molecules. This is done in Chapters 2, 3, and 4. We shall also give considerable attention to what was left out, or what refinements are needed for meaningful descriptions. We should not forget, however, that these bonds and antibonds which we construct are not real energy eigenstates. They are coupled one to another, one such coupling being the negative of the metallic energy $-V_1 = (\varepsilon_s - \varepsilon_p)/4$ which we shall introduce. We shall explore the resulting energy bands in Chapter 5, and the electronic properties which are understandable in terms of them in Chapters 6 and 7. In Chapter 8 we discuss impurities and defects in semiconductors in terms of the same outlook on the electronic structure. Throughout the analysis we included only s- and p-states in the electronic structure. The small role which d-states play is postponed to Section 17-5 in the discussion of transition-metal compounds.

E. Overall Organization of Solid-State Systems

We have discussed briefly the electronic structure of metals and semiconductors. It may be helpful before proceeding to see how these, as well as insulators, fit in the scheme of the periodic table of the elements.

In Figure 1-11 we show the periodic table, rearranged as the *Solid-State Table* (Pantelides and Harrison (1975)). This again reads like a book in order of increasing atomic number, as in the usual arrangement, but

	I	II	III	IV	V	VI	VII	VIII	IX	X	
							H	**He**	Li		
		Be	B	**C**	N	O	F	**Ne**	Na		
		Mg	Al	**Si**	P	S	Cl	**Ar**	K		
Trans-ition Metals	Cu	Zn	Ga	**Ge**	As	Se	Br	**Kr**	Rb	Ca	
	Ag	Cd	In	**Sn**	Sb	Te	I	**Xe**	Cs	Ba	f-Shell Metals
	Au	Hg	Tl	**Pb**	Bi	Po	At	**Rn**	Fr	Ra	

Type B Metals   Nonmetals   TypeA Metals

Fig. 1-11 The Solid-State Table with the nonmetals at the center, between the tetrahedral semiconductors in column IV and the inert gases in column VIII.

now has the metals which are sometimes referred to as of type A moved to the right of the inert gases, and with the transition metals placed to the left and the f-shell metals placed to the right, rather than written as an insert below the table. Column IV in the center contains the elemental semiconductors. Because they have four electrons per atom they can place one electron in each of four two-center bonds with four nearest neighbors, as just described. This is the reason they form in a tetrahedral structure and our understanding of the electronic structure is based upon the nearly-independent, two-center bonds. To the right of this, in Column VIII, are the inert gas atoms. Their eight valence electrons are just enough to fill the valence s- and p-states. Their occupied states are strongly bound and their empty states are high in energy so that they tend to be chemically inert. Inert-gas atoms are only weakly bound to each other in their condensed state since the mechanisms which we find to bind metallic, covalent, and ionic solids do not apply. We no not return to the source of this other bonding, the van-der-Waals interaction, until Chapter 20. All of the other elemental nonmetals lie between these two columns, IV and VIII.

To the right of Column VIII lie the alkali metals and alkaline earths, the simple metals of type A. To the left of the Column IV elements lie the simple metals of type B. Both types have free-electron-like bands and our understanding of them is based upon such a free-electron basis, with pseudopotentials which we shall introduce in Section 1-6 to represent the effects of the lattice atoms.

When a nonmetal (Column V through VII) combines, in a one-to-one ratio, with a simple metal to the left, a tetrahedral semiconducting compound is formed. There are again just enough electrons to doubly occupy all of the bonds, but the bonds have now become polar, describable by the analog of Eq. (1-27), and have the charge shifted towards the nonmetallic atom. It is a slight generalization of the simple two-center σ-bonds we have described for N_2 and the elemental semiconductors.

An isoelectronic series such as Ge, GaAs, ZnSe, and CuBr has two-center bonds successively increasing in their polarity. Sometimes a useful way of thinking about the electronic structure of such a series is to start with the germanium, and our simple description in terms of two-electron bonds. We then freeze the electron cloud and remove a proton from one germanium nucleus, making it a gallium nucleus, and insert it in a neighboring nucleus, making it an arsenic nucleus. Then the electron cloud is allowed to relax toward the arsenic nucleus because of its extra proton. This "theoretical alchemy", converting one element to another, has produced gallium arsenide without any fundamental change in the electronic structure. The transfer of an additional proton changes the first nucleus to a zinc nucleus and the second to a selenium nucleus, and transferring a third changes the first to a copper nucleus and the second to a bromine nucleus. The theoretical alchemy has allowed us to keep track of the nuclear charges as we change the system. This will be of particular interest in Section 19-3-C when we discuss polar surfaces. It will allow us to see if any charge accumulation arises at a polar surface, assuming that the bonds deform but no electrons are transferred in or out of the bonding levels when the interface is formed.

If instead of combining a nonmetal with a simple metal to the left, we combine a nonmetal with a simple metal on the right, it forms a structure with more than four neighbors. We shall see in the following section that electrons are transferred from the metal atoms to the nonmetal atoms to form closed shells, just as in the inert-gas atoms which they surround in the Table. In fact, we may understand the electronic structure again using theoretical alchemy, as we just did for polar semiconductors, but now starting with inert-gas atoms. For example we may make up a crystal of argon atoms. We freeze the electron clouds on each atom and remove a proton from one nucleus, making a chlorine nucleus. The proton is then inserted in the nucleus of a neighboring atom to make it a potassium nucleus. The cloud of electrons on each ion thus formed is then allowed to relax, shrinking around the potassium to become a real potassium ion and expanding around the chlorine to become a real chlorine ion. Thus theoretical alchemy has allowed us to construct the ionic crystal without any transfer of electrons among different states, only deformation of the states. In this case the ion which receives the proton is the metal, whereas for semiconductors the atom which gave up the proton was the metal. Transferring a second proton produces CaS, and a third would produce ScP, though such a compound seems not to exist. ScN, from the same columns does form.

The ions formed in this process are electronically inert, but are charged, that is *ions* rather than *atoms* . We may associate their more tightly-packed structures with the tendency to bring more oppositely-charged neighboring ions closer to them; we shall see that this lowers the electrostatic energy, though we prefer to understand the cohesion in terms of the electron transfers.

It is elements with the largest difference in term values which form in these ionic structures, but it is the structure which is the feature determining how we should understand them. Magnesium sulfide apparently forms in both a tetrahedral structure and in an ionic structure and we should base our understanding upon the structure it forms, not on the basis of its atomic term values. Similarly, silicon can form in metallic structures, and is a simple metal when melted; in this case it should be understood as we understand magnesium and aluminum.

We have placed the nonmetals at the center of the Solid-State Table in Fig. 1-11, where they seem to belong, and have moved the special cases: d-shell and f-shell metals to the edges of the Table and we shall discuss them in the Section 1-7, explicitly adding the effects of these additional orbitals of higher angular momentum. Admittedly, if one is describing the electronic structure of atoms, the more traditional arrangement is preferable, with Column VIII forming the right border. It is with them that we complete the filling of a shell, make a typewriter carriage return and begin with a new set. For organizing solids, and for suggesting their electronic structure, the arrangement of Fig. 1-11 is best. At the back of the book we have repeated this Solid-State Table with the parameters for each element which are most essential for an understanding of the solids of which they are a part.

5. Ionic Insulators

Compounds such as rock-salt form in structures with more than four neighbors, six in the case of rock-salt. It is then not possible to make independent hybrids directed at each neighbor from the four valence states on each atom. The tight-binding states, and the tight-binding parameters, remain appropriate, but the simplification provided by the independent bonds is not applicable. Fortunately in the compounds which form in these structures with many neighbors, the atomic energies on the metallic component (sodium) are very much higher than those on the nonmetallic component (chlorine). Then we may understand the electronic structure as a transfer of electrons from the metallic component (the sodium 3s-electron) to the states on the nonmetallic component (filling the last 3p-chlorine state) and the corrections due to the relatively small coupling can be treated as a correction.

A. Independent Ions

If we neglect this coupling altogether we would say that the band gap for NaCl would be $\varepsilon_s(Na) - \varepsilon_p(Cl) = 8.83\ eV$, in good accord with the experimental gap of 8.5 eV. This is a very large separation in energy between the highest-energy occupied state and the lowest-energy empty state. The probability of any electron being in one of the upper (sodium-like) states will be of order $e^{-E_g/(2k_BT)} \approx 10^{-74}$ in thermal equilibrium. There is a negligible chance of finding a single electron in this band in a cubic *meter* of rock-salt. Since the states in the lower (chlorine-like) band are all occupied, for any current carried by a state there is an electron going in the opposite direction to cancel its effect. This crystal, and other ionic crystals with gaps greater than a few volts are electrical *insulators*.

We would similarly say that the cohesive energy would be the energy gained in the transfer, this same 8.83 eV per atom pair, in reasonable agreement with the experimental 8.04 eV cohesive energy. These are very large cohesive energies, and the reason why atoms of column I combine to form compounds with atoms of column VII. The same driving force which makes the cohesion large also makes them insulators. Given enough options nature is driven to form insulating compounds.

Probably the most important insulators, silicon dioxide or glass, have electronic structure intermediate between that of the semiconductors and ionic crystals, but the concepts used for semiconductors are more useful. Their discussion is given in Chapter 12, after the discussion of ionic solids.

Concerning ionic crystals, we might worry that we have neglected the effect of the Coulomb repulsion between the extra electrons on the chlorine, but we shall see in Chapter 9 that this is largely canceled by an interionic Coulomb attraction.

B. Couplings and Perturbation Theory

We have also neglected the effect of the coupling $V_{sp\sigma}$ in the calculation of the cohesive energy, and correspondingly the effect of the overlap repulsion which would cancel part of the resulting contribution to the cohesion. The small effect of this couplings will be included in Chapters 9 through 11 when we discuss these ionic crystals. This is done in Chapter 9 using the familiar second-order perturbation theory of quantum mechanics. It is such an important technique that we should pause a moment to introduce it.

We may do this by considering the solution, Eq. (1-27) , for a carbon and an oxygen hybrid separated in energy by $2V_3$ and coupled by a small matrix element V_2 . Then we may expand the square root in Eq. (1-27) to obtain the approximate result for the lower state as

$$\varepsilon \approx \varepsilon_h(O) - \frac{V_2{}^2}{2V_3} \; . \tag{1-29}$$

The energy is lowered due to the interaction by the coupling squared, divided by the energy difference between the coupled states. Similarly, the upper state is raised above the $\varepsilon_h(C)$ by the same amount. If a state is coupled to a number of other states the shifts due to coupling with each may be added. Thus in general, to second order in the coupling V_{ij} between states $|i\rangle$ and $|j\rangle$, which in the absence of this coupling have energies ε_i^0 and ε_j^0 , the energy of the i'th state is given by *second-order perturbation theory* as

$$\varepsilon_i = \varepsilon_i^0 + \sum_j \frac{V_{ij}V_{ji}}{\varepsilon_i^0 - \varepsilon_j^0} \quad . \tag{1-30}$$

We will not need it so often, but it is also easy to show that the state itself, to first order in V_{ij} , is given by $|i\rangle + \Sigma'_j V_{ij}|j\rangle / (\varepsilon_i^0 - \varepsilon_j^0)$.

It has not been necessary thus far to talk about how the wavefunction changes with time, but it is sometimes needed and these changes also can be treated in perturbation theory. We gave the time dependence of the wavefunction in Eq. (1-8), the Schroedinger Equation,

$$-\frac{\hbar}{i}\frac{\partial\psi(\mathbf{r},t)}{\partial t} = -\frac{\hbar^2\nabla^2}{2m}\psi(\mathbf{r},t) + V(\mathbf{r})\,\psi(\mathbf{r},t) \quad . \tag{1-31}$$

We went on to seek approximate energy eigenstates of the Hamiltonian, which vary in time as $e^{-i\varepsilon_j t/\hbar}$ according to the Schroedinger Equation, with ε_j the energy eigenvalue. The time dependence is then not important for our purposes. However, frequently we will add an additional term to the Hamiltonian, such as the influence of the electric field of a light wave, and the effect can be to cause transitions of an electron from one eigenstate to another. Such transitions only occur if there are a continuous, or closely-spaced, set of states into which the electron may transfer. (Otherwise the electron simply oscillates back and forth between the two states.) The rate of such transitions is given by the *Golden Rule* which is derived in every quantum-mechanics text. The electron state is initially written as one of the eigenstates $|i\rangle$ of the original Hamiltonian. When the perturbing term, H_1 , which might be from a light wave, is introduced the state is written as a linear combination of the eigenstates of the original Hamiltonian, $\Sigma_j u_j(t) |j\rangle$, which is substituted in the Schroedinger Equation. If H_1 is independent of time, it is found that the coefficients of the states $|j\rangle$ which have energy very close to that of the initial state $|i\rangle$ grow with time such that the total probability of occupation of one of these final states grows at a rate, called the transition rate,

$$P_{ij} = \frac{2\pi}{\hbar}\,|\langle j|H_1|i\rangle|^2\,\delta(E_j - E_i)\,. \tag{1-32}$$

The meaning of the delta function is that one is to sum over these closely spaced final states which are coupled by similar matrix elements

$<j/H_1/i>$ and write the sum as an integral over energy, in the form we shall obtain in Section 6-6. Our principal aim in this text will be to obtain matrix elements to be used in such calculations, but we will actually carry out evaluations for absorption of light in Section 6-6, for scattering due to electron-phonon interactions in Section 7-5-H and for scattering due to impurities semiconductors in Section 8-1-D and in metals in Section 13-8. In each case it is the $<j/H_1/i>$ which cause the transitions.

If the perturbation varies sinusoidally with time, so that there are matrix elements $<j/H_1/i>\, e^{i\omega t}$ and $<j/H_1/i>\, e^{-i\omega t}$, Eq. (1-32) applies for one set of final states with $\delta(E_j - E_i)$ replaced by $\delta(E_j - E_i + \hbar\omega)$ (from the matrix element with $e^{i\omega t}$) and for another set of states with $\delta(E_j - E_i)$ replaced by $^1/_4\delta(E_j - E_i - \hbar\omega)$ (from $e^{-i\omega t}$). These correspond to the electron emitting, or absorbing, respectively, one quantum of energy.

6. Simple Metals and Pseudopotentials

We noted in Section 1-3 that the electrons in a solid, or in the lithium chain discussed there, are quite free-electron-like. At first this is very surprising in view of the strong potentials arising from the atomic nucleus and core electrons. Indeed this potential is strong enough to produce all of the core states lying below the free-electron valence states. For the same reason, the periodic table of the elements itself is surprising. Adding an additional eight protons to the nucleus and eight bound states of one sp-shell brings us back to very similar energies. One way of understanding both surprises is the idea of a pseudopotential, apparently introduced first by Enrico Fermi to treat the interaction between protons and neutrons. (An early, but rather comprehensive, exposition of its use in solids is given in *Pseudopotentials in the Theory of Metals* (Harrison, 1966).)

A. Empty-Core Pseudopotentials

One of the simplest views of the pseudopotential, and the one we shall use, is illustrated in Fig. 1-11. The atomic potential $v(r)$ is replaced by a pseudopotential $w(r) = 0$ for r less than a *pseudopotential core radius* r_c and equal to the real potential (usually approximated by $-Ze^2/r$ for an element in the column Z of the periodic table) for $r > r_c$. This is called an *empty-core pseudopotential* and was first introduced by Ashcroft and Langreth (1967). The radius r_c is chosen such that the potential gives its lowest-energy s-state, called a *pseudowavefunction* at the real energy ε_s of the valence s-state for that atom. This pseudowavefunction, $\phi(r)$ in Fig. 1-11, has no node (no zero in the function of r), though the real atomic

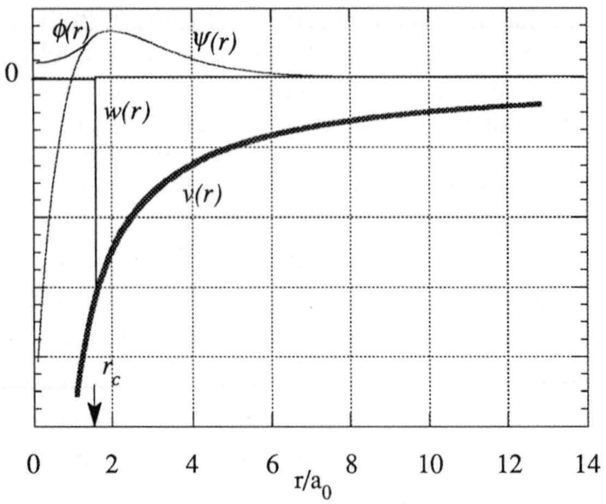

Fig. 1-12. The heavy line $v(r)$ shows the potential for a lithium atom and the thin line $\psi(r)$ shows the corresponding valence 2s-state with energy ε_s. If the potential is replaced by the empty-core pseudopotential $w(r)$, the lowest-energy s-state becomes $\phi(r)$ again with energy ε_s. $\phi(r)$ is the pseudowavefunction representing the valence state. Radii are shown in units of the Bohr radius a_0. For lithium the core radius is $1.73\ a_0$.

wavefunction must have, as seen, or it cannot be orthogonal to the core 1s-state. Since both the pseudowavefunction and the true wavefunction could be obtained by integrating the eigenvalue equation at the same energy from large r, the two functions must be the same outside the core radius, as shown, except possibly for a normalization factor.

Fig. 1-12 shows the pseudopotential for lithium, but the same construction can be made for heavier atoms where there are more core states and more nodes in the valence wavefunctions. There is also a 2p-state solution for the same atomic pseudopotential, higher in energy than the s-state shown, which turns out to be rather close to the true ε_p for the atom, approximately one half ε_s as we have seen. Thus this pseudopotential can be used as the potential determining all valence states for the atom and in fact for a solid made up of these atoms.

Pseudopotential core radii for the elements will be given in Chapter 13. They are typically equal to a quarter of the internuclear distance, or half the distance to the midpoint between atoms. Thus the empty-core region, $r < r_c$, eliminates all but the tails of the atomic potential and a superposition of pseudopotentials on the individual atoms leaves a small net solid-state pseudopotential. To a first approximation this potential is flat and the electrons perfectly free, which is why we could take the bands of the lithium chain as free-electron-like, providing us with values for the tight-binding parameters. We add the effects of the small pseudopotential as a perturbation in the theory of the simple metals.

Because the pseudopotential core radii are similar from row to row in the periodic table, the theory of each row becomes very much the same and the periodicity of the properties follows immediately. A special feature of this similarity of pseudopotentials is that, for example, a sodium atom substituted in a lithium crystal really makes a very major change in the wavefunction ψ in introducing an additional node near the sodium, but because the *pseudopotential* is so similar, we may expect the small difference to produce only weak scattering which we can treat in perturbation theory as discussed in the preceding section.

B. The Effective Interaction Between Ions

One other aspect of this theory which we should note here is the interatomic interaction which occurs through the pseudopotential. We saw in Section 1-3 that the interatomic couplings $V_{ss\sigma}$ led to a lowering of the energy of the metal proportional to $1/d^2$, or $1/\Omega^{2/3}$ in terms of the total volume Ω. Part of this was canceled by the electron kinetic energy, also proportional to $1/d^2$. There was in addition an overlap repulsion which we took to vary as $1/d^4$. This repulsion gives a change in energy as the atoms in the metal move with respect to each other even at constant total volume. We shall see in Chapter 14, by treating the pseudopotential in second-order perturbation theory, that there is also a change in energy in this context due to atomic motion under fixed total volume that corresponds to a purely repulsive interaction between atoms a distance d apart of the form

$$V_0(d) = \frac{Z^2 e^2 \cosh^2 \kappa r_c \, e^{-\kappa d}}{d} , \tag{1-33}$$

with r_c again the pseudopotential core radius. We shall see that the screening parameter κ is defined by $\kappa^2 = 4e^2 mk_F/(\pi\hbar^2)$ for a three-dimensional free-electron gas. Both equations appear in the Solid-State Table at the back of this book. We might ask whether this repulsion can be identified with the $1/d^4$ overlap interaction which we saw in Section 1-2 arose from the nonorthogonality of the valence states on neighboring atoms, and which is required in tight-binding theory to prevent the molecule or lattice from collapsing.

The comparison would suggest so. We do this for diamond, silicon, germanium, and tin since we need a three-dimensional system and we shall use the overlap repulsion most in the semiconductors. We may first see if the dependence upon d near the equilibrium spacing is appropriate, $d\partial V_0/\partial d = -4V_0$, corresponding to a C/d^4 dependence, as we suggested in Sections 1-2 and 1-3. k_F is related the valence electron density and for the tetrahedral semiconductor is given by $(k_F d)^3 = 9\sqrt{3}\pi^2/2$. From Eq. (1-33) we have $d\partial V_0/\partial d = -(1+\kappa d)V_0$ and evaluating kd leads to $1 + \kappa d = 4.97, 5.90. 6.00,$ *and* 6.35 for C, Si, Ge, and Sn, respectively. These correspond to a form near equilibrium of

$$V_0(d) \approx \frac{C}{d^n} \tag{1-34}$$

with n equal to the $1 + \kappa d$ just given. This only crudely accords with C/d^4 which we have assumed, and this is the first of a number of indications that higher values of the exponent may be appropriate, increasing as we move down in the periodic table. Two others will be in the discussions of the bulk modulus and of structural stability in Chapter 2. For uniformity and simplicity, we retain the exponent four in our discussions.

We may discuss the magnitude, as well as the variation, of the overlap interaction from Eq. (1-33) for semiconductors. In Chapter 2 we shall discuss the bonding in semiconductors in terms of two-electron bonds between hybrids, as we did for N_2 in Section 1-4, with a covalent energy as in Eq. (1-25) varying as $1/d^2$. With two electrons in the bond, the tension is $-4V_2/d$ which we can tabulate given the observed spacing d_0 (with the coefficient 3.22 rather than the 3.19 of Eq. (1-25), as we shall see). We can then ask what value of r_c/d_0 gives a repulsion equal to this tension. We find $r_c/d_0 = 0.30, 0.27, 0.27$, and 0.27 for C, Si, Ge, and Sn, respectively. Indeed these are not so far from the values $r_c/d_0 = 0.24, 0.24, 0.22$, and 0.21 obtained from our table of r_c values obtained from pseudopotential theory and given in Chapter 13 and the Solid State Table.

It is somewhat gratifying that the overlap repulsions arising from the free-electron and pseudopotential concepts are in such close accord with the tight-binding picture. It may also be useful to note that the ratio r_c/d_0 is so independent of system, certainly on the scale of the variation of d_0 . We could redefine the constant C in our $V_0(d) = C/d^4$ which we used in Section 1-2 so that

$$V_0(d) = 16\,V_2\,\frac{r_c^2}{d^2}\,, \tag{1-35}$$

where the constant 16 has been chosen such that $-2V_2 + V_0(d)$ will be minimum at the equilibrium space if $r_c/d_0 = 0.25$. There is no real advantage in this form since in most cases we adjust the scale factor to give the observed spacing in any case. However, it is interesting that Eq. (1-35) corresponds in tight-binding terms to a nonorthogonality $\langle h|h'\rangle = 8r_c^2/d^2$ giving additional meaning to the pseudopotential core radius. Further, if we are seeking the interaction between two different elements, we may expect Eq. (1-35) will simply have r_c^2 replaced by the product of the core radii for the two atoms.

C. Partly-Filled Bands

We saw in Section 1-3-B that the energy bands could be understood both in terms of tight-binding theory or in terms of free-electrons, supported now by the pseudopotential concept. When the bands are completely full or empty, as in semiconductors, the tight-binding view is much more convenient for obtaining the total energy. We simply sum bond energies, which we shall see represent the center of gravity of the bands. This will also prove to be the best way to treat the dielectric and bonding properties. In metals, by definition, the bands are only partly occupied, knowledge of the center of gravity is not adequate and an integration over free-electron bands is much easier and more appropriate. We in fact saw that already in summing the energies for the lithium chain in Section 1-3-C to obtain the cohesion. Thus the electronic structure of the simple metals is studied in Chapter 13 beginning with the free-electron view and adding the corrections for a nonzero pseudopotential. This is again the appropriate approach for the bonding properties of the simple metals discussed in Chapter 14, though for most of those properties the interatomic interaction, given in Eq. (1-33) and based upon the same pseudopotential, provides the simplest way.

8. Transition-Metal and f-Shell Systems

We noted in Section 1-1 that the atomic d-states are partially occupied in the transition metal atoms; they become valence states and must be included in the description of the electronic structure. These d-states have energy similar to that of the s-states, but have two units of angular momentum. It is easy to see that because of this they are much more strongly localized near the nucleus. A classical argument suffices: we imagine an electron moving in a circular orbit with a particular angular momentum, and with a kinetic energy equal to half the magnitude of the potential at that radius if it is held by a Coulomb potential. (This is from the virial theorem.) If now the energy is kept the same, but the orbital motion is made to be parallel to the radius, corresponding to zero angular momentum, this kinetic energy will carry the electron to a radius twice as large, again for a Coulomb potential. The s-state orbit therefore extends much further from the atom than the d-state orbit. The argument will be illustrated by Fig. 15-1 when we return to transition metals. The same qualitative effect occurs for other forms of potential and the d-states are generally much more confined than s-states of comparable energy. Because of this, they are much more weakly coupled to states on the neighboring atoms than are the s-states and they form much narrower energy bands. It remains appropriate to treat the s-like levels in terms of free electrons and pseudopotentials, but the d-like levels are more appropriately treated in tight-binding terms. However, the d-bands are not full, as in the semiconductors and ionic solids, and the coupling between s- and d-states mixes them up in any case.

A. Coupling of d-States.

In Chapter 15 we see how this can be unscrambled. The d-states are taken not to overlap d-states on neighboring atoms, but a d-state having angular dependence $Y_2^m(\theta,\phi)$, on an atom at $\boldsymbol{r}_j$, is coupled to a free-electron state $|\boldsymbol{k}\rangle$ by a matrix element which is found to be of the form (Eq. (15-11))

$$\langle k|H|d\rangle = \sqrt{\frac{4\pi r_d^3}{3\Omega}}\,\frac{\hbar^2 k^2}{m} Y_2^m(\theta_k,\phi_k) e^{-i\boldsymbol{k}\cdot\boldsymbol{r}_j} . \tag{1-36}$$

Thus it is characterized by a d-state radius r_d appropriate to the element in question. These d-state radii can be obtained from an elementary calculation, called the Atomic Surface Method, from the tail of the free-atom d-state wavefunctions and will be tabulated for all transition metals. We call such matrix elements "hybridization terms" because, as in the hybrids discussed in Section 1-4, they admix states of different angular momentum (taking free states to be basically s- and p-like) on the same site. It seems bad form to use this same term also for interatomic matrix elements, which are qualitatively different.

Then the only coupling between neighboring d-states arises indirectly through the free-electron states. It is evaluated in second-order perturbation theory and found to be of the form

$$V_{ddm} = \eta_{ddm} \frac{\hbar^2 r_d^3}{m d^5} \tag{1-37}$$

with $\eta_{dd\sigma} = -45/\pi$, $\eta_{dd\pi} = 30/\pi$, and $\eta_{dd\delta} = -15/2\pi$. This is remarkably similar to the coupling between s- and p-states, given in Eq. (1-14). However, because the states are so strongly localized, the interaction drops more rapidly with distance as $1/d^5$, and because of this a length cubed must enter to obtain the units of energy. This length r_d is different for each transition metal, and is tabulated in Chapter 15 and the Solid-State Table. We may imagine that when the matrix elements were proportional to $1/d^2$ (because of the free-electron nature, as we have seen), the lengths characterizing the states enter to the zero power, so the matrix elements were independent of which element was involved, as in Eq. (1-21).

B. The Coulomb U_d

A second important feature arises because of the localization of the d-states near the nuclei: the Coulomb interaction between two electrons in a d-state will be quite large, and in fact it will be larger by a quantity called U_d , some five or six eV for the 3d-states, than the interaction if one of the electrons is in an s- or p-state. We noted at the end of Section 1-5 that the Coulomb interaction between p-states in ionic crystals is nearly equal to the Coulomb interactions from neighboring sites so that the shift of electrons between atoms did not greatly affect the atomic term values ε_p. This is no longer true for transition-metal atoms. Thus the coupling in Eq. (1-36), which transfers electrons from the d-state to s-like (or free) states, also shifts the ε_d for these d-states in an important way.

In the metal, the atomic d-level for the usual $d^{Z-2}s^2$ configuration tends to lie below the Fermi energy, and the coupling to the s-like states transfers charge into the d-state from the s-like states, raising the ε_d . This shift in turn tends to reduce the transfer. The problem needs to be solved self-consistently, but when that is done in full band calculations it is found in the end that there are some 1.5 s-electrons per atom. In some sense, because of the large U_d , the d-level "floats" in the gas of free-electrons leaving about the same free-electron occupation, and therefore Fermi wavenumber, in all the transition metals.

C. The Energy Bands and the Friedel Model

Having the coupling between d-states, given by Eq. (1-37), the hybridization of these d-bands with the free-electron states, given in Eq.

(1-36), and ε_d fixed relative to ε_s by the self-consistent calculation just discussed, we have everything needed to make an elementary band calculation, which we do in Section 15-2-C. This calculation leads to a set of five bands, which could accommodate ten electrons over a narrow energy range, threaded through by the free-electron band. The Fermi energy comes in the midst of the set of d-like bands, corresponding to the partial filling of the d-like states which we associate with a transition metal. This leaves a difficulty in calculating total energies from the sum of energies over occupied states; we cannot use the center of gravity of the bands because they are only partly full but we don't have a simple free-electron form over which to integrate. The difficulty is avoided here by using the Friedel Model.

In the Friedel Model, the density of states per unit energy per atom $n(\varepsilon)$ arising from these d-bands is approximated by a constant value $10/W_d$ (since ten electrons per atom could be accommodated) over an energy range W_d , centered at ε_d , and with W_d adjusted so that this density of states has the same second moment as the true density of states. The value for the true density of states can be estimated directly from the tight-binding matrix elements. With this square density of states the needed integrals can be done simply. Thus the parameters for the electronic structure lead us to a model which can be used directly for estimating total energies and properties.

In particular we use this model to make the self-consistent determination of the d-state energy ε_d in Section 15-4-B. We even see in Section 15-4-D that the use of the 1.5 s-electrons per atom has included the effects of hybridization between d-like and s-like bands in the context of the model. We also use it to estimate the total energy and various bonding properties in the remainder of Chapter 15. We see that an important consequence of the partly- filled d-bands is to compress transition metals in the middle of each series, and increase their cohesion and elastic constants. The same model is also used to discuss Pettifor's (1979, 1995) theory of transition-metal alloys. Adding an exchange interaction, we also discuss the origin of ferromagnetism in the context of the Friedel Model.

D. Correlated States in f-Shell Metals

As we might expect, many features of the d-states in transition metals extend to f-states in the rare-earth (4f-shell) metals and actinides (5f-shell). The matrix elements with f-states, $\langle k|H|f\rangle$, are direct generalizations of Eq. (1-36) but because they vary as k^3 rather than k^2 , the element-dependent factor $r_d^{3/2}k^2$ must be replaced by $r_f^{5/2}k^3$. Further the coupling between neighboring f-states becomes

$$V_{ffm} = \eta_{ffm} \frac{\hbar^2 r_f^5}{md^7}, \tag{1-38}$$

with geometric coefficients η_{ffm} given in Eq. (16-9). In addition, the extra Coulomb repulsion associated with f-states, U_f, is even larger than that for d-states and makes a qualitative change in the behavior of these states.

In the rare earths, in particular, if the isolated atom contains Z_f f-electrons, in the metal we find the removal energy for the Z_f th electron well below the Fermi energy, but the removal energy for the $Z_f + 1$ st electron (U_f higher) well above the Fermi energy. They do not form a partially-filled band crossing the Fermi energy as in the transition metal. There remain effects of the coupled f-orbitals, but it is a fundamentally different system, which we call a *correlated state*. The nature of such a system, and intermediate systems, is explored for a two-level system (the Heitler-London system) for which an exact solution is possible, as well as an approximate (Unrestricted Hartree-Fock) solution. For treating the metals we actually generalize the exact solution using a Friedel-like model, which incorporates the Coulomb U_d or U_f, to study cohesion and volume-dependent properties for both the transition and f-shell metals. In most f-shell metals all but three electrons are found to be in f-states. This leaves three (or in some cases two) electrons in the free-electron bands, rather than the 1.5 in free-electron bands as in the transition metals.

One feature of the strongly correlated states in the rare earths is that the Z_f occupied states are so much like atomic states that, as according to Hund's rule for the atom, the spins align to the extent possible, producing a magnetic moment, localized on the atom, which is free to rotate on each atom in the presence of applied fields. This is also true in the heavier actinides, curium and beyond, but the lighter actinides are more analogous to transition metals. In Section 16-5 we treat these states as scattering resonances, and then apply the Unrestricted Hartree-Fock method to obtain a criterion for the formation of the local moment.

E. Transition-Metal and f-Shell-Metal Compounds

The important interatomic coupling in transition-metal compounds is found to be between the metallic d-states and the valence p-states on the nonmetals. In Chapter 17 this coupling is formulated in terms of matrix elements analogous to those in Eqs. (1-37) and (1-38) as

$$V_{pdm} = \eta_{pd\sigma}\frac{\hbar^2\sqrt{r_p r_d^3}}{md^4} , \qquad (1\text{-}39)$$

with the geometric coefficients η_{pdm} and the additional parameters r_p for each of the nonmetals given. Again ε_d is shifted by the sharing of occupation of the transition-metal d-state and the nonmetal p-states, and must be calculated self-consistently. Then all parameters are available for calculation of approximate bands for the compounds.

For AB-compounds (one metal atom to one nonmetal atom) three bonding bands, usually extending down from ε_p , are formed, as well as three antibonding bands, extending up from ε_d , and two nonbonding bands at ε_d . When only the bonding bands are occupied, the compounds are insulators. When there are electrons in the nonbonding bands, the compounds are poorly conducting metals. We only find a count indicating electrons in the nonbonding bands, when the effects of pd-coupling are extremely weak, as indicated by our generalization of the exact correlated-state theory. In that case local moments are expected to form. They are a result, not the cause, of the weak coupling.

In all of these cases the total energy is again formulated with separate Friedel-like densities of states for the bonding band and for the antibonding band. Again the widths are fit to the second-moment which is obtainable from tight-binding theory. This scheme is used both to make the self-consistent calculation of ε_d and to estimate the total energy for the bonding properties, the prediction of the equilibrium spacing in particular The predicted spacings do not accord well with experiment, nor do the effects of U_d correct the discrepancies.

In Chapter 18 the approach which was used for transition-metal compounds is applied to compounds of the f-shell metals. The coupling V_{pfm} , analogous to Eq. (1-39), is written and used with the r_p and r_f which had already been given. The resulting pf-bonding terms are of little consequence in the rare-earth compounds, nor at the beginning of the actinide series, nor beyond plutonium. However, their predicted effect is sizable in compounds of uranium, neptunium and plutonium. It is affected little by inclusion of the Coulomb U_f , nor does the combination account for an observed increasing spacing of the nitrides, phosphides, and arsenides in going from plutonium to americium to curium.

Also in Chapter 18 the approach for the monoxides is applied to transition-metal dioxides in the rutile structure. It is also applied to the trioxides in the perovskite structure, and important class of materials. For these trioxides the d-state energy is determined self-consistently, and the bands calculated. The stability of the perovskite structure is discussed in terms of the pd-bonding and the Madelung energies, but the general bonding properties are not addressed.

Finally the cuprates, in which high-temperature superconductivity was discovered, are studied. Parameters for the electronic structure are the same as for the other transition-metal compounds, and energy bands are obtained, corresponding to a metallic compound. In terms of these, the instability against formation of an antiferromagnetic insulating state is analyzed, as well as the excitations which have been considered sources for the electron-electron attraction responsible for superconductivity. These are phonons, paramagnons, and excitons.

8. Surfaces and Other Bonding Types

We chose to discuss surfaces and interfaces for all different systems in a separate discussion in Chapter 19, partly to show the great contrast between the surface energies of different types of solids.

A. Surface Energies

There is a natural quasi-chemical view of surface energies, as the energy arising from the breaking of interatomic bonds. Thus one can define a bond energy as the cohesive energy, listed for the elements in Table 1-3, divided by the number of nearest-neighbor contacts per atom (six in metals with twelve nearest neighbors since each "bond" is shared by two atoms). Then the formation of the ideal surface is imagined as the removal of all atoms on one side of a surface plane, leaving all remaining atoms in their original positions, and the surface energy is predicted to be this bond energy times half the number of such bonds broken per unit area (half of the energy is associated with each of the two surfaces created).

A consideration of the electronic structure shows that this is wrong in serious ways in most solid types. The concept would seem most plausible for semiconductors which we have discussed in terms of two-electron bonds, but we may see from our discussion of hybrid states in Section 1-4 that it greatly underestimates the surface energy. We may imagine forming the electronic structure for the semiconductor silicon in steps. We first form sp^3 hybrids, each with energy $(\varepsilon_s + 3\varepsilon_p)/4$, and each with one electron in it, at a cost in energy per atom of $\varepsilon_p - \varepsilon_s$, called the *promotion energy*, in comparison to the s^2p^2 configuration of the free atom. However, we gain that energy back, and a little more, when we form the bonds with each electron in a bond dropping in energy by V_2 , a covalent energy analogous to that given in Eq. (1-25). If we perform the same calculation for a silicon crystal with an ideal surface, we still promote every electron to a hybrid state, *including those hybrids pointing out of the surface* since the hybrids pointing into the solid are needed to form bonds, and those that point out must also be sp^3 hybrids if they are to be orthogonal to the others. Thus the same promotion energy is required for the crystal with surfaces and we lose the entire covalent energy for each bond, not just the small difference between the covalent energy and the promotion energy.

The resulting surface energy far exceeding the cohesive energy per broken bond has important consequences for semiconductor surfaces. The atoms near the surface almost always find a way to rebond to lower the energy. This ordinarily lowers the symmetry in comparison to the ideal surface and is called *reconstruction* , discussed in Section 19-3 where a few important and illustrative cases are considered.

The situation is quite the opposite in ionic crystals, where we saw that the cohesion could be understood as the transfer of electrons from metallic to the nonmetallic atom. If we form a crystal with a surface, we still gain the principal contribution to the cohesion from every electronic state and the surface energy is *very small* compared to the cohesive energy per broken bond. We see in Section 19-1-B that this is true only if the geometry corresponds to neutral surfaces, but such surfaces naturally are the cleavage planes and growth surfaces of low energy so this is the case. The small surface energy can be estimated in terms of the small extra electrostatic energies arising at such surfaces.

For simple metals also, the theory of the surface energies, based upon a combination of tight-binding theory and free-electron theory described in Section 19-1-C, is based upon quite different terms in the energy from those for the theory of cohesion. However, both that theory, and the quasi-chemical bond picture, predict surface energies of the order of one or two percent of $E_F k_F^2$. Thus in this one case, seemingly the least likely, the quasi-chemical theory gives reasonable estimates. In Section 19-1-D we include still higher terms in this surface energy to obtain the interaction between different surfaces, and to see its relevance to Giant Magnetoresistance.

In Section 19-2 we find that photothresholds and work functions, the energy required to remove an electron from a semiconductor or a metal, cannot be taken directly from the energy eigenvalues of the electronic states, as they could for the free atoms, but have Coulomb corrections of around four electron volts. In Section 19-4 we discuss interfaces between different solids, and see that the band structure retains meaning right up to a surface, or for very thin layers. We also see how the wavefunctions are matched from one solid to another and how the band line-ups are determined, again requiring in some cases large Coulomb corrections.

B. Other Systems

Finally in Chapter 20 we discuss the electronic structure of a number of solid types which do not fit well into the categories discussed before. Covalent sulfur, selenium and tellurium, with fewer than four neighbors, cause no problem, but boron with more neighbors requires a reformulation of a multicenter bond and the explicit inclusion of nonorthogonality as we did in Section 1-2-A. This also arises in a number of compounds.

Inert-gas solids are in some ways simpler than the other solids we have treated since they are described so well as independent atoms, with respect to electronic and dielectric properties. The bonding, which arises from quantum fluctuations of the atomic dipoles, is called the van-der-Waals interaction and is fundamentally different from the mechanisms treated earlier in the text. It is formulated in terms of the polarizability, which is in turn estimated from the electronic structure. Combining it with the overlap repulsion leads to the familiar Lennard-Jones interaction in terms of which the bonding properties are readily calculated.

This same van-der-Waals interaction gives intermolecular bonding in molecular crystals and the bonding of different graphite planes to each other. The bonding within individual graphite planes contains π-bonding, which is seen to be a special case of resonant bonds. Using a moments method it is seen that the contribution of a resonant π-bond to the energy is equal to the π-bond energy (as in N_2, discussed in Section 1-4-B), scaled by the square root of the number of sites among which it resonates. This is illustrated for benzene, graphite, and the fullerenes, and for the important biological molecules containing the heme group.

Finally we turn to bonding involving hydrogen, the only systems for which the atomic positions are dominated by electrostatic interactions. Both in H_2, and the central hydrides HF, H_2O, NH_3 and CH_4, we may begin with an inert-gas atom, freeze the electronic charge density, and pluck protons one by one from the nucleus as in the theoretical alchemy described in Section 1-4-E. Allowing them to fall to their positions of lowest electrostatic energy gives reasonable spacings, but to obtain the appropriate angles between protons, bonding terms must be included. A particularly important special case arises when a proton rests between two molecules, providing a hydrogen bond, or bridge, between them. The stability of such an arrangement is not as obvious as it first seems. However, proceeding from the central hybrids, it is seen how these weak bonds can be formed in systems such as ice.

Problem 1-1 Li_2

Li_2 is an extremely simple molecule; the distance d between the two nuclei is 2.67 Å.

a) Look up the energy of the one valence electron in each free atom from Table 1-1.

b) Obtain the coupling energy between the valence states, evaluated at the observed spacing (Eq. 1-14)

c) What is the energy of the bond state and of the antibonding state?

d) What is the total change in energy of occupied one-electron states in the formation of the molecule. This is a contribution to the formation energy of Li_2. We shall see that an overlap repulsion cancels about half of this. Even then, cohesions are typically overestimated by a factor of two for first-row-element systems, though not for elements from lower rows. The experimental value of cohesion is 1.07 eV for Li_2.

e) What energy of photon is required to excite an electron into an excited electronic state of the molecule?

Problem 1-2 O_2

Redo the calculations for O_2 (which has spacing d equal to 1.22 Å) that we did for N_2 in Section 1-4-B:

a) Obtain the energies of the eight molecular orbitals. [We note that there are errors in the sign of these levels in parentheses in the book, *Electronic Structure and the Properties of Solids,* Harrison (1980) for O_2 and F_2 as taken from the literature.]

b) Note which are occupied in the neutral molecule.

c) The fact that not all levels of the same energy are occupied means that, by Hund's rule, those will be occupied with parallel spin as far as possible. What net spin do you expect for the molecule? (It is true that O_2 is paramagnetic.)

d) Estimate the cohesive energy from these results, dividing by two to take account of the repulsion. The experimental cohesion is 5.1 eV and is considerably smaller than your estimate, as in Problem 1-1.

e) Redo the energies of the σ-orbitals using sp-hybrids and the *Bond-Orbital Approximation*, as we did for nitrogen to see how well they agree with the full diagonalization.

Problem 1-3 Size of neglected matrix elements.

a) Obtain the formula for the matrix element between one inward- and one outward-pointing sp-hybrid, in analogy with Eq. (1-25) for two inward-pointing hybrids.

b) Obtain the corresponding formula for the matrix element between two outward-pointing hybrids.

CHAPTER 2

Bonding in Tetrahedral Semiconductors

We saw in Section 1-4 that the electronic structure of tetrahedral semiconductors could be represented in terms of two-center, two-electron bonds. All bonds are identical. This makes them one of the nicest systems to discuss theoretically in an organized way, as well as one of the most important categories of solids. We begin with a description of the crystal structure.

1. The Crystal Structure and Notation

We have indicated that in the tetrahedral structure each atom is surrounded by four neighbors. The zincblende structure of zinc sulfide is such a structure and is shown in Fig. 2-1 and in Fig. 2-2; gallium arsenide is also in this structure. We may think of each dark circle as a gallium atom. In Fig. 2-1 the atoms are drawn as contacting spheres. Each atom in the forward-most plane is in contact with two atoms in the plane below, and each of these with two in the plane below it. It may be easier to understand the structure in Fig. 2-2 where the spheres are too small to contact, and lines are drawn from each gallium atom to the four nearest -neighbor arsenic atoms. These four arsenic atoms are the corners of a

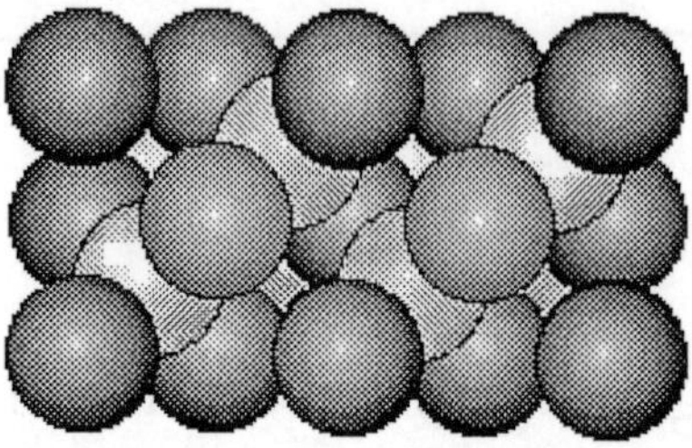

Fig. 2-1. A model of the zincblende crystal structure of most tetrahedral semiconductors, such as GaAs, drawn as contacting spheres. The darker atoms could be Ga, the lighter As, or all could be silicon. The cube axes are horizontal, vertical and perpendicular to the figure.

regular tetrahedron with the gallium atom at the center. Similarly each arsenic atom is surrounded by four gallium atoms which are the corners of a regular tetrahedron with the arsenic in the center. This second tetrahedron is congruent with the first, but is inverted; its corners are oppositely oriented. In the diamond structure, the structure of silicon, all

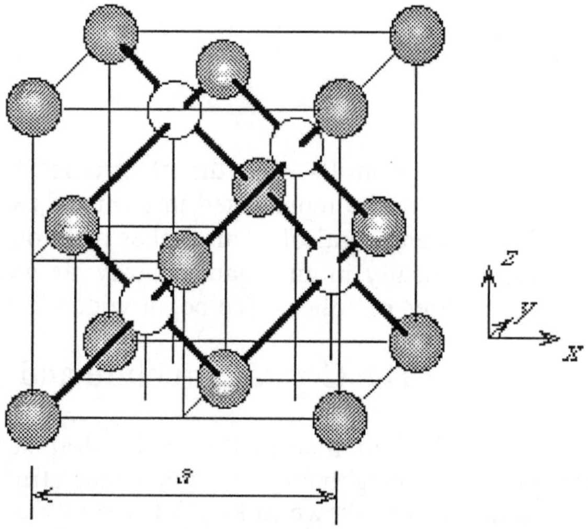

Figure 2-2. The zincblende structure organized for calculations. Shaded atoms are metallic (e.g. gallium),unshaded nonmetallic (e.g. arsenic). If all atoms were of the same element (e.g. silicon), it would be the diamond structure. The lines connecting the metallic atoms form a cube with a metallic atom also at each face-centered position. For the conventional coordinate system shown, a nonmetallic atoms lies at the position *[a/4, a/4, a/4]* with respect to each metallic atom forming a displaced face-centered cubic lattice.

atoms are of the same element but otherwise the structure is the same. Though all atoms in this case are the same we can distinguish two types of lattice sites by the orientation of the tetrahedra formed by the surrounding atoms. Such elemental semiconductors are from the column IV in the periodic table and are called Column IV semiconductors. Those such as gallium arsenide with one constituent from column III and one from column V are called III-V semiconductors. Similarly zinc selenide and copper bromide are II-VI and I-VII semiconductors, respectively.

For the purposes of making calculations, the structure is most conveniently viewed as in Fig. 2-2. There it is seen that the gallium atoms form a cube of edge a with an additional gallium atom at the center of each cube face. If the cube is divided into eight "cubies" (one is drawn in the lower left front corner in Fig. 2-2) it is seen that half have arsenic atoms at the center. Each is drawn with a line down to the bottom face in Fig. 2-2. The structure then has the symmetry of a regular tetrahedron, half the symmetry of a cube since, for example, the ninety-degree rotations through the cube center which take a cube into itself, place the arsenic atoms in different cubies. This is sufficiently high symmetry to make the conductivity and any other second-rank physical tensor isotropic, but we shall see in Chapter 3 that it allows nonzero third-rank tensors though these would vanish for full cubic symmetry.

In organizing calculations it will be convenient to take a rectangular coordinate system along the cube edges, as shown in Fig. 2-2. Then directions and position vectors are specified by the x-, y-, and z-components of the vector with square brackets; in Figure 2-2 there is a nonmetallic (arsenic) atom at a position $[a/4, a/4, a/4]$ relative to each metallic (gallium) atom. The nearest-neighbor distance d is equal to $\sqrt{3}a/4$. For specifying directions it is customary to use unit components; the two neighbors lie in a [111] direction relative to each other. The direction [111] can be used to refer to any of the eight cube-diagonal directions or one can distinguish directions such as [1-11]. The other frequently discussed directions are the cube-edge direction [100] and the face diagonal [110]. We shall wish to use this coordinate system immediately in the next section as we construct hybrid states on the atoms.

The gallium arsenide crystal is made up of a large number of such cubes repeated in all three dimensions. The gallium atoms are said to form a face-centered-cubic lattice because of their arrangement on cubes such as that drawn in Fig. 2-2. The environment of every gallium atom is identical to that of every other, whether we think of it as a cube-face atom or a cube-corner atom; we could have drawn the cubes with any one atom chosen as a cube corner. This face-centered-cubic lattice is said to make up the translational symmetry (or Bravais Lattice) for the crystal since translating the entire crystal to take one atom to the site of another face-centered-cubic site replaces every atom by an identical atom within the interior of the crystal. One also says that each gallium atom is *translationally equivalent*. Obviously the arsenic atoms also form a face-centered-cubic lattice and have the same translational symmetry. If one divides the entire lattice into identical cells, each containing one gallium atom, the cells are called *primitive cells*. The primitive cell here contains one gallium atom and one arsenic atom.

Having two atoms in the primitive cell makes an important difference when we construct electron states, or lattice vibrations, in crystals. When we constructed electron states for a lithium chain in Chapter 1, the solutions of our tight-binding equations, Eq. (1-19), were Bloch sums, with coefficients $u_j = e^{ikdj}/\sqrt{N}$, which satisfied all of the Eqs. (1-19) because every site was translationally equivalent. When we construct electron states for semiconductor crystals, which have two atoms per cell, we will need to construct separate Bloch sums for states on each of the two atom types.

In silicon, both the gallium and the arsenic sites are occupied by silicon atoms but the two types of sites are not translationally equivalent, thus we will again need to construct separate Bloch sums of atomic states for the two types of sites. Sometimes it will be useful to notice the additional symmetry when the types of sites are occupied by the same atom. For example, in the silicon structure an inversion of the entire crystal through the point in the center of one of the nearest-neighbor bonds takes every atom into a site where an atom had previously been (within the bulk of the crystal). There is a related symmetry, called a *screw-axis symmetry* , which is present in a silicon lattice. If we construct an axis parallel to the x-axis (in a [100] direction) through the point $[0,a/8,a/8]$ relative to the lower left front cube corner, we may confirm that a translation of $[a/4,0,0]$ accompanied by a 90° clockwise rotation around that axis again takes every silicon atom into the previous position of a silicon atom. Indeed it interchanges what were gallium and what were arsenic sites. For silicon it could be used to construct special Bloch sums, $u_j = e^{ikaj/4}/\sqrt{N}$, where j is an index numbering (100) atomic planes (perpendicular to [100]) along the x-axis. For electrons propagating along this direction (or a [010] or [001] direction) electron states can be constructed for the homopolar (e. g., silicon) lattice as if there were only one atom per primitive cell. They are special solutions, but they will sometimes be useful. The same simplification occurs for longitudinal vibrational modes, which will be discussed in Section 3-3, for wavenumbers along a [100] direction.

The wurtzite structure has the same tetrahedral nearest-neighbor coordination as the zincblende structure but the translational symmetry is different. Note that we could construct a (111) plane through the center of the cube in Fig. 2-2, perpendicular to the three bonds which run upward and to the right. Note that they form rows of equally spaced parallel bonds. The portion of the crystal below and to the left of these bonds could be rotated 180° around one of these bonds so that again there was perfect tetrahedral coordination but the structure below would have changed orientation. Such a change in the structure is called a *stacking fault* . Making such a fault at alternate sheets of bonds throughout the crystal converts it to a wurtzite structure. All of our calculations here will be done for the zincblende structure, but many results are unaffected by the difference

2. Hybrids, and Their Coupling

In the tetrahedral structure each atomic s-state is coupled equally to the s-state on each of its four neighbors and these s-states would broaden out into bands, as in the lithium chain discussed in Section 1-3, not the bonding and antibonding states as in Li_2, discussed in Section 1-2, yet we have indicated that the electronic structure is understandable in terms of individual two-center bonds. This is accomplished by an old trick, the formation of tetrahedral hybrids in close analogy to the sp-hybrids we introduced for the N_2 molecule in Section 1-4-C.

A. sp^3-Hybrids

It is not really *necessary* to form hybrids. We could simply expand the states in the solid in terms of s- and p-states on each atom and obtain the equations, analogous to Eq. (1-19), which minimize the energy. We could then solve the equations without ever explicitly introducing hybrids. We shall do just this, and in some detail, in Chapter 4 when we obtain the energy bands. The hybrids we introduce here are simply a way of understanding the results of such an exact solution, just as were the hybrids for N_2 in Section 1-4-C. They also, more importantly, will provide us a way of obtaining approximate values for the total energy and a wide range of semiconductor properties.

The hybrids are four different linear combinations of the s- and p-states on a single atom.

$$|h_i\rangle = u_{i1}|s\rangle + u_{i2}|p_x\rangle + u_{i3}|p_y\rangle + u_{i4}|p_z\rangle \;. \tag{2-1}$$

It is necessary that the hybrids be constructed to be orthogonal and normalized,

$$\langle h_i|h_j\rangle = u_{i1}^*u_{j1} + u_{i2}^*u_{j2} + u_{i3}^*u_{j3} + u_{i4}^*u_{j4} = \delta_{ij}. \tag{2-2}$$

Thus the four amplitudes for a hybrid may be considered as components of a four dimensional vector orthogonal to the four-dimensional vectors corresponding to each of the other hybrids on the same atom. We make this requirement since the true electronic states in quantum-mechanics are orthogonal and any effect we might introduce by using nonorthogonal hybrids would be expected to go away when we went to a more accurate solution. Expanding the electronic states for the crystal in terms of such hybrids on the atoms is entirely equivalent to expanding them in terms of the atomic states which constitute the hybrids. The desired hybrids for a shaded atom in Fig. 2-1 or 2-2 have the form

$$\begin{aligned}
|h_1\rangle &= {}^1/_2(|s\rangle + |p_x\rangle + |p_y\rangle + |p_z\rangle)\,,\\
|h_2\rangle &= {}^1/_2(|s\rangle + |p_x\rangle - |p_y\rangle - |p_z\rangle)\,,\\
|h_3\rangle &= {}^1/_2(|s\rangle - |p_x\rangle + |p_y\rangle - |p_z\rangle)\,,\\
|h_4\rangle &= {}^1/_2(|s\rangle - |p_x\rangle - |p_y\rangle + |p_z\rangle)\,,
\end{aligned} \tag{2-3}$$

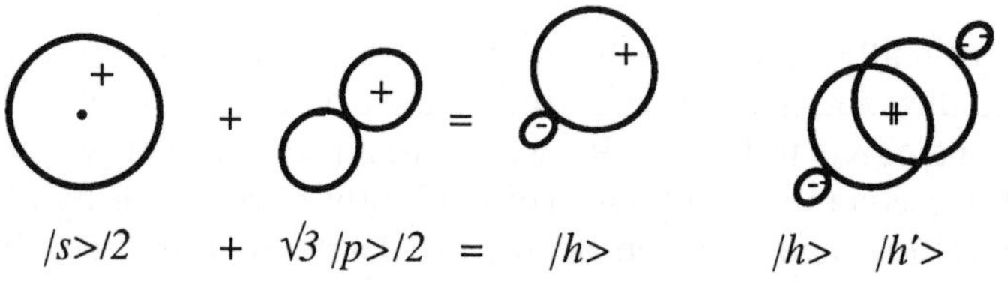

Figure 2-3. Representation of a hybrid state. In a hybrid sp-state, the amplitude is enhanced in the direction where the p-state is positive and suppressed on the opposite side; thus, it "leans" in one direction. Two hybrids directed into the same bond, $|h\rangle$ and $|h'\rangle$, have a large coupling energy (called the *covalent energy*); each is only weakly coupled to those on the other atom that are directed into other bonds.

where the individual p-states are written in terms of the Cartesian coordinate system of Fig. 2-2. Note that the first combination of p-states is a p-state $\sqrt{3}|p\rangle/2$ oriented toward the $[1,1,1]\ a/4$ neighbor, which can be seen by decomposing the p-states into σ- and π-oriented components as we did in Section 1-4. Such hybrids are illustrated in Figure 2-3.

The orthogonality and normalization, Eq. (2-2), of these four hybrid-states can be immediately verified. Thus the hybrids on one atom form an orthonormal set, as do the s- and p-states on that atom. The occupation probability on the s-state is $1/4$ and that on p-states is $3/4$, based upon the squared amplitudes from Eq. (2-3). It is for this reason they are called *sp^3-hybrids*. The energy of each hybrid $\varepsilon_h = \langle h_i|H|h_i\rangle$ is the weighted average of the s- and p-state energies,

$$\varepsilon_h = \frac{\varepsilon_s + 3\varepsilon_p}{4} \tag{2-4}$$

For silicon we use Table 1-1 to obtain $\varepsilon_h = (-14.79 - 3\times 7.58)/4 = -9.38$ eV, which was also listed in Table 1-1, along with the corresponding values for the other elements which form semiconductors.

B. The Covalent Energy, V_2

We can see that the coupling between two hybrids on neighboring atoms directed into the same bond is very large, because the various matrix elements from Figure 2-3 all add up, just as they did for the N_2 inward-directed hybrids discussed in Section 1-4. This was the reason for the choice of hybrid in both cases. Here we may note again that the hybrid $|h_1\rangle$ consists of one half times an s-state plus $\sqrt{3}/2$ times a p-state which is σ-oriented with respect to the neighbor in the $[111]$ direction. $|h_1'\rangle$ on the $[1,1,1]$ neighbor will be of the form of $|h_1\rangle$, but with the opposite sign for the amplitude of the p-states. The net coupling between these two hybrids, $\langle h_1|\,H\,|h_1'\rangle$, can be written in terms of the σ-coupling matrix elements from Eq. (1-8) by collecting the contributions for the coupled pairs of s- and pσ-states. It is negative and has magnitude (called the *covalent energy* V_2) equal to

$$V_2 = -\,{}^1/_4(V_{ss\sigma} - 2\sqrt{3}\,V_{sp\sigma} - 3V_{pp\sigma})/4 = 3.22\hbar^2/md^2\,, \tag{2-5}$$

Table 2-1. Metallic, covalent, and polar energies (eV) for the tetrahedral semiconductors (based upon Table 1-1). The subscript ± designates the metallic (+ for Ga in GaAs) and nonmetallic (- for As in GaAs) constituent. Also listed is the equilibrium internuclear distance d and the polarity from Eq.(2-12).

	d (Å)	α_p	V_{1+}	V_{1-}	V_2	V_3
C	1.54	0	2.08		10.35	0
Si	2.35	0	1.80		4.44	0
Ge	2.44	0	1.96		4.12	0
Sn	2.80	0	1.57		3.13	0
SiC	1.88	0.26	1.80	2.08	6.94	1.88
BN	1.57	0.34	1.26	3.10	9.95	3.62
BP	1.97	0.18	1.26	2.42	6.32	1.14
BAs	2.07	0.15	1.26	2.48	5.73	0.89
AlN++	1.89	0.59	1.25	3.10	6.87	4.99
AlP	2.36	0.49	1.25	2.42	4.41	2.50
AlAs	2.43	0.48	1.25	2.48	4.16	2.25
AlSb	2.66	0.41	1.25	1.97	3.47	1.58
GaN*	1.94	0.60	1.47	3.10	6.52	4.90
GaP	2.36	0.48	1.47	2.42	4.41	2.41
GaAs	2.45	0.47	1.47	2.48	4.09	2.16
GaSb	2.65	0.39	1.47	1.97	3.49	1.49
InN*	2.15	0.70	1.19	3.10	5.31	5.19
InP	2.54	0.58	1.19	2.42	3.80	2.70
InAs	2.61	0.56	1.19	2.48	3.60	2.45
InSb	2.81	0.50	1.19	1.97	3.11	1.77
BeO*	1.65	0.63	0.66	4.33	9.01	7.30
BeS	2.10	0.60	0.66	3.10	5.56	4.13
BeSe	2.20	0.58	0.66	3.05	5.07	3.64
BeTe	2.40	0.54	0.66	2.40	4.26	2.75
MgTe*	2.76	0.75	0.76	2.40	3.22	3.67
ZnO*	1.98	0.79	0.99	4.33	6.26	8.02
ZnS	2.34	0.73	0.99	3.10	4.48	4.85
ZnSe	2.45	0.73	0.99	3.05	4.09	4.36
ZnTe	2.64	0.70	0.99	2.40	3.52	3.47
CdS	2.53	0.79	0.81	3.10	3.83	4.95
CdSe	2.63	0.78	0.81	3.05	3.55	4.47
CdTe+	2.81	0.75(0.73)	0.81(1.32)	2.40(2.46)	3.11	3.57(3.36)
HgTe+	2.81	0.76(0.67)	0.79(1.41)	2.39(2.46)	3.11	3.60(2.82)
CuF	1.84	0.83	1.34	5.73	7.25	10.94
CuCl	2.34	0.84	1.34	3.85	4.48	6.96
CuBr	2.49	0.84	1.34	3.64	3.96	6.18
CuI	2.62	0.82	1.34	2.84	3.57	5.05
AgI	2.80	0.85	1.11	2.84	3.13	5.05

*Wurtzite structures, + Relativistic values in parentheses (Harrison, 1983f).

quite close to the value of $V_2 = 3.19\hbar^2/md^2$ obtained in Section 1-4 for sp-hybrids with equal weighting of the s- and p-states. Values for these, obtained from Eq. (2-5), are listed in Table 2-1.

This covalent energy is much larger than the coupling $-V_1^x$ between this same orbital $|h_1'>$ on the neighbor and $|h_2>$, $|h_3>$, or $|h_4>$ on the first atom, obtained in the same way,

$$V_1^x = -\,^1/_4(V_{ss\sigma} - 2\sqrt{3}\,V_{sp\sigma}/3 + V_{pp\sigma}) = 0.18\ \hbar^2/md^2\ . \tag{2-6}$$

Note that in neither case is there a contribution from $V_{pp\pi}$. The effect of this smaller matrix element will be of interest later (Section 3-2-C), but a very good approximation for the moment is to include only the coupling energy $-V_2$ for two hybrids directed into the same bond. We can then again treat the bonds independently as in Li_2 and other diatomic molecules. The internuclear distance for silicon is 2.35 Å; hence, the covalent energy becomes 4.44 eV. An analysis exactly like the Li_2 analysis gives a bond energy at -13.82 eV and an antibonding energy at -4.94 eV. The gain in energy -8.88 eV for the two electrons in the bond relative to the uncoupled hybrids is the principal contributor to the cohesive energy of silicon.

The individual bonds are not truly independent. We have noted in Eq. (2-6) a coupling $-V_1^x$ between a hybrid in one bond and a hybrid in another. We will discuss an even more important one in Section 2-D. Before proceeding to that, we should carefully specify the parameters that characterize the bond in a compound semiconductor.

C. The Polar Energy, V_3

We use here the terminology given in Section 1-4 for the polar bond in a diatomic molecule. We have already given the magnitude of the energy of coupling between two hybrids forming a bond, the $-V_2$ given in Eq. (2-5). Note that V_2 depends only on internuclear distance and is applicable to a compound semiconductor (e.g. GaAs) where the two hybrids come from different elements. However, in such a case, the two hybrid energies given in Eq. (2-4) are different. The polarity of a bond is characterized by this energy difference; a *polar energy* V_3 is defined to be half the difference,

$$V_3 = \frac{\varepsilon_{h+} - \varepsilon_{h-}}{2}\ , \tag{2-7}$$

where the plus indicates the higher-energy hybrid, or the positively charged atom, the metallic atom, the cation. The negative indicates the lower-energy hybrid on the negatively charged atom, the anion. For purposes of a computer program, for example, interchanging the atoms changes the sign of the polar energy and the formulae which depend upon it remain appropriate. Values are listed in Table 2-1.

Repeating the principal steps from Section 1-4, the bond orbital is written with amplitudes u_+ and u_- for the two hybrids, (analogous to the u_1 and u_2 in Section 1-4) and the energy [written just before Eq. (1-26)] is given by

$$\varepsilon = u_+^2\varepsilon_{h+} - 2u_+ u_- V_2 + u_-^2\varepsilon_{h-} \ . \qquad (2\text{-}8)$$

The minimization conditions become

$$\begin{aligned} \varepsilon_{h+}u_+ - V_2u_- &= \varepsilon u_+ \ , \\ -V_2u_+ + \varepsilon_{h-}u_- &= \varepsilon u_- \ . \end{aligned} \qquad (2\text{-}9)$$

Solution of these two equations together gives

$$\varepsilon_b = \frac{\varepsilon_{h+}+\varepsilon_{h-}}{2} - \sqrt{V_2^2 + V_3^2} \ , \qquad (2\text{-}10)$$

and a second solution ,ε_a, with the sign before the $\sqrt{V_2^2 + V_3^2}$ changed. Equation (2-10) is the bond state; the other is the antibonding state orthogonal to it. Equation (2-10) can be substituted back into Eq. (2-9) to obtain values for the coefficients u_+ and u_- , which we designate by a super b for the bonding state,

$$u_-^b = \sqrt{\frac{1+\alpha_p}{2}} \ , \quad u_+^b = \sqrt{\frac{1-\alpha_p}{2}} \ , \qquad (2\text{-}11)$$

where α_p is the *polarity* defined by

$$\alpha_p = \frac{V_3}{\sqrt{V_2^2 + V_3^2}} \ . \qquad (2\text{-}12)$$

That these are correct can be checked, with a little algebra, by substituting Eqs. (2-10), (2-11), and (2-12) into Eqs. (2-9), The parameter complementary to the polarity, called *covalency* , is frequently useful and is defined by

$$\alpha_c = \frac{V_2}{\sqrt{V_2^2 + V_3^2}} \; . \tag{2-13}$$

We have indicated that the antibonding energy is given by

$$\varepsilon_a = \frac{\varepsilon_{h+} + \varepsilon_{h-}}{2} + \sqrt{V_2^2 + V_3^2} \; , \tag{2-14}$$

and the corresponding amplitudes are

$$u_-^a = \sqrt{\frac{1-\alpha_p}{2}} \; , \qquad u_+^a = -\sqrt{\frac{1+\alpha_p}{2}} \; . \tag{2-15}$$

It is readily confirmed that the bonding and antibonding states are orthogonal and normalized, which is sufficient to confirm the correctness of the antibonding state if the bonding state is correct. They approach the even and odd bonding and antibonding states discussed in Section 1-2 as the polarity approaches zero. For the polar case the probability density of the bond shifts to the lower-energy hybrid, and the probability density for the antibond must correspondingly shift to the higher-energy hybrid. In the limit of high polarity ($\alpha_p = 1$), where the coupling V_2 is negligible compared to the energy difference $2V_3$, the bond state becomes simply the low-energy hybrid state and the antibond the high-energy hybrid state.

D. The Metallic Energy, V_1

Another important coupling is that between two hybrids on the same atom. This coupling is evaluated just as was the weighted hybrid energy of Eq. (2-4), but using one set of amplitudes from one hybrid and one from the other, rather than both sets the same. [ε_h was a sum over atomic states i as $\Sigma_i u_i^2 \varepsilon_i$; here one u_i is taken from the first hybrid, one from the second.] We obtain that coupling from Eq. (2-3) noting that by symmetry the coupling between two *atomic* states on the same atom is zero. The coupling between two hybrids is negative; its magnitude, the *metallic energy* , is

$$V_1 = \frac{\varepsilon_p - \varepsilon_s}{4} \; . \tag{2-16}$$

Note that in a compound it will be different for the two constituent elements. Both are listed for the tetrahedral semiconductors in Table 2-1.

This metallic energy is the principal coupling between bonds; the next most important one is the $V_1{}^x$ defined by Eq. (2-6). These couplings between the bond orbitals broadens those levels into a *valence energy band.* They also broaden the antibonding levels into a *conduction band.* The simplest theory of the bands in a homopolar semiconductor (e.g., silicon), to be described in Section 5-2-A, corresponds to a lattice of bonds, of identical energy, each coupled by $-V_1/2$ to six neighboring bonds, a straightforward generalization of the calculation of a row of lithium atoms given in Section 1-3. It leads to valence bands stretching from $\varepsilon_b - 3V_1$ to $\varepsilon_b + V_1$ and the corresponding calculation for the antibonds leads to conduction bands stretching from $\varepsilon_a - 3V_1$ to $\varepsilon_a + V_1$. Then the gap in energy between the two bands is

$$E_g = 2V_2 - 4V_1 = 2V_2\,(1 - \alpha_m) \tag{2-17}$$

where we have defined a *metallicity* by

$$\alpha_m = 2V_1/V_2\ . \tag{2-18}$$

The conclusion that the band gap goes to zero as the metallicity defined this way goes to one is also very nearly true with much more complete descriptions of the bands (Harrison, 1981b). We have seen that $\varepsilon_p - \varepsilon_s$ depends principally on column number and does not change greatly as we move down a column in the periodic table; e.g. carbon, silicon, germanium, tin, lead. On the other hand, because V_2 decreases as $1/d^2$ for the heavier, larger atoms, metallicity increases through such a series. Metallicity characterizes the trends of all properties as a function of their position vertically in the periodic table.

It would be easy to generalize this metallicity to polar semiconductors [in analogy with Eqs. (2-12) and (2-13)]; however, a different V_1 [from Eq. (2-16)] arises for the two atoms and the particular combination that enters the theory depends upon the property being considered (Harrison, 1983a, b). We therefore make the quantitative definition [Eq. (2-18)] only for the homopolar semiconductor, but use the qualitative term to describe the increased effect of the metallic energy for semiconductors made from the heavier elements. Of particular interest for the bonding properties is the effect of this metallic energy in coupling the individual bond to its environment, an effect that we call *metallization.*

We might note, for completeness, that there is one further attribute which characterizes a semiconductor: a *skewness,* which can be specified by the row number for the nonmetallic element minus the row number of the metallic element (in Table 1-1) divided by four. Then, by specifying the metallicity (depending upon average row number), polarity (depending upon column difference), and skewness (depending on row difference), we specify the semiconducting compound uniquely. Skewness is not frequently discussed because most properties are rather insensitive to it. Increasing skewness does tend to decrease the electronegativity difference between the metallic and nonmetallic elements and therefore reduce the polarity, but that effect is already included in the polar energy of Table 2-1. Increasing skewness also increases the metallic energy V_{1-} more than it decreases the metallic energy V_{2+} and therefore increases the effect of the metallization (described in Section 2-3-B). Such effects are automatically included when we use the parameters of Table 2-1.

E. The Overlap Repulsion, $V_0(d)$

There will be enough electrons in the semiconductor to fill all of the bond states, with energy given by Eq. (2-10), $\varepsilon_b = (\varepsilon_{h+} + \varepsilon_{h-})/2 - \sqrt{V_2^2 + V_3^2}$, with no electrons left over to fill the antibonding levels; this is, of course, because there are exactly eight valence electrons per pair of atoms and, in the tetrahedral structure, exactly four bonds per pair of atoms. The bond energy of Eq. (2-10) decreases as the atoms are brought closer together because V_2 is proportional to d^{-2} and increases as d decreases. This corresponds to a bond tension given by the derivative of the energy $2\varepsilon_b$ associated with the bond with respect to d , a tension of $(4/d)V_2^2/\sqrt{V_2^2 + V_3^2}$. Of course, this bond tension must be balanced by a repulsion if the crystal is not to collapse, the overlap repulsion discussed for Li_2 in Section 1-2.

Ultimately that repulsion comes predominantly from the increased electronic kinetic energy that accompanies the decreased volume available to the electrons. In Chapter 1 we associated it with the nonorthogonality, $S_{ij} = \langle i|j\rangle$, of atomic orbitals on adjacent atoms. If we attempt to place two atoms at the same position, we would be putting *two* electrons of each spin in the atomic states, violating the exclusion principle. The nonorthogonality is the first indication of this need to promote the electrons to higher states and increase the kinetic energy as the atoms are brought together. Interestingly enough, there are three ways of thinking of this repulsion to estimate its form, and they all suggest - though none very convincingly - the same dependence, C/d^4 , of the repulsion for covalent solids. There is a fourth way which suggests slightly higher powers.

The first way is to regard the *lowering* in energy from the coupling V_2 as arising from Coulombic potential energy, and the repulsive term as arising from kinetic energy. For such a system the virial theorem states that the potential energy should have twice the magnitude of the kinetic energy. (See, for example, Schiff (1968), p. 180, if the potentials are Coulombic; i. e., vary as $1/r$.). It then follows that if the lowering in energy is proportional to $1/d^2$ and we are to fit the value and slope of the overlap repulsion at the equilibrium spacing to C/d^n , that n must be four. This follows from minimizing $-A/d^2 + C/d^n$, giving $2A/d^3 - nC/d^{n+1} = 0$ at $d = d_0$. Then the energy at the minimum is $-A/d_0^2 + (2A/n)/d_0^2$ with the second term half the first only if $n = 4$.

The second way is the analysis we gave Section 1-2 for Li_2. We saw that the effect of nonorthogonality is to shift the average of the bond and the antibond energy upward by the nonorthogonality $<s1|s2>$ times the magnitude of the coupling $V_{ss\sigma}$ which produced the bonding-antibonding splitting. Further, the Extended Hückel Theory of Roald Hoffmann (1963) plausibly suggested that interatomic couplings should be proportional to the nonorthogonality. Thus if $<\alpha|H|\beta>$ varies as $1/d^2$, so must $<\alpha|\beta>$ and the product, which is the repulsion, should be proportional to $1/d^4$.

The corresponding analysis of nonorthogonality for polar bonds was carried out in Appendix B of Harrison (1980) leading to bonding and antibonding energies of the form of Eqs (2-10) (with a correction due to nonorthogonality absorbed also in V_3) and (2-14) and again a shift of the average of the two upward by the nonorthogonality times the effective interorbital coupling, V_2, suggesting applicability of the $1/d^4$ dependence also to polar semiconductors.

An additional point may be made concerning these nonorthogonalities. The conclusion that the repulsion for a tetrahedral bond, $2<h|h'>V_2 \equiv 2SV_2$ for the two electrons in the bond, is half the gain in energy $2V_2$ means that the nonorthogonality must be $S \approx 1/2$ at equilibrium. This is consistent with values obtained by fitting the couplings we use with the formula given in Section 1-2 from Extended Hückel Theory and by detailed calculations of the overlap using known atomic wavefuntions, as described following Eq. (1-16). This rather large value is comforting for its consistency, but means that in our analysis where we largely lump all of the effects of this large nonorthogonality into an overlap repulsion, we may be making significant errors in our calculations. It is a compromise we make in order to gain the extraordinary simplicity of the theory which follows.

The third way is based upon local-density theory of many-electron systems, which will be described in Section 10-1, when we discuss ionic solids. It will lead to an exponential overlap repulsion, $e^{-\mu d}$, with the coefficient μ depending upon the energy of the atomic states involved. Fitting this to a form C/d^n will lead to n near eight for ionic solids, but values close to four for covalent solids, largely due to their smaller equilibrium spacings.

The fourth way is from pseudopotential theory and the theory of metals, described briefly in Section 1-7. There also we obtained an exponential repulsion, which when fit to a form C/d^n gave an n varying from 5 to 6.4 from C to Sn. This also led to a specific form for the coefficient in the overlap repulsion, Eq. (1-28), which for sp^3-hybrids becomes

$$V_0(d) = \frac{C}{d^4} \equiv 16 \times 3.22 \frac{\hbar^2 r_c^2}{m d^4}. \tag{2-19}$$

Here r_c is the pseudopotential empty-core radius, introduced in Section 1-7 and listed for each element in Chapter 13. We shall generally adjust the value of the constant C to give the observed equilibrium spacing and will find then that predicted properties then depend just on the parameters V_1, V_2, and V_3, just given. Then the more complicated final form in Eq. (2-19) is just an inconvenience. This is of course taking the bond length from experiment rather than prediction. When we discuss trends in bond length from a theoretical point of view in Section 2-4-A, and for SiO_2 in Chapter 12, we shall make a little use of this form.

We indicated in Chapter 1 that the use of this overlap repulsion appears to be the weakest aspect of this tight-binding theory of electronic structure. When one seeks more accurate estimates of bonding properties, it is possible to use a form with *two* adjustable parameters and fit not only the experimental bond length but also the experimental bulk modulus. This will obviously improve at least the description of the elastic constants. Such an analysis was made by Bechstedt and Harrison (1989). Bechstedt noted a finding by Van Schilfgaarde and Sher (1987) that calculated nonorthogonalities went more closely with $1/d$ than $1/d^2$, as assumed above. Thus the overlap repulsion arising from the nonorthogonality, $2SV_2$, should vary as d^{-3}. He then included an additional term, ΔE_{rep}, varying as $1/d^{12}$, as in the traditional Lennard-Jones model, to describe the repulsion between valence orbitals and neighboring core states. This may also be thought of as the effect of the repulsive term in the pseudopotential for the neighboring atom. Bechstedt's overlap repulsion takes the form

Table 2-2. Parameters for the Bechstedt overlap repulsion, Eq. (2-20), obtained by fitting the equilibrium spacing and the experimental bulk modulus.

Semiconductor	$S(d_0)$	ΔE_{rep} (eV)	$V_0(d_0)$ (eV)
C	0.583	0.326	12.40
Si	0.384	0.446	3.86
Ge	0.351	0.418	3.31
Sn	0.247	0.460	2.01
GaAs	0.389	0.348	3.53
ZnSe	0.332	0.313	3.03

$$V_0(d) = 2S(d_0)V_2(d_0)(d_0/d)^3 + \Delta E_{rep}(d_0)(d_0/d)^{12} \ , \qquad (2\text{-}20)$$

with the various parameters, $S(d_0)$, etc., evaluated at the equilibrium spacing. d_0 . We have added this, in place of the form Eq. (2-19), to the total energy per bond, calculated in the following section. Using these two adjustable parameters, $S(d_0)$ and ΔE_{rep}, we have fit the observed equilibrium spacing and the experimental bulk modulus (the experimental κ_0 to be given in Table 2-5) to obtain the values for a sample of materials listed in Table 2-2. (They differ slightly from those listed by Bechstedt and Harrison (1989); in particular, our ΔE_{rep} is the contribution per *bond* rather than per electron.)

These values make seem sensible and lend some confidence to the use of this form for making numerical estimates. First, the nonorthogonalities $S(d_0)$ are not far from the value 0.5 which was suggested above. Second, it is not very different in the isoelectronic series, Ge, GaAs, and ZnSe. Third, the magnitude of the term ΔE_{rep} representing the repulsion between the valence electrons and the neighboring cores is very nearly the same for all materials, in spite of its very rapid variation with d for a given material. Finally, the total repulsion $V_0(d_0)$ is not far from the covalent energy V_2 so that again half of the bonding energy is canceled by the overlap repulsion.

The fact that the parameters do not vary greatly with polarity suggests that one could use the values for the homopolar semiconductors, C, Si, Ge, and Sn, for polar systems isoelectronic with them to predict the spacing, as well as bulk modulus for polar systems. Bechstedt did just that, and used geometric means (e. g., $\sqrt{S(Ga)S(P)}$) for compounds (e. g., GaP) with constituents from two rows, and confirmed that the estimates are quite good. He then proceeded to use these parameters to predict the distortion of the lattice around a wide variety of substitutional impurities in a wide variety of semiconductors.

Use of the Bechstedt form would appear to be a significant numerical improvement, but also a significant complication of the resulting formulae. Further, even it becomes quite inaccurate when we go to different structures, with different numbers of neighbors. These systems have been treated by van Schilfgaarde and Harrison (1986) by a more complete analysis of the effect of nonorthogonalities, which will be discussed in Section 20-1-B. A more recent study of this has been made by Dorantes-Dávila and Pastor (1995). We choose to continue the analysis using the simpler form, Eq. (2-19), but will at some points note the effects of replacing that repulsion by the Bechstedt form.

F. Total Energies and Coulomb Effects

We noted in Section 1-1 that the atomic term values, ε_s and ε_p may be regarded as the removal energies for electrons from the atom. We call these one-electron energies. We noted at the end of Section 1-1 that the total energy of the atom, or the energy required to remove *all* of the electrons, cannot be taken as the sum of these one-electron energies since the energy to remove the second electron exceeds the energy to remove the first by approximately the Coulomb interaction U , listed in Table 1-1, between the two electrons and the third requires even greater energy. As we noted in Chapter 1 this lack of equality of the total energy and the sum of one-electron energies is a universal aspect of calculations of electronic structure, where the one-electron energy contains the interaction of that electron with all nuclei, and with all other electrons. Then when we add the one-electron energy for a second electron it includes again the Coulomb interaction with the first so that in the end all electron-electron interactions are counted twice and must be subtracted. Further, we must add the Coulomb repulsion between all nuclei to the total energy.

Fortunately, if the atoms *remain rather neutral* as they are rearranged, the *change* in the total energy as the atoms are rearranged will be approximately equal to the change in the sum of the one-electron eigenvalues since the Coulomb interactions within the atom change little and the subtracted inter-atom electron terms just cancel the inter-atom nuclear terms. (This fact that the change in energy is given by the change in the sum of one-electron energies of occupied states for neutral atoms was derived, for example, in Harrison (1985) and follows directly from the evaluation of an expression for the pressure by Pettifor (1976) using the Atomic Sphere Approximation of Andersen (1973). It has been widely used as an approximation for many years.) We find also that the use of free-atom term values, appropriate to neutral atoms, generally is quite adequate for the estimates of energy bands and properties, even in the ionic solids, where we *formally* think of electrons as being completely transferred between atoms. This also is consistent with the neglect of Coulomb corrections to the total energies. A similar point was made by Pauling (1960) in his "electroneutrality" concept. It will always be the change in total energy which is of interest here so we are able to proceed with the extraordinary simplification of regarding the total energy as simply a sum of the energies, such as we calculated in Eq. (2-10), of the occupied electronic states.

3. The Cohesive Energy

It is convenient to develop the electronic structure of the semiconductor step by step from the atoms using the concepts we have given. This becomes particularly useful when we discuss the binding energy for the solid, which is then given by the total change in the energy of the occupied states as we move from the free atom to the solid.

A. The Bond-Orbital Approximation and Matrices

We begin with the approximation which we used for N_2 in Section 1-4 where we included only the coupling, V_2, between two hybrids directed into the same bond. This is called the *Bond-Orbital Approximation* (Pantelides and Harrison, 1975). After making that approximation here, we return at the end of the section to a more formal mathematical statement of what this has done.

We can follow the energy change step-by-step using an energy-level diagram, such as those in Figs 1-3, 1-7, and 1-9 for molecules. We begin here with a homopolar semiconductor, silicon. The atomic energy levels ε_s and ε_p are shown to the left in Figure 2-4. The free atom has two electrons in the valence 3s-state and two in the 3p-state. When we form sp-hybrids in the next step, the energy of these four electrons becomes $4\varepsilon_h = \varepsilon_s + 3\varepsilon_p$ (rather than $2\varepsilon_s + 2\varepsilon_p$), for an increase in energy of $(\varepsilon_p - \varepsilon_s)/4 = V_1$ per electron. This *promotion energy* is $2V_1$ per bond, because there are two electrons per bond. In the next step, we form bond orbitals and antibonding orbitals of energy $\varepsilon_h \pm V_2$, and gain a bonding energy, $-2V_2$ per bond. We saw in Part E of Section 2-2 that the overlap interaction cancels half of this binding energy, $V_0 \approx V_2$. This leaves us with a net change in energy per bond of

$$-E_{coh} = -2V_2 + 2V_1 + V_2 = -V_2(1 - \alpha_m), \tag{2-21}$$

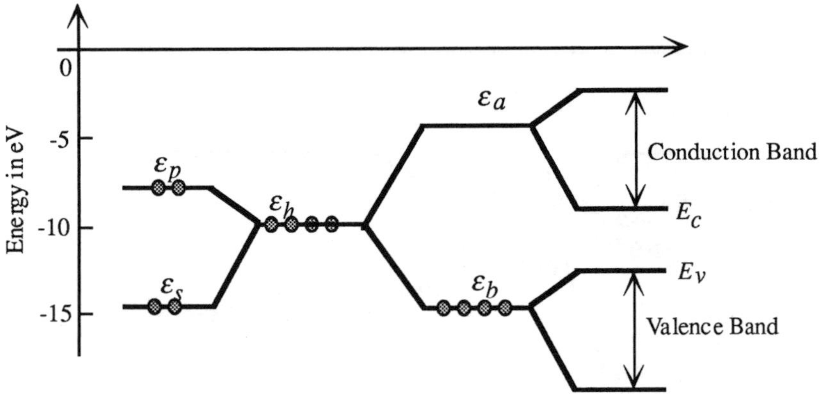

Fig. 2-4. Development of the electronic structure of silicon. The single s-states and p-states on each atom are transformed to four hybrid states, which are combined with neighboring hybrids to form bonds. The Bond-Orbital Approximation stops at that point and neglects additional couplings which finally broaden the bonding and antibonding levels into energy bands in the crystal.

where we have used the definition of metallicity α_m from Eq. (2-18). This is a drop in energy (if $\alpha_m < 1$) and we note that the cohesive energy E_{coh} is taken as the magnitude. If we could stop at this step, we would have a very simple theory of the binding in the solid. The bond energy would be the covalent energy, reduced by the effect of metallicity and going to zero when the gap goes to zero, according to Eq. (2-17). It is basically true that the gain in energy from bond formation is largely canceled by the repulsion and the promotion energy, and because of the cancellation the much smaller terms in the energy - metallization, which we consider in the following section - become very important to the net cohesion. They arise from couplings which we neglect in the Bond-Orbital Approximation. We note also that this is our first case in which, as mentioned earlier, eliminating the coefficient of the overlap repulsion, C , with the equilibrium condition leads us to a result in terms only of our parameters V_1, V_2, etc.

The development of the electronic structure for a polar semiconductor proceeds in the same way, as illustrated in Figure 2-5. Hybrids are formed on both atom types. The promotion energy is perhaps most easily calculated by first transferring electrons between atoms so that there are four electrons on each atom and the rest of the promotion energy for each is calculated exactly as in the homopolar case. For a III-V semiconductor such as GaAs the transfer is of a single electron from the p-state on the nonmetallic atom (indicated by a "-") to the p-state on the metallic one

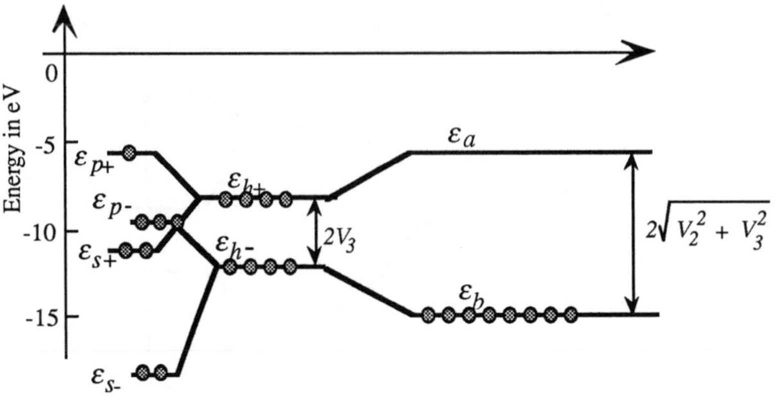

Fig. 2-5. Development of the electronic structure of gallium arsenide. The s-state and p-state energies given in Table 1-1 are transformed to individual hybrids, each with energy $(\varepsilon_s + 3\varepsilon_p)/4$ for the atom in question, which then form bonding and antibonding states with the other hybrid in the bond. We have shown all eight electrons for an atom pair for gallium arsenide. In the promotion of electrons to hybrids we left half the electrons in gallium hybrids and half in arsenic hybrids costing $4E_{pro}$ for each pair of atoms, with E_{pro} the energy per bond given in the text.

(indicated by a "+") and a promotion energy *per bond* of $E_{pro} = (\varepsilon_{p+} - \varepsilon_p)/4 + V_{1+} + V_{1-}$. For a II-VI semiconductor such as ZnSe, two electrons must be transferred for a promotion energy per bond of $E_{pro} = (\varepsilon_{p+} - \varepsilon_p)/2 + V_{1+} + V_{1-}$. For a I-VIII semiconductor such as CuBr, an electron must also be transferred to the s-state of the copper atom, giving a promotion energy per bond of $E_{pro} = (\varepsilon_{s+} - \varepsilon_{p-})/4 + (\varepsilon_{p+} - \varepsilon_{p-})/2 + V_{1+} + V_{1-}$.

The formation of bonds then gains an energy $2\sqrt{V_2{}^2 + V_3{}^2}$ per bond, deduced from Eq. (2-10) since the average energy of the electrons in hybrids before that, as in Fig. 2-5, is $(\varepsilon_{h+} + \varepsilon_{h-})/2$. If we add the overlap repulsion at this point, $V_0 = C/d^4$, and minimize with respect to d at the equilibrium spacing, we find $V_0 = \alpha_c V_2$, evaluated at the equilibrium spacing. This reduces to the V_2 for the homopolar case as we found before. These bonds are finally broadened into bands, as in silicon; that step is not shown in Figure 2-5.

It is helpful to state in mathematical terms what we have been doing here. Our basic assumption is that the eigenstates could be written as an expansion in terms of the valence s- and p-states $|i\rangle$ on all of the atoms. We number the states of the solid by k, which may be a wavenumber for a perfect lattice but in general may just be an index.

$$/\psi_k> = \Sigma_i u_i(k)\ /i> \ . \tag{2-22}$$

This was already done in Sections 1-2, 3, and 4 for simple systems, but the sum now contains $4N$ terms for N atoms. We may construct the energy $\varepsilon_k = <k/H/k>/<k/k>$ which is $\Sigma_{i,j} u_i^* u_j H_{ij} / \Sigma_i u_i^* u_i$ if we replace the effects of nonorthogonality $<i/j>$ by an overlap repulsion $V_0(d)$ as we did for Li_2 in Section 1-2. Minimizing this with respect to the u_i gives

$$\Sigma_j H_{ij} u_j = \varepsilon_k u_i \qquad \text{for } i = 1, 2, ...4N. \tag{2-23}$$

This is a direct generalization of Eq. (1-19) which we wrote for the lithium chain.

In Chapter 1 following Eq. (1-12) we saw that the elements of the Hamiltonian matrix $H_{ij} \equiv <i/H/j>$ could be taken as the free-atom term value for $i = j$, as universal parameters given in Eq. (1-21) for nearest-neighbor atomic states, and as zero otherwise. Each step we took in this chapter replaced this orthonormal set by a new orthonormal set, $/j> = \Sigma_i U_{ij}/i>$. One such transformation was to hybrid states on each atom, so that for each hybrid $/j>$ there were four terms in the sum $\Sigma_i U_{ij}/i>$ and these were given in Eq. (2-3). We shall see explicitly in Section 2-5-A that such a transformation from one orthonormal set to another is a *unitary* transformation, meaning that the inverse of the matrix is equal to the conjugate transpose, $U^{-1}{}_{ij} = U_{ji}^*$. The unitary transformation changes the Hamiltonian matrix to $H_{kl} = \Sigma_{i,j} U_{ki}^{-1} H_{ij} U_{jl} = \Sigma_{i,j} U_{ik}^* H_{ij} U_{jl}$. In the case of the transformation to hybrids the *diagonal* matrix elements H_{kk} become simply the hybrid energies ε_h . In fact, the diagrams which we drew in Figs. 2-4 and 2-5 simply trace out the changes in diagonal matrix elements of the Hamiltonian matrix with each transformation. Of course the transformations also change the off-diagonal matrix elements, some of which were $V_{ss\sigma}$ in the atomic basis but became perhaps V_2 or V_1 or V_{1x} in terms of the hybrid basis.

An important aspect of any such unitary transformation is that it leaves the average of the diagonal matrix elements unchanged. We see that explicitly for the transformations to hybrids and to bonds and antibonds in Figs. 2-4 and 2-5, and it follows generally from the definition of unitarity we gave above, and we shall show it explicitly (Eq. (2-51)) in Section 2-5-A. This is also true of the unitary transformation $U_{jk} = u(k)_j$ which gives the exact solutions of Eq. (2-23), the transformation which *diagonalizes* the Hamiltonian matrix to the form $H_{kk'} = \varepsilon_k \delta_{kk'}$. The average energy of all of the electronic states of the solid obtained from Eq. (2-23) is equal to the average energy of the atomic states which entered as diagonal elements of the Hamiltonian matrix. However, it will prove particularly useful for clarifying the meaning of our approximate diagonalizations.

We imagine the $4N\times4N$ Hamiltonian matrix for a homopolar system of N atoms. There will be 4×4 blocks along the diagonal which in the transformation to hybrids become

$$H = \begin{pmatrix} \varepsilon_s & 0 & 0 & 0 \\ 0 & \varepsilon_p & 0 & 0 \\ 0 & 0 & \varepsilon_p & 0 \\ 0 & 0 & 0 & \varepsilon_p \end{pmatrix} \rightarrow \begin{pmatrix} \varepsilon_h & -V_1 & -V_1 & -V_1 \\ -V_1 & \varepsilon_h & -V_1 & -V_1 \\ -V_1 & -V_1 & \varepsilon_h & -V_1 \\ -V_1 & -V_1 & -V_1 & \varepsilon_h \end{pmatrix} . \quad (2\text{-}24)$$

Diagonalization of this matrix on the right would lead back to that on the left. However, proceeding from the $4N\times4N$ matrix with hybrids along the diagonal to a matrix with bond and antibond energies along the diagonal gives

$$\begin{pmatrix} \varepsilon_b & V & V & V & 0 & V' & V' & V' \\ V & \varepsilon_b & V & V & V' & 0 & V' & V' \\ V & V & \varepsilon_b & V & V' & V' & 0 & V' \\ V & V & V & \varepsilon_b & V' & V' & V' & 0 \\ 0 & V' & V' & V' & \varepsilon_a & V & V & V \\ V' & 0 & V' & V' & V & \varepsilon_a & V & V \\ V' & V' & 0 & V' & V & V & \varepsilon_a & V \\ V' & V' & V' & 0 & V & V & V & \varepsilon_a \end{pmatrix} \quad (2\text{-}25)$$

This is written illustratively as an 8×8 array rather than the full $4N\times4N$ matrix for a crystal of N atoms. It shows bond and antibond energies along the diagonal. The off-diagonal matrix element for each bond with the antibond in the same site is zero. We have used a V to represent any matrix elements V_1 or V_{1x}, or other matrix elements, between pairs of bonds or pairs of antibonds and have used a V' to represent any of the remaining matrix elements between a bond and neighboring antibond.

The Bond-Orbital Approximation (Pantelides and Harrison (1975)) is exactly the neglect of the matrix elements of the type V'. If these are neglected we are left with two quadrants, a bonding quadrant in the upper left and an antibonding quadrant in the lower right and we include no coupling between them. We could then diagonalize each of these quadrants independently and the upper-left quadrant would contain all of the energies in the valence bands and in the pure system these would all be occupied. The important point is that the sum of the energies of all of these occupied states is exactly the sum of the energies ε_b since, as we indicated, the sum of the diagonal elements is not changed by the diagonalization. We can obtain the total energy without this last diagonalization, and this remains true even if this crystal has defects, or is even amorphous, as long as we are able to transform first to one set of orbitals, bond orbitals, which are occupied and another set, antibonding orbitals, which are empty. Thus, we are not neglecting the matrix elements V which broaden the real electronic states into bands, as represented to the far right in Fig. 2-4; it is just that that formation of bands has not affected the total energy. We are neglecting the one set of matrix elements indicated by V' above. When we can do that, the simple theory based upon independent two-electron bonds will be valid.

The effect of these neglected couplings tends to be small since, as we have seen, they are small compared to the V_2 coupling the two hybrids in the same bond. Further the neglected couplings are between states which are considerably different in energy, $2V_2$. Nonetheless in a case such as the cohesion, where the promotion energy may very nearly cancel the entire gain from bond formation (V_2 per electron), these small additional terms can be important.

These couplings can actually be included approximately in the matrix given above before the final diagonalization of the bonding quadrant to form band states. This approximate contribution, which we call metallization will be considered next.

B. Metallization of the Bonds.

We use perturbation theory, which was derived in Eq. (1-30) for the case in which the coupling between levels is small compared to their energy difference, exactly the situation we anticipate here. We found in Eq. (1-30) that an energy level initially at ε_i^0, coupled by V_{ij} to states initially at energy ε_j^0, is shifted to an energy

$$\varepsilon_i = \varepsilon_i^0 + \sum_j \frac{V_{ij}V_{ji}}{\varepsilon_i^0 - \varepsilon_j^0} \quad . \tag{2-26}$$

We use this to obtain the shift of the bond levels initially at ε_b due to their coupling with the neighboring antibonding levels initially at ε_a. The energy difference $\varepsilon_b - \varepsilon_a$ in the denominator is large compared to the couplings so this formula is applicable. For each term j there is an upward shift of the antibonding state, equal and opposite to the downward shift of the bonding level, but since only the bonding levels are occupied, we gain energy by this shift, which arises from the V' in the matrix, Eq.

(2-25). The formula is not applicable to the matrix elements V from Eq. (2-25) since the energy denominator would be zero. However, we have seen that those couplings broaden the band without shifting the average so they did not affect the energy. Every upward shift of an occupied state was canceled by a downward shift of another occupied state.

We have all of the parameters needed to evaluate from Eq. (2-26) the shift in energy of each bond state due to its coupling with the neighboring antibonding states. Eq. (2-16) gives the metallic energy V_1 , the negative of which is the coupling between two *hybrids* on the same atom. To obtain the coupling between a bond orbital and an antibonding orbital sharing the same atom, we must weight $-V_1$ by the amplitudes that each orbital has on that atom. For a homopolar crystal we have seen that each amplitude is $\pm 1/\sqrt{2}$ so the coupling has magnitude $V_1/2$. For a polar crystal the amplitudes are obtained from Eqs. (2-11) and (2-15). For a bond and an antibond sharing a metallic atom, the coupling energy is

$$V_{ab+} = -u_+^b u_+^a V_{1+} = \sqrt{1 - \alpha_p^2}\, V_{1+}/2 = \alpha_c V_{1+}/2 \ . \qquad (2\text{-}27)$$

The coupling through a nonmetallic atom is the same, but with V_{1+} replaced by V_{1-} and the sign changed because u_+^a as given in Eq. (2-15) is negative. This sign does not matter because it enters Eq. (2-26) squared.

The energy shift of a bond resulting from metallization can be immediately obtained by substituting these coupling energies into the perturbation-theory formula, Eq. (2-26). The energy difference between the two coupled states is $2\sqrt{V_2^2 + V_3^2}$ and the bond orbital is coupled to three antibonding orbitals at each end as seen in Figure 2-6; accordingly, the total shift for the two electrons in the bond is the metallization energy

$$E_{met} = -\frac{3\alpha_c^2}{4}\frac{V_{1+}^2 + V_{1-}^2}{\sqrt{V_2^2 + V_3^2}} \ , \qquad (2\text{-}28)$$

which for the homopolar case becomes simply $-3V_1^2/(2V_2)$.

Before continuing the calculation, we may note the change in the bond states themselves due to this metallization. We noted following Eq. (1-30) that this same coupling which gave the shift in Eq. (2-26) admixes the coupled state to the starting state, in the homopolar case here adding a term to the bond state as $|b\rangle = |b\rangle^0 + (V_1/2)/(2V_2)\,|a\rangle$ for each antibonding state coupled by $V_1/2$. In this modified bond state, the probability of finding an electron occupying the state on a neighboring antibond is the coefficient squared, $V_1^2/(16V_2^2) = \alpha_m^2/64$, to second order in the metallicity. This is illustrated in Fig. 2-6, showing the bond orbital with

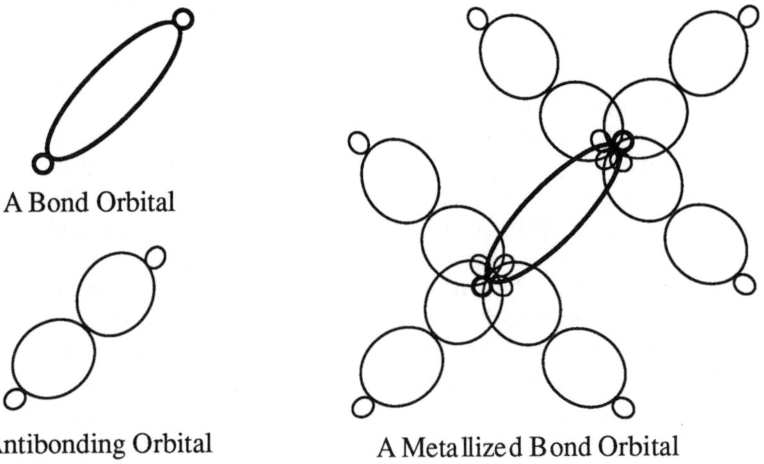

Fig. 2-6. Metallization of a bond arises from the coupling of the bond state, shown in the center, to antibonding states in the six neighboring bond sites. In the homopolar case, the coupling energy is $-V_1/2$; in the polar case, it is different at the metallic end from that at the non-metallic end (Problem 2-4).

admixed probability density on the six neighboring antibonds. This is a relatively small total probability, $6\alpha_m^2/64$, equal to 0.06 for silicon, but we may think of this "relaxation" of the bond into the neighboring bond sites as the origin of the metallization energy.

When we evaluate the total energy of the system, we must add this shift in energy, Eq. (2-28), for each bond. This contributes a negative term to the bond tension, which can be obtained from Eq. (2-28) by taking a derivative with respect to d noting that V_2 enters α_c , as well as the denominator. We obtain

$$\frac{\partial E_{met}}{\partial d} = - \frac{3(V_{1+}^2 + V_{1-}^2)\alpha_c^2(3\alpha_c^2 - 2)}{2d\sqrt{V_2^2 + V_3^2}} , \qquad (2\text{-}29)$$

This contributes to the tension in the bonds and will modify the equilibrium condition, which we shall obtain shortly.

In computing such a contribution to a system in which all bonds do not remain identical (as in an inhomogeneous distortion or an alloy), we note that there is a contribution to a particular bond tension, owing both to the terms arising from terms such as Eq. (2-29), where the bond in question lies in the site where the tension is being computed, and from such terms where the bond lies in a neighboring site and the antibond lies in the site where the tension is being computed. We automatically included both for a homogeneous distortion when we obtained Eq. (2-29) by taking the derivative of Eq. (2-28) with respect to d.

The qualitative effect of bond metallization is now quite clear: It always enhances the bond energy because [as seen from Eq. (2-28)] the energy of the occupied bond states is always lowered. We may envision this as a lowering of the energy of the system as the bond is allowed to relax out into its environment. We note further that, in the homopolar semiconductor, metallization *reduces* the bond tension and allows the bond length to increase, because Eq. (2-29) gives a negative tension when the covalency is one. This may also be thought of as an effect of allowing the bond to expand into its environment. In the polar semiconductor, on the other hand, $3\alpha_c^2 - 2$ becomes negative (crossing zero roughly at III-V semiconductors) and metallization increases the tension; this has important effects on the bond length in polar semiconductors. Once the bond is sufficiently well localized on the nonmetallic atom, the effect of relaxing into the environment is to contract the lattice. These ways of thinking about the bond are not explanations of anything; rather, they are simply ways of remembering a complicated theoretical result [Eqs. (2-28) and (2-29)].

C. Homopolar Systems

We may now easily reevaluate the energy per bond of the system, relative to that of free atoms, including metallization which we obtain from Eq. (2-28). We do this first for elemental ($\alpha_p = 0$, $\alpha_c = 1$) semiconductors. This contains the terms obtained as in Fig. 2-4 and the overlap repulsion, which can be conveniently rewritten as AV_2^2 which is equivalent to C/d^4.

$$E_{tot}/bond = 2V_1 - 2V_2 - 3V_1^2/(2V_2) + AV_2^2 \; . \qquad (2\text{-}30)$$

A is to be adjusted such that this energy is minimum at the equilibrium spacing. As Eq. (2-30) is written the spacing enters only through V_2 so we may minimize with respect to V_2 to obtain

$$-2 + 3V_1^2/(2V_2^2) + 2AV_2 = 0 \; , \qquad (2\text{-}31)$$

at the equilibrium spacing. It is of interest to use this value of A in Eq. (2-30) and plot the resulting $E_{tot}/bond$ as a function of d , as in Fig. 2-7.

A has been chosen so the minimum comes at $d = d_0$, the equilibrium spacing. We see also that the theoretical curve drops at large distances, due to the metallization, which increases as d^2 due to the decreasing energy denominator. This reminds us that all expressions, and the metallization in particular, are only good near the equilibrium spacing and we have carried the plot well beyond that. We may nevertheless expect that the expression and the curve is appropriate very near equilibrium. This drop-off is of no concern unless we wish to consider such large spacings, and then the metallization and all other terms will need to be modified to eliminate such unrealistic behavior.

Fig. 2-7 also shows the total energy without metallization, which will be used later for some properties. Note that it approaches a positive value, $2V_1$, at large d. This is the promotion energy from the starting atomic states.

Of more interest are the values at equilibrium, which are obtained by solving Eq. (2-31) to obtain $AV_2^2 = V_2 - 3V_1^2/(4V_2)$ at equilibrium, which may be substituted back into Eq. (2-30) to obtain

$$E_{tot}/bond = 2V_1 - V_2 - 9V_1^2/(4V_2) \qquad (2\text{-}32)$$

at the equilibrium spacing.

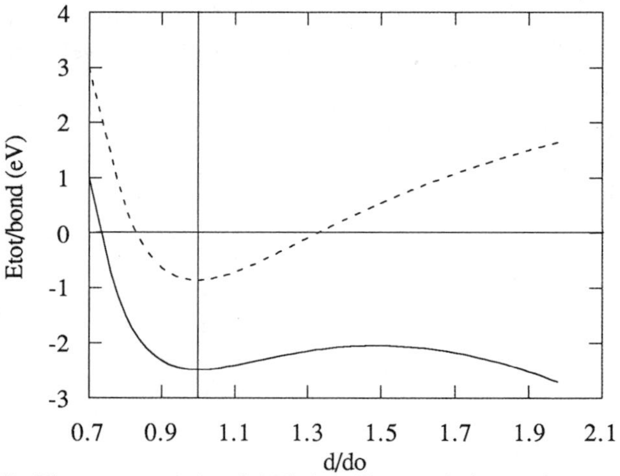

Fig. 2-7. The energy per bond (eV) for silicon, relative to the energy of free atoms, as a function of atom spacing, the solid line is from Eq. (2-30). The dashed line is without metallization, and with the overlap interaction suitably scaled to give the correct equilibrium spacing. All contributions are really meaningful only near the equilibrium spacing. The metallization term becomes particularly inappropriate at large spacing.

It is convenient also in this case to write the magnitude of this negative energy as the cohesive energy per bond, E_{coh} , and to use the definition of metallicity, $\alpha_m = 2V_1/V_2$ for homopolar materials, to write

$$E_{coh} = V_2(1 - \alpha_m + \frac{9}{16}\alpha_m^2) . \qquad (2\text{-}33)$$

The cohesive energy appears as an expansion in metallicity, with a third term added to Eq. (2-21). The leading term is the covalent energy from the bond (reduced by an overlap repulsion). The next is the promotion energy and the third term is the metallization (enhanced by the reduction in the overlap repulsion). As the metallicity rises towards one (at tin as we move through C, Si, Ge, Sn), the metallization term more and more dominates the cohesion.

The value E_{coh} is our estimate of the cohesive energy, the energy required to separate the solid into atoms. The predicted values are listed in Table 2-3 and compared to the corresponding experimental values. For silicon, germanium, and tin, the agreement is remarkably good, given the simplicity of the theory, the lack of any parameters fitted to cohesion, and the rather considerable cancellation owing to promotion. Note in particular that in tin, with $\alpha_m = 1$, Eq. (2-33) shows that the entire cohesion arises from metallization. This fits well intuitively with the

Table 2-3. The cohesive energy E_{coh} , or heat of atomization, in eV per bond, compared with experiment.

Semiconductor	α_m	Theory	Experiment
C	0.40	7.13	3.68
BN		7.04	3.34
BeO		6.40	3.06
Si	0.81	2.48	2.32
AlP		2.62	2.13
Ge	0.95	2.30	1.94
GaAs		2.09	1.63
ZnSe		2.19	1.29
CuBr		1.71	1.45
Sn	1.00	1.76	1.56
InSb		1.55	1.40
CdTe		1.68	1.03
AgI		1.33	1.18

vanishing band gap for tetrahedrally coordinated tin (gray tin) and the almost equal cohesion of metallic tin (white tin). The next in the series, lead, is stable only as a metal.

In the case of diamond, we have underestimated the cohesion by a factor of nearly two. This was also true of the estimates by Hoffmann (1963) using the Extended Hückel Theory. It is generally true that this simple theory works best for silicon-row semiconductors and for heavier ones. Although it is poorest for carbon-row systems, even there the theory is meaningful. For cohesion it appears to be a good rule of thumb that for the first row we underestimate by a factor of two; for compounds half from the first row (SiO_2) , a factor of $\sqrt{2}$.

Use of the Bechstedt form for the overlap repulsion, Eq. (2-20), makes only a small change in the predictions. It replaces the overlap repulsion $V_2 - 3V_1^2/(4V_2)$ given above Eq. (2-32) by the values listed as $V_0(d_0)$ in Table 2-2. This lowers the prediction in Table 2-3 by 2.37 eV for C, reducing considerable the discrepancy for this first-row material. It increases the predictions for Si, Ge, and Sn by 0.03 eV, 0.11 eV, and 0.56 eV, respectively, slightly worsening agreement with experiment. The general scale of agreement for cohesion is not greatly affected by the difference.

D. Polar Systems

The evaluation of the bond energy for polar semiconductors is also straightforward, and was outlined in Section 2-3-A. It contains a promotion energy E_{pro}, which includes transfer of electrons between atoms as well as promotion to p-states . A gain of $-2\sqrt{V_2^2 + V_3^2}$ is obtained in the bond formation. The metallization energy is given in Eq. (2-28). The term in the bond tension from the derivative of the bond-formation energy with respect to d in the polar case is $4V_2\alpha_c/d$ and that from the metallization is given in Eqs. (2-29). They are equated to $4C/d^5$ to obtain a value of C and C/d^4 which is the overlap repulsion. With this procedure the cohesive energy per bond becomes

$$E_{coh} = (2 - \alpha_c^2)\sqrt{V_2^2 + V_3^2} + \frac{9\alpha_c^4(V_{1+}^2 + V_{1-}^2)}{8\sqrt{V_2^2 + V_3^2}} - E_{pro} , \qquad (2\text{-}34)$$

with the E_{pro} described in Section 2-3-A. This reduces to Eq. (2-33), with E_{pro} equal to $2V_1$, for the homopolar case, and $\alpha_c = 1$. Eq. (2-34) was evaluated using Tables 1-1 and 2-1 for the compounds isoelectronic with carbon, silicon, germanium and tin and the results were listed with the homopolar values in Table 2-3. The accuracy is comparable to that for the homopolar semiconductors. The factor-of-two overestimate from diamond carries over to BN and BeO. The trend of decreasing cohesion with decreasing covalency (empirically E_{coh} is approximately proportional to α_c) was noted earlier by Kraut and Harrison (1984).

This is not consistent with a familiar concept of additive ionic and covalent contributions to the bond energy (Pauling(1960), Phillips (1973)), which arose in connection with *formation energies*, the energy of a compound relative to its isolated constituent elements in the standard condensed state. Formation energies would seem a poor guide to our intuition since they depend so much on the particular state formed by the element (e. g., Cl_2 for CuCl). Further, as we extrapolate back from polar crystals, ZnSe, GaAs to Ge, the limiting case is separating GeGe into two Ge crystals at no cost in energy. Thus almost automatically the formation energy increases with polarity, independent of the nature of the bonding.

It may not be so important to have such an approximate theory of a well-known property, but just the same procedure is applicable to the heat of solution and other properties of imperfect semiconductors. We return to that in Chapter 8.

E. Coulomb Corrections.

One might ask if we have not made a serious error in discussing polar bonds by ignoring the repulsion U between electrons occupying the same atom. We found in Section 2-2-C that a bond electron had a probability $(u_-^b)^2 = (1 + \alpha_p)/2$ of lying on the nonmetallic atom. Thus the total number of electrons on that atom, from the two electrons in each bond, is $4(1 + \alpha_p)$ or about six for a III-V semiconductor, with polarity of 0.5. Thus there is one extra electronic charge, relative to the neutral atom, on the arsenic in GaAs, an effective charge Z^* of one as defined in Section 4-1. We might therefore expect a shift of UZ^* in the atomic energies ε_i which we used in our calculations. However, there is also a shift due to the neighboring atoms which have charges. This shift is called the Madelung potential and will be discussed in the chapter on ionic crystals. It is the sum of the shift due to *all* neighbors, assuming that their charge distributions due not overlap the atom in question. It is $\alpha Z^* e^2/d$, with $\alpha = 1.64$ for zincblende and wurtzite structures. For GaAs this is $9.64Z^*$ eV, which more than cancels the average of the shifts $\frac{1}{2}(U_{Ga} + U_{As})Z^*$ obtained from Table 1-1 as $7.46Z^*$ eV. The effects of neighbors came out larger than the intraatomic term because in fact charge distributions on one atom do overlap those on the neighbors and reduce the correction (Harrison and Klepeis (1988)). In any case, the corrections are much smaller than one would at first have thought and our procedure of using free-atom term values in the polar solid does quite well. This was learned early in fitting full band calculations with tight-binding parameters and noting that the fit energy-level differences been the two atomic species were roughly consistent with the free-atom term values.

The appropriateness of neglecting Coulomb shifts of the term values also bears on the question of total-energy calculations discussed in Section 2-2-F. We noted there that if the atoms remain neutral as they are rearranged, the total change in energy is equal to the change of the sum of the one-electron eigenvalues. Here we see that even though charges shift between atoms, the Coulomb shifts may cancel, and again the use of a sum of one-electron energies for the total change in energy becomes appropriate. A more complete discussion of the appropriateness of this method for calculation of the total energy is given by Weinert, Watson and Davenport (1985).

It is of course no accident that these cancellations occur. When charges are transferred, giving nonzero Z^*'s, the atoms tend to arrange themselves so that the missing electrons are made up for by the distributions on the neighboring atoms so that the final charge distribution looks much like a superposition of free-atom densities, a frequent approximation used in band calculations. There will be many cases where we force a different charge distribution, by exciting an electron from the bonding band and placing it a distance from the site with the missing electron. The energy change is the experimental band gap, and we will see that Coulomb corrections should be made to the values which come from the band calculation. There are also cases, as at semiconductor surfaces, where we would predict a reconstruction based upon doubly occupying half the surface sites and leaving the other empty unless we take into account the Coulomb corrections which in fact prevent it. Another approach to this question was given by Pettifor (1976) in the context of the total energy of transition metals, using the Atomic Sphere Approximation of Andersen (1973) for evaluation of an expression for the pressure due to Liberman (1971). This has been discussed also by Mackintosh and Andersen (1980).

The corrections discussed above can all be made within Hartree-Fock theory. There are also Coulomb corrections to Hartree-Fock theory itself, or density-functional theory, which lie outside those theories. We present an argument here that the errors in Hartree-Fock theory itself are small, but it is not a simple subject and is not essential to most of our considerations. The discussion is given in more detail in Harrison (1985b).

Corrections to the energy beyond Hartree-Fock are generally called *configuration interactions* by chemists and *correlation energies* by physicists. It is only possible to calculate them approximately using many-body theory; e.g. , Horsch, Horsch, and Fulde (1984). They can also be understood rather simply and estimated in terms of tight-binding theory using the intraatomic coulomb repulsion U (Harrison, 1985a). The most important physical effect can be understood in terms of a simple two-center nonpolar bond, such as the Li_2 molecule discussed in Section 1-2. If only a single electron were involved, its energy would be lowered by the coupling energy V_2. If two electrons were involved, one from each atom, we gain $2V_2$, but now there is a fifty-percent probability of both simultaneously occupying the spin-up and spin down state on one atom, raising the energy by U when it occurs, a quantity which had no influence while the atoms were separate. This would be included in the Hartree-Fock theory of the bond itself and not in our free-atom Hartree-Fock values. This is reduced by the positive charge on the neighboring atom when this happens, so U should be replaced by $U^* = U - e^2/d$. For the Li_2 molecule discussed in Section 1-2, $U = 8.16$ eV and $d = 2.67$ Å, so $U^* = 2.77$ eV, now to be divided by two since there is also a fifty-percent probability that the electrons are found on different atoms. This is a reduction in the cohesive energy included in a full Hartree-Fock theory. A lower-energy state can be constructed by admixing some antibonding state (or configuration) to produce a two-electron state in which the probability of two electrons occupying the same state is reduced - that is, the electron motion is correlated. We will see how this is accomplished in Unrestricted Hartree-Fock theory when we discuss correlated states in f-shell metals in Section 16-1. We ignore these corrections here, feeling that they are small compared to other errors in the tight-binding theory.

4. The Dependence of Energy upon Volume

Our analysis above was directed at obtaining the energy at the equilibrium spacing, but resulted in a form for the energy, Eq. (2-30) for homopolar systems which was plotted as Fig. 2-7. That form should be valid near the equilibrium spacing. Thus a number of properties, involving the various derivatives of the energy with respect to spacing d, can be addressed using that curve.

The algebra involving these derivatives can be confusing since we often are interested in the variation of properties with changes in volume. Then, for example, the derivative of the energy per unit volume, with respect to volume, contains terms from the change in volume as well as from the change in each bond energy with volume. We seek to minimize this confusion by working always in terms of a curve such as those in Fig. 2-7 which gave a bond energy as a function of bond length, and we can clearly discuss properties in terms of the various derivatives of this energy with respect to d, evaluated exactly at the equilibrium bond length d_0. We shall use this framework also when we discuss volume-dependent properties of ionic and metallic crystals.

A. Trends in Bond Length

Of course we fit the coefficient in the overlap repulsion to obtain the correct bond length so the position of the minimum automatically agrees. Nevertheless, we can obtain some insight into the dependence of bond length on material. This was done earlier (Bechstedt and Harrison (1989)) with the Bechstedt overlap interaction discussed at the end of Section 2-2-E, but here we use the simpler C/d^4 form. The increase in bond length with metallicity (the increase as we move down the rows of the Periodic Table) can be associated with the increased core radius r_c entering Eq. (2-19) for atoms with more electrons in the cores, expanding the electron cloud increases the overlap repulsion.

We might imagine that the coefficient $C = 16{\times}3.22(\hbar^2 r_c^2/m)$ would not vary much with the polarity, as one atom shrinks (the nonmetallic one) and the other expands, and indeed the average r_c obtained from Table 13-5, or from the Solid-State Table, is relatively constant. At first that would seem to explain the familiar nearly-constant bond length in a series such as Ge, GaAs, ZnSe, and CuBr. However, the bond tension arising from the bonding term, $-2\sqrt{V_2{}^3 + V_3{}^2}$, is $-2\alpha_c\ \partial V_2/\partial d$. The equilibrium condition, without metallization and with C fixed at the germanium value, then gives the result that the equilibrium d should vary as $1/\sqrt{\alpha_c}$, giving bond lengths for these systems, and a number of others, listed in Table 2-4.

This expansion does not generally occur; it is canceled by the decreased metallization, and correspondingly decreased outward relaxation of the bond due to it. We must add the decreased bond tension from metallization, Eq. (2-28), in determining the equilibrium spacing. This leads, if we hold the C of the overlap interaction at the value for homopolar systems, to the predictions given in the third column in Table 2-4. It was necessary to obtain these numerically by iterating the equation for the equilibrium condition. The corresponding calculation carried out by Bechstedt and Harrison (1989) gave the values listed in parentheses, in slightly better accord with experiment.

Metallization has largely eliminated the discrepancy. (Note the 2.59 Å and 2.95 Å predictions for GaAs and ZnSe without metallization.) The reason for the indifferent predictions (particularly for the II-VI compounds) is quite clear. The effects of polarity alone are very large, but are then canceled by the effects of metallization. These effects are of quite

Table 2-4. Bond lengths (Å) for compounds, predicted assuming that the overlap repulsion is the same as for the elemental semiconductor isoelectronic with that compound. The first value $d(homo)/\sqrt{\alpha_c}$ is the prediction without metallization and the C / d^4 overlap repulsion. $d(theory)$ includes metallization but the same form of repulsion. The value in parentheses also includes metallization, but the Bechstedt overlap repulsion (Bechstedt and Harrison(1989)).

Semiconductor	$d(homo)/\sqrt{\alpha_c}$	$d(theory)$	$d(experiment)$
C	1.54		1.54
BN	1.59	1.58(1.58)	1.57
BeO	1.75	1.74	1.65
Si	2.35		2.35
AlP	2.52	2.40(2.38)	2.36
Ge	2.44		2.44
GaAs	2.59	2.42(2.43)	2.45
ZnSe	2.95	2.70(2.54)	2.45
Sn	2.80		2.80
InSb	3.01	2.76(2.79)	2.81
CdTe	3.44	3.23(2.88)	2.81

different origin and are not given individually with sufficient accuracy that the small difference is reliably predicted. The divergence from experiment is even greater in the I-VII compounds, not included in Table 2-4. However, the qualitative result appears to be correct: In the heavier compounds the effect of polarity is largely canceled by metallization. Only for the carbon series, where the metallization is small, do we find experimentally the increase in bond length with increasing polarity expected in the simplest theory.

B. The Radial Force Constant

The value of the curve in Fig. 2-7 gives the cohesive energy, setting the first derivative equal to zero gives the equilibrium spacing, and the second-derivative gives the force constant and the bulk modulus. This property is particularly sensitive to the exact form of the overlap repulsion, C/d^4 , and we therefore cannot expect it to be so reliably given. We proceed with that form to see both how good the predictions are and what the effect from metallization is.

It may be helpful to carry out the calculation without metallization first. The energy becomes

$$E_{tot}/bond = E_{pro} - 2\sqrt{V_2^3 + V_3^2} + C/d^4 \qquad (2\text{-}35)$$

for polar systems, without metallization. The equilibrium condition becomes

$$\partial(E_{tot}/bond)/\partial d = \frac{4V_2^2}{\sqrt{V_2^3 + V_3^2}\,d} - \frac{4C}{d^5} = 0 \ , \qquad (2\text{-}36)$$

which gives

$$C = \alpha_c V_2 d_0^4 \ , \qquad (2\text{-}37)$$

with all quantities evaluated at the equilibrium spacing d_0. We also obtain the second derivative, which we call a "spring constant" $\kappa_0 = \partial^2(E_{tot}/bond)/\partial d^2$, from Eq. (2-36) and it is of course important that we only substitute C from the equilibrium condition after performing the derivative,

$$\kappa_0 = -\frac{20V_2^2}{\sqrt{V_2^3 + V_3^2}\,d^2} + \frac{8V_2^4}{(V_2^3 + V_3^2)^{3/2}d^2} + \frac{20C}{d^6} = \frac{8V_2\alpha_c^3}{d_0^2} \ . \qquad (2\text{-}38)$$

We compare the corresponding predictions with experiment in Table 2-5. We have chosen to tabulate $\kappa_0 d_0^2$, which has the units of energy and therefore may give a better sense of the magnitudes, which are quite large on the scale of V_2. The change in energy per bond due to a change in bond length δd is $^1/_2\kappa_0 d_0^2(\delta d/d_0)^2$. Experimental values were obtained from the bulk modulus, $B \equiv \Omega\partial^2 E_{tot}/\partial\Omega^2 = \kappa_0/(4\sqrt{3}d_0)$, which gives the change in energy per unit volume $^1/_2 B(\delta\Omega/\Omega)^2$ due to a fractional change in volume $\delta\Omega/\Omega = 3\delta d/d$ with the volume per bond given by $4d^3/(3\sqrt{3})$. The general magnitudes and trends with polarity and metallicity are correct, but the values are generally too small and the variation with α_c and α_m are much smaller experimentally than predicted, effects we shall return to after including metallization.

We next carry out the calculation including metallization, Eq. (2-28). The second derivative of the total energy then includes a derivative of Eq. (2-29) with respect to d. Only then can we substitute the equilibrium condition to eliminate C. The resulting force constant is given by

$$\kappa_0 = \frac{8V_2\alpha_c^3}{d_0^2} - \frac{9(V_{1+}^2 + V_{1-}^2)}{V_2 d_0^2}(5\alpha_c^2 - 4)\alpha_c^5 \ . \qquad (2\text{-}39)$$

Table 2-5. $\kappa_0 d_0^2$ (eV), with κ_0 the predicted spring constant of Eq. (2-38) and (2-39), and d_0 the equilibrium spacing. Experimental values are the values which would lead to the observed bulk modulus for the crystal.

Semiconductor	*No Metallization* Eq. (2-38)	*With Metallization* Eq. (2-39)	*Experiment**
C	82.8	75.3	70.4
BN	66.1	63.0	
BeO	33.8	39.0	
Si	35.5	22.4	55.0
AlP	23.2	24.8	48.9
Ge	33.0	16.2	47.7
GaAs	22.6	23.5	47.2
ZnSe	10.5	16.1	38.0
Sn	25.0	10.8	50.4
InSb	16.3	18.0	44.4
CdTe	7.1	11.3	40.0
CuCl	3.7	7.6	21.8

*Compiled in Harrison (1980), p. 196.

The final term includes both the derivative of the metallization term and the change in the overlap repulsion due to metallization. These results are also shown in Table 2-5.

The comparison is quite informative. The addition of metallization has largely eliminated the variation of the force constant with polarity, or covalency, and it is indeed correct that the force constant is rather insensitive to polarity. This reduction in dependence has come about by a decrease in the estimate for the homopolar semiconductors (where $5\alpha_c^2 - 4$ in Eq. (2-39) is positive) and increasing the estimate in the most polar semiconductors (where $5\alpha_c^2 - 4$ in Eq. (2-39) is negative). As in the estimates of bond length, metallization is necessary to describe the observed insensitivity to polarity since a strong dependence on polarity would be predicted without it.

In spite of rectifying the error in this predicted trend, the magnitudes of the force constant predicted for the homopolar semiconductors are still too small by a factor of two to three, except for the carbon-row materials. This appears to be from the inaccuracy of the overlap repulsion. We may redo the analysis which led to Eq. (2-38), but with a $V_0(d) = C/d^n$. For the homopolar case it yields $\kappa_0 = 4(n-2)V_2/d_0^2$, reducing to Eq. (2-38) for $n = 4$. Using the larger values for n suggested by pseudopotential theory just above Eq. (1-34) replaces the first column in Table 2-5 by 123, 69, 66, and

54 for C, Si, Ge, and Sn, respectively. This more than eliminates the underestimate compared to experiment suggesting that indeed the use of the $1/d^4$ repulsion is responsible for much of the discrepancy. Of course the two-term form proposed by Bechstedt and Harrison (1989) eliminates the discrepancy entirely by fitting the experimental value of the bulk modulus, as discussed in Section 2-2-E.

It may be useful to think then of an increasing *abruptness* n of the overlap repulsion with increasing metallicity, relative to our simple estimate of $n = 4$. We distinguish this "abruptness" from a "stiffness" or "hardness" which may better be associated with the value of κ_0 itself. κ_0 , varying as $1/d^4$ for homopolar materials ($\alpha_c = 1$ in Eq. (2-38)), decreases as we move down in the periodic table, while the abruptness of the potential as defined above increases.

C. The Grüneisen Constants

The third derivative of the energy, Fig. 2-7, with respect to bond length is also of interest, and the corresponding derivative will be of interest in all of the systems we consider in this book. The third derivative determines the Grüneisen constant γ which, for a particular vibrational mode, is defined as the logarithmic derivative of the frequency ω of the mode with respect to changes in volume Ω of the system, $\gamma = -\partial ln(\omega)/\partial ln(\Omega) = -(\Omega/\omega)\partial\omega/\partial\Omega$. When we consider thermal expansion we shall see that it defines the real momentum of a vibrational mode in that the force the phonon exerts upon reflection is as if it had a momentum $\gamma\hbar q$.

This dimensionless parameter depends very much upon which mode we consider - having different sign for different modes in semiconductors - and depends upon exactly how it is defined. In most cases we will be interested in exactly the third-derivative with respect to d of a curve such as that in Fig. 2-7, at the point where the first derivative is zero, and the second derivative is defined as κ_0 as in Eq. (2-38). We try to avoid ambiguity by defining a Grüneisen constant in terms of that third derivative. If the nearest-neighbor radial force constant $\kappa_0 = \partial^2 E_{bond}/\partial d^2$ dominates the rigidity, the frequency of a given mode, such as the $q = 0$ optical mode we shall discuss in Section 3-3-B, will be given by a geometrical constant times $\sqrt{\kappa_0/M}$, with M the atomic mass. Then the Grüneisen constant $\gamma = -(\Omega/\omega)\partial\omega/\partial\Omega = -{}^1/_3\ (d/\omega)\partial\omega/\partial d = -{}^1/_6\ (d/\kappa_0)\partial\kappa_0/\partial d$. Thus we shall define a Grüneisen constant

$$\gamma \equiv -\frac{d}{6}\frac{\partial^3 E_{bond}/\partial d^3}{\partial^2 E_{bond}/\partial d^2} , \qquad (2\text{-}40)$$

Table 2-6. Predicted Grüneisen constant, γ, first neglecting metallization and using $V_0 = C/d^4$, second including metallization and C/d^4 and third including metallization and using the Bechstedt form for V_0. Experimental values are also listed, derived from the observed shifts in optical-mode frequency ω_{TO} and in bulk modulus B with volume.

Semiconductor	Eq. (2-41) *(no metallization)*	Eq. (2-43) *(with metallization)*	*Bechstedt* V_0 *(with metallization)*	*Experiment** ω_{TO}	*Experiment** B
C	1.5	1.57	2.25	1.19	1.85
Si	1.5	1.89	2.93	0.98	1.90
Ge	1.5	2.19	3.07	1.12	2.11
Sn	1.5	2.36	3.11		
GaAs	1.72		2.56	1.39	2.08
ZnSe	2.03		2.79	1.45	2.19

*Weinstein and Zallen (1982). See the discussion following Eq. (2-42).

evaluated at the equilibrium spacing $d = d_0$. We have written the $E_{tot/bond}$ of Eq. (2-35) as E_{bond}. First neglecting metallization, we take the derivatives of $E_{bond} = E_{pro} -2\sqrt{V_2^2 + V_3^2} + C/d^4$ *before* substituting the value for C from the equilibrium condition. After some algebra we obtain simply

$$\gamma = \frac{5}{2} - \alpha_c^2 = \frac{3}{2} + \alpha_p^2. \tag{2-41}$$

These are listed in Table 2-6. Experimental values obtained from the optical-mode frequency, which motivated our derivation of the form Eq. (2-40) are also listed. We see that we have slightly overestimated the shift in these vibrational frequencies.

The experimental value with which we generally will wish to compare is the logarithmic derivative of the bulk modulus with respect to volume, $-\partial ln(B)/\partial ln(\Omega) = -(\Omega/B)\partial B/\partial\Omega$ which is also equal to the change in bulk modulus divided by a hydrostatic applied pressure causing the change. There seems to be ambiguity in this quantity associated with the fact that $\partial E_{tot}/\partial\Omega$ is no longer zero at the changed volume. We noted following Eq. (2-38) that $B = \Omega\partial^2 E_{tot}/\partial\Omega^2$. It takes some algebra to identify with the Grüneisen constant defined in Eq. (2-41). We write out $\partial^3 E_{tot}/\partial\Omega^3$ and $\partial^2 E_{tot}/\partial\Omega^2$ in terms of $\partial^3 E_{bond}/\partial d^3$, $\partial^2 E_{bond}/\partial d^2$, and $\partial E_{bond}/\partial d$. We then write out $(\Omega/B)\partial B/\partial\Omega$ and evaluate it at the equilibrium spacing taking $\partial E_{bond}/\partial d = 0$ at the equilibrium volume and obtain $(\Omega/B)\partial B/\partial\Omega = -2\gamma - 1$. On the other hand if we define the bulk modulus by $B = \kappa_0/(4\sqrt{3}d_0)$ with $\kappa_0 = \partial^2 E_{bond}/\partial d^2$,which we obtained following Eq. (2-38) we obtain the traditional result (see for example Weinstein and Zallen (1982))

$$\frac{\Omega}{B}\frac{\partial B}{\partial\Omega} = -2\gamma - \frac{1}{3} \tag{2-42}$$

which we shall use since the data were reduced using that expression.

We obtained experimental values from the Grüneisen constants γ_{ave} given by Weinstein and Zallen (1982). They gave values with the B in Eq. (2-42) replaced by the elastic constants c_{11} and by a shear constant $c = (c_{11}-c_{12})/2$ which we discuss in Chapter 3, in both cases with the $-{}^1/_3$ term in Eq. (2-42). (The definitions of elastic constants will be detailed at the beginning of Chapter 3.) The first, γ_1 for c_{11} was obtained from the longitudinal acoustic mode at Γ. The second, γ_2 for $c = (c_{11}-c_{12})/2$ was obtained from the transverse acoustic mode. Writing $B = (c_{11}+2c_{12})/3 = c_{11}-4c/3$, we obtain using their definition of γ_{ave} the result $(\Omega/B)\partial B/\partial\Omega = -2(c_{11}\gamma_1 - 4c\gamma_2/3)/(c_{11} - 4c/3) - {}^1/_3$. We use these resulting values of $(\Omega/B)\partial B/\partial\Omega$ in Eq. (2-42) to obtain values of γ, listed as the experimental value under B in Table 2-6, to compare with our prediction.

There is a sizable difference in the Grüneisen constants deduced from the optical-mode vibrations and that from the bulk modulus, indicating an inaccuracy in associating all elastic properties with the same interatomic forces. The discrepancies between our estimates and these experimental estimates are similar to those between the two experimental estimates. Over all, the predictions are in rough accord with experiment, for γ values obtained from either source, and show correct trends.

We have reevaluated the Grüneisen constant for the nonpolar systems adding the metallization term $-3V_1{}^2/(2V_2)$ to the energy, but otherwise proceeding as indicated above. We obtain

$$\gamma = \frac{3}{2}\,\frac{1 - {}^5/_4\, V_1{}^2/V_2{}^2}{1 - {}^9/_4\, V_1{}^2/V_2{}^2} \;. \tag{2-43}$$

This is compared with the simpler estimate and experiment also in Table 2-6. It is not difficult to obtain the corresponding expression for polar crystals using Mathematica, or some such program, but we do not reproduce the expression here.

We have also recomputed the Grüneisen constant including metallization and the Bechstedt form, Eq. (2-20) for the overlap repulsion.

That was done numerically, leading to the values labeled *Bechstedt* V_0. in Table 2-6. They are increased due to the increase in abruptness of the repulsion which increased the force constants.

The linear thermal expansion coefficient α (defined by $\delta d/d = \alpha\delta T$) is directly related to an appropriate average of the Grüneisen constants over the vibrational modes of the system. (See again Weinstein and Zallen (1982).) Frequently all of these constants are of similar magnitude in metals and we can use the Grüneisen constant for the radial force constant to estimate the linear coefficient, $\alpha = \gamma C_V/3B$, where C_V is the heat capacity (at constant volume) per unit volume.

There is an interesting way of interpreting this formula for α. It follows if we associate a momentum $\gamma\hbar q$ with the mode of wavenumber q. The thermal pressure which expands the lattice then comes from the thermal phonons reflecting from the surface. Thus the formula $\alpha = \gamma C_V/3B$ is particularly reasonable at high temperatures where all modes are weighted equally. Then also C_V equals three times the Boltzmann constant divided by the volume per atom or, $9\sqrt{3}k_B/8d^3$. Combining with our expression (without metallization) for γ and B we obtain

$$\alpha = \frac{27k_B(1 + {}^2/_3\alpha_p^2)}{32V_2\alpha_c^3} \quad . \tag{2-44}$$

For a homopolar semiconductor $\alpha = 27k_B/32V_2$, or 14×10^{-6} per degree Kelvin for silicon. Weinstein and Zallen (1982) give a curve for α as a function of temperature for silicon giving a value of 3×10^{-6} /°K for the linear coefficient at room temperature and rising with increasing temperature. At lower temperatures the thermal expansion coefficient for most semiconductors, including silicon, is negative. The simple theory gives a mixed description at best.

This negative expansion coefficient may be qualitatively understandable in terms of a view given by Xu, Wang, Chen, and Ho (1991) which compares effects of angular and radial forces. When radial forces are dominant it is easy to imagine that in a *close-packed* structure, any vibrational distortion will tend to increase the volume and correspondingly pressure will tend to raise the frequency. This is usual for a metal. In an *open* structure, such as a tetrahedral semiconductor, there should be some distortions which lead towards closer packing and thus to a decrease in volume. In this case we expect a decrease in frequency with pressure. If these are low-frequency modes they may cause negative thermal expansion at low temperature as in silicon. In diamond, with larger relative importance of angular forces, the negative coefficients do not occur.

These complications, and very likely the apparent discrepancy between our prediction for silicon and experiment, arise from the strong variation of γ from mode to mode. It does not appear that our simple theory will do well with these variations. The shear constant $(c_{11}-c_{12})/2$ calculated in Chapter 3 varies as $1/d^5$ for homopolar semiconductors and would give a Grüneisen constant of 2/3 for all homopolar semiconductors. This agrees poorly with the experimental values of 0.51, -0.05, and 0.17 for carbon, silicon, and germanium, respectively, given by Weinstein and Zallen (1982) and suggests that the thermal expansion coefficient would not be well predicted.

5. Structural Stability and Moments

We turn next to the relative energies of different structures, which requires treatments of systems for which sp^3-hybrids are no longer appropriate and we must depart from the approach we have been using. It has become clear that full density-functional calculations of the electronic structure, such as those by Cohen and coworkers (e. g., Froyen and Cohen (1983)), are adequate to distinguish which of alternative structures are stable. This became particularly apparent when plots of the energy as a function of volume, as in Fig. 2-8, were given, rather than simply confirming which structure is stable at standard pressure and temperature. We may for example draw a common tangent to the curves for the NiAs structure and the zincblende structure in the figure to the left. The negative of the slope of that tangent is the predicted pressure necessary to transform AlAs to the NiAs structure. We also see how the different structures shift in relative energy as we go from compound to compound. It seems likely also that a full tight-binding calculation, using universal parameters, would be adequate but that has not really been established. Here we are one step down in seeking an approximation to that method, the moments method. We shall find that as directly applied it is not successful in predicting correct structures, but it provides some insights into structural stability, and into the use of hybrids for tetrahedral structures, which were not otherwise available.

The moments method is a mathematical manipulation which provides a short-cut for estimating total energies from the electronic structure, used for example by Lannoo and Decarpigny (1974). It is also a systematic method for accurate calculations if carried to sufficiently high orders, but it seems never to be carried very far. We shall find it very useful in the study of transition metals and their compounds but make only limited use here. In this discussion we follow the analysis given in Harrison, (1990).

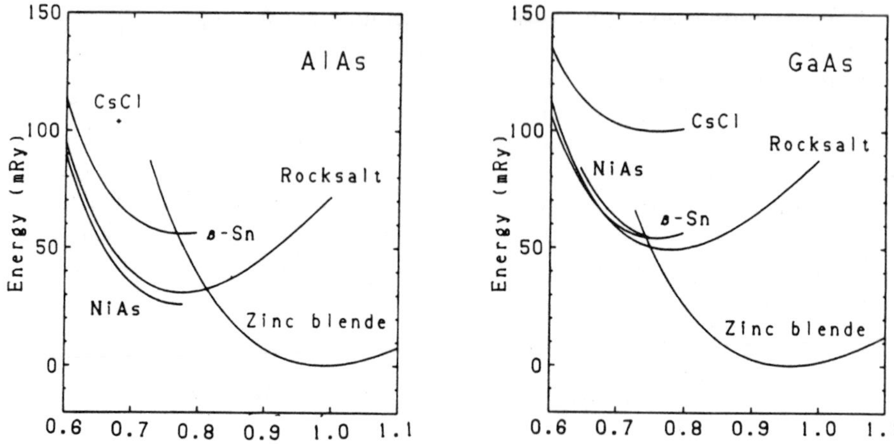

Fig. 2-8. Calculated total energies per molecule, as a function of volume, for AlAs and GaAs from Froyen and Cohen (1983).

We use moments for discussing the relative energies of different crystal structures since, in contrast to the Bond-Orbital Approximation, it can equally well be applied to structures with different numbers of nearest neighbors. The point is that the moments of the bands can be used for estimates of the average energies of occupied states and that they can be directly computed without diagonalizing the Hamiltonian matrix to obtain the energies of the states themselves. The results are surprising and interesting. We shall find that in the first approximation the nearest-neighbor distances are expected to vary strongly with variation of the number of nearest neighbors, but that the energies of these structures are expected to be nearly the same. This does not do well for predicting structures, but both conclusions are basically correct. Improvements over these approximations suggest correct trends, but do not yield reliable predictions. The moments method will also be informative with respect to interatomic interactions. It seems to be a valuable conceptual tool, but in the end we will conclude that it does not seem a promising quantitative approach.

A. The Second Moment

We return again to the Eq. (2-23),

$$\Sigma_j H_{ij} u_j = \varepsilon u_i \qquad \text{for } i = 1, 2, \ldots 4N, \qquad (2\text{-}45)$$

which if solved exactly would give us a set of $4N$ eigenvalues ε_k representing the energies of the $4N$ electronic states based upon the s- and the three p-states on each of the N atoms. We designate the states by k which would include a wavenumber for a periodic structure and the spin of the electron and a band index since there are more than one band, but we now let it be simply an index designating the state. For each eigenvalue ε_k the $u_i(k)$ are the coefficients in the expansion of the corresponding state,

$$/\psi_k> = \Sigma_i\, u_i(k)\, /i> , \tag{2-46}$$

and the H_{ij} in Eq. (2-45) are the term values and couplings which we have introduced. To obtain the total energy we seek a sum over the energies of the occupied states. For convenience we label the states such that those with $k < 0$ are occupied and those with $k \geq 0$ are empty. Thus we seek

$$E_{tot} = \Sigma_{k<0}\, \varepsilon_k\ . \tag{2-47}$$

In order to evaluate it we must, as we indicated in Section 2-3-A, diagonalize this Hamiltonian matrix, H_{ij}, which can only be done for small systems or numerically for ideal crystalline systems.

To proceed we need to organize the mathematics. The set of numbers $U_{ik} = u_i(k)$ is also a matrix. It has the feature that its columns are orthonormal, that is, that

$$\Sigma_i U_{ik'}{}^* U_{ik} = \delta_{k'k}\ . \tag{2-48}$$

This is simply the statement that the states $/\psi_k>$ are normalized and orthogonal to each other, but Eq. (2-48) is also the mathematical statement that the U_{ik} are *unitary* matrices. It is also the statement that the conjugate transpose of a unitary matrix, $U^+{}_{ik} \equiv U^*{}_{ki}$, is its inverse. We may substitute Eq. (2-46) into Eq. (2-45) to obtain

$$\Sigma_j\, H_{ij} U_{jk} = \varepsilon_k U_{ik}\,. \tag{2-49}$$

We multiply both sides by $U_{ik'}{}^*$ and sum over i, using Eq. 2-48, to obtain

$$\Sigma_{ij}\, U_{ik'}{}^* H_{ij} Ujk = \Sigma_{ij}\, U^+{}_{k'i}\, H_{ij} Ujk = \varepsilon_k \delta_{k'k}\,. \tag{2-50}$$

This is the source of the mathematical statement that the matrix H_{ij} is *diagonalized* by multiplying on the right by the unitary matrix U and on the left by U^+ since the matrix to the right of the = in Eq. (2-50) has only diagonal elements, which are the eigenvalues we seek.

We may now see explicitly that the sum of the diagonal elements $\Sigma_i H_{ii}$ is not changed by a unitary transformation. We do this by setting $k' = k$ in Eq. (2-50) and summing over k using the unitarity in the form $\Sigma_k U_{ik}{}^* U_{jk} = \delta_{ij}$, to obtain

$$\Sigma_k \Sigma_{ij} U_{ik}{}^* H_{ij} U_{jk} = \Sigma_i H_{ii} = \Sigma_k \varepsilon_k \ . \tag{2-51}$$

This is the relation which enabled us to use the Bond-Orbital Approximation to replace the sum over the valence bands by the sum over bond energies. Here we use it differently.

We now seek the second moment, multiplying Eq. (2-49) on the left by H_{gi} and summing over i.

$$\Sigma_{ji} H_{gi} H_{ij} U_{jk} = \varepsilon_k \Sigma_i H_{gi} U_{ik} = \varepsilon_k^2 U_{gk} \ . \tag{2-52}$$

In the last step we used Eq. (2-49). This says that the eigenvalues of the matrix $\Sigma_i H_{gi} H_{ij}$, which is the square of the Hamiltonian matrix, are ε_k^2. Using again the fact, Eq. (2-51), that the sum of the diagonal elements of a matrix is unchanged by a unitary transformation, we find that

$$\Sigma_k \varepsilon_k^2 = \Sigma_{ij} H_{ij} H_{ji} \ . \tag{2-53}$$

This is the essential result. It tells us that the second moment of the electronic states,

$$M_2 \equiv \frac{1}{4N} \Sigma_k \varepsilon_k^2 = \frac{1}{4N} \Sigma_{ij} H_{ij} H_{ji} \ , \tag{2-54}$$

can be calculated from the starting Hamiltonian matrix *without carrying out the diagonalization.* Here again $4N$ is the number of orbitals in the calculation. Furthermore if we measure all energies from the average of all of the atomic-orbital energies which appear on the diagonal of the Hamiltonian matrix, $(\varepsilon_{s+} + 3\varepsilon_{p+} + \varepsilon_{s-} + 3\varepsilon_{p-})/8$, we might take $-\sqrt{M_2}$ as an estimate of the average energy ε_b of the occupied orbitals, as illustrated in Fig. 2-9. In semiconductors and insulators the occupied and empty states are separated, and in our analysis equal in number. If the widths of the individual bands, indicated by W in the figure, are small $\sqrt{M_2}$ approaches the energy of each, measured from the average. We shall return to estimate corrections to this using the fourth moment, which can also be calculated without diagonalization.

It is in fact quite easy to calculate this second moment. The average energy of the atomic orbitals is also the average hybrid energy. For a homopolar system, then, one quarter of the diagonal elements are ε_s, which is $-3V_1$ relative to the hybrid energy, and three quarters are ε_p, which is V_1 relative to the hybrid energy and the sum of $H_{ii}{}^2/(4N)$ is $N[(3V_1)^2 + 3V_1{}^2]/(4N) = 3V_1{}^2$. For the polar case we must add or subtract V_3 to each before squaring and we obtain $3(V_{1+}{}^2 + V_{1-}{}^2)/2 + V_3{}^2$.

For the sum of the off-diagonal terms, the only terms are from coupling between the orbitals on an atom and those on its nearest neighbor. If we consider one particular atom and a particular neighbor we can sum the couplings $H_{ij}H_{ji}$ In fact, we may transform the orbitals to be σ- and

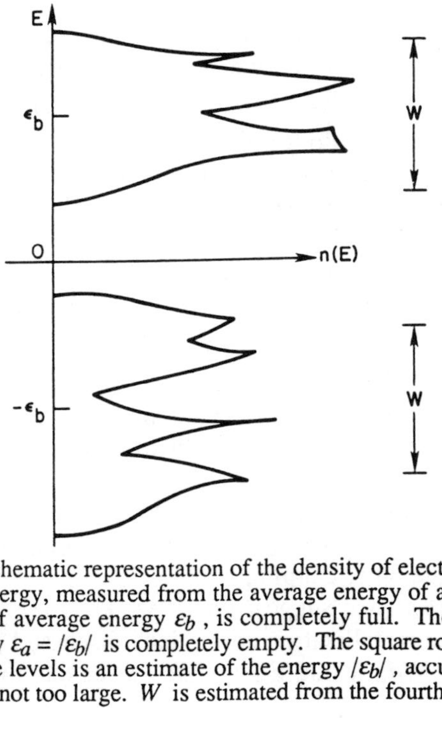

Fig. 2-9. A schematic representation of the density of electronic states as a function of energy, measured from the average energy of all orbitals. The lower band, of average energy ε_b, is completely full. The upper band of average energy $\varepsilon_a = |\varepsilon_b|$ is completely empty. The square root of the second moment of the levels is an estimate of the energy $|\varepsilon_b|$, accurate if the band widths W are not too large. W is estimated from the fourth moment.

π-oriented with respect to those neighbors since that is a unitary transformation upon this four-by-four matrix representing the squared Hamiltonian, $\Sigma_j H_{ij} H_{jk}$, and the sum of the diagonal elements is unchanged. For reasons to be seen shortly we call that sum for a pair of atoms $V_2{}^{*2} = \Sigma_{ij} H_{ij} H_{ji}$ (with i running over the orbitals on the first atom and j running over those on the second) and see that $V_2{}^*$ is given by

$$V_2{}^* = \sqrt{V_{ss\sigma}{}^2 + 2V_{sp\sigma}{}^2 + V_{pp\sigma}{}^2 + 2V_{pp\pi}{}^2} = 3.39\,\frac{\hbar^2}{md^2}\,. \qquad (2\text{-}55)$$

The contribution to the second moment is obtained by multiplying $V_2{}^{*2}$ by the number X of nearest neighbors to each atom (which is four for the tetrahedral structure, but will be different for others), and by the number N of atoms in the system, and dividing, according to Eq. (2-54) by $4N$. These are combined with the diagonal contributions to obtain

$$M_2 = \frac{3}{2}(V_{1+}{}^2 + V_{1-}{}^2) + \frac{X}{4}V_2{}^{*2} + V_3{}^2\,. \qquad (2\text{-}56)$$

If we neglect the effects of the $V_{1\pm}$, as in the Bond-Orbital Approximation, and take $X = 4$, we see that our estimate of the energy gain per electron becomes $\sqrt{M_2} = \sqrt{V_2{}^{*2} + V_3{}^2}$. That is of course the reason we defined $V_2{}^*$ as in Eq. (2-55); we have obtained exactly the same form for the energy per electron gained which we obtained in the Bond-Orbital Approximation, Eq. (2-10), but now with the covalent energy V_2 replaced by $V_2{}^*$. Remarkably, the values for the two covalent energies are almost the same $3.39\ \hbar^2/(md^2)$ for $V_2{}^*$ rather than $3.22\ \hbar^2/(md^2)$ for V_2 . This is of course no accident. Another unitary transformation which we could have made on the H^2 matrix, $\Sigma_j H_{ij} H_{jk}$, would have been a transformation to sp^3-hybrids and one contribution would have been $[3.22\ \hbar^2/(md^2)]^2$ for the two hybrids directed into the bond. The sums of squares of all the other couplings of hybrids must exactly make up the small difference from $[3.39\hbar^2/(md^2)]^2$. Indeed, the coupling V_2 is completely dominant, making up 90% of the contribution to the second moment. This is strong support for our Bond-Orbital Approximation in which we neglected the other couplings.

One might at first glance suggest that if we keep the terms in $V_{1\pm}$ we will have included the metallization corrections discussed in Section 2-3-B. Indeed, if we expand $\sqrt{M_2}$ using Eq. (2-56) we see that the leading term is proportional to $(V_{1+}{}^2 + V_{1-}{}^2) / \sqrt{V_2{}^{*2} + V_3{}^2}$ for tetrahedral semiconductors, as was the metallization energy. However, at a closer look one sees that it gives twice as large a correction as it should since it includes contributions to M_2 not only from the lowering of the energy of the bonding band, but also from the broadening of the bonding and antibonding bands, which do not contribute to the cohesion. We proceed with the full form of Eq. (2-56), though it is not clear this aspect is an improvement.

B. The Dependence Upon Structure

As we noted, an important feature of using $-\sqrt{M_2}$ as an estimate of the bonding energy is that it can be used for different structures and larger coordination X, whereas that could not be done with the Bond-Orbital Approximation. Thus it gives us a way to compare the energies of different structures. We write the energy per atom-pair since we have different numbers X of neighbors for different structures, but we always have eight electrons per atom-pair and $-8\sqrt{M_2}$ as the gain in energy from bond formation. We add the overlap energy per atom-pair as $XV_0(d) = XC/d^4 = AXV_2{}^{*2}$, but note that this is a much more serious approximation for higher coordinations. As we shall discuss in Section 20-2, it is really necessary to incorporate the nonorthogonality of atomic orbitals explicitly in order to have matrix elements which transfer well from one structure to another. However, the overlap energy used here seems to do well enough to be informative and we proceed with it. We also introduce the promotion energy (required to take all electrons to the average hybrid energy). We obtain the energy per atom-pair in analogy with Eq. (2-30) as

$$E_{tot}/\,\text{atom-pair} = E_{pro} + AXV_2{}^{*2} - 8\sqrt{\tfrac{3}{2}(V_{1+}{}^2 + V_{1-}{}^2) + \tfrac{X}{4}V_2{}^{*2} + V_3{}^2} \,. \tag{2-57}$$

Our first step would be to adjust A such that the minimum for some known structure comes at the observed spacing. Then we could use the same A to determine the spacing and energy of another structure. Then Eq. (2-57) provides a series of parabolas, one for each structure, analogous to Fig. 2-8.

A remarkable thing is that the only place where the spacing d enters is as $XV_2{}^{*2}$ in the repulsion and in the square root. Thus the same value of

$XV_2{}^{*2}$ gives the minimum energy, independent of X. It follows that we predict that the equilibrium d varies with coordination X as

$$d \propto X^{1/4} . \tag{2-58}$$

A still more remarkable feature is that it follows that all structures have the same energy at their own equilibrium spacing, where the values of $XV_2{}^{*2}$ are identical. In the context of Fig. 2-8 we obtain a series of parabolas with minima at the same energy but with different positions of the minimum. In the comparison of rock-salt and zincblende structure, the volume per atom pair for rock-salt is $(2d_{rs})^3/4$ and for zincblende is $(4d_{zb}/\sqrt{3})^3/4$ so with $d_{zb} = (4/6)^{1/4}d_{rs}$ the volume at the minimum for zincblende is 1.14 times that for rock-salt, smaller but comparable to the ratio found by Froyen and Cohen (1983).

The theory has failed to provide a widely recognized experimental trend for the more polar systems (large V_3) to form in the structures with higher coordination. That trend is predicted by the more complete calculation (e. g., Froyen and Cohen (1983)) and it is easy to rationalize, but its origin is subtler than one might have guessed and it does not show up in this theory.

It might seem disappointing at first that we have not distinguished which structure is stable, though we in fact correctly predict an increase in nearest-neighbor distance d with coordination. This last is a quite universal trend in solids and molecules, which now becomes a simple quantitative estimate, Eq. (2-58). It in fact predicts a ratio of the spacing in graphite to that in diamond of $(3/4)^{1/4} = 0.93$, to be compared with the observed 0.92 . Further, it is true that though the couplings change significantly, $V_{ss\sigma}$ changing from 4.24 eV to 4.99 eV from diamond to graphite, the different structures do in fact differ in energy by a small amount, typically of order k_BT for T equal to a transformation temperature or perhaps a tenth of an eV per atom.

We note that this exact independence of the energy upon coordination arose because we assumed that the repulsion was proportional to $1/d^4$ or $V_2{}^{*2}$. If the overlap repulsion were more abrupt, varying say as $1/d^6$ consistent with the discussion in Section 1-7, we would find that the energy decreased with increasing abruptness and the structures of higher coordination would be favored. We can understand this by noting that for a *very* abrupt repulsion, there would be negligible energy stored in it and the larger coordination X is the more we would gain from the square-root term in Eq. (2-57).

We noted at the end of Section 2-4-B on force constants, that there is evidence that the exponent n in the repulsion C/d^n should be taken larger than 4 for systems in the lower rows of the periodic table. This is in qualitative accord with the tendency to increasing coordination with metallicity, that diamond is stablest with $X = 3$, silicon and germanium with $X = 4$, and tin and lead with larger coordination. It is interesting that this trend arises from the changing overlap repulsion rather than from the increase in metallization energy

C. The Fourth Moment

A further improvement can be made by using the calculated fourth moment to correct for our estimate of $-\sqrt{M_2}$ as the bonding energy per electron. This was carried out in detail by Harrison (1990) and we summarize briefly here.

It is not surprising that the fourth moment can be obtain in exact analogy to our evaluation of the second moment in Eq. (2-54),

$$M_4 \equiv \frac{1}{4N}\Sigma_k \varepsilon_k{}^4 = \frac{1}{4N}\Sigma_{ijkl} H_{ij}H_{jk}H_{kl}H_{li} \ . \tag{2-59}$$

Each term in the sum can be thought of as a path from the *ith* orbital to the *jth* (which may or may not be the same as the *ith*) to the *kth* to the *lth* and back to the *ith*. Thus we may evaluate it for the tight-binding Hamiltonian as we did the second moment in Eq. (2-56), though it is very much more intricate.

We noted that if the energy bands in the crystal were a very narrow occupied valence band and a very narrow empty conduction band, the average energy in the occupied bands, relative to the average of all energies, would be $\varepsilon_b = -\sqrt{M_2}$. Now let the bands be broadened out such that the second moment within the valence band is $W^2 = (1/2N)\Sigma_{j<0}(\varepsilon_j - \varepsilon_b)^2$ and within the conduction band is the corresponding form. We use the average for the two when they are different. We have already obtained the energy to zero-order in $W^2/\varepsilon_b{}^2$ by taking it to be the square root of the second moment, as if W were zero. We now seek the energy to the next order. It is not difficult to expand both the second and fourth moments in W and solve for the average over the valence band as

$$\varepsilon_b{}^2 = M_2 - \frac{M_4 - M_2{}^2}{4M_2} \ . \tag{2-60}$$

This gives us a correction to our initial estimate of the bonding energy based upon the fourth moment which we can also evaluate without diagonalizing the Hamiltonian matrix. Tests of this formula for sample distributions of states (Harrison, 1990) suggested that the correction removed much of the lowest-order error.

D. Structural Stability

We then included the fourth moment in the total energy to compare the total energy of a number of structures, a zig-zag chain ($X = 2$), a graphite structure ($X = 3$), the tetrahedral structure ($X = 4$) and the simple cubic structure ($X = 6$). In the evaluation of M_4 there are contributions which arise entirely from orbitals on a single atom, terms which involve a pair of nearest-neighbor atoms, and terms which involve three atoms. For the simple cubic structure there were also terms involving four atoms. We included a repulsion of the Bechstedt form, Eq. (2-20), fit to the observed spacing and bulk modulus for the stable form of the compound, using the ε_b from Eq. (2-60) (with ε_b expanded to first order in $M_4 - M_2{}^2$), rather than $-\sqrt{M_2}$ as in the preceding section. We then calculated the equilibrium spacing for each structure and the cohesive energy. The latter is reproduced in Table 2-7. The largest cohesive energy, corresponding to the predicted stable structure, is shown bold face.

We see that the tetrahedral structure is correctly predicted for the semiconductors except for carbon, and that for carbon the graphite structure is correctly found more stable (though the energy difference is certainly too large). However, if we apply it to systems which are in fact in the rock-salt structure (LiF, NaCl, and CaO), only calcium oxide is

Table 2-7. Predicted energy, relative to isolated atoms, in eV per atom pair, for the carbyne structure with $\phi = 126^o$ ($X = 2$), the graphite structure ($X = 3$), the tetrahedral structure ($X = 4$), and the simple-cubic structure ($X = 6$). The overlap repulsion, of the Bechstedt form Eq. (2-20), was fit to the spacing and bulk modulus of the observed stable structure. The energy for the predicted stable structure is given bold face.

	$X = 2$	$X = 3$	$X = 4$	$X = 6$
C	-16.38	**-18.74**	-17.62	-8.40
Si	-7.04	-8.34	**-8.84**	-6.56
Ge	-6.94	-7.94	**-8.42**	-6.68
Sn	-6.28	-7.38	**-8.18**	-7.66
GaAs	-7.50	-8.39	**-8.83**	-7.53
ZnSe	-8.69	-9.23	**-9.51**	-8.74
CuBr	-7.08	-7.31	**-7.41**	-6.85
LiF	-15.71	**-15.80**	-15.63	-13.92
NaCl	-9.82	-9.93	**- 9.99**	-9.76
CaO	-13.11	-13.67	-14.16	**-14.54**

correctly predicted. It appears that this fourth-moment theory is not sufficient to reliably predict structures.

It does suggest an additional trend. We noted in Part B that the increasing abruptness of the overlap repulsion with metallicity favors higher coordinations far down in the periodic table, with tin having a metallic structure at very nearly the energy of the tetrahedral structure, and lead being stable with a coordination of twelve. This trend is reinforced by the fourth moment which always increases the energy and since it arises from pairs of neighbors it tends to vary with the square of the coordination. Thus it always favors low coordination and since it varies as the fourth power of the coupling it drops rapidly with increasing spacing and is thus suppressed at high metallicity. It also drops rapidly with covalency (as α_c^4) so that it correctly suggests that the more polar systems will tend to have higher coordination than the less polar systems. The trend is correct, but Table 2-7 suggests that the fourth moment underestimates it. It may well be that our dismissal of Coulomb effects in these polar systems is not justified, which we shall return to in Chapter 7.

E. Angular Rigidity

We saw in Part C that in the tetrahedral structure the fourth moment contained three-atom contributions, as well as contributions from atom pairs and single atoms. Thus it is the leading contribution which might provide angular rigidity and stability of the tetrahedral structure, which it would not have with purely two-body, central-force interactions.

The evaluation of these three-atom contributions, $M_4(3)$, to the fourth moment is quite intricate, but was carried out in Harrison (1990). In the tetrahedral structure the two nearest neighbors to any atom are second-neighbors to each other (at a distance $1.633d$) so their coupling is neglected but each three-atom set contains many nearest-neighbor "paths", $ijkli$ in Eq. (2-59). They can begin on any of the three atoms and can proceed in either direction, but each is divided by $4N$. We associate all with the central atom and for each pair of neighbors (e. g., the neighbors in directions [111] and [1-1-1]; we only count that pair once) we obtain a contribution to the moment of

$$M_4(3) = (V_{ss\sigma}^2 + V_{sp\sigma}^2)^2 + 2(V_{sp\sigma}^2 + V_{pp\sigma}^2)V_{pp\pi}^2 + V_{pp\pi}^4 +$$

$$+ 2V_{sp\sigma}^2(V_{ss\sigma} - V_{pp\sigma})^2\cos\phi + (V_{sp\sigma}^2 + V_{pp\sigma}^2 - V_{pp\pi}^2)^2\cos^2\phi$$

$$= (0.150 + 0.382\cos\phi + 0.324\cos^2\phi)V_2^{*2}(d_1)V_2^{*2}(d_2)\,, \qquad (2\text{-}61)$$

for a pair of neighbors at distances d_1 and d_2 with an angle ϕ between them. To obtain the final form we substituted universal matrix elements (Eq. (1-8)) and wrote the result in terms of the $V_2{}^*$ of Eq. (2-55). It is interesting that this contribution leads to an energy which is minimum at 126°, not at the tetrahedral angle of 109.5°. This was not anticipated, but is not unreasonable. When these are added for an atom with four neighbors, the total energy minimum still comes with the neighbors at the tetrahedral angle, and there is no reason why the individual angular forces should correspond to a minimum at the tetrahedral angle. Further, since 126° exceeds 120°, this angular interaction correctly favors a planar graphite structure for three-fold coordination, which an angular interaction with minimum at 109.5° would not.

In Section 3-2-A we shall discuss an elastic distortion which modifies the angles between neighbors without modifying the bond lengths, and can therefore directly test Eq. (2-61) quantitatively. We note here the features relevant to the moments method. Osgood and Harrison (1991) corrected a 13% error in the Harrison (1990) estimate of the elastic constant $C = (c_{11} - c_{12})/2$ to obtain 1.45×10^{11} ergs/cm^3, compared to the experimental 5.14 ergs/cm^3, an underestimate of a factor of 3.5. (The estimates made in Section 3-2-A in the Bond-Orbital Approximation are much closer. The error here is specifically that of the moments method.) They noted, however, that the rigidity associated with another property (ω_{TA} , discussed in Section 3-4) was *overestimated* by a factor of 1.2, Thus they argued that the discrepancy likely comes from longer-range forces which the fourth-moment theory omits. Our analysis in Section 3-4 supports that view. Though the needed long-range terms may involve only four atoms, they require a sixth moment to describe them and very possibly even higher moments contribute appreciably to this force as well as to the three-body force.

We have found that there are serious quantitative problems with the use of the moments method for structure prediction or for elastic distortions. The calculation of the fourth moment, leading to Eq. (2-61), was already so complicated that moving to higher moments seems impractical. A more direct approach, begun in Section 3-4, seems much more promising for improving upon the Bond-Orbital Approximation with respect to interatomic forces. Another approach to this has been recently described by Horsfield, Bratkovsky, Fearn, Pettifor, and Aoki, (1996). A quantitative analysis of structural stability may require much more computationally intensive study than that used in this book.

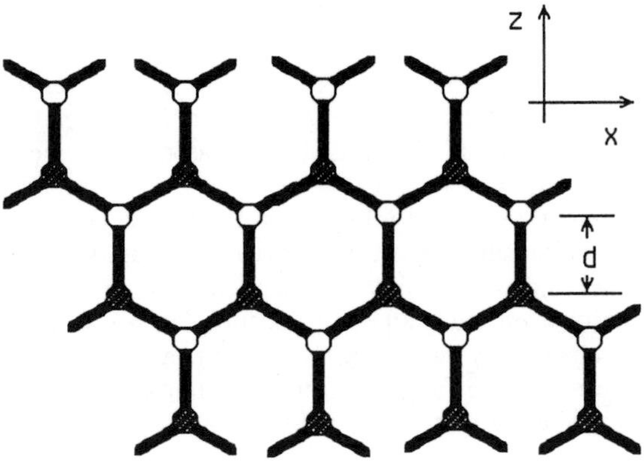

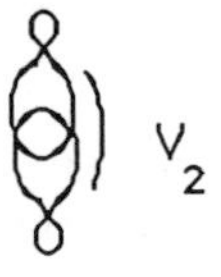

Problem 2-1 sp²-hybrids

Graphite is a simple planar structure with carbon atoms arranged as shown above (Fig 3-30 in *Electronic Structure*), with $d = 1.42$ Å.. The p_y -states, oriented perpendicular to the planes, are uncoupled to the others and we discard them for the time. The remaining s-, p_x- and p_z-states can be combined to form orthogonal and normalized sp^2-hybrids directed toward the nearest neighbors in the plane in exact analogy to the sp^3-hybrids of the diamond structure. [You can, for example, pick the relative value of the s- and p-terms such that two hybrids separated by 120° are orthogonal.]

a) Find the hybrid energy (formula and value).

b) What is the covalent energy V_2 coupling two hybrids directed into the same bond (formula and value)?

c) What is the metallic energy V_1 of coupling between two hybrids on the same atom?

Problem 2-2 Polar crystals.

The figure above could as well have been hexagonal boron nitride which has a bond length close to the graphite $d = 1.42$ Å.

a) Obtain values for the polar energy and the polarity of the sp^2-bonds for this BN, using the results from Problem 2-1. (Note that the ε_h values given for *tetrahedral* semiconductors do not apply to sp^2-hybrids.)

b) It is of some conceptual interest to define an effective charge Z^* on each boron and each nitrogen. It is obtained by adding for each atom the fraction , $u_{\pm}^{b\,2}$, of each of the 2-electron bonds it shares. The π-electrons form bands which we shall describe in Section 20-3 and a moments argument indicates that 20% of the total π-charge lies on the borons and 80% on the nitrogen. We may add the appropriate share and note that the boron atom would be neutral if it had three electrons and the nitrogen if it had five. What is the discrepancy, Z^*, for each atom type and which atom is positive?

Problem 2-3 Bonding properties of graphite

Consider graphite, beginning with diamond and correcting for the differences. *We neglect metallization* in this problem so that the energy per bond in diamond is $2V_1 - 2V_2 + C/d^4$, with of course $V_2 = 3.22\, \hbar^2/md^2$. Evaluate C for carbon using the 1.54 Å equilibrium spacing of diamond.

In graphite V_2 has a different coefficient (obtained in Problem 2-1) and there is additional attraction between atoms from the π-bond. If we assumed it was like the π-bond in N_2, the contribution would be $V_{pp\pi} = -0.63\hbar^2/md^2$ per π-electron, or $^2/_3 V_{pp\pi}$ for each nearest-neighbor bond. However, we shall see in Section 20-3-A that for such resonating bonds the energy is enhanced by the square root of the number of sites among which it resonates ($\sqrt{3}$) so we must add $(2\sqrt{3}/3)V_{pp\pi}$ to the σ-bonding expression, $-2V_2$.

a) Predict the equilibrium spacing in graphite by minimizing this energy, using the same C.

b) Predict the force constant κ_0 from this same expression, evaluated at the predicted equilibrium spacing. (Give the result as $\kappa_0 d_0^2$ in eV so it can be compared with diamond from Table 2-5 .)

Problem 2-4 Metallization

Find the probability an electron, initially occupying a bond orbital in GaAs, lies because of metallization in one of the three neighboring antibonds at the As end. Redo this for the neighbors at the Ga end. This is in close analogy with the 0.01 for each of the neighboring antibonding orbitals in silicon, found in the text and illustrated in Fig. 2-6.

CHAPTER 3

Elastic Properties of Semiconductors

The elastic constants of the solid describe the rigidity of the lattice against uniform distortions. We first give the definition of those constants and then proceed to their calculation in terms of the tight-binding electronic structure. This calculation will also give us the dependence of the elastic constants on material, on polarity in particular, and on pressure. We go on to study the complete vibration spectrum in terms of the same theory.

A central finding is that it appears that the elastic properties are quite well described by the tight-binding theory and universal parameters we use and that most discrepancies in results come from the approximations made upon that theory. We shall find that matrix elements we neglect in the Bond-Orbital Approximation become important for a shear constant in polar crystals and we find that the suppression of the electronic degrees of freedom inherent in force-constant models can lead to important errors in the vibration spectrum.

For elasticity we shall see that the effect of metallization arising from the metallic energy V_1 is quite small and we obtain the energy near equilibrium neglecting it, as in Section 2-3-A. We shall, however, note the magnitude of the corrections which arise if this metallization is included.

1. Elastic Constants and Bulk Modulus

Strains are defined in terms of the variations of the material displacements $\boldsymbol{u}$ with position. The uniaxial strains are

$$\varepsilon_{xx} = \partial u_x/\partial x, \qquad \varepsilon_{yy} = \partial u_y/\partial y, \qquad \varepsilon_{zz} = \partial u_z/\partial z. \quad (3\text{-}1)$$

The shear strains are

$$\begin{aligned}\varepsilon_{yz} = \varepsilon_{zy} &= {}^1/_2(\partial u_y/\partial z + \partial u_z/\partial y),\\ \varepsilon_{zx} = \varepsilon_{xz} &= {}^1/_2(\partial u_z/\partial x + \partial u_x/\partial z),\\ \varepsilon_{xy} = \varepsilon_{yx} &= {}^1/_2(\partial u_x/\partial y + \partial u_y/\partial x).\end{aligned} \quad (3\text{-}2)$$

These are the six independent strains. The combinations such as $\partial u_y/\partial z$ - $\partial u_z/\partial y$ give rotations of the system rather than strains.

The usual contracted notation for strains is

$$\begin{aligned}e_1 &= \varepsilon_{xx}, & e_2 &= \varepsilon_{yy}, & e_3 &= \varepsilon_{zz},\\ e_4 &= 2\varepsilon_{yz}, & e_5 &= 2\varepsilon_{zx}, & e_6 &= 2\varepsilon_{xy}\,.\end{aligned} \quad (3\text{-}3)$$

The elastic energy per unit volume can be written in terms of these strains as

$$E_{elast} = {}^1/_2 \textstyle\sum_{i,j} c_{ij}\, e_i e_j\,, \quad (3\text{-}4)$$

with the sum over all i and all j so that each cross term appears twice. For semiconductors in the zincblende structure the elastic tensor c_{ij} has the symmetry appropriate to a cubic crystal and Eq. (3-4) may be rewritten as

$$\begin{aligned}E_{elast} = {}^1/_2\, c_{11}\,(\, e_1^2 + e_2^2 + e_3^2\,) \;+\; c_{12}\,(\, e_2e_3 + e_3e_1 + e_1e_2\,)\\ + {}^1/_2\, c_{44}\,(\, e_4^2 + e_5^2 + e_6^2\,)\end{aligned} \quad (3\text{-}5)$$

following Voigt's notation (1910) as given by Seitz (1940). Our aim, then is to estimate the three independent elastic constants, c_{11} , c_{12} , and c_{44} . In fact it will be preferable to evaluate three particular combinations of them which are most directly obtainable theoretically since different theoretical approximations are required for the different combinations and comparison with experiment tests the different approximations. In particular, we have already considered the energy change due to a fractional change in volume (with no distortion) , a dilatation, $\Delta = \delta\Omega/\Omega = e_1 + e_2 + e_3$ with $e_1 = e_2 = e_3$. Then the energy change, per unit volume, is written $\frac{1}{2}B(\delta\Omega/\Omega)^2 = \frac{9}{2}Be_1{}^2$. Using Eq. (3-5) this can also be written $(\frac{3}{2}c_{11} + 3c_{12})e_1{}^2$, so the bulk modulus is given by $B = (c_{11} + 2c_{12})/3$. The energy per bond, with metallization dropped (the Bond-Orbital Approximation), relative to that of isolated atoms, was given in Chapter 2, Eq. (2-35), as $E_{tot/bond} = E_{pro} - 2\sqrt{V_2{}^2 + V_3{}^2} + C/d^4$. From this we evaluated $\kappa_0 = \partial^2 E_{tot/bond}/\partial d^2$, obtaining the value of C from $\partial E_{tot/bond}/\partial d|_{d_o} = 0$, and can obtain $B = \kappa_0/(4\sqrt{3}d)$ from the expressions there, or from Table 2-5. This was obtained using the relations $d\partial/\partial d = \frac{1}{3}\Omega\partial/\partial\Omega$ and the volume per bond as

$$\Omega_b = \frac{4d^3}{3\sqrt{3}} . \tag{3-6}$$

The result, without metallization, is

$$B = \frac{c_{11} + 2c_{12}}{3} = \frac{2}{\sqrt{3}}\frac{V_2}{d^3}\alpha_c{}^3 . \tag{3-7}$$

2. Shear Constants

In all of the analysis to this point we have retained the geometry of a regular tetrahedron, with four identical sp^3-hybrids. When the geometry is changed, we must consider the possibility of changing our starting tetrahedral hybrids. In the trivial case in which the lattice is simply rotated, we should of course rotate the hybrids [Eq. (2-3)] so that they still are directed at the neighboring atoms. We then obtain (as we should) an identical energy for the system. In a less trivial case, hybrids can be constructed such that they are no longer oriented toward the corners of a regular tetrahedron. In that case, the individual hybrids must change their sp-composition because we require that the four hybrids be mutually orthogonal. Carlsson (1996) made an alternative approach by constructing nonorthogonal sp^3-hybrids directed at the neighbors, and corrected afterward for the nonorthogonality. It is a valid and interesting approach which leads to a description including "bond-twisting", or torsional, interatomic forces in addition to the angular forces discussed here.

A particularly interesting and simple case of modified *orthogonal* hybrids is in the graphite structure in which each carbon atom is surrounded by *three* neighbors, arranged as an equilateral triangle with the

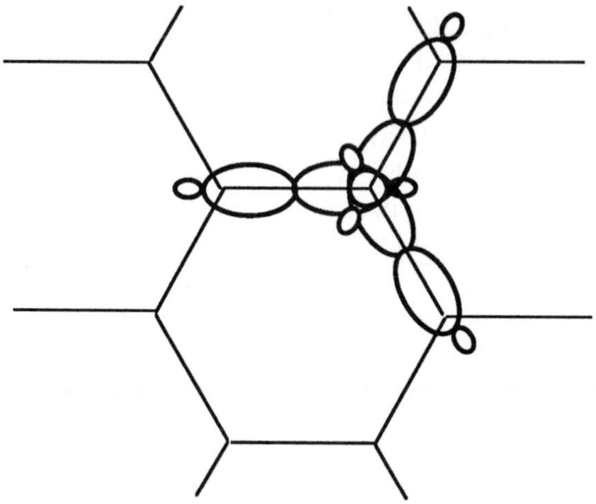

Fig. 3-1. A plane of atoms in the graphite structure. sp^2-hybrids are constructed (Problem 2-1) to lie in the direction of the nearest-neighbor atoms in the plane of the figure. Their composition (squared amplitudes) is one part |s> and two parts |p>, in contrast to the one and three for the sp^3 hybrids for the tetrahedral case.

carbon atom in question at the center. This is illustrated in Figure 3-1. The distance to the other carbon atoms out of this plane is large enough that they may be ignored for the moment. We can construct a hybrid (Problem 2-1) as shown in Fig. 3-1, oriented from a particular atom toward its neighbor to the right, and two more rotated 120° toward the other two neighbors. We require, as after Eq. (2-3), that these are normalized and that they are orthogonal [note that $\langle p'|p\rangle = cos\theta$ for two p-states making an angle θ with each other); the ratio for the coefficients for the s- and for the p-states is fixed by the orthogonality. There is a fourth "hybrid", a p-state oriented perpendicular to the plane of the figure and orthogonal to the others. The first three are often called *sp^2-hybrids* (constructed from an s-state and the two p-states in the plane of the figure), to distinguish them from the *sp^3-hybrids* of tetrahedral structure (constructed from an s-state and the three p-states).

In Problem 2-1 we constructed the covalent energy for such hybrids, in analogy with Eq. (2-5), and found a slightly larger coupling than that given in Eq. (2-5). We also obtained a modified hybrid energy and metallic energy V_1 . We shall not carry out the analysis of the bond orbitals in graphite except in the problems at the ends of the chapters. There it provides a good exercise in precisely the techniques which we have used for the tetrahedral semiconductors. We also return to a study of the states based upon the fourth state on each atom, the p-state - or π-state - oriented perpendicular to the plane in Section 20-3. The distortion of the tetrahedral structure (which, if carried far enough, would lead to the graphite structure) will be important. We call this the *graphitic distortion,* and will see that we can associate with it the elastic rigidity c_{44}.

Distortions to which the hybrids cannot accommodate and still remain orthogonal also exist; in this case the covalent energy is modified because of a misalignment of the two hybrids forming the bond. Such a distortion—one that would try to twist the tetrahedral hybrids into one plane—is the *twist distortion,* with which we associate an elastic rigidity which turns out to be the combination of elastic constants $(c_{11} - c_{12})/2$. This particular combination is chosen since it is exactly $(c_{11} - c_{12})/2$ which becomes equal to the shear constant c_{44} in an isotropic elastic system, but not generally in a cubic system.

Combined with the bulk modulus, these two types of rigidity against shear distortion correspond to the three independent elastic constants of the cubic crystal. The corresponding three rigidities, $B = (c_{11} + 2c_{12})/3$, $(c_{11} - c_{12})/2$, and c_{44} more clearly distinguish the physics of different types of distortion than do the three elastic constants c_{11}, c_{12}, and c_{44} by themselves. There are also nonuniform distortions of the crystal; the complete set of vibrational modes for the lattice constitutes the complete set of small distortions. The rigidity against such distortions can be calculated on just the basis we use for the uniform distortions; however, the calculation is more complex. We consider enough limiting cases to provide understanding of the full spectrum. In most cases, we proceed without including metallization for simplicity and clarity, but note its effect on occasion.

A. Twist Distortion

To evaluate the change in energy under a uniform distortion, we need consider only a single tetrahedron. Such a tetrahedron is illustrated in Figure 3-2a, as is the distortion we call a twist. It may take some concentration to see that this distortion should be called a twist and that the hybrids cannot accommodate to it and still remain orthogonal. Each displacement, indicated by an arrow and a u, is perpendicular to the bond to the central atom. The angular rotation of each internuclear vector is therefore $\theta = u/d$ and the two hybrids making up a bond are therefore misaligned as shown in Figure 3-2b. We must reevaluate the covalent energy as in Eq. (2-5), which now becomes

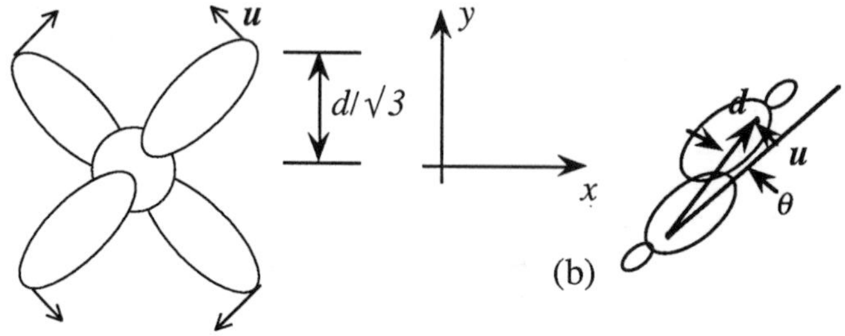

Fig. 3-2. (a) A single tetrahedron under a twist distortion. The displacements are in the plane of the figure, and relative to the central atom. (b) The misalignment of the two hybrids forming a bond, with $\theta \approx tan\theta = u/d$.

$$V_2 = -(V_{ss\sigma} - 2\sqrt{3}V_{sp\sigma}\cos\theta - 3V_{pp\sigma}\cos^2\theta - 3V_{pp\pi}\sin^2\theta)/4$$
$$= \frac{3.22\hbar^2}{md^2}(1 - \lambda\theta^2 + ...) \qquad (3\text{-}8)$$

with the omitted terms being of higher order in θ . Here λ is readily obtained as

$$\lambda = \frac{-\sqrt{3}V_{sp\sigma} - 3V_{pp\sigma} + 3V_{pp\pi}}{V_{ss\sigma} - 2\sqrt{3}V_{sp\sigma} - 3V_{pp\pi}} = 0.854 . \qquad (3\text{-}9)$$

This analysis is carried out in somewhat more detail in Harrison(1980), p. 187. The bond length is not changed to first order in θ and therefore the change in bond length does affect the energy in second order. Hence, only the change in the bond energy $-2\sqrt{V_2^2 + V_3^2}$ arising from the correction in Eq. (3-8) contributes to the elastic energy from the distortion. It is $-2V_2\delta V_2/\sqrt{V_2^2 + V_3^2} = -2\alpha_c\delta V_2$.

To write this in convenient form, we can relate the distortion to the elastic strains. [Harrison (1980)] Taking the x- and y-axes as shown in Fig. 3-2, the strain $e_1 = \partial u_x/\partial x = \sqrt{3/2}\, u/d = \sqrt{3/2}\, \theta$. Furthermore, $e_2 = -e_1$ and the strain energy is

$$E_{strain} = (c_{11} - c_{12})\, e_1^2 = \frac{3}{2}(c_{11} - c_{12})\theta^2 , \qquad (3\text{-}10)$$

per unit volume. Combining these results and noting that the bond density from Eq. (3-6) is $3\sqrt{3}/(4d^3)$, we obtain

Table 3-1. Elastic shear constants in units of 10^{11} ergs/cm^3. (10^{11} ergs/cm^3 = 0.0624 eV/Å^3).

Semiconductor	$(c_{11} - c_{12})/2$ Theory		$(c_{11} - c_{12})/2$ Experiment*	c_{44} Theory		c_{44} Experiment*
C	33.54		47.55	54.3		57.7
Si	4.05		5.09	6.56		7.96
Ge	3.36		4.03	5.44		6.71
Sn	1.69			2.73		
Ge	3.36†	3.36‡	4.03	5.44#	5.44§	6.71
GaAs	2.91†	2.28‡	3.25	4.80#	3.74§	5.29
ZnSe	2.25†	1.05‡	1.61	3.72#	1.74§	4.41
CuBr	1.64†	0.48‡		2.95#	0.87§	

*Compiled by Harrison(1980) p. 196. † From Eq. (3-11). ‡ From Eq. (3-20). # $\alpha_c c_{44}$(Ge). § $\alpha_c^3 c_{44}$(Ge).

$$\frac{c_{11} - c_{12}}{2} = \frac{\sqrt{3}\lambda V_2 \alpha_c}{2d^3} = 2.38 \frac{\hbar^2 \alpha_c}{md^5} . \qquad (3\text{-}11)$$

This can be evaluated immediately using Table 2-1; the results for the homopolar semiconductors and for the polar series isoelectronic with germanium are compared with experiment in Table 3-1. The magnitudes are rather good, as is the variation with metallicity. However, we considerably underestimate the variation with covalency. This has been studied by Sokel (1978), who found that the error arises from the neglect of interatomic metallization, to which we return in Section 3-2-C. The study of elastic constants by Kitamura, Muramatsu, and Harrison (1992) did not make the Bond-Orbital Approximation we have used here (and used Extended-Hückel Theory, rather than our tight-binding theory). The better agreement with experiment for all of the elastic constants supports the thought that the relatively small errors arise from the Bond-Orbital Approximation, not the tight-binding theory itself.

B. Graphitic Distortion and Internal Displacements

For the second independent shear distortion, we again consider a single tetrahedron, but now with a different geometry. We choose one bond along the vertical axis as illustrated in Fig. 3-3. The elastic shear displaces the upper atom in the upward direction, as indicated, by u . The central atom also tends to be displaced upward; however, the magnitude of its

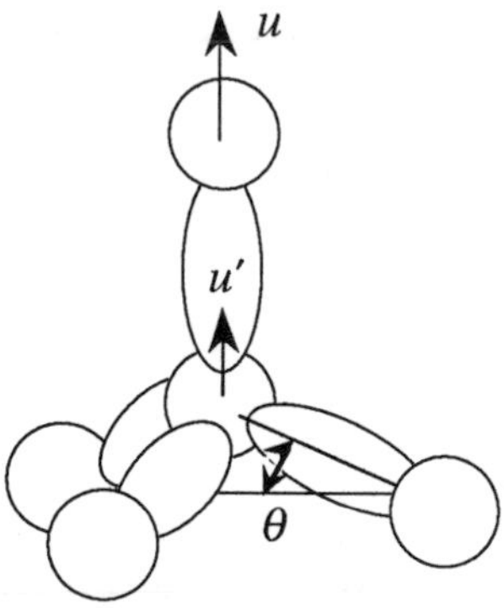

Fig. 3-3. A single tetrahedron under a graphitic distortion. A negative u' large enough to make $\theta = 0$ would produce the graphite structure, Fig. 3-1. An applied strain fixes the displacement u ; then u' adjusts to minimize the energy.

displacement u' is not determined by symmetry. (Note, for example, that if the angular forces were strong enough, the atom would remain fixed, so that no angles change.) We must let u' be arbitrary and at some later stage in the calculation minimize the energy with respect to it. A uniform expansion of the lattice in the vertical direction would correspond to $u'/u = 1/4$. Deviations in u' from that value are called *internal displacements;* our calculation will give us an estimate of an internal displacement parameter. This is a complication in comparison to the twist distortion.

A second complication, as indicated in Section 3-1, is that for this distortion we can allow the hybrid orientation to accommodate to the distorted structure. This could be done by decreasing the p-content of the upper hybrid and increasing the contribution of the same p-state to the three hybrids below, so that they were better aligned with the internuclear vector. In fact, we should allow the orientation to be a variable, as we allow the internal displacements to be a variable, and minimize the energy with respect to that choice. This was the procedure carried out by Sokel (1978) for a different geometry, and we shall give his results. However, we shall first carry out the derivation without allowing relaxation of the hybrids. This is very much simpler and the results of the more complete calculation are easily understood in terms of the simpler form. This *rigid-hybrid* model for the elastic properties may not be much less accurate than the theory with relaxation and could be used as a first approach for treating, for example, lattice vibrations. All parameters which enter the calculations are obtained from the theory but could be adjusted to fit experimental elastic properties for greater accuracy.

The analysis is straight-forward but somewhat more intricate than that for the twist distortion. We carry it out only for the homopolar case; the only difference in the analysis for polar systems is the use of a bonding energy $-2\sqrt{V_2{}^2 + V_3{}^2}$ rather than simply $-2V_2$, which introduces corrections depending upon α_c in the formulae. These complicate the results and introduce uncertainties concerning the overlap interaction and metallization which we discussed for radial forces in Section 2-4. For the present it seems best to stay with the homopolar case without metallization, though we shall discuss briefly the influence of polarity and metallization for this distortion in Part 3-2-C.

For the distortion illustrated in Fig. 3-3 we see that a displacement u' changes the angle θ of the internuclear vector for the lower bonds, which we call transverse bonds. If the hybrids are held rigidly at tetrahedral angles, there will be a misalignment of the hybrids, just as in Fig. 3-2b, of $\delta\theta = \sqrt{8/9}\, u'/d$ and, by Eq. (3-8), a change in the covalent energy due to this change in angle of $\delta V_2 = -(8V_2/9)\lambda u'^2/d^2$ plus terms of higher order in u'.

There is also a change in the length, d_t, of these bonds given by $d_t{}^2 = (8/9)d_0{}^2 + (u'+d_0/3)^2$ and we must keep both first- and second-order terms in evaluating the change in the covalent energy and the change in the overlap interaction (which we again take to vary as $1/d^4$ with a coefficient chosen so that the derivative of the total energy with respect to d_t vanishes prior to shearing the lattice). For the longitudinal (the vertical) bond, there is no angular misalignment of the hybrids but there is a change in bond length of $\delta d_l = u - u'$. The energy in these four bonds is obtained by summing $-2V_2$ over the three transverse and one longitudinal bond in the distorted structure, and subtracting the value for the undistorted structure. Terms are kept up to second order in u and u' but the coefficient of the overlap interaction is taken so that the first-order terms vanish. The resulting change in energy of the four bonds if the sp^3-hybrids remain rigid is found to be

$$\delta E_{tot} = \frac{4V_2\,[u^2 - 2uu' + {}^4/_3(1 + \lambda)u'^2]}{d^2} . \tag{3-12}$$

This is a quadratic form in u and u' and any other theoretical analysis must also come up with the same quadratic form but with coefficients which may differ. In particular, we have calculated the change in energy taking the opposite extreme assumption, that the hybrids reorient completely to the direction of the neighbors (changing their sp-composition in the process) (Harrison, 1983a). Then the expression in square brackets, $[u^2 - 2uu'$

$+{}^4/_3(1 + \lambda)u'^2] = [u^2 - 2uu' + 2.47u'^2]$ above, becomes $[u^2 - 1.66uu' + 2.91u'^2]$. Sokel's analysis (1978) in which the orientation of the hybrids was optimized gave $[u^2 - 2uu' + 2.06u'^2]$. (It is surprising that the coefficient of uu' in this case is the same as in Eq. (3-12), but we have not redone the calculation to check it.) We shall compare these different results shortly.

The next step for all three approximations is the minimization of the energy with respect to u' to obtain the internal displacements. By setting $\partial\delta E_{tot}/\partial u' = 0$ from Eq. (3-12) we obtain immediately

$$u' = \frac{3u}{4(1+\lambda)} = 0.40u \ . \tag{3-13}$$

It is convenient to write this in terms of the *internal displacement parameter* ζ introduced by Kleinman (1962). In his notation the internal displacement u_x arising from a simple shear in the yz-plane, $e_4 = e_{yz} = \partial u_y/\partial z + \partial u_z/\partial y$, is written $u_x = -\zeta a e_{yz}/4 = -\zeta d e_{yz}/\sqrt{3}$. This may be understood in terms of Fig. 2-1 or 2-2 by imagining displacement of two neighbors above a white atom (in the positive z-direction from it) displaced in the positive y-direction and two neighbors below displaced in the negative y-direction; thinking of the bonds as radial springs, all four neighbors push or pull the white atom in the same negative-x direction. Had we chosen a shaded atom the sign would have been reversed. For this structure we have chosen our definition of ζ above to correspond to the white atom. It is readily seen that with this definition of ζ no internal displacement, as defined at the beginning of this subsection, corresponds to $\zeta = 0$ and a displacement which keeps all bond lengths equal corresponds to $\zeta = 1$. In our calculation for Fig. 3-3 a displacement $u' = 3u/4$ would correspond to $\zeta = 1$ and a displacement of $u' = u/4$ would correspond to $\zeta = 0$, so a calculated displacement of u' corresponds to a ζ of $(u' - u/4)/(u/2) = 2u'/u - 1/2$. The result from Eq. (3-13) corresponds to a

$$\zeta = \frac{2 - \lambda}{2(1+\lambda)} = 0.31 \ . \tag{3-14}$$

The calculation mentioned after Eq. (3-12) in which the hybrids were rotated to point at the nearest neighbors gives in the same way a ζ of 0.07 and Sokel's optimized calculation gives 0.47. These compare very poorly with the experimental values of 0.63 ± 0.04 given by Segmuller and Neyer (1965) for both silicon and germanium and a more recent experimental value for silicon of 0.73 ± 0.03 given by d'Amour, Denner, Schulz, and Cardona (1982).

The difficulty appears to have to do with the overlap repulsion. Metallization and optimization of the hybrid orientation as carried out by Sokel (1978) has limited effect. (He found the angle between hybrids changed only by about a third as much (0.36 with our parameters) as the angle between internuclear vectors.) We have seen that the internal displacement is a compromise between radial forces which cause the central atom in Fig. 3-3 to move upward and angular forces - attempting to keep the hybrids aligned - which would favor no motion ($u' = 0$); this competition is reflected in the proportionality of ζ to $2 - \lambda$ in Eq. (3-14). If we were to make the repulsion stiffer, for example by using A/d^n with $n > 4$,we would increase ζ .

It is not difficult to redo the analysis, for a homopolar system without metallization, and with just the V_2 in each bond, with its d-dependence and its variation with misorientation, and a repulsion A/d^n . This leads to

$$\zeta = \frac{n - 2 - \lambda}{n - 2 + 2\lambda} , \qquad (3\text{-}15)$$

which of course reduces to Eq. (3-14) for $n = 4$. This gives values for ζ of 0.31, 0.55, and 0.67 for $n = 4, 6,$ and 8 . A similar model gives the force constant of Eq. (2-38) as $4(n - 2)V_2$, equal to the $8V_2$ of Eq. (2-38) for our choice of $n = 4$. A value in better agreement with experiment is obtained for $n = 6$, at least for the Si, Ge, and Sn. This would presumably also be true for Bechstedt's overlap repulsion.

The internal displacement parameter turns out to be sensitive to the details of the theory and even the best calculations have limited accuracy, as found by Sánchez-Dehesa, Tejedor, and Verges (1982) and by Nielsen and Martin (1983). Thus we cannot be certain of the accuracy of even very complete calculations such as those by Kim, Lambrecht, and Segall (1996) on the nitrides, who obtained $\zeta = 0.1$ for BN, 0.6 for AlN, 0.5 for GaN, and 0.7 for InN.

We learn from our calculation that the rigid-hybrid approximation is better than a full-alignment approximation. We learn also that the overlap repulsion we use is too soft, but different values of n would be obtained from different properties and we continue with the value four.

We may substitute Eq. (3-13) for the internal displacement back into Eq. (3-12) to obtain

$$\delta E_{tot} = \frac{1+4\lambda}{1+\lambda}\frac{V_2 u^2}{d^2} = -2.38\frac{V_2 u^2}{d^2} . \qquad (3\text{-}16)$$

with d again the equilibrium spacing. We may multiply by the density of pairs of atoms (δE_{tot} is the change in energy per atom pair) of $3\sqrt{3}/16d^3$ to obtain the energy density.

This can be directly related to the elasticity by noting that the distortion shown in Fig. 3-3 corresponds to $e_1 = e_2 = e_3 = u/4d$ and $e_4 = e_5 = e_6 = u/2d$. Then the energy density, written in terms of the elastic constants, becomes, by Eq. (3-5),

$$^3/_2 c_{11} e_1^2 + 2c_{12} e_1^2 + {^3/_2} c_{44} e_4^2$$
$$= {^1/_2}\left(\frac{c_{11} + 2c_{12}}{2}\right)\left(\frac{3u}{4d}\right)^2 + \frac{2}{3} c_{44}\left(\frac{3u}{4d}\right)^2 . \tag{3-17}$$

The first term is simply half the bulk modulus times the dilatation squared. Using the bulk modulus from Eq. (3-7) (without metallization in accord with our analysis here), the first term becomes $V_2(3u/4d)^2/(\sqrt{3}d^3)$. We can then combine the elastic expression with the energy density obtained from Eq. (3-16) to obtain the elastic constant c_{44} as

$$c_{44} = \frac{3\sqrt{3}\lambda V_2}{2(1+\lambda)d^3} = 3.85\,\frac{\hbar^2}{md^5} . \tag{3-18}$$

This is evaluated for the homopolar semiconductors and compared with experiment in Table 3-1. The formula again describes well the variation with bond length and predicts that c_{44} is larger than $(c_{11} - c_{12})/2$ by a factor of $3/(1+\lambda) = 1.62$ (they would be equal in an isotropic system) in rough accord with experiment; the agreement is comparable for the two independent shear constants. Even the best theories do not yield accurate values for c_{44} [see Sánchez-Dehesa, et al. (1982)].

Relaxing the hybrids, as done by Sokel (1978), of course lowers the energy of the distorted structure and reduces the predicted c_{44} . It leads to $c_{44} = 2.96\hbar^2/(md^5)$, in poorer agreement with experiment than Eq. (3-18), but not in an important way; a stiffer overlap repulsion raises it again. This may suggest, as it did for the internal displacements, that the extra complexity of relaxing hybrids may not be justified.

C. Interatomic Metallization and Effects of Polarity

In Section 3-2-A we saw that that dependence of $(c_{11} - c_{12})/2$ on polarity was given very poorly. It is not difficult to see that this discrepancy arises from the neglect of the coupling between a hybrid directed into a bond and a hybrid from the neighbor directed into another bond. This couples a bond and neighboring antibonds and therefore contributes to the metallization. However, in this case it comes from *inter*atomic coupling rather than the *intra*atomic coupling V_1 ; hence, we call it *interatomic metallization.* We noted in Section 2-2-B that this coupling is very small in comparison to V_2 ; however, when we make a distortion of the lattice as in this chapter such that the covalent energy decreases, the coupling with the other hybrids increases and the change cannot always be neglected. In particular, it becomes important in polar semiconductors. It is rather simple to calculate in the linear-chain model treated in Problem 3-2, but becomes somewhat complicated in tetrahedral solids.

Although the calculation of this interatomic metallization is intricate, the result is quite simple, and very much the same as in the linear-chain model, which is carried out in Problem 3-2. We consider two neighboring bond sites in a polar semiconductor, as illustrated in Fig. 3-4. We can construct a bond state on the left and an antibonding state on the right. The interatomic coupling between hybrids is indicated by $-V_1{}^x$ as given in Eq. (2-6); it is the same for both sets. To compute the shift in energy of the bond orbital owing to this coupling, we must multiply each $-V_1{}^x$ by the appropriate coefficients indicated, giving a total coupling between the bond and the antibond of $-(u_-{}^a u_+{}^b + u_-{}^b u_+{}^a)V_1{}^x = -\alpha_p V_1{}^x$ [using the coefficients from Eqs. (2-11) and (2-15)]. It is interesting that there is only metallization from this coupling for polar semiconductors;

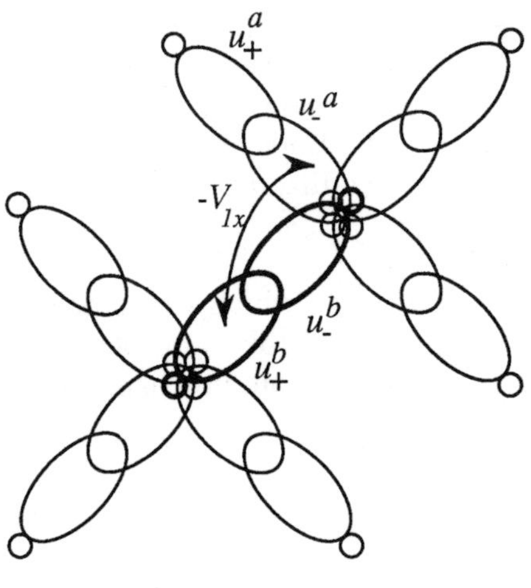

Fig. 3-4. Interatomic coupling between a bond and a neighboring antibond in a tetrahedral semiconductor due to the interatomic coupling $-V_1{}^x$.

because there is no effect for carbon, silicon, germanium, and tin, our comparisons with experiment for them are unaffected. The total shift in energy per bond will be

$$E_{met} = -2\alpha_p^2 \sum \frac{V_1^{x2}}{2\sqrt{V_2^3 + V_3^2}} \tag{3-19}$$

obtained from the perturbation-theory formula, Eq. (2-26). The leading factor of two is for spin and the sum is over the six neighboring antibond sites. $2\sqrt{V_2^3 + V_3^2}$ is the energy difference between the coupled states. In the undistorted crystal, the sum over neighbors would simply give a factor of six; however, we are interested in the effect for the distorted case.

Let us focus upon the hybrid with coefficient u_+^b to the left in Fig. 3-4. It has coupling energy V_2 with the hybrid with coefficient u_-^b and the coupling $-V_1^x$ with each of the three hybrids with coefficient u_-^a (only one is labeled in Fig. 3-4), all on the same neighboring atom. The sum of the squares of these four coupling matrix elements does not change if the four hybrids on that atom are rotated. (This follows from the mathematical fact that such a rotation of hybrids is a unitary transformation on them as discussed in Section 2-5-A.) Thus any decrease in the squared covalent energy, must be accompanied by an equal increase in the sum of the three terms $V_1^{x\,2}$. Thus the sum of $V_1^{x\,2}$ in Eq. (3-19) must increase by $-2V_2\delta V_2$, owing to a change in covalent energy due to the rotation of the hybrids to the right, relative to the internuclear axis. There is also a change in V_2 resulting from the rotation of the hybrids on the atom to the left in Fig. 3-4, which enters the expression for the metallization of the central bond with antibonds on the left. Both terms were included in the δV_2 in Eq. (3-8) so the change in metallization, including both sides, becomes $\delta E_{met} = 2\alpha_p^2 V_2 \delta V_2 / \sqrt{V_2^3 + V_3^2} = 2\alpha_p^2 \alpha_c\, \delta V_2$, to be added to the shift $-2\delta\sqrt{V_2^2 + V_3^2} = -2\alpha_c\, \delta V_2$, which we computed in Section 3-2-A. This gives $-2\alpha_c\, \delta V_2\, (1 - \alpha_p^2) = -2\alpha_c{}^3 \delta V_2$. (Note this calculation assumed no change in bond length and therefore no change in overlap interaction.) Eq. (3-11) is therefore replaced by

$$\frac{c_{11} - c_{12}}{2} = \frac{\sqrt{3}\lambda V_2 \alpha_c^3}{2d^3} = 2.38\, \frac{\hbar^2 \alpha_c^3}{md^5} \,. \tag{3-20}$$

This is equivalent to a result obtained for a simple linear chain model (Harrison (1980) p. 189, Section 3-4-C, and Problem 3-2) and is very close to the result obtained by Sokel (1978) for the full three-dimensional calculation.

Table 3-1 lists the values for $(c_{11} - c_{12})/2$ obtained from Eq. (3-20). We see that the trend with polarity is now very well described (compare ratios of $(c_{11} - c_{12})/2$ to the homopolar value); indeed, the neglect of interatomic metallization for distorted structures was responsible for the error in the trend from Eq. (3-11).

We cannot directly make this correction for c_{44}, even in the rigid-hybrid approximation, because the distortions change the bond length. However, we might try applying the factor α_c^3 to the homopolar formula, Eq. (3-18), for c_{44}. The result, $c_{44}(Ge)\alpha_c^3$ for GaAs, ZnSe, and CuBr (with $c_{44}(Ge)$ the value from Eq. (3-18) for germanium), is the second theoretical value listed for c_{44} for the polar semiconductors, and compared with experiment, in Table 3-1. We see that it considerably overestimates the dependence of c_{44} upon polarity and in fact the simple dependence upon α_c, listed as the first theoretical value, is in much better accord with experiment. Clearly the dependence of elasticity upon polarity is more complicated than we might have hoped, but the full analysis of this problem including both polarity and metallization has not yet been carried out. Until it is, we may regard the dependence of $(c_{11} - c_{12})/2$ as α_c^3 and c_{44} as α_c as appropriate.

3. Vibrational Frequencies

The calculation of normal modes in crystals is vastly simplified by the translation periodicity of the lattice in just the way we saw that the electronic states in a chain of N atoms was greatly simplified in Section 1-3. It may in fact be useful to carry out the calculation of the vibration frequencies for such a chain, introducing the notation which we shall use in three dimensions, and including the complex vibrational amplitudes which we shall use. We then write down the corresponding expressions for three-dimensional crystals.

We consider a chain of atoms as in Fig. 1-4, each with a mass M and each coupled to its nearest neighbor by a spring, relaxed at spacing d and with a spring constant κ_0. For displacements δx_j of the atom at the positions $x_j = jd$, equating force to mass times acceleration gives

$$M\partial^2\delta x_j/\partial t^2 = \kappa_0(\delta x_{j+1} + \delta x_{j-1} - 2\delta x_j) \tag{3-21}$$

and if we seek normal modes, for which every atom moves together with time dependence $\cos\omega t$, we have

$$-M\omega^2\delta x_j = \kappa_0(\delta x_{j+1} + \delta x_{j-1} - 2\delta x_j)\,, \tag{3-22}$$

the counterpart of Eq. (1-19) for the electronic states of the chain. For solving that set of N equations (one for each j) we used the complete form $u_j = e^{ikdj}/\sqrt{N}$. That solved all N Eqs. (1-19) with $\varepsilon_k = \varepsilon_s + 2V_{ss\sigma}\cos kd$. It is customary, and very convenient, to describe the vibrations also in terms of complex amplitudes.

We do this by expanding the displacements of the atoms in complex form,

$$\delta x_j = \Sigma_q\, u_q e^{iqx_j}/\sqrt{N}. \tag{3-23}$$

We have chosen q rather than k to distinguish the vibrational wavenumbers from the electron wavenumbers, important when we treat the electron-phonon interaction in Chapter 7. Since the displacements are real we must require that $u_{-q} = u_q^*$ so there is just one displacement parameter (rather than independent real and imaginary parts) for each q. As for electronic states following Eq. (1-19) we apply periodic boundary conditions and find discrete wavenumbers $q = 2\pi n/(Nd)$ and $-\pi/d < q \leq \pi/d$ and there are just as many wavenumbers allowed as atoms, or degrees of freedom, for the chain.

Substituting Eq. (3-23) into Eq. (3-21) we obtain

$$-\frac{M\omega^2}{\sqrt{N}}\Sigma_q\, u_q\, e^{iqx_j} = \frac{\kappa_0}{\sqrt{N}}\,\Sigma_q\, u_q e^{iqx_j}(e^{iqd} + e^{-iqd} - 2)\,. \tag{3-24}$$

We may select some particular q, say q_0, multiply Eq. (3-24) through by $e^{-iq_0x_j}$, and sum over j. All terms give zero except for $q = q_0$, for which every term is identical. Canceling the factor $\sqrt{N}$ and rewriting q_0 as q we have

$$-M\omega^2 u_q = \kappa_0\,(e^{iqd} + e^{-iqd} - 2)\,u_q = -\,4\kappa_0\sin^2(qd/2)\,, \tag{3-25}$$

an independent harmonic-oscillator equation for each value of q, corresponding to a frequency

$$\omega_q = \sqrt{\frac{4\kappa_0}{M}}\,\sin\frac{qd}{2}\,. \tag{3-26}$$

For small positive q (so $\sin(qd/2) \approx qd/2$) it corresponds to a sound wave propagating in the positive x-direction with phase velocity $\omega_q/q = \sqrt{\kappa_0/M}$.

The only complication from use of complex amplitudes comes when we calculate, for example, a total kinetic energy. The velocity of the j th atom is $\Sigma_q \partial u_q/\partial t \; e^{iqx_j}/\sqrt{N}$, which is squared by multiplying by a corresponding sum over q', multiplied by $M/2$, and summed over q and q'. The sum over j is performed before the sum over wavenumbers, giving a contribution only when $q' = -q$ and leading to a total kinetic energy

$$K.E. = \frac{M}{2} \Sigma_q \frac{\partial u_q}{\partial t} \frac{\partial u_{-q}}{\partial t} \; . \tag{3-27}$$

One must take particular care with these cross terms in the quantum-mechanical formulation of the vibrational states. (See, for example, Harrison (1970), 407 ff.)

We proceed directly to vibrational modes of three-dimensional crystals with the displacement $\delta \boldsymbol{r}_j$ of an atom at a position $\boldsymbol{r}_j$. We note first that for a system such as gallium arsenide the equations analogous to Eq. (3-21) for the gallium atoms differ from those for the arsenic atoms, so we have two sets of equations to satisfy. The equations in each set are identical, because of the translational equivalence of each atom of one type, but we must introduce two amplitudes, each in analogy with Eq. (3-23), and these will be vectors,

$$\delta \boldsymbol{r}_j = \sum_q \frac{\boldsymbol{u}_q{}^i}{\sqrt{N_c}} \; e^{i\boldsymbol{q}\cdot\boldsymbol{r}_j} \; . \tag{3-28}$$

with $i = 1$ for gallium, and $i = 2$ for arsenic. N_c is the number of atoms of each type, or the number of cells in the crystal if the crystal is divided into primitive cells each containing one of each atom type. When the geometry associated with the wavenumbers is treated in detail in Chapter 5 it will be seen that two $\boldsymbol{u}_q{}^i$ are needed even in silicon because half the atoms have different environments than the other so there are again two atoms per primitive cell. The factor $1/\sqrt{N_c}$ was introduced so that quantities like the total energy of the system, Eq. (3-27), in terms of the $\boldsymbol{u}_q{}^i$ do not depend upon the size of the system. That kinetic energy, now, becomes $1/_2 M_1 \omega^2 u_q{}^1 u_{-q}{}^1 + 1/_2 M_2 \omega^2 u_q{}^2 u_{-q}{}^2$.

Here it will be sufficient to limit ourselves to wavenumbers $\boldsymbol{q}$ along a [100] direction. We will see in Chapter 5 that a range of wavenumbers along this direction from $-2\pi/a$ to $2\pi/a$ (with a the cube edge) will generate all of the $\boldsymbol{q} \,//\, [100]$ normal modes. Any wavenumber outside this *Brillouin Zone* will reproduce a mode already included within it.

For each wavenumber $\boldsymbol{q}$ in the Brillouin Zone for the diamond or zincblende structure there are six vibrational modes, each with amplitude vectors u_q^1 and u_q^2 for the two distinguishable types of atomic sites (the dark and light spheres in Figs. 2-1 and 2-2). We may think of the six components for each mode as components of a six-dimensional vector. The set of six amplitude vectors at each wavenumber form a set of six orthogonal vectors, one for each of the normal modes of that wavenumber. The amplitudes for each mode could be decorated with further subscripts or superscripts indicating which of the six modes for each wavenumber is under discussion, but we shall not complicate the notation with these indices. We discuss the frequency spectrum of these modes in detail for the homopolar semiconductor, silicon, first and return to polar semiconductors after.

A. Acoustic Modes

The measured vibrational frequencies (from Dolling (1962)) for silicon for wavenumbers along a [100] direction in the Brillouin Zone are shown in Fig. 3-5, along with theoretical curves which we shall obtain from the limiting cases we treat first. Such spectra are given for silicon, and also

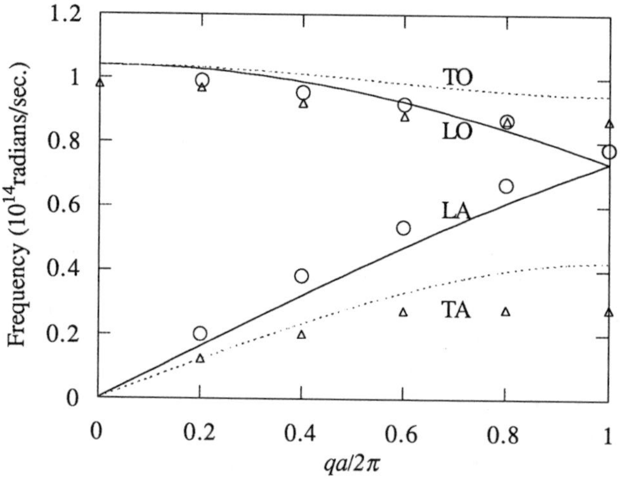

Fig. 3-5. Vibrational frequencies (radians per second) for silicon with wavenumber $\boldsymbol{q}$ in a [100] direction. Experimental points are from Dolling (1962); curves are from Eqs. (3-33) and (3-34). The linear portion of the dispersion curve ω_{LA} is described by Eq. (3-29), that of ω_{TA} is described by Eq. (3-30).

for a wide range of other systems in Bilz and Kress (1979). The low-frequency modes for small q are easily understandable as simple acoustic waves. There is a longitudinal mode for which $\boldsymbol{u_q}^1$ and $\boldsymbol{u_q}^2$ (and therefore all displacements) are parallel to $\boldsymbol{q}$. (At wavenumbers which do not lie in a high-symmetry direction such as [100] the modes may not be purely longitudinal.) The frequency ω_{LA} of this longitudinal-acoustic mode is readily calculated using elasticity theory, giving $\omega_{LA}{}^2 = c_{11}q^2/\rho$, with the density given in terms of the atomic mass M by $\rho = 3\sqrt{3}M/(8d^3)$. We may obtain c_{11} from our expressions for the bulk modulus $B = (c_{11} + 2c_{12})/3$ and$(c_{11} - c_{22})/2$ giving $c_{11} = 2\sqrt{3}(1+\lambda)\,V_2/(3d^3)$. We then obtain

$$\omega_{LA}{}^2 = \frac{16(1+\lambda)V_2q^2}{9M} \quad . \tag{3-29}$$

with ω_{LA} proportional to q . This corresponds to a speed of sound of $[16(1+\lambda)\,V_2/(9M)]^{1/2}$.

For small q there are also two transverse modes of equal frequency ($\omega_{TA}{}^2 = c_{44}q^2/\rho$ from elasticity theory) with $\boldsymbol{u_q}^1$ and $\boldsymbol{u_q}^2$ perpendicular to $\boldsymbol{q}$. Using Eq. (3-18) we obtain

$$\omega_{TA}{}^2 = \frac{4\lambda V_2q^2}{(1+\lambda)M} \quad . \tag{3-30}$$

We found in Eq. (3-26) that a chain of masses M , each coupled to its neighbors at a distance $s = d$ by a spring of constant κ_0 , has longitudinal-acoustic modes with frequency given by $\omega_q = 2\sqrt{\kappa_0/M}\,\sin(qs/2)$. It is the discreteness of the lattice which causes the curves for ω_q to turn flat at the Brillouin-Zone boundary, $qs = \pi$ in that formula and $qa = 2\pi$ in Fig. 3-5. This form could in fact be used as an approximation to extend our small-q equations to larger wavenumbers. For the transverse modes a detailed calculation for the diamond structure, even using simple nearest-neighbor spring constants, shows that the effective spacing is $s = a/2$, the spacing in a [100] direction between equivalent (same color) atoms shown in Fig. 2-2. Thus we extrapolate Eq. (3-30) to $\omega_{TA}{}^2 = 64\lambda V_2 \sin^2(qa/4)/[(1 + \lambda)Ma^2]$, which is shown as the dotted line labeled TA in Fig. 3-5. We return to the discrepancy with experiment at large q in Section 3-4.

We might think that the same extrapolation would apply to longitudinal waves but we see from the experimental LA curve in Fig. 3-5 that it does not appear to. The reason is the subtle symmetry, screw-axis symmetry, which we discussed in Section 2-1. We saw there that Bloch sums for states propagating along a [100] direction could be used with factors $e^{ikaj/4}/\sqrt{N}$, as if the spacing were $s = a/4$ and the appropriate extrapolation of Eq. (3-29) is $\omega_{LA}{}^2 = 1024(1 + \lambda)V_2 \sin^2(qa/8)/(9Ma^2)$, which is plotted as the lower solid line in Fig. 3-5.

This form continues to higher frequencies beyond the Brillouin Zone face at $q = 2\pi/a$, in the negative as well as the positive directions. We have translated the curve for $q < -2\pi/a$ by $4\pi/a$ to the right, to bring it into the Brillouin Zone, and plotted it as the upper solid curve in Fig. 3-5. This second longitudinal mode at each wavenumber in the Brillouin Zone is called an "optical mode" because its frequency is that of infrared light. We ordinarily think of these optical modes differently than we thing of acoustic modes, though we see that in silicon at $q = 2\pi/a$ they have the same frequency. We next consider in general this second set of modes.

B. Optical modes.

All three branches of modes at higher frequency are called *optical modes*. The modes at $q = 0$ are modes in which all atoms of a particular type move exactly in phase with the others (this follows from the form proportional to $e^{i\mathbf{q}\cdot\mathbf{r}_j}$ given in Eq. (3-28).), but the two types have equal and opposite amplitude vectors, $\mathbf{u}_q{}^2 = -\mathbf{u}_q{}^1$, for elemental semiconductors. Thus this is a mode in which the two sublattices vibrate uniformly against each other. The modes with three perpendicular directions of displacements have the same frequency in silicon, as seen in Fig. 3-5. Such modes are important in the infrared spectrum for the solid. We may immediately predict their frequency using Fig. 3-3 and Eq. (3-12). The potential energy in such a mode is obtained by taking u equal to zero and u' is the relative displacement of the two sublattices. (In terms of Eq. (3-28), $u' = 2u_q{}^1 e^{-i\omega t}/\sqrt{N_c}$. We derive the frequency first using u' as a real displacement in an atomic cell and then note the same result is obtain using complex amplitudes.) In terms of u' the elastic energy per atom pair is obtained from Eq. (3-12) as

$$\delta E_{tot} = \frac{16(1+\lambda)V_2 u'^2}{3d^2} = 31.84\frac{\hbar^2 u'^2}{md^4}. \tag{3-31}$$

In the motion, the center of gravity of the crystal remains fixed and the two atoms in the primitive cell have displacements in the laboratory frame of $\pm u'/2$ (for the homopolar case where both masses M are the same). We may multiply by a factor $\cos\omega_O t$ to obtain the time dependence and the kinetic energy per atom pair becomes $^1/_4 Mu'^2\omega_O{}^2\sin^2\omega_O t$, which in any vibrational problem will have the same time average as the elastic energy of $16(1+\lambda)V_2u'^2\cos^2\omega_O t\,/(3d^2)$ per atom pair. (In terms of the complex amplitudes this kinetic energy would have been $Mu_q^{1*}\cdot u_q^1 \omega_O^2 / N_c$ per atom pair and the potential energy $64(1+\lambda)V_2 u_q^{1*}\cdot u_q^1/(3Nd^2)$ per atom pair.) We solve for ω_O^2 (using either notation) and have immediately

$$\omega_O^2 = \frac{64(1+\lambda)V_2}{3Md^2} = \frac{127.\hbar^2}{mMd^4} \quad . \tag{3-32}$$

This is in fact exactly what we obtained by extrapolating the longitudinal acoustic frequencies as $sin(qs/2)$ with $s = a/4$ above, and in a detailed calculation in Problem 3-1. That is because all three calculations are based upon the same rigid-hybrid approximation . In fact

$$\omega_{LO} = \omega_O \sin(qa/8) \tag{3-33}$$

is exactly the prediction of the rigid-hybrid approximation for the entire longitudinal spectrum before folding back into the Brillouin Zone. As we indicated, these are plotted along with the experimental spectrum in Fig. 3-5.

The full transverse-mode calculation is not this simple. It becomes relatively simple if we not only hold the angles between hybrids on each atom fixed, but we do not allow free rotation of the hybrids. We have carried out such a calculation obtaining

$$\omega_{TO,TA}^2 = \frac{\omega_0}{2}\left(1 \pm \sqrt{\cos^2(qa/4) + 0.441\sin^2(qa/4)}\,\right) , \tag{3-34}$$

which were also plotted in Fig. 3-5. We return to a discussion of these curves after first comparing the $q = 0$ optical-mode frequencies with experiment.

The values of ω_O from Eq. (3-32) are listed, along with experimental values, in Table 3-2. The agreement is remarkably good, especially for the heavier elements. Use of optimized hybrids (Sokel (1978)) leads to

Table 3-2. Optical-mode frequencies for the homopolar semiconductors (in units of 10^{14} radians/sec.)

Element	Theory Eq. (3-32)	Experiment*
C	3.71	2.46
Si	1.04	0.98
Ge	0.60	0.57
Sn	0.36	0.38

* Compiled by Harrison (1980, p. 208).

predictions 9% smaller, in comparable agreement with experiment. Both the longitudinal and transverse modes have this frequency in homopolar materials, written $\omega_{TO} = \omega_{LO} = \omega_O$.

We may also compare the full spectrum in Fig. 3-5. It was anticipated that the spectra would be in reasonable agreement with experiment since the predicted elastic constants and the optical-mode frequencies were in good accord with experiment. However, we see that there is a noticeable discrepancy in that the observed transverse acoustic frequencies are lower than predicted. One reason for this is our failure to allow free rotation of the hybrids in deriving Eq. (3-34). In our calculation of the elastic constant c_{44}, one hybrid remained oriented in a [111] direction, corresponding to a rotation of the hybrids with the crystal for an $e_4 = 2\varepsilon_{yz}$ distortion. Thus since free rotation *was allowed* in the calculation of the elastic constant, we obtained a lower prediction (Table 3-1). In holding the hybrid orientations fixed in calculating the transverse vibrational frequencies, we overestimated the energy of distortion, bringing the predicted TA modes at small q into better accord with experiment but raising all other modes too high. These are rather small differences in any case. It is rather impressive that we do this well without *any* parameters adjusted to the elastic or vibrational properties of the solid.

None of these approximations to the tight-binding theory (rigid hybrids, etc.) are really necessary. Chadi and Martin (1976b) have started with essentially the same tight-binding theory and made an evaluation numerically for the Zone-boundary modes in C, Si, and Ge, without any further approximations. They found some, but less than observed, lowering of the transverse spectrum at the Zone boundary, but the numerical results did not shed much light on it. We return to this feature, beginning in Section 3-4-B.

C. Effects of Polarity

We may briefly consider the application to polar semiconductors. As we indicated in discussing the graphitic distortion, the expressions become much more complex. However, for the longitudinal modes with wavenumbers in a [100] direction, for which there were only compressional and twist distortions, it is quite simple. We noted, above Eq. (3-29), that the longitudinal acoustic frequencies, $\omega_{LA}{}^2 = c_{11}q^2/\rho$, were proportional to $\sqrt{c_{11}}$. We saw in Eq. (3-7) that $B = (c_{11} + 2c_{12})/3$ was proportional to $\alpha_c{}^3$ and that when metallization was included, Eq. (3-20), so was $(c_{11} - c_{12})/2$. Thus we may expect the entire longitudinal branch of frequencies, Eq. (3-26), to scale as $\alpha_c{}^{3/2}$. In addition, it is not difficult to see that in a compound the mass M in the denominator is replaced by twice the reduced mass, $2M_r = 2M_+M_-/(M_+ + M_-)$. We shall see in Section 4-3-B that there is also an extra rigidity for the longitudinal optical modes due to electrostatic interactions, with the transverse optical mode not showing that shift. Thus it is preferable to test the $\alpha_c^{3/2}$ dependence by looking at the transverse optical frequency which we then expect to be generalized from Eq. (3-32) as

$$\omega_{TO} = \sqrt{\frac{32(1+\lambda)V_2\alpha_c^3}{3M_r d^2}} = \sqrt{\frac{127.\hbar^2\alpha_c^3}{2M_r md^4}} . \qquad (3\text{-}35)$$

These values are compared with experiment in Table 3-3 for all the polar semiconductors for which we have experimental values. This can be taken as a test of the predicted variation with polarity since the predictions for the homopolar semiconductors Si, Ge, and Sn were seen in Table 2-7 to be within about 5%. The dependence is seen here to be even closer, in strong support of the $\alpha_c^{3/2}$ scaling for this optical branch.

This dependence upon $\alpha_c^{3/2}$ will certainly not be appropriate for the transverse *acoustic* modes, which depend upon the elastic constant c_{44} which we saw at the end of Part 3-2-C was proportional to α_c . This was really an empirical finding and tells us that these transverse-acoustic frequencies vary as $\alpha_c^{1/2}$.

We turn finally to the vibrational modes at the Brillouin-Zone edge. These modes illustrate another indirect effect of polarity. We saw in Fig. 3-5 that the longitudinal optical and acoustic modes for silicon have the same frequency at the Brillouin-Zone edge. It is not difficult to see that in fact the two modes can be chosen such that $\mathbf{u}_q^1$ is zero for one mode and $\mathbf{u}_q^2$ is zero for the other. They are then two modes in which the atoms of only one type vibrate in alternating directions and the others remain

Table 3-3. Transverse optical-mode frequencies at $\mathbf{q} = 0$ for the polar semiconductors (in units of 10^{14} radians/sec.)

Compound	Theory+ Eq. (3-35)	Experiment*
GaP	0.69	0.69
GaAs	0.50	0.51
InSb	0.29	0.35
ZnS	0.48	0.52
CuCl	0.33	0.34
CuBr	0.23	0.24
CuI	0.21	0.24

* Compiled by Harrison (1980, p. 208).
+ A much more extensive list is given in Table 5-4.

stationary. Since the two atoms were identical, the two frequencies were also. The modes are of the same form in a polar crystal, but then the masses are different. The optical and acoustical branches therefore split in polar crystals, as the transverse modes already did in the homopolar crystal. The longitudinal optical mode is shifted up by a factor equal to the square root of the ratio of the average mass to the lighter atomic mass; the acoustical frequency is shifted down by the corresponding factor containing the heavier mass.

In summary, we have seen that the vibration spectrum of the *homopolar* semiconductors is predicted to be just that shown in Fig. 3-5, but scaled from element to element as $\sqrt{\hbar^2/(mMd^4)}$. For polar semiconductors here is an additional scaling of the longitudinal modes by $\alpha_c^{3/2}$, a replacement of the atom mass M by twice the reduced mass $M_r = M_+M_-/(M_+ + M_-)$, a splitting of the longitudinal mode at the Zone face proportional to the mass difference (and an additional spitting of the longitudinal and transverse optical modes at $q = 0$ to be calculated in Section 4-3-B.) The transverse optical modes at small q also scale as $\alpha_c^{3/2}$, but the transverse acoustic modes scale as $\alpha_c^{1/2}$. The transverse modes are already split at the Zone face, and the splitting is increased by the mass difference. The frequencies could be interpolated between these limits with formulae such as Eq. (3-34), with appropriate masses and covalency factors inserted in such a way to give the correct limits. This is usually done with force-constant models, which we discuss in the next section. The vibration spectrum provides a complete specification of the elastic properties of the perfect crystal. We shall return in a later chapter to further calculations of elastic properties in connection with defects and impurities.

4. Interatomic-Interaction Models

Before there was a sufficient understanding of electronic structure to allow calculation of vibration frequencies, models were proposed, parameters fit to elastic constants, and vibrational spectra calculated. Even now such models are frequently used for the calculation for modes other than those with wavenumbers at symmetry points, or to study amorphous or other complicated systems. Our effort here is to try to understand the range of validity and the short-comings of such models generally, not to compare different models nor to propose a new one. There are also many continuing efforts to understand the interatomic forces from a fundamental basis. One by Carlsson (1996) is very much in the spirit of the approach we use here.

A. Traditional Models

For the calculation of vibrational spectra (Born and Huang (1954) is the classic reference in this field), the first were based upon the derivatives of the total energy of the lattice as a function of the atomic positions, $W(\mathbf{r}_1, \mathbf{r}_2, ...)$. The derivatives $\kappa_{ij}{}^{\mu\nu} = \partial^2 W(\mathbf{r}_1, \mathbf{r}_2, ...) / \partial r_i{}^\mu \partial r_j{}^\nu$ (μ and ν indicate components of the vector) enter the dynamics as force constants, which can be fit to experimental elastic constants, or other properties. This is rigorously correct if one keeps enough terms, but it was found that as more neighbors were added to the fit, it did not converge well and all previously fit force constants changed. There seemed always to be long-range forces spoiling the fit. The simplest fits such as a radial interatomic interaction $^1/_2\ \kappa_0(d-d_0)^2$ for nearest neighbor interactions, where κ_0 could be estimated from Eq. (2-38), and an angular interaction $^1/_2\ \kappa_1(\phi-\phi_0)^2$, for changes in angle $\delta\phi$ from the tetrahedral angle $\phi_0 = 109.5\ ^o$ between adjacent internuclear vectors, are the most widely used and seem to be useful. The two constants can be fit to c_{11} and c_{12} and then c_{44} can be predicted. It is not a very accurate prediction and adding complexity by fitting an additional parameter to fit c_{44} then doesn't do well on the next prediction.

A number of interatomic-force models have been proposed recently which have more parameters, fit to such properties as full calculations of the energy of small clusters. The best known of these are by Keating (1966), Stillinger and Weber (1985), Biswas and Hamann (1985, 1987), Khor and Das Sarma (1988), and Tersoff (1986a, 1988). We shall not survey such approaches in detail, but we can use our understanding of the electronic structure to shed some light on the applicability of such models.

First we mention that it seems that whenever one derives an interatomic interaction from the electronic structure, it is much different than the plausible guesses one usually uses in generating models. We shall see that in simple metals the interatomic interaction is best thought of as a purely repulsive interaction, with the balancing attraction being so featureless that it is thought of as an external pressure on the system. This is much different than an almost universally assumed repulsion (as B/d^{12} in the Lennard-Jones model discussed in Section 20-2-E) and a more-slowly-varying attraction ($-A/d^6$ in the Lennard-Jones model). We saw in Section 2-5 that when we attempted to derive an angular force for tetrahedral semiconductors using the moments method, we found it minimum at a 126° angle rather than at the tetrahedral angle of 109.5^o usually assumed. (A plausible, but unanticipated finding.)

Second, we note that since the energy is obtained from the sum of the energies of occupied levels it, and the derived forces, will certainly be changed if different levels are occupied. For example the equilibrium positions at a vacancy in a semiconductor will depend upon the charge state, whereas no interatomic-interaction models we know have such dependence. Similarly the structure of molecules, such as ammonia, depends upon their charge state.

Third, It may be essential to have an electronic degree of freedom in the model, rather than just nuclear coordinates. This is in fact accomplished in "shell" models (Dick (1965)) in which a shell surrounding each atom is coupled to the nucleus and also shells to each other. This has the feature that if one atom in a crystal is moved, neighboring shells are displaced and their displacement extends to large distances, giving forces on atoms a large distance away. The feature arose in our calculations in this chapter in that for our calculation of $(c_{11} - c_{12})/2$ the hybrids could not rotate; however, when we went on to calculate c_{44} it was possible to allow them to accommodate. In not allowing it in our estimates we made an error. We were ignoring one aspect of the electronic degrees of freedom.

B. The Angular Force Puzzle

Weber (1974) had proposed that such an electronic effect was responsible for the low frequencies of the transverse-acoustic mode for large wavenumbers, relative to the slope at small wavenumbers, apparent in Fig. 3-5. He called his model a "bond-charge model", but it did as well with a charge of zero. To us this is an important distinction: there are long-range forces of electrostatic origin, but they seem to ordinarily be quite negligible, in the case of silicon in particular. There are also long-range forces of electronic origin, represented by the shell interactions in the shell model and the spring constants between the "bond charges" in Weber's model, and they were the dominant effect. We supported that explanation (Harrison (1980, p. 212)) using derived forms (though not derived magnitudes) of the long-range interaction. There is no question that such forces could remove the discrepancy if they were large enough. We now return to that question and argue that the cause is different.

We consider the transverse acoustic mode discussed by Weber, which provided the only significant discrepancy between the observed vibration spectrum of silicon and that which we derived from elementary theory and plotted in Fig. 3-5. There are uncertainties in how close the simple theory will be for the magnitudes for any property, but the error in the ratio for these two measures of the angular rigidity, the slope of $\omega_{TA}(q)$ at small q and the value at $q = 2\pi/a$, seems more fundamental.

Osgood and Harrison (1991) examined these two properties using the second-moment calculation of the energy discussed in Section 2-5 and found that the Zone-face transverse mode frequency was well given, but that c_{44} elastic-constant prediction was too small by a factor of four. This is the same qualitative discrepancy, but both magnitudes are smaller than in the theory given here. They agreed with the earlier conclusion (Harrison (1980)) that there must be long-range forces contributing to the elastic constants, but not strongly affecting the Zone-boundary modes.

Kitamura and Harrison (1991) sought to determine the range of the important forces using a full tight-binding calculation of the elastic constants (c_{44} in particular, which determines the [100] transverse sound speed). Kitamura indeed found that it was in reasonable accord with experiment, the predicted value being about 0.71 times the experimental value, suggesting that tight-binding theory itself, with the universal parameters we are using, contains any important long-range forces.

His calculation included a full tight-binding band calculation, as in Chapter 5, and a sum of energies over a grid in the Brillouin Zone, as an approximation to the appropriate integral over the Zone. We were able to argue that replacing the integral by a sum over a grid eliminated interatomic forces of longer range than a distance of the order of the reciprocal of the grid. One way of arguing this is to note that a grid in wavenumber space can be thought of as the wavenumber lattice for a larger primitive cell in the real lattice. Then any interatomic forces between one atom and an atom outside the cell can be treated as an interaction with the equivalent one inside. Some analysis of simple systems supported this idea. Kitamura found (in the usual terms of band structure, to be outlined in Chapter 5) that taking a grid in the wavenumber lattice, corresponding to a primitive cell of one cube of edge a (See Fig. 2-2) had not converged and predicted a value of c_{44} only half of the experimental value. (In band notation, this included the energies at Γ and at X .) On the other hand, a primitive cell twice that large (including also the point L in the Brillouin Zone) had converged and gave essentially the same result as for much finer grids, both for c_{44} and $(c_{11} - c_{12})/2$. This suggested that there were important forces longer range that first-neighbor like atoms (second-neighbors in the tetrahedral structure) but not of much longer range. We turned to examination of these relatively short-range interactions.

In calculating the elastic constant c_{44} we noted that there were changes in bond length so that the overlap repulsion enters, but we shall see (Fig. 3-8) that this is not the case for the transverse acoustic mode at the Zone face. That could give deviations from the $sin(qa/4)$ extrapolation, but the approximate agreement for the longitudinal, as well as the transverse, speed of sound would suggest that it is not the cause for the discrepancy. It is interesting, however, that this Zone-boundary mode - like the elastic constant $(c_{11} - c_{12})/2$ - does not involve changes in d so that if we take the simplest model with radial and angular energies, $^1/_2\kappa_0(d-d_0)^2$ and $^1/_2\kappa_1(\phi-\phi_0)^2$, the two experimental values give independent fits to κ_1 without any uncertainties arising from the overlap repulsion. We noted earlier (Harrison, 1980) that the two determinations differ generally by more than a factor of two and for silicon in particular are $\kappa_1 = 3.2$ eV if fit to $(c_{11} - c_{12})/2$ and $\kappa_1 = 1.07$ eV if fit to the transverse-acoustic Zone-face mode. It will be simplest to address this discrepancy between the angular force constant values in the context of the distortions which do not involve any changes in distance.

C. A Solution for Carbyne

The missing force seems to have been found in studying the same chain of atoms, carbyne, which showed that interatomic metallization was needed to understand the polarity dependence of $(c_{11} - c_{12})/2$ (Section 3-2-C, and discussed in Problem 3-2). We consider the zig-zag chain of carbon atoms lying in a plane, separated by d , and illustrated in Fig. 3-6. We let two electrons from each atom remain in the valence s-state, and ignore them. We then construct tight-binding states based only upon $|p_x\rangle$ and $|p_y\rangle$ states in the plane. The second two electrons per atom fill a bonding band, leaving the antibonding band empty, and the $|p_z\rangle$ states are empty. The physics of the bonding bands is sufficiently similar to the bonding bands in semiconductors that calculations on the chain have provided real insight into the electronic structure of semiconductors.

In particular, if we include only $V_{pp\sigma}$ coupling between the p-states we may orient them as shown for a chain with 90° angles between neighboring

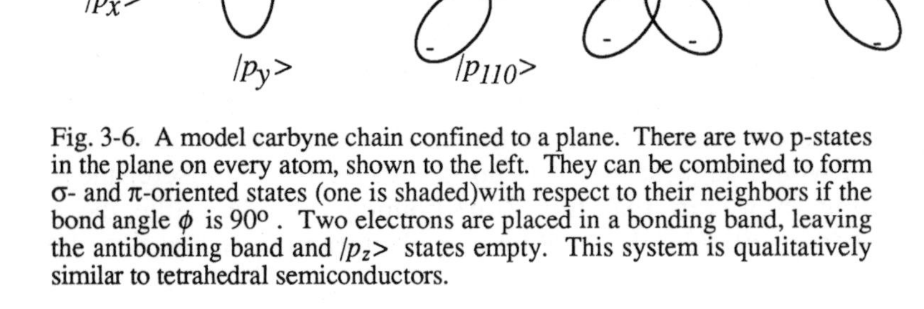

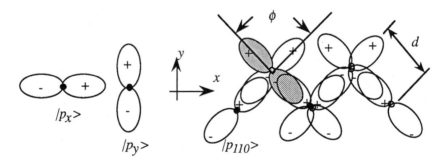

Fig. 3-6. A model carbyne chain confined to a plane. There are two p-states in the plane on every atom, shown to the left. They can be combined to form σ- and π-oriented states (one is shaded)with respect to their neighbors if the bond angle ϕ is 90° . Two electrons are placed in a bonding band, leaving the antibonding band and $|p_z\rangle$ states empty. This system is qualitatively similar to tetrahedral semiconductors.

bonds, and the only coupling for each p-state is the p-state on one neighbor, yielding bond energies of ε_p - $V_{pp\sigma}$ and antibond energies of $\varepsilon + V_{pp\sigma}$. Adding the smaller $V_{pp\pi}$ couples bonds with (second-neighbor) bonds and broadens the levels into bonding and antibonding bands. $V_{pp\pi}$ plays the role of the metallic energy in the semiconductor in producing the bands, and also gives an interatomic metallization. Such approximations can be made, but it is also possible to solve for the states and energy exactly to test these approximations. Here we discuss some exact results, but will then understand them in terms of reasonable approximations.

In the present context we wish to study deviations of the bond angles from 90° and we keep the spacing at the equilibrium d, so the overlap repulsion is irrelevant. From the exact solution we find that the total energy is minimum with all adjacent bonds at 90°. We also calculate the change in energy under two small uniform distortions, one for which every bond angle is changed by the same $\delta\phi$, corresponding to a uniform stretch of the chains. For the second we let alternate bond angles change by $+\delta\phi$ and $-\delta\phi$, corresponding to bending the chain in a circular arc. Of interest is the fact that both could be characterized by a change in angle, $\delta\phi$, and therefore two independent fits could be made to an angular force constant κ_1 representing the change in energy per atom by $^1/_2\kappa_1(\phi-\pi/2)^2$. Indeed we find a result strongly analogous to that in tetrahedral semiconductors, if we fit the angular force constant to an elastic constant $(c_{11} - c_{12})/2$ we obtain a much larger value than if we fit it to the transverse mode at the Zone face.

For the stretch distortion ($\delta\phi_i = \delta\phi_{i-1}$) in the carbyne chain we find a change in energy per bond of

$$\delta E_{angle} = 2(V_{pp\sigma} + V_{pp\pi})A_2(R)\,\delta\phi^2\,, \quad \text{(stretch)} \tag{3-36}$$

including a factor of two for two electrons per bond, where

$$A_2(R) = \frac{1 + R^2}{2}\frac{2}{\pi}\int_{0,\pi/2}\frac{\sin^2 x\,dx}{\sqrt{1 + R^2\sin^2 x}}\,, \tag{3-37}$$

and

$$R^2 = -\frac{4V_{pp\sigma}V_{pp\pi}}{(V_{pp\sigma} + V_{pp\pi})^2}\,. \tag{3-38}$$

This corresponds to an exact calculation of the $\delta\phi^2$ term in the energy. We give these in detail since they have not been published elsewhere. Note $V_{pp\sigma}$ is positive and $V_{pp\pi}$ is negative in Eqs. (3-36) and (3-38).

For the bend distortion ($\delta\phi_i = -\delta\phi_{i-1}$)we obtain

$$\delta E_{angle} = 2(V_{pp\sigma} + V_{pp\pi})A_3(R)\ \delta\phi^2 , \qquad \text{(bend)} \qquad (3\text{-}39)$$

where

$$A_3(R) = \frac{1}{2}\frac{2}{\pi}\int_{0,\pi/2} \frac{\cos^2 x\, dx}{\sqrt{1 + R^2 \sin^2 x}} \quad . \qquad (3\text{-}40)$$

These integrals required numerical evaluation. For our universal parameters Eq. (3-38) leads to $R = 1.48$ and Eqs. (3-37) and (3-40) to $A_2(R) = 0.511$ and $A_3(R) = 0.210$. The stretch rigidity is greater than the bend rigidity by a factor of 2.4. This is closely analogous to the peculiarity we observed in the tetrahedral structures.

For this simple system it is also not difficult to see exactly how this arose. We note first that for $V_{pp\pi}$ equal to zero, the two rigidities are identical. (Because R is zero, $A_3(R) = A_2(R)$.) They may be understood in terms of a misalignment of the p-states on adjacent atoms just as for the misalignment of hybrids we calculated in Section 2-2-A. In this case the p-states *must* remain rigidly oriented 90° from each other to remain orthogonal. For both deformations the misalignment is $\theta = \delta\phi/2$ so each electron in the bond has energy $\varepsilon_p - V_{pp\sigma}\cos^2\theta$ so the shift in energy for the two electrons is $2V_{pp\sigma}\delta\phi^2/4$ in accord with both Eq. (3-38) and (3-39) for $V_{pp\pi} = 0$. This corresponds to an angular force constant $\kappa_1 = V_{pp\sigma}$.

If now $V_{pp\pi}$ is not zero, we see that in the stretch distortion there is an additional contribution to the energy of each bond electron of $-V_{pp\pi}\sin^2\theta$, with the minus sign because the positive lobes of the inward-pointing p-states are tilted in opposite directions; this contributes $-2V_{pp\pi}\delta\phi^2/4$ to the bond energy, stiffening the lattice since $V_{pp\pi}$ is negative. On the other hand, in the bend distortion the two p-state lobes are tilted in the same direction and the contributions is $2V_{pp\pi}\delta\phi^2/4$, which softens the lattice. The ratio of the angular rigidity against stretch to that against bending becomes $(V_{pp\sigma} - V_{pp\pi})/(V_{pp\sigma} + V_{pp\pi}) = 1.8$. This is of the same order as the ratio 2.4 which came from the complete analysis. The discrepancy appear to come from taking terms only to lowest order in $V_{pp\pi}/V_{pp\sigma}$ in this illustrative analysis.

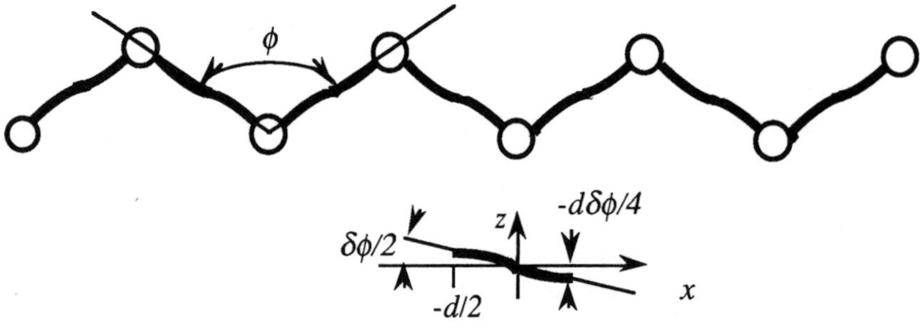

Fig. 3-7. A ball-and-stick model of the carbyne chain. The two sticks at each ball are required to emerge from the balls at a 90° angle from each other, but are flexible. Here the chain is stretched so each is bent into an s-shape. Below, one is rotated and a coordinate systems set up for calculating the energy of distortion.

Before going on to the tetrahedral structure, we may note a remarkably simple mechanical system which behaves in very much the same way. It is the "ball and stick" model in which each atom is represented by a ball with two elastic sticks emerging from the ball at 90° angles as in Fig. 3-7. The length of the sticks is not allowed to change, but there is elastic energy proportional to the square of the curvature of the stick during distortion. [We are not allowing shear of the stick which would allow the angle of emergence of the stick to differ from 90°.] In a uniform stretch, the bond ends remain at 45° from the chain axis but the two ends of each bond are displaced perpendicular to the stick by $\pm(\delta\phi/2)d/2$ since each internuclear axis is rotated by $\delta\phi/2$ to change each bond angle by $\delta\phi$. This is illustrated in the lower part of Fig. 3-7, which is drawn with the initial stick axis horizontal. The curvature of the bond is readily calculated to be $d^2z/dx^2 = 6\delta\phi x/d^2$. The elastic energy of the stick is proportional to the average value of $(d^2z/dx^2)^2$ which is $3\delta\phi^2/d^2$. For the bend distortion, each bond is bent into a single arc with a change in angle of $\delta\phi$ over the entire length of the bond and the curvature is $d^2z/dx^2 = \delta\phi/d$. The elastic energy of the stick is proportional to the average value of $(d^2z/dx^2)^2$ which is $\delta\phi^2/d^2$, one third that occurring in the uniform stretch with the same change in angle $\delta\phi$ at each atom. This is remarkably close to the value of 2.4 which we found for carbyne. This could be studied more completely by incorporating an increase in energy for shearing the stick along its axis. The extra parameter would enable one to fit both $V_{pp\sigma}$ and $V_{pp\pi}$, but it would seem not to have any advantage over using the electronic structure directly and may even be more complicated.

It seems quite remarkable that this simple and familiar ball-and-stick model describes so well the subtlety of the bond in such a covalent system.

It seems quite remarkable that this simple and familiar ball-and-stick model describes so well the subtlety of the bond in such a covalent system. It is, however, not a model that would be easy to use. For an arbitrary movement of the atoms, it would be necessary to minimize the energy with respect to the orientation of each ball in order to obtain the distortion energy. This is the counterpart of including an "electronic" degree of freedom in the real electronic structure which we discussed at the beginning of this section. It is for this reason that essentially all force-constant models (with the exception of the shell model discussed above) have written the energy entirely in terms of atomic positions.

There is such an option in the carbyne system which we have studied here. We might add a term in the energy $^1/_2\Sigma_i\, \kappa_2 \delta\phi_i \delta\phi_{i+1}$ to the $^1/_2\Sigma_i\, \kappa_1 \delta\phi_i^2$ which we introduced before. This would add to the rigidity under a stretch (because $\delta\phi$ is of the same sign at adjacent atoms) and subtract for a bend (because $\delta\phi$ is of opposite sign at adjacent atoms). The results we gave above for the distortion energies, which were to lowest order in $V_{pp\pi}/V_{pp\sigma}$, correspond to

$$\kappa_1 = V_{pp\sigma}\text{, and} \qquad\qquad (3\text{-}41)$$
$$\kappa_2 = -V_{pp\pi}$$

for the carbyne chain. In this particular pair of distortions it seems possible to model the important electronic effect by a force constant κ_2 but that will not generally be the case. In particular, making a bend in the lattice at one atom only will distort the sticks many atoms away (though with distortions dropping exponentially with distance) though none of the intervening atoms have moved. Such long-range forces are omitted in the κ_1 , κ_2 description.

Finally, it is interesting to relate this additional interaction to the moments method which we described in Section 2-5. We note that the $\delta\phi_i\delta\phi_{i+1}$ terms can only be defined by specifying the positions of four neighboring atoms. Thus in the moments method if this correction is to be obtained we need products $H_{ij}H_{jk}H_{kl}...H_{ni}$ which span these four atoms and this requires six steps in the round trip diagram. The correction can be obtained in the moments method only if we include the *sixth* moment. We actually calculated the corresponding term in the sixth moment for this model of carbyne and found that it was indeed of the right form, but much too small. Thus many other higher-moment terms involving these four neighboring atoms are necessary. This suggests that the moments method may be too complicated, and require moments of too high an order, to be very useful at this elementary level of property calculation. (See also Kress and Voter (1995).)

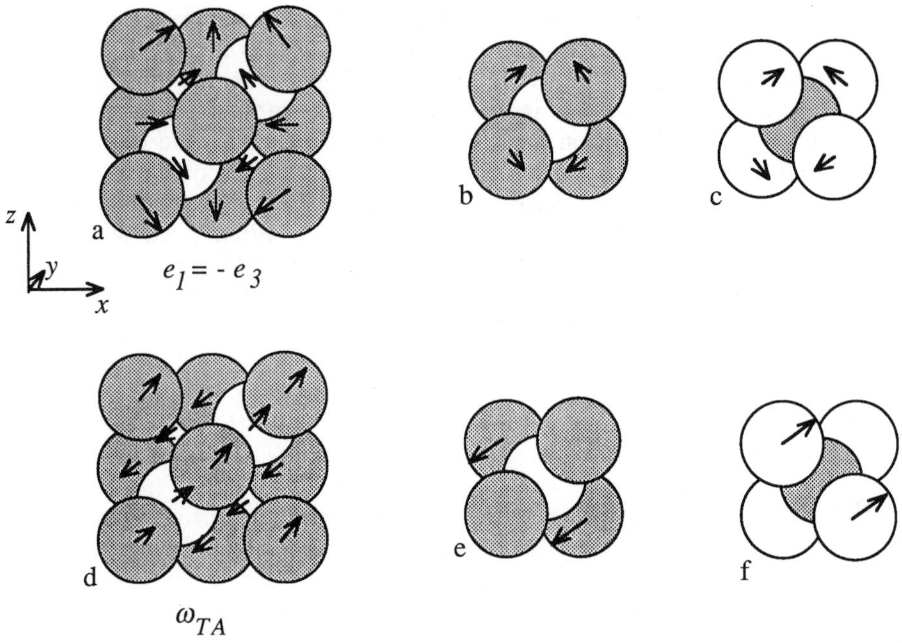

Fig. 3-8. The atomic displacements, above, corresponding to an elastic shear and, below, corresponding to the transverse acoustic mode at the (100) Brillouin Zone face. In both cases all displacements are perpendicular to the nearest-neighbor vectors so that only angular forces enter if the radial forces are for nearest neighbors only.

D. Application to Tetrahedral Semiconductors

We return to the tetrahedral structure and to the two properties which do not entail changes in bond distance, an elastic shear $e_2 = -e_1$, and the transverse-acoustic mode at the Zone face. We sought the counterpart of the mechanism which we found for the carbyne chain, the possibility that there was again a term $^1/_2\kappa_2\delta\phi_i\delta\phi_{i+1}$ for successive atoms along zig-zag chains in the three-dimensional structure. The calculation is too intricate to be usefully given in detail here, but we may note the gist of the argument, the results, and what we concluded from them.

It was necessary first to understand these two distortions, which we attempted in Fig. 3-8. Above is an elastic distortion which, with our choice of coordinate system is $e_1 = -e_3$ with e_3 positive. It is a little difficult to see the relative motion of nearest neighbors since all but the atom at the center of the front face have been displaced. We sketched to the right the motion of the four shaded neighbors relative to a white one and the motion of the four white neighbors relative to a shaded one. For this distortion the displacement of the neighbors is the same for every atom of the same type. Below we gave the distortion for the Zone-boundary mode, and in this case some atoms of the same type had opposite displacement of neighbors.

We then sought the sequence in angle changes along the various [110] zig-zag chains in the three-dimensional structure. Finally we wrote the energy changes due to misalignment of the hybrids for these various sequences exactly as for the p-states in carbyne, and could represent them as $1/2\Sigma_i\,(\kappa_1\delta\phi_i^2 + \kappa_2\delta\phi_i\delta\phi_{i+1})$, with

$$\kappa_1 = 0.708V_2\,, \qquad \kappa_2 = 0.147V_2\,. \tag{3-42}$$

This is qualitatively similar to the effect we found in carbyne, but is *not* large enough to explain the angular-force discrepancy of a factor of three. For the elastic constant the net angular force, as for the stretch in carbyne, was the total, $0.855V_2$. For the transverse mode we showed in Fig. 3-8, half of the contributions $\delta\phi_i\delta\phi_{i+1}$ were positive and half negative so κ_2 does not contribute. We are left with $0.708V_2$, hardly the factor of three difference we sought.

It seemed likely that the shortcoming of the theory was isolation of effects of individual zig-zag chains, and treating them as if all displacements were coplanar with the chain, as we did for carbyne. It may be necessary to perform the variational calculation for the hybrids on each atom, which would be more intricate than justified by the approximate tight-binding representation of the electronic structure.

The central point may be that the electronic degrees of freedom are essential here, as they were for carbyne, and it may not be possible to account for them with third-neighbor interatomic forces as seemed possible for carbyne. In any case we have not succeeded in doing that and we do not believe any existing atomic-force models do. The errors in force-constant models would seem to remain on the scale of this factor of three discrepancy for the angular-force puzzle.

Problem 3-1 Optical modes in silicon

We may estimate the separate contributions to the restoring force $\boldsymbol{F} = -\kappa_{eff}\boldsymbol{u}$ on an atom in silicon which is displaced by $\boldsymbol{u}$ from its equilibrium position while all other atoms remain fixed.

a) Find the contribution to κ_{eff} from the change in bond length with the nearest neighbors, which can be obtained from $\delta E/bond = {}^1/_2\ \kappa_0\ \delta d^2$. This arises from the overlap repulsion and from the change in V_2 with bond length. κ_0 was obtained in Eq. (2-38) as $\kappa_0 = 8V_2/d_0^2$. (You can let the displacement $\boldsymbol{u}$ be in a [100] direction, and calculate δd for each of the four bonds.)

b) There is also a contribution from the resulting misalignment of the hybrids. Obtain that contribution to the energy using the rigid-hybrid approximation just as we found it for the twist distortion. [It is only the angular change you calculate since the radial change in V_2 is already included in Part a). You need θ^2 which can be used with Eqs. (3-8) and (3-9).] By what factor does this change the effective spring constant κ_{eff} obtained from part a)?

Problem 3-2 Elasticity and interatomic metallization.

Consider the carbyne chain of Fig. 3-6, with ϕ the same at every atom and near 90°. We neglect $V_{pp\pi}$ but keep $V_{pp\sigma}$. In the Bond-Orbital Approximation the only coupling included in the calculation of energy is that between two p-states directed into each bond.

a) Confirm that because of their misalignment when ϕ is not 90^0, there is a change in energy of ${}^1/_2\kappa_1\delta\phi^2$ per bond, with $\kappa_1 = V_{pp\sigma}$.

b) Let alternate atoms have p-state energies $\varepsilon_p \pm V_3$. What is now the rigidity κ_1 in the Bond-Orbital Approximation.

c) Now include interatomic metallization (again with $V_{pp\pi} = 0$), which arises from the coupling between each bond and the nearest-neighbor antibonds at each end, in the polar case ($V_3 \neq 0$).

[This calculation is closely parallel to that leading to Eq. (3-19), and in this case we have only a single antibond at each end, so we do not have the problem of the summation over three antibonds.]

d) Add this lowering in energy to the increase in energy found in part c), to obtain the formula for κ_1 including interatomic metallization, analogous to Eq. (3-20), which applied to tetrahedral solids.

CHAPTER 4

The Dielectric Properties of Semiconductors

It is indeed remarkable that aspects of solids as diverse as the bonding and the dielectric properties should be described by the same elementary theory. Phillips (1973) noted such a connection and defined an *ionicity* of the bonds in semiconductors in terms of the dielectric constant, rather than in terms of the bond energy as had been done by Pauling (1960). They gave remarkably similar ionicity scales, but Phillips argued that the dielectric basis was better because of its more direct relation to the electronic structure. We would agree with this, noting that the bond energy (cf. Eq. (2-34)) contains a term E_{pro} and a complicated dependence upon polarity, while we shall find a rather simple dependence (proportionality to $\alpha_c^3 = (1- \alpha_p^2)^{3/2}$) for the susceptibility and approximately the same for the dielectric constant.

The defining of an ionicity to be used to scale properties from one system to another is of course a much less ambitious undertaking than a derivation of the properties in terms of the electronic structure. On the other hand, because of its empirical content it can be a more accurate predictor of experimental values. It can be used, for example, to accurately predict the elastic constant of GaAs by interpolating between Ge and ZnSe which are isoelectronic with it. The result will be much more accurate than our prediction of the value from Eq. (3-11). On the other hand, because our value does not depend on empirical values for similar systems we can make predictions for totally different systems and obtain the dependence upon other features, such as the bond length or metallicity. The goals are completely different. The fact that our polarity is a similar concept to ionicity (in fact related more closely to the square root of the ionicity of Pauling and Phillips than to the ionicity itself (Harrison(1980), p. 190, Christensen, Satpathy, and Pawlowska(1987)), should not be allowed to confuse the totally different goals and methods used in the two approaches. Our covalent, polar and metallic energies have been obtained directly from the free-atom term values and the simultaneous validity of free-electron and tight-binding concepts (with a slight tuning of coefficients to fit the germanium energy bands) and we may proceed to direct predictions of the dielectric properties in terms of them without further parameters or approximations.

1. Bond Dipoles and Effective Charges

The basis aspect of the bond which determines all of the dielectric properties, is the bond dipole, $\boldsymbol{p}$. In the presence of a uniform electric field $\boldsymbol{E}$, the bond energy is lowered by $-\boldsymbol{p}\cdot\boldsymbol{E}$, which represents the interaction energy of the bond with the electric field. This interaction deforms the bond, resulting in a modified dipole moment and therefore an electric polarization of the system that is describable by a dielectric susceptibility. We return in Part 4-2 to a more careful definition and interpretation of such an electric field in a dielectric solid.

Figure 4-1 depicts a bond in a polar semiconductor. The figure lies in an xz-plane and the internuclear distance d shown has equal components in the x-, y-, and z-directions. This is the most convenient geometry with a field in the x-direction, because every bond in the system makes the same angle (the cosine of which is $1/\sqrt{3}$) with the electric field. We have chosen a bond with the negatively charged atom to the right and the positively charged atom to the left; equal numbers of bonds exist with the negative atom to the left and the positive atom to the right.

A bond such as that shown in Figure 4-1 was constructed in Section 2-2-C as a mixture of a hybrid $|h\text{-}\rangle$ on the right atom and a hybrid $|h+\rangle$ on the left:

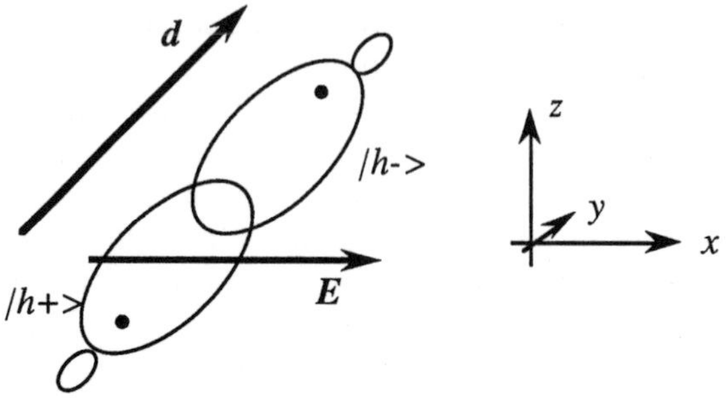

Fig. 4-1. A bond dipole formed by a hybrid on a metallic atom (+) and a hybrid on a nonmetallic atom (-) . The bond dipole is antiparallel to ***d***. An electric field ***E*** as shown will make the bond dipole less negative.

$$|b\rangle = \sqrt{\frac{1+\alpha_p}{2}}\,|h-\rangle + \sqrt{\frac{1-\alpha_p}{2}}\,|h+\rangle \;, \tag{4-1}$$

where the polarity α_p was related to the covalent energy

$$V_2 = 3.22\,\hbar^2/md^2 \tag{4-2}$$

and the polar energy

$$V_3 = \frac{\varepsilon_{h+} - \varepsilon_{h-}}{2} \tag{4-3}$$

by

$$\alpha_p = \frac{V_3}{\sqrt{V_2^2 + V_3^2}} \;. \tag{4-4}$$

The hybrid energies for each atom are $\varepsilon_h = (\varepsilon_s + 3\varepsilon_p)/4$, evaluated from Table 1-1.

The fractional occupation of each constituent of a mixed state is the square of the amplitude, the square of the coefficient in Eq. (4-1). Thus, if the center of gravity of each hybrid were at the nucleus of its atom, the center of the gravity of the bond in Figure 4-1, measured from the bond center, would be

$$r = -\left(\frac{1 - \alpha_p}{2}\right)\frac{d}{2} + \left(\frac{1 + \alpha_p}{2}\right)\frac{d}{2} = \frac{\alpha_p\, d}{2} . \tag{4-5}$$

It might seem a poor approximation to assume that the hybrids are centered on their nuclei, because they were specifically designed to "lean" toward the neighboring site. However, the shift in center of gravity because of this lean is small (see Fig. 1-10) compared to Eq. (4-5), and we neglect it here. We will find that we underestimate our dipoles in any case, so this small correction would worsen that underestimate slightly. Thus, Eq. (4-5) is not quantitatively justified on fundamental grounds; in fact the earlier analysis (Harrison, 1980) introduced an empirical scale factor γ to correct for the approximation in its use. We do not use any such correction here and proceed with Eq. (4-5) as a clear starting point of our treatment of dielectric properties and shall discuss the errors it produces.

Because there are two electrons in the bond, it has a net charge of $-2e$; therefore, unless we include a compensating positive charge from the nuclei of neighboring atoms, the dipole associated with the bond depends on the origin of coordinates. For the analysis here, it is best to treat the nuclear charges separately and to define the electronic dipole, measured relative to the bond center, using Eq. (4-5). The bond dipole becomes

$$p = -\alpha_p e \boldsymbol{d} , \tag{4-6}$$

with $\boldsymbol{d}$ the vector from the metallic (e. g., Ga) to the nonmetallic (e. g., As) nucleus.

The bond dipole is zero in the elemental semiconductor, with $\alpha_p = 0$, and the four bond dipoles surrounding a metallic atom in the polar crystal sum to zero; accordingly, the crystal has no net dipole when no field is applied. However, the formulation given above suggests an effective charge on each atom in a polar crystal. For example, the four doubly occupied bonds described by Eq. (4-1) place an electronic charge of $-4e(1 - \alpha_p)$ on the metallic atom in a polar semiconductor. If the valence charge of that atom is Z (equal to the column number in Table 1-1), the effective charge on the atom becomes

$$Z^* = Z - 4 + 4\alpha_p , \tag{4-7}$$

with an equal and opposite charge on the nonmetallic atom. Using polarities from Table 2-1, we see that these charges are approximately $+1$ for the metallic atom in III-V and II-VI semiconductors and somewhat less for the I-

VII semiconductors. The value of Z^* has no direct quantitative consequence, but provides a starting point for a number of analyses. We might fear that the electron energies for these charged atoms would be shifted from their values of Table 1-1, but we saw in Section 2-3-E that the shifts are largely canceled by the charges on neighboring atoms in the bulk semiconductors.

One intuitive feature can be noted, that the effective charges of the atoms are not directly associated with the polarity of the bonding, which for example reduced the angular force constants by $\alpha_c^3 = (1 - \alpha_p^2)^{3/2}$. That polarity continues to grow as we take constituent atoms further and further from column IV in the periodic table, though Z^* grows and then wanes. In fact, Z^* could have remained as zero if the polarity had grown as 0.25, 0.5, 0.75 in successive compounds but the trends in properties with polarity would have been similar to what we find here.

We have noted in Section 2-2-D that each bond state is coupled to the neighboring antibond state by a coupling energy proportional to the metallic energy V_1 (listed in Table 2-1). The effect of this coupling, which we called metallization , shifted the bond energy and it will also modify the bond dipole. We consider this effect of metallization in Section 4-2-B, but proceed now without it.

2. Electronic Susceptibilities

A. The Linear Response.

At optical frequencies (well above the frequencies of the lattice vibrations, discussed in Section 3-3), any displacement of the atoms themselves by the electric field becomes negligible and the only electric polarization comes from the electronic bond dipoles, Eq. (4-6). It becomes trivial to calculate that dipole if we note that α_p in Eq. (4-6) depends on the polar energy, Eq. (4-3), and that an electric field shifts the polar energy by

$$\delta V_3 = {}^1/_2\, \delta(\varepsilon_{h+} - \varepsilon_{h-}) = -\, {}^1/_2\, \mathbf{E} \cdot \mathbf{d} \quad . \tag{4-8}$$

This follows from exactly the same approximation that led to Eq. (4-5). We let the electric field lie in the x-direction as indicated in Figure 4-1 and compute the polarization in the x-direction from Eq. (4-6); the cubic symmetry of the zincblende or diamond structure guarantees that the *net* polarization proportional to the field (when summed over bonds) will be parallel to the field with magnitude independent of the field direction. In particular, the change in polarization δp_x for this one bond resulting from the field is

$$\delta p_x = -\, e d_x \delta\alpha_p = \frac{-e d_x(1-\alpha_p^2)\,\delta V_3}{\sqrt{V_2^2+V_3^2}} = \frac{e^2 d_x^2(1-\alpha_p^2)\,E_x}{2\sqrt{V_2^2+V_3^2}}$$

$$= \frac{e^2 d^2(1-\alpha_p^2)E_x}{6\sqrt{V_2^2+V_3^2}}, \qquad (4\text{-}9)$$

using in the last step $d_x^2 = d^2/3$ for bonds in a [111] direction. We can multiply by the number of bonds per unit volume, $3\sqrt{3}/(4d^3)$, to obtain the polarization density $\boldsymbol{P}$; it is related to the field by the linear susceptibility, $\boldsymbol{P} = \chi^{(1)}\boldsymbol{E}$. We obtain

$$\chi^{(1)} = \frac{\sqrt{3}\, e^2(1-\alpha_p^2)}{8\sqrt{V_2^2+V_3^2}\; d} = \frac{\sqrt{3}\, e^2\alpha_c^3}{8\, V_2\, d}\;, \qquad (4\text{-}10)$$

where α_c is the covalency which we defined by

$$\alpha_c = \frac{V_2}{\sqrt{V_2^2+V_3^2}}\;. \qquad (4\text{-}11)$$

This susceptibility is related to the dielectric constant ε by

$$\varepsilon = 1 + 4\pi\,\chi^{(1)}\;, \qquad (4\text{-}12)$$

and the refractive index is just the square root of this dielectric constant; accordingly, the susceptibility can be directly determined from experiment and compared with the prediction in Eq. (4-10).

The theoretical and experimental values for the homopolar semiconductors, silicon, germanium, and tin, listed in Table 4-1, show that we have considerably underestimated the values, *and* the ratio by which the susceptibilities increase from row to row in the periodic table. When we include metallization (Section 4-2-B) the latter discrepancy is largely removed; however, there remains a factor of about two between theory and experiment. We prefer to leave that as a discrepancy, rather than introduce an empirical scale correction factor as in Harrison (1980).

Table 4-1. Optical dielectric susceptibilities, $\chi^{(1)}$, for elemental semiconductors.

Element	Theory $\frac{\sqrt{3}\, e^2}{8V_2 d}$	Theory $\frac{\sqrt{3} e^2 (1+11\alpha_m{}^2/16)}{8V_2 d}$	Experiment*
Carbon	0.20	0.22	0.37
Silicon	0.30	0.43	0.88
Germanium	0.31	0.50	1.19
Tin	0.36	0.60	1.83

* From Harrison (1980), p. 114.

We can also examine the effect of polarity on the susceptibility. To separate out the absolute discrepancy in the homopolar semiconductor, Table 4-2 gives the ratio of the polar semiconductor susceptibility to the susceptibility of the corresponding elemental semiconductor, which we shall list in Table 5-4. In fact, the trend with polarity is extremely well described. This is an interesting result, particularly because the parameters used here are quite different from those used in Harrison (1980) and because these parameters are the same as those used for the bonding properties in Chapter 2. In fact, this dependence on α_c^3 is exactly the same as the dependence of the elastic rigidity constant $(c_{11} - c_{12})/2$ against twist distortions.

Our calculation of the polarization assumes a uniform electric field $\boldsymbol{E}$ within the semiconductor. Although this approach differs from many traditional analyses, it is appropriate in the context of the present analysis. The applied field has polarized each bond and therefore has increased the atomic charge Z^* at one end of the bond and decreased it at the other. However, each *atom* has lost charge from the bonds on one side and gained an equal amount from the bonds on the other; thus, the net atomic charge on any atom in the interior does not change: charge is simply passed through the crystal without any net local distortion of the electronic charge distribution. For this reason it is not appropriate to construct a local electric field arising from the neighboring bond dipoles, nor to cut a spherical hole in the dielectric medium and examine charges on the hole surface. We shall discuss this point in more detail in Section 4-2-C when we discuss higher-order susceptibilities.

B. The Effect of Metallization.

We saw in Section 2-3-B that each bond state is coupled to three neighboring antibonding states sharing the metallic atom by a coupling

Table 4-2. Effect of polarity on the optical dielectric susceptibility, $\chi^{(1)}$.

Semiconductor	Isoelectronic Element	Ratio of Susceptibility Theory, α_c^2	Experiment*
BN	C	0.85	0.74
BeO	C	0.50	0.43
A1P	Si	0.66	0.64
GaAs	Ge	0.69	0.66
ZnSe	Ge	0.32	0.33
CuBr	Ge	0.17	0.23
InSb	Sn	0.66	0.64
CdTe	Sn	0.28	0.27
AgI	Sn	0.15	0.17

* Taken from the $\varepsilon = 1 + 4\pi\chi^{(1)}$ in Table 5-4.

energy $-\alpha_c V_{1+}/2$ and to the three antibonding states sharing the nonmetallic atom by $-\alpha_c V_{1-}/2$. We found (in perturbation theory) that this shifted the bond energy by $-{}^3/_8\alpha_c^2(V_{1+}^2 + V_{1-}^2))/\sqrt{V_2^2 + V_3^2}$ per electron, an effect we called metallization. An applied electric field shifts the energies of the coupled levels and the matrix elements (through the factor α_c) and therefore the metallization energy. We see in Section 4-2-C that the shift in the total energy per unit volume due to an electric field$E$ is ${}^1/_2\chi^{(1)}E^2$, so the shift in the metallization energy proportional to the square of the field directly contributes to the dielectric susceptibility. The analysis becomes quite intricate for the polar semiconductor and has only been carried out here for the case of elemental semiconductors.

Consider a bond and its six neighboring antibonds (Figure 4-2). Without a field the energy of the bond state is $-V_2$; however, the effect of the electric field is to shift the polar energy (zero in this case with no field) by $\delta V_3 = -eEd/(2\sqrt{3})$. This give a bond energy (without metallization) of

$$\varepsilon_b = \varepsilon_h - \sqrt{V_2^2 + \delta V_3^2} \approx \varepsilon_h - V_2 - {}^1/_2\, \delta V_3^2/V_2 \qquad (4\text{-}13)$$

to second order in the field.

Similarly, each antibonding state energy is shifted by the positive quantity ${}^1/_2\delta V_3^2/V_2$. In addition, the center of gravity of the antibonds labeled (1) and (2) in Figure 4-2 lies to the right of the central bond by $d/\sqrt{3}$; therefore, the average hybrid energy (and the antibonding energy) is lower by an additional $-eEd/\sqrt{3} = +2\delta V_3$. Antibond (4) does not have this

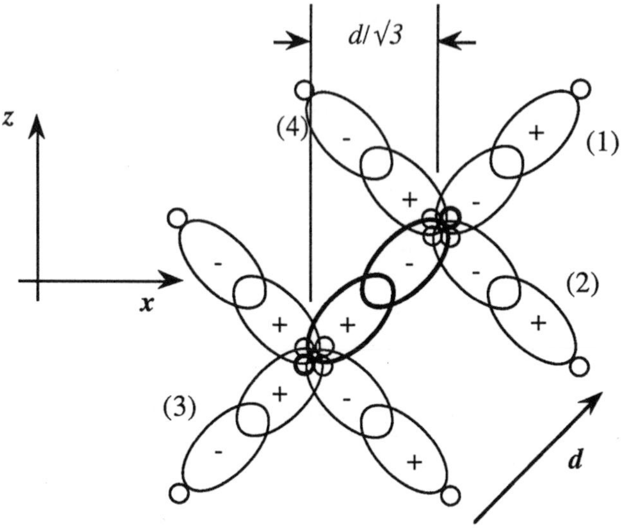

Fig. 4-2. A bond, as in Fig. 4-1, and the six neighboring antibonds with which it shares an atom. The coupling to these antibonds gives a metallization energy, which depends on the electric field and therefore contributes to the dielectric susceptibility. The numbering of antibonds and choice of axes are used in the analysis.

shift and antibond (3) has a shift in the opposite direction. The coupling between the bond and antibond (1) is $-V_1$ times the coefficient of the hybrid forming the right end of the bond, $\backslash r(1 + \alpha_p) / \sqrt{2}$, times the coefficient, $\sqrt{1 + \alpha_p} / \sqrt{2}$, of the hybrid forming the left end of the antibond (1). With the polarity $\delta V_3/\sqrt{V_2{}^3 + \delta V_3{}^2}$, the coupling becomes $-(1 + \delta V_3/V_2)V_1/2$, plus terms of third order in the field. The coupling to antibond (2) is the same; in antibond (4), the coefficient of the hybrid to the right is $\sqrt{1 - \alpha_p} / \sqrt{2}$ and the coupling becomes $-\sqrt{1 - \delta V_3{}^2 / V_3{}^2}\, V_1/2$. The coupling for the three antibonding states to the left is the same as the corresponding ones to the right, but with the sign of δV_3 changed.

The bond energy shift resulting from these six terms is obtained by squaring the coupling , dividing by the energy difference for each set, and summing the shifts. We obtain the shift in bond energy from metallization as

$$\delta\varepsilon_b{}^{met} = -\frac{2(1 + \delta V_3/V_2)^2\, V_1{}^2/4}{2V_2 - 2\delta V_3 + \delta V_3{}^2/V_2} - \frac{(1 - \delta V_3{}^2/V_2{}^2)V_1{}^2/4}{2V_2 + \delta V_3{}^2/V_2} + \ldots \quad (4\text{-}14)$$

The first term represents the effect of the coupling with antibonds (1) and (2) and the second term represents the effect of the coupling with antibond (4). Two more terms are obtained by changing the sign of δV_3, giving the effect of the coupling with the antibonds to the left. We can expand this result to obtain the metalization shift in bond energy to second order in δV_3,

$$\delta\varepsilon_b^{met} = -\frac{3V_1^2}{4V_2} - \frac{11}{8}\frac{V_1^2}{V_2}\frac{\delta V_3^2}{V_2^2} \quad . \tag{4-15}$$

We had in Eq. (4-13) a bond energy shift of $-{}^1/_2\delta V_3^2/V_2$. Due to the second term in Eq. (4-15), this is increased now by a factor $1 + (11/4)(V_1/V_2)^2$; the susceptibility is increased by the same factor. It is convenient to write this in terms of the metallicity, defined in Eq. 2-18 by

$$\alpha_m = 2V_1/V_2 \quad . \tag{4-16}$$

We then combine this scale factor with Eq. (4-10), for $\alpha_c = 1$, to obtain

$$\chi^{(1)} = \frac{\sqrt{3}\,e^2}{8\,V_2\,d}\left(1 + \frac{11\alpha_m^2}{16}\right) \tag{4-17}$$

for elemental semiconductors. These values, including the metallization correction factor $1 + 11\alpha_m^2/16$, are listed in Table 4-1.

Although the corrected values are still about a factor of two too small, the ratios of the values for elements of different metallicities are now quite well given. In fact, the predicted values are all close to those obtained by Ren (Ren and Harrison (1981)) using the same formulation of the electronic structure (although slightly different parameters) but making a full calculation of the energy bands and of the dielectric susceptibility in terms of them. Residual errors seem to result from the approximate basic description of the electronic structure we are using and maybe from the one-electron description itself. One error inherent in Ren's calculation is the use of our parameters which were fit to the energy bands, as described in Section 5-3-D. These bands (Chelikowsky and Cohen (1976)) were initially adjusted empirically to give the observed band gaps. It would have been preferable not to have adjusted the bands, giving band gaps smaller than experiment but appropriate to the susceptibility calculation (ultimately because this is a ground-state, rather than excited-state calculation), and then correcting afterward, according to Section 5-5-A for Coulomb effects, if one really wanted to know the gaps associated with excited states. This more correct procedure would have enhanced the calculated susceptibility, but we do not know by how much. Fortunately, these residual errors seem to be describable by a universal factor of about two and the trends from system to system are well described, provided that metallization is included.

A correction similar to, but more complicated in form than, that in Eq. (4-17), could be derived for the susceptibility of polar semiconductors and for the other dielectric properties that we will discuss. This has not yet been done. The description without metallization, as in Section 4-2-A, is much simpler and suffices to describe the main features of the dielectric properties.

C. Higher-Order Susceptibilities.

The linear susceptibility derived in Section 2-2-A is a second-rank tensor and the symmetry of the zincblende structure of the polar semiconductors (or the symmetry of the diamond structure) is sufficient that it is diagonal with each diagonal element the same; that is, it is a numerical constant, independent of field orientation. However, there exist polarizations that are not parallel to the field in these structures that are of higher order in the electric field. They are described by a higher-order susceptibility.

The definition of dielectric quantities can be confusing and ambiguous, and in most physics applications it is best not to introduce such quantities as the displacement $\boldsymbol{D}$ which may simplify analysis when one is doing many repeated problems on similar systems. It is best to think in more physical terms, in this case of individual dipoles associated with each bond, as in the preceding sections of this chapter. Then there is an electric field in the region of each bond due to any applied field and due to other dipoles in the system. We have approximated this by a uniform field $\boldsymbol{E}$ though of course neighboring dipoles will produce nonuniform fields and modifications of the polarization from such nonuniformities are called *local-field corrections.* We have neglected them because our view of the bond polarization is the transfer of charge from one atom to the next so that in a uniform field each atom receives as much charge as it gives up, remains neutral, and leaves the field uniform. The atoms at the surface of the crystal do *not* remain neutral. However, if the surface is distant we may represent their effect as a surface charge numerically equal to the normal component of polarization, which leads to a field within the crystal reduced relative to an applied uniform field by a factor of the dielectric constant given in Eq. (4-12). Our estimate of the energy, $-2\sqrt{V_2^2 + V_3^2}$ without metallization, includes any electrostatic interactions between the two electrons in that bond. Our treatment is internally consistent, and makes each aspect of the system unambiguous, but such neglect of local-field effects may be partly responsible for the inaccuracy of our predictions given in Tables 4-1.

For our definition of higher-order susceptibilities we may define the total energy of the bonds E_{tot} in the presence of the uniform electric field. This is then expanded in the components of the electric field,

$$E_{tot} = -{}^1/_2 \sum_{ij} \chi_{ij}^{(1)} E_i E_j - {}^1/_3 \sum_{ijk} \chi_{ijk}^{(2)} E_i E_j E_k$$
$$- {}^1/_4 \sum_{ijkl} \chi_{ijkl}^{(3)} E_i E_j E_k E_l - \ldots \qquad (4\text{-}18)$$

Here the indices $i = 1$ corresponds to the component of the field in the x-direction, 2 to the y-direction and 3 to the z-direction. Note that in the sum over indices, all combinations appear such that there are $3^2 = 9$ terms in the first sum; e.g. $i = 1, j = 2$ as well as $i = 2, j = 1$. The electric polarization density in the i th direction is given by the negative of the derivative of the energy with respect to the field in the i th direction,

$$P_i = \sum_j \chi_{ij}^{(1)} E_j + \sum_{jk} \chi_{ijk}^{(2)} E_j E_k + \sum_{ikl} \chi_{ijkl}^{(3)} E_j E_k E_l + \ldots \qquad (4\text{-}19)$$

Note that the two terms (for example) in the first sum of Eq. (4-18), $\chi_{12}^{(1)} E_1 E_2 + \chi_{21}^{(1)} E_2 E_1$, can be combined and we did that, taking $\chi_{21}^{(1)} = \chi_{12}^{(1)}$; both terms contributed when we took the derivative of Eq. (4-18) with respect to E_1 so that the leading factor of ${}^1/_2$ was canceled. If we had not done that, the $\chi_{ij}^{(1)}$ would have been left as ${}^1/_2\,(\chi_{ij} + \chi_{ji})$.

We should pause to see what can be learned about these susceptibility tensors from symmetry and exactly the same arguments will apply to other high-rank tensors that we will consider in the subsequent analysis. We make the symmetry arguments for the zincblende or diamond structure, not the wurtzite structure (semiconductors with that structure are indicated in Table 2-1). We therefore include the most important semiconductors and the arguments would need to be slightly modified for the hexagonal symmetry of the wurtzite systems.

Figure 2-2 depicted the zincblende structure and is redrawn as Fig. 4-3. A cube was drawn through eight of the metallic atoms to help show the overall symmetry. A dashed line has now been added through the center of

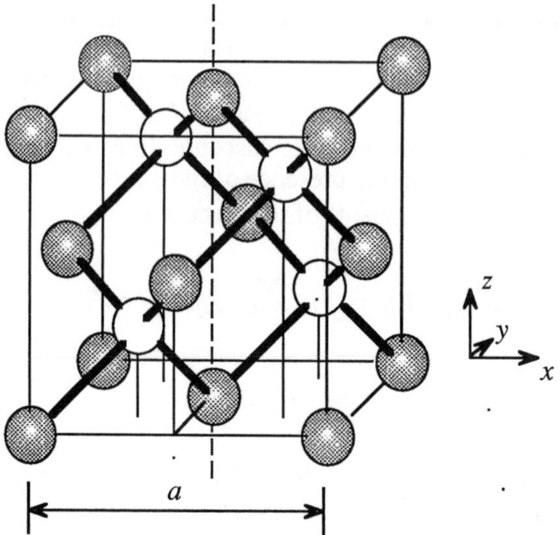

Fig. 4-3. The zincblende crystal structure, from Fig. 2-2. Shaded atoms are metallic, unshaded atoms are nonmetallic. A rotation of 180^0 around the dashed axis drawn through the top and bottom face-centered atoms is a symmetry operation, taking the crystal into itself.

the cube parallel to the z-axis; it is not difficult to see that rotating the crystal 180° around that line will leave every position originally occupied by a metallic atom still occupied by a metallic atom, and each nonmetallic atom by a nonmetallic atom. This rotation - said to "take the crystal into itself" - is a symmetry operation of the crystal and can be used to learn about the susceptibility tensor. We can illustrate the procedure by asking if a field in the y-direction can give a polarization in the x-direction proportional to the square of that field. From Eq. (4-19), we see that such a polarization would be given by $P_1 = \chi_{122}^{(2)} E_2^2$. Let us now rotate the crystal *and* the fields by 180° around the z-axis. Because of the rotation, the polarization will also be rotated and will now be oppositely directed. With the new crystal orientation, we have $P_1 = -\chi_{122}^{(2)} E_2^2$. (Note E_2^2 did not change sign.) However, the crystal symmetry is such that this orientation is equivalent; thus, we could start over and write $P_1 = \chi_{122}^{(2)} E_2^2$. Combining the two, we have $\chi_{122}^{(2)} = -\chi_{122}^{(2)}$ and both must be identical to zero; there can be no P_1 proportional to E_2^2. We can use similar rotations of 180^0 around the x- and y-axes to show that any component of $\chi_{ijk}^{(2)}$ with two indices the same must vanish and the only allowed components have all indices different. Similarly, a symmetry rotation of 120^0 around a cube diagonal parallel to $x = y = z$ can be used to show that $\chi_{123}^{(2)} = \chi_{312}^{(2)} = \chi_{231}^{(2)}$.

Similar symmetry arguments lead to the conclusion given earlier that $\chi_{ij}^{(1)}$ is diagonal, with all three diagonal elements equal. Furthermore, the same kind of argument can be used to show that all components of $\chi_{ijkl}^{(3)}$ vanish unless the indices appear in pairs. With all fields equivalent in Eq. (4-18), there are only two independent indices, $\chi_{1111}^{(3)}$ and $\chi_{1122}^{(3)}$. [Note that $\chi_{1212}^{(3)}$ was taken equal to $\chi_{1122}^{(3)}$ as we took $\chi_{ij}^{(1)} = \chi_{ji}^{(1)}$.]

There are cases where different electric fields and different polarizations are distinguishable (e.g., the electrooptic effect discussed in Section 4-3-F) We apply a dc field E_j and ask for the optical polarization P_i resulting from an optical field E_k . We shall see that because the system responds differently to the static field (the atoms are displaced by it), we cannot immediately deduce that (for example) $\chi_{213}^{(2)} = \chi_{123}^{(2)}$. In this particular case, a reflection of the crystal through the plane x = y guarantees that even for the electrooptic effect they are equal and there is only one independent element $\chi_{123}^{(2)}$ in the zincblende structure.

The same arguments can be applied to any second-rank physical tensor (e.g., conductivity σ_{ij}) or third-rank tensor (e.g., the piezoelectric constant e_{ijk}). For the fourth-rank tensor we must be careful about reshuffling indices: There are three independent elastic constants (c_{1111}, c_{1122}, and c_{1212}); similarly there are three independent elastooptic coefficients in the zincblende structure.

We turn to the second-order susceptibility $\chi_{ijk}^{(2)}$, which must be zero if all three indices are not different. In the diamond structure (nonpolar), even those elements are zero (deduced from an inversion through a bond center), but for zincblende structures we cannot prove them to be zero. Symmetry arguments are one-sided in this sense; we can prove a parameter is zero, but we cannot prove by symmetry that it will be nonzero.

A nonzero $\chi_{123}^{(2)}$ is allowed by symmetry, and we proceed to estimate it. We begin by expanding Eq. (4-6) for the bond dipole to second order in the electric field, which contributes $-e\mathbf{E}\cdot\mathbf{d}/2$ to V_3 . The second-order term is

$$p^{(2)} = \frac{3e^3\,\alpha_p(1-\alpha_p^2)\,(\mathbf{E}\cdot\mathbf{d})^2}{8\sqrt{V_2^2+V_3^2}}\,\mathbf{d}\,. \tag{4-20}$$

For any set of applied fields, we can simply substitute in Eq. (4-20) and add up the dipoles from all the bonds.

To be specific, consider the four bonds surrounding a metallic atom as illustrated in Figure 4-4. Let the field have components in both the x- and z-

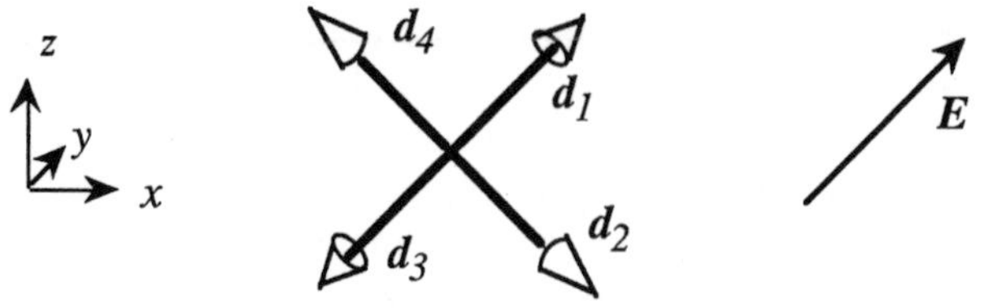

Fig. 4-4. Four bond dipoles surrounding a metallic atom With electric field components in both the z- and x-directions, a net dipole arises in the y-direction, corresponding to a second-order susceptibility.

directions. We can immediately evaluate the relevant geometrical quantities in Eq. (4-20) for each of these bonds as

Internuclear vector	$\mathbf{d}\cdot\mathbf{E}$	$\frac{(\mathbf{d}\cdot\mathbf{E})^2\, 3\sqrt{3}}{d^3}\,\mathbf{d}$
$\mathbf{d}_1 = [1\ 1\ 1]d/\sqrt{3}$	$(E_x + E_z)d/\sqrt{3}$	$(E_x^2 + 2E_xE_z + E_z^2)[1\ 1\ 1]$
$\mathbf{d}_2 = [1\text{-}1\text{-}1]d/\sqrt{3}$	$(E_x - E_z)d/\sqrt{3}$	$(E_x^2 - 2E_xE_z + E_z^2)[1\text{-}1\text{-}1]$
$\mathbf{d}_3 = [\text{-}1 1\text{-}1]d/\sqrt{3}$	$(-E_x - E_z)d/\sqrt{3}$	$(E_x^2 + 2E_xE_z + E_z^2)[\text{-}11\text{-}1]$
$\mathbf{d}_4 = [\text{-}1\text{-}1 1]d/\sqrt{3}$	$(-E_x + E_z)d/\sqrt{3}$	$(E_x^2 - 2E_xE_z + E_z^2)[\text{-}1\text{-}1\ 1]$

(4-21)

By summing over the four bonds, we see that only the y-component of the polarization gives a nonzero value (as the symmetry argument tells it should). We obtain a net dipole for the four bonds surrounding this atom of

$$\Sigma p_y = \sqrt{3}\, e^3 d^3 \alpha_p (1 - \alpha_p^2) \frac{E_x E_z}{3(V_2^2 + V_3^2)} \quad . \tag{4-22}$$

We can multiply by the density of nonmetallic atoms, $3\sqrt{3}/(16d^3)$, to obtain the polarization density; that same polarization density is written using Eq. (4-19) as $(\chi_{213}{}^{(2)} + \chi_{231}{}^{(2)})E_xE_z$. Matching the polarization densities and taking the two susceptibility components equal to $\chi_{123}{}^{(2)}$ we obtain

$$\chi_{123}{}^{(2)} = \frac{3}{32} \frac{e^3 \alpha_p (1 - \alpha_p{}^2)^2}{V_2{}^2} . \tag{4-23}$$

This is not dimensionless, as was $\chi^{(1)}$, and it is customarily given in electrostatic units by substituting e as 4.8×10^{-10} esu and V_2 in ergs, giving a value of order 10^{-7} esu. Microscopic units may be preferable here.

We can write $e^2 = 14.4\ eV\text{-}\AA$ and the other factor of e as one eV per volt. Then $\chi_{123}^{(2)}$ has units of reciprocal potential gradient, Å per volt; in these units it will be of order one.

Such formulae can be extremely useful. If, for example, we wished to find a material with as large a $\chi_{123}^{(2)}$ as possible, for use in generation of higher harmonics of a light beam, we would note first that the form, Eq. (4-23), vanishes for α_p equal to zero or one, and reaches a maximum as a function of polarity at $\alpha_p = \sqrt{1/5} = 0.45$, near the values for III-V semiconductors. Then we note that at fixed α_p it varies as $1/V_2^2 \propto d^4$ so we would select the III-V semiconductor with the greatest bond length d , which is indium antimonide. That is indeed the best choice among the tetrahedral semiconductors.

We expect significant numerical error because of the neglect of metallization. To factor these out, we could compute the ratio of $\chi_{123}^{(2)}$ to $\chi^{(1)}$ (both without metallization) because similar corrections might be expected for each. (The use of a scale factor γ in Harrison (1980) was analogous to comparing the ratios of $\chi_{123}^{(2)}$ to $\chi^{(1)}$.) Then, combining Eq. (4-10) and (4-23), we have

$$\frac{\chi_{123}^{(2)}}{\chi^{(1)}} = \frac{\sqrt{3}}{4}\frac{ed\alpha_p\alpha_c}{V_2} \tag{4-24}$$

These are compared with experiment in Table 4-3. The predicted trends are correct and the order of magnitude right; however, the error is larger than that in the first-order susceptibility. The origin of the error was discussed in Harrison (1980): the analysis in terms of bond states treats all occupied

Table 4-3. Ratio of second-order susceptibility to first-order susceptibility, $\chi_{123}^{(2)}/\chi^{(1)}$, in units of Å/volt.

	Theory Eq.(4-24)	Experiment*
Ge	0.	0.
GaAs	0.11	0.27
ZnSe	0.13	0.32
CuBr	0.12	0.05
Sn	0.	0.
InSb	0.17	0.71
CdTe	0.19	0.54

* Collected, with other values, in Harrison (1980) p. 122.

states as having the same energy (and all empty states another), whereas in the real bands the energy difference between coupled full and empty states varies widely. In successively higher-order susceptibilities, those with smallest energy differences become more and more dominant and more seriously underestimated in our simple approach. The predictions are still meaningful for the second-order susceptibility given by Eq. (4-24), but the theory is not useful for the third-order susceptibility. It was carried out in Harrison (1980) giving $\chi_{1111}^{(3)}/\chi^{(1)} = e^2d^2(1 - \alpha_p^2)(5\alpha_p^2 - 1)/(24V_2^2)$. The sign is not correct for homopolar semiconductors (note it is nonzero for this case) and the values are in error by orders of magnitude.

D. Magnetic Susceptibility

The analysis of the response to a magnetic field is rather intricate, somewhat specialized, and of less practical importance than the response to an electric field. We give here a brief account and refer to the full analysis (Harrison (1980) p. 131), based upon Chadi, White and Harrison (1975).

In quantum (as well as in classical) mechanics, a magnetic field can be introduced into the dynamics of the electrons (charge $-e$) by replacing the momentum $\boldsymbol{p}$ by $\boldsymbol{p} - (-e)\boldsymbol{A}/c$, where $\boldsymbol{A}$ is the vector potential, related (in one choice of gauge) to a uniform magnetic field $\boldsymbol{H}$ by $\boldsymbol{A} = -\boldsymbol{r} \times \overline{\boldsymbol{H}}/2$. The Hamiltonian (the energy) contains a term $p^2/2m$, which becomes

$$H = \frac{p^2}{2m} - \frac{e}{2mc}\,\boldsymbol{r} \times \boldsymbol{H} \cdot \boldsymbol{p} + \frac{e^2}{8mc^2}\,|\boldsymbol{r} \times \boldsymbol{H}|^2 . \qquad (4\text{-}25)$$

There are two effects on the energy of a bond. The final term shifts the energy of the bond just as an added potential energy $V(\boldsymbol{r})$ would, by an amount proportional to the square of the magnetic field. The shift is also proportional to r^2 averaged over the bond. The value for one direction (e.g. the x-direction) can be estimated as if the charge were concentrated at the atomic sites $((d/(2\sqrt{3}))^2 = d^2/12$. Both directions perpendicular to $\boldsymbol{H}$ contribute, giving a factor of two. This gives an energy per electron of $e^2d^2H^2/(48\ mc^2)$. We can multiply this by the electron density of $3\sqrt{3}/(2d^3)$ and equate it to the energy density written in terms of magnetic susceptibility $-{}^1/_2\chi_L H^2$ to obtain

$$\chi_L = -\frac{\sqrt{3}\,e^2}{16mc^2d} . \qquad (4\text{-}26)$$

This is called the *Langevin term* in the susceptibility. It is diamagnetic (negative) because it represents an increase in the energy in the field. It is very small, because e^2/d is of the order of electron volts while mc^2 is half a million electron volts; this is partly why the effect is of limited interest. It seems plausible that the energy of an orbiting electron should be shifted by a magnetic field that tends to deflect it, but it is actually quite a subtle effect, occurring only in a quantum-mechanical system; the diamagnetism of a classical charged gas is zero. Another diamagnetic term χ_c arises in just the same way from the core electrons (e. g. $1s^22s^23p^6$ for silicon); however, for that term the r^2 average is of the order of the square of a core radius, very much smaller.

There is also a contribution to the energy of a bond from the second term in Eq. (4-25). Although the average of that term over a bond is zero, it gives a coupling energy between the bond orbital and the neighboring antibonding orbitals (although not between the bond and antibond on the same bond site) very much like the metallic energy coupling gives the metallization energy that was calculated in Section 4-2-B. A vector identity allows us to write $\boldsymbol{r} \times \boldsymbol{H} \cdot \boldsymbol{p} = \boldsymbol{p} \times \boldsymbol{r} \cdot \boldsymbol{H}$ and $\boldsymbol{p} \times \boldsymbol{r} = -\boldsymbol{L}$ is the negative of the angular momentum. There is an angular-momentum matrix element between p-states on the same atom given by $\langle p_y|L_z|p_x\rangle = i\hbar$, and corresponding matrix elements obtained by rotating indices, giving matrix elements between bonds and antibonds in neighboring bond sites (described in some detail in Harrison (1980) p. 131). For states on an atom the angular average of the square gives a factor of $2/3$ as did the average over r^2 in the Langevin term. There is also a factor of $\alpha_c/2$ in the coupling energy because of the coefficients of the hybrids on the atoms shared by the bonding and antibonding states. The shift in energy of the bond state is obtained by squaring the coupling [the averaged square is $e^2\hbar^2H^2\alpha_c^2/(24m^2c^2)$], dividing by the energy difference , [$2\sqrt{V_2^2 + V_3^2}$], and multiplying by the number [six] of antibonding states coupled. This is then multiplied by the electron density and related to the susceptibility as was the Langevin term. We obtain

$$\chi_{VV} = \frac{3\sqrt{3}\, e^2\, \hbar^2 \alpha_c^3}{8m^2c^2d^3V_2} \,. \tag{4-27}$$

This is a prediction of the *Van Vleck* term in the susceptibility; it is paramagnetic (positive) because it is a lowering in the energy. (Coupling of low-energy occupied states to high-energy full states must lower the energy.)

It is very small because of the same factors that made the Langevin term small. We can in fact write $V_2 = 3.22\ \hbar^2/md^2$ to obtain $\chi_{VV} = -1.85\ \chi_L\ \alpha_c^3$.

It is interesting that the diamagnetic term is predicted not to depend on polarity: the bond shifts toward one atom but does not change its size. The paramagnetic term is expected to decrease with polarity, partly from the increase in energy gap separating occupied and empty states, but more strongly from the separation of bonding and antibonding states to separate atoms and the corresponding reduction in the coupling between them. This was also exactly the origin of the α_c^3 factor in the dielectric susceptibility, Eq. (4-10).

These contributions have been evaluated for a sample of semiconductors and compared with experiment in Table 4-4. The individual contributions are quite well given, considering the approximate estimates of r^2 and $(\boldsymbol{p} \times \boldsymbol{r})^2$

Table 4-4. Magnetic susceptibilities (multiplied by 10^6): χ_c for the cores*, χ_L , χ_{VV} , and total χ from theory and experiment*.

Semi-conductor	χ_c Core	χ_L Eq. (4-26)	χ_L Exp.	χ_{VV} Eq. (4-27)	χ_{VV} Exp.	χ Theory	χ Exp.
C	-0.03	-1.98	-3.56	3.69	1.91	1.68	-1.70
BN	-0.04	-1.94	-	3.00	-	1.02	-
BeO	-0.05	-1.85	-	1.61	-	-028	-1.43
Si	-0.19	-1.30	-1.63	2.42	1.56	0.93	-0.26
Ge	-0.51	-1.25	-1.84	2.33	1.87	0.57	-0.58
GaAs	-0.51	-1.24	-1.73	1.60	1.12	-0.15	-1.22
ZnSe	-0.55	-1.24	-	0.74	-	-1.05	-1.70
CuBr	-0.49	-1.22	-	0.36	-	-1.36	-1.96
Sn	-0.78	-1.09	-	2.03	-	0.16	-1.55
InSb	-0.80	-1.08	-	1.33	-	-0.56	-1.60
CdTe	-0.82	-1.08	-	0.57	-	-1.33	-2.04
AgI	-0.83	-1.09	-	0.30	-	-1.62	-2.22

* Compiled in Harrison (1980), p. 135, from Selwood (1956), Hudgens, Kastner, and Fritzsche (1974), and Bailley and Manca (1972).

that were made; however, because they cancel against each other in the net susceptibility χ (also listed), that value does not agree at all well with experiment, even with respect to the sign. The earlier analysis (Harrison, 1980, p. 131) introduced universal empirical scale factors for the Langevin and Van Vleck terms, thereby bringing the predictions into quite good accord with experiment.

3. Internal Displacement Properties

A. The Transverse Charge.

In all the preceding analyses the nuclear positions remained at the perfect-crystal sites and only the electrons responded. There is a range of interesting properties that involve both electromagnetic fields and atomic displacements and we begin considering them now.

We defined (in Section 4-1) an effective charge Z^* and indicated that it had no direct experimental meaning. Even if we knew precisely the density of charge as a function of position in the crystal, the allocation of it to individual atoms would not be unique. However, we are always interested in some experimental effect, and frequently a natural effective charge arises for that experiment. Different experiments have different effective charges and none, except possibly the Coulomb shifts discussed in Section 2-2-F and 2-3-E, seem to correspond to Z^*. One of the most important physical effects of this kind is the coupling energy between an electric field and the displacement of an individual atom with respect to its neighbors. For displacement $\boldsymbol{u}$ and field $\boldsymbol{E}$ that coupling energy is written

$$\delta H = -e\, e_T^* \, \boldsymbol{u} \cdot \boldsymbol{E} \quad , \qquad (4\text{-}28)$$

which becomes the definition of the *transverse charge* e_T^*. In a crystal with cubic symmetry (as is zincblende), it is a scalar quantity equal to the local electric dipole induced by a displacement u divided by eu; that is a more convenient definition for purposes of calculation.

If the atomic charge Z^* moved rigidly with the atom as it was displaced, the transverse charge would be Z^*. However, the displacement modifies the covalent energy in each of its bonds; therefore, the polarity of the bond's changes and the corresponding charge transfer also contributes to the dipole. Note that even if the theory contained electrostatic shifts in the energy levels owing to the charges on neighbors (as discussed in Section 2-3-E), that shift would be *quadratic* in the displacement for the tetrahedral symmetry of a semiconductor and thus would not contribute to the dipole linear in displacement.

Consider a single metallic atom (say Ga in GaAs) being displaced in an x-direction as in Figure 4-5 (a). The bond length for either bond to the right becomes $d' = d - u/\sqrt{3}$ plus terms of order u^2. Then the covalent energy becomes $(1 + 2u/(\sqrt{3}d))V_2$, because $\partial V_2/\partial d = -2V_2/d$. The charge that the two electrons in that bond places on the atom to the right is $-2e(1 + \alpha_p)/2 = -e(1 + \alpha_p - 2\alpha_p\alpha_c^2 u/(\sqrt{3}d))$. The same form obtains for atoms to the left, with the sign of the last term changed. The net dipole (proportional to u) as a result of the shift of these charges is the charge times $\pm d/\sqrt{3}$ summed over the four atoms, or $+8\alpha_p\alpha_c^2 u/3$. This is to be added to the dipole from the charge $Z^*e = Ze - 4(1 - \alpha_p)e$, which is displaced by u. Here Z is the column number of the element in Table 1-1. (Again corrections to the central charge only contribute to the dipole in higher order in u.) We obtain a transverse charge of

$$e_T^* = Z - 4 + 4\alpha_p + \frac{8}{3}\alpha_p\alpha_c^2 \ . \qquad (4\text{-}29)$$

The transverse charge for the nonmetallic atom (As in GaAs), calculated similarly, is equal and opposite.

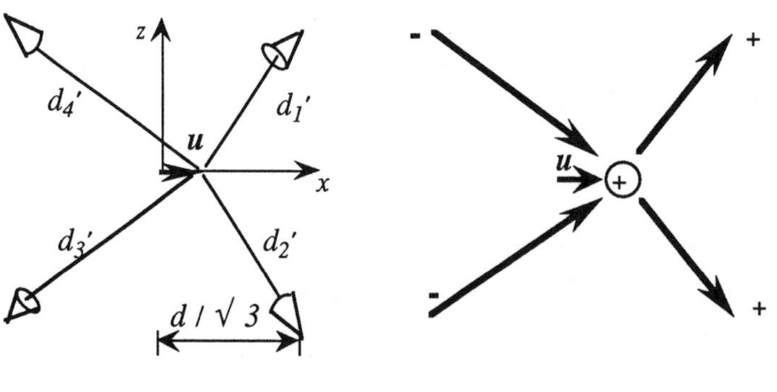

(a) Displacement of atom (b) Charge transfer

Fig. 4-5. Calculation of the transverse charge. A single metallic atom is displaced by u in the x-direction, as in (a). The right-hand bonds have modified lengths $d_1' = d_2' = d - u/\sqrt{3}$, to first order in u; the left-hand bond lengths d_3' and d_4' are increased by an equal amount. This causes transfer of electrons from the two atoms on the left to the two atoms on the right, adding to the dipole moment and therefore to e_T^*.

Table 4-5. Effective transverse charges for compound semiconductors.

Semiconductor	Theory [Eq. (4-29)]	Experiment*
BN	1.17	2.47
BeO	1.53	1.83
AlP	1.97	2.28
GaAs	1.84	2.16
ZnSe	1.83	2.03
CuBr	1.02	1.49
InSb	1.97	2.42
CdTe	1.88	2.35
AgI	1.03	1.40

*Excerpted from a large table given by Lucovsky, Martin, and Burstein (1971) and quoted in Harrison (1980, p220).

It is interesting to note what has occurred: The metallic atom is generally positive ($4\alpha_p$ is greater than $4\text{-}Z$), giving a positive dipole. In addition, the shortening of the right-hand bonds reduces their polarity and transfers electronic charge backward. An equal charge is transferred backward by the increased polarity of the left-hand bonds, as illustrated in Figure 4-5(b). Thus, the additional dipole [the final term in Eq. (4-29)] has arisen from transfer of electrons from the right neighboring atoms to those on the left. Eq. (4-29) is evaluated for a series of polar semiconductors (note e_T^* is zero for the elemental semiconductors) and compared with experiment in Table 4-5. The significant enhancement over the value of Z^* is well described and the general magnitudes and trends are given. Bennetto and Vanderbilt (1996) have carefully explored and corrected for the principle approximations in this analysis, finding that most corrections reduce the values, worsening the agreement with experiment. The theory has been extended to the effects of strain on e_T^* by Anastassakis and Cardona (1985).

B. Optical Mode Splitting

In this analysis, we have displaced a single nonmetallic atom. We can similarly displace all nonmetallic atoms by the same amount and directly add the effects of each. The transfer charge now cancels on every atom in the interior of the crystal (just as it canceled for the central atom in Figure 4-5), but the surface atoms become charged. The effect is conveniently calculated by adding the dipoles to obtain a polarization density. Let the displacement in the x-direction on the metallic atom be u_+ and that of the nonmetallic atom be u_-. Then the polarization density for the $3\sqrt{3}/(16d^3)$ atoms pairs per unit volume becomes

$$P = \frac{3\sqrt{3}\, e\, e_T^*(u_+ - u_-)}{16d^3} . \tag{4-30}$$

If these displacements are modulated in the x-direction, corresponding to a longitudinal optical mode, for which the two atoms move opposite to each other as described in Section 3-3-B, there will be charge density accumulation $\rho = -\partial P/\partial x$ and an electric field E_x given by Poisson's equation, $\partial E_x/\partial x = 4\pi\rho$. This field is reduced by the optical dielectric constant ε (a result of bond polarization, as discussed in Section 4-2) and the resulting field gives a force on the metallic atom (as follows from Eq. (4-28)) of $ee_T^*E_x$. Combining these, we obtain the force on the metallic atom,

$$F = -3\sqrt{3}\,\pi e^2 e_T^{*2}(u_+ - u_-)/(4\varepsilon d^3) . \tag{4-31}$$

This force adds directly to the force arising from the bonds which led to Eq. (3-32) for the optical-mode frequency, increasing the longitudinal optical mode frequency ω_{LO} above the transverse optical frequency ω_{TO}, since for the transverse mode there is no charge accumulation and no stiffening.

The formula, Eq. (3-32), was actually obtained by equating the elastic and kinetic energies, but could equally as well be derived by equating mass times acceleration with the applied force. We redo this derivation in terms of forces and then include the force from Eq. (4-31). The displacement of the metallic atom is written u_+ and its mass M_+. Then the acceleration times the mass is $-M_+\omega^2 u_+$ and the force F from bonding is obtained as the negative of the derivative of Eq. (3-12) with respect to $u' = u_+ - u_-$ for the homopolar case, or α_c^3 times it for the polar case. Setting mass times acceleration equal to force for the transverse optical mode (for which there are no effects of e_T^*), we obtain $-M_+\omega_{TO}^2u_+ = F = -32(1+\lambda)V_2\alpha_c^3(u_+ - u_-)/3d^2$. A similar equation is written for u_- with M_- and the two solved together to obtain, as in Eq. (3-35),

$$\omega_{TO}^2 = \frac{32(1+\lambda)V_2\alpha_c^3}{3M_r\,d^2} \tag{4-32}$$

with the reduced mass $M_r = M_+M_-/(M_+ + M_-)$. This reduces to Eq. (3-32) for $M_r = M/2$ and $\alpha_c = 1$.

For the longitudinal mode the force given in Eq. (4-31) adds directly to the force F (also a negative number times $u_+ - u_-$) given above leading to a longitudinal-mode frequency given by

$$\omega_{LO}^2 = \omega_{TO}^2 + \frac{3\sqrt{3}\pi e^2\,e_T^{*2}}{4M_r\,d^3\varepsilon} \quad . \tag{4-33}$$

This formula does not depend upon our use of the Eq. (4-32) for ω_{TO}^2 . Eq. (4-33) is called the *Lyddane-Sachs-Teller relation* (Kittel, 1976). The experimental values for e_T^* in Table 4-5 were obtained from the measured difference in the two frequencies.

The shifts tend to be rather small. Substituting Eq. (4-32) into Eq. (4-33) allows us to obtain

$$\omega_{LO}^2 = \omega_{TO}^2\left(1 + \frac{0.064e^2md\,e_T^{*2}}{\hbar^2\alpha_c^3\varepsilon}\right) \quad . \tag{4-34}$$

For gallium arsenide this corresponds approximately to a 12% difference in frequency based upon all theoretical parameters, comparable to the observed 8%. Because the experimental e_T^* values in Table 4-5 came from the measured splittings, that table gives a comparison of Eq. (4-33) or (4-34) with experiment for other systems.

C. The Static Dielectric Constant.

The transverse charge also enters the static dielectric constant describing the response of the system to a static electric field. In fact, the Lyddane-Sachs-Teller relation (Kittel, 1976) can be written $\omega_{LO}^2/\omega_{TO}^2 = \varepsilon_s/\varepsilon$ where $\varepsilon_s = 1 + 4\pi\chi_s$ is the static dielectric constant and ε is the optical dielectric constant of Eq. (4-12). Thus the static dielectric constant could be directly obtained from the results of the preceding section, Section 4-3-B. However, it is helpful to treat the problem directly, giving intermediate results, which we shall need.

An electric field E_x within the material exerts opposite forces on the metallic and nonmetallic atoms, each with magnitude $ee_T{}^*E_x$. This must be balanced by an elastic force proportional to the relative displacement of the atoms and a coefficient related to the transverse-optical-mode frequency, because the same kinds of relative displacements are involved in the two cases.

This field will also tend to shear the lattice, just as a shear in the lattice tends to produce a polarization in the piezoelectric effect, discussed in the following part. In fact, the shearing stress can be written in terms of the field and the piezoelectric constant. We do not include that effect here. We imagine that the surfaces of the crystal are clamped and look only at the internal displacements. Any effects of shear in a real system have this added effect which must be included using the piezoelectric theory.

We note that the force on the vibrating metallic atom was equal to the mass times its acceleration $-\omega_{TO}{}^2 M_+u_+ = \omega_{TO}{}^2 M_r(u_+ - u_-)$. Equating this relation between force and displacement to $-ee_T{}^*E_x$ we obtain $u_+ - u_- = ee_T{}^*E_x/(M\omega_{TO}{}^2)$ and [using Eq. (4-30)] a polarization density of

$$P = \frac{3\sqrt{3}e^2 e_T{}^{*2} E_x}{16\, M_r \omega_{TO}{}^2\, d^3} \, . \tag{4-35}$$

This adds directly to the electronic polarization discussed in Sections 4-2-A. We write the static susceptibility as χ_s and the polarization density in Eq. (4-35) corresponds to a difference between the static and optical [from Eq. (4-10)] susceptibility of

$$\chi_s - \chi^{(1)} = \frac{3\sqrt{3}e^2\, e_T{}^{*2}}{16 M_r \omega_{TO}{}^2 d^3} \, . \tag{4-36}$$

To compare with experiment, we again use the formula Eq. (4-32) for $\omega_{TO}{}^2$ to rewrite Eq. (4-36) as

$$\chi_s - \chi^{(1)} = 0.0051\, \frac{e^2 md}{\hbar^2 \alpha_c{}^3}\, e_T{}^{*2} \, . \tag{4-37}$$

This same result could be obtained using Eq. (4-33) and the relation $\omega_{LO}^2/\omega_{TO}^2 = \varepsilon_s/\varepsilon$.

Equation (4-37) was evaluated for the compounds for which experimental values were available; the results are listed in Table 4-6. The agreement is rather good, both with respect to absolute values and trends.

Table 4-6. The lattice-displacive contribution to the static susceptibility for compound semiconductors.

Semiconductor	d(Å)	e_T^*	$\chi_s - \chi^{(1)}$	
		[Eq. (4-29)]	[Eq. (4-37)]	Experiment*
A1Sb	2.66	1.57	0.084	0.10
GaAs	2.45	1.84	0.116	0.10
GaSb	2.65	1.46	0.070	0.09
InP	2.54	2.34	0.247	0.10
InAs	2.61	2.28	0.231	0.18
InSb	2.81	1.97	0.160	0.15
ZnO	1.98	2.95	0.713	0.33
ZnS	2.34	2.84	0.582	0.49
CdS	2.53	2.95	0.926	0.32

*Aigran and Balkansky (1961)

D. The Piezoelectric Effect

We noted in Section 3-2-B that a shear distortion of a lattice can cause a displacement of one sublattice (the metallic atoms) with respect to the other sublattice (the nonmetallic atoms). In a polar semiconductor, the resulting electrical polarization is called the *piezoelectric effect.* It is an important property of tetrahedral semiconductors and one easily treated by the techniques used here.

Perhaps the simplest geometry from which to understand the displacements of the two sublattices is that shown in Figure 4-6, where a shear strain (see Eqs. (3-2) and (3-3))

$$e_5 = 2\varepsilon_{zx} = \frac{\partial u_z}{\partial x} + \frac{\partial u_x}{\partial z} \ . \tag{4-38}$$

is applied, with the two terms taken equal, $\partial u_z/\partial x = \partial u_x/\partial z = e_5/2$, for the figure. Let the central atom be metallic (positively charged). This strain displaces the four nonmetallic atoms tetrahedrally surrounding the central atom as shown. An identical set of strains applies to the metallic sublattice; however, we have taken our origin of coordinates in the metallic atom at the center. The strongest interatomic forces are the radial ones and it is not difficult to see that the displacements shown for all four of the neighbors tend to pull or push the central atom into the plane of the figure, in the positive y-direction. The individual displacements tend also to drive a displacement in other directions but the x- and z-tendencies both sum to zero. The net effect is a displacement of the nonmetallic sublattice out of the plane of the figure and of the metallic sublattice into the plane of the figure. Calling the difference between the two displacements the *internal displacement* u_y, the Kleinman internal displacement parameter (Kleinman (1962))ζ, which we defined after Eq. (3-13), corresponds to

$$u_y = \zeta e_5 d/\sqrt{3} \quad . \tag{4-39}$$

Similarly we would have, $u_z = \zeta e_6 d/\sqrt{3}$ and $u_x = \zeta e_4 d/\sqrt{3}$ if those strains had been applied. The sign of ζ depends upon the choice of coordinate system; the choice here [and in Figure 4-3] with the nonmetallic atom at *[a/4,a/4,a/4]* from the metallic one seems to be customary. For that choice, ζ is positive if the internal displacements are in the direction favored by the radial forces. Note that if the internal displacement were made to minimize the change in angle between bonds, a displacement of the central atom of Figure 4-6 in the *negative* y-direction would be favored.

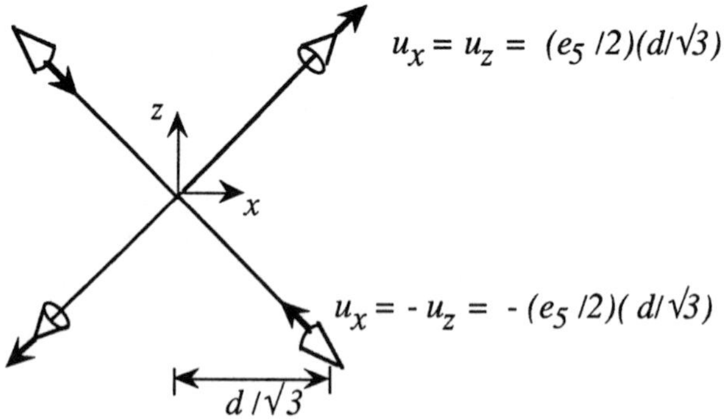

Fig. 4-6. Displacement of nonmetallic (As) neighbors to a metallic (Ga) atom resulting from a shear strain e_5. The corresponding pushes and pulls on the central atom cancel except for the y-direction, in which they all add, causing an internal displacement in the positive y-direction (away from the observer).

We can now estimate the local dipole introduced by these displacements. If the charges Z^* moved rigidly with each atom, the polarization would of course be $u_y Z^* e = \zeta Z^* e d e_5 / \sqrt{3}$ per atom pair. We must add to this the contribution of the charge redistribution, as we did for the transverse charge. We can take components of each displacement along the bond to see that the change in bond length for the bond to the upper right is

$$\delta d = (u_x + u_z - u_y)/\sqrt{3} = (1 - \zeta)\, e_5\, d/3 \ . \tag{4-40}$$

That change is equal for the bond at the lower left; those for the other two are equal and opposite. The change in charge on the atom to the upper right owing to this δd is $-e\delta\alpha_p = -2e\alpha_p\alpha_c^2\delta d/d$. The four contributions to the dipole from the four bonds add in the y-direction contributing $-8ed\alpha_p\alpha_c^2(1 - \zeta)\, e_5 /(3\sqrt{3})$ to the dipole per atom pair, in addition to the $\zeta Z^* ed\ e_5 / \sqrt{3}$ arising from the displacement of the charge Z^*e . We multiply by the density of atom pairs, $3\sqrt{3}/(16d^3)$ To obtain the polarization density per unit volume, P_y ,which is related to the strain by

$$P_y = e_{231}\, e_5\ , \tag{4-41}$$

with the piezoelectric constant then given by

$$e_{231} = e_{123} = \frac{3e}{16d^2}\, [Z^* \zeta - \frac{8}{3}\alpha_p\alpha_c^2(1 - \zeta)] \ . \tag{4-42}$$

In discussing the result, it may be helpful to define the piezoelectric charge e_p^* that would give the same piezoelectric constant if it moved rigidly with the atom. That charge is

$$e_p^* = Z^* - \frac{8(1 - \zeta)}{3\zeta}\, \alpha_p\alpha_c^2 \ . \tag{4-43}$$

The remarkable part of this result is that the charge redistribution *reduces* the effective charge below the atomic charge Z^* , in contrast to the transverse charge where the atomic charge was enhanced. The difference comes from the fact that for e_T^* the metallic atom moved against its neighbors and the reduced polarity of the bonds in front transferred electrons in the direction opposite to the displacement, adding to the dipole. In the piezoelectric effect, the metallic atom is pulled by the atoms in front of it so that the increased polarity transfers electrons in the same direction as the displacement, reducing the dipole.

The comparison with experiment is complicated by uncertainty in the internal displacement parameter ζ. Our analysis in Section 3-2-B gave a value of 0.31 for all homopolar semiconductors and the suggestion that it does not depend strongly upon polarity. Experimental values were higher, as was Sokel's (1978) tight-binding estimate of 0.47. If we use this value of 0.47 for all systems, we obtain the predictions given in Table 4-7, in quite good accord with the experimental values deduced by Martin (1970). The cancellation is clearly demonstrated (although the sign is not reliably predicted with the result near zero). The trends are also rather well described. On the other hand, the value of ζ that we used is quite different from the values estimated by Martin (1970) and use of his values worsens the agreement with experiment.

An important point, apparent in Table 4-7, is that this experimentally observable e_P^* is not necessarily even of the same sign as e_T^*. The intuitive idea of well-defined atomic charges is a very misleading one. It is essential to work with the observable property directly.

Table 4-7. Piezoelectric charges predicted from Eq. (4-43), compared with experimental values calculated using the internal displacement parameter $\zeta = 0.47$ from Sokel (1978), see Section 3-2-B.

Semiconductor	Theory	Experiment*
AlP	-0.15	-
GaAs	-0.23	-0.47
ZnSe	-0.11	+0.13
CuBr	-0.37	-
InSb	-0.14	-0.24
CdTe	+0.04	+0.09
AgI	-0.31	-

*From Martin (1970)

E. Elastooptic Effect

The variation of the optical dielectric response (discussed in Sections 4-2-A and B) resulting from elastic strains is called the *elastooptic effect,* and is customarily written in terms of the dielectric tensor that is related to the susceptibility as $1 + 4\pi \chi_{ij}^{(1)}$. It is preferable to write the elastooptic effect in terms of the $\chi_{ij}^{(1)}$ with the customary elastooptic constants q_{ijkl} defined by

$$\delta\chi_{ij}{}^{(1)} = \frac{1}{4\pi} \Sigma_{kl}\ q_{ijkl}\ \varepsilon_{kl} \quad . \tag{4-44}$$

in terms of the strains ε_{kl} defined in Eqs. (3-1) and (3-2). We can then use the these elastooptic constants to write the change in polarization density resulting from an elastic strain,

$$\delta P_i = \frac{1}{4\pi}\ \Sigma_{jkl}\ q_{ijkl}\ \varepsilon_{kl} E_j \quad . \tag{4-45}$$

These were treated using essentially the approach we use here by Ren and Harrison (1981). We have seen (Section 4-2-C) that the components of such a fourth-rank tensor vanish unless the indices appear in pairs. Thus, there may be elements of the type q_{1111}, q_{1122} , and q_{1212} . [The latter two are not equivalent because the last two indices refer to the strain and are not interchangeable with the first two in the counterpart of Eq. (4-18)], so there are three independent components. Ren and Harrison (1981) did not consider the element q_{1212} .

We have one combination immediately: A uniform distortion corresponds to $\varepsilon_{11}\ \varepsilon_{22} = \varepsilon_{33} = \delta d/d$ and other strains zero. Under such a dilation, the polarization remains parallel to the field and

$$\delta P = \frac{1}{4\pi}(q_{1111} + 2q_{1122})\varepsilon_{11}E \ . \tag{4-46}$$

We can also obtain the change in polarization directly from Eq. (4-10), the susceptibility (calculated without metallization), by differentiating with respect to d. Then $\delta P = d\ (\partial\chi^{(1)}/\partial d\ )\varepsilon_{11}E$. Combining with Eq. (4-46) and the derivative, $(d/\chi^{(1)})\partial\chi^{(1)}/\partial d = 6\alpha_c^2 - 5$ obtained from Eq. (4-10), we have

$$\frac{q_{1111} + q_{1122}}{3} = \frac{4\pi d}{3} \frac{\partial \chi^{(1)}}{\partial d} = \frac{4\pi}{3} \chi^{(1)} (6\alpha_c^2 - 5) . \qquad (4\text{-}47)$$

A second combination can be obtained by applying the twist distortion, $\varepsilon_{22} = - \varepsilon_{11}$, discussed in Section 3-2-A. This distortion does not change any bond lengths to first order in the strain, but simply rotates the internuclear vectors. We found there a change in V_2, but only to second order in the strains. For fields and polarizations in the x-direction, the only change in the susceptibility comes from the scaling of the two factors of d_x in the third form in Eq. (4-9), here by $1 + \varepsilon_{11}$ each. Thus, $\delta\chi^{(1)} = (q_{1111} - q_{1122}) \varepsilon_{11}/(4\pi)$, we obtain

$$\frac{q_{1111} - q_{1122}}{2} = 4\pi \chi^{(1)} . \qquad (4\text{-}48)$$

This is the elastooptic constant for the twist distortion, analogous to the shear constant $(c_{11} - c_{12})/2$ for rigidity against the twist distortion.

Finally there is the graphitic shear distortion illustrated by $\varepsilon_{31} = \varepsilon_{13} = e_5/2$ in Figure 4-6. The corresponding elastooptic elements are $q_{3131} = q_{1331} = q_{3113} = q_{1313}$. We can evaluate them by imagining a field applied in the x-direction and seeking the polarization in the z-direction. In direct analogy with Eq. (4-10), we obtain the z-component of the dipole in each bond that is proportional to E_x as

$$\delta p_z = \frac{e^2 V_2^2 d_z d_x E_x}{2(V_2^2 + V_3^2)^{3/2}} . \qquad (4\text{-}49)$$

For the bonds to the upper right and lower left in Figure 4-6 we have $d_z d_x = d^2[1 + \varepsilon_{31}]^2/3$, or $d^2/3 + 2d^2\varepsilon_{31}/3$ to first order in the strain. We found in Eq. (4-40) that the change in bond length for these two bonds is $(1 - \zeta) 2\varepsilon_{31}d/3$ to lowest order in the strain. The change in $V_2^2/(V_2^2 + V_3^2)^{3/2}$ is $(-4 + 6\alpha_c^2)[V_2^2/(V_2^2 + V_3^2)^{3/2}]\delta d/d = 2(1 - 3\alpha_p^2)[V_2^2/(V_2^2 + V_3^2)^{3/2}] (1 - \zeta) 2\varepsilon_{31}/3$. Accordingly, to first-order in the strain, Eq. (4-49) becomes

$$\delta p_z = \frac{e^2 V_2^2 d^2 E_x}{6(V_2^2 + V_3^2)^{3/2}}[1 + 2\varepsilon_{31} + {}^4/_3(1 - 3\alpha_p^2)(1 - \zeta)\varepsilon_{31}]. \qquad (4\text{-}50)$$

For the upper left and lower right bonds in Figure 4-6, $d_z d_x$ is negative. The strain decreases its magnitude; thus, although the zero-order terms form

Eq. (4-50) cancel, the first order terms add. This is also true of the final term in Eq. (4-50). We can add the contributions from the four bonds and sum over atom pairs to obtain a polarization density

$$P_z = \frac{\sqrt{3}e^2V_2^2}{4(V_2^2 + V_3^2)^{3/2}d}[1 + {}^2/_3(1 - 3\alpha_p^2)(1 - \zeta)]\, \varepsilon_{31} E_x . \quad (4\text{-}51)$$

From Eq. (4-45), we note that polarization density is also equal to $(q_{3131} + q_{3113})\, \varepsilon_{31} E_x/(4\pi)$. We solve for $q_{3131} + q_{3113}$ but write the result in terms of the more usual and equal $q_{1212} + q_{1221}$,

$$q_{1212} + q_{1221} = 8\pi\chi^{(1)}[1 + {}^2/_3(1 - 3\alpha_p^2)(1 - \zeta)] \; . \quad (4\text{-}52)$$

We note that all components are proportional to $4\pi\chi^{(1)} = \varepsilon - 1$, with ε the dielectric constant for these cubic systems. It is therefore appropriate to tabulate the ratio of the elastooptic coefficients to $4\pi\chi^{(1)}$. Values obtained from Eqs. (4-47), (4-48), and (4-52) for germanium and the semiconductors isoelectronic with it are listed in Table 4-8. As in the piezoelectric calculation the theoretical value of $\zeta = 0.47$ from Section 3-2-BC was used in Eq. (4-52). Because the experimental values assembled by Ren and Harrison (1981) did not show consistency, we selected here a single set obtained from a single experimental source for comparison.

The theory predicts that the ratios given in Table 4-8 would be the same for all homopolar semiconductors to those given for germanium. These

Table 4-8. Elastooptic coefficients, divided by $\varepsilon - 1$, for germanium and compounds isoelectronic with it.

		Ge	GaAs	ZnSe	CuBr
$\frac{q_{1111}+2q_{1122}}{12\pi\chi^{(1)}}$	Eq. (4-47)	0.33	-0.11	-0.73	-1.08
	Experiment*	2.41	1.20	-	-1.13
$\frac{q_{1111}-q_{1122}}{8\pi\chi^{(1)}}$	Eq. (4-48)	1	1	1	1
	Experiment*	0.22	0.40	-	0.42
$\frac{q_{1212}+q_{1221}}{4\pi\chi^{(1)}}$	Eq. (4-52)	2.70	2.24	1.58	1.22
	Experiment*	2.52	1.92	-	0.76

*Biegelsen, Zesch, and Schwab (1976). Note that their p_{44} relates $\Delta(1/\varepsilon)_{23}$ to $e_4 = 2\varepsilon_{23}$.

were in fact in better accord with experiment for carbon and silicon than for germanium: Experimental values for $(q_{1111} + 2q_{1122})/(12\pi\chi^{(1)})$ were 0.48 and 0.31 for carbon and silicon, respectively, to be compared with the predicted 0.33. For $(q_{1111} - q_{1122})/(8\pi\chi^{(1)})$ they were 1.5 and 0.75, to be compared with the predicted 1. Germanium was not so well given, but the trends with polarity in the isoelectronic series did reasonably well. Ren and Harrison (1981) found that agreement was improved in a full tight-binding analysis, analogous to including metallization.

F. Electrooptic Effect

We have derived (Section 4-2-C) a formula for the second-order susceptibility in which a polarization in the y-direction arose that was proportional to E_x and E_z. These fields were assumed to be of optical frequencies; hence, there was negligible motion of the atoms and the polarization was purely electronic. On the other hand, if the field E_x is static, there will be internal displacements of the atoms, as we found in our discussion of the static dielectric susceptibility. These displacements will modify the optical polarization in the y-direction due to an optical electric field in the x-direction. Thus, through the internal displacements, a static field has caused a shift in the optical susceptibility. This is a contribution to the *electrooptic effect,* in addition to the purely electronic contribution obtained earlier.

We have all of the ingredients of the analysis for this electrooptic effect from our previous studies. It will be more convenient in using these results to let the static field be in the y-direction. We found [just above Eq. (4-35)] that such a field will cause internal displacements parallel to the static field E_y^{stat} given by $u_+ - u_- = ee_T^*E_y^{stat.}/(M_r\omega_{TO}^2)$. This arose in the analysis of the static dielectric constant, in which the surfaces of the crystal were clamped so that there was no net shear of the lattice, and we again assume that here.

We found [in Eq. (4-51)] in studying the elastooptic effect that an optical-frequency polarization density in the z-direction arose from an optical-frequency field in the x-direction as a result of the strains. The final term in that expression, proportional to ζ, arose from internal displacements. That final term was

$$P_z = -\frac{\sqrt{3}e^2V_2^2}{6(V_2^2 + V_3^2)^{3/2}d}(1 - 3\alpha_p^2)\,\zeta\,\varepsilon_{31}E\,. \tag{4-53}$$

The corresponding internal displacements were $u_y = u_+ - u_- = 2\zeta\, \varepsilon_{31}\, d/\sqrt{3}$. We can thus rewrite Eq. (4-53) in terms of the u_y and hence for this case in terms of the static field that causes the displacements here:

$$P_z = -\ \frac{e^3 e_T^*(1 - \alpha_p^2)^{3/2}(1 - 3\alpha_p^2)}{2d^2(2M_r\omega_{TO}^2)V_2}\ E_x E_y^{stat}\ . \tag{4-54}$$

The coefficient of $E_xE_y^{stat}$ on the right might be thought of as a static contribution to $\chi_{312}^{(2)} = \chi_{123}^{(2)} = \chi_{231}^{(2)}$, with the third index referring to the static field. When we evaluated the electronic polarization in Section 3-2-C, there were two equal contributions: $\chi_{312}^{(2)} + \chi_{321}^{(2)}$. We should add both to the static contribution, Eq. (4-54)[e.g. we add $2\chi_{312}^{(2)}$ from Section 3-2-C to the coefficient in Eq. (4-54)]. It is convenient for comparison to use the theoretical value [Eq. (4-32)] of $M_r\omega_{TO}^2$ for homopolar semiconductors in Eq. (4-54), so that the two contributions can be directly compared. We obtain (rotating indices)

$$\chi_{123}^{(2)\ stat} = \frac{3e^3}{16V_2^2}\ [\alpha_p(1 - \alpha_p^2)^2 - 0.081(1 - \alpha_p^2)^{3/2}(1 - 3\alpha_p^2)e_T^*]\ . \tag{4-55}$$

It is interesting first that the two contributions are of opposite sign. This followed from systematic use of the geometry shown in Figures 4-3 through 4-6, but can be verified by thinking it through physically. In Figure 4-4, for example, with a positive atom in the center, a dc field E_y will move the central atom away from the viewer, increasing the covalent energy for d_1 and reducing its polarizability. Thus, an optical field in the x-direction will give reduced dipole in the z-direction and a negative contribution to $\chi_{312}^{(2)}$. For the electronic effect, the same dc field raises the electronic energy on the nonmetallic atom at d_1, reduces the polarity of the bond, and enhances its polarizability; it gives a positive contribution to $\chi_{312}^{(2)}$. In contrast, the two terms added in the static dielectric susceptibility, $\chi^{(1)}$.

Secondly we note that the displacive contribution [the final term in Eq. (4-55)] tends to be much smaller than the electronic part. This is not because of a fundamental difference in parameters that enter but because the numerical coefficient is smaller. Perhaps one could say that the smallness came from the large coefficient, $32(1 + \lambda)/3 = 19.8$, in the expression [Eq. (4-32)] giving the ω_{TO}^2. This argument can also be made for the static susceptibility.

The experimental data on the static electrooptic effect are somewhat sparse and show considerable variation. Therefore, we have not attempted a tabulated comparison with experiment.

G. Dielectric Response in the Infrared

We have discussed the optical susceptibility for which it was assumed that the ions were not displaced by the field and the static dielectric response in which they were assumed to be in elastic equilibrium at all times. The intermediate region comes of course where the optical frequency is comparable to the vibrational frequency of the lattice, which we have seen to be of order 10^{14} cycles per second, in the infrared.

Imagine an infrared wave, polarized in the x-direction and propagating in the z-direction. The field is given by $E_x \cos(qz - \omega t)$. We evaluate the force on a metallic atom at $z = 0$. It is $ee_T^* E_x \cos\omega t$, where ω is 2π times the frequency of the light. We saw in deriving Eq. (4-35), for the polarization owing to a static field, that there is also an elastic restoring force on the metallic atom of $-\omega_{TO}^2 M_r(u_+ - u_-)\cos\omega t$, where we have written the displacements as $u_+\cos\omega t$ and $u_-\cos\omega t$. We can set these equal to the mass times the acceleration of the "+" atom,

$$ee_T^* E_x \cos\omega t - {}^1/_2\, \omega_{TO}^2 M_r(u_+ - u_-)\cos\omega t = M_+ u_+ \partial^2 \cos\omega t / \partial t^2 =$$
$$= -\, {}^1/_2 M_r \omega^2 (u_+ - u_-)\cos\omega t\,, \qquad (4\text{-}56)$$

and the corresponding expression for the "-" atom. We can solve for $u_+ - u_-$, multiply by ee_T^* and the density of ion pairs, $3\sqrt{3}/16d^3$, to obtain the displacive part of the frequency-dependent susceptibility,

$$\delta\chi(\omega) = \chi(\omega) - \chi^{(1)} = \frac{3\sqrt{3}e^2 e_T^{*2}}{16M_r(\omega_{TO}^2 - \omega^2)d^3}\,. \qquad (4\text{-}57)$$

This reduces to Eq. (4-36) for $\omega = 0$, the static case. It increases as the frequency is raised toward ω_{TO}, as illustrated in Figure 4-7. Above that frequency, it becomes negative and finally approaches zero at very high frequency; the susceptibility approaches the optical susceptibility (Harrison, 1980), calculated in Sections 2-2-A and B.

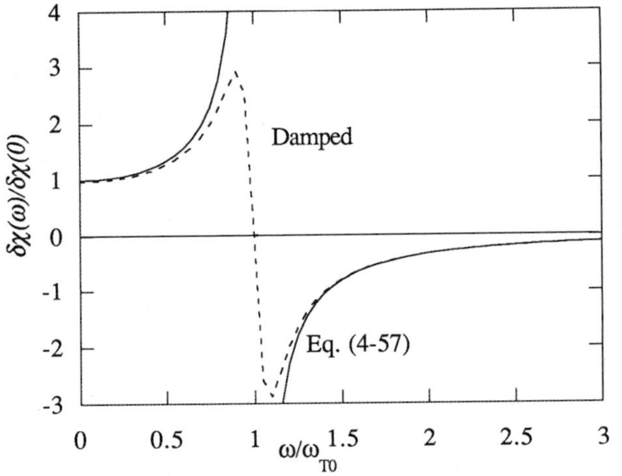

Fig. 4-7. The displacive part of the dielectric susceptibility as a function of the frequency. The "damped" curve corresponds to a relaxation time τ such that $\omega_{TO}\tau = 0.17$.

It might at first seen peculiar that the displacements are in the opposite direction to the applied force for ω above ω_{TO} , but that is in accord with everyday experience. If we slowly push a pendulum, the displacements are at any time parallel to the force; however, if we shake it at a high frequency, the required force is maximum when the displacement is oppositely directed. In both of these limits the displacive amplitudes may be small, the amplitudes do not grow with time, and no energy is fed into the system. Near the resonant frequency, ω_{TO} , the amplitudes become very large and any loss mechanism (e.g. damping or scattering of the vibrations) becomes important. Such a loss mechanism reduces the amplitude as illustrated by the dashed line in Figure 4-7 and also gives an absorptive contribution (frequently treated as an imaginary term in the dielectric susceptibility), which has a peak at the resonant frequency.

Many different vibrational modes exist in the crystal (three times as many modes as atoms in the crystal) and they have a variety of frequencies. However, the infrared wave is directly coupled only to the transverse optical mode of the same wavenumber. Thus only a single mode enters and its wavenumber is so small that the frequency is essentially that of the mode of infinite wavelength. Thus, our treatment of a single mode above was appropriate.

When the frequency of the light becomes high enough, electrons can be excited into empty states in a way similar to that described for vibrational modes, giving a resonant absorption of the same form. For our description of the electronic states with empty antibonding states higher in energy than the filled bonding states by $2\sqrt{V_2^2 + V_3^2}$, we would deduce a susceptibility curve with a singularity of just the form given in Figure 4-7, but with the resonance occurring at $\hbar\omega = 2\sqrt{V_2^2 + V_3^2}$. Our treatment of the optical susceptibility assumed that the frequencies were well below this resonance where the susceptibility approaches a constant value, as did $\chi(\omega) - \chi^{(1)}$.

We can now consider the behavior of the optical susceptibility at higher frequencies. In this case, finally, our treatment of the electronic states as simple bonding and antibonding levels is inadequate. We noted in Section 2-2-D that each of these levels is in fact broadened out into bands, just as there are broad bands of vibrational frequencies for the lattice. In the absorption of light by electrons, transitions between many different sets of electronic states occur and the electronic states cannot be treated as individual levels. We need now to describe in some detail the actual energy-band structure of the electronic states. This will also be necessary for the discussion of conduction and other transport properties.

Problem 4-1. Dielectric susceptibility of hexagonal BN

Obtain a formula for the electric polarization per unit area ($^1/_A\Sigma_i p_i$) from the σ-bonds in boron nitride due to a field in the plane of the crystal. In terms of the figure for Problems 2-1 and 2-2, this is probably easiest for a field in the x-direction. This is essentially redoing the derivation of Eqs. (4-9) and (4-10) with different geometry.

This could be redone for fields in the z-direction to confirm that the same result is obtained.

Problem 4-2. Second-order susceptibility

Carry the derivation in Problem 4-1 to one higher order in E_x to obtain the polarization per unit area in the z-direction proportional to E_x^2. This corresponds to a second-order susceptibility. You need to be careful with the relative contributions for the two bonds, but it is less important to keep the absolute sign. For an input field containing a term $E_x cos\omega t$ this will produce a polarization proportional to $E_x^2 cos^2\omega t = E_x^2(1 + cos(2\omega t))/2$ and can reradiate at twice the frequency of the driving field.

[The correct absolute sign can be obtained by taking the shaded atoms as nitrogen and focusing on the bond to its left. An electric field to the right pushes electrons to the left, decreasing the polarity of that bond and increasing its polarizability so the electric polarization due to a second factor of field is to the right and up. That from the other bond is to the left and up so the net polarization is in the positive z-direction - again if the shaded atoms are nitrogen.]

CHAPTER 5

Semiconductor Energy Bands

Our analysis in Chapters 2, 3 and 4 has been based upon bond orbitals and antibonding orbitals. We made a Bond-Orbital Approximation in neglecting all interatomic couplings except those between hybrids directed into the same bond. This Bond-Orbital Approximation is appropriate when we consider properties such as the total energy or the electric polarization density which depend upon sums over completely filled bands since, as we noted in Section 2-3-A, the sum over the entire band (the sum of the diagonal elements of the Hamiltonian matrix) is not affected by the coupling between bond levels which broadens them into bands. We now consider properties of individual electronic levels and this Bond-Orbital Approximation does not apply.

We shall introduce couplings between neighboring bonds and between neighboring antibonds to see the general form of the bands which arise, as we did for the lithium s-bands in Section 1-3. This will gain us insight into the bands which is more difficult to obtain from more accurate solutions. We shall then return to the expansion of the electronic states as a full linear combinations of the s- and p-states on each atom, to obtain a more accurate description. In all cases we shall write the states in terms of wavenumber, as when we wrote the $u_j = e^{ikdj}/\sqrt{N}$ to obtain a solution of Eq. (1-19) for the lithium chain. We should first generalize this to the three-dimensional structure of the diamond or zincblende lattice.

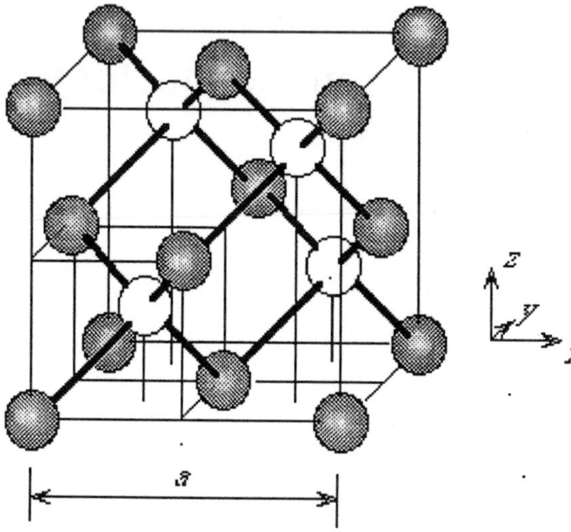

Fig. 5-1. The tetrahedral crystal structure. The shaded and white atoms are distinguishable, as at right or left ends of the [111]-oriented bonds, even if both are silicon. The shaded atoms form a face-centered-cubic lattice, the white atoms another.

1. Wavenumbers and Brillouin Zones

We take a moment to organize the postulated states in a way applicable also to more accurate representations of the bands. Most of our analysis will not depend upon this detail, which is given in any solid-state text. We redraw the crystal structure from Fig. 2-2 in Fig. 5-1 and distinguish the atom (shaded) lying to the lower left of a [111] bond from that (white) at the upper right. (These might both be silicon, or one might be gallium and the other arsenic.) The atoms of the first type form a lattice called *face-centered cubic*, because of their arrangement on a cube as shown with identical atoms also at the center of each face.

The *primitive translations* $\tau_1 = [011]a/2$, $\tau_2 = [101]a/2$, and $\tau_3 = [110]a/2$, or any combination of an integral number of each, take every atom to a position previously occupied by an identical atom. A *primitive cell*, which may be taken as a parallelepiped with edges given by τ_1, τ_2 and τ_3, contains one atom of the first type (an eighth of an atom at each of the eight corners) and one of the second (at a position *[111]a/4* from the first. We imagine the crystal made up of N_1 primitive cells along the τ_1 direction N_2 along τ_2 and N_3 along τ_3 and we may take periodic boundary conditions on each of the parallel surfaces, just as we took periodic boundary conditions on the lithium chain. We might then ask for a *Bloch sum* of s-states on the atoms of type one, the three-dimensional counterpart of our lithium-chain state for which we wrote $u_j = e^{ikdj}/\sqrt{N}$ with N values of j in the sum. We do this now in terms of *primitive lattice wavenumbers*,

$$\mathbf{k}_1 = 2\pi\frac{\tau_2\times\tau_3}{\tau_1\cdot\tau_2\times\tau_3},$$

$$\mathbf{k}_2 = 2\pi\frac{\tau_3\times\tau_1}{\tau_2\cdot\tau_3\times\tau_1}, \qquad (5\text{-}1)$$

$$\mathbf{k}_3 = 2\pi\frac{\tau_1\times\tau_2}{\tau_3\cdot\tau_1\times\tau_2},$$

chosen such that $\mathbf{k}_1$ is perpendicular to τ_2 and τ_3, etc. Then the coefficients for our Bloch sum can be written $u_j = e^{i\mathbf{k}\cdot\mathbf{r}_j}/\sqrt{N_c}$ where $N_c = N_1N_2N_3$ is the number primitive cells, or of atom pairs, in the crystal. We include the subscript "c "so that we can continue to use N for the number of atoms; in this case $N_c = N/2$. The wavenumber $\mathbf{k}$ is given by

$$\mathbf{k} = \frac{n_1\mathbf{k}_1}{N_1} + \frac{n_2\mathbf{k}_2}{N_2} + \frac{n_3\mathbf{k}_3}{N_3}, \qquad (5\text{-}2)$$

where n_1, n_2, and n_3 are integers. This is the appropriate generalization of the one-dimensional case and a normalized sum; note that if we change $\mathbf{r}_j$ by τ_1 then u_j changes by $e^{2\pi i\, n_1/N_1}$ just as a change in the index j by one gave a change in u_j by a factor $e^{2\pi i n/N}$ in the lithium chain. Periodic boundary conditions are satisfied since adding $N_1\tau_1$ to $\mathbf{r}_j$ gives a factor of $e^{2\pi i n_1} = 1$.

The wavenumbers of Eq. (5-2) are very closely spaced for a large system, for which N_1, N_2, and N_3 are large numbers, and we can almost think of them as a continuous set of wavenumbers. In some cases, particularly in the evaluation of total scattering rates from the Golden Rule of quantum mechanics, Eq. (1-32) we will wish to sum over states and it will be convenient to replace the sum by an integral. It is not difficult to see how to do this. With the N_i large the allowed $\mathbf{k}$ form a fine grid with spacing k_i/N_i along the $\mathbf{k}_i$ axis, so that they are uniformly distributed in wavenumber space. It is very easy to write down the density of states in this wavenumber space if the τ_i, and therefore the $\boldsymbol{k}_i$, are perpendicular to each other. Then $k_i = 2\pi/\tau_i$ and the cell associated with each state has volume $(2\pi)^3/(N_1\tau_1 N_2\tau_2 N_3\tau_3) = (2\pi)^3/\Omega$ with Ω the volume of the system, and the density of states in wavenumber space is the reciprocal of this. The same final result, in terms of Ω, follows for general τ_i. We may therefore replace any sum over wavenumbers, of some function of wavenumber, by an integral over that function,

$$\sum_{\mathbf{k}} \rightarrow \frac{\Omega}{(2\pi)^3}\iiint d^3k . \tag{5-3}$$

If we are summing over electrons occupying these states there will be an additional factor of two for the two spin states. The result is independent of the geometry of the lattice or the number of atoms in the primitive cell and depends only on having boundary conditions on a large volume Ω.

If we restrict the integers, $-N_1/2 \leq n_1 < N_1/2$, etc., there will be just as many wavenumbers $\boldsymbol{k}$ as there are atom pairs in the crystal. Use of wavenumbers outside this region would simply reproduce the same solutions as a wavenumber inside. It is actually more convenient to use another limitation on the wavenumbers included, one which is a different generalization of the one-dimensional case. In that case the primitive lattice wavenumber, Eq. (5-1), was $2\pi/d$ or $-2\pi/d$ and we could

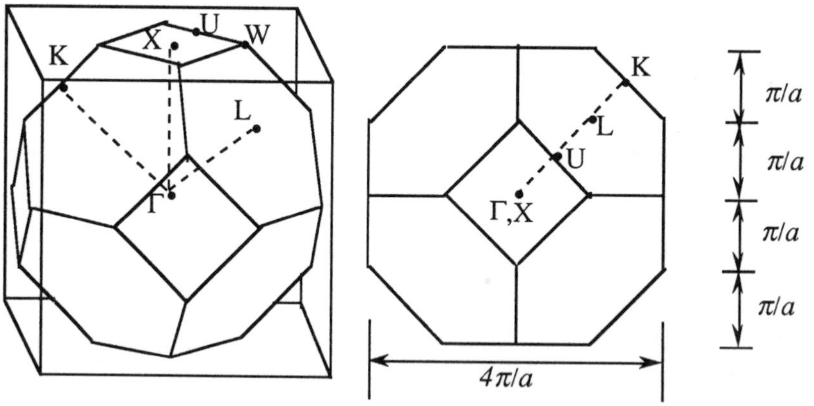

Fig. 5-2. The Brillouin Zone for the face-centered cubic crystal lattice, and therefore also for the diamond or zincblende lattices which have the same translational periodicity. To the right it is observed along a [100] direction. Symmetry points of the Brillouin Zone are indicated.

define a wavenumber lattice of values $k_m = 2\pi m/d$ with m any integer such that any two Bloch sums, $u_j = e^{ikd_j}/\sqrt{N}$, differing in wavenumber k by a lattice wavenumber yield identical coefficients u_j. For that construction our Brillouin Zone, $-\pi/(2d) \leq k < \pi/(2d)$ was a restriction to the wavenumbers, allowed by periodic boundary conditions, which were closer to $k = 0$ than to any other lattice wavenumber. The Brillouin Zone generally chosen for a three-dimensional solid is all of those wavenumbers closer to $\mathbf{k} = 0$ than to any other lattice wavenumber, $\mathbf{k_m} = m_1\mathbf{k_1} + m_2\mathbf{k_2} + m_3\mathbf{k_3}$. That wavenumber lattice can be seen, using Eq. (5-1), to be a *body-centered cubic* lattice in wavenumber space, a simple-cubic lattice with edge $4\pi/a$ with lattice wavenumbers also at all body-center positions, such as $[111]2\pi/a$. The corresponding Brillouin Zone is shown in Fig. 5-2.

This procedure and these definitions also reduce to the wavenumbers for the simple cubic lattice discussed in Problem 1-5 for the case where τ_1, τ_2 and τ_3 are perpendicular to each other and of length d. Then the primitive lattice wavenumbers are along the three cubic axes with length $2\pi/d$ and the Brillouin Zone is a simple cube of edge $2\pi/d$.

2. Bonding and Antibonding Bands

In Section 2-2 we formed a bond orbital $|b_j>$ and an antibonding orbital $|a_j>$ in each bond site, where we have now numbered the bond sites by the index j. We noted that there was a coupling $-V_1$ between two hybrids on the same atom which were directed into different bond sites and

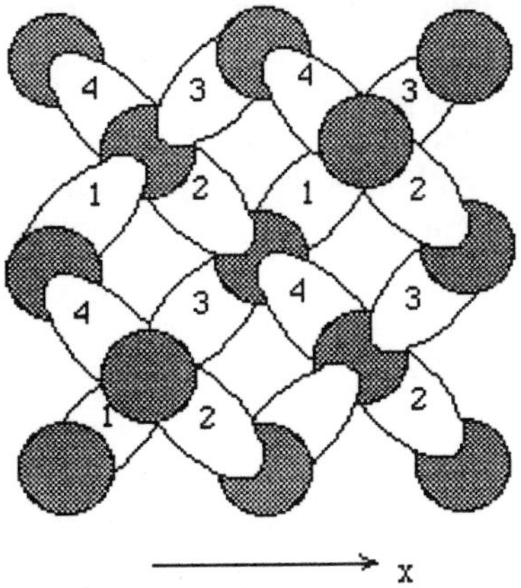

Fig. 5-3. A lattice of bonds in a tetrahedral lattice. Note that there are four types of bonds distinguished by their orientation. We have numbered the four types.

correspondingly there is a coupling between neighboring bonds, each of which is given by $(|h> + |h'>)/\sqrt{2}$ in a homopolar crystal, of $-V_1/2$. In just this way we found in Section 2-3-B that there was a coupling $\pm V_1/2$ between a bond and each neighboring antibond in the homopolar crystal. Assuming that those bond-antibond couplings were small with respect to the energy difference between them, we could treat them in perturbation theory as the metallization shift of the energy of each bond. We now neglect those couplings but look at the effects of the coupling between adjacent bonds. These cannot be treated in perturbation theory since the coupled states have the *same* energy. Their effect is to broaden the identical bond states into a band of states, just as in the lithium chain discussed in Section 1-3. Again we have states of the same energy, ε_b rather than ε_σ , with each coupled to its nearest-neighbor states, here by $-V_1/2$ rather than $V_{ss\sigma}$.

There are two immediate distinctions. First, this is a three-dimensional lattice, rather than a one-dimensional one and each bond is coupled to six nearest neighbors, as illustrated in Fig. 5-3. Second, we may distinguish four different kinds of bonds, oriented along four different [111] directions. We can still take a propagating-wave kind of solution for the coefficients for the bonds which are translationally equivalent (oriented for example along the same [111] direction), $u_j \propto e^{i\mathbf{k}\cdot\mathbf{r}_j}$, but the coefficients can be different for the different orientations, just as if the lithium chain were replaced by two distinguishable atoms, such as alternate lithium and sodium atoms.

A. Bonding Bands from Coupling V_1 .

We return to the construction of bonding bands. It follows from the translational symmetry that the states can be chosen such that each has a wavenumber satisfying periodic boundary conditions as described above. Thus each bonding-band state will be written as a linear combination of Bloch sums for the four bond-orbital types indicated in Fig. 5-3. Each Bloch sum is of the form we also introduced as a solution of Eq. (1-19) and may be written

$$|b_{\mathbf{k}}^{(i)}> = \frac{1}{\sqrt{N_c}}\Sigma_{j(i)}e^{i\mathbf{k}\cdot\mathbf{r}_j}\,|b_j> , \qquad (5\text{-}4)$$

where $i = 1, 2, 3, 4$ indicates the bond type from Fig. 5-3 and j is summed over bond orbitals $|b_j>$ centered at the site $\mathbf{r}_j$ for this type of bond. There is one bond orbital of each type in each primitive cell so N_c is also the number of bonds of each type. We choose the bond sites $\mathbf{r}_j$ to lie at the center of each bond, midway between the two atoms forming the bond. These bond sites are displaced from the *atom* positions just discussed in

Section 5-1, but that does not cause any problem. Then the state of wavenumber $\boldsymbol{k}$ can be written

$$|\psi_k\rangle = u_1|b_k^{(1)}\rangle + u_2|b_k^{(2)}\rangle + u_3|b_k^{(3)}\rangle + u_4|b_k^{(4)}\rangle. \qquad (5\text{-}5)$$

This has reduced the problem to a problem of four coupled states. We may minimize with respect to the coefficients u_i to obtain, as in going from Eq. (1-18) to Eq. (1-19),

$$\varepsilon_b\, u_i + \Sigma_{j=1\text{-}4}\, V_{ij}\, u_j = \varepsilon_k\, u_i \qquad (5\text{-}6)$$

for $i = 1, 2, 3,$ and 4. Here $\varepsilon_b = \varepsilon_h - V_2$. The V_{ij} is the matrix element between the two Bloch sums, and will be different for different wavenumbers, as will the u_i. However, we may select a $\boldsymbol{k}$ in the Brillouin Zone, evaluate the V_{ij} and solve for the four energy eigenvalues for that wavenumber. Here we do this just for wavenumbers in a [100] direction; that is, along the line Γ to X in the Brillouin Zone shown in Fig. 5-2.

To evaluate the matrix element between, for example, type 1 and type 2 Bloch sums, we need to write out the Bloch sums.

$$V_{12} = \langle b_k^{(1)}|H|b_k^{(2)}\rangle = \frac{1}{N_c}\Sigma_{j(1)}\Sigma_{i(2)}e^{-i\boldsymbol{k}\cdot\boldsymbol{r}_i}\, e^{i\boldsymbol{k}\cdot\boldsymbol{r}_j}\, \langle b_i|H|b_j\rangle. \qquad (5\text{-}7)$$

Each term j corresponds to a particular bond orbital of type 1 and there will be contributions from each i corresponding to a bond orbital of type 2 which neighbors it. We may see from Fig. 5-3 that there are two orbitals of type 2, displaced relative to each orbital of type 1 along the x-axis by $\pm a/4$; only the x-component is of concern since $\boldsymbol{k}$ is taken to lie along the x-axis. The phase factors for these two neighbors become $e^{-i\boldsymbol{k}\cdot\boldsymbol{r}_i}\, e^{i\boldsymbol{k}\cdot\boldsymbol{r}_j} = e^{\pm ika/4}$. The coupling between two bond orbitals is $-V_1/2$ in both cases. There are N_c such terms for the N_c different bond orbitals of type one, so their sum cancels the factor $1/N_c$. We obtain simply $V_{12} = -V_1 cos(ka/4)$. We similarly obtain the other matrix elements to obtain the four equations, Eq. (5-6), as

$$\begin{aligned}
\varepsilon_b u_1 - V_1 cos(ka/4)u_2 - V_1 cos(ka/4)u_3 - V_1 u_4 &= \varepsilon_k u_1\,, \\
-V_1 cos(ka/4)u_1 + \varepsilon_b u_2 - V_1 u_3 - V_1 cos(ka/4)u_4 &= \varepsilon_k u_2\,, \\
-V_1 cos(ka/4)u_1 - V_1 u_2 + \varepsilon_b u_3 - V_1 cos(ka/4)u_4 &= \varepsilon_k u_3\,, \\
-V_1 u_1 - V_1 cos(ka/4)u_2 - V_1 cos(ka/4)u_3 + \varepsilon_b u_4 &= \varepsilon_k u_4\,.
\end{aligned} \qquad (5\text{-}8)$$

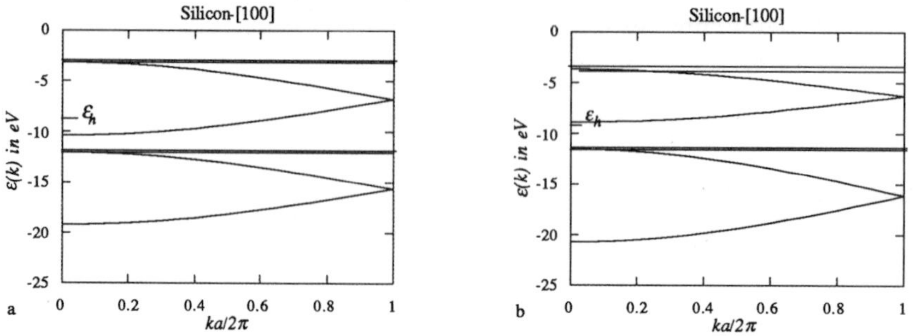

Fig. 5-4. In Part a are the energy bands for silicon obtained by including the intraatomic coupling $-V_1/2$ between each bond and its neighboring bonds, and the same coupling between each antibond and its neighboring antibonds. In Part b we have included also the interatomic coupling $\pm V_1{}^x$ between neighboring bonds and neighboring antibonds. In both cases $ka/2\pi = 0$ corresponds to the center of the Brillouin Zone, Γ, of Fig. 5-2 and $ka/2\pi = 1$ corresponds to the point X at the center of the square face of the Brillouin Zone.

For this symmetric choice of wavenumber it is not difficult to guess solutions to these equations. For example, the states will be even or odd under a 180° rotation around the x-axis. One odd state is given by $u_1 = -u_4$ and $u_2 = u_3 = 0$. Substituting these in, we see that they satisfy all equations with $\varepsilon_k = \varepsilon_b + V_1$, independent of k. The other odd state is given by $u_3 = -u_2$ and $u_1 = u_4 = 0$ and has the same energy. Other orthogonal combinations of these two would serve equally well. The even states are obtained by substituting $u_4 = u_1$ and $u_3 = u_2$, to obtain two sets of two identical equations, which can be solved together to obtain $\varepsilon_k = \varepsilon_b - V_1 \pm 2V_1 \cos(ka/4)$. These are shown for silicon in Part a of Fig. 5-4. Also shown are the antibonding bands found in the same way. For that case, between each pair of neighbors the antibond is given by $(|h\rangle - |h'\rangle)/\sqrt{2}$ and we may take the hybrid on the white atoms of Fig. 5-3 as the one with the minus sign in every antibond. Then the coupling between all neighboring antibonds is again $-V_1/2$ in all cases and we obtain exactly Eqs. (5-8) with ε_b replaced by ε_a. The four antibonding, or conduction, bands are identical in form to the bonding, or valence, bands. This is the simplest description of the bands in silicon. We have evaluated only the bands along the [100] direction, between Γ and X. They are of course defined over the entire Brillouin Zone, and we shall show bands along other symmetry lines presently, but this one set of curves will be sufficient here.

The band picture which this provides is in fact qualitatively correct. The electronic states do not have a single bonding energy, but form a *band* of energies. There are exactly as many states in the bonding band as there were bond orbitals, and with four electrons per atom they are exactly filled with an electron of each of the two spin states in each band state. There is a gap between the highest-energy occupied state and the lowest empty conduction-band state, characteristic of an insulator. In this case the gap is not large and electrons can be thermally excited across the gap providing electrical conduction both from these electrons and from the equal number of *holes* left in the valence band. Thus it is called a semiconductor; electrons can also be introduced by *doping* , substituting for example phosphorus for some silicon atoms. This does very little to the energy bands, but the extra electron which the phosphorus has is put in the conduction band, again providing conductivity.

Note that the total width of the conduction bands, and of the valence bands, is given by $4V_1 = \varepsilon_p - \varepsilon_s$. Note also that the band gap is given by $2V_2 - 4V_1 = 2(1-\alpha_m)V_2$ as indicated in Eq. (2-17) in Section 2-2-D; the band gap goes to zero as the metallicity $\alpha_m = 2V_1/V_2$ goes to one. This correctly gives a vanishing gap with our parameters for tin. Otherwise, we shall see that these bands only qualitatively resemble the true bands for silicon.

B. Bands for Polar Crystals.

For polar crystals the antibonding and bonding energies are replaced by $(\varepsilon_{h+} + \varepsilon_{h-})/2 \pm \sqrt{V_2^2 + V_3^2}$ and the coupling between neighboring bonds is

$$V_{bond\text{-}bond} = -(1 \mp \alpha_p)\frac{V_{1\pm}}{2} \qquad (5\text{-}9)$$

depending upon whether they share a metallic atom (e. g., gallium, with the minus before α_p and V_{1+} evaluated from gallium term values) or a nonmetallic atom (e. g., arsenic, with the plus before α_p and V_{1-} evaluated from arsenic term values). This opens a gap at X (at the Brillouin Zone face) between the two lowest valence bands, given by $2(1 + \alpha_p)V_{1-} - 2(1 - \alpha_p)V_{1+}$ (Harrison (1980), p. 147), which vanishes for the homopolar case. The coupling between neighboring antibonds is the same as Eq. (5-9), but with the sign changed in each case before the α_p . This opens the corresponding gap at X between the two lowest conduction bands. It also changes the gap between valence and conduction bands, by is a complicated form depending upon polarity and V_{1+} and V_{1-} , which we do not write down. We note that the metallic energies do not enter as the sum of squares as they did for the cohesive energy in Eq. (2-28). There is not a single metallicity describing all properties for polar crystals.

C. The Effects of V_1^x.

These bands do not accord well with the true bands because we have omitted a number of couplings. One is the interatomic coupling $-V_1^x$ between a hybrid directed into a bond and the hybrids on the other atom which are directed into different bonds. This was found to be $-0.18\ \hbar^2/md^2$ in Eq. (2-6). Note that it also contributes to the coupling between a bond and its neighboring bonds. In the homopolar case it in fact adds twice $-V_1^x/2$ to the intraatomic coupling of $-V_1/2$ which we included before, twice since each of the hybrids in one bond are coupled by $-V_1^x$ to one of the hybrids in the other. This simply expands the valence bands as seen in Part b of Fig. 5-4. It, however, enters with opposite sign to the coupling between neighboring antibonds since one of the hybrids which enters has opposite sign to the first. Thus the conduction bands are reduced in width by a corresponding amount, also seen in Part b of Fig. 5-4. This correction is qualitatively correct; the conduction bands are somewhat more compressed than the valence bands. This may also be associated with the fact that for free-electron bands, folded back into the Brillouin Zone, the number of bands per unit energy must increase with energy since the density of states increases with energy (as the square root of energy).

There are other couplings we have neglected, $V_{pp\pi}$, for example, which contributes (along with other couplings) a curvature to the bands which are completely independent of k in Fig. 5-4. We shall not add these one-by-one, but return to the expansion of the states in the full set of atomic orbitals. Seeing the bands as arising from bonds and antibonds may have been an informative intermediate step. Our purpose now is to fit the tight-binding parameters, $V_{ss\sigma}$, etc., to the true energy bands, as we fit the free-electron bands for the lithium chain in Section 1-3.

3. Full Tight-Binding Bands

A. sp^3-bands.

Rather than expanding the states as Bloch sums of the four types of bond orbitals, as in Eq. (5-4), we may more generally expand in terms of the *eight* types of atomic orbitals. There are now eight independent coefficients for each wavenumber, those for the s- and p-states on the two atom types in GaAs or on the two distinguishable atoms in silicon. (Again, two nearest-neighbor silicon atoms are not translationally equivalent. The neighbors to each are tetrahedrally arranged, but with differently oriented tetrahedra.) The Bloch sum of s-states on one of the two types of atoms ($i = 1$ or 2) is

$$|s_k^{(i)}\rangle = \frac{1}{\sqrt{N_c}}\Sigma_{j(i)} e^{ik\cdot r_j} |s_j\rangle . \qquad (5\text{-}10)$$

with N_c being the number of terms in the sum since there is one s-state of each type per primitive cell. The other seven Bloch sums are written similarly for the seven values of i corresponding to the other s-state and the three p-states on each atom type. The coupling is only between nearest neighbors, and therefore only between states of different atom types, and the diagonal matrix elements are, for example, $\langle s_k^{(i)}|H|s_k^{(i)}\rangle = \varepsilon_s^{(i)}$, simply the appropriate atomic term value from our Table 1-1. The off-diagonal elements are also simple, for example $\langle s_k^{(1)}|H|p_k^{(i)}\rangle = (V_{sp\sigma}/\sqrt{3})\, 4i\sin(ka/4)$ for the $|p_k^{(i)}\rangle$ based on a p_x-state on the second type of atom and for $\boldsymbol{k}$ along a [100] direction. The $1/\sqrt{3}$ came from the cosine in the matrix element as in Fig. 1-6 and the $4i\sin(ka/4)$ from summing the phase factors over the four neighboring atoms. We obtain eight equations analogous to the four given in Eq. (5-8). That means that the states can in general be obtained by solving eight simultaneous equations in eight unknowns, a simple task for a small computer, for each wavenumber.

The solution is even simpler for wavenumbers in special directions such as the [100] direction which we considered in Section 5-2. We make the symmetry argument a little more completely. If the crystal were rotated by 180° around an x-axis drawn through an atom, every atom would be taken to the position previously occupied by an atom of the same type, as can be seen from Fig. 5-1. The interior of the crystal is unchanged. Correspondingly, if an electronic energy eigenstate $|\boldsymbol{k}\rangle$ of wavenumber $\boldsymbol{k}$ in the x-direction is rotated by 180° around this axis, it remain an energy eigenstate in the crystal with the same wavenumber $\boldsymbol{k}$ and the same energy. It could be exactly the same wavefunction or it could be the negative of the same wavefunction $-|\boldsymbol{k}\rangle$, or it could be a different one of the same energy $|\boldsymbol{k}\rangle'$. In this last case, we could construct an even normalized combination $(|\boldsymbol{k}\rangle + |\boldsymbol{k}\rangle')/\sqrt{2}$ which is identical if rotated 180° and which is again an eigenstate of the same wavenumber and energy. We could also construct an odd combination$(|\boldsymbol{k}\rangle - |\boldsymbol{k}\rangle')/\sqrt{2}$ which goes into its negative if rotated 180°. Thus in every case, the eigenstates can be chosen to be either even or odd under this rotation.

For $\boldsymbol{k}$ in the x-direction, only the $|p_y\rangle$ and $|p_z\rangle$ states are odd with respect to this rotation, so the odd states can be constructed using only these two states on the two atoms. These states are analogous to the π-states which we found for N_2 in Section 1-4, but each of the bands (except at $\boldsymbol{k} = 0$) involves a mixture of both $|p_y\rangle$ and $|p_z\rangle$ states. The four simultaneous equations can be reduced to two for the case of homopolar semiconductors, (making use of a screw-axis symmetry, by which the state is translated along $\boldsymbol{k}$ by $a/4$ and then rotated 90° around $\boldsymbol{k}$) giving two-fold degenerate bands (as for the highest valence and highest conduction bands in Fig. 5-4). They were obtained by Chadi and Cohen (1975) as

$$\Delta_5 = \varepsilon_p \pm \sqrt{[{}^4/_3(V_{pp\sigma}+2V_{pp\pi})\cos(ka/4)]^2 + [{}^4/_3(V_{pp\sigma}-V_{pp\pi})\sin(ka/4)]^2}\ . \tag{5-11}$$

The notation Δ_5 for this state derives from the symmetry of the state. Similarly the even states can be reduced to the solution of quadratic equations. Correcting a misprint in the expression given by Chadi and Cohen (1975) we obtain four bands given by

$$\Delta_1 = {}^1/_2[\varepsilon_s + \varepsilon_p \pm (4V_{ss\sigma} + {}^4/_3V_{pp\sigma} + {}^8/_3V_{pp\pi})\cos(ka/4)] \ \mp \tag{5-12}$$

$$\sqrt{{}^1/_4[\varepsilon_p - \varepsilon_s \pm ({}^4/_3V_{pp\sigma} + {}^8/_3V_{pp\pi} - 4V_{ss\sigma})\cos(ka/4)]^2 + {}^{16}/_3V_{sp\sigma}^2\sin^2(ka/4)}$$

Here the meaning of the $\mp$ notation is that four combinations of the signs $\pm$ $\mp$ $\pm$ in the equation arise; they are + + +, + - +, - + -, and - - -. These bands are shown in the first panel of Fig. 5-5. The wavenumber is zero at the left and $2\pi/a = \sqrt{3}\pi/2d$ at the right. For larger wavenumbers, the coefficients of the atomic orbitals become identical to those obtained with a smaller wavenumber and are not independent states. The lines from $\Gamma_{25'}$ to X_4 contains two states. They are in quite good accord with the true valence bands shown in the third panel but differ considerably for the conduction bands. Discrepancies in the conduction bands would not necessarily cause difficulties for the properties which we have discussed in the preceding chapters since we sought the sum of the energies of the *occupied* states to obtain the total energy. However, it causes difficulty here since we wish to adjust our parameters, $V_{ll'm}$, to accurate band structures to obtain the universal set which we have used throughout. The flaw which is intrinsic to the calculation as we have carried it out is that the conduction bands are inevitably predicted to rise in going from Γ to X whereas the real bands drop (as do the free electron bands to the far right, which we shall discuss).

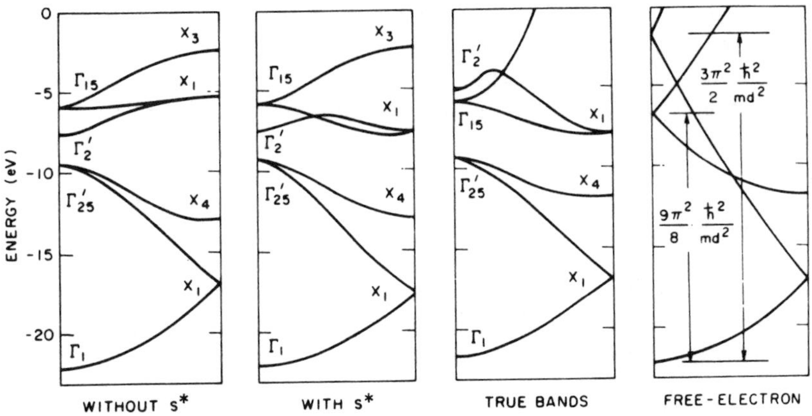

Fig. 5-5. In the first panel are the bands for silicon obtained from Eqs. (5-11) and (5-12). In the second panel the effects of a peripheral excited s-state are included. In the third are more accurate bands obtained by Chelikowsky and Cohen (1976), and in the final panel are free-electron bands. The zero of energy in the third and fourth panels was selected so that the valence-band maximum comes at the same energy. The figure is taken from Harrison (1981b).

B. A Peripheral s*-State.

There are two ways in which this discrepancy can be rectified, the addition of interatomic interactions beyond the nearest neighbors or the addition of higher atomic states on each atom. We are convinced that in semiconductors the latter is the most appropriate physically and the most useful practically. Adding distant couplings introduces parameters for which we do not know the variations with distance, (they presumably drop much more rapidly than $1/d^2$) while the inclusion of higher states proceeds naturally and has led us to parameters which have worked well in a wide variety of circumstances.

Vogl, Hjalmarson, and Dow(1983) had found that supplementing the sp^3 basis which we have been using by a single excited s-state on each atom, called the s*-state, was sufficient to rectify this shortcoming of the conduction bands. This increased the basis to ten orbitals per primitive cell, but if the corresponding Hamiltonian matrix was diagonalized by computer it was not a serious problem. For our attempts to make all evaluations analytically, it *is* a problem. A way around the expansion of the matrix had been suggested earlier by Louie (1980). He added an s*-state on each atom and also five d-states on each atom. This is a more serious complication but he noted that the d-states could be added in perturbation theory so that the matrix actually diagonalized remained a ten-by-ten (or a five-by-five for homopolar semiconductors with $\boldsymbol{k}$ along [100]). He noted that part of the reason the perturbation theory was valid was that even though the atomic d-state is not much higher in energy than the others, orthogonalizing it to the neighboring atomic orbitals would raise its energy considerably, making the perturbation theory more accurate.

We return to the addition of only the s*-state, but use Louie's perturbation approach for adding the effects of this "peripheral" state (Harrison, 1981b). Writing this state $|s^*\rangle$, we let it be coupled to a neighboring p-state by $V_{s^*p\sigma}$; the effect should be larger than that of the coupling to the starting s-states, which are much lower in energy, and we include only the coupling to the p-states. We are then adding a Bloch sum $|s^{*(i)}{}_{\boldsymbol{k}}\rangle = 1/\sqrt{N_c}\Sigma_{j(i)}e^{i\boldsymbol{k}\cdot\boldsymbol{r}_j}\,|s^{*(i)}{}_j\rangle$, for each of the two atom types. For wavenumbers along the x-direction, each will be coupled to the Bloch sum of $|p_x\rangle$ states on the other sublattice, and that coupling will be $4i/\sqrt{3}V_{s^*p\sigma}sin(ka/4)$. The Δ_5 states of Eq. (5-11) are unaffected since the sum of the couplings to the four neighboring $|p_y\rangle$ and $|p_z\rangle$ states is zero. In perturbation theory the effect is to shift the ε_p value entering Eq. (5-12) to ε_p' given by

$$\varepsilon_p' = \varepsilon_p + \frac{16}{3}\frac{V_{s^*p\sigma}^2}{\varepsilon_p - \varepsilon_{s^*}}\, sin^2(\frac{ka}{4}) . \tag{5-13}$$

In the polar case, the shift could be different on the two atoms in the primitive cell, but Eqs. (5-11) and (5-12) are for homopolar systems.

C. Fitting the Free-Electron Bands.

Although we have used this tight-binding description, we have noted that at the same time the bands are quite free-electron like, as seen in Fig. 5-5, and as we noted for the lithium chain in Chapter 1. We may legitimately think of these bands as arising from electrons moving through the crystal, only rather weakly interacting with the lattice through a pseudopotential (discussed in Section 13-4) describing that interaction. We see that the effect is weak by imagining following the true bands shown in Fig. 5-5 as that pseudopotential is gradually set to zero. The resulting free-electron bands were shown in Figure 5-5d . This free electron parabola, $E = \hbar^2k^2/2m$, appears in pieces because of our convention to compress all states to a limited wavenumber range; parts of that parabola have simply been translated into the Brillouin Zone. The resemblance to the true bands is indeed striking.

We are now in a position to choose the various tight-binding parameters to fit either the true bands of silicon or the free-electron bands, both shown in Fig. 5-5. We shall make the fit first to the free-electron bands, fitting values for the bands at the end points as we did for the lithium chain in Chapter 1. Values there can be obtained directly from Eqs. (5-11) and (5-12), taking $ka/4 = 0$ at the points labeled Γ in Fig. 5-5 and $ka/4 = \pi/2$ at points designated X . Note that the effect of the peripheral state, Eq. (5-13), enters the states labeled X_1 but not those at Γ where $\sin ka/4 = 0$. The corresponding formulae will be given later more generally for polar semiconductors. This constitutes eight band energies (rather than sixteen, because of degeneracies). Our parameters for homopolar materials are ε_s , ε_p , $V_{ss\sigma}$, $V_{sp\sigma}$, $V_{pp\sigma}$, $V_{pp\pi}$, and ${}^{16}/_{3}V_{s^*p\sigma}{}^2/(\varepsilon_p - \varepsilon_{s^*})$. This enables us to fit all eigenvalues except the highest one at X which is very poorly given by this tight-binding theory which omits the many excited atomic states which occur at these high energies. Mattheiss (1980) pointed out that all of the upper eigenvalues at $k = 0$ ($\Gamma_{2'}$, $\Gamma_{25'}$ and Γ_{15}) arise from the one (eightfold degenerate) free-electron band at $(9\pi^2/8)\hbar^2/(md^2)$ (See Fig. 5-5.). The lower X_1 energy arises from the free-electron band at $(3\pi^2/8)\hbar^2/(md^2)$ and the X_4 and upper X_1 energy arise from the free-electron bands at $(3\pi^2/4)\hbar^2/(md^2)$. Matching these up we obtain (Harrison, 1981b)

$$V_{ll'm} = \eta_{ll'm}\frac{\hbar^2}{md^2} , \tag{5-14}$$

$$\frac{V_{s^*p\sigma}{}^2}{\varepsilon_p - \varepsilon_{s^*}} = \lambda_{sp\sigma}\frac{\hbar^2}{md^2} , \tag{5-15}$$

and

$$\varepsilon_p - \varepsilon_s = \frac{9\pi^2}{16}\frac{\hbar^2}{md^2} , \tag{5-16}$$

with

$$\begin{aligned} \eta_{ss\sigma} &= -9\pi^2/64 , \\ \eta_{sp\sigma} &= 3\sqrt{3}\pi^2/64 , \\ \eta_{pp\sigma} &= 3\pi^2/16 , \\ \eta_{pp\pi} &= -3\pi^2/32 , \\ \lambda_{sp\sigma} &= -27\pi^2/256 . \end{aligned} \tag{5-17}$$

These parameters are quite close to what we shall shortly choose for universal parameters simply because the true bands so closely resemble free-electron bands, and they do produce the drop in the conduction band in going from Γ to X . However, they lead to a vanishing band gap at Γ and they scale the bands exactly from one system to another, not showing the trends with metallicity and with polarity which dominated our discussion in the preceding chapters. The principal difficulty is Eq. (5-16) for ε_p - ε_s , which is about right for silicon, but badly overestimates the real term-value difference for diamond and underestimates it for germanium and tin. Thus we return to free-atom term values instead of Eq. (5-16) and this will also allow us to treat the effects of polarity without further adjusted parameters. With this change, the other parameters given in Eqs. (5-14) through (5-17) would probably serve well for the treatment of the many properties which we considered in the preceding chapters. However, there is really no practical advantage in using these numbers for the $\eta_{ll'm}$, which have analytic formulas, instead of some other set of numbers and we have chosen to use values which give generally a better description of the energy bands from more complete band calculations. It is clear now that the use of the empirical-pseudopotential bands includes in the band calculation the enhancement of the gaps due to Coulomb effects, described in Section 5-5-A. The bands corresponding to correct total-energy calculations are the "unenhanced" bands and it would have been better to remove those enhancements before fitting. If we had it to do over, we would certainly do that and would probably choose the free-electron $\eta_{ll'm}$ just for the sake of conceptual purity. However, it did not seem worth changing parameters for such a minor point.

D. Obtaining Universal Parameters.

We repeated the above fit (Harrison (1981b)) using Eqs. (5-14) and (5-15) for the couplings, free-atom term values from Table 1-1, and the bands calculated by Chelikowsky and Cohen (1976) for Si, Ge, Sn, GaAs, InSb, ZnSe, and CdTe rather than free-electron bands. Parameters $\eta_{ll'm}$ and $\lambda_{sp\sigma}$ so obtained indeed did not vary greatly from material to material, showing the largest deviations for silicon. We then chose the values for germanium, listed in Table 5-1, and used throughout this text. They are of course particularly appropriate for tetrahedral systems to which they were fit, and any study suggests that there should be dependence upon the number neighbors (Froyen and Harrison (1979), Tang, Wang, Chan, and Ho (1996)). The parameters of Eq. (5-17) do differ significantly from the earlier attempt (Froyen and Harrison (1979), Harrison (1980)) to fit the bands using only

Table 5-1. Parameters obtained from the free-electron fit (Eq. (5-17)) and obtained from a fit to the germanium bands. In this text the latter are taken to be "universal", and used with the free-atom term values from Table 1-1.

	Free-electron	Universal
$\eta_{ss\sigma}$	-1.39	-1.32
$\eta_{sp\sigma}$	0.80	1.43
$\eta_{pp\sigma}$	1.85	2.22
$\eta_{pp\pi}$	-0.93	-0.63
$\lambda_{sp\sigma}$	-1.04	-0.40

the sp^3-basis with no peripheral state, leading to values $\eta_{ss\sigma} = -1.40$, $\eta_{sp\sigma} = 1.84$, $\eta_{pp\sigma} = 3.24$, $\eta_{sp\sigma} = -0.81$.

Earlier in the development of the theory we used the Herman-Skillman values of the free-atom term values, which arose from the same local-density approximation as was used in contemporary band calculations. This seemed appropriate since our principle comparisons were with such calculations. We now prefer the Hartree-Fock values which more closely reproduce for example the measured free-atom ionization energies, as seen in Table 1-2. These, with the couplings given in Table 5-1 constitute what we call *universal tight-binding parameters*.

The energy bands for silicon, obtained using these universal parameters, are shown in Fig. 5-5. The valence band is clearly very well given. The inverted order of the Γ_{15} and $\Gamma_{2'}$ levels in the conduction band is a conspicuous, but not surprising, error. The first of these is determined entirely by ε_p , $V_{pp\sigma}$ and $V_{pp\pi}$ while the latter by ε_s and $V_{ss\sigma}$. There is strong cancellation between the effects of the term values and the coupling and the small difference is in error. Indeed the ordering predicted, with $\Gamma_{2'}$ lower, is correct for germanium and tin, but not for silicon and carbon where the spacing is small enough that the effects of the interatomic couplings exceeds the ε_p - ε_s which is relatively independent of material. The addition of peripheral d-states (Louie (1980)) can readily remove that discrepancy for silicon and carbon. There is also a noticeable discrepancy for the highest band energy X_3 which was why it was not used in the fit. Recognizing that the discrepancies are worst for silicon so all other materials are even closer, one may conclude that the universal parameters do remarkably well for such simple calculations and demonstrate the proper trends among the homopolar semiconductors.

E. Application to Polar Systems.

The solution for the band energies is particularly simple at the Γ and X points even for polar semiconductors. Using symmetry as we did above, the eight simultaneous equations, of the form of Eq. (5-8), reduce to the solution of quadratic equations. The solutions have been given by Chadi and Cohen (1975) who did not include a peripheral state. Their forms for these levels remain correct if we use the primed values for the p-states, including the correction from Eq. (5-13). Their results for Γ (for which $\varepsilon_{p'} = \varepsilon_p$), X_1 and X_3 (for which ε_p' is given by Eq. (5-13) with $\sin ka/4 = 1$), and one set at L (where the correction corresponding to Eq. (5-13) is almost the same as at X, $\varepsilon_p' = \varepsilon_p + 4V_{s*p\sigma}^2/(\varepsilon_p - \varepsilon_{s*})$ at L rather than the $\varepsilon_p' = \varepsilon_p + (^{16}/_3\, V_{s*p\sigma}^2/(\varepsilon_p - \varepsilon_{s*})$) at X) are

$$E(\Gamma_1) = \frac{\varepsilon_{s+} + \varepsilon_{s-}}{2} \pm \sqrt{\left(\frac{\varepsilon_{s+} - \varepsilon_{s-}}{2}\right)^2 + (4V_{ss\sigma})^2}\ . \qquad (5\text{-}18)$$

$$E(\Gamma_{15}) = \frac{\varepsilon_{p+} + \varepsilon_{p-}}{2} \pm \sqrt{\left(\frac{\varepsilon_{p+} - \varepsilon_{p-}}{2}\right)^2 + \left(\frac{4}{3}V_{pp\sigma} + \frac{8}{3}V_{pp\pi}\right)^2}\ , \qquad (5\text{-}19)$$

$$E(X_1) = \frac{\varepsilon_{s+} + \varepsilon_{p-}'}{2} \pm \sqrt{\left(\frac{\varepsilon_{s+} - \varepsilon_{p-}'}{2}\right)^2 + \frac{16}{3}V_{sp\sigma}^2}\ , \qquad (5\text{-}20)$$

$$E(X_3) = \frac{\varepsilon_{s-} + \varepsilon_{p+}'}{2} \pm \sqrt{\left(\frac{\varepsilon_{s-} - \varepsilon_{p+}'}{2}\right)^2 + \frac{16}{3}V_{sp\sigma}^2}\ , \qquad (5\text{-}21)$$

$$E(X_5) = \frac{\varepsilon_{p+} + \varepsilon_{p-}}{2} \pm \sqrt{\left(\frac{\varepsilon_{p+} - \varepsilon_{p-}}{2}\right)^2 + \left(\frac{4}{3}V_{pp\sigma} - \frac{4}{3}V_{pp\pi}\right)^2}\ , \qquad (5\text{-}22)$$

$$E(L_3) = \frac{\varepsilon_{p+}' + \varepsilon_{p-}'}{2} \pm \sqrt{\left(\frac{\varepsilon_{p+}' - \varepsilon_{p-}'}{2}\right)^2 + \left(\frac{4}{3}V_{pp\sigma} + \frac{2}{3}V_{pp\pi}\right)^2}\ . \qquad (5\text{-}23)$$

The notation for the states differs from that for nonpolar materials because the symmetry is different. (There is no screw-axis symmetry of the type we mentioned before for homopolar crystals.) In Harrison (1981b) the values predicted by these formulae for Si, Ge, Sn, GaAs, InSb, ZnSe, and

CdTe are compared with values obtained from full band calculations by Chelikowsky and Cohen (1976). [The corresponding bands will be displayed in Fig. 5-6.] We illustrate the comparison by giving the case, CdTe, which is the greatest extrapolation from the germanium fit. We list the values (in eV) obtained with Eqs. (5-18) through (5-23) and a corresponding evaluation at L, with the values from the full band calculation in parentheses (indicated by a "v" if it is for a valence band and a "c" if it is for a conduction band): $\Gamma_1{}^v$: -22.00 (-20.46), $\Gamma_{15}{}^v$: -9.80 (-9.69), $\Gamma_1{}^c$: -5.33 (-7.80), $\Gamma_{15}{}^c$: -3.73 (-3.86), $X_3{}^v$: -19.85 (-18.51), $X_1{}^v$: -13.27 (-14.44), $X_5{}^v$: -11.36 (-11.24), $X_1{}^c$: -5.56 (-5.91), $X_3{}^c$: -5.34 (-5.44), $L_1{}^v$: -20.14 (-19.03), $L_2{}^v$: -13.47 (-14.12), $L_3{}^v$: -10.47 (-10.22), $L_1{}^c$: -5.75 (-6.57), and $L_3{}^c$: -3.61 (-3.10). The agreement is rather extraordinary considering such a band calculation can readily be made for any semiconductor and may be expected to have accuracy on this scale. The most serious discrepancy is in the value for $\Gamma_1{}^c$ which is too high by more than two volts, making the corresponding overestimate in the band gap itself. Such discrepancies arise for the other polar semiconductors and could be rectified by reducing the value of $\eta_{ss\sigma}$ for these systems. We shall see in Section 5-5-A that there are Coulomb enhancements of this general size of the band gap obtained from full local-density-theory band calculations. As we discussed at the end of Section 5-3-C, these corrections were included in the bands used to obtain our universal parameters so it would be inappropriate to introduce them again here.

F. The Band States

Of course once we have calculated the energies of the states for some wavenumber, whether with tight-binding theory or some other way, we may substitute the energy back into the equations from which we determined the energy in order to obtain the states themselves, just as we did for the polar bonding and antibonding states in Section 2-2-C. We obtain the coefficients of each of the basis states which were used in expanding the band state. For arbitrary wavenumbers these were the eight states, s-states on each atom type and three p-states on each atom type (plus the peripheral s*-states if we include that). For each of the states at symmetry points with energies given in Eqs. (5-18) to (5-23), only two atomic states entered and we may tell which by seeing which atomic energies entered the band energy. For example, for the states Γ_1 given in Eq. (5-18) only the s-states entered. Then the calculation of states is almost* exactly parallel to that for the polar

*The one complication is that what enters as $-V_2$ in Eqs. (2-9) for the coupling between two s-Bloch sums is actually $\Sigma_j V_{ss\sigma} e^{i\mathbf{k}\cdot\mathbf{d}_j} = 4V_{ss\sigma}(\cos k_x a/4 \cos k_y a/4 \cos k_z a/4 - i \sin k_x a/4 \sin k_y a/4 \sin k_z a/4)$. For $\mathbf{k} = 0$ this is exactly equivalent to the bond problem with $-V_2 = 4V_{ss\sigma}$, but for other wavenumbers complex couplings can be obtained and the coefficients are found to have a phase difference equal to the phase of the new $-V_2$. The formulae here give the magnitudes of the coefficients correctly in any case.

bond in Section 2-2-C with $-V_2$ now replaced by $4V_{ss\sigma}$ and V_3 now replaced by $(\varepsilon_{s+} - \varepsilon_{s-})/2$. Let us write this more generally, taking the $i = 1$ for the lower-energy s-state (or p-state) and $i = 2$ for the upper one, so that V_3 is positive. That is,

$$V_3 = \frac{\varepsilon_2 - \varepsilon_1}{2} \tag{5-24}$$

with ε_2 being the higher of the two atomic energies which appear in Eqs. (5-18) through (5-23) and ε_1 being the lower. V_2 is the square root of the second term under the square root in each case, $4/V_{ss\sigma}/$ for Eq. (5-18). We define a polarity for this band state from these coefficients obtained from Eqs. (5-18) through (5-23) as

$$\alpha_p = \frac{V_3}{\sqrt{V_2^2 + V_3^2}} \tag{5-25}$$

which is positive. Then the lower-energy Bloch sum of Eq. (5-10) is written $/\psi_k^1>$ and the higher-energy one $/\psi_k^2>$. The band states are written as combinations $u_1/\psi_k^1> + u_2/\psi_k^2>$ of these Bloch sums. For the lower of the two states, given in Eqs. (5-18) through (5-23) with the minus sign, the coefficients are obtained in analogy with Section 2-2-C and are

$$u_1^- = \sqrt{\frac{1+\alpha_p}{2}}\,, \qquad u_2^- = \sqrt{\frac{1-\alpha_p}{2}}\,. \tag{5-26}$$

For the upper states the coefficients are

$$u_1^+ = \sqrt{\frac{1-\alpha_p}{2}}\,, \qquad u_2^+ = -\sqrt{\frac{1+\alpha_p}{2}}\,. \tag{5-27}$$

We will actually only need the states at these symmetry points in our study of properties, but the more general states at lower symmetry, and containing more terms, could be similarly obtained. If we use these factors with the form of the Bloch sums from Eq. (5-10) we see that the coefficient on each atomic state has a magnitude $\sqrt{(1 \pm \alpha_p)/N}$ since $N = 2N_c$ for this structure. For some cases there may be a phase difference for the states, as indicated in the footnote on the preceding page.

G. Spin-Orbit Coupling

It is familiar that in atoms there is a splitting of the three p-states due to the interaction of the magnetic moment of the electron with the magnetism of the electron motion around the nucleus. This same effect modifies the states in the solid. The interaction is a consequence of the Dirac relativistic theory of the electron and is a term in the Hamiltonian of the form (see, for example Schiff (1968), p. 433) we might anticipate for such a magnetic interaction,

$$H_{SO} = \xi(r)\, \mathbf{L}\cdot \sigma \tag{5-28}$$

where $\xi(r)$ depends upon the spherically-symmetric potential as $\xi(r) = (1/(2m^2c^2r))\partial V(r)/\partial r$ and $\mathbf{L}$ is the angular momentum of the electron and σ is its spin. We do not have other occasion to use these angular-momentum operators and we simply outline the analysis, first for the free atom.

We select a z-axis and the state $|p_z\rangle$ has zero angular momentum around that axis. This will be a slight inconvenience when we construct the bands away from the center of the Brillouin Zone in the next chapter because it will then be best to take the wavenumbers also in the z-direction rather that along the x-direction as we have before. However, angular momentum is customarily quantized along the z-axis. The $|p_x\rangle$ and $|p_y\rangle$ states are written in terms of states with one unit $(\hbar)$ of angular momentum around the z-axis $((|p_x\rangle + i|p_y\rangle)/\sqrt{2}\,)$ and states with minus one unit of angular momentum (the complex conjugate of the other). Similarly the spin states have components of angular momentum of $\pm {}^1/_2\,\hbar$ · The dot-product in Eq. (5-28) can be written out in components, $L_x\sigma_x + L_y\sigma_y + L_z\sigma_z$. The x- and y-components are written in terms of the raising and lowering operators, much as the $|p_x\rangle$ and $|p_y\rangle$ states were written terms of angular-momentum eigenstates. $L_x = {}^1/_2(L^+ + L^-)$ and $L_y = {}^1/_{2i}\,(L^+ - L^-)$, so that $\mathbf{L}\cdot\sigma$ becomes ${}^1/_2(L^+\sigma^- + L^-\sigma^+) + L_z\sigma_z$. These angular momentum operators have the properties that L_z operating on one of our states gives back the state, multiplied by its component of angular momentum along the z-axis. Similarly σ_z operating on the spin state gives $\pm\hbar/2$. The raising operator, L^+, operating on a state raises its component of angular momentum along the z-axis by one unit (or gives zero if it is already at the maximum) and multiplies it by $\sqrt{2}\hbar$. [The formula for this factor, e. g., Schiff (1968) p. 204, is actually $\sqrt{l(l+1) - m(m+1)}\;\hbar$ where l is the total angular momentum quantum number ($l = 1$) and m is the quantum number for the component.] The lowering operator, L^-, lowers it by one and multiplies by $\sqrt{2}\hbar$. [The formula is $\sqrt{l(l+1)-m(m-1)}\,\hbar$.] The spin raising and lowering operators do the same, but multiply simply by $\hbar$. [The same formulae.]

With this combination of operators we can find the matrix elements of the spin-orbit coupling of Eq. (5-28) between any pair of atomic p-states. In particular we find that a state $|p_z+\rangle$, the plus indicating a spin component of $+{}^1/_2\hbar$ along the z-axis, is coupled to a state $|p_x-\rangle$. The matrix element is real and has a magnitude which depends upon the appropriate integral involving the $\xi(r)$ in Eq. (5-28). That matrix element is called λ and the others are related to it. [We note that Phillips (1973), p. 178, defined a spin-orbit λ which was larger by a factor of two.] There are in fact matrix elements coupling the states $|p_x+\rangle$, $|p_y+\rangle$, and $|p_z-\rangle$ but no coupling of these to the other three. They are

$$\langle p_x+|H_{SO}|p_z-\rangle = \lambda \ ,$$

$$\langle p_y+|H_{SO}|p_z-\rangle = -i\,\lambda \ , \tag{5-29}$$

$$\langle p_x+|H_{SO}|p_y+\rangle = -i\,\lambda \ .$$

The matrix elements with the two states interchanged are the complex conjugate and the diagonal ones, $\langle p_x+|H_{SO}|p_x+\rangle$, etc., are zero. Of course the other set of three states with reversed spins from these are also coupled among each other. The matrix elements, with all spins flipped are the complex conjugate of those given above; e. g., $\langle p_y-|H_{SO}|p_z+\rangle = i\lambda$.

For each set we may readily diagonalize the three-by-three matrix for the free atom to obtain eigenvalues λ , λ , and -2λ . An identical set of eigenvalues comes from the other set of three. We can see from the definition of $\xi(r)$ that λ is positive; we also expect this physically since with parallel spin and orbital angular momentum the magnetic fields add and the energy is higher. Thus we have found four states with higher energy and parallel spin, and two with lower energy and antiparallel spin. These are more usually derived for atoms directly in terms of the quantum theory of angular-momentum eigenstates, writing the total angular momentum $\boldsymbol{J} = \boldsymbol{L} + \boldsymbol{\sigma}$ and evaluating $J^2 = L^2 + 2\boldsymbol{L}\cdot\boldsymbol{\sigma} + \sigma^2$. J can be ${}^1/_2$ or ${}^3/_2$ and writing $J^2 = J(J+1)\hbar^2$, etc., we obtain $\boldsymbol{L}\cdot\boldsymbol{\sigma}$ equal to ${}^1/_2\ \hbar^2$ for $J = {}^1/_2$ with two possible components of angular momentum ($J_z = \pm{}^1/_2$)around the z-axis and -1 for $J = {}^3/_2$ with four components ($J_z = \pm{}^1/_2 , \pm{}^3/_2$). We need the set of matrix elements, Eq. (5-29) for solids.

These are intraatomic matrix elements, as were the $-V_1$ couplings between hybrids. We may add them directly in the solid to the Hamiltonian from which we obtain energy bands. Our Hamiltonian becomes a sixteen-

Table 5-2. Spin-orbit coupling parameters λ (in eV) for valence p-states, renormalized for use in solids, compiled by Chadi (1977). The spin-orbit splitting at the top of the valence band is 3λ.

	Al 0.008	Si 0.015	P 0.022	S 0.025
Zn 0.025	Ga 0.058	Ge 0.097	As 0.140	Se 0.160
Cd 0.076	In 0.131	Sn 0.267	Sb 0.324	Te 0.367

by-sixteen matrix since we must distinguish spin states for each orbital. One might hope to divide this into two submatrices, as we could for the atom, and that can be done at wavenumbers of high symmetry in the Brillouin Zone, but in general the full matrix is needed. A renormalization of the atomic λ values is needed in the solid, typically by a factor of 1.5, due to the change in environment of the atom (Elliot (1954), Phillips (1973), Braunstein and Kane (1962), Chadi (1977). A set of such renormalized values is presented in Table 5-2. They grow rapidly with increasing atomic number, so that they are really important only in heavier semiconductors, as with other relativistic effects. They also increase across each row of the periodic table. Both trends may be associated with the $\partial V/\partial r$ in the $\xi(r)$ of Eq. (5-28) which has the same trends.

Spin-orbit coupling splits the valence bands, as will be seen in the heavier elements in Fig. 5-6. The splitting at the top of the valence band can be immediately obtained and we do that here for polar, as well as homopolar semiconductors. (We will only discuss homopolar semiconductors in Chapter 6 when we obtain more complete bands.) We include the three states from the cation (Ga in GaAs) which enter Eq.(5-29) and their coupling, λ_+ . Each contributes a Bloch sum of zero wavenumber to the expansion of the state. The sum based on the $|p_x+\rangle$ on the cations is coupled to the sum based on the $|p_x+\rangle$ on the anions, etc., and each of those couplings is

$$V_{xx} = {}^{4}/_{3}V_{pp\sigma} + {}^{8}/_{3}V_{pp\pi} . \tag{5-30}$$

It is also coupled to the other cation states by λ_- , but to no other anion states. Thus these six states are coupled only to each other and the Hamiltonian matrix becomes

$$H = \begin{pmatrix} \varepsilon_{p+} & -i\lambda_+ & \lambda_+ & V_{xx} & 0 & 0 \\ i\lambda_+ & \varepsilon_{p+} & -i\lambda_+ & 0 & V_{xx} & 0 \\ \lambda_+ & i\lambda_+ & \varepsilon_{p+} & 0 & 0 & V_{xx} \\ V_{xx} & 0 & 0 & \varepsilon_{p-} & i\lambda_- & \lambda_- \\ 0 & V_{xx} & 0 & -i\lambda_- & \varepsilon_{p-} & i\lambda_- \\ 0 & 0 & V_{xx} & \lambda_- & -i\lambda_- & \varepsilon_{p-} \end{pmatrix} \qquad (5\text{-}31)$$

Here the first column (and row) represents the state $/p_x+>$, the second $/p_y+>$, the third $/p_z+>$, the fourth $/p_x ->$, etc.

This matrix is easily diagonalized. The transformation which diagonalizes the upper left three-by-three (by taking orthogonal linear combinations of the p_+-states) gives

$$H_{3\times 3} = \begin{pmatrix} \varepsilon_{p+}+\lambda_+ & 0 & 0 \\ 0 & \varepsilon_{p+}+\lambda_+ & 0 \\ 0 & 0 & \varepsilon_{p+}-2\lambda_+ \end{pmatrix}, \qquad (5\text{-}32)$$

in direct analogy with the free-atom calculation and leaves the upper-right and lower-right three-by-threes unchanged. Similarly the lower-right three-by-three can be diagonalized and finally the first and fourth states, coupled by V_{xx} , yield

$$\Gamma_8 = \frac{\varepsilon_{p+} + \lambda_+ + \varepsilon_{p-} + \lambda_-}{2} \pm \sqrt{\left(\frac{\varepsilon_{p+} + \lambda_+ - \varepsilon_{p-} - \lambda_-}{2}\right)^2 + V_{xx}^2} , \qquad (5\text{-}33)$$

with the minus sign of course giving the valence-band maximum. For the other states, designated by their symmetry as Γ_7 , the same expression is obtained but with each $\lambda_\pm$ replaced by $-2\lambda_\pm$. For the homopolar case the energies are simply $\Gamma_8 = \varepsilon_p + \lambda \pm V_{xx}$ and $\Gamma_7 = \varepsilon_p - 2\lambda \pm V_{xx}$.

The weighting of the λ_+ and λ_- for the polar case is a little more complex than the $(1 \pm \alpha_p)/2$ suggested in Harrison (1980), p. 161. However, if we expand Eq. (5-33) for small $\lambda_\pm$ we obtain such a weighting with the $\alpha_p(p)$ from Eq. (5-25) based upon $(\varepsilon_{p+} - \varepsilon_{p-})/2$ rather than V_3 and V_{xx} rather than V_2 . It can in fact be seen from the start that this weighting must be appropriate for small λ by treating the spin-orbit coupling in first-order perturbation theory using the starting states without spin-orbit coupling. We shall see how the splittings vary with wavenumber as we consider the results of full band calculations next and in Chapter 6 when we consider the form of the valence bands near Γ. For that analysis it will be necessary to restrict consideration to bands for homopolar systems, but we shall use the approximate $(1 \pm \alpha_p)/2$ weighting to extend the results to polar-semiconductor bands and write out the explicit form for the effective λ in Eq. (6-28).

4. More Accurate Energy Bands

There have of course been very many full calculations of semiconductor energy bands and semiconductor properties, almost all of which provide higher accuracy than the elementary methods used here. For our purposes they are like experiment in guiding our understanding and testing our approximations. A review of such *ab initio* calculations has been given by Chelikowsky and Cohen (1992) in the *Handbook on Semiconductors*. Another very extensive study is the *Handbook of the Band Structure of Elemental Solids* by Papaconstantopoulos (1986). Chelikowsky, Troullier, and Saad (1994) have also provided a new method, which seems promising to us, for calculating states on a real-space grid rather than expanding them in plane waves. There are even more accurate tight-binding approaches, such as that by Mercer (1996). We here have been making comparisons with bands which have been tuned to experiment, which has difficulties which we discuss in Section 5-5.

The energy bands obtained by Chelikowsky and Cohen (1976), which were used for the fit in Table 5-1, are displayed in Figure 5-6. The bands of silicon from Γ to X are the same as in Fig. 5-5, but additional lines in the Brillouin Zone are included here. Note that additional splittings have occurred in the bands of heavier semiconductors, arising from the spin-orbit splitting discussed in the preceding section. That also is the reason different notation is used for the symmetry points. The additional splittings are of importance in some electronic properties, but are generally not large enough to be of importance in the bonding and dielectric properties of semiconductors.

Trends in the band structure are apparent in this set of bands. Considering the series of homopolar semiconductors at the left, we see that the conduction-band level $\Gamma_{2'}$ (or Γ_1 in the notation appropriate to heavier elements) drops with metallicity. We have seen in tight-binding terms that this is because as we move down in the table the coupling ($V_{ss\sigma}$) which pushed this level so high is decreasing due to the increase in bond length, while the sp-splitting which would place such s-like states below the p-like

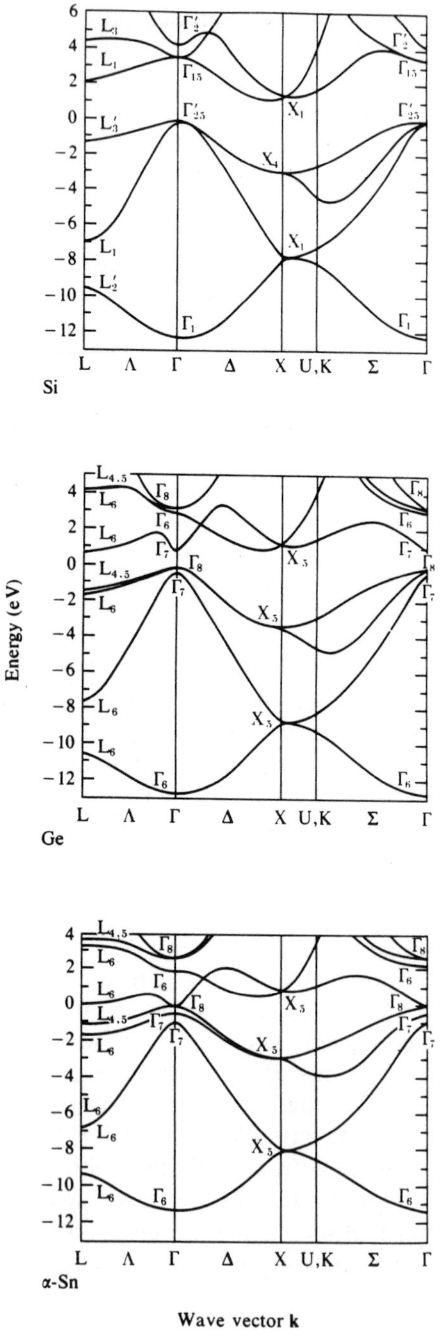

Fig. 5-6. The energy bands of silicon, germanium, and tin, and polar semiconductors isoelectronic with them, as calculated by Chelikowsky and Cohen (1976) using an empirical pseudopotential method. (From Harrison, 1980.)

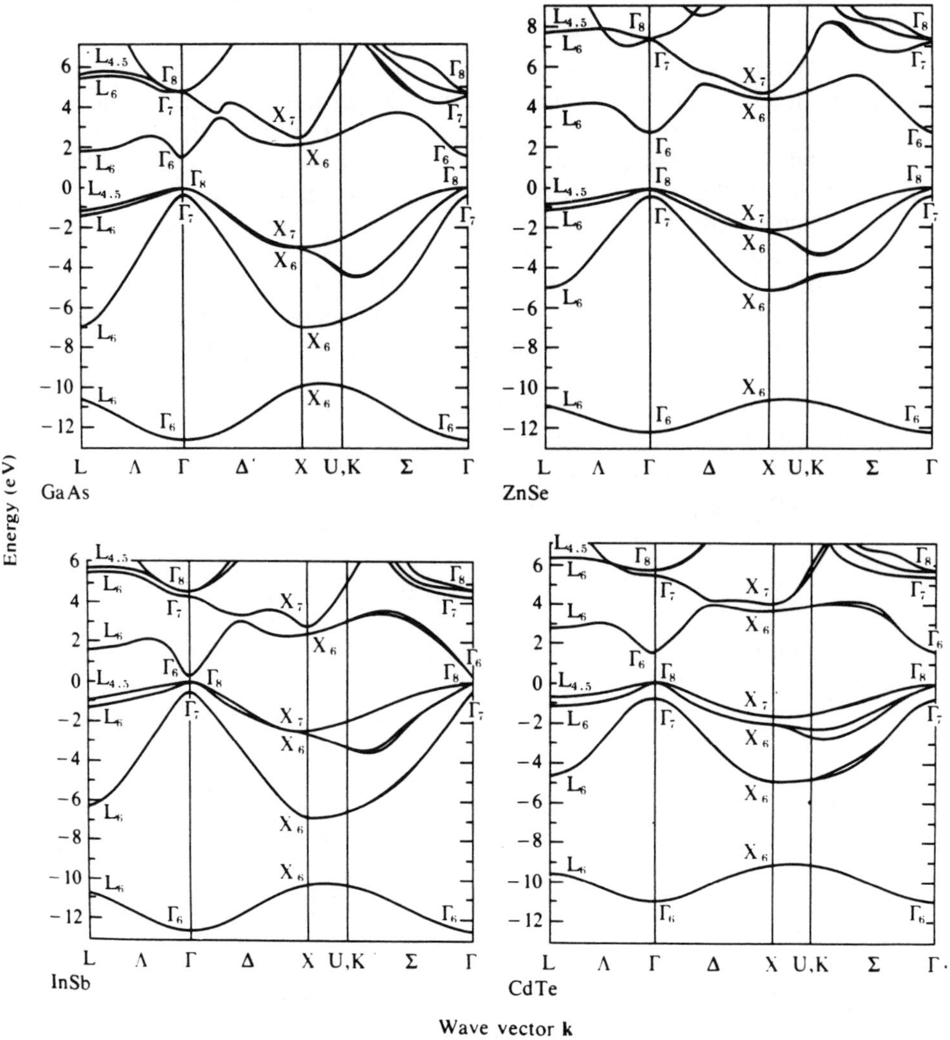
Energy (eV)
GaAs
ZnSe
InSb
CdTe
L Λ Γ Δ X U,K Σ Γ
Wave vector k

states does not change nearly as much. This variation with metallicity produces a reduction in the gap, and finally its disappearance at tin.

As we move to the right in the figure, increasing polarity, we see that the band gaps tend to increase, in tight-binding terms because the splitting between bonding and antibonding states increases as $2\sqrt{V_2^2 + V_3^2}$, with V_2 remaining quite constant. There also arises a gap between the lower conduction bands which we noted for the simpler representation of the bands in Section 5-2-B. Here we see from Eqs. (5-20) and (5-21) that at X each of these states is a mixture of an s-state on one atom in the cell and a p-state on the other. In the homopolar case it did not change the energy when the s- and p-states were interchanged, but it does in the polar case. Such a splitting also seems to be responsible for the shifting upward of the conduction bands away from Γ which favors the conduction-band minimum occurring at the center of the Brillouin Zone, Γ . This is an important feature of gallium arsenide, in contrast to silicon and germanium where the minimum occurs elsewhere (near X for silicon and at L for germanium). There are a number of features of the band structure which may be discerned from such sets of curves, and which we may understand in terms of tight-binding theory. We do that in the following chapter as we consider the relation of these bands to the electronic properties. Before doing that, we should look carefully at important effects that bear on the meaning of the bands themselves.

5. Corrections to the Bands

Our outlook from the very start of Chapter 1 has been based upon a one-electron approximation, in which each electron is imagined to be moving in an effective potential due to the nuclei and all other electrons present. We noted in Section 1-2-D that if we simply add the corresponding electron energies to obtain the total energy, we have counted each electron-electron interaction twice and must subtract one contribution This was not an error being corrected, but the subtraction is an inherent part of the one-electron approximation. These one-electron energy states in the atom remain the proper estimates of the removal energy for an electron from the atom, without these corrections.

Similarly, our energy bands provide meaningful electron energy levels, but for particular properties they cannot be used directly. The same corrections for doubly-counted Coulomb interactions must be included unless we can see that they are canceled by other terms. We shall also see in Section 19-2-A that there are corrections to the removal energy of an electron from the solid due to the image potential produced by the solid. Also in semiconductors the calculated energy gap is not exactly what is measured in any experiment. There are two important effects which we discuss here. The first is a Coulomb correction, closely related to the U parameter for the atom, listed in Table 1-1, which distinguishes the electron affinity of the atom from the ionization energy. It is important that this not be incorporated as a correction to the band calculation since it enters only some properties. As we have indicated, it would have been better had the energy bands given in Fig. 5-6 not been adjusted to empirical gaps. We would then add a correction afterward to obtain the energy required to excite electrons. We discuss these corrections in Part A. The second important effect is for our assumption - called the *Born-Oppenheimer Approximation* - that the nuclei are classical charges at positions $\boldsymbol{r}_j$ rather than dynamical particles described by quantum mechanics, just as the electrons are. Corrections to this assumption are discussed in Part B.

A. Coulomb Enhancement of the Gaps

The band gap in the spectrum of electronic states is the first quantity we have considered which has major Coulomb corrections to the simple one-electron theory. Attention was drawn to this fact by Sham and Schlüter (1983) and by Perdew and Levy (1983) who indicated that the density-functional theory used in most theoretical calculations will significantly underestimate the gap. Corrections to these band calculations have been estimated by Yin(1985), by Carlsson (1985), and by Hybertsen and Louie (1985) in the context of many-body theory, where they are thought of as "correlation corrections" due to the correlated motion of the electrons, or as a "discontinuity in the energy-dependent correlation energy". They are also very directly understandable, and easily calculated, in terms of parameters available here (Harrison (1985)).

When we treated the bond polarizability in Section 3-1, our analysis consisted of admixing the antibonding to the bonding state in each bond due to an applied electric field. This left two electrons in each bond and the polarization of the bond did not appreciably affect the Coulomb interaction energy of those two electrons. In contrast, when we excite an electron optically or thermally and remove it from its initial site, it is placed in an antibonding state where the two bonding states are already occupied. It is analogous to adding an electron to a neutral atom. We noted at the end of Section 1-1 that the corresponding electron affinity is less than the ionization energy, represented by our atomic term values, by a Coulomb energy U , listed as 7.64 eV for silicon in Table 1-1. Thus we might at first expect the gap energy to be enhanced by some such amount.

It is not difficult to see that this value is greatly reduced by the polarization of the medium. For this purpose it is very helpful to think of the electron state, here a bond, as a spherical shell with radius r_s given by

$$\frac{e^2}{r_s} = U \tag{5-34}$$

For such a shell a second electron in the state has its energy raised by the potential at that shell e^2/r_s so this is consistent. The band gaps which the band calculation, or local density theory, provides correspond to transfer of an electron from a bonding state to an antibonding state within the same bond site, as for the calculation of bond polarizability. Thus the parameters from the band calculation are appropriate to the calculation of dielectric properties. A measured band gap corresponds to transferring an electron from a bond state to an antibonding state at *another* site, just as it would if we transferred an electron between two isolated atoms and therefore the energy required contains the additional U , the difference between the electron affinity and the ionization potential for the atom. If the second atom is imbedded in a dielectric medium, the potential is reduced to $e^2/(\varepsilon r_s) = U/\varepsilon$ *due to the polarization of the bonds in the surrounding medium.* It is reasonable to use the dielectric constant for this estimate because we are using it outside the atom, where the fields are slowly varying, not within the atom itself.

We must decide whether we should use the static dielectric constant ε_s of Section 4-3-C or the optical dielectric constant ε of Section 4-2-A and B. This turns out to be a question with important consequences, which we return to in the next Part, B, on polarons. For the present we discuss only electronic effects and thus use the optical dielectric constant as if the atoms did not have time to move during the absorption. We saw in Table 4-6, and will see more completely in Table 5-4, that the differences between the two dielectric constants are not large, so that we are obtaining the principal correction to the band gap in any case. Thus the extra energy required to make the electron-hole pair is U/ε giving an enhancement of the gap as

$$E_c - E_v = (E_c - E_v)_{band\ calc.} + U/\varepsilon . \tag{5-35}$$

This is in fact a raising of the entire antibonding, or conduction, band over that predicted by tight-binding theory, or by more accurate forms of density-functional theory [Harrison, (1985)] . It applies not only to semiconductors but to insulators with much larger gaps.

The estimate may be made directly using the values of U from Table 1-1 [average values for the two elements in a compound are to be used since the bonding orbitals and antibonding orbitals enter symmetrically in moving an electron from the valence band to the conduction band]. For the dielectric constant $\varepsilon = 1+4\pi\chi^{(1)}$ we use experimental values of $\chi^{(1)}$ from Tables 4-1

Table 5-3 Predicted enhancement U/ε of the band gaps in semiconductors and insulators, compared with experimental values, obtained from the difference between experimental gaps and those from band calculations (all in eV).

System	U/ε	Experimental
C	2.06	1.37[a]
Si	0.65	0.69[a]
Ge	0.47	0.27[a]
Sn	0.28	
AlP	0.95	0.92[b]
GaAs	0.69	
ZnSe	1.43	
CuBr	1.92	
LiF	6.1	4.7[b]
NaCl	3.5	3.7[b]
CsCl	2.8	3.3[b]
MgO	3.6	2.5[b]
CaS	1.7	2.8[b]
BaS	1.6	2.1[b]
Ne	13.8	9.9[b]
Ar	6.9	6.1[b]
Kr	5.8	4.6[b]

[a] M. T. Yin (1985)
[b] A. E. Carlsson (1985)

and 4-2, or the more complete set of experimental values which we shall list in Table 5-4. This would seem more relevant than using our Bond-Orbital-Approximation estimates which were a factor of order two too small. We similarly use experimental values of ε and average U 's in making the estimate for insulators to obtain the predictions given in Table 5-3.

The results may be compared with the difference between band gaps obtained from experiment and those derived from full band calculations. Such "experimental" differences have been given by Carlsson (1985) and Yin (1985) for a number of semiconductors and are listed in Table 5-3 along with our estimates. The agreement is remarkable indeed, and certainly much better than the agreement between tight-binding estimates of semiconductor band gaps which we shall present in Table 6-1 and the corresponding values from band calculations. Thus we may say that this simple theory gives a very good account of the enhancement, but that the tight-binding description of the conduction band energies themselves is not very accurate.

We noted in Section 3-D that it would have been more consistent, and preferable, to fit the universal parameters to band calculations which were not fit to give the observed band gaps. That would in fact have made only small changes in the parameters and would not have eliminated the large errors in the conduction bands. In that sense the distinction is not so important.

We noted that we predict the same shift for the entire conduction band relative to the valence band; more detailed calculations confirm this to be approximately true (e. g., Del Sole and Girlanda (1993)). A reasonably accurate computational method for obtaining these corrections is the *GW Method* (Hedin (1965)). The G represents a "dressed Green's function" and the W a "dynamical screened Coulomb interaction". Hybertsen and Louie (1985) used such a method to compute the corrections. A recent application of this method to AlN and GaN has been made by Rubio, Corkill, Cohen, Shirley, and Louie(1993), which includes references to a number of other works.

We may note also that the principal trends arise from the trends in the dielectric constant from one system to another, making the enhancements larger for polar semiconductors since, as we have seen, the dielectric constant decreases with polarity approximately as $\alpha_c^3 = (1 - \alpha_p^2)^{3/2}$. For this same reason the corrections are still larger in ionic crystals, as can be seen for those included in Table 5-3. It will be necessary to watch for their effects when we treat those systems and we shall reconsider the enhancement again in Section 9-1-F.

B. Polarons

In deciding whether to use the high-frequency or low-frequency dielectric constant for the gap enhancement discussed in part A, the intuitive question one might be inclined to ask is whether the atoms "have time to displace" during the optical absorption or, alternatively, whether the Franck-Condon principle applies and the atomic motion is much slower than the electronic motion. The answer to that question [see, for example, Harrison (1970), 333 ff.] is that the atoms should be said to have time to displace only if the corresponding lowering in energy which would arise from the relaxation is small compared to phonon energies for the vibrational modes relevant to the relaxation, $\hbar\omega_{LO}$. A way of understanding this is to note that the final state in the absorption will be the relaxed state, with or without integral numbers of phonons. If the phonon energy is very large, we will not expect phonons and the relaxed state is appropriate, shifted by U divided by the static dielectric constant. If the phonon energy is small, we will expect the large number of phonons required to make up the difference between the

Table 5-4. Experimental optical dielectric constant (compiled by Harrison (1980), pp. 114, 115), static dielectric constant, obtained by adding $4\pi\delta\chi^{(1)}$ from Eq. (4-37), the polaron energy shift from Eq. (5-36), and the transverse optical phonon energy from Eq. (3-35).

	$d(\text{Å})$	$\varepsilon(optical)$	$\varepsilon_s(static)$	$\delta\varepsilon_{pol}(eV)$	$\hbar\omega_0(eV)$
C	1.54	5.70	5.70	0.00	0.239
Si	2.35	12.0	12.0	0.00	0.068
Ge	2.44	16.0	16.0	0.00	0.039
Sn	2.80	24.0	24.0	0.00	0.023
SiC	1.88	6.7	7.42	0.14	0.131
BN	1.57	4.50	4.81	0.16	0.212
BP	1.97	8.02*	8.02	0.00	0.126
BAs	2.07	9.49*	9.49	0.00	0.105
AlN	1.89	4.8	7.23	0.69	0.095
AlP	2.36	8.0	9.67	0.16	0.054
AlAs	2.43	9.1	10.6	0.12	0.044
AlSb	2.66	10.2	11.2	0.06	0.037
GaN	1.94	5.23*	7.92	0.64	0.079
GaP	2.36	9.1	10.6	0.12	0.045
GaAs	2.45	10.9	12.3	0.08	0.033
GaSb	2.65	14.4	15.2	0.03	0.027
InN	2.15	5.03*	10.3	0.98	0.052
InP	2.54	9.6	12.7	0.24	0.033
InAs	2.61	12.3	15.1	0.11	0.023
InSb	2.81	15.7	17.7	0.05	0.019
BeO	1.65	3.00	3.99	1.03	0.148
BeS	2.10	4.70*	5.67	0.36	0.087
BeSe	2.20	5.58*	6.49	0.24	0.075
BeTe	2.40	6.90*	7.59	0.12	0.065
MgTe	2.76	4.83*	8.92	0.72	0.022
ZnO	1.98	4.0	7.89	1.38	0.049
ZnS	2.34	5.2	8.25	0.67	0.031
ZnSe	2.45	5.9	8.98	0.49	0.022
ZnTe	2.64	7.3	10.0	0.29	0.019
CdS	2.53	5.2	10.2	0.78	0.021
CdSe	2.63	5.8	10.7	0.64	0.015
CdTe	2.81	7.2	11.4	0.38	0.012
HgTe	2.81	7.2*	11.4	0.38	0.011
CuF	1.84	2.24*	3.59	1.97	0.045
CuCl	2.34	5.7	7.55	0.37	0.022
CuBr	2.49	4.4	6.39	0.60	0.015
Cul	2.62	5.5	7.11	0.32	0.014
AgI	2.80	4.9	7.34	0.51	0.009

*Interpolated in Harrison (1980)

relaxed and unrelaxed state, so the unrelaxed state is appropriate, shifted by U divided by the optical dielectric constant.

The lowering in energy of an electron - or a hole - excitation due to the displacement of the atoms, or polarization of the lattice as in the static dielectric susceptibility, is called the *polaron* energy, and the moving carrier, with its accompanying polarization-displacements of the lattice, is called a polaron. A brief but clear analysis of polaron effects has been given by Callaway (1976). The polaron energy is negative and its magnitude is the difference between the energy estimated with the two different dielectric constants,

$$\delta\varepsilon_{pol} = \frac{U}{\varepsilon} - \frac{U}{\varepsilon_s} \qquad (5\text{-}36).$$

It is evaluated in Table 5-4, where we list the experimental optical dielectric constants (or estimates of them, since our direct predictions for the $\chi^{(1)}$ were considerably too small) and the static dielectric constants where we have added the displacive correction from Eq. (4-37), which were seen in Table 4-6 to give a good account of the difference between optical and static dielectric susceptibilities. We used the average U for the two constituents from Table 1-1 to obtain the polaron energy from Eq. (5-36).

These shifts $\delta\varepsilon_{pol}$ correspond to a reduction in the gap enhancement of Table 5-3, if they apply, and we confirm as indicated that they are ordinarily much smaller and the total gap enhancement as predicted in Part A is appropriate. $\delta\varepsilon_{pol}$ is of course zero for the homopolar semiconductors. For the polar semiconductors we may compare the shifts with the optical phonon energy, which we obtain from Eq. (3-35) and list as the final column in Table 5-4. We saw in Table 3-3 that this formula gave a good account of those frequencies.

For the most familiar semiconductors, such as GaAs and InSb, we see that the shifts are large compared to the phonon energies, which means that the gap observed in optical absorption will be that enhanced by U/ε and listed in Table 5-3, as if the atoms did not move when the carriers were produced. Vibrational energy will be left behind as the polarons subsequently form. In other systems, notably nitrides, which have high vibrational frequencies, the shifts are much smaller than the phonon energies and the enhancement of the gap will be reduced by lattice polarization. These reductions are mostly not so large. Then electrons are formed as polarons and there are no phonons generated in the transition.

The dynamics of these polarons is quite a separate question. (See Callaway (1976)). Already in the bands the electrons may move with very small effective masses, as small as $0.0134m$ for InSb as we shall see in Table

6-2, and this dramatically affects the polaron properties though it did not enter the shift itself given in Eq. (5-36).

C. Wannier Functions

Before leaving the accurate bands, which we have understood in terms of tight-binding representations, we should mention a rigorously accurate parallel formulation by Wannier (1937). Just as we formed approximate band states as sums over j of $e^{i\boldsymbol{k}\cdot\boldsymbol{r}_j}$ times bond orbitals, we could transform back as sums over $\boldsymbol{k}$ of $e^{i\boldsymbol{k}\cdot\boldsymbol{r}_j}$ times these Bloch states. This is simply a unitary transformation and its inverse based on these approximate states. It is also possible to make the same unitary transformation on the exact Bloch state, summed over $\boldsymbol{k}$ in the Brillouin Zone, to obtain localized orbitals $w(\boldsymbol{r} - \boldsymbol{r}_j)$, called Wannier functions, in terms of which the sum over j would yield the *exact* band states. This has proven a useful basis for making formal treatments analogous to our approximate tight-binding treatment, but has not been widely used for computation, partly because of limited knowledge of the exact Bloch states. Although it could be done in principle with one Wannier function for each band, it really only makes sense for semiconductors to construct four *bond* Wannier functions representing the four valence bands.

Such calculations of bond Wannier functions have recently been made by Marzari and Vanderbilt (1997), which may prove useful for real numerical calculations. Their paper also contains references to many other discussions of Wannier functions.

Problem 5-1. Energy bands in carbyne

Consider the carbon chain of Fig. 3-6, with ϕ exactly 90°. You can form bonds and antibonds with energies $\varepsilon_p \pm V_{pp\sigma}$. Now introduce $V_{pp\pi}$ which couples second-neighbor bonds (of parallel orientation, or if every bond is numbered, odd-numbered bonds are coupled to odd-numbered bonds). In the Bond-Orbital Approximation (neglecting coupling between bonds and antibonds) there are then two identical bonding bands, one for even-numbered and one for odd-numbered bonds. Obtain and sketch these bond bands, $-\pi/(d\sqrt{2}) < k < \pi/(d\sqrt{2})$. Similarly obtain and sketch the antibonding bands on the same plot. Use the matrix elements of Eq. (1-21) with $d = 1.30$ Å.

Recalculate the bands without the Bond-Orbital Approximation (now including coupling between bonds and antibonds, but again, only odd-numbered orbitals couple with odd-numbered orbitals.

Problem 5-2. σ-bands in graphite

Calculate the energy bands for graphite arising from the bond orbitals, in direct analogy with the calculation for the bands of tetrahedral semiconductors in Section 5-2-A. Use wavenumbers along a bond, in the z-direction in the figure for Problem 2-1. Then the states can be chosen to be either even or odd with respect to reflection in a z-axis through a bond. This allows the three equations, analogous to the four of Eq. (5-8), to be reduced to two and easily solved.

CHAPTER 6

Electronic Properties

In Chapter 5 we described the results of full band calculations for semiconductors and various levels of understanding them in terms of a tight-binding representation. In earlier chapters we had seen how to use bond orbitals to evaluate properties, such as the total energy, which depended upon integrals over the bands. The bands themselves are required for electronic properties, but in most cases only near the top of the occupied valence bands and the bottom of the mostly empty conduction bands. We proceed now to provide an understanding of the parameters for these states.

1. The Valence-Band Maximum.

Note that the zero of energy in the bands of Fig. 5-6 was taken at the valence-band maximum. In many band-calculational techniques the zero of energy is arbitrary; the states are expanded in a Fourier series, the Fourier components of the potential (or pseudopotential) are used, but the constant term is ill-defined. In such circumstances, the valence-band maximum is a good choice of zero. It is of similar form in all semiconductors, occurring at the center of the Brillouin Zone, Γ , and triply degenerate (except for spin-orbit splitting). The choice does not affect the dynamical properties of the electrons nor the predicted energies required to excite an electron from one state to another. On the other hand, it prevents one from guessing how the bands would line up if two crystals were brought together, or from estimating the energy required to remove an electron from the crystal.

In contrast, the tight-binding bands of Eqs. (5-18) through (5-23) come automatically on the same scale as the free-atom term values of Table 1-1. This will provide a starting point for suggesting natural band line-ups in heterojunctions between semiconductors in Section 19-4-B and predicting photoelectric thresholds for electron emission in Section 19-2-B, though we shall see that in both cases there are important corrections to be made. Having the valence-band maximum on the scale of atomic term values is also extremely valuable in discussing adsorption of atoms on semiconductor surfaces. It puts the semiconductors on an absolute electronegativity scale with atoms , molecules and with other solids.

Note that the Coulomb corrections which we discussed in Section 5-5-A do not come up for the discussion of the valence-band maximum. Our energies in some sense represent a removal energy for an electron and the corrections come, as we saw, when we add electrons as in a conduction band. All such corrections have to do with shifts of the conduction bands relative to this scale based upon the valence band.

A convenient way of formalizing this scale is to list the energy of the valence-band maximum E_v obtained as the lower $E(\Gamma_{15})$ of Eq. (5-19) ,

$$E_v = \frac{\varepsilon_{p+} + \varepsilon_{p-}}{2} - \sqrt{\left(\frac{\varepsilon_{p+} - \varepsilon_{p-}}{2}\right)^2 + \left(1.28\,\frac{\hbar^2}{md^2}\right)^2} \quad , \qquad (6\text{-}1)$$

where we have substituted formulae from Eq. (5-14) and universal values from Table 5-1 for the couplings $V_{pp\sigma}$ and $V_{pp\pi}$. The resulting values are given in Table 6-1. These will be useful in a number of our discussions. They nominally represent the energy of the highest occupied states relative to an electron at rest a large distance away (an electron at the *vacuum* level), but we shall see that the real removal energy for an electron (the photothreshold) is reduced by three or four volts due to image potentials near a semiconductor surface.

The analyses in Chapters 2 through 4 incorporated approximations that reduced these valence bands to a single bonding level and the conduction bands to a single antibonding level. When we treated metallization, we included coupling between bond orbitals and neighboring antibonding orbitals in such a way that it left the energy levels for all bonds identical, though shifted. The principal cause of the broadening into bands shown in Figure 5-6 is the coupling between neighboring bond orbitals for the valence bands and coupling between neighboring antibonding orbitals for

Table 6-1. Internuclear distance, d , energy of the valence-band maximum E_V (in eV) on the scale of the atomic term values of Table 1-1. Also the predicted and experimental optical threshold E_0 (in eV) and the experimental indirect gap E_g for the indirect-gap semiconductors. Experimental values taken from Harrison (1980), mostly from a compilation by Lawaetz (1971).

	d (Å)	E_V	E_0 (Eq. (6-5))	E_0 (Exp.)	E_g
C	1.54	-15.18	12.78	5.5	
Si	2.35	-9.35	1.85	4.18	1.13
Ge	2.44	-8.97	0.58	0.89	0.76
Sn	2.80-	-8.00	0.09		
SiC	1.88	-12.59	7.12	7.75	2.3
BN	1.57	-15.93	13.61		
BP	1.97	-11.56	5.98		
BAs	2.07	-11.00	4.59		
AlN	1.89	-14.67	9.89		
AlP	2.36	-10.22	3.65		
AlAs	2.43	-9.67	2.82	2.77	2.10*
AlSb	2.66	-8.76	1.68	2.5	1.87
GaN	1.94	-14.59	8.67		
GaP	2.36	-10.21	3.01	2.77	2.38
GaAs	2.45	-9.64	2.06	1.52	
GaSb	2.65	-8.76	1.13	0.81	
InN	2.15	-14.34	8.01		
InP	2.54	-10.03	3.06	1.37	
InAs	2.61	-9.48	2.31	0.42	
InSb	2.81	-8.61	1.41	0.24	
BeO	1.65	-17.79	12.32		
BeS	2.10	-12.35	4.54		
BeSe	2.20	-11.40	3.25		
BeTe	2.40	-10.19	2.01		
MgTe	2.76	-9.81	4.90		
ZnO	1.98	-17.19	12.79	3.4	
ZnS	2.34	-12.00	6.89	3.80	
ZnSe	2.45	-11.06	5.67	2.82	
ZnTe	2.64	-9.87	4.36	2.39	
CdS	2.53	-11.89	6.78	2.56	
CdSe	2.63	-10.97	5.68	1.84	
CdTe	2.81	-9.80	4.47	1.60	
CuF	1.84	-20.32	16.25		
CuCl	2.34	-14.05	8.61		
CuBr	2.49	-12.67	6.93		
CuI	2.62	-11.20	5.54		
AgI	2.80	-11.15	5.65		

* Kamoto, Wood, and Eastman (1981)

the conduction bands. We noted in Section 2-3-A that such a coupling among bond levels does not change the average energy (excluding the overlap interaction energy, which was treated separately). Thus, when we computed the total energy (for bonding properties, but also for dielectric properties since they can be formulated as field-dependent shifts in energy), it did not matter that we had neglected the coupling. Our success in those calculations was not because the true electron energies were single bond energies; Figure 5-6 shows that they form very broad bands. The success was because we could approximately calculate the average energies directly. We now turn to properties where the occupation of individual levels must be considered and this average-energy approach is not relevant.

2. Optical Absorption

Optical absorption through electronic transitions is ordinarily proportional to the intensity of the light and can be calculated just as well at arbitrarily low intensities where the time-dependent perturbation theory described in Section 1-5-B is well justified. In doing this, we neglect peculiarities that arise at extraordinarily high intensities. Our principle goal in this section is to calculate the matrix elements between electronic states which arise from the light wave. These can be used directly to calculate the absorption rates as described in Section 1-5-B.

A. Vertical Transitions

In a single atom, transitions can occur between occupied and empty electronic states, seen in Table 1-1 to be separated by energies of the order of electron volts. As in Section 4-3-G, we can consider light of frequency ω propagating in a z-direction and polarized in an y-direction. The corresponding field is written $\boldsymbol{E}\cos(\boldsymbol{q}\cdot\boldsymbol{r} - \omega t) = \boldsymbol{E}\,(e^{i\boldsymbol{q}\cdot\boldsymbol{r}} + e^{-i\boldsymbol{q}\cdot\boldsymbol{r}})/2$ with $\boldsymbol{q}$ in a z-direction and $\boldsymbol{E}$ a constant vector amplitude in the y-direction. The photons of such energy have very small wavenumbers, corresponding to wavelengths of thousands of Angstroms. Thus the matrix elements between atomic states for an atom at $\boldsymbol{r}_j$ can be calculated as if the field from the light were uniform over the atom, $\boldsymbol{E}\cos(\boldsymbol{q}\cdot\boldsymbol{r}_j - \omega t)$. Similarly the matrix element between a bond and an antibond in the same site at $\boldsymbol{r}_j$ can be calculated as if the field were uniform. However, if we construct a *band* state, $\sqrt{1/N_b}\,\Sigma_j\, e^{i\boldsymbol{k}\cdot\boldsymbol{r}_j}|b_j\rangle$, with N_b the number of bonds, the matrix element between it and an antibonding state of wavenumber $\boldsymbol{k}' = \boldsymbol{k} + \boldsymbol{q}$ or wavenumber $\boldsymbol{k}' = \boldsymbol{k} - \boldsymbol{q}$ will $1/N_b\Sigma_j 1 = 1$ times a bond-antibond matrix element for a field $\boldsymbol{E}/2$.

We only have matrix elements between electronic states which differ in wavenumber by the tiny $\boldsymbol{q}$ of the light (conserving momentum). On the scale of the bands shown in Fig. 5-6, these are between states of essentially the same wavenumber, or *vertical* transitions in the band diagram, with energy differences equal to the photon energy. $\boldsymbol{k}$ and $\boldsymbol{k} \pm \boldsymbol{q}$ could also differ by a lattice wavenumber, but such states are replotted in the Brillouin Zone so they are of the *same* wavenumber, and so transitions are still considered vertical.

B. The Optical Matrix Elements

For a more general representation of the band states the matrix element between two band states depends on the constitution of those band states (in terms of atomic orbitals) and on the coupling between individual atomic states. This is customarily calculated (Harrison (1980), p. 98) by introducing the field through a vector potential $\boldsymbol{A}(\boldsymbol{r},t)$, with the electric field given by $\boldsymbol{E}(\boldsymbol{r},t) = -(1/c)\, \partial\boldsymbol{A}(\boldsymbol{r},t)/\partial t$ as we introduced the magnetic field in Section 4-2-D. In analogy with Eq. (4-25) the terms in the Hamiltonian due to the vector potential are $e\boldsymbol{A}(\boldsymbol{r},t)\cdot\boldsymbol{p}/(mc) + (e^2\boldsymbol{A}(\boldsymbol{r},t)\cdot\boldsymbol{A}(\boldsymbol{r},t)/(2mc^2))$. The first term is responsible for absorption and emission of light, the second only for scattering of light (with or without loss of energy) . If the electric field is given by $\boldsymbol{E}\cos(\boldsymbol{q}\cdot\boldsymbol{r} - \omega t)$, the vector potential is $c\boldsymbol{E}/\omega \sin(\boldsymbol{q}\cdot\boldsymbol{r} - \omega t)$ and the coupling between the light and the electron becomes

$$H_{le} = \frac{e\boldsymbol{E}\cdot\boldsymbol{p}}{m\omega} \sin(\boldsymbol{q}\cdot\boldsymbol{r} - \omega t) . \qquad (6\text{-}2)$$

It is 90° out of phase with the electric field. This is more appropriate than a form $H_{le} = e\boldsymbol{E}\cdot\boldsymbol{r}\cos(\boldsymbol{q}\cdot\boldsymbol{r} - \omega t)$ for solids where $\boldsymbol{r}$ extends over large distances and therefore the leading factor $\boldsymbol{E}\cdot\boldsymbol{r}$ becomes very large. $\boldsymbol{E}\cdot\boldsymbol{r}$ is also a form which is inconsistent with the periodic boundary conditions which we ordinarily use in treating solids. For small systems the two forms must be equivalent and we shall see that this is the case when we obtain matrix elements of H_{le} for atomic states in Eq. (6-4).

We are writing our band states as linear combinations of atomic orbitals, so the matrix element between two band states becomes the sum of such matrix elements between atomic states, both between states on the same atom and states on different atoms. In semiconductors it turns out that the *interatomic* terms are the largest by far. This may be seen from the study of susceptibilities (Chapter 4) which could also be calculated in perturbation theory using Eq. (6-2) and the results are the same if we include only coupling between states on different atoms. The intraatomic matrix elements give very small susceptibilities, and correspondingly dielectric constants near one for inert-gas solids (Section 20-2) and ionic solids (Chapter 11) where interatomic couplings are small. Their contribution will be small also in semiconductors. The semiconductor dielectric constants are very large, as we saw in Chapter 4. This comes from the interatomic coupling, and the associated highly polarizable bonds. Thus for semiconductors we include only these interatomic terms and we shall see that it is possible to obtain the matrix elements in terms of our universal interatomic couplings between neighboring atomic states. We shall also use the result to see that the neglect of intraatomic terms is related to our taking the hybrids to be centered at the atoms in treating the susceptibility in Chapter 4.

We begin with a rigorous relation between matrix elements of the momentum and the position operators between two eigenstates, which we call $|b\rangle$ and $|a\rangle$ (Schiff (1967), p. 404). The essential idea is two evaluations of the "commutator", $C = \langle a|H\boldsymbol{r} - \boldsymbol{r}H|b\rangle$, first using the Hamiltonian of Eq. (1-7). We note that only the ∇^2 in the Hamiltonian gives extra terms, $\nabla^2 \boldsymbol{r} - \boldsymbol{r}\nabla^2 = 2\nabla$, giving $C = \hbar\langle a|\boldsymbol{p}|b\rangle/(im)$. For the second evaluation we use the definition of energy eigenstates, Eq. (1-9), $H|b\rangle = \varepsilon_b|b\rangle$ and $\langle a|H = \varepsilon_a\langle a|$ to obtain $C = (\varepsilon_a - \varepsilon_b)\langle a|\boldsymbol{r}|b\rangle$. The combination gives $\hbar\langle a|\boldsymbol{p}|b\rangle = im\,(\varepsilon_a - \varepsilon_b)\langle a|\boldsymbol{r}|b\rangle$. We then apply this rigorous result to our approximate description of tight-binding states, taking the atomic state $|i\rangle$ at a nucleus a distance $\boldsymbol{d}$ from an atomic state $|j\rangle$. We write $V_3 = (\varepsilon_i - \varepsilon_j)/2$ and the coupling between them $V_2 = \langle i|H|j\rangle$. From these two states, we construct a bonding eigenstate $|b\rangle = \sqrt{(1 - \alpha_p)/2}\,|i\rangle + \sqrt{(1 + \alpha_p)/2}\,|j\rangle$, with α_p of course given by Eq. (2-12), and an antibonding state $|a\rangle = \sqrt{(1 + \alpha_p)/2}\,|i\rangle - \sqrt{(1 - \alpha_p)/2}\,|j\rangle$. Then we may substitute these into the rigorous relation given above, taking of course $\langle j|\boldsymbol{p}|i\rangle = -\langle i|\boldsymbol{p}|j\rangle$ and $\langle i|\boldsymbol{p}|i\rangle = \langle j|\boldsymbol{p}|j\rangle = 0$ and $\langle i|\,\boldsymbol{r}\,|j\rangle = \langle j|\,\boldsymbol{r}\,|i\rangle = 0$. We obtain $\hbar\langle i|\boldsymbol{p}|j\rangle = im\,\alpha_c\,(\varepsilon_a - \varepsilon_b)\,\boldsymbol{d}/2$, with $\alpha_c = \sqrt{1 - \alpha_p^2} = 2\langle i|H|j\rangle/(\varepsilon_a - \varepsilon_b)$, giving a direct relation between the matrix elements, $\hbar\langle i|\boldsymbol{p}|j\rangle = im\langle i|H|j\rangle\,\boldsymbol{d}$. Writing the momentum operator as $\boldsymbol{p} = (\hbar/i)\nabla$, we write this for atomic states of specific angular-momentum quantum numbers in order to relate the matrix element of the gradient to our universal parameters,

$$\langle l,m\,|\,\nabla\,|\,l',m\rangle = -\frac{m\boldsymbol{d}V_{ll'm}}{\hbar^2} = -\eta_{ll'm}\frac{\boldsymbol{d}}{d^2}\,. \qquad (6\text{-}3)$$

Note that the sign is as expected, as can be seen by imagining coupling between identical wavefunctions representing two neighboring s-states, with $\boldsymbol{d}$ the vector distance from the first (l) to the second (l'), with $\eta_{ss\sigma}$

negative. This form will be useful in Section 6-3 for obtaining effective masses.

This important connection between the tight-binding matrix elements and the optical matrix elements was apparently first made by Dresselhaus and Dresselhaus (1967), and then applied to universal parameters by Harrison (1980), p. 116. More recently Yew Yan Voon and Ram-Mohan (1993) have made a more complete study of such optical matrix elements in the context of tight-binding theory.

We may also use the first form $\hbar\langle i|\boldsymbol{p}|j\rangle = im\,\langle i|H|j\rangle\,\boldsymbol{d}$ to write the matrix elements of the electron-light interaction from Eq. (6-2) in more convenient form. We obtain $\langle i|H_{le}|j\rangle$ using this matrix element for $\langle i|\boldsymbol{p}|j\rangle$ and writing out $sin(\boldsymbol{q}\cdot\boldsymbol{r}_j - \omega t)$ as a sum of complex exponentials. We note further that the exponential $e^{-i(\boldsymbol{q}\cdot\boldsymbol{r}_j - \omega t)}$ will only enter when the state $\varepsilon_i < \varepsilon_j$, the opposite of what we assumed above and the ω in the denominator should be changed in sign, leading to

$$\langle j|H_{le}|k\rangle = -\frac{eV_{jk}\,\boldsymbol{E}\cdot\boldsymbol{d}}{\sqrt{(\varepsilon_k - \varepsilon_j)^2/4 + V_{jk}^2}}\,cos(\boldsymbol{q}\cdot\boldsymbol{r}_j - \omega t). \tag{6-4}$$

For two s-orbitals of the same energy $(V_{jk} < 0)$ the coupling is simply $e\boldsymbol{E}\cdot\boldsymbol{d}cos(\boldsymbol{q}\cdot\boldsymbol{r}_j - \omega t)$ as we would expect, and for states of different energy we obtain a factor of a covalency based upon the parameters for the two coupled states. This may be combined with Eq. (6-3) to obtain the needed optical matrix elements.

The rigorous quantum-mechanical relation, $\hbar\langle a|\boldsymbol{p}|b\rangle = im\,(\varepsilon_a - \varepsilon_b)\langle a|\boldsymbol{r}|b\rangle$, upon which these approximate relations are based, can also be applied to intraatomic matrix elements to learn something about hybrids. We note that $\langle s|\boldsymbol{p}|p\rangle$ is proportional to $\langle s|\boldsymbol{r}|p\rangle$. The center of gravity of an sp-hybrid is $\langle h|\boldsymbol{r}|h\rangle = {}^1/_2\,(\langle s|\boldsymbol{r}|s\rangle + 2\langle s|\boldsymbol{r}|p\rangle + \langle p|\boldsymbol{r}|p\rangle)$ so taking a hybrid to be centered at the atom, as we did in calculating susceptibilities in Chapter 2, corresponds to assuming $\langle s|\boldsymbol{r}|p\rangle$ and $\langle s|\boldsymbol{p}|p\rangle$ are zero. Other intraatomic matrix elements, $\langle s|\boldsymbol{p}|s\rangle$, $\langle p_i|\boldsymbol{p}|p_j\rangle$, are zero by symmetry. Thus our neglect of intraatomic terms is completely consistent with the bond-orbital approach we used in Chapter 2.

The actual calculation of the optical absorption spectrum in terms of these matrix elements is straightforward and does not require the introduction of any further parameters but it is complicated enough, because it requires summations over the band states of the solid, to require significant computer calculations (Ren and Harrison (1981)). However, the energies of various features of the absorption spectrum are perhaps the quantities of most interest and we can directly predict these using our tight-binding formulation of the energy bands.

Except for special points of high symmetry, each occupied state in Figure 5-6 is coupled by light to each empty state at the same wavenumber. An exception, for example, is that there is no coupling between Γ_1 and $\Gamma_{2'}$ in silicon; however, when the wavenumber differs from that point, there is coupling from fields polarized along $\boldsymbol{k}$. One might have carelessly guessed this on the grounds that both states consist entirely of s-states and there is no absorption between s-states on the free atom. However, one cannot make the argument on the basis of atomic states alone. One must utilize the full symmetry of the state. After all, there is absorption from the bonding to the antibonding state of the Li_2 molecule discussed in Section 1-2, though both may be regarded as consisting entirely of s-states. Similarly, there is optical coupling between different p-like bands which does not depend upon the presence of any s-state contribution.

When the light frequency is equal to the energy difference between those two states, divided by $\hbar$, absorption will occur, just as there is absorption when the optical frequency equals the vibrational mode frequencies as described in Section 4-3-G. It is convenient to represent the absorption by an imaginary term in the susceptibility. Note that the susceptibility relates the polarization density to the electric field, $\boldsymbol{P} = \chi \boldsymbol{E}$, so if the field varies as $\boldsymbol{E}_0\, e^{-i\omega t}$ (or the real part of that expression) the current density $\boldsymbol{j} = \partial \boldsymbol{P}/\partial t = -i\omega\chi \boldsymbol{E}_0\, e^{-i\omega t}$ is 90° out of phase with the field and no work is done by the field. If χ has an imaginary part there will be absorption of energy from the light field in proportion to that part. This $Im\chi$ gives a dimensionless representation of the absorption. It can be directly calculated in terms of the energy bands and the coupling (Harrison (1980), p. 98; Ren and Harrison (1981)). It can also be determined experimentally from the optical reflectivity. Such an experimental curve for germanium is given in Fig. 6-1. The real part of χ goes through zero near the peak, much as in Fig. 4-7.

C. The Absorption Threshold

This experimental spectrum can be directly related to the energy bands for germanium given also in Fig 6-1 without spin-orbit coupling. We note first that the highest occupied state in the inset (and in fact also for all wavenumbers) is $\Gamma_{25'}$ and the lowest-energy empty state of the same wavenumber is $\Gamma_{2'}$, slightly less than one volt above it. This is called the *direct optical threshold* E_0 and it is indicated in Figure 6-1; it is the lowest-energy for direct optical absorption (vertical transitions in the figure). Experimental values for a number of systems were listed in Table 6-1.

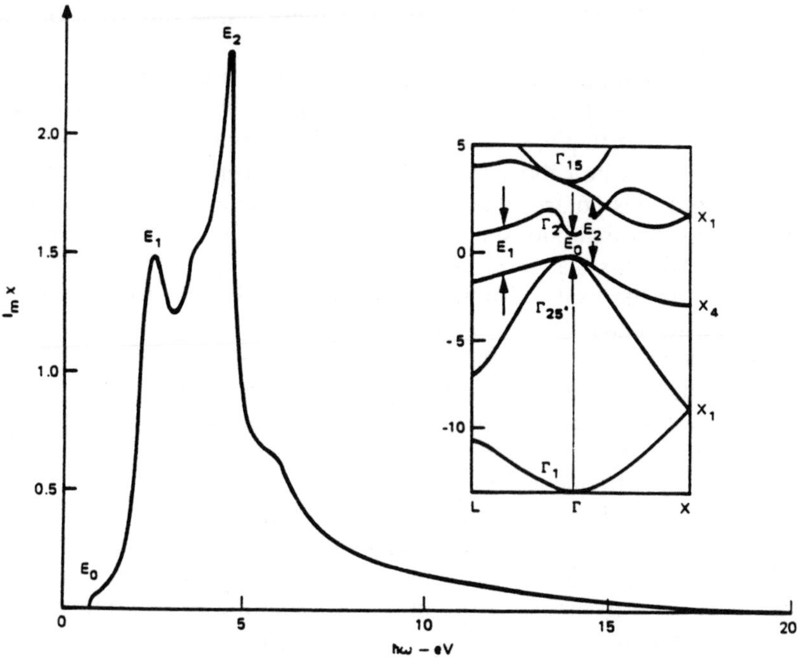

Fig. 6-1. The imaginary part of the optical susceptibility, representing the experimental absorption spectrum for light in germanium, from Phillip and Ehrenreich (1963). The energies of the three principal features are indicated and can be identified with features of the energy bands, as shown in the inset.

We can independently estimate E_0 , from our own calculations by subtracting the lower-energy state of Eq. (5-19) (with the minus sign) from the higher energy state of Eq. (5-18) (with the plus sign) .

$$E_0 = \frac{\varepsilon_{s+} + \varepsilon_{s-}}{2} - \frac{\varepsilon_{p+} + \varepsilon_{p-}}{2} + \sqrt{\left(\frac{\varepsilon_{s+} - \varepsilon_{s-}}{2}\right)^2 + (4V_{ss\sigma})^2}$$

$$- \sqrt{\left(\frac{\varepsilon_{p+} - \varepsilon_{p-}}{2}\right)^2 + \left(\frac{4}{3}V_{pp\sigma} + \frac{8}{3}V_{pp\pi}\right)^2} , \qquad (6\text{-}5)$$

The corresponding theoretical values are listed in Table 6-1. These thresholds are the differences in large quantities and the error is appreciable, particularly for the more polar semiconductors; however, the trends are right and the values close enough to be of some use. We may note, as indicated in the preceding chapter, that the errors are large compared to the Coulomb corrections and we should regard them as the result of the limited-basis tight-binding theory not giving good descriptions of the conduction bands. In particular, we should not make Coulomb corrections to our predictions since our predictions were based on parameters which ultimately were fit to the observed optical gaps. This does not preclude our using the predicted Coulomb corrections with the *experimental* E_0 to obtain the unenhanced gap for use in formulas for which the unenhanced gap is appropriate.

Eq. (6-5) is quite simple for the case of homopolar semiconductors,

$$E_0 = \varepsilon_s - \varepsilon_p - 4V_{ss\sigma} + {}^4\!/_3\, V_{pp\sigma} + {}^8\!/_3\, V_{pp\pi}$$
$$= -4V_1 + 6.56\hbar^2/(md^2) \approx 2V_2\,(1 - \alpha_m) \quad . \qquad (6\text{-}6)$$

The final form was obtained noting that $V_2 = 3.22\hbar^2/md^2$ and is only approximate. The same formula for the gap was obtained as the exact result of the simpler analysis of the bands in terms of bond orbitals. [Obtained at the end of Section 5-2-A.] It seems remarkable that the results are so close since Eq. (6-6) has nothing to do with hybrids, and in fact represents coupling between purely p-like and between purely s-like states. As we noted, the dependence on $(1 - \alpha_m)$ is indicative of the decrease in photothreshold with increasing metallicity, reaching zero at tin.

We may observe in Figure 5-6 that the conduction-band energy Γ_{15} in silicon lies below the $\Gamma_{2'}$ energy [that is, the upper energy Γ_1 in the notation of Eq. (5-18)]; thus, the real optical threshold arises from a different set of states in silicon from that in germanium, but has very similar values to Eq. (6-5). This inversion occurs also in diamond. This particular inversion was not reproduced by our band calculation, as we indicated in Section 5-3-D.

As we go to polar semiconductors, Eq. (6-5), the leading terms giving the average s- and p-state energies do not change appreciably (we noted in Section 1-1 that the atomic energies vary approximately linearly with column number in the periodic table), nor do the coupling terms under the square roots (since the bond lengths remain approximately constant) but the terms in the difference between s-state energies and between p-state energies do grow directly with polarity. Thus the direct optical threshold increases with polarity. This is apparent also in the bands shown in Fig. 5-6.

D. Indirect Gaps

We note another aspect of the bands of silicon and germanium, as shown in Fig. 5-6 and in Fig. 6-1. Although the highest energy in the valence band comes at Γ ; that is, $\boldsymbol{k} = 0$, the lowest-lying conduction-band lies lower elsewhere, near X in silicon and at L in germanium. Aluminum arsenide is another important semiconductor where this occurs. These are called indirect-gap semiconductors, and the gaps, equal to the difference between Γ_{15} and the lowest conduction-band state, for such systems are also indicated in Table 6-1. (The conduction-band minimum also occurs away from Γ in tin, and is in fact lower in energy than the valence-band maximum.) When electrons are present in the conduction band, from thermal excitation or from doping, they will predominantly lie near these conduction-band minima (within energy of order k_BT of the minimum). Thus concerning properties such as transport this is the true gap. Because of the restriction to vertical transitions for optical absorption, we do not expect absorption at the indirect-gap energy and indeed none was indicated in the insert in Fig. 6-1. Neither can systems which have an indirect gap be good photoemitters, semiconductor light sources, nor lasers. One may introduce a population inversion by injecting many electrons and many holes but we do not expect them to readily recombine producing photons because of this restriction. Their recombination is accomplished by more complex, and radiationless, processes. (See Section 8-3-C.) A very important feature of gallium arsenide is that it is a *direct-gap* semiconductors, as can be seen from Fig. 5-6, and it is an excellent choice for construction of semiconductor lasers.

There can in fact be some absorption, or emission, at the indirect gap if something in the crystal makes up for the needed momentum change, $\hbar$ times the difference in wavenumber. A phonon of that wavenumber can provide it. Even with no phonons present, a photon can be absorbed and a phonon emitted as the electron is taken to the indirect minimum. The argument we gave for vertical transitions in Section 6-2-A breaks down since the Hamiltonian contains a term varying as $e^{i\boldsymbol{q}'\cdot\boldsymbol{r}_j}$, with $\boldsymbol{q}'$ the phonon wavenumber, so the transitions can occur with a large change in wavenumber, analogous to the small change in wavenumber we indicated there for the photon involved. Such processes with spontaneous emission of phonons are of low probability, but as the temperature increases the stimulated emission and the absorption of phonons becomes more likely and a weak absorption below the direct threshold is observed.

It is also possible for transitions to occur where other features, such as impurities or defects, take up the momentum. Such defects could provide mechanisms for radiative transitions but optical transitions arising in this way seem not to be of importance. One exception is in "porous silicon" which is a strong photoemitter and this is attributed to the complex inner surfaces. A simple planar surface does not ordinarily lead to optical transitions. For example, a (111) surface of a germanium crystal will reflect the electrons and produce states proportional to sin(**k**·**r**), coupling a wavenumber and a value reflected through the (111) plane, but this does not couple a state at Γ to a state at L.

Another important feature of such indirect-gap semiconductors is that the variation of the energy with wavenumber is very much different with change in wavenumber parallel to the position of the minimum than that perpendicular to it. We shall return to these effects in Section 6-3-D.

E. The Optical Absorption Peaks

The large peak in absorption at energy E_2 in Figure 6-1 arises from the nearly-parallel bands through $\Gamma_{25'}$ and Γ_{15} as indicated in the inset of Fig. 6-1. It can be estimated from the difference in those two energies obtained from Eq. (5-19). For homopolar semiconductors it is

$$E_2 = |\, {}^{8}/_{3}\, V_{pp\sigma} + {}^{16}/_{3}\, V_{pp\pi} | = 2.56\hbar^2/md^2 \quad . \tag{6-7}$$

For polar semiconductors, E_2 is the square root of the sum of this value squared and $(\varepsilon_{p+} - \varepsilon_{p-})^2$, corresponding to twice the square root from $E(\Gamma_{15})$ in Eq. (5-19); this has a dependence on polarity similar to that of the bonding-antibonding splitting. The value from Eq. (6-7) for germanium is 3.28eV. There is no coupling between these two bands exactly at Γ, but the states *near* Γ contribute a part of the absorption peak indicated by E_2, and there are many other contributions of similar energy and the maximum absorption is near 4.2 eV. This transition is dominated by coupling between p-like states in the conduction and valence bands.

The smaller peak indicated by E_1 has large contributions from the nearly parallel bands between Γ and L indicated in the inset of Fig. 6-1. Ren and Harrison (1981) noted that these are between largely p-like states in the valence bands and s-like states in the conduction bands. For this reason they move further below the E_2 peak at greater metallicities where the sp-splitting is more significant in comparison to the interatomic matrix elements. In silicon, the E_1 and E_2 peaks merge to a very narrow absorption peak.

Our treatment of the electronic structure in Chapter 2, 3, and 4 as a bonding and antibonding level separated by $2V_2$ would replace this absorption peak in silicon by a single line at 8.24 eV. This is nearly twice E_2 but represents some average of the band energy whereas the optical couplings heavily weights the p-like states separated by approximately the E_2 of Eq. (6-7).

3. Effective Masses

We have noted that in electrical conduction and other transport properties, the band states of minimum energy in the conduction bands and of maximum energy in the valence bands are of most importance. Conduction is obtained only by adding electrons (thermally or by doping with impurities), which go to the bottom of the conduction band, or by removing electrons, leaving empty states (or holes), which rise to the top of the valence band. In indirect-gap semiconductors, it is the actual conduction-band minimum - not the local minimum at Γ - which is of interest. For most transport properties, the energy bands away from these band extrema are quite irrelevant. Near a band extremum at $\boldsymbol{k}_0$ the energy varies in proportion to the square of $\boldsymbol{k} - \boldsymbol{k}_0$ and all we need are the coefficients that give the curvature of the band for different directions of $\boldsymbol{k} - \boldsymbol{k}_0$.

There is a very useful and simple method for calculating the curvature of a band, called the $\boldsymbol{k}\cdot\boldsymbol{p}$ method ("kay-dot-pea"). In particular it provides a convenient basis for discussing trends in effective masses, and it will provide us a check on the optical matrix elements which we derived in Section 6-2-B. The $\boldsymbol{k}\cdot\boldsymbol{p}$ method is based on perturbation theory, treating the deviation of the wavenumber from $\boldsymbol{k}_0$ as a perturbation. We write the result for an extremum at $\boldsymbol{k}_0 = 0$, the point Γ (Harrison (1970) p. 141)

$$\varepsilon_k = \varepsilon_0 + \frac{\hbar^2 k^2}{2m} + \sum_n \frac{\hbar^2}{m^2} \frac{|\boldsymbol{k}\cdot <n|\boldsymbol{p}|0>|^2}{\varepsilon_0 - \varepsilon_n} . \qquad (6\text{-}8)$$

where the sum is over all of the bands at $\boldsymbol{k} = 0$; these bands are numbered by n. $<n|\boldsymbol{p}|0>$ is the matrix element of the momentum operator between the two band states. These are also called optical matrix elements since they enter directly the formulae for optical absorption, as discussed in Section 6-6. When these states are written in tight-binding theory it is a sum of matrix elements between atomic states. We saw in Section 6-2-B that only *inter*-atomic couplings are important and from Eq. (6-4) that they are given in terms of universal parameters by $-\eta_{ll'm}\hbar\boldsymbol{d}/d^2$.

In Eq. (6-8) the energy differences which appear in the denominator should represent, in our opinion, the differences in energy unenhanced by the Coulomb effects discussed in Section 5-5-A. (There are widely held opinions to the contrary.) As in the polarization of a bond by an electric field, it corresponds to admixture of conduction-bands states at the same bond site, rather than the carrying of an electron to a distant bond site. We shall make note of that in interpreting the results of our estimates.

A. Conduction-Band Mass at Γ

The simplest case is the conduction band edge for direct-gap semiconductors, where the $\mathbf{k}_0 = 0$ is appropriate, or for the secondary minimum at Γ in the homopolar semiconductors.. Then for the zincblende structure, with cubic symmetry, the same dependence must obtain in the x- y-, and z-directions and the energy can be written $\varepsilon_k = \varepsilon_0 + \hbar^2k^2/(2m_c)$ with a unique conduction band *effective mass* m_c. From Eq. (6-8) we obtain

$$\frac{m}{m_c} = 1 + \frac{2}{m}\sum_n \frac{|\langle n|p_x|0\rangle|^2}{\varepsilon_0 - \varepsilon_n} . \qquad (6\text{-}9)$$

The dominant term comes from the valence-band maximum at energy E_0 below the conduction-band minimum. It was designated by Γ_7 and Γ_8 for the compound semiconductors in Fig. 5-6 because spin-orbit coupling was included in the calculation. Without spin-orbit it would be Γ_{25}' for homopolar semiconductors and Γ_{15} for polar semiconductors. Here it will be best to use the notation "lower Γ_{15}" for the valence band and "upper Γ_{15}" for the p-like conduction bands at $\mathbf{k} = 0$. We shall see that the lower Γ_{15} band gives the *only* contributing term among the bands we include in homopolar crystals.

The conduction-band minimum consists purely of atomic s-states and the valence-band maximum purely of p-states (we consider first those which are oriented along the y-axis, making an angle, the cosine of which is $\pm 1/\sqrt{3}$, with the internuclear separation for each of its four neighbors). Then for a homopolar semiconductor, the conduction-band state is a sum of $\pm 1/\sqrt{2N_c}$ times an s-state, with the sign opposite for the two atoms in each cell. Each of the $2N_c$ s-states (One is illustrated in Fig. 6-2) is coupled to four neighboring p-states, all of which are on the same sublattice and therefore have the same orientation of p-state (for $\mathbf{k} = 0$) . Because these p-states have the same orientation the sp-coupling to the two neighbors in the positive y-direction is of opposite sign to those lying in the negative y-direction, which was why there was no net coupling at Γ in the band calculation, and no contribution of $V_{sp\sigma}$ to the bands at that point. However, the matrix elements of p_y which enter here also contain a factor of $\mathbf{d}$ to these atoms (from $\eta_{ll'm}\hbar\mathbf{d} / d^2$), which also has opposite sign for atoms in the positive and negative y-direction, so the four matrix elements of

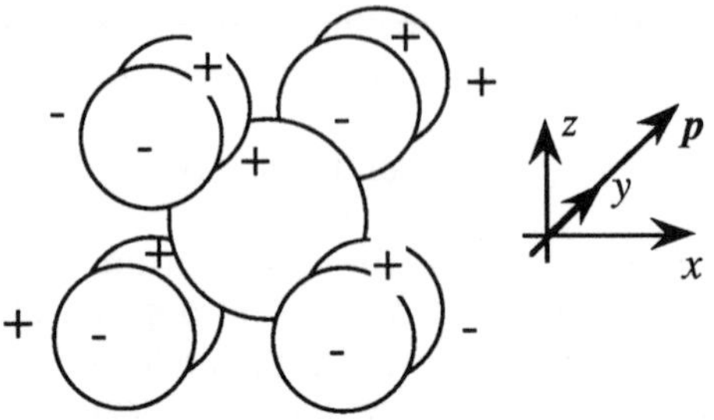

Fig. 6-2. A diagram to help obtain optical matrix elements, matrix elements of the momentum operator. An s-state at the center is coupled to four neighboring p-states, here taken as p_y-states. The sign of each coupling depends on whether the s-state overlaps a negative (indicated by the sign on the corresponding circle) p-state lobe ($V_{sp\sigma}/\sqrt{3}$) or a positive p-state lobe ($-V_{sp\sigma}/\sqrt{3}$). That sign is indicated to the right or left of each p-state. For a field in the y-direction as shown the factor $\boldsymbol{d}\cdot\boldsymbol{p}$ also is different for each p-state and is indicated to the left or right of each p-state. The coupling times $\boldsymbol{d}\cdot\boldsymbol{p}$ is negative for all four p-states giving a net nonvanishing matrix element.

the p_y operator add. (Again, see Fig. 6-2.) Furthermore, because the p-states are oppositely oriented on the two sublattices (putting them in a bonding relationship to each other in the valence band, the lower Γ_{15} band), and the s-states also appear with opposite sign on the two sublattices (putting them in an antibonding relationship to each other in the conduction band, the upper Γ_{15}), the couplings from the s-states on two sublattices add. We obtain

$$<n/\boldsymbol{p}/0> \;=\; 4\,\frac{\hbar}{i}\,\frac{\eta_{sp\sigma}}{\sqrt{3}}\,\frac{d}{\sqrt{3}d^2} \;=\; -\frac{4i\,\hbar\eta_{sp\sigma}}{3d}\,, \tag{6-10}$$

with $/0>$ corresponding to the upper Γ_1 state (conduction band) and $/n>$ representing the lower Γ_{15} state (valence band) .

If we follow the same path in constructing matrix elements of p_x between this Γ_1 state and the Γ_{15} state *both in the conduction band*, the argument remains the same up until this last point, where the contribution from the s-states on the two sublattices cancels. Further, had we selected p-states oriented along the x-axis, the net matrix elements of p_y would have canceled. The matrix element given in Eq. (6-10) is the only one contributing to the conduction-band effective mass at Γ in homopolar semiconductors.

For a polar semiconductor, there are factors containing s-like polarities and p-like polarities (each based on half the energy difference and the coupling between them) and so there is no longer exact cancellation for the matrix element of p_x between the Γ_1 state and the upper Γ_{15} state. In fact this difference in symmetry is related to the difference in symmetry symbols, and the fact that both upper and lower p-like states are called Γ_{15} in polar semiconductors is related to the fact that both are coupled by matrix elements of $\boldsymbol{p}$. It is not difficult to include the correction, $1/2\sqrt{(1+\alpha_p(s))(1-\alpha_p(p))} \pm 1/2\sqrt{(1+\alpha_p(p))(1-\alpha_p(s))}$, to Eq. (6-10) with the plus obtaining for coupling to the lower p-like state and the minus for the upper. However, the effects are not so large and we use Eq. (6-10) also for polar semiconductors. The reciprocal of Eq. (6-9) becomes

$$\frac{m_c}{m} = \frac{1}{1 + \dfrac{32\eta_{sp\sigma}^2 \hbar^2}{9md^2E_0}} = \frac{1}{1 + \dfrac{2.23\,V_2}{E_0}} \quad . \tag{6-11}$$

The last form is convenient because of our frequent use of V_2. We could use the last form in Eq. (6-6) for E_0 to write this $m/m_c = 1 + 1.11/(1-\alpha_m)$ for homopolar systems. However, because of the inaccuracy of Eq. (6-6) in predicting the direct gap, it seemed more informative to evaluate Eq. (6-11) using the experimental values of E_0 from Table 6-1. Such values for a series of semiconductors are compared with experiment in Table 6-2. The experimental trends are quite well given, the decrease of m_c with metallicity, and the decrease with increasing polarity, both arise from the corresponding trends in E_0. The predicted magnitudes are typically a factor of two too large.

We indicated that really the unenhanced E_0 should be used in such evaluations so it is of interest to see whether the agreement is improved when that is done. This is accomplished simply by subtracting the Coulomb enhancement from Table 5-3 (or calculating the corresponding corrections which were not included) from the experimental E_0 of Table 6-1 before substituting in Eq. (6-11). The effects are sizable and in general the agreement with experiment is significantly improved. This supports our view that the Coulomb enhancements should be removed from the experimental E_0 in predicting conduction-band energies, but it is not terribly convincing since simply dividing the estimates from the first column in Table 6-2 by a factor of two would do as well; it is a test on the scale of one number, and the fact that it fits many semiconductors is not compelling. We should mention in passing that in two of the estimates, InAs and InSb, the Coulomb enhancement exceeded the observed gap, suggesting that the unenhanced gap is negative. We shall discuss this aspect in Section 6-4

Table 6-2. Effective masses at Γ for systems for which experimental values of E_0 are given in Table 6-1, compared with experiment. Experimental m_c were compiled by Harrison (1980), experimental m_{hh} compiled by Lawaetz (1971). The experimental values (Dresselhaus, Kip, and Kittel (1953)) for m_c/m listed for silicon and germanium are transverse masses at the conduction-band minimum, discussed in Section 6-3-D.

Material	m_c/m from Eq. (6-11)			m_{hh}/m	
	Using E_0	Using E_0-U/ε	Exper.	Eq. (6-17)	Exper.
Si	0.30	0.26	0.19(X)	0.30	0.53
Ge	0.09	0.04	0.082(L)	0.30	0.35
Sn	0.33	0.29		0.30	0.29
AlAs	0.23	0.17		0.42	
AlSb	0.24	0.11		0.40	
GaP	0.22	0.17		0.45	
GaAs	0.14	0.084	0.066	0.42	0.62
GaSb	0.09	0.040	0.045	0.40	0.49
InP	0.14	0.067	0.077	0.51	0.85
InAs	0.05	0.020	0.024	0.48	0.60
InSb	0.03	0.026	0.0137	0.45	0.47
ZnO	0.28	0.04	0.19	0.82	
ZnS	0.24	0.18	0.28	0.71	1.76
ZnSe	0.24	0.13	0.17	0.68	1.44
ZnTe	0.23	0.14	0.16	0.66	1.27
CdS	0.23	0.10	0.20	0.81	
CdSe	0.19	0.06	0.13	0.77	
CdTe	0.19	0.08	0.096	0.74	1.38

where we can see that we should simply take the absolute value for the calculated gap in estimating the effective mass. It is related to the fact that the light-hole mass and the electron mass are approximately equal as we shall see.

It is also quite interesting to compare this $\boldsymbol{k}\cdot\boldsymbol{p}$ result, Eq. (6-11), with what we would obtain for the mass directly from the tight-binding bands as given in Eq. (5-12). The appropriate choice of signs is - - - . We may expand all terms to second order in k to obtain

$$\Delta_1 = \varepsilon_s - 4V_{ss\sigma} + \left(- \frac{4\eta_{ss\sigma}}{3} + \frac{32\eta_{sp\sigma}^2 \hbar^2}{9md^2E_0} \right) \frac{\hbar^2k^2}{2m} + \ldots \qquad (6\text{-}12)$$

The factor in brackets preceding the $\hbar^2k^2/2m$ is the predicted m/m_c. The second term is identical to the second term in Eq. (6-11), indicating that our intuitive derivation of the matrix elements of the momentum and gradient operators was correct. The first term represents the dependence on energy without interband coupling. It is negative here since without interband coupling this s-band would peak at $\boldsymbol{k} = 0$, whereas in Eq. (6-11) the term without coupling was effectively the free-electron mass corresponding to taking V_2 equal to zero in that expression. Using this form increases our predicted m_c which were already too large and we have not tabulated them. Perhaps one should not be concerned with the meaning of this correction to the large term, represented by the coupling term in both Eqs. (6-11) and (6-12).

B Light-Hole Mass.

We may note from Figure 5-6 that in every system there is one valence band that is very nearly a mirror image of the conduction band near Γ. It is called the *light-hole* band, in contrast to the pair of valence bands with much smaller curvature, the *heavy holes*. We must be cautions in using the $\boldsymbol{k}\cdot\boldsymbol{p}$ method, Eq. (6-8), for obtaining the hole masses since it is based upon perturbation theory and at $\boldsymbol{k} = \boldsymbol{0}$ there are three hole bands of the same energy. This requires a special treatment, degenerate perturbation theory (see, for example, Schiff (1968), p. 248). The difficulty comes from the arbitrariness in the choice of basis states in this case. The problem actually goes away if $\boldsymbol{k}$ is taken to lie along a [100] direction. Then only one band couples to the conduction-band state and the others, heavy-hole bands, become irrelevant. Then with the same approximations we obtain the same formula, Eq. (6-11), but with the second term replaced by its negative. [We could of course also readily calculate the curvature by expanding Eq. (5-12), which is for wavenumbers in a [100] direction, as we did in obtaining Eq. (6-12). This leads to Eq. (6-12) with ε_s replaced by ε_p', and $4V_{ss\sigma}$ replaced by $4V_{pp\sigma}/3 + 8V_{pp\pi}/3$, and the corresponding replacement for $\eta_{ss\sigma}$. We stay with the simpler $\boldsymbol{k}\cdot\boldsymbol{p}$ form.] The curvature of the bands is negative and is associated with a negative effective mass.

This light-hole mass also has a significant contribution from the peripheral s-state introduced in Eq. (5-13). We may expand that equation for small k to obtain

$$\varepsilon_p' = \varepsilon_p + \frac{32}{9}\frac{V_{s^*p\sigma}^2}{(\varepsilon_p - \varepsilon_{s^*})\hbar^2/(md^2)}\frac{\hbar^2k^2}{2m} + \ldots \qquad (6\text{-}13)$$

Recalling that we took $V_{s^*p\sigma}^2/(\varepsilon_p - \varepsilon_{s^*}) = \lambda_{sp\sigma} = -0.40\hbar^2/(md^2)$, we obtain a contribution of -32×0.40/9 = -1.42 to the coefficient of $\hbar^2k/2m$, and therefore the corresponding contribution to the hole mass. Because only coupling of p-states with the peripheral state was included, this contribution does not occur for the electron mass m_c .

For the hole bands, the coupling terms (the second term in the denominators of Eq. (6-11)) are negative, and we add this additional negative term -1.42 in the denominators. The entire expression is negative, and it is well known that the negative curvature of a valence-band maximum can be described in terms of a positive hole mass (see, for example, Harrison (1970) p. 149). Therefore, we reverse the sign of the result, and obtain a positive light-hole mass of

$$\frac{m_{lh}}{m} = \frac{1}{0.42 + \dfrac{32\eta_{sp\sigma}^2\hbar^2}{9md^2E_0}} = \frac{1}{0.42 + \dfrac{2.23\,V_2}{E_0}} . \tag{6-14}$$

This was derived for wavenumbers in a [100] direction but frequently the bands are approximated as isotropic, as we discuss in Section 6-5. These light-hole masses are not as well known experimentally as the conduction-band mass because of the larger density of states of the heavy holes. It is clear from band calculations that the result is essentially correct that the light-hole and conduction-band masses are approximately equal (though we shall see in Eq. (6-32) that even in the limit of very small masses the effect of spin-orbit coupling is to increase the light-hole mass to $^3/_2$ the conduction-electron mass).

C. Heavy-Hole Mass.

The heavy-hole masses for motion of electrons in a [100] direction for the homopolar semiconductors can be obtained by expanding Eq. (5-11) for small k , leading to

$$\Delta_5 = E_v - \frac{8V_{pp\pi}(2V_{pp\sigma} + V_{pp\pi})}{9\sqrt{4V_{pp\sigma}/3 + 8V_{pp\pi}/3}}\, k^2d^2 + \ldots \tag{6-15}$$

This is understandable in terms of the $\boldsymbol{k \cdot p}$ method; the square root in the denominator is half the energy difference between the lower Γ_{15} state and the upper Γ_{15} state, the only state coupled to the lower by the matrix elements of $\boldsymbol{p}$. One effect of going to a polar crystal would be to modify that square root to $\sqrt{(4V_{pp\sigma}/3 + 8V_{pp\pi}/3)^2 + (\varepsilon_{p+} - \varepsilon_{p-})^2/4}$. We might also expect some modification of the numerator, as we noted for the conduction band mass following Eq. (6-10). The effects there were not large and we assume they are not large here either. Eq. (6-15) contains any effects of the curvature of the band without coupling, the counterparts of the $-4\eta_{ss\sigma}/3$ of Eq. (6-12) or the 1 in the denominators in Eq. (6-11) so we may obtain an approximate expression for the heavy-hole mass from Eq. (6-15). We multiply and divide the second term by $\hbar^2/2m$ and extract the ratio of the curvature to that of a free-electron. This corresponds to a negative curvature but we again think of it as a positive hole mass which is given by

$$\frac{m_{hh}}{m} = \frac{9\hbar^2\sqrt{(4V_{pp\sigma}/3 + 8V_{pp\pi}/3)^2 + (\varepsilon_{p+} - \varepsilon_{p-})^2/4}}{16md^2V_{pp\pi}(2V_{pp\sigma} + V_{pp\pi})} . \tag{6-16}$$

The result for homopolar semiconductors is an exact result for the tight-binding theory and can be evaluated using our universal parameters as $m_{hh} = 0.30m$. We see in Table 6-2 that it is in agreement for tin but the experimental values are larger for germanium and silicon. The experimental values in fact must represent some average which we should not really compare with ours evaluated for [100] electron motion. The values for the polar semiconductors can also be obtained from Eq. (6-16) using universal parameters as

$$\frac{m_{hh}}{m} = \frac{\sqrt{(2.56\hbar^2/(md^2))^2 + (\varepsilon_{p+} - \varepsilon_{p-})^2}}{2.56\hbar^2/(md^2)}\,0.30 . \tag{6-17}$$

The resulting values are compared with experiment in Table 6-2. As with germanium and silicon we have generally underestimated the effective masses but not by a large amount. It does not seem that we should be correcting the energy difference, which appears in the numerator, for gap enhancement since we did not use experimental gaps here as we did for m_c/m and it is not clear to what extent our parameters incorporate empirical values for the Γ_{15} to Γ_{15} gap. Making such corrections would reduce our predicted mass even further, worsening the agreement with experiment.

D. Masses for Indirect Minima

The same $\boldsymbol{k \cdot p}$ theory also applies to the indirect minima in silicon and germanium. In Eq. (6-8) we replace $\boldsymbol{k}$ by $\boldsymbol{k} - \boldsymbol{k}_0$. We let this deviation in wavenumber lie in a particular direction, along $\boldsymbol{k}_0$ or perpendicular to it; then only that component p_{k-k_0} of the momentum operator enters, and we perform the same sum over states at the same wavenumber, but in other bands, to obtain the energy dependence for that direction.

$$\frac{m}{m^*} = 1 + \frac{2}{m} \sum_n \frac{|\langle n | p_{k-k_0} | 0 \rangle|^2}{\varepsilon_0 - \varepsilon_n} \quad . \qquad (6\text{-}18)$$

In the case of silicon (and also for germanium), the coupling for wavenumbers changing parallel to the $\boldsymbol{k}_0$ of the conduction-band minimum is not with the nearest bands below which are doubly degenerate, but with the lowest two bands. [For reference, see the bands for silicon in Fig. 5-6, or the very similar germanium bands in the inset of Fig. 6-1. The conduction-band minimum of interest comes most of the way from Γ to X.] The reason is that, as we have seen, the top valence-band states (from $\Gamma_{25'}$ to X_4) are odd under 180° rotation around $\boldsymbol{k}_0$ while the conduction-band-minimum state is even under that rotation. The effects of the counterpart (for nonzero $\boldsymbol{k}_0$) of the last term in Eq. (6-18) are just from the bands near the lower X_1 and are even smaller than the effects of the higher conduction bands and the peripheral state. The contribution from the upper bands is negative so the net effect of the sum is to give a value for the right side of Eq. (6-18) less than one and the corresponding *longitudinal effective masses* can be greater than the electron mass. From cyclotron resonance measurements (Dresselhaus, Kip, and Kittel (1955)) they are found to be $0.98m$ for silicon and $1.64m$ for germanium (at the point L in the Brillouin Zone).

For variation transverse to the direction of $\boldsymbol{k}_0$, on the other hand, the coupling is with the doubly degenerate band approximately E_2 below the conduction-band minimum. In fact, E_2 is quite comparable to E_0 , especially in silicon, and the interatomic matrix elements of the momentum are very similar in magnitude (although different in detail) to those that entered the light-hole and electron bands at Γ. Thus, the transverse mass (applicable to both transverse directions) is expected to be comparable to those of Eq. (6-11) and that is the case. Experimental values were included in Table 6-1 for germanium and silicon and are seen to be in as good accord with Eq. (6-11) as are the direct-gap masses m_c .

4. Dynamics of Band States

We have introduced the effective masses as a parameterization of the energy of the bands near the band minima or maxima. They give the coefficients in a Taylor expansion of the ε_k in the deviations of the three-dimensional wavenumber $\boldsymbol{k}$ from the band extremum. Their role in the dynamics of electrons in the presence of fields is an even more important role. It is not difficult to show (e. g., Harrison (1970) p. 64-71) that electrons in bands are describable by Hamiltonian dynamics with the momentum given by $\hbar\boldsymbol{k}$ and the Hamiltonian given by the band energy as a function of $\boldsymbol{k}$ and any added potential as a function of $\boldsymbol{r}$. Correspondingly, if the bands are describable by an effective-mass expression, $\hbar^2k^2/2m^*$, the dynamics of the electrons are just those of particles of charge e and mass m^* . This is so central to semiconductor physics, and it is so important to understand the assumptions - and therefore the limitations - that we describe the derivation briefly here.

A. The Dynamical Equations

In order to discuss dynamics it is preferable to consider wave packets made up of the band states which we have constructed. For example, in one dimension a gaussian packet of free-electron states $\int dk\ e^{ikx}\ e^{-\alpha(k-k_0)^2}$ is centered at a position $x = 0$, though spread over a distance of order $\sqrt{\alpha}$, and has an energy of approximately $\varepsilon(k_0)$. This remains true if the packet is made up of band states in a solid. Further, each different band state making up the packet varies with a different frequency as $e^{-i\,\varepsilon(k)t/\hbar}$ and it is easy to show that the center of the packet moves with velocity

$$\boldsymbol{v} = (1/\hbar)\partial\varepsilon_k/\partial\boldsymbol{k} . \tag{6-19}$$

The vector notation means that we have one equation with v_x and k_x , one with v_y and k_y and one in v_z and k_z . This is the counterpart of the first Hamilton's equation, $\boldsymbol{v} = \partial H/\partial\boldsymbol{p}$.

We may also add a static potential $V(\boldsymbol{r})$ to the Hamiltonian, *which must be slowly varying with distance on the scale of our wave packet.* Then as the potential energy of the packet changes with time, $\boldsymbol{v}\cdot\partial V/\partial\boldsymbol{r}$, the band energy must change $(\partial\varepsilon_k/\partial\boldsymbol{k})\cdot\partial\boldsymbol{k}/\partial t$ such that the total energy is conserved. Combining this with Eq. (6-19) we have $\boldsymbol{v}\cdot(\partial V/\partial\boldsymbol{r} + \partial\boldsymbol{p}/\partial t) = 0$. This provides the second of Hamilton's equations, $\partial\boldsymbol{p}/\partial t = -\partial H/\partial\boldsymbol{r}$, at least for components along the velocity; it turns out to apply also for forces such as those from a magnetic field which are perpendicular to the velocity. This is perhaps most usefully written in terms of the force $\boldsymbol{F}$ itself and in terms of the wavenumber for the packet,

$$\hbar \frac{d\boldsymbol{k}}{dt} = \boldsymbol{F} = -e(\boldsymbol{E} + \frac{1}{c}\boldsymbol{v} \times \boldsymbol{H}) . \qquad (6\text{-}20)$$

In the last form we wrote out the force in terms of electric and magnetic fields.

For our purposes the most important point is that for simple bands, $\varepsilon_k = \varepsilon_0 + \hbar^2k^2/(2m^*)$, the dynamics are exactly those of a particle of charge $-e$ and mass m^*. Eqs. (6-19) and (6-20) however remain valid whatever the band structure is. We will see that in metals they can be used to trace the motion of electrons near complicated Fermi surfaces in the presence of applied fields.

B. Limitations of the Theory

In all cases, however, we must remember that the formulae apply to wave packets and are meaningful only if the fields vary slowly over a reasonable packet. We cannot use them to calculate deflection of an electron by an impurity atom which introduces a potential varying on the atomic scale. That would require constructing a packet on an atomic scale, which would require use of states over entire band widths and a corresponding uncertainty in energy so large that the problem was not meaningful. For applied fields in semiconductors, or for "band-bending" fields arising from densities of charged impurity atoms, Eqs. (6-19) and (6-20) can be used with confidence. We can even use the fact that Hamilton's equations apply, with $H(\boldsymbol{p},\boldsymbol{r}) = \boldsymbol{p}^2/(2m^*) + V(\boldsymbol{r})$, to write a Schroedinger Equation for the electron packets and calculate quantum-mechanical aspects of the electron behavior, such as the trapping of electrons in quantum wells or the binding of electrons to charged donor atoms.

We should mention a second limitation upon these dynamical equations which is less familiar. If an electron is accelerated very rapidly within a band, it may "jump" to a neighboring band. As the packet moves through the band, another packet of the same central wavenumber may begin to develop in the other band. The required energy is supplied by the applied field. When the acceleration is caused by a magnetic field it is called "magnetic breakdown", a term coined by Cohen and Falicov (1961); see also Blount (1962). This effect may best be understood in terms of the free-electron bands such as those in Fig. 5-5. We have noted that the effect of the atoms in the lattice can be described in terms of a pseudopotential which diffracts the electrons and opens up gaps such as those shown in the central or left panels of Fig. 5-5. Then when an electron packet moving through one band reaches the Zone face at X it proceeds from the equivalent point at the opposite Zone face and stays always within the same band, tracing out a path determined by Eq. (6-20). Obviously this rule breaks down if the pseudopotential is zero, or sufficiently weak. Then the electron is free and it follows simply $\partial k/\partial t = \boldsymbol{F}$ along the free-electron parabola, corresponding to a jumping from band to band in our Brillouin Zone picture. The criterion for when this breakdown occurs can be obtained by writing a state which is a linear combination of the two states $/k_1>$ and $/k_2>$, which are coupled by a pseudopotential W. Then the energy difference between the two states is written $\Delta\varepsilon(t)$ and the combined state substituted into the time-dependent Schroedinger Equation. It is then seen that transitions between bands begin to become important when $\hbar\partial\Delta\varepsilon(t)/\partial t \approx W^2$. Translated into energy bands, the gaps are of the order of the pseudopotential so breakdown occurs when there are rates of change in the energy of the electrons larger than the gap squared, divided by $\hbar$. Translated into magnetic breakdown it is when $\hbar\omega_c \times E_F$ exceeds the squared gap.

C. The Dynamics of Holes

The dynamics of holes, missing electrons in the valence band, are also described by Eqs. (6-19) and (6-20) since the surrounding electrons are. We may make a packet for a missing electron, like a bubble, and it will move according to these equations. This does not directly correspond to the simple behavior we associate with holes until we define a hole energy ε_k which increases downward in the bands, since it requires more energy to remove the electron from deeper in the band. We must also associate a hole wavenumber which is the negative of the wavenumber at the center of the packet. Then in terms of this energy and this momentum the velocity of the hole, its change in "wavenumber" or momentum with time in the presence of electric and magnetic fields, the current it carries, and the fields it introduces in the system are those for a positively charged particle with positive mass determined by the curvature of the band as we obtained in Sections 6-3-B and 6-3-C. (See, for example, Harrison (1970), 147ff).

These various effective masses are a major ingredient in the theory of transport properties. All such theories begin with these as given parameters. An additional important ingredient for transport theory is the scattering rate for the electrons, arising either from vibrations of the lattice or from defects or impurities in the semiconductor. We shall calculate the scattering by lattice vibrations in Section 7-4-H and the scattering from impurities in Section 8-1-D.

5. The Form of the Hole Bands

The conduction bands, near the minimum for direct-gap semiconductors, are simple isotropic parabolic bands and there is no difficulty in treating the electron dynamics. Even for the conduction bands in indirect-gap semiconductors, for which the masses are anisotropic, the dynamics is straightforward, though complicated. As indicated following Eq. (6-19), the vector equations can be separated into components, each treated separately. For the holes we have the additional complication of three bands coming together at Γ, in the absence of spin-orbit splitting, and the separation of those bands by spin-orbit splitting. We should describe them more closely.

A. Light- and Heavy-Hole Bands

If we consider only states with wavenumbers in a [100] direction there is no complication in the presence of the additional hole bands. The $\boldsymbol{k} \cdot \boldsymbol{p}$ formula, Eq. (6-8) is a rigorously correct formula, making only the approximation that k is small, if accurate band states are used in the evaluation. The behavior for motion in other directions is more complicated.

Symmetry does not require that the hole-band masses are isotropic, even though they must be the same in all six [100] directions This is because of the degenerate bands at Γ. We may see why this is so for a simple model system of a square lattice with two in-plane p-states on each atom and with nearest-neighbor coupling. A $/p_x>$ state on one atom is coupled to the $/p_x>$ states in the positive and negative x-directions by $V_{pp\sigma}$ and to the $/p_x>$ states in the positive and negative y-directions by $V_{pp\pi}$ but we see immediately that it is not coupled to the $/p_y>$ state on any of these neighbors, nor of course to the $/p_y>$ state on the same atom. Thus they form independent bands, with the $/p_x>$ band given by $\varepsilon_k = 2V_{pp\sigma}cos(k_xd) + 2V_{pp\pi}cos(k_yd)$ and the $/p_y>$ band given by $\varepsilon_k = 2V_{pp\sigma}cos(k_yd) + 2V_{pp\pi}cos(k_xd)$. If we assume the usual negative $V_{pp\pi}$ and positive $V_{pp\sigma}$ then each band has negative curvature in one direction and positive curvature in the other. In order to illustrate the valence bands we take both positive, with $V_{pp\pi}$ smaller. We can then expand for small $\boldsymbol{k}$ and see that a constant-energy surface near the band edge for each is an ellipse, as illustrated by the two ellipses to the left in Fig. 6-3, and all that the square symmetry guarantees is that the combination of the two ellipses has square symmetry. Similarly for the generalization to a simple-cubic lattice of p-states, we obtain constant-energy surfaces which are three disk-like

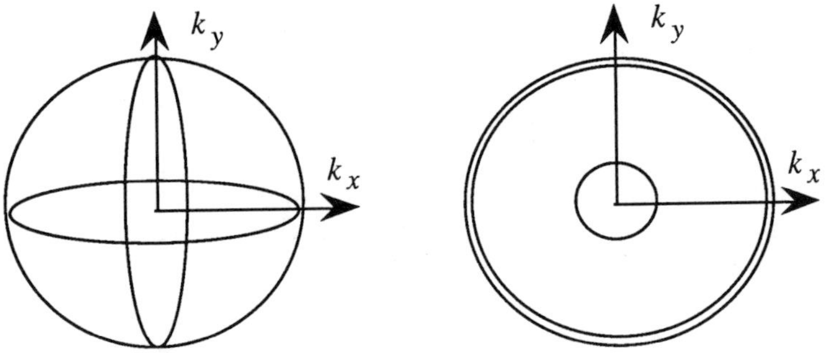

Cubic Lattice, $0 < V_{pp\pi} < V_{pp\sigma}$ Isotropic Approximation

Fig. 6-3. Constant energy surfaces, near the band edge, for p-like bands with nearest-neighbor interactions for a square or cubic structure. Valence bands for tetrahedral structures, without spin-orbit coupling, are analogous. An isotropic approximation, shown to the right, yields much different dynamical properties and may be appropriate when spin-orbit coupling is included. Note that the bands along the cube directions could be the same.

ellipsoids. Along each cube direction we have a light-hole band (corresponding to half the thickness of a constant-energy ellipsoid) and two degenerate heavy-hole bands (corresponding to the large radii of two constant-energy ellipsoids) but they really correspond to three highly anisotropic bands as far as the dynamics is concerned. The left part of Fig. 6-3 represents the xy-plane in wavenumber space for bands in such a simple-cubic structure (again with positive $V_{pp\pi}$). In such a system the dynamics within each band is complicated, and there is scattering between bands as well as within bands. It is more usual to treat semiconductor valence bands as a single isotropic light-hole band and two isotropic heavy-hole bands which, in the case of the simple-cubic p-bands, would correspond to the constant energy surfaces shown to the right in Fig. 6-3. This isotropic approximation is inappropriate for this simple-cubic case and we shall see that there are anisotropies for tetrahedral semiconductors, similar in size to those illustrated to the left in Fig. 6-3.

We could do the analysis for our tight-binding bands, keeping at first only the p-states. The coupling between each $|p_x\rangle$ state and its four neighbor $|p_x\rangle$ states is $V_{pp\sigma}/3 + 2V_{pp\pi}/3$ and of course the same coupling applies between neighboring $|p_y\rangle$ states and between neighboring $|p_z\rangle$ states. However, now there is also coupling between a $|p_x\rangle$ state and its neighboring $|p_y\rangle$ states, given by $\pm(V_{pp\sigma} - V_{pp\pi})/3$. We may add all of these with the appropriate phase factors $e^{i\mathbf{k}\cdot(\mathbf{r}_i - \mathbf{r}_j)}$ and expand for small $\mathbf{k}$ to see that the situation is quite analogous to that for the simple-cubic system of Fig. 6-3. We again have a degenerate heavy-hole band for $\boldsymbol{k}$ in a [100] direction. In this case there are no degenerate bands in the [110] direction, but there are in the [111] direction and the anisotropies are comparable to those shown to the left in Fig. 6-3.

There are other complications in the hole bands, the principal one being the spin-orbit splitting of the bands which we discuss in part 6-5-C. We shall postpone quantifying the anisotropies until we have done that. Kane (1956) has also noted that in compound semiconductors the heavy-hole bands are degenerate only at Γ (as can be seen in Fig. 5-6)) and that the maxima for the two heavy-hole bands are shifted slightly from Γ, in a symmetric way. Boykin (1997) has pointed out that, in terms of tight-binding theory, this cannot occur unless spin-orbit coupling matrix elements between neighboring orbitals are included and it would not occur with the on-site spin-orbit coupling which we introduced in Section 5-3-G.

B. Negative Gaps

The nature of the bands as the gaps get small, or disappear, as in tin, is important to understand. We found in Chapter 5 four bonding bands per atom pair and four antibonding bands. There would be a qualitative change in the bands if we were to increase the spacing of the atoms, decreasing the interatomic coupling, which eventually would lead to two low-energy s-bands and six, partly occupied, higher-energy p-bands. The system would have become metallic. The manner in which this happens can be seen already in the simple bond-orbital bands of Fig. 5-4. The gap there is quite small and as V_2/V_1 is decreased it would go to zero and then it would open up again, but with the doubly degenerate bands attached to the bottom of the upper set of bands. All of our bond-orbital analysis was taking advantage of the fact that V_1 is small compared to V_2 and we were essentially expanding in the ratio. When the gap has closed and the system has become metallic because of the partly occupied upper bands, we are approaching the regime where the opposite expansion is appropriate, an expansion in the ratio V_2/V_1. When we discuss metals we will see that the V_1 which has provided the band width here corresponds to the electron kinetic energy in the free-electron picture and the V_2 which produced the gap is the counterpart of the pseudopotential. Thus indeed when we treat metals by including the pseudopotential as a perturbation, we are making exactly that expansion.

It is interesting to see how this happens in more accurate representations of the bands. What has occurred as we cross this region of vanishing gap is that the upper Γ_1 energy crosses the lower Γ_{15} energy. (That notation is for polar semiconductors. For homopolar systems they were labeled $\Gamma_{2'}$ and $\Gamma_{25'}$ and with spin-orbit splitting still different symbols.) We may see from

Eqs. (5-18) and (5-19) that these two levels depend upon totally different parameters so it is not difficult to imagine a crossing. However, as we move from Γ the coupling between these levels changes the bands dramatically. This is illustrated, for states near the gap in Fig. 6-4. The bands to the left correspond to germanium, those to the right correspond to tin, and the center is close to that for a compound SnGe, which however at 50% is expected to have a small positive gap (Brudevoll, Citrin, Christensen, and Cardona (1994)). Such a crossing occurs also in the semiconducting alloy $Hg_xCd_{1-x}Te$ as one increases x. CdTe is a semiconductor as in Part a, but HgTe is metallic, as in Part c.

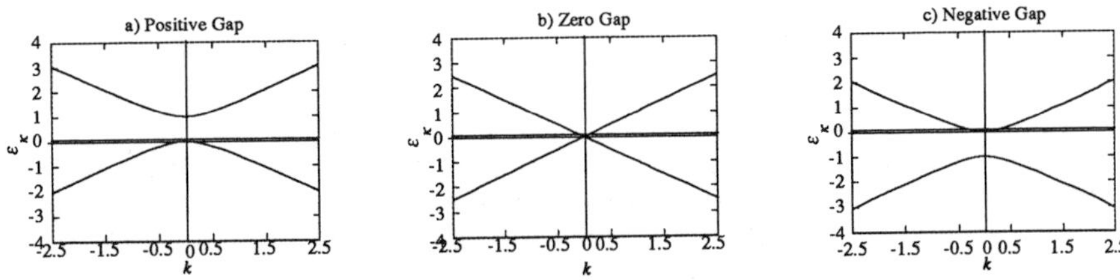

Fig. 6-4. The energy bands near Γ as the gap goes through zero. The horizontal lines at zero represent the heavy-hole bands; the two curves are the light holes and electrons. Part a shows a normal semiconductor, Part b a vanishing gap, and Part c a negative gap as in a metal.

C. The Spin-Orbit Split Valence Bands

We again consider bands with wavenumbers along a [100] direction in a homopolar semiconductor, and add spin-orbit coupling. We add the effects of polarity after. We follow the construction of bands which we carried out for the valence-band maximum in Section 5-3-G. We rewrite the six-by-six Hamiltonian matrix of Eq. (5-31), taking a wavenumber now along the z-direction, which is more convenient than the x-direction since the spin is quantized along the z-direction. We use the same ordering of states as in Eq. (5-31) with the first three columns (and rows) representing the states on the first atom in the cell which are coupled to each other by spin-orbit coupling. One such set is $/p_x +>$, $/p_y +>$, and $/p_z - >$ (with the plus now indicating spin up and the minus, spin down, rather than cation and anion as in Section 5-3-G). The fourth through the sixth column (and row) represent the corresponding states on the second atom ($/p_x +>$, $/p_y +>$, and $/p_z - >$) to which they are coupled by interatomic matrix elements $V_{pp\sigma}$ and $V_{pp\pi}$. For homopolar systems all ε_p are the same. We include at first only the p-states, but will need to add the effect of the coupling $V_{sp\sigma}$ which produces the small light-hole mass. The three-by-three in the upper left and lower right corners of Eq. (5-31), which provide spin-orbit coupling, are unchanged by the introduction of a nonzero wavenumber, but the upper right and lower left are modified. The coupling V_{xx} becomes

$$V_{xx}cos(ka/4) = \left(\frac{4V_{pp\sigma}}{3} + \frac{8V_{pp\pi}}{3}\right)cos(ka/4) = 1.28\frac{\hbar^2}{md^2}cos(ka/4)\,. \quad (6\text{-}21)$$

A coupling also arises between the $/p_x>$ and the $/p_y>$ states which we may write

$$iV_{xy}\,sin(ka/4) = i\left(\frac{4V_{pp\sigma}}{3} - \frac{4V_{pp\pi}}{3}\right)sin(ka/4) = i3.80\,\frac{\hbar^2}{md^2}sin(ka/4)\,. \quad (6\text{-}22)$$

There is no coupling between the other states when ***k*** is along the z-direction so there is no coupling between these six states and the complementary set made up of $/p_x\,->$, $/p_y\,->$, and $/p_z\,+>$ states on each atom. We need to be careful to obtain the correct sign of the matrix elements by drawing the orbitals and the structure. The 3×3 matrix in the upper right of Eq. (5-27) becomes

$$H_{3\times3} = \begin{pmatrix} V_{xx}cos(ka/4) & iV_{xy}\,sin(ka/4) & 0 \\ iV_{xy}\,sin(ka/4) & V_{xx}cos(ka/4) & 0 \\ 0 & 0 & V_{xx}cos(ka/4) \end{pmatrix}, \quad (6\text{-}23)$$

and the 3×3 matrix to the lower left of Eq. (5-31) becomes the complex conjugate of this. (These submatrices are not Hermitian by themselves, $H_{3\times3}^{+} \neq H_{3\times3}$, but the full matrix is.) We did two unitary transformations which eliminated the elements in these submatrices by mixing the states on the two atoms. We were left with a matrix in the upper left given by

$$H_{3\times3} = \begin{pmatrix} \varepsilon_p - R & -i\lambda & \lambda \\ i\lambda & \varepsilon_p - R & -i\lambda \\ \lambda & i\lambda & \varepsilon_p - V_{xx}cos(ka/4) \end{pmatrix}, \quad (6\text{-}24)$$

where

$$R = \sqrt{V_{xx}{}^2 cos^2(ka/4) + V_{xy}{}^2 sin^2(ka/4)} \ . \tag{6-25}$$

At this stage the first and second columns (and rows) represent bonding states which could be combinations of $/ p_x+ >$ and $/ p_y +>$ on each atom, but they can again be transformed to make them no longer mixtures; it is not surprising that the matrix elements of the spin-orbit coupling are not changed.

The matrix to the lower right is the same, but with the negative sign in each of the diagonal matrix elements replaced by a plus sign. It is the matrix of Eq. (6-24) which leads to the valence bands in which we are interested. Of course there would be an identical matrix leading to the same energies, with all spins reversed, so that there are two states at each energy. By writing out the secular equation for Eq. (6-24) it may be confirmed that the eigenvalues are

$$\varepsilon_k = \varepsilon_p + \lambda - R,$$

$$\varepsilon_k = \varepsilon_p + \frac{-\lambda - R - V_{xx}cos(ka/4) \pm \sqrt{(\lambda + R - V_{xx}cos(ka/4))^2 + 8\lambda^2}}{2} \ . \tag{6-26}$$

The second expression with the plus sign corresponds to the light-hole band, but we have not included the $V_{sp\sigma}$ coupling which makes it light. We may see how that coupling affects the results by taking $\lambda = 0$, which leads

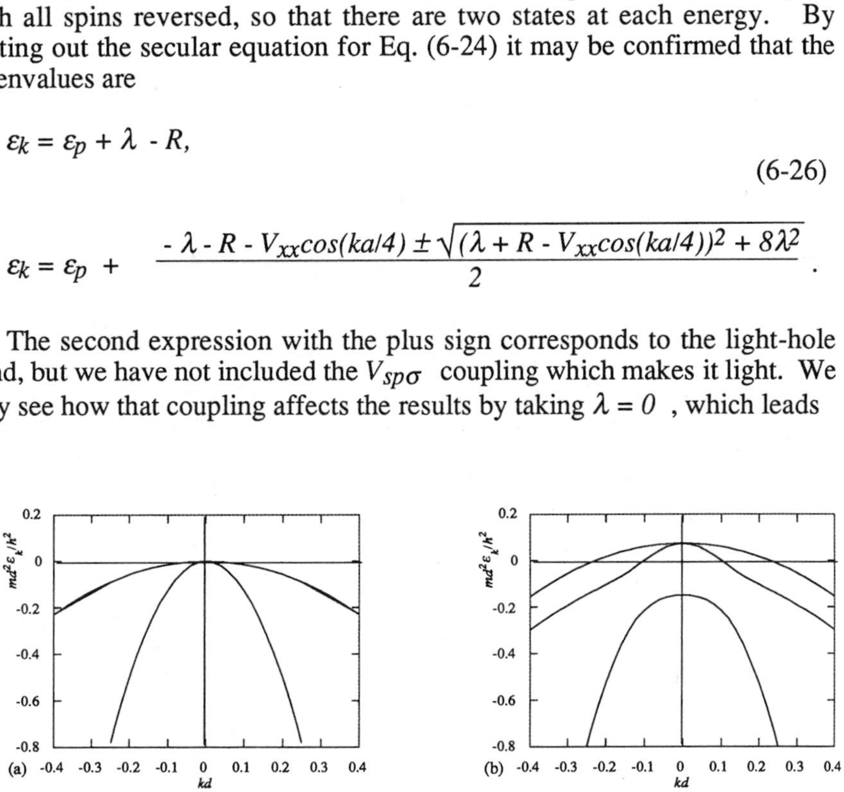

Fig. 6-5. The top of the valence bands of germanium (a) without spin-orbit coupling and (b) with spin-orbit coupling from Eq. (6-27) using effective-mass bands for ε_{lh} and ε_{hh} . The energies are in units of $\hbar^2/md^2$. Note that the light-hole band is at the upper energy at $\boldsymbol{k} = 0$.

to the final two bands given by $e_{hh} = \varepsilon_p - R$ and $\varepsilon_{lh} = \varepsilon_p - V_{xx}cos(ka/4)$. We may note that the second of these, the light-hole band, does not even have the appropriate negative curvature. However, we see that Eq. (6-26) may be rewritten in terms of these light- and heavy-hole bands without spin-orbit coupling as

$$\varepsilon_K = \varepsilon_{hh} + \lambda \ ,$$

$$\varepsilon_k = \frac{\varepsilon_{hh} + \varepsilon_{lh} - \lambda \pm \sqrt{(\varepsilon_{hh} - \varepsilon_{lh} - \lambda)^2 + 8\lambda^2}}{2} \ . \tag{6-27}$$

In a similar way we could rewrite the matrix of Eq. (6-24) with the upper two diagonal elements replaced by ε_{hh} and the last by ε_{lh} .

We may use these also with effective-mass expressions for ε_{hh} and ε_{lh} . For germanium we take our estimate of the electron mass, $0.04m$, as an estimate of the light-hole mass, use the experimental heavy-hole mass in germanium of 0.35 (both from Table 6-2)), and take $\lambda = 0.095$ eV from Table 5-2 to obtain the bands shown in Fig. 6-5. It is interesting that although the splitting at Γ is 3λ the splitting of the heavy-hole bands at large k is only 2λ . (This can be verified by expanding the second of Eq. (6-27) with the plus sign for large ε_{lh} and large ε_{hh} . It is necessary to keep two terms in the expansion of the square root.) This form is quite different from Eq. (5-29), and the corresponding equation for Γ_7, which gave the splitting at Γ, but those equations, if evaluated for homopolar semiconductors, are the same as Eqs. (6-27) evaluated at Γ .

Fig. 6-5 gave the bands for wavenumbers in a [100] direction. We saw in Section 6-5-A that without spin-orbit splitting there could be major anisotropy of the bands. The split-off band in Part b of Fig. 6-5 is not degenerate and the $\boldsymbol{k \cdot p}$ arguments which we gave earlier for an isotropic band apply. The two upper bands however remain anisotropic as illustrated to the left in Fig. 6-3. The effective masses for the holes given by Woerner and Elsaesser (1995), for example, for germanium are $m_{lh}/m = 0.042$ for the light hole at $\boldsymbol{k} = \boldsymbol{0}$, but a heavy-hole mass m_{hh}/m varying with direction as 0.21 for [100], 0.37 for [110], and 0.49 for [111].

We may expect the same formula for the bands to be approximately correct for polar semiconductors. There are some additional complications such as a shift of the valence-band maximum slightly off the $\boldsymbol{k} = \boldsymbol{0}$ point due to the lower symmetry, noted at the end of Part 6-5-A, and similar shifts specifically for wurtzite structures discussed by Lew Yan Voon, Willatzen,

Cardona, and Christensen (1996), but these are small effects. For polar semiconductors the ε_{lh} is obtained from Eq. (5-12) and the ε_{hh} from Eq. (5-11) or we may use the effective mass expansions of those bands. There are in fact anisotropies similar to those we gave before for germanium. The effective masses for the holes for gallium arsenide, given also by Woerner and Elsaesser (1995), are $m_{lh}/m = 0.074$ for the light hole , and a heavy-hole mass varying with direction as 0.36 for [100], 0.66 for [110], and 0.9 for [111]

We saw at the end of Section 5-3-G that it should be a good approximation to use an average λ for the compound semiconductors, given by

$$\lambda = \frac{1 + \alpha_p(p)}{2}\lambda_- + \frac{1 - \alpha_p(p)}{2}\lambda_+ \qquad (6\text{-}28)$$

with

$$\alpha_p(p) = \frac{(\varepsilon_{p+} - \varepsilon_{p-})/2}{\sqrt{(\varepsilon_{p+} - \varepsilon_{p-})^2/4 + V_{xx}^2}} \qquad (6\text{-}29)$$

for the valence band, with the λ_- and λ_+ interchanged for the conduction band. These formulae, with universal parameters and the spin-orbit parameters from Table 5-2, provide approximate bands and band states, near the band edges, for all of the semiconductors.

Spin-orbit coupling affects other properties, such as the somewhat complicated effects on defect levels which are pulled out of the spin-orbit-split valence band, discussed by Rebane (1993).

D. Density of States

A particularly important aspect of the bands is the density of states, per unit energy, as a function of energy. It enters directly the electronic specific heat and, as we shall see, the optical absorption near threshold. For a parabolic band the density of states varies as the square root of the energy measured from the band edge. It is readily calculated for a band with $\varepsilon = \hbar^2k^2/(2m^*)$ as

$$n(\varepsilon) = \frac{\Omega_0}{2\pi^2}\left(\frac{2m^*}{\hbar^2}\right)^{3/2}\sqrt{\varepsilon} \qquad (6\text{-}30)$$

states per atom per unit energy, where Ω_0 is the atomic volume, $8d^3/3\sqrt{3}$ for tetrahedral solids. This comes from writing the number of states of each spin in a spherical shell in wavenumber space of radius k and thickness dk as $(\Omega/(2\pi)^3)4\pi k^2 dk$ if the system has volume Ω. We may easily redo this for an isotropic band of arbitrary dispersion finding $n(\varepsilon)$ at any energy given by

$$n(\varepsilon) = \frac{8d^3}{3\sqrt{3}\pi^2}\frac{k^2}{d\varepsilon/dk}\cdot , \tag{6-31}$$

of which Eq. (6-30) is a special case. This evaluation was carried out taking the bands of Eq. (6-27) as isotropic and substituting the effective-mass forms for ε_{lh} and ε_{hh}, solving for k for each band, evaluating the derivative and substituting in Eq. (6-31) The results are given in Fig. 6-6.

The top band provides the upper parabola which looks much like the free-electron density of state of Eq. (6-30). The light-hole band beginning at the same energy of 0.095 eV initially gives a tiny parabolic contribution, corresponding to the fact the light-hole mass is only about a ninth of the heavy-hole mass and the density of state is proportional to $m^{*3/2}$ as seen in Eq. (6-30). It soon switches over to a curve parallel to the top band, as we

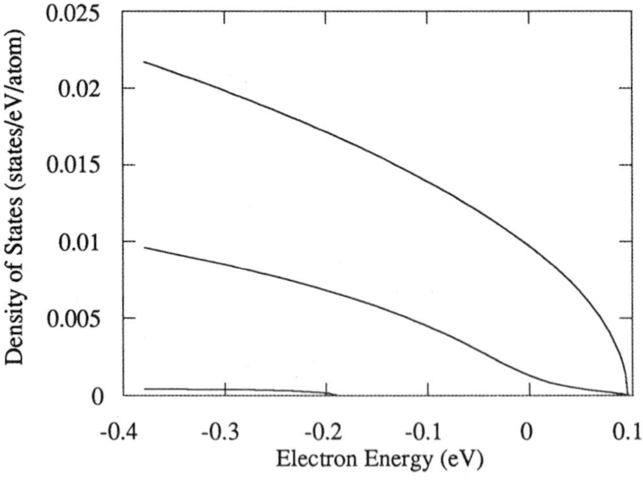

Fig. 6-6. The density of states for the three spin-orbit-split valence bands of Eq. (6-27), taken to be isotropic, and evaluated for parameters appropriate to germanium. The energy is measured from the unsplit band maximum at $\varepsilon_p - V_{xx}$.

might anticipate from the bands as seen in Fig. 6-5. Somewhat more surprising is how small is the density of states arising from the lowest band which begins at -0.19 eV. Nominally, by its symmetry as $\boldsymbol{k}$ leaves zero, it is a heavy-hole band, but its spin-orbit coupling with the light-hole band increases its curvature and makes its mass and density of states very small.

We may see this in detail by expanding the second of the Eqs. (6-27) for small ε_{lh} and ε_{hh} to find them varying as $\varepsilon \approx \varepsilon_{lh} - \lambda \pm (3\lambda + (\varepsilon_{lh} - \varepsilon_{hh})/3)$. Correspondingly the upper and lower effective masses at the band edges are

$$\frac{1}{m^*} = \frac{2}{3m_{lh}} + \frac{1}{3m_{hh}} \quad \text{(upper band)},$$
$$\frac{1}{m^*} = \frac{1}{3m_{lh}} + \frac{2}{3m_{hh}} \quad \text{(lower band)}. \qquad (6\text{-}32)$$

Since m_{lh} is so much smaller it dominates *both* masses and they differ only by a factor of two. This only describes the bands within energies of order of λ of the band edge but in these heavier semiconductors these are

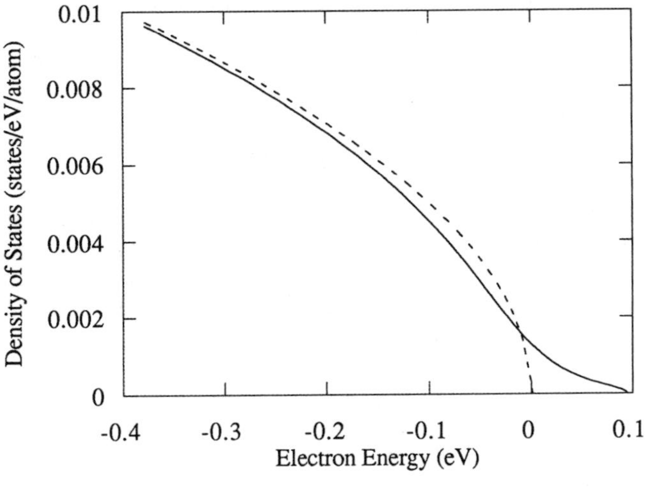

Fig. 6-7. The density of states (dashed) from a free-electron band, corresponding to $m^* = 0.22m$ and a band edge at the unsplit value, compared to the middle valence band for germanium (solid).

the important energies. Table 5-2 giving λ can be used to decide in each case.

The picture which this provides may be useful. The top band is a true heavy-hole band with m_{hh} unaffected by spin-orbit splitting. The split-off band may be thought of as a light-hole band and its contribution to the density of states tiny. So also is the contribution of the middle band near the band maximum, but at deep enough energy it begins contributing to the density of states, much like another heavy-hole band, but originating 2λ below the first. We tried modeling its contribution by its asymptotic form, which is equal to the upper curve in Fig. 6-6, but with its onset shifted to -0.095 eV. That is much too large. If one wished to model it by a parabola, something such as shown in Fig. 6-7 would be much more accurate, but has no other justification, nor would it be accurate at much deeper energies.

6. The Optical Absorption Rate

Two of the most important uses of the direct-gap semiconductors are for light-emitting diodes (LED's) and for semiconductor lasers. Because only nearly vertical transitions are allowed (Section 6-2-A), it is important that the electrons which are to produce the light accumulate in states near Γ where the holes also are. That is, we must use a direct-gap semiconductor such as gallium arsenide. It is then important to understand the coupling between the conduction-band states and the various hole bands. We do that for polar semiconductors, making use of the matrix elements we have obtained in Section 6-2-B.

A. The Matrix elements

We wrote the optical matrix element between two neighboring atomic states in Eq. (6-4). The conduction-band states at small wavenumbers (near Γ) are s-like and the valence-band states are p-like so we obtain optical matrix elements, based upon a field $\boldsymbol{E}\, cos(\boldsymbol{q}\cdot \boldsymbol{r}_j - \omega t)$ with $\boldsymbol{E}$ in a y-direction for a pair of neighbors (with σ-oriented p-states) of

$$<s|H_{le}|p> = -\frac{eV_{sp\sigma}\boldsymbol{E}\cdot\boldsymbol{d}}{\sqrt{(\varepsilon_s - \varepsilon_p)^2/4 + V_{sp\sigma}^2}}\, cos(\boldsymbol{q}\cdot \boldsymbol{r}_j - \omega t). \tag{6-33}$$

We are interested here in polar semiconductors, such as GaAs, so we might construct a Bloch sum of p-states $|p_{y-}>$ on arsenic atoms, $|p_{y-},\boldsymbol{k}> = \sqrt{1/N_b}\,\Sigma_{j(As)}\, e^{i\boldsymbol{k}\cdot \boldsymbol{r}_j}\, |p_{yj}>$, and a Bloch sum of s-states on gallium atoms $|s_+, \boldsymbol{k}'> =$

$\sqrt{1/N_b}\,\Sigma_{i(Ga)}\,e^{i\mathbf{k}'\cdot r_i}\,|s_i>$. In evaluating the matrix element between such Bloch sums, using Eq. (6-33) with its $cos(\mathbf{q}\cdot r_j - \omega t) = [e^{i(\mathbf{q}\cdot r_j - \omega t)} + e^{-i(\mathbf{q}\cdot r_j - \omega t)}]/2$, we will obtain matrix elements for $\mathbf{k}' = \mathbf{k} + \mathbf{q}$ and for $\mathbf{k}' = \mathbf{k} - \mathbf{q}$, each with a factor $^1/_2$ and the first with a factor $e^{-i\omega t}$ and the second with a factor $e^{i\omega t}$. There will be factors $e^{\pm i\mathbf{q}\cdot(r_i - r_j)}$ which we can set equal to one (small q). For p-states oriented along a y-direction there will be a factor $1/\sqrt{3}$ and we need add the contributions from the four neighboring atoms to each s-state, giving a factor 4 . The resulting matrix elements between Bloch sums are

$$<s_+, \mathbf{k}'|H_{le}|p_{y-}, \mathbf{k}> = \frac{2edE_yV_{sp\sigma}/3}{\sqrt{(\varepsilon_{s+} - \varepsilon_{p-})^2/4 + V_{sp\sigma}^2/3}}\, e^{\pm i\omega t} \tag{6-34}$$

for $\mathbf{k}' = \mathbf{k} \mp \mathbf{q}$ with the $e^{-i\omega t}$ applying for $\mathbf{k}' = \mathbf{k} + \mathbf{q}$ and the $e^{i\omega t}$ applying for $\mathbf{k}' = \mathbf{k} - \mathbf{q}$.

Of course we would also obtain such nonzero matrix elements between s- and p_x-states with $\mathbf{E}$ in the x-direction and between the s- and p_z-states with $\mathbf{E}$ in the z-direction. There is also a nonzero matrix element (large because $V_{sp\sigma}/\sqrt{3}$ is replaced by V_{xy}) between the p_x-state and the p_y-state for fields in the z-direction, but it does not couple to the conduction-band minimum where the states are s-like.

The conduction-band state is the antibonding counterpart of the Γ_1 bonding state at the bottom of the valence band and therefore has the form, in analogy with Eq. (2-15), of

$$\sqrt{\frac{1-\alpha_p(s)}{2}}\,|s_-, \mathbf{k}'> - \sqrt{\frac{1+\alpha_p(s)}{2}}\,|s_+, \mathbf{k}'>. \tag{6-35}$$

Now the polarity depends only on s-state parameters,

$$\alpha_p(s) = \frac{(\varepsilon_{s+} - \varepsilon_{s-})/2}{\sqrt{(\varepsilon_{s+} - \varepsilon_{s-})^2/4 + (4V_{ss\sigma})^2}} . \tag{6-36}$$

The top of the valence band is a bonding combination of p-states and without spin-orbit splitting could be chosen with one as p_x-states, one p_y, and one p_z . We use that form now and will make the appropriate combination for spin-orbit coupling after. Thus our Bloch sums for the valence-band maximum are of the form

$$\sqrt{\frac{1+\alpha_p(p)}{2}}\,/p_-,\mathbf{k}> \;+\; \sqrt{\frac{1-\alpha_p(p)}{2}}\,/p_+,\mathbf{k}> \qquad (6\text{-}37)$$

with a polarity $\alpha_p(p)$ given in Eq. (6-29).

We now seek the matrix element between the two band states. In each cell there is a contribution from the state $/s_+>$, which appeared in Eq. (6-34), and which is weighted by the second square root in Eq. (6-35) and the first square root in Eq. (6-37). There is also a contribution from the state $/s_->$, given by Eq. (6-34) but with ε_{s+} replaced by ε_{s-} and with ε_{p-} replaced by ε_{p+}, and it is weighted by the first square root in Eq. (6-35) and the second square root in Eq. (6-37). We add the two terms to obtain the matrix element between s-like (upper Γ_1) and p_y-like (lower Γ_{15}) band states of

$$<\Gamma_1/H_{le}/\Gamma_{15}> = \frac{edE_yV_{sp\sigma}}{3}\times$$

$$\left(\sqrt{\frac{(1+\alpha_p(p))(1+\alpha_p(s))}{(\varepsilon_{s+}-\varepsilon_{p-})^2/4+V_{sp\sigma}^2/3}}+\sqrt{\frac{(1-\alpha_p(p))(1-\alpha_p(s))}{(\varepsilon_{s-}-\varepsilon_{p+})^2/4+V_{sp\sigma}^2/3}}\right)e^{\pm i\omega t} \qquad (6\text{-}38)$$

,

again for $\mathbf{k'} = \mathbf{k} \mp \mathbf{q}$. For homopolar semiconductors both α_p's are zero and the denominators in the two roots become the same so that this becomes the same as the right side of Eq. (6-34). The complexity for polar systems seems essential to obtaining a reasonable estimate. It need only be evaluated once for each semiconductor. This coupling between the two band states for the same spin can be written simply as

$$<\Gamma_1/H_{le}/\Gamma_{15}> = edE_i\,\eta\,e^{\pm i\omega t} \qquad (6\text{-}39)$$

for the ith component of the field with η given for a number of semiconductors in Table 6-3. If we are willing to neglect spin-orbit coupling this gives directly the needed optical matrix elements.

If we are to include spin-orbit coupling, we must return to the matrix in Eq. (6-24). The corresponding eigenstates on each atom, or on each Bloch sum, and their energies are

$$\begin{aligned} &(/p_x+> + i/p_y+>)/\sqrt{2}, && \varepsilon=\varepsilon_p+\lambda, \\ &(/p_x+> - i/p_y+> +2/p_z->)/\sqrt{6}, && \varepsilon=\varepsilon_p+\lambda, \\ &(/p_x+> - i/p_y+> - /p_z->)/\sqrt{3}, && \varepsilon=\varepsilon_p-2\lambda, \end{aligned} \qquad (6\text{-}40)$$

Table 6-3. The parameter η determining the optical matrix elements between Bloch sums at $\boldsymbol{k} = 0$ in Eq. (6-39).

C 0.62	AlN 0.76		
Si 0.35	AlP 0.87	InP 0.96	AlAs 0.76
Ge 0.30	GaAs 0.61	ZnSe 0.61	CuBr 0.44
Sn 0.29	InSb 0.61	CdTe 0.61	AgI 0.42

with the superscript $\pm$ indicating spin. These combinations are the same on both atoms, though λ is different on the two. We have all of the needed pieces, with the couplings between Bloch sums which produce the bands, as discussed in Section 6-5-C, and the optical matrix elements between each Bloch sum deduced from Eq. (6-34). The task is intricate in general, particularly as we move away from $\boldsymbol{k} = 0$, with possibilities for various approximations, but we shall not carry it further. We proceed to the absorption rate, using Eq. (6-39), which is appropriate to the band edges in the absence of spin-orbit coupling.

B. The Rate

It may be helpful to complete the calculation, using these matrix elements, for one case, transitions between a heavy-hole band (we neglect spin-orbit splitting) and a conduction-band state near the minimum at $\boldsymbol{k} = 0$. We use the matrix elements $iedE\,\eta/\sqrt{3}$ from each initial state of wavenumber $\boldsymbol{k}$ to a single final state of wavenumber $\boldsymbol{k}'$ which differs by a negligible amount from $\boldsymbol{k}$. This is a particularly simple case with a matrix element independent of the wavenumbers of the states and we can use simple parabolic bands. The generalization to other circumstances is direct.

We gave in Eq. (1-32) the transition rates due to a perturbation, the Golden Rule of quantum mechanics, and noted that if the perturbation varied with a frequency ω, leading to matrix elements $\langle j|H_1|i\rangle e^{\pm i\omega t}$ as in Eqs. (6-34), (6-38), and (6-39), we could write rates for both losses and gains in energy as

$$P_{ij} = \frac{2\pi}{\hbar} |\langle j|H_1|i\rangle|^2 [\delta(E_j - E_i - \hbar\omega) + \delta(E_j - E_i + \hbar\omega)] , \qquad (6\text{-}41)$$

with the first term representing absorption of light; the energy of the final state E_j is higher than that of the initial state E_i. We keep only the first term and $E_j - E_i - \hbar\omega$ becomes $E_0 + \hbar^2k^2/2m_{hh} + \hbar^2k^2/2m_c - \hbar\omega$, with E_0 of course the energy from the upper valence band to the conduction-band minimum at $\boldsymbol{k} = 0$ and we neglect the difference between initial and final wavenumbers. We take all states in the valence band to be occupied and all states in the conduction band to be empty. The sum over final states is a sum over the conduction-band state $\boldsymbol{k'}$ which becomes occupied, leaving an empty valence-band state $\boldsymbol{k} = \boldsymbol{k'} - \boldsymbol{q} \approx \boldsymbol{k}$. We multiply by the energy lost to the system in the process, $\hbar\omega$, to obtain the loss rate,

$$\frac{dE_{light}}{dt} = -2\pi\omega \sum_{k,\sigma} \frac{e^2d^2E^2\eta^2}{3} \delta(\hbar\omega - E_0 - \hbar^2k^2/2m_{hh} - \hbar^2k^2/2m_c). \quad (6\text{-}42)$$

The sum over σ is a sum over spin orientations, which in this case is simply a factor of two. Under the sum only the delta function depends upon $\boldsymbol{k}$ so if we replace the sum over $\boldsymbol{k}$ by an integral, according to Eq. (5-3), we can write it $\Sigma_k = \Omega/(2\pi)^3 \int 4\pi k^2 dk$. The integral over the delta function gives a factor $1/[\hbar^2k/m_{hh} + \hbar^2k/m_c]$ and Eq. (6-42) becomes

$$\frac{dE_{light}}{dt} = -\frac{2\eta^2e^2d^2E^2\Omega\omega}{3\pi\hbar^2} \frac{m_{hh}m_c}{m_{hh} + m_c} k \quad (6\text{-}43)$$

with k evaluated from $\hbar\omega - E_0 - \hbar^2k^2/2m_{hh} - \hbar^2k^2/2m_c = 0$. We may relate E^2 to the energy flux associated with the light, $\Pi = cE^2/8\pi$ (e. g., Richtmeyer and Kennard (1947), p. 58, with E the electric-field amplitude) with c the velocity of light in vacuum. Also the density of energy of the light is $E_{light}/\Omega = \varepsilon E^2/8\pi$. It is convenient to write the absorption in terms of an attenuation rate α for the light energy density,

$$\frac{1}{\Omega}\frac{dE_{light}}{dt} = -\alpha \frac{E_{light}}{\Omega}. \quad (6\text{-}44)$$

Then we obtain

$$\alpha = \frac{8\eta^2e^2d^2}{3\hbar^3\varepsilon}\left(\frac{m_{hh}m_c}{m_{hh} + m_c}\right)^{3/2} \omega\sqrt{\hbar\omega - E_0} \quad (6\text{-}45)$$

with the k-factor written in terms of $\hbar\omega$, for $\hbar\omega > E_0$ and zero otherwise. The η are given in Table 6-3. We have attempted to keep all factors of two correct, but it is not easy. α is plotted as a function of $\hbar\omega$ in Fig. 6-8, illustrating the dependence above threshold.

This can also be written in terms of the imaginary term in the susceptibility. For an isotropic system the polarization density is proportional to the electric field, $P = \chi^{(1)}E$ so if the field is proportional to $e^{-i\omega t}$ the current density is $j = dP/dt = -i\omega\chi^{(1)}E$ and the rate work is done on the carrier system is $E^*j = -i\omega\chi^{(1)}E^*E$ per unit volume. As above, $E^*E = 4\pi E_{light}/\varepsilon$ and $dE_{light}/dt = -4\pi\omega Im(\chi^{(1)})E_{light}/\varepsilon$ so $Im(\chi^{(1)}) = \alpha/4\pi\omega$.

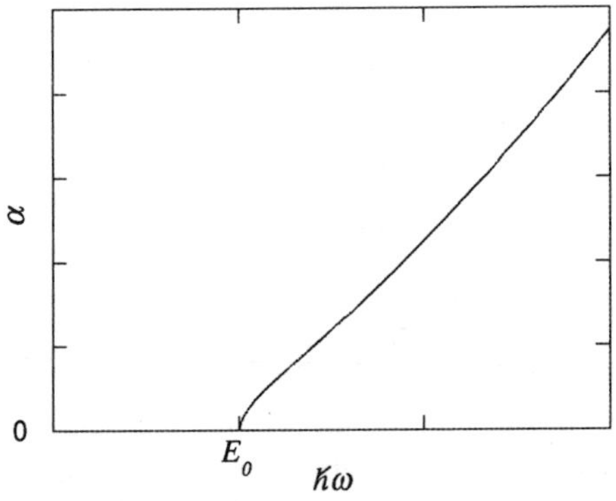

Fig. 6-8. A plot of the optical absorption coefficient α from Eq. (6-45) as a function of energy $\hbar\omega$ above the optical threshold, E_0.

Problem 6-1 Optical Threshold

Eq. (6-5) gives the optical threshold for tetrahedral semiconductors.

a) In general do you expect it to decrease or increase with pressure?

b) What spacing d would be necessary in InSb for E_0 to be equal to zero.

Problem 6-2. Optical Matrix Elements

Eqs. (6-38) and (6-39) give the dominant terms in the matrix element between an s-like and a p-like band at $\boldsymbol{k} = 0$. This was obtained from the *interatomic* matrix elements. There is also an *intraatomic* contribution arising from the matrix element between an s-state and a p-state on the same atom. Estimate the magnitude of that contribution using an $H_{1e} = eEx$, and model wavefunctions $\psi_s = Ae^{-\mu r}$ and $\psi_p = Bxe^{-\mu' r}$, with μ and μ' chosen to fit the silicon s- and p-term values from Table 1-1, and A and B adjusted so that each wavefunction is normalized, $\int\psi(r)^2\, 2\pi \sin\theta\, d\theta\, r^2\, dr = 1$. Write the result as $\eta' eEd$, with $d = 2.35$ Å for silicon, so its square can be directly compared numerically with the squared interatomic contribution from Eq. (6-39). The two contributions obtained, for example, from Eq. (6-42) directly add. This may be a meaningful estimate of the relative importance of the two contributions.

CHAPTER 7

Effects of Lattice Distortions

There are a number of circumstances under which the changes in the bands when the lattice is distorted become important. Of course one such was for the change of the total energy under a lattice shear, which is described by the elastic constants which we calculated in Chapter 3, but there we needed only the average energy of the bands and we obtained it in a Bond-Orbital Approximation without considering the band structure. However, such a shear will have other effects such as making the x- and the y-directions in the lattice inequivalent and lifting the degeneracy of the three bands at the top of the valence bands. This arises for example when a semiconductor of one equilibrium lattice distance is grown epitaxially on a semiconductor with another lattice distance. The added layer takes on the lateral spacing of the substrate and is correspondingly distorted, producing this split valence-band maximum at Γ. We shall calculate just what these shifts are for a strain-layer heterostructure in Section 19-4-D.

Another important case is when there are lattice vibrations present which locally shift the energy bands because of the local distortion. For each state this shift is analogous to a shift in its potential energy and these shifts are called "deformation potentials". When the vibrations are of long wavelength, the deformation-potential shifts can be calculated as for uniform strains, so the effects on the bands we calculate in this section will be directly usable when we treat this interaction between electrons and lattice vibrations, called the electron-phonon interaction, in Section 7-3. However, we shall see that there can also be electron-phonon interactions arising from local *rotations* of the lattice and these contributions are missed in the uniform-distortion calculations of deformation potentials. Fortunately, for the most important couplings in tetrahedral semiconductors these additional effects do not enter and deformation potentials can be used.

These detailed shifts in the bands under distortion are rather well given by our simple tight-binding theory, except in the case of the indirect minimum in the conduction bands. The analysis may be useful in understanding the nature of the shifts as well as the general magnitudes. The reasons why they come out as they do are much more difficult to discern from the full accurate density-functional calculations which we shall also discuss. The electron-phonon interaction is extremely complicated in principle because of the many different cases which may be distinguished, coupling of light-hole states to light-hole states by longitudinal modes, by transverse modes of each polarization or coupling of light-hole states to states in each of the heavy-hole bands by each type of mode, etc. Tight-binding theory can yield all of these, reducing all to a few parameters. It can also suggest approximations which reduce the complexity.

1. Deformation Potentials for Dilatations

A uniform strain will modify the energy bands which we have calculated in Chapters 5 and 6. In particular, a uniform dilatation Δ (equal to $\delta\Omega/\Omega$) will modify the bond lengths by $\delta d/d = \Delta/3$ and shift the energy of, for example, the $E(\Gamma_1)$ of Eq. (5-18) just as if a potential had been applied. The shift will of course be proportional to the dilatation and if the dilatation varies from place to place, so also with the shift in the band due to it. This effective potential was called the *deformation potential* by Bardeen and Shockley (1950). It has been treated very carefully by Cardona and Christensen (1987) and we make use of their results, though carrying out the formulation in terms of our tight-binding parameters.

For the particular case of the minimum in the conduction band Γ_1 in direct-gap semiconductors a shear of the lattice can only shift the energy to second order in the shear (since symmetry requires that the negative of the same shear gives an equal shift) so a single deformation potential constant D_c relates the deformation potential $V(\mathbf{r})$ to the dilatation $\Delta(\mathbf{r})$; that is, $V(\mathbf{r}) = D_c\Delta(\mathbf{r})$, with D_c given by

$$D_c = \Omega\frac{\partial E(\Gamma_1)}{\partial\Omega} = \frac{d}{3}\frac{\partial E(\Gamma_1)}{\partial d} . \tag{7-1}$$

A. Nonorthogonality and Electrostatic Contributions

Two problems arise immediately concerning the zero of energy and screening of effective potentials in the crystal. If we use Eq. (5-18) directly, all energies are measured relative to the zero of energy for our free-atom term values. However, we have noted in Section 1-2 that the nonorthogonality of orbitals on adjacent atoms shifts the average energy of all the states upward, and in Section 2-2-E we represented that effect by an overlap repulsion $V_0(d)$ per bond, or $V_0(d)/2$ per electron, neglecting any effect of the nonorthogonality on the relative energy of different (e. g., s- or p-) levels. That means that we have assumed a shift of $V_0(d)/2$ in every free atom term value. It is an approximate representation of a real shift in the levels and as such should be added to the $E(\Gamma_1)$ of Eq. (5-18) when evaluating Eq. (7-1), giving an additional term in D_c of $(d/3)\partial(V_0/2)/\partial d = -(2/3)V_0(d)$ if V_0 varies as $1/d^4$.

It is in fact convenient to rewrite this shift from orthogonality using the theory of the total energy which led to Eq. (2-21). That was for the homopolar case for which we found $V_0(d) = V_2$ at equilibrium. The same calculation for the polar case (with bond energy $\bar{\varepsilon}_h - \sqrt{V_2^2 + V_3^2}$) gives $V_0(d) = \alpha_c V_2 = V_2^2/\sqrt{V_2^2 + V_3^2}$ at the equilibrium spacing, to be evaluated using the V_2 and V_3 from Table 2-1. There were additional smaller terms if we included metallization, but these did not prove helpful for the elastic properties and we do not include them here. Thus we incorporate shifts in all term values of $\alpha_c V_2/2$ and contributions to all deformation potentials for dilatations of $-(2/3)V_0(d) = -2\alpha_c V_2/3 = -{}^2/_3 V_2^2/\sqrt{V_2^2 + V_3^2}$.

The second question which arises concerns any charge redistributions due to the distortions and resulting electrostatic contributions to the deformation potentials. We are thinking here of the polarization of the bonds by such effective potentials, as we discussed in Chapter 4, and a resulting electrostatic shift in the term values. It is best to think of this in terms of susceptibilities rather than dielectric constants. Thus the relevant effective potentials are those associated with the polarizable bonds, or the transfer of charge between bonds.

A local compression of the lattice shifts the energies of the hybrids which form each bond, as we have just discussed, and that nonorthogonality shift will be dominated by the nonorthogonality of the two hybrids making up the bond. It will shift each the same, so that in a homopolar crystal it does not make a polar bond, nor any bond dipole. Similarly, the local compression will shift the covalent energy V_2 without introducing bond dipoles. Further, since the bond energy was a minimum at the equilibrium spacing, under the influence of these two energies, the two shifts cancel. Not only are no bond dipoles introduced, but there are no shifts in the energies of the bonds to first order in the local compression and no induced transfer of charge between bonds. Thus there are no electrostatic contributions to the deformation potential of Eq. (7-1) for homopolar crystals. There will be electrostatic contributions in polar crystals, to which we shall return. This same argument ruling out electrostatic contributions to the deformation potentials applies also when metallization is included since the argument depends only upon the equilibrium condition. For a crystal under hydrostatic pressure there presumably would be such contributions.

B. Predicted Deformation Potentials

Thus we obtain the deformation potential directly by substituting Eq. (5-18), with the plus sign for the conduction band, in Eq. (7-1), using the $1/d^2$ dependence of $V_{ss\sigma}$ upon bond length to obtain the deformation-potential constant for the shift relative to the average atomic (or hybrid) energies, and adding the contribution $-2\alpha_c V_2/3$ for the shift in the hybrid energies due to nonorthogonality. We obtain $\delta E_c = D_c \Delta$ with

$$D_c = \frac{-32 V_{ss\sigma}^2}{3\sqrt{\left(\frac{\varepsilon_{s+} - \varepsilon_{s-}}{2}\right)^2 + (4V_{ss\sigma})^2}} - \frac{2\alpha_c V_2}{3}. \tag{7-2}$$

Similarly for the valence-band maximum ($\delta E_v = D_v \Delta$) we use the band energy from Eq. (5-19), with the minus sign for the valence band, again adding the shift of the hybrid energies, and obtain a deformation potential constant

$$D_v = \frac{2 V_{xx}^2}{3\sqrt{\left(\frac{\varepsilon_{p+} - \varepsilon_{p-}}{2}\right)^2 + V_{xx}^2}} - \frac{2\alpha_c V_2}{3}. \tag{7-3}$$

with again $V_{xx} = {}^4/_3 V_{pp\sigma} + {}^8/_3 V_{pp\pi} = 1.28\hbar^2/md^2$.

For indirect-gap semiconductors the deformation potential for the conduction band is obtained from the corresponding derivative of the band energy. For silicon it would be obtained from the energy $E(X_1)$ of Eq. (5-20) (with the plus sign) since the conduction-band minimum in silicon is quite close to the point X . Note that only in that case the energy ε_p' enters and $d\partial\varepsilon_p'/\partial d = -(32/3)\lambda_{sp\sigma}\hbar^2/md^2$ is +5.88 eV for silicon. We must include again the contribution $-2\alpha_c V_2/3$ for the shift in the hybrid energies due to nonorthogonality. Including all three terms we have $\delta E_c = D_{X1}\Delta$, with

$$D_{X1} = \frac{-32V_{sp\sigma}^2}{9\sqrt{\left(\frac{\varepsilon_{p-} - \varepsilon_{s+}}{2}\right)^2 + \frac{16V_{sp\sigma}^2}{3}}} - \frac{2\alpha_c V_2}{3}$$

$$+ \left(1 + \frac{\frac{\varepsilon_{p-} - \varepsilon_{s+}}{2}}{\sqrt{\left(\frac{\varepsilon_{p-} - \varepsilon_{s+}}{2}\right)^2 + \frac{16V_{sp\sigma}^2}{3}}}\right)\frac{d}{6}\frac{\partial\varepsilon_p'}{\partial d}\,. \qquad (7\text{-}4)$$

We have evaluated it for the lower of the two X_1 conduction bands for the polar case; for silicon the $\pm$ subscripts on the atomic term values disappear and the two conduction bands at X_1 are degenerate. Values for these constants for a number of systems are given in Table 7-1.

We find the deformation-potential constants for all conduction-band states to be large and negative because they are antibonding states, which drop in energy with increasing spacing and therefore decreasing coupling. The second contribution, due to nonorthogonality, is also negative and slightly smaller. For the conduction-band states at Γ the magnitude of the constants is very much larger than those for the valence bands, which are bonding states. Because of the smaller bonding antibonding splitting the corresponding *positive* contribution is smaller and cancels part of the negative nonorthogonality contribution.

Both of these shifts of the bands at Γ are roughly in accord with the results of Cardona and Christensen (1987). Their calculations are intrinsically much more accurate than ours but because of the large effect of the screening fields in the context of their calculation, typically six eV, and their uncertainty, it is not clear how much more reliable theirs are. They obtain typically -7 eV for D_c and -1.5 eV for D_v for a similar set of semiconductors, very similar to our values.

They typically obtain 0.7 eV for D_{X1}, the deformation potential constant for the conduction-band values at X, quite different from our values near -4. eV. It seems likely that we have considerably underestimated the final term in Eq. (7-4), which is positive and equal to 1.60 eV for silicon. However, it would require quadrupling that term to obtain agreement with Cardona and Christensen's value for the deformation-potential constant, corresponding to a variation as $1/d^8$ rather than $1/d^2$. Furthermore we shall see that such a scaling up seems not necessary for the germanium value for the L-point. It would appear that correcting the conduction bands away from the Γ-point

Table 7-1. Deformation potentials, in eV, obtained from Eqs. (7-2), (7-3), (7-4), and (7-6) for the conduction band at Γ, the conduction-band at X for indirect-gap semiconductors, the valence band at Γ, and the valence-band shift under shear, respectively. The constant $D_v(shear)$ should be reduced by a factor $1/\sqrt{3}$ due to spin-orbit coupling, as indicated in Eq. (7-11)

	D_c	D_{X1}	D_v	$D_v(shear)$
Si	-7.81	-3.73	-1.78	-6.75
Ge	-7.25	-3.27	-1.65	-6.26
AlP	-6.71	-4.43	-1.77	-3.31
AlAs	-6.34	-4.10	-1.66	-3.47
GaP	-6.83	-4.31	-1.80	-3.27
GaAs	-6.34		-1.65	-3.32
GaSb	-5.70		-1.45	-3.23
InP	-5.43		-1.48	-2.07
InAs	-5.15		-1.43	-2.26
InSb	-4.74		-1.25	-2.26
ZnSe	-4.83		-1.39	-0.79
CeTe	-3.62		-1.08	-0.36
CuBr	-3.83		-1.11	+0.54

requires something more complicated than the $\lambda_{sp\sigma}$ term which we introduced in Eq. (5-15).

C. Experimental Values

Cardona and Christensen (1987) have also discussed the *experimental* values, and found them somewhat uncertain, largely because deformation potentials must be obtained through their effect on electron scattering. They noted, for example, values of 4.5 eV and 8.2 eV from two different experiments for the *magnitude* of the conduction-band deformation potential for indium antimonide, to be compared with their theoretical estimate of $D_c = -6.6\ eV$ and our estimate of $D_c = -4.74\ eV$.

Measurements of changes in band *gaps* under pressure can be more reliably obtained theoretically because the effects of the screening fields and of the nonorthogonality terms $-2\alpha_c V_2/3$ cancel out and experimentally since they are obtained from variations of optical-absorption peaks under pressure rather than for example fitting to electron-mobility data. Paul and Waschauer (1963) compiled a series of optically determined values for the change in E_0, which are compared with our values of $D_c - D_v$ in Table 7-2. They found considerable variation from one experiment to another even in these values. There are certainly more recent and accurate values, which we have not sought out, but they probably do not change the picture greatly. We seem to have underestimated the values somewhat, but the trends we find are similar to those observed.

It is of interest also to compare our estimates of the direct-gap deformation potentials with the much more accurate calculations by Cardona and Christensen (1987) for a range of systems, which is also done in Table 7-2. Theirs would seem to be in accord with experiment within the accuracy of the experiment. The full calculations for E_0 by Cardona and Christensen *should* be in accord with experiment since they come simply from comparing band calculations for different volumes, both of which are expected to accurately represent the bands. The most serious concern is with the Coulomb enhancements of the gap discussed in Section 5-5-A, but they are quite small. For silicon, for example, the conduction-band is raised by these effects by U/ε equal to about 0.63 eV according to Table 5-3. $\varepsilon = 1+4\pi\chi^{(1)}$ varies approximately in proportion to d (Eq. (4-10) for $\alpha_c = 1$ and neglecting the 1 in comparison to $4\pi\chi^{(1)}$) so the resulting $\Omega\partial E_c/\partial\Omega$ contribution is only $^1/_3$ 0.63 eV and quite negligible. This contribution would not be included in the Cardona-Christensen calculation.

Our predictions of band-gap deformation potentials are further off for the indirect band gaps of silicon as expected, because of our apparent errors in the $D_c(X)$. We find a $D_c(X) - D_v$ which is negative for both silicon and germanium (-1.95 and -1.61 eV, respectively) while Van de Walle and Martin(1986) find positive values for both (1.72 eV and 1.31 eV), in agreement with experiment (Laude, Pollak, and Cardona (1971) give a value 1.50±0.3 eV for silicon, no value for germanium). Similarly we obtain -3.10 eV for the germanium conduction-band minimum at L, but quite surprisingly this appears to be approximately correct. Van de Walle and Martin obtain -2.78 eV for the L-point in germanium in accord with experiment (Balslev (1966) gives -2.0±0.5 eV). We nevertheless conclude

Table 7-2. Comparison of the calculated combined deformation potentials, $D_c - D_v$ from Table 7-1, in eV, theoretical estimates (C&C) by Cardona and Christensen (1987) with the experimental $\Omega\partial E_0/\partial\Omega$ compiled by Paul and Waschauer (1963).

	$D_c - D_v$ (Table 7-1)	$D_c - D_v$ (C&C)	$D_c - D_v$ (Experiment)
Ge	-5.6	-	-9.
GaAs	-4.7	-7.2	-8.
GaSb	-4.3	-8.4	-7.9
InP	-4.0	-5.5	-6.2
InAs	-3.7	-6.7	-3.8
InSb	-3.5	-5.8	-6.4

that our treatment of the conduction bands at the Brillouin Zone faces is not reliable.

It is interesting that our fitting tight-binding parameters to the observed bands was similar to fitting them to the free-electron bands, and free-electron bands *do* lead to an X_1 level rising relative to the $\Gamma_{25'}$ level with increasing volume. The free-electron energies (as can be seen in Fig. 5-5) for $\Gamma_{25'}$ and X_1 are $3(\hbar^2/(2m))(2\pi/a)^2 = (9\pi^2/8)\hbar^2(md^2)$ and $(3\pi^2/4)(\hbar^2/(2md^2))$, respectively, and the difference $(3\pi^2/8)\hbar^2/(md^2)$ decreases with increased volume, corresponding to a $D_c(X) - D_v = 3.40\ eV$ for silicon, positive as is experiment but of opposite sign to what we obtained. The free-electron L-point comes at $(33\pi^2/32)\hbar^2/(md^2)$ so the difference from $\Gamma_{25'}$ and the corresponding deformation potential are only a quarter as big. This partially explains the large difference between the observed $D_c(X) - D_v$ and the $D_c(L) - D_v$ values.

2. Band Shifts Due to Shear Distortions

Strictly speaking a shear distortion cannot give rise to a deformation potential $V(\mathbf{r})$ since, by symmetry, an elongation in the x-direction and contraction in the y-direction ($e_2 = -e_1$) must give the same potential as with the directions interchanged, and therefore the sign reversed. However, such a shear could raise the energy of electrons near a silicon conduction-band minimum in the x-direction and lower those at the minimum in the y-direction, reversing the shifts if the strains were reversed. Thus we can introduce deformation-potential-like shifts, which depend upon the wavenumber of the state under consideration. Such shifts have been treated carefully for silicon and germanium by Van de Walle and Martin (1986) and we shall compare our tight-binding predictions with their results.

Shear distortions can not give rise to screening potentials for the same reason that they cannot produce scalar deformation potentials. A similar argument indicates that the nonorthogonality terms cannot lead to a linear shift in the average term values due to shear. However, shear can lead to shifts of some p-states relative to others and these enter our deformation potentials for shear. We have ignored such effects of different shifts for different states up to this point, as when in Section 2-2 we replaced all of the effects of nonorthogonality by a single overlap repulsion in the treatment of bonding properties. Their contributions to the deformation potentials are significant and we consider them here.

A. The Effects of Nonorthogonality

The shift of each level, due to interaction with a neighboring orbital, is given by $-S_{ij}V_{ij}$, the negative of the product of the overlap S_{ij} and the coupling V_{ij} , as we indicated in Section 1-2 and in 2-2-E. V_{ij} and S_{ij} will have the same dependence upon orientation of the orbitals relative to the internuclear separation and in Extended Hückel Theory (R. Hoffmann (1963)) it is assumed that V_{ij} is given by a universal constant times $(\varepsilon_i + \varepsilon_j)S_{ij}/2$. Actually the ratios of these interatomic matrix elements, $V_{ll'm}$, and nonorthogonalities, $S_{ll'm}$, from one m- value to another (at fixed l and l') have been obtained by Wills (Wills and Harrison (1983)) using a muffin-tin-orbital approach. It was seen that V_{ij} does not vary from one to the other as S_{ij} does, but it is close enough to make a meaningful approximation. Thus we take the shift of the two levels i and j as proportional to $V_{ij}^2/(\varepsilon_i + \varepsilon_j)$ with the proportionality constant C chosen to give the correct average shift associated with the overlap repulsion. The total shift in the state with initial energy ε_i is estimated as

$$\delta\varepsilon_i = C\sum_j \frac{V_{ij}^2}{\varepsilon_i + \varepsilon_j} , \tag{7-5}$$

with the sum over all neighboring states. (Again, this is from Extended Hückel Theory, not second-order perturbation theory.) We may write this for a homopolar system (so only one ε_s and one ε_p enter) for a single neighbor, for an s-state and for σ- and a π-oriented p-states as

$$\delta\varepsilon_s = C\left(\frac{V_{ss\sigma}^2}{2\varepsilon_\sigma} + \frac{V_{sp\sigma}^2}{\varepsilon_s + \varepsilon_p}\right),$$

$$\delta\varepsilon_{p\sigma} = C\left(\frac{V_{sp\sigma}^2}{\varepsilon_s + \varepsilon_p} + \frac{V_{pp\sigma}^2}{2\varepsilon_p}\right), \tag{7-6}$$

$$\delta\varepsilon_{p\pi} = C\frac{V_{pp\pi}^2}{2\varepsilon_p} .$$

We evaluate the constant C by noting that the overlap repulsion is obtained by adding the shift of all eight electrons in hybrid states, or equivalently the sum of shifts for all eight atomic states, $V_0 = 2(\delta\varepsilon_s + \delta\varepsilon_{p\sigma} + 2\delta\varepsilon_{p\pi})$, and equating them at the equilibrium spacing d_0 where $V_0 = V_2$ It is convenient in doing this to take $\varepsilon_p = \varepsilon_s/2$, which we saw in Chapter 1 to be an excellent approximation. This leads to $C = 1.12\varepsilon_p/V_2(d_0)$. We then rewrite the above shifts, letting the p-state have an axis with an angle θ with the internuclear distance, in terms of the $\eta_{ll'm}$ rather than the $V_{ll'm}$ to obtain the convenient forms

$$\delta\varepsilon_s = 0.119 V_2 \frac{d_0^2}{d^2},$$
$$\delta\varepsilon_p = (0.316\cos^2\theta + 0.021)V_2 \frac{d_0^2}{d^2}, \tag{7-7}$$

both varying as $1/d^4$. The average over angle of the p-state shift is $0.127V_2d_0^2/d^2$, only 6% greater than that for the s-state so it was a perfectly good approximation to take a single shift in all term values when we looked at the effects of pure dilatations, contributing the term $-2\alpha_c V_2/3$ in Eqs. (7-2), (7-3) and (7-4). We need these individual shifts in the energies of orbitals due to nonorthogonality with neighboring orbitals when we construct deformation potentials for the effects of shear.

B. Indirect-Gap Conduction Bands

The simplest case is the twist distortion which we discussed for elasticity theory, $e_2 = -e_1$, $e_3 = 0$. We have seen by symmetry that this cannot shift the conduction-band minimum at Γ to first order in the strain. However, it can shift the conduction band of silicon, near X , and the valence bands. With this distortion, the bond lengths do not change to first order but the angle θ between, for example, $|p_x\rangle$ states on neighboring atoms and the direction of their internuclear separation does change. We shall make our task easier if we are careful to define our terms for convenient analysis of general strains. We allow strains e_1, e_2, and e_3 and we may readily obtain the corresponding angle θ between the p_x-state and the vector distance to the neighbors, to first order in e_1 , as given by

$$\cos\theta = \frac{1}{\sqrt{3}}(1 + e_1 - {}^1/_3(e_1 + e_2 + e_3)). \tag{7-8}$$

For volume-conserving strains the final term does not contribute. Angles for the p_y- and p_z-states are obtained by rotating indices. We have already obtained the contributions from a dilatation, $e_1 + e_2 + e_3 = \Delta$, and it is not difficult to see that a shear strain, e_4, e_5, or e_6 , does not shift the energy of the conduction-band minimum in a [100] direction to first order in the strain.

This follows from the screw-axis symmetry discussed in Section 2-1. This is a rotation of 90° around the x-axis and a displacement of $a/4$ in the x-direction which takes the undistorted crystal into itself, and leaves the same minimum in the [100] direction, but changes the sign of the strain e_4 and, if performed twice, changes the sign of e_5 and e_6 .

The energy of the conduction-band minimum in silicon, near X, depended upon $V_{sp\sigma}\cos\theta$, so that in Eq. (5-20) the $16V_{sp\sigma}^2/3$ becomes $16V_{sp\sigma}^2(1+e_1 - \Delta/3)^2/3$. In just the same way, the correction in ε_p' becomes $16\lambda_{sp\sigma}\hbar^2(1 + e_1 - \Delta/3)^2/(3md^2)$. Finally, the p-state shift due to nonorthogonality is obtained from Eq. (7-7) with $\cos^2\theta = {}^1/_3(1 + 2e_1 + ...)$, giving an additional shift in ε_p' of $0.316V_2(8e_1/3)$ from nonorthogonalities with the four neighbors. All three effects contribute giving a deformation-potential constant $D_{X1}(shear)$, which we define such that the effective potential shift of the minimum in the x-direction is $D_{X1}(shear)(e_1 - \Delta/3)$. Here again $\Delta/3 = (e_1 + e_2 + e_3)/3$ is the average strain, zero in the case of $e_2 = - e_1$. We have taken care to make this definition systematic and easy to use. The effect of the shears e_1 , e_2 , and e_3 on the [100] conduction bands are described by this constant. With this definition, $\delta E_c = D_{X1}(shear)(e_1 - \Delta/3)$, and

$$D_{X1}(shear) = \frac{16V_{sp\sigma}^2}{3\sqrt{\left(\frac{\varepsilon_{p-} - \varepsilon_{s+}}{2}\right)^2 + 16V_{sp\sigma}^2/3}} + \left(1 + \frac{\frac{\varepsilon_{p-} - \varepsilon_{s+}}{2}}{\sqrt{\left(\frac{\varepsilon_{p-} - \varepsilon_{s+}}{2}\right)^2 + \frac{16V_{sp\sigma}^2}{3}}}\right)\left(\frac{16\lambda_{sp\sigma}\hbar^2}{3md^2} + 0.842V_2\right) \tag{7-9}$$

Of course for the minima in the y-direction the shift is $D_X(shear)(e_2 - \Delta/3)$. None of these minima, as we indicated, are shifted to first order by a strain e_4 , e_5 , or e_6 . The first term in Eq. (7-9) is -3/2 times the first term in Eq. (7-4). The second term is -3 times the third term in Eq. (7-4). The term $0.842V_2$ replaces the second term in Eq. (7-4). The factors of three arise because of the proportionality of the shift to e_1 rather than the dilatation Δ . For Eq. (7-9) both terms were taken proportional to the same $(1 + e_1 - \Delta/3)^2$. The first term is positive because the distortion enhances the coupling $V_{sp\sigma}\cos\theta$ for p-states oriented along the x-direction so that the coupling with the states in the valence band at lower energy raises the energy more.

The term in $\lambda_{sp\sigma}$ is negative ($\lambda_{sp\sigma} = -0.40$) because it arises from coupling with the peripheral s-states above. The nonorthogonality term, $0.842V_2$ is positive because the shear increases the overlap of the p_x-states with their neighbors.

The values of $D_{X1}(shear)$ from Eq. (7-9) for Si, Ge, AlP, AlAs, and GaP in eV are 4.83, 4.31, 5.34, 5.04, and 5.34, respectively. This result for silicon differs from that given by Van de Walle and Martin by an amount comparable to the discrepancies we found for the D_{X1} for dilatation. They find a value of +9.16 eV for this constant for silicon (which is $\Xi_u{}^{\Delta}$ in their notation), close to the experimental value of 8.6±0.4 eV from Laude, Pollak, and Cardona (1971).

For strains $e_2 = -e_1$, and $e_3 = 0$, the shift for the conduction-band minima in the y-direction, [010] , is the negative of that in the x-direction and there is no shift of the minimum in the z-direction. Such shear strains have the interesting effect of lifting the degeneracy of the six conduction-band minima. This enters a number of properties, such as the conductivity in a particular direction.

C. Valence Bands

The coupling determining the valence band maxima ($E_v = \varepsilon_p - V_{xx}$ for a homopolar semiconductor) for this same strain is

$$V_{xx} = 4(V_{pp\sigma}\cos^2\theta + V_{pp\pi}\sin^2\theta)$$

$$= {}^4/_3V_{pp\sigma} + {}^8/_3V_{pp\pi} + {}^8/_3(V_{pp\sigma} - V_{pp\pi})e_1 \qquad (7\text{-}10)$$

$$= 1.28\frac{\hbar^2}{md^2} + 7.60\frac{\hbar^2}{md^2}e_1 \,.$$

There is in addition the shift of $0.316V_2(8e_1/3) = 0.842V_2\, e_1$ for every p-state due to the nonorthogonality with its neighbors which we obtained following Eq. (7-8). Thus the deformation potential constant $D_v(shear)$ is defined such that the shift in the energy of the states based upon p_x-states is $\delta E_v = D_v(shear)\,(e_1 - \Delta/3)$. It is

$$D_v(shear) = -\frac{5.94V_{xx}^2}{\sqrt{\left(\frac{\varepsilon_{p+} - \varepsilon_{p-}}{2}\right)^2 + V_{xx}^2}} + 0.842V_2\,, \qquad (7\text{-}11)$$

where the 5.94 was from 7.60/1.28 . For the uniform strain $e_2 = -e_1$, $e_3 = 0$, with e_1 positive the band based upon p_x-states drops by $|D_v(shear)\, e_1|$ because $D_v(shear)$ is negative, that based upon p_y-states rises by that amount and the third remains unshifted to first order. Values for this parameter were listed in the final column of Table 7-2. The deformation potential parameter b defined by Pollak and Cardona (1968) is related by $D_v(shear) = 3b$. Using this conversion we see that the theoretical estimate of $D_v(shear)$ by Van de Walle and Martin (1986) for silicon was -7.05 eV, and the experimental value they quote (Laude, Pollak, and Cardona (1971)) is -6.3 eV, in comparison to our -6.75 eV. The corresponding values for germanium are -7.65 eV and -8.58 eV, in comparison to our -6.26 eV. These give some support to the nonorthogonality corrections which we gave in Eq. (7-7), but we shall find that with spin-orbit coupling included the deformation potential $D_v(shear)$ is reduced by a factor $1/\sqrt{3}$ so these values which appear to be in good accord with experiment are in fact quite a bit too small.

Van de Walle and Martin also predict the shift in eigenvalues if the crystal is elongated in a [111] direction and contracted in a lateral direction . The corresponding shift of the p-band based on a p-state along that same $[111]$ direction is written $\delta E_v(\mathbf{r}) = D_v'(shear)\,(e_{111} - \Delta/3)$. The deformation potential constant d which they evaluate is related to ours by $D_v'(shear) = \sqrt{3}d$. The value they predict for silicon corresponds to $D_v'(shear) = -9.21$ eV, to be compared with an experimental -8.40 eV. For germanium they predict -9.52 eV, compared to an experimental -9.14 eV. Values for these from tight-binding theory are not immediate because they correspond to strains e_4 and, as we have seen, internal displacements which affect the results. The values for the two directions are sufficiently close that for many purposes it will be adequate to take a single value of $D_v'(shear) = D_v(shear)$ for all orientations of shear strain.

For the particular distortion with e_1 positive and $e_2 = -e_1$, we may see what happens to the bands near Γ , looking first at the bands without spin-orbit coupling. There is a shift in the bands at and near Γ and for long-wavelength vibrational modes we may take that shift as independent of wavenumber. The light-hole band in the [100] direction is based upon p_x-states and is lowered in energy, pulling the constant-energy surface at a particular energy closer to the origin in the x-direction. The light-hole band in the y-direction, based upon p_y, is similarly raised and the constant-energy surface moved out, while it is unshifted in the z-direction. The constant-energy surface for the light-hole band is thus sheared in the opposite sense from the shear of the lattice. The constant-energy surface for one heavy-hole band is sheared in the same sense as the lattice, and by a much larger amount than for the light-hole band since the magnitude of the energy shift is the same but $\partial E/\partial k$ is much smaller. The other heavy-hole band is unshifted. These shifts are only estimated for cube directions, [100], etc.. In the isotropic approximation of Fig. 6-3 we would distort two of the spheres into ellipsoids.

Furthermore, the graphitic distortion discussed for the elastic constant c_{44} corresponds to an elongation of the lattice in the [110] direction, or in a [111] direction, and contracting it in lateral directions. It is much more complicated to calculate in detail because of the internal displacements. In the approximation that the internal displacements hold the bond lengths fixed (only the angles change) the distortion has the effect of distorting the light-hole constant-energy surfaces in the opposite sense to the distortion of the lattice, and the heavy-hole constant-energy surfaces in the same sense as the lattice. It would be possible to proceed in more detail, but the result is inevitably complicated. It seems most reasonable at the start to approximate the effects of lattice distortion on the heavy-hole band by the shift given by Eq. (7-3) for dilatations of the lattice, and a shift given by Eq. (7-11) for all shear distortions.

D. Shifts of the Spin-Orbit Split Bands

It is straightforward to estimate the shifts for the band including the spin-orbit splitting. This was done earlier by Pollak and Cardona (1968). For dilatations the theory is essentially unchanged since the spin-orbit splitting at $k = 0$ is an additive effect (as can be seen from Eq. (6-26) for $k = 0$) and the same deformation-potential constants are obtained for the split bands. That is not true in the polar case (as can be seen from Eq. (5-33) the spin-orbit splitting enters in a more complicated way), but the same deformation potential constants are probably a good approximation for the polar case, at least on the scale of the accuracy of the theory.

The effects of shear distortions are more complicated, even in the homopolar case. The analysis is straightforward but it requires more algebra than one might expect and the results are a little surprising. We proceed as in Section 6-5-C for including spin-orbit coupling for homopolar semiconductors, but now with $k = 0$. The diagonal elements will be different for the three states because the V_{xx} matrix elements are different for p-states oriented along different axes if the lattice is distorted(the V_{xy}-matrix elements do not contribute at $k = 0$). We choose the distortion $e_2 = -e_1$ which we treated in detail above. The transformation which led to Eq. (6-24) can still be made, leading to

$$H_{3\times3} = \begin{pmatrix} \varepsilon_p - V_{xx} & -i\lambda & \lambda \\ i\lambda & \varepsilon_p - V_{yy} & -i\lambda \\ \lambda & i\lambda & \varepsilon_p - V_{zz} \end{pmatrix} = \begin{pmatrix} E_v + De_1 & -i\lambda & \lambda \\ i\lambda & E_v - De_1 & -i\lambda \\ \lambda & i\lambda & E_v \end{pmatrix} . \tag{7-12}$$

In the last form we have written the unsplit valence-band maximum as $E_v = \varepsilon_p - V_{xx}$, where the term value ε_p includes the shift $\alpha_c V_2/2$ due to nonorthogonality in the absence of shear, and used the deformation-potential constant $D = D_v(shear)$ of Eq. (7-11). The secular equation can readily be obtained from the last form. Measuring energies from E_v it is

$$-\varepsilon(\varepsilon^2 - D^2 e_1^2) + 3\lambda^2\varepsilon - 2\lambda^3 = 0 . \tag{7-13}$$

We require a solution for small strains e_1 but they do not seem to come easily from Eq. (7-13). We may, however, obtain the formal solution of cubic equations (the first time the author has ever found these solutions useful) and expanding for small e_1 to obtain

$$\varepsilon = E_v + \lambda + \frac{D_v(shear)\, e_1}{\sqrt{3}} ,$$

$$\varepsilon = E_v + \lambda - \frac{D_v(shear)\, e_1}{\sqrt{3}} , \tag{7-14}$$

$$\varepsilon = E_v - 2\lambda ,$$

plus terms of order e_1^2 . This is quite similar, but not identical, to the result obtained by Pollak and Cardona (1968). Their Eq. (6) was equivalent to this but with the $\sqrt{3}$ replaced by 2 . We believe that they made an error in neglecting the coupling between states of positive and negative m_J (the quantum number for the z-component of the total angular momentum J) on the grounds that the Kramers (spin) degeneracy was not lifted. It would appear that the neglected coupling modifies the results without, of course, eliminating the Kramers degeneracy.

A second way of obtaining this result, Eq. (7-14), is to solve for the modes, $u_x|p_x\rangle + u_y|p_y\rangle + u_z|p_z\rangle$, without the De_1 terms in Eq. (7-13). The corresponding three equations for the energy $E_v + \lambda$ are all $u_z = u_x + iu_y$. Two orthonormal solutions are $|1\rangle = [u_x, u_y, u_z] = [1, i, 0]/\sqrt{2}$ and $|2\rangle = [1, -i, 2]/\sqrt{6}$. We may seek the normalized combinations, $\cos\theta|1\rangle + \sin\theta|2\rangle$, which have the maximum and minimum expectation value of the

De_1 terms in the Hamiltonian, $2cos\theta sin\theta De_1/\sqrt{3}$. This becomes the shift in first-order perturbation theory and selects the right combination. The solutions are at $\theta = \pm \pi/4$ and are those given in Eq. (7-14). This will be the most convenient way to obtain the solutions in the next section for a different set of strains.

In any case, the spin-orbit coupling has done more than split the levels; it has forced a mixture of the two-upper bands such that the deformation potential is considerably reduced. In this same way we found that spin-orbit splitting mixed the light- and heavy-hole bands and gave intermediate masses. Again constant-energy surfaces in the heavy-hole band will be distorted in the same sense as the distortion of the lattice but the amount of the deformation is reduced by a factor of $1/\sqrt{3}$. We saw in Part C that our predictions of the deformation potential were in rather good accord with experiment before this correction was made, so this factor spoils the good agreement.

3. The Electron-Phonon Interaction

In Section 3-3 we calculated the frequencies of lattice vibrations as a function of wavenumber $\boldsymbol{q}$. For the zincblende or diamond structure, with two atoms per primitive cell, there were six modes for each wavenumber within the Brillouin Zone (acoustic and optical, longitudinal and transverse for wavenumbers in symmetry directions). Each of these is a harmonic oscillator for which we calculated the frequency. In quantum theory the excitations of a harmonic oscillator of frequency ω are quantized and given by $\hbar\omega(n + {}^1/_2)$, with n an integer, so the vibrational energy in each vibrational mode of the lattice is quantized in this way. It is customary in solids to say that a mode in such an excitation state contains n *phonons*, and we shall use that terminology here though the quantization is not essential to our discussion. For a simple lattice, with every atom identical, we found in Eq. (3-27) the kinetic energy, equal also to the potential energy, to be given by ${}^1/_2 M u_q u_{-q} \omega^2$ in terms of the amplitude vectors of Eq. (3-23).

The presence of such phonons can give an interaction between electron states, called the electron-phonon interaction, just as the presence of light gave a term in the Hamiltonian, Eq. (6-2), which provided a coupling, Eq. (6-4), between electronic states. An extensive discussion of this interaction for semiconductors has been given by Cardona (1982) and a recent study, with further references, has been given by Shields, Popovíc, Cardona, Spitzer, Nötzel, and Ploog (1994). Here we give a discussion in the context of tight-binding theory.

The difference in wavenumber for the coupled electronic states will equal the wavenumber of the phonon (sometimes plus or minus a lattice wavenumber) as we shall see in Section 7-3-B. In semiconductors the occupied electron states (or empty hole states) are ordinarily within k_BT of the band edges so their wavenumbers tend to be quite small (or at least the differences between electron wavenumbers within the same valley if the band minimum occurs far from $\mathbf{k} = 0$ as in silicon). In this circumstance the most important phonons are those of small wavenumber, or long wavelength. Further, at low temperatures the long-wavelength modes have lower frequencies and are more strongly excited. We shall focus our attention on long-wavelength acoustical phonons, though tight-binding theory - in contrast to effective-mass theory - is just as appropriate to the treatment of modes at wavelengths comparable to the lattice spacing. At various points we shall note features of short-wavelength systems and shall return later to optical-mode phonons.

When the wavelengths of acoustic modes are long compared to a bond length it has seemed natural to try to understand the interaction with the electrons by looking at local regions where the lattice strain due to the vibrations is quite uniform and the electron energies are shifted locally by the energies we estimated in the preceding section. This is the basic idea of the *deformation potentials* , which were postulated by Bardeen and Shockley (1950), and which we introduced in Section 7-1. We generalized these earlier to optical-mode vibrations (Harrison (1956b)). We shall show in Section 7-3-A that the matrix elements of the electron-phonon interaction can be correctly calculated, in a long-wavelength limit, from such deformation potentials calculated from uniform distortions. This may be misleading since we shall also see that although uniform rotations of an entire system do not change the energy of the eigenstates, local rotations can contribute to the electron-phonon matrix elements. Thus the deformation concept can lead one to omit an important part of the interaction arising from phonons. It is still of course possible to define deformation potentials which correspond to the full electron-phonon interaction (Harrison (1956b), Resta (1991)). However, calculations on uniform systems such as we have described do not give all of the corresponding deformation-potential constants.

There is a second complication which we shall return to in Section 7-3-G. When strains are nonuniform they may lead to charge accumulation, either from the polarization of bonds or from the displacement of charged carriers, and the electrostatic potentials arising from these charges are part of the interaction. They do not arise for uniform strains so they did not enter before and here we postpone their discussion to the end, except for the electron-phonon interaction in polar crystals. In that case we also include a reduction of the resulting electrostatic potentials from the lattice charges by a factor of the dielectric constant.

A. Interactions for Simple Systems

The formulation of the electron-phonon interaction, when the principle shifts arise from interatomic couplings, is not as obvious as it might seem. It may be useful first to go through a series of simple systems to understand the physics of the complications.

We begin with atoms and work up. For a single atom with an electron in an atomic s-state, we may shake the nucleus back and forth and the electron must follow in that atomic s-state if only that state is included in our representation of the states. If on the other hand we include also an atomic p-state, it is possible that the electron is excited into that state. Ultimately the corresponding rate is obtained from the Schroedinger Equation, Eq. (1-31), which tells how the wavefunction changes with time. The rate can in fact be calculated using perturbation theory, Eq. (1-32), where the matrix element arises from the change in potential due to displacing the atom. This matrix element, like the matrix element of an electric field perturbation $-e\boldsymbol{E}\cdot\boldsymbol{r}$, is not directly obtainable from tight-binding theory and we assume that it is small in comparison to interatomic terms in semiconductors. We neglect the corresponding contribution to the electron-phonon interaction, though there is in fact such a contribution.

If two atoms with single electrons in atomic s-states are combined to form a molecule, the two electrons occupy a bonding state as discussed in Section 1-2. If the two atoms are made to vibrate against each other, the coupling $V_{ss\sigma}$ varies with time and the bond level rises and drops in energy as if a potential were applied. This is describable as a "deformation potential" but it does not cause transitions to the antibonding state because the antibonding state is odd around the molecular center while the bonding state and deformation potential are even. The integral which is the matrix element is zero in the case of this symmetric molecule. If the molecule were polar, this matrix element between the bonding and antibonding state would not be zero and there could be electronic excitation from such molecular vibrations if the frequency of the vibration, times $\hbar$, were comparable to the energy difference. The same features occur for molecular orbitals based upon p-states when the two atoms vibrate against each other.

However, if we apply a rocking motion to the molecule, with the atoms moving perpendicular to their internuclear separation, a new feature arises. A $p\sigma$-bonding state may consist of $|p_z\rangle$ states and these retain that orientation as the atoms are displaced in the x-direction in the absence of any changed matrix elements, which follows from Eq. (1-31). However, a $p\sigma$-

matrix element and a $p\pi$-matrix element, proportional to the magnitude of the rocking displacement, arise between the this p_z-state and a p_x-state on the other atom. The contributions from the two cross terms add for the two atoms for coupling between a $p\sigma$-bonding state and the $p\pi$-antibonding state and can cause a transition between these states, calculated with the perturbation formula of Eq. (1-32). (Again the time-dependent perturbation provides a term $\hbar\omega$ in the energy delta-function allowing a transition if this $\hbar\omega$ equals the energy difference for the states.) Physically this transition makes sense: by rocking the molecule we may transmit angular momentum to the electron taking it to another molecular state. It is *not* a transition describable by a deformation potential in the sense that a static rotation of the molecule leaves the electron in a bonding state of the same energy without a deformation-potential shift, and yet the dynamic rocking can cause a transition. When we constructed deformation potentials for a silicon lattice, we considered the effects of shear strains $e_4 = \partial\delta z/\partial y + \partial\delta y/\partial z$ (with for example δz the displacement of atoms in the z-direction) but not rotations $\theta_1 = \partial\delta z/\partial y - \partial\delta y\partial z$, which do not shift the energy. For the electron-phonon interaction, both can play a role and we should not work exclusively in terms of strains.

We proceed next to the simple case of the chain of lithium atoms discussed in Section 1-3 to see the tight-binding representation of the electron-phonon interaction. It will then be apparent how the result is generalized to each three dimensional case

We constructed electronic states for the lithium chain given by

$$|k\rangle = \Sigma_j e^{ikjd}|s_j\rangle/\sqrt{N} \quad , \tag{7-15}$$

with any k satisfying periodic boundary conditions in the length Nd of the chain. Each s-state was coupled to its neighboring s-state by the same $V_{ss\sigma}$ producing an energy band. If we now introduce a long-wavelength compressional mode (displacements along the chain) as we did for the crystals in Section 3-3, these have the effect of modifying the couplings by $-2V_{ss\sigma}\delta d/d$ for neighbors which change their separation by δd , taking of course $V_{ss\sigma}$ as proportional to $1/d^2$. We take the displacements in the form given by Eq. (3-23),

$$\delta x_j = \Sigma_q u_q e^{iqjd} / \sqrt{N} . \tag{7-16}$$

We thought of this as a transformation of coordinates from δx_j to u_q in which we allow the u_q to be complex. To first order in the u_q each term will contribute separately and we consider the effect of one such term.

At long wavelengths jd may be regarded as a continuous coordinate x and the change in bond length locally becomes $\delta d = d\partial(\delta x)/\partial x = iqdu_q e^{iqx}/\sqrt{N}$. Thus the change in our Hamiltonian due to this term in the lattice distortion is a change in local interatomic matrix elements of $\delta H_{m,m+1} = -i2V_{ss\sigma}\, qu_q e^{iqx_m}/\sqrt{N}$, where in the exponents we do not need to distinguish between the m and the $m+1$ position. (The difference gives a term in $\delta H_{m,m+1}$ of higher order in q.) The matrix element between two states becomes

$$<k'|\Sigma_m \delta H_{m,m+1}|k> = \frac{1}{N^{3/2}}\Sigma_{n,j,m} e^{-ik'nd} <s_n|\delta H_{m,m+1}|s_j> e^{ikjd}\,. \qquad (7\text{-}17)$$

Now for each m in the Hamiltonian sum there will be two contributions from the sum over indices n and j for the electron states, one when $n = m$, $j = m+1$, and one when $n = m+1, j = m$. Thus

$$<k'|\Sigma_m \delta H_{m,m+1}|k> = \frac{-i2V_{ss\sigma}qu_q}{N^{3/2}}\Sigma_m[e^{-ik'md}e^{iqmd}e^{ik(m+1)d} + e^{ik'(m+1)d}e^{iqmd}e^{ikmd}] \qquad (7\text{-}18)$$

$$= \frac{-i2V_{ss\sigma}qu_q}{N^{3/2}}[e^{ikd} + e^{-ik'd}]\Sigma_m e^{i(k+q-k')md}\,.$$

With each wavenumber satisfying periodic boundary conditions on the chain the sum will vanish unless $k' = k + q$ (or $(k' - k - q)d$ is an integral multiple of 2π, which we shall come back to). If these conditions are satisfied, every term is unity and the sum is N. At long wavelengths we can take the quantities e^{ikd} and $e^{-ik'd}$ equal to one. We have the electron-phonon matrix element

$$<k+q|H_{e\phi}|k> = <k+q|\Sigma_m \delta H_{m,m+1}|k> = -i4V_{ss\sigma}qu_q/\sqrt{N}\,. \qquad (7\text{-}19)$$

Matrix elements for all other combinations of wavenumbers vanish. These are the matrix elements of the electron-phonon interaction for this simple system. In the full quantum treatment (see, for example, Harrison (1970), 414ff) this matrix element gives the absorption of a quantum of vibrational energy taking the electron from the state $|k>$ to the state $|k+q>$, conserving momentum $\hbar k + \hbar q = \hbar(k+q)$. It turns out (*ibid*) that the complex conjugate of this term gives the emission of a phonon of wavenumber $-q$, also conserving momentum. In the time-dependent perturbation theory of such processes the conservation of energy is enforced by the energy delta function of Eq. (1-32).

We noted above that a nonzero matrix element would also be obtained if, for example, $(k' - k - q)d$ were equal to 2π. Then each term in the sum would be $e^{-i2\pi m} = 1$. Mathematically this is because, as we have indicated, the wavenumbers which differ by a lattice wavenumber are equivalent and correspond to the same states. Physically we could say that an electron can be diffracted by the lattice (changing its wavenumber by a lattice wavenumber) as it absorbs or emits a phonon. It does not come up here because we were discussing electron states with small wavenumbers and phonons with small wavenumbers so a lattice-wavenumber change takes us out of the range of wavenumbers we are considering.

For this case, as for the case of bonding s-states in a diatomic molecule discussed above, we may match our result, Eq. (7-19), with deformation-potential theory. By changing the atom spacings locally the lattice vibration has shifted the band energy, $\varepsilon_s + 2V_{ss\sigma}\cos kd$, locally. For long wavelengths the shift was $2\delta V_{ss\sigma} = -4V_{ss\sigma}\delta d/d$. In a continuum approximation this is like a potential $V(x) = -4iV_{ss\sigma}qu_qe^{iqx}/\sqrt{N}$, obtained using the continuum form of Eq. (7-16) which we gave following that equation. Then free electrons would have matrix elements of this perturbation given by $\langle k'|V(x)|k\rangle = (1/L)\int dxV(x)e^{i(k-k')x}$. If $k' = k + q$ we obtain again $-4iV_{ss\sigma}qu_q/\sqrt{N}$. This confirms the plausible deformation-potential view for this case. If we extended this to p-like states in the chain and transverse vibrations we would find other cases, analogous to the rocking motion in the diatomic molecule, where deformation potentials did not describe the matrix elements. As we proceed to the three-dimensional case for tetrahedral semiconductors we will see that the matrix elements of the electron-phonon interaction for interaction between conduction-band states within a single valley and for hole scattering near the valence-band maximum can be calculated from the deformation potentials we obtained in Sections 7-1 and 7-2, but some other matrix elements can not.

B. Acoustic Modes and Conduction Electrons

We first note that ordinarily when a phonon is absorbed or emitted the change in energy of the electron is negligible compared to the electrons initial energy. The phonon energy is $\hbar\omega = \hbar qv_s$ where v_s the speed of sound, typically 10^5 cm/sec. The change in energy of the electron is $\hbar^2(\boldsymbol{k} + \boldsymbol{q})^2 - k^2)/2m^* = \hbar^2(2\boldsymbol{k}\cdot\boldsymbol{q} + q^2)/2m^* \approx \hbar qv$ with v the velocity of the electron, $\hbar k/m^*$, equal to $\sqrt{m/m^*}\ 10^7$ cm/sec. for a thermal electron at room temperature. Thus for scattering of an electron the change in energy is typically only one one-hundredth of the electron energy. For the scattering from the state $\boldsymbol{k}$ to $\boldsymbol{k}' = \boldsymbol{k} + \boldsymbol{q}$ we may take $k' \approx k$.

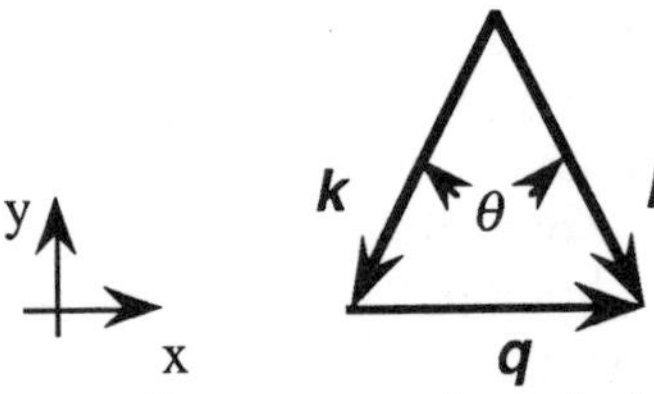

Fig. 7-1. The coordinate system for calculating the scattering of an electron of wavenumber $\boldsymbol{k}$ to a state $\boldsymbol{k'}$ by absorption of a phonon of wavenumber $\boldsymbol{q}$ for the case treated in the text, with a longitudinal phonon wavenumber (and x-axis) along the crystal axis [100], and electron wavenumbers in the x-y plane.

We first consider one case in detail, as we did with the one-dimensional chain, and see that for it the deformation potential calculated from the corresponding uniform distortion gives the entire coupling matrix element. We let $\boldsymbol{k}$, $\boldsymbol{k'}$, and $\boldsymbol{q}$ all lie in the xy-plane of the crystal, with $\boldsymbol{q}$ along a [100] direction as illustrated in Fig. 7-1. Then the displacements arising from the phonon vary only in the x-direction.

We consider first a longitudinal wave propagating in this [100] direction and s-like conduction-band states for $\boldsymbol{k}$ near zero. Because of the screw axis symmetry along this [100] axis (for elemental semiconductors) discussed in Section 2-1, and its effect on the longitudinal modes discussed in Section 3-3-A, $\boldsymbol{u}_q^2 = \boldsymbol{u}_q^1$, both parallel to $\boldsymbol{q}$ and all displacements can be written, following Eq. (3-28), as $\delta x_j = u_q e^{iqx_j}/\sqrt{N_c}$ with N_c again the number of atomic cells, half the number of atoms. Thus the displacements of the four neighbors, each at a distance $\pm d/\sqrt{3}$ along the x-axis relative to the central atom, are given by $u_q e^{iqx_j}(e^{\pm iqd/\sqrt{3}} - 1)/\sqrt{N_c}$ or $\pm iqu_q de^{iqx_j}/\sqrt{3N_c}$ for small wavenumbers, with the + for the atoms to the right and the - for the atoms to the left. If the states of interest were near a conduction-band minimum at $\boldsymbol{k} = 0$, where the energy is $\varepsilon_s - 4V_{ss\sigma}$ for a homopolar semiconductor, the change in each of the four matrix elements $V_{ss\sigma}$ would be $-2iqu_qV_{ss\sigma}e^{iqx_j}/(3\sqrt{N_c})$ due to the same change $iqu_qde^{iqx_j}/(3\sqrt{N_c})$ in each internuclear distance. (The displacement $-\delta x$ of atoms to the left makes the same change in spacing as δx of atoms to the right.) Then just as in the one-dimensional case, Eq. (7-19), (except for four neighbors and two factors of $1/\sqrt{3}$) for the case of a homopolar semiconductor with a conduction-band minimum at the zone center we obtain a matrix element

$$\langle \boldsymbol{k}+\boldsymbol{q}|H_{e\phi}|\boldsymbol{k}\rangle = \frac{-8iqu_qV_{ss\sigma}}{3\sqrt{N_c}} \rightarrow \frac{iqu_q}{\sqrt{N_c}} D_c \ , \qquad (7\text{-}20)$$

with again N_c the number of atom pairs in the crystal. In calculations of physical properties based upon this interaction, the matrix element will enter squared and will be summed over the number N_c of longitudinal-mode wavenumbers in the Brillouin Zone so this factor disappears in the end. In the last form we have used the homopolar form of the deformation-potential constant of Eq. (7-2). (Note that $-V_{ss\sigma}^2/|V_{ss\sigma}| = V_{ss\sigma}$.) As in Eq. (7-19) we have taken the small-k limit in setting factors such as $e^{i\mathbf{k}\cdot\mathbf{d}}$ equal to one. For the polar case, u_q^2 is not exactly equal to u_q^1, but for long wavelengths they are the same to lowest-order in q and we take them equal. The matrix element is multiplied by the coefficient of the conduction-band wavefunction for the s-state on the metallic (Ga) atom and the coefficient on the nonmetallic (As) atom (at the conduction-band minimum), giving exactly the final form of Eq. (7-20) with D_c given by Eq. (7-2). The polar case is important; most heteropolar semiconductors have just such conduction-band minima at $\mathbf{k} = 0$. The deformation-potential constant D_c gives the correct result for this case.

For coupling between two states in the same valley ($\mathbf{k}$ near the same [100] direction, as required if $\mathbf{q}$ is small) in indirect-gap semiconductors the deformation potential also gives a correct description (to lowest order in wavenumber so that the combination of s- and p-states entering the band states can be taken as the same for initial and final states). We found two contributions to the deformation-potential, one from the coupling with the peripheral s*-state, through the $\lambda_{sp\sigma}$ discussed in Section 5-3-B, in addition to the $V_{sp\sigma}$ coupling between valence states. It is interesting that for the strain e_1 associated with the longitudinal mode propagating in a [100] direction, $V_{sp\sigma}$ with the p-state oriented along the [100] direction does not change to first-order in the strain. This is because the $\cos\theta$ in $V_{sp\sigma}\cos\theta$ changes as a factor $(1 + 2e_1/3)$ according to Eq. (7-8). Because d varies as $(1 + e_1/3)$, and $V_{sp\sigma}$ as $1/d^2$, the $V_{sp\sigma}$ varies as $(1 - 2e_1/3)$. There is no net change in $V_{sp\sigma}\cos\theta$ to first order in e_1. In the absence of the $\lambda_{sp\sigma}$ contribution there would be no net electron-phonon interaction for this mode for an electrons in the [100] valley. The two contributions would *add* for the two transverse valleys giving a large electron-phonon interaction. In discussing the deformation potentials for these valleys we found that the uncertainties in the form of the $\lambda_{sp\sigma}$ term made the predictions for these off-center minima very inaccurate so that these features may not be so interesting here.

We consider next a transverse mode, again with $\mathbf{q}$ along a [100] direction as in Fig. 7-1 and again s-like conduction-band states. We first imagine that $u_q^2 = u_q^1 = u_q$ along [010], though the screw axis symmetry does not require this and we shall correct the error after. For the approximate form, the local strain is $e_4 = \partial u_y / \partial x = iqu_q e^{iqx} / \sqrt{N_c}$, and there is a local rotation around the z-axis, also equal to this value. Both are included. For an s-like conduction-band minimum at Γ only $V_{ss\sigma}$ enters the coupling between neighboring atomic states and there are no angular factors. Then, for each atom, two bond lengths increase and two others decrease by the same amount to first order in $\boldsymbol{u_q}$ so no electron-phonon interaction arises. The main error in the assumption that $\boldsymbol{u_q}^2 = \boldsymbol{u_q}^1$ is from the internal displacements (other deviations are of higher order in q) which we described in Section 3-2-B. These also provide canceling increasing and decreasing spacings (of the same order as those from $\boldsymbol{u_q}$) so there is no electron-phonon matrix element with these transverse modes, consistent also with the lack of a deformation-potential shift of the conduction band with shear. Our calculated deformation-potentials have described this case of long-wavelength acoustic modes propagating along a [100] direction. (It would appear that symmetry guarantees also a vanishing for interaction between shear modes propagating along a [100] direction and two states near any one of the conduction-band minima near X .)

If we were to consider wavenumbers $\boldsymbol{q}$ which are not along a [100] direction (and not along another symmetry direction, [110] or [111]) we would find that unless the system is elastically isotropic the distortions were not purely longitudinal nor purely transverse and there would be matrix elements with all three acoustic modes. (Elastic isotropy in this case means $(c_{11} - c_{12})/(2c_{44}) = 1$. That ratio is typically 0.6 for the semiconductors.) It seems clear that the interaction for conduction-band minima at Γ would still be correctly described by deformation potential theory, but it would require analysis of the modes to determine the dilatation associated with each.

The deformation potentials which we have obtained for conduction-band minima near X appear also to fully describe the coupling by modes with wavenumbers in arbitrary directions for electronic states in one valley. For each mode we need determine the local strain components e_1 , e_2 , *and* e_3 , giving the deformation potentials for electrons in a particular valley near X. Each mode will also give rise to strains e_4 , e_5 , *and* e_6 , and local rotations but these do not shift the energy of the states (to first order in $\boldsymbol{u_q}$) and do not couple states within the same valley. (We will note in Section 7-3-F that local rotations do in fact couple states in different valleys.) The interactions we obtain for indirect-gap semiconductors are not simple. Note that though a longitudinal mode in a [100] direction couples states within each of the six valleys, a longitudinal mode in the [110] direction produces only the strain e_6 locally and gives no scattering within any valley.

There are intrinsically many complications, but for scattering of conduction-band electrons by long-wavelength acoustic vibrational modes they are describable in terms of the three deformation-potential constants D_c,

D_{X1}, and $D_{X1(shear)}$ listed in Table 7-1, except for electrostatic contributions to be discussed next. The analysis using these deformation potentials can be greatly simplified by making the approximation of elastic isotropy. To lowest order in q the internal displacements do not enter except, as we shall see, through the piezoelectric constant. There they contribute because of an additional factor of $1/q^2$ arising from the Coulomb interaction. We turn to that next.

C. Electrostatic Contributions in Polar Crystals

For polar crystals, such as gallium arsenide, there is an additional contribution to the electron-phonon interaction which is not included in the deformation-potential interaction. The internal displacements, mentioned for the shear modes above, give a volume electric polarization in polar crystals. These are describable in terms of the piezoelectric effect discussed in Section 4-3-D. For the particular modes treated above, propagating along a [100] direction, these polarizations for longitudinal modes are zero and for transverse modes are perpendicular to $\boldsymbol{q}$ and there is no charge accumulation (equal to the divergence of the polarization, zero in this case). However, for a longitudinal mode propagating along a [111] direction, the internal displacements are parallel to $\boldsymbol{q}$ and there *is* charge accumulation and a resultant electrostatic potential.

The potential is readily calculated using the piezoelectric constant e_{123} of Eqs. (4-41) and (4-42). We again take $\boldsymbol{u_q}^2 = \boldsymbol{u_q}^1 = \boldsymbol{u_q}$ except for internal displacements which give rise to the piezoelectric polarization. For $\boldsymbol{u_q}$ and $\boldsymbol{q}$ along a [111] direction (longitudinal modes), we find $e_4 = e_5 = e_6 = 2iqu_q e^{i\boldsymbol{q}\cdot\boldsymbol{r}}/(3\sqrt{N_c})$. The net polarization is in the [111] direction and is $\sqrt{3}e_{123}$ times one of these strains. The negative of the divergence of this polarization is $-iq$ times this and is the corresponding local charge density change. The potential energy of an electron is $-4\pi e/q^2$ times this charge density, according to Poisson's Equation, and relaxation of the bonds in the presence of these fields reduces the potential by a factor of the reciprocal of the optical dielectric constant. This yields an electrostatic potential energy for a conduction electron, in the presence of a longitudinal phonon with wavenumber $\boldsymbol{q}$ in the [111] direction of

$$V(\boldsymbol{r}) = -\frac{8\pi e e_{123}}{\varepsilon\sqrt{3N_c}} u_q e^{i\boldsymbol{q}\cdot\boldsymbol{r}} . \tag{7-21}$$

This gives a contribution to the electron-phonon matrix element of $\langle \boldsymbol{k}+\boldsymbol{q}|H_{e\phi}|\boldsymbol{k}\rangle = -8\pi e e_{123} u_q / (\varepsilon\sqrt{3N_c})$. Comparing with the deformation-potential matrix element in Eq. (7-20) we note that this matrix element is real while the deformation-potential matrix element is imaginary. Physically this means that the electrostatic potential has maximum value where the displacements are maximum while the deformation potential is maximum where the displacements are zero (and the strains are maximum). This also has the consequence that when we calculate a scattering rate from $|\langle \boldsymbol{k}+\boldsymbol{q}|H_{e\phi}|\boldsymbol{k}\rangle|^2$ we obtain the sum of the absolute values squared of the two terms; the two mechanisms can be thought to scatter independently.

A second interesting difference in the two mechanisms is that though both are proportional to u_q, the piezoelectric contribution is independent of q while the deformation-potential contribution is proportional to q. In fact, from Eq. (4-42) we see that e_{123} is of order e/d^2 so the ratio of the deformation- potential matrix element to the piezoelectric matrix element is of order $V_{ss\sigma}qd^2\varepsilon/e^2$ which in turn is of order q times a Bohr radius times the dielectric constant. It will always be true for sufficiently long-wavelength phonons that the piezoelectric contribution will dominate. This is known to be the case for ultrasonic sound-wave experiments. Various numerical factors, including the dielectric-constant factor, seem to make the deformation-potential mechanism dominant in electron scattering at room temperatures.

Working out the geometry associated with the piezoelectric effect for longitudinal and transverse modes propagating in arbitrary directions is complicated, but straightforward. It was first worked out by Meijer and Polder (1953) and more completely by Harrison (1956a). We shall not carry it further here.

D. Interaction with Optical Modes

We should discuss briefly the interactions of electrons with optical-mode phonons before proceeding to holes. We have noted that these phonons have high enough values of $\hbar\omega$ that they are not so strongly excited thermally and tend not be so important in limiting electron mobilities as are the acoustic phonons. However, for the same reason they can be dominant in the relaxation of the energy of hot electrons, electrons with energy greater than the $\hbar\omega$ for the optical phonons. These electrons may still have wavenumbers near the band minimum and phonons of importance also will have small wavenumber, where the two atoms in each primitive cell move in opposite directions, in contrast to acoustic phonons where they move approximately together. If the two atoms in each primitive cell move exactly out of phase, the shift in energy of a conduction-band minimum at Γ must be proportional to the square of the displacements since, by symmetry, reversing the direction of a very small displacement must give the same shift. It follows that deformation-potential-like shifts will need to be proportional to the wavenumber of the phonon, or of the electron, which breaks the symmetry. Thus we may expect matrix elements of the same general magnitude as those given in Eq. (7-20). They can be thought of as arising from an optical-mode deformation potential (Harrison, 1956b) which can be calculated in tight-binding theory from the shift in the conduction-band minimum due to this motion of alternate atoms in opposite directions.

For polar crystals the optical modes have particularly strong electrostatic coupling with the carriers. For $\boldsymbol{q}$ near zero the positively and negatively charged atoms move in opposite directions producing a local electric polarization. In Chapter 4 we obtained the resulting potential, $\phi(x)$, for a longitudinal optical mode of wavenumber $\boldsymbol{q}$ leading to Eq. (4-31) for a force on the atom $F = -3\sqrt{3}\,\pi e^2 e_T{}^{*2}(u_+ - u_-)/(4\varepsilon d^3)$ equal to $-iqee_T{}^*\phi(x)$. Here the $u_\pm$ were the displacements of the positive and negative atoms. We may write these in terms of an amplitude u_q as $u_+ - u_- = u_q e^{iqx}/\sqrt{N_c}$ with N_c again the number of cells, or atom pairs, in the system. Then this same potential is seen by an electron giving a potential energy

$$V(x) = \frac{3\sqrt{3}\,\pi e^2 e_T{}^* u_q\, i}{4\varepsilon d^3 q\sqrt{N_c}}\; e^{iqx}\ . \qquad (7\text{-}22)$$

The resulting electron-phonon matrix element is

$$\langle \boldsymbol{k}+\boldsymbol{q}|H_{e\phi}|\boldsymbol{k}\rangle = \frac{3\sqrt{3}\,\pi e^2 e_T{}^* u_q\, i}{4\varepsilon d^3 q\sqrt{N_c}}\ . \qquad (7\text{-}23)$$

In this form we may compare with the deformation-potential-like matrix element arising from the interatomic couplings, of order of Eq. (7-20). We see that the deformation-potential matrix elements tend to be smaller by a factor q^2d^2 since $e_T{}^*e^2/(\varepsilon d)$ is similar in magnitude to $V_{ss\sigma}$. The electrostatic potential contribution, Eq. (7-23), will be the dominant effect. In contrast to interaction with acoustic modes these contributions are not out of phase with each other. The same electrostatic potential is of course seen by holes. The matrix elements for holes, however, are reduced by the overlap factors which will be discussed in Section 7-3-F.

E. Holes and Local Lattice Rotations

The valence-band maximum is always at Γ and the states at the maximum are based entirely upon atomic p-states. The three degenerate (in the absence of spin-orbit coupling) bands arise from the three orthogonal orientations of p-states. For wavenumbers $\boldsymbol{k}$ along symmetry directions the light holes correspond to a p-state parallel to $\boldsymbol{k}$. There are many more cases to consider than for electrons since for the matrix element $<\boldsymbol{k}+\boldsymbol{q}|H_{e\phi}|\boldsymbol{k}>$, the state $|\boldsymbol{k}>$ can come from any of the three bands, the state $|\boldsymbol{k}+\boldsymbol{q}>$ can come from any band, and the mode could be of any of the three polarizations so there are twenty-seven different cases even after the wavenumbers have been specified. Before proceeding in detail we note a simplification that the effects of local rotations can be ignored, again at small wavenumbers.

The fact that local rotations of the crystal do not produce coupling at the band maximum may be obvious if one looks at it the right way, but we found it necessary to treat each case to make certain and it is only true because of the high symmetry of the states at the top of the band, Γ. One state at Γ has p-states oriented in a particular direction, say the x-direction. Another might have orientation in the z-direction. For the undistorted lattice there is no coupling between these perpendicularly oriented band states since the couplings with the four nearest neighbors to any one orbital are of the same magnitude and half positive and half negative. This is best seen by sketching the geometry as in Fig. 7-2(a). The p_x-band state has p_x-orbitals on every atom but we show only one. The p_z-band state has p_z-orbitals on every atom but we show only those on the four neighbor atoms. If we define a vector internuclear distance, as shown, and define the angle θ_i as the angle between it and one p-state (defined by the direction of the positive lobe), and θ_j as the corresponding angle for the other, the coupling between each of these nearest-neighbor pairs can be seen to be

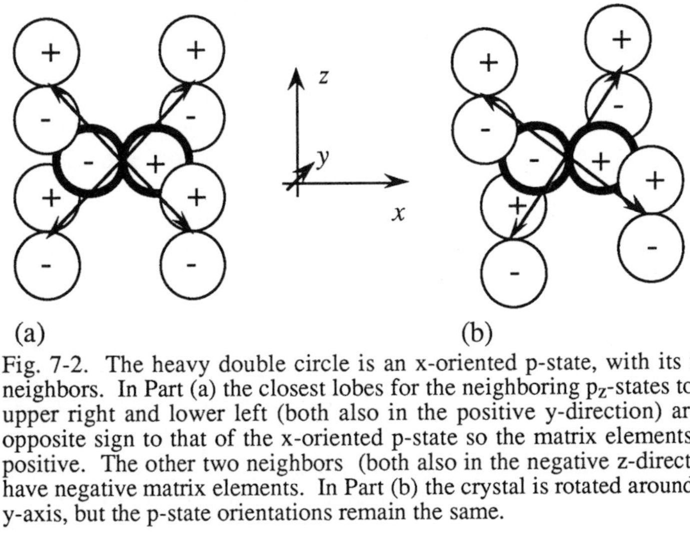

Fig. 7-2. The heavy double circle is an x-oriented p-state, with its four neighbors. In Part (a) the closest lobes for the neighboring p_z-states to the upper right and lower left (both also in the positive y-direction) are of opposite sign to that of the x-oriented p-state so the matrix elements are positive. The other two neighbors (both also in the negative z-direction) have negative matrix elements. In Part (b) the crystal is rotated around the y-axis, but the p-state orientations remain the same.

$$V_{ij} = \cos\theta_i \cos\theta_j (V_{pp\sigma} - V_{pp\pi}) - V_{pp\pi} \cos\theta_{ij} \tag{7-24}$$

where the $\cos\theta_{ij}$ is the cosine of the angle between the two p-state axes, one if the two are parallel or zero if they are perpendicular as in this case.

The coefficients for the four neighboring p-states are related to each other as $e^{i\mathbf{k}\cdot\mathbf{r}_j}$ and therefore are equal if $\mathbf{k} = 0$. We see that V_{ij} is $+(V_{pp\sigma} - V_{pp\pi})/3$ for the neighbors to the upper left and lower right and the $-(V_{pp\sigma} - V_{pp\pi})/3$ for the other two. The sum is zero at $\mathbf{k} = 0$ and there is no coupling between the two band-states, as must be since these are eigenstates. If $\mathbf{k}$ were different from zero, the coefficients on the four states would be different and the coupling would not be zero. Correspondingly, the three band states at this wavenumber will be slightly different in energy. The same is true for all combinations of perpendicular p-states ($\cos\theta_{ij} = 0$). For parallel p-states, the coupling is $(V_{pp\sigma} + 2V_{pp\pi})/3$ for all four neighbors, as we obtained in our study of energy bands, leading to the band energies for homopolar systems of $\varepsilon_p \pm 4(V_{pp\sigma} + 2V_{pp\pi})/3$, again at $\mathbf{k} = 0$.

If we locally rotate the crystal, the neighbors are shifted, as in Part (b) of Fig. 7-2, but the orientation of the p-states remains the same. There could now be coupling between the two modified states as there was in the case of bonding and antibonding p-states in the diatomic molecule, but we can see that there is not in this case for $\mathbf{k} = 0$. Before rotation, the axis for each of the p-states shown made an angle θ with its own internuclear distance shown with $\cos\theta = \pm 1/\sqrt{3}$. (When added, the net coupling was zero.) We focus on the coupling between the central p-state and that to the upper right in Part (b). After the rotation the $\cos\theta$ for the central p-state is decreased by a term linear in the angle of rotation, but the $\cos\theta$ for the upper-right p-state is *increased* by an equal term linear in the angle of rotation so the coupling only changes to second order in the angle. The same is true for each of the other neighbors so there is no coupling between the two band states linear in the rotation and therefore no contribution to the first-order electron-phonon interaction at $\mathbf{k} = 0$.

The same result obtains of course for any combination of p-states and rotation axes all perpendicular to each other. For perpendicular p-states and a rotation parallel to one axis, it is seen that one positive matrix element, and one negative matrix element, increases linearly and the other decreases linearly so the net linear change is again zero. If the two p-states are parallel, all matrix elements add before rotation, but with a rotation around an axis perpendicular to the p-states, half increase and half decrease. For a rotation parallel to the p-states, there is no change in any matrix elements. This exhausts the possibilities and the high symmetry has eliminated all rotational contributions to the linear electron-phonon interaction at $\mathbf{k} = 0$.

Again for wavenumbers *near* zero there will be a small coupling arising from local rotations of the lattice. It becomes necessary to include terms not only in the strains such as $e_4 = \partial\delta z/\partial y + \partial\delta y/\partial z$ but also rotations $\theta_1 = \partial\delta z/\partial y - \partial\delta y\partial z$, or equivalently to write the matrix elements in terms of the *unsymmetrized strains* such as $\partial\delta z/\partial y = iq_y u_z e^{i\boldsymbol{q}\cdot\boldsymbol{r}}/\sqrt{N}$. We may think of these as additional deformation potentials, and in fact they are obtained automatically in some formulations of the electron-phonon interaction such as the rigid-ion model (Seitz (1948), Harrison (1956b), Resta (1991)). However, if one begins with deformation potentials as calculated for uniformly strained crystals it would be easy to overlook them. For electron states of large wavenumber the rotational contributions are of the same general magnitude as the strain contributions but near the valence-band maximum they are smaller by a factor of order kd and we shall not include them here.

F. Acoustic Modes and Holes

Even though we may restrict the discussion to symmetrized strains we cannot *directly* use the deformation potentials which we discussed in Sections 7-1 and 7-2. A convenient way into this complex set of couplings involving holes and phonons is again to treat the simplest case first, and then to discuss increasingly complicated systems, seeing ways to approximate them.

a. Light Holes and Longitudinal Modes. The simplest case would seem to be the geometry of Fig. 7-1 in Section 7-3-B with a longitudinal phonon in the *[100]* direction and light holes in the xy-plane, all in a homopolar crystal such as silicon.

The only strain present for this mode is

$$e_1 = iqu_q e^{iqx}/\sqrt{N_c}\,, \tag{7-25}$$

obtained from the displacements $\delta\boldsymbol{r}_j = \boldsymbol{u}_q e^{i\boldsymbol{q}\cdot\boldsymbol{r}_j}/\sqrt{N_c}$. We will need the shift of all three p-states due to this strain so we may combine the deformation potentials we introduced in Sections 7-1 and 7-2 to obtain the shift in the energy of the p_x-band at $\boldsymbol{k} = 0$ due to shears e_1, e_2, and e_3 as

$$\begin{aligned}\delta\varepsilon_1 &= D_v(e_1 + e_2 + e_3) + D_v(shear)(e_1 - (e_1 + e_2 + e_3)/3) \\ &= (D_v + {}^2\!/_3 D_v(shear))e_1 + (D_v - {}^1\!/_3 D_v(shear))(e_2 + e_3)\,,\end{aligned} \tag{7-26}$$

and the shifts for the other two p-states are obtained by rotating indices. For only e_1 this gives

$$\delta\varepsilon_1 = (D_v + {}^2/_3 D_v(shear))\, e_1\,,$$
$$\delta\varepsilon_2 = \delta\varepsilon_3 = (D_v - {}^1/_3 D_v(shear))\, e_1\,. \qquad (7\text{-}27)$$

In Fig. 7-1 of Section 7-3-B for both the states $/\boldsymbol{k}>$ and $/\boldsymbol{k}'>$ the wavenumber makes an angle $\theta/2$ with the y-axis. As an approximation we take every p-state entering a particular $/\boldsymbol{k}>$ to be parallel to every other and oriented along the wavenumber $\boldsymbol{k}$ for light holes. This is true with wavenumbers along any [100] direction, and may be approximately true for small wavenumbers in other directions. Then in a homopolar semiconductor the coefficients for each of these p-states is given by $e^{i\boldsymbol{k}\cdot \boldsymbol{r}_j}/\sqrt{(2N_c)}$, with $2N_c$ the number of atoms in the system . This is a computational approximation which is not necessary, but is helpful.

We may use deformation potentials to calculate the local shift for each band state, $/\boldsymbol{k}>$, and the matrix element with some final state $<\boldsymbol{k}'/H_{e\phi}/\boldsymbol{k}>$ will come from that shift, summing over atoms just as in the coupling between s-like states, but here it will be reduced because the p-states for the two band states will be differently oriented. If the p-states for the state $/\boldsymbol{k}>$ all have magnitude $\Sigma_{j=1,2,3}\cos\alpha_j/p_j>/\sqrt{(2N_c)}$ and those for the state $/\boldsymbol{k}'>$ all have magnitude $\Sigma_{j=1,2,3}\cos\beta_j/p_j>/\sqrt{(2N_c)}$ the matrix element will contain a factor $\Sigma_{j=1,2,3}\cos\alpha_j \cos\beta_j\ \delta\varepsilon_j$, in contrast to the $\delta\varepsilon_s = D_c(e_1 + e_2 + e_3)$ which entered for a conduction-band minimum at Γ. All cross terms in $\cos\alpha_j\cos\beta_i$ with $i \neq j$ vanish because there is no coupling between p-band states oriented along different axes, as we showed in Section 7-3-E.

For the geometry of Fig. 7-1, with our assumption of p-states parallel to $\boldsymbol{k}$ for light holes, $\cos\alpha_1 = -\sin(\theta/2)$, $\cos\alpha_2 = -\cos(\theta/2)$, $\cos\beta_1 = \sin(\theta/2)$, $\cos\beta_2 = -\cos(\theta/2)$, and $\cos\alpha_3 = \cos\beta_3 = 0$. For this vibrational mode with $\delta\varepsilon_j$ given by Eq. (7-27), this leads to

$$\begin{aligned}\Sigma_{j=1,2,3}\cos\alpha_j \cos\beta_j\ \delta\varepsilon_j &= (\cos^2(\theta/2) - \sin^2(\theta/2))D_v e_1 \\ &\quad + ({}^2/_3\cos^2(\theta/2) + {}^1/_3\sin^2(\theta/2))D_v(shear)e_1 \qquad (7\text{-}28)\\ &= \cos\theta\, D_v e_1 + ({}^1/_2 + {}^1/_6\cos\theta)D_v(shear)e_1.\end{aligned}$$

The matrix element contains a sum over the $2N_c$ atoms in the crystal, with the phase-factors canceling and leads to an electron-phonon matrix element of

$$<k+q|H_{e\phi}|k> = \frac{iqu_q}{\sqrt{N_c}}\,[\cos\theta\, D_v + ({}^1/_2 + {}^1/_6 \cos\theta) D_v(shear)], \qquad (7\text{-}29)$$

for a longitudinal mode and light holes with θ the angle between $\boldsymbol{k}$ and $\boldsymbol{k} + \boldsymbol{q}$. The only change for a polar semiconductor is that the coefficient of the p-state on one atom type has a factor $\sqrt{1 + \alpha_p}$ and the other a factor $\sqrt{1 - \alpha_p}$ where the polarity is appropriate to the states at the valence-band maximum, (Eq. (5-19)),

$$\alpha_p = \frac{\dfrac{\varepsilon_{p+} - \varepsilon_{p-}}{2}}{\sqrt{\left(\dfrac{\varepsilon_{p+} - \varepsilon_{p-}}{2}\right)^2 + \left(\dfrac{4}{3}V_{pp\sigma} + \dfrac{8}{3}V_{pp\pi}\right)^2}} . \qquad (7\text{-}30)$$

(The factor $1/\sqrt{2}$ from these coefficients is already included in the coefficient $e^{i\boldsymbol{k}\cdot\boldsymbol{r}_j}/\sqrt{(2N_c)}$ used for the states.) The expression on the right in Eq. (7-29) is to be multiplied by $\alpha_c = \sqrt{1 - \alpha_p^2}$ for polar crystals.

Eq. (7-29) is closely parallel to Eq. (7-20) for conduction electrons arising from the same vibrational mode. Here, however, even with elastic isotropy, this matrix element does not apply to longitudinal modes propagating in other directions because of the geometry associated with the hole states on the tetrahedral lattice. The shear term in Eq. (7-29) arises from an e_6 rather than an $(e_1 - e_2)/2$. However, it may not be unreasonable as an approximation to use this form, which depends only upon the angle difference θ between $\boldsymbol{k}$ and $\boldsymbol{k}'$, as a general interaction between light-hole states arising from longitudinal modes.

We note that the first term in Eq. (7-29) is equivalent to the response to a simple potential $D_v\Delta(\boldsymbol{r})$ with a reduction factor $\cos\theta$ from the misalignment of p-states. The matrix element of the electrostatic potential discussed in Section 7-3-C enters in just this way: it is the same matrix element as for conduction-band electrons, but for light holes is reduced by the factor $\cos\theta$. For a longitudinal mode propagating along a [100] direction, there is no piezoelectric potential (it requires a strain e_4 , e_5 , or e_6 as in Eq. (4-41) which contained $e_5 = 2\varepsilon_{zx}$) but for a longitudinal mode propagating in a [110] direction it would. The geometry of the crystal would need to be considered to obtain the electrostatic potentials, but then the matrix elements could be obtained from those for conduction electrons at Γ by multiplying by $\cos\theta$.

This complication of reduction factors due to different atomic character of the band states arises also for matrix elements between conduction-band states in different conduction-band minima, giving rise to intervalley scattering. For example, a conduction-band state at X in the Brillouin Zone ($\boldsymbol{k} = [2\pi/a,0,0]$) consists of an s-state on one atom in the primitive cell and a p_x- state on the other, while at another X-point, $\boldsymbol{k} = [0, 2\pi/a,0,]$, it consists of an s-state on the one atom and a p_y-state on the other. The interatomic matrix element for the electron-phonon interaction between states at these two minima contains a factor of the shift $\delta\varepsilon_s$ and the fractional coefficient (the same for both valleys) for the s-state for each of these states.

b. Light Holes and Transverse Modes. If we were to retain the same geometry, Fig. 7-1, for transverse modes, the strains would be of the more complicated type, e_5 , and e_6 . We therefore choose a similar geometry but with the phonon wavenumber along a [110] direction as in Fig. 7-3. For this geometry the displacements for a transverse wave polarized in the plane of the figure are given by

$$u_x = -\sqrt{\frac{1}{2N_c}}\, u_q e^{iq(x+y)/\sqrt{2}} , \qquad u_y = \sqrt{\frac{1}{2N_c}}\, u_q e^{iq(x+y)/\sqrt{2}} . \tag{7-31}$$

This leads to strains $e_2 = \partial u_y / \partial y = -e_1 = {}^1/_2\, iqu_q\, e^{iq(x+y)/\sqrt{2}}/\sqrt{N_c}$ and rotations, which do not concern us, but no other strains (such as $e_6 = \partial u_y/\partial x + \partial u_x/\partial y$). We again write the coefficients of the p_x- and p_y-states for each of the light-hole states as $cos\alpha_i$ or $cos\beta_i$ as for Eq. (7-28). For example, the coefficient of the p_x state for the state of wavenumber $\boldsymbol{k}$ is seen from Fig. 7-3 to be $cos({}^{\pi}/_4 + {}^{\theta}/_2)$. There is no dilatation so the shifts are simply $\delta\varepsilon_i = D_v(shear)e_1$. This leads to

$$\Sigma_{j=1,2,3} cos\alpha_j\, cos\beta_j\, \delta\varepsilon_j = D_v(shear)e_2(-cos({}^{\pi}/_4+{}^{\theta}/_2)cos({}^{\pi}/_4-{}^{\theta}/_2) + \tag{7-32}$$
$$sin({}^{\pi}/_4+{}^{\theta}/_2)sin({}^{\pi}/_4-{}^{\theta}/_2) = -D_v(shear)e_2 cos({}^{\pi}/_2) = 0 .$$

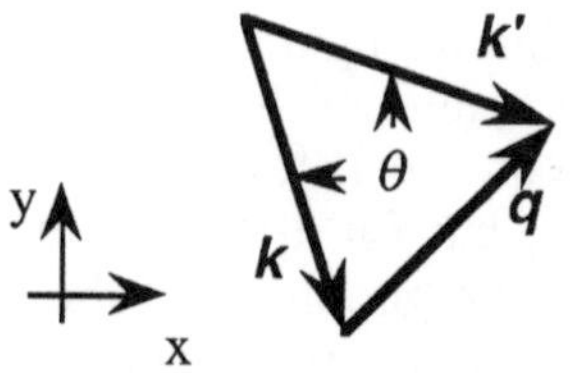

Fig. 7-3. The coordinate system for calculating the scattering of an electron of wavenumber $\boldsymbol{k}$ to a state $\boldsymbol{k'}$ by absorption of a transverse phonon of wavenumber $\boldsymbol{q}$ in a [110] direction. This choice leads to simpler strains than would the geometry of Fig. 7-1.

The electron-phonon matrix elements between light-hole states for these transverse modes are zero, as were the matrix elements for the conduction-electron states. It does not seem an unreasonable approximation to take this to be true for all transverse modes for the light-hole states.

c. Heavy Holes and Light-Heavy Matrix Elements . When we include heavy-hole states we have many more possibilities. We may use again the simple geometries of Figs. 7-1 and 7-3 for longitudinal and transverse modes, respectively, and it is again simply a matter of computing $\Sigma_{j=1,2,3}\cos\alpha_j \cos\beta_j \,\delta\varepsilon_j$ for each case.

G. Electrostatic Potentials and Screening

The term *screening* of a potential has come to mean the additional electrostatic potentials which arise from redistribution of charged carriers, potentials which add to the applied potential. That applied potential may itself be an electrostatic potential or a deformation potential arising from phonons. There are also contributions to the electrostatic potential from polarization of the bonds, which often enter as a reduction of potentials by a factor of the reciprocal optical dielectric constant. We consider first the effect of polarization of the bonds. We included the corresponding dielectric response for the electrostatic potentials which arose from piezoelectricity for acoustic modes in polar crystals and from the transverse charge of optical modes in polar crystals. These were a straightforward division by the dielectric constant, but somewhat surprisingly when deformation potentials are involved the dielectric response seems to require a deep understanding of the electronic structure. It becomes clear again as long as we think of the dielectric response as a polarization of the individual bonds.

In the case of the deformation potentials which we use to describe the interaction between electrons and acoustic lattice vibrations, different effective potentials are seen for different bands (equal to the shift in energy of that particular band due to the local elastic strain). The relevant one for producing dipoles in the bonds is the effective potential seen by the *bonds* . We may argue that the term in the bond energy linear in dilatation, the deformation potential, vanishes because the bond energies are minimum at the equilibrium spacing. This is only approximately true because it depends on our tight-binding view of the total energy as a sum of individual bond energies. However, for the approximations which we are using it is true and with bond deformation potentials equal to zero, there is no dielectric reduction in the deformation potentials we have introduced in either polar or nonpolar crystals. Further, any shift in bond energies due to shear of the lattice will average to zero when bonds of all four orientations are considered. Thus the deformation potentials we have calculated for uniform strains apply directly to nonuniform systems.

Strains do modify bond *dipoles* in polar crystals by modifying the polarities of the bonds. We have calculated the resulting electrostatic fields, and the effects of the dipoles they produce by using the factor $1/\varepsilon$ in Eqs. (7-21), (7-22), and (7-23). These electrostatic potentials are distinct from what we have defined as deformation potentials, and these deformation potentials are not reduced by any dielectric-constant factors .

Screening of the electron-phonon interaction by carriers which are present is a much more intricate problem. We have treated it in some detail (Harrison (1995)) and give only the results here, which are simple and quite plausible.

The basis for such screening analyses is given for example in Harrison (1970), 280ff. The response of the carriers to any potentials and deformation potentials is treated only to first order in the potentials. Because the response is linear in the potentials, we may make a Fourier expansion of any potential in position and time and calculate the response to each term, proportional to $V(\mathbf{q},\omega)\, exp(i\mathbf{q}\cdot\mathbf{r} - i\omega t)$, separately. The majority carriers will see an electrostatic potential and some deformation potential with this dependence and will respond by providing a density fluctuation, $n(\mathbf{q},\omega)\, exp(i\mathbf{q}\cdot\mathbf{r} - i\omega t)$, with $n(\mathbf{q},\omega) = X(\mathbf{q},\omega)V(\mathbf{q},\omega)$ and the response function will depend upon the carrier density in the absence of the perturbation, which we take to be an equilibrium distribution. $V(\mathbf{q},\omega)$ contains any applied potential *and* the screening potential arising from $n(\mathbf{q},\omega)$. It will also contain a deformation potential $D\Delta(\mathbf{q},\omega)$ due to local dilatations $\Delta(\mathbf{r})$, but not due to shear strains since such anisotropic terms may increase the number in some directions but decrease it in others and the average must be zero. (By symmetry, reversing the shear must give the same density shift so the linear term must vanish.) For any vibrational mode, we have found in the preceding sections the magnitude of the deformation term and any electrostatic term in the polar crystals but we do not yet know the screening potential, $V_{sc}(\mathbf{q},\omega)$ arising from the carriers. It is related to $n(\mathbf{q},\omega)$ by Poison's Equation, $-q^2V_{sc}(\mathbf{q},\omega) = -4\pi e^2 n(\mathbf{q},\omega)/\varepsilon$. (Note that here we included dielectric screening also of the screening potential itself.) We substitute for $n(\mathbf{q},\omega)$ in terms of the total potential and solve for the total potential in terms of the applied potential plus deformation potential to obtain $V(\mathbf{q},\omega) = V_0(\mathbf{q},\omega)/\varepsilon(\mathbf{q},\omega)$ with (Harrison (1970), p. 282)

$$\varepsilon(\mathbf{q},\omega) = 1 - \frac{4\pi e^2 X(\mathbf{q},\omega)}{\varepsilon\, q^2} \quad . \tag{7-33}$$

The whole problem is the solution of the constitutive equation which gives $n(\mathbf{q},\omega)$ in terms of the total potential, finding $X(\mathbf{q},\omega)$. It can be done classically with the Boltzmann Equation or quantum mechanically using a density-matrix. This is a substantial problem. The screening process cannot really be separated from the scattering process, for example electron scattering by phonons. We need to include the response of the electrons, including the compressional-wave aspects called *plasma oscillations*, and that solution will give imaginary and real contributions to $\varepsilon(\mathbf{q},\omega)$. The imaginary contributions are energy-loss terms which will include not only phonon emission and absorption but also plasma emission and absorption and the excitation of carrier-like modes which are called quasiparticle excitations. To the extent that we can separate out scattering by screened phonons it is found (Harrison (1996)) that we can approximately represent them in terms of two limits of $\varepsilon(\mathbf{q},\omega)$ to be used with the matrix elements, both limits being the same classically and quantum-mechanically. One is the familiar static dielectric constant at long wavelengths,

$$\varepsilon(q\ small, 0) = 1 + \frac{\kappa^2}{q^2} \tag{7-34}$$

For Maxwell-Boltzmann statistics κ is called the Debye-Hückel screening parameter given by

$$\kappa^2 = -\frac{4\pi e^2 X(0,0)}{\varepsilon} = \frac{4\pi e^2 N}{\varepsilon\, k_B T} \tag{7-35}$$

with N the density of carriers and of course k_B is the Boltzmann constant and T the temperature. For a degenerate gas of free carriers, with states filled to the Fermi energy ε_F , the corresponding Fermi-Thomas screening parameter is

$$\kappa^2 = \frac{3}{2}\,\frac{4\pi e^2 N}{\varepsilon\, \varepsilon_F}\ . \tag{7-36}$$

For intermediate degeneracy it appears quite accurate to interpolate $1/\kappa^2 = 1/\kappa^2_{DH} + 1/\kappa^2_{FT}$ between Eqs. (7-35) and (7-36) with ε_F determined for a degenerate gas at the density N .

The other limit needed is the high-frequency limit at long wavelengths,

$$\varepsilon(0,\omega) = 1 - \frac{\omega_p{}^2}{\omega^2} \tag{7-37}$$

with ω_p the plasma frequency,

$$\omega_p{}^2 = \frac{4\pi Ne^2}{\varepsilon\, m^*} \tag{7-38}$$

m^* is the effective mass of the majority carriers. If there are more than one type of carrier present, such as light and heavy holes, we may add the contributions from Eq. (7-38) for each carrier and the formulae for κ^2 would remain appropriate except that the carriers see different deformation potentials and one must redo the calculation to obtain the screening from the two carrier types.

Our recent study indicated that the static dielectric constant (Eq. (7-34)) is appropriate for longitudinal acoustic modes, though it is in error when the phonon wavenumber q becomes greater than the typical electron wavenumbers (obtained for an electron of energy k_BT or ε_F). Then $\varepsilon(q,\omega)$ is near one in any case so it does not matter. For transverse acoustic phonons, we have found here that there is no screening of the deformation potential, but if there is a piezoelectric potential (polar crystals) it is screened, as are the longitudinal acoustic phonons, by the static dielectric constant of Eq. (7-34). For longitudinal optical phonons, the high-frequency form, Eq. (7-37), was found (Harrison (1996)) to be appropriate if the phonon frequency is much greater than the plasma frequency and the static form, Eq. (7-34), is appropriate if it is much less than the plasma frequency. If the two frequencies are comparable, we should include scattering both by the optical phonons and by the plasma modes and it is a good approximation to treat both as unscreened by any carrier dielectric function.

These are approximate results, but appropriate on the scale of accuracy of the tight-binding theory of the interaction. In particular, they eliminate divergences which can occur in summing over scattering events. They are quite similar to what is commonly assumed, except possibly for the last case in which the phonon and plasma frequencies are comparable. All of these results were obtained assuming a majority carrier dominated the screening and we were calculating the screened electron-phonon interaction seen by the dominant carrier. Other cases are more complicated, but are straightforward. For example, for a particular phonon the screening may be provided by the majority carrier, but then a minority carrier sees the same electrostatic (including screening) potential, but a different deformation potential.

H. Electron Scattering by Acoustic Phonons

It may be desirable to carry out one calculation of scattering based upon these matrix elements to make it clear how they enter. We treat the scattering of electrons in an isotropic band due to acoustical phonons. We found in Section 7-1 that only longitudinal modes are coupled to such electrons. The calculation is similar to our calculation of photon absorption by electron-hole creation in Section 6-6. We use the Golden Rule, Eq. (1-32) and will treat the phonons classically, as we treated the light wave. This is appropriate at high temperatures where the phonon amplitudes are such that the kinetic energy given in Eq. (3-27) as $^1/_2 M\omega^2 u_q u_{-q}$ is equal to $^1/_2 k_B T$. We noted just before Eq. (3-28) that for acoustic modes in a crystal with two atoms per cell, M is equal to the average mass. It is not difficult to generalize this calculation to other cases. From Eq. (6-41) we have the rate for transitions due to time-dependent potentials, with matrix elements varying as $e^{\pm i\omega t}$ as do ours since the $u_{\pm q}$ have that variation,

$$P_{ij} = \frac{2\pi}{\hbar} |\langle j|H_1|i\rangle|^2 \,[\delta(E_j - E_i - \hbar\omega) + \delta(E_j - E_i + \hbar\omega)\,] . \qquad (7\text{-}39)$$

We saw at the beginning of Section 7-3-B that the change in energy $\hbar\omega$ is negligible compared to the electron energies, so the two terms can be combined, as in the static case, Eq. (1-32). $E_j - E_i$ becomes $\hbar^2(\mathbf{k} + \mathbf{q})^2/2m_c - \hbar^2k^2/2m_c = \hbar^2kq\,cos\theta/2m_c + \hbar^2q^2/2m_c$ for scattering of an electron from a state $\mathbf{k}$ to a state $\mathbf{k} + \mathbf{q}$, with θ the angle between $\mathbf{q}$ and $\mathbf{k}$. The matrix element from Eq. (7-20) is $iqu_q D_c/\sqrt{N_c}$, this time depending upon q. We do not include the piezoelectric contribution, which turns out to have a small effect for scattering by thermal electrons. As we have indicated in the preceding subsection, the deformation-potential matrix elements would be screened by the Debye-Hückel dielectric function, $\varepsilon(\mathbf{q},\omega) = 1 + \kappa^2/q^2$, for longitudinal acoustic modes if only electrons were present. (Holes see a different deformation potential and their response gives electrostatic potentials also seen by the electrons.) We obtain the total scattering rate by summing over all final states, designated by $\mathbf{q}$ as

$$P_{tot} = \frac{4\pi D_c^2 k_B T}{\hbar M N_c} \sum_q \frac{q^2}{\omega^2(1 + \kappa^2/q^2)^2}\, \delta\left(\frac{\hbar^2}{2m_c}(2kq\cos\theta + q^2)\right), \quad (7\text{-}40)$$

where we have also substituted for $u_q u_{-q} = k_B T/(M\omega^2)$ (which follows from setting the kinetic energy of Eq. (3-27) equal to the classical $^1/_2 k_B T$) with M the average atomic mass. For acoustic modes $\omega = v_s q$ so the leading factor in the sum is $1/\,[v_s^2(1 + \kappa^2/q^2)^2]$ We again write the sum as an integral according to Eq. (5-3), this time as $\Sigma_q = \Omega/(2\pi)^3 \int 2\pi q^2 dq\, sin\theta\, d\theta$. For the scattering rate appropriate to transport properties we would weight each event by the fraction of initial momentum lost, $1 - cos\theta$, (e. g., Harrison (1970) p. 194ff) but we do not do that here. We perform the integral over θ first, using the delta function, to obtain

$$P_{tot} = \frac{D_c^2 m_c k_B T \Omega}{\pi \hbar^3 k M v_s^2 N_c} \int \frac{q^5 dq}{(q^2 + \kappa^2)^2} \ . \tag{7-41}$$

Had we inserted the factor $1-cos\theta$ there would have been an additional factor $q^2/(2k^2)$ in the integrand. The integral extends from zero to $2k$, beyond which the delta-function does not contribute. For simplicity we may evaluate this for such small electron densities that κ^2 can be neglected and obtain

$$P_{tot} \approx \frac{2D_c^2 m_c k_B T \Omega_c k}{\pi \hbar^3 M v_s^2} \ . \tag{7-42}$$

Ω_c is the volume of the primitive cell. There would be a weak dependence upon the screening length if we had retained the dependence upon κ^2. Had we been treating scattering of holes by shear waves, there would be no dependence upon κ^2 since we have seen that shear wave potentials are not screened. The size of the system has of course canceled out. We may verify that with D_c, Mv_s^2, and $\hbar^2/(m_c\Omega_c k)$ all having units of energy, P_{tot} has units of reciprocal time. The scattering rate is proportional to the speed of the electron through k and it is proportional to temperature, since the squared amplitudes of the modes were proportional to temperature.

Problem 7-1. Deformation Potentials

Use the minimum in the σ-conduction band in graphite, obtained in Problem 5-2, to define a deformation potential. In this case the deformation potential constant would be the ratio of the shift in band energy to a fractional change in area.

CHAPTER 8

Impurities and Defects

Almost all of our analysis has been directed at translationally symmetric systems, for which the states can be selected to have well-defined wavenumbers and for which we may construct energy bands. However, for our bond-orbital analysis we were able to construct local states, which lead to the same total energy when summed over occupied states, and these did not require translational symmetry of the lattice. This same approach enables us to obtain approximate total energies for systems with impurities or defects though in fact there may be no real localized states associated with the defect. In some cases there will be localized states and we shall see how they may also be approximated. When we treat metals we will again be able to treat systems with defects, and even liquid metals, using the fact that the electron states are free-electron-like, a fundamentally different approach.

We shall proceed by treating heterovalent semiconductor solutions, such as gallium arsenide in germanium, for which the change in bond energy dominates the energy change. We then go on to isovalent solutions, such as InAs in GaAs, for which the misfit energy associated with different natural bond lengths dominates the energy. We then return to a more complete analysis of energies of substitution, for which we include also effects of doping, or carrier creation, and changes in metallization. We then return to point defects, vacancies and interstitials in pure materials.

1. Semiconductor Alloys

In most cases an impurity in a semiconductor sits on a site previously occupied by a host atom, rather than taking an interstitial position between host atoms. We shall return to interstitial defects. Thus a phosphorus atom in silicon will sit on a silicon site. It will cause some distortion of the surrounding lattice, but we ignore that initially. We often associate an enthalpy of solution with the change in energy from such a replacement, and we shall treat such a case immediately. We quickly see that we need to state more exactly just what change in energy we mean. We then return to define a convenient quantity, the energy of substitution, in terms of which properties such as the enthalpy of solution may be obtained.

A. Enthalpies of Solution

One of the simplest cases is the solution of one semiconductor into another, which we illustrate with the solution of gallium arsenide in germanium. Considering it will also introduce a series of problems which must be addressed. We shall in fact conclude that to obtain meaningful numbers it will often be necessary to go beyond the approximations which we actually carry out here.

We imagine a system consisting of the two semiconductors such as gallium arsenide and germanium, as illustrated in Fig. 8-1. The structure is continuous across the interface, in which case the interface is called a heterojunction, as we shall discuss in Section 19-4 . In the case drawn, we have two planar $(1\bar{1}0)$ heterojunctions, separated by a step running parallel to the z-axis perpendicular to plane of the figure. [The $\bar{1}$ indicates a -1. A plane $(1\bar{1}0)$ is perpendicular to a line $[1\bar{1}0]$, which is perpendicular to a line [110].]

If we were to interchange a gallium atom at one planar heterojunction with a germanium atom from the bulk, as indicated by arrow *a* , we would have dissolved a gallium atom in germanium, but would have also introduced a defect in the interface and part of the change in energy should be associated with that defect; the change in energy cannot be associated just with an enthalpy of solution. That was the reason for introducing the step, because we could now interchange a gallium at the step with a germanium atom in the bulk, arrow *b* , (really every gallium atom along the step but we don't see the

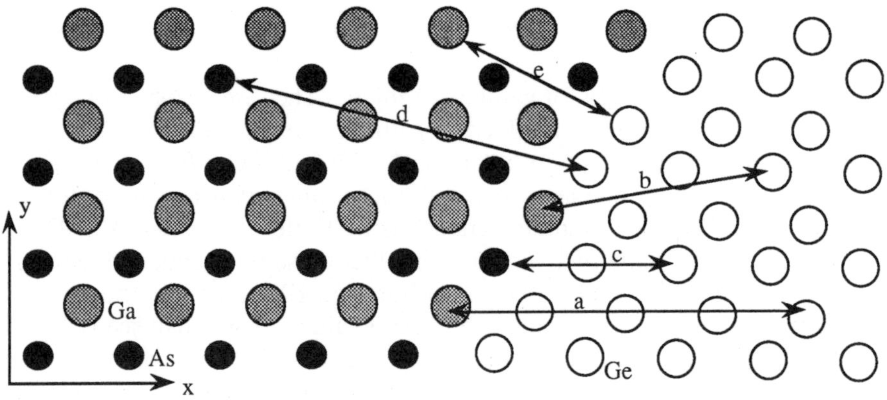

Fig. 8-1. Heterojunctions separating a gallium arsenide crystal from a germanium crystal. A step was inserted between the two $(1\bar{1}1)$ heterojunctions. Interchanges discussed in the text are indicated.

others in the figure) and the step has moved. In fact it is now a step terminating in an arsenic atom rather than a gallium atom, so the step is different. However, we can also interchange that arsenic atom from the new step with a bulk germanium, arrow c , and the step becomes the same as initially. We have now dissolved a gallium arsenide molecule (really a row of molecules along the edge) in the germanium, and moved the step without changing the step energy.

It is interesting that in order even to define a heat of solution we find it necessary specify so completely the system we will treat. Other energies, such as those leaving a modified interface, may also be of interest and their energies could be calculated. Sometimes the exercise of defining the problem is as useful as the actual calculation of the energy.

For this particular case we see that the insertion of the gallium and arsenic into the germanium produces four gallium-germanium bonds and four germanium-arsenic bonds while eliminating four germanium-germanium bonds and four gallium-arsenic bonds. (This is a little tricky but we may obtain the result by counting the fifteen bonds involving the four sites which are modified, before and after interchange.) Each bond energy is $-2\sqrt{V_2^2 + V_3^2}$ with the V_3 in each case evaluated from the hybrid energies entering that bond. If we leave all bond lengths the same, all V_2 values are the same and since the polar energies for the germanium-arsenic and germanium-gallium bonds are very nearly half that for gallium arsenide, we have a change in bond energy of

$$\delta E_{bond} \approx 8\sqrt{V_2^2 + V_3^2} + 8V_2 - 16\sqrt{V_2^2 + V_3^2/4}\ , \tag{8-1}$$

with V_3 the polar energy for gallium arsenide. The same expression applies to two germanium atoms dissolved in gallium arsenide (Problem 8-1). We may expand for small V_3 to obtain $2V_3^2/V_2 = 2.3$ eV for the change in energy in dissolving one atom pair. Without this expansion the difference obtained from Eq. (8-1) and the parameters of Table 2-1 is 2.04 eV. This is quite large on the scale of k_BT indicating that these two semiconductors are quite insoluble in each other. Other contributions to the energy, metallization energy in particular, shift these values, as we shall see in Part 8-1-C, to 0.76 eV per GaAs atom pair in Ge and to 1.04 eV per Ge pair in GaAs . There may be further shifts if other corrections are included, as we shall see.

It *is* generally true that heterovalent solutions, those in which the two compounds mixed come from different columns of the periodic table (e. g., III-V's in column-IV systems) have large positive enthalpies of solution. The simple estimates based only on the bond energy are qualitatively correct for such heterovalent solutions, but there is very little experimental information except that the energies are large The corrections to these simple estimates which we make in Section 8-1-C are significant, but it will appear there that even full tight-binding results may only be semiquantitatively correct.

In contrast to the estimates for heterovalent solutions, if we look at isovalent solutions such as silicon in germanium, the predicted enthalpies of solution are very small. For a solution of InAs in GaAs the approach given above gives an energy change of exactly *zero* ; we are simply moving four InAs bonds into the GaAs and moving four GaAs bonds to the interface. For such cases energy associated with the different equilibrium bond lengths - the misfit energy - becomes important. This was estimated many years ago by Stringfellow (1972, 1973) and we make a similar analysis next.

B. Lattice Distortions

We could proceed, as we did with cohesion, with an overlap repulsion fit to the observed spacing, and interpolate to obtain repulsions between, for example, germanium and silicon. However, the central results can be obtained more directly by denoting a natural spacing d_0 for each set of neighbors and a bond-distortion energy $^1/_2\,\kappa_0(d-d_0)^2$ with the corresponding bond. We take as an example InAs dissolved in GaAs and write the d_0 as $d_H = 2.45$ Å for the host GaAs bonds and $d_0 = 2.61$ Å for InAs bonds, both obtained from Table 2-1 for the bulk systems.

Harrison and Kraut (1988) estimated natural bond lengths for this case, and for cases such as Zn-Ge bonds where no corresponding tetrahedral bulk system exists, by minimizing the tight-binding energy with respect to spacing in the alloy. These spacings did not accord well with experiment where values existed. For example, they predicted that the natural bond length for arsenic dissolved in silicon is 2.31 Å, yielding a relaxed bond length (from Eq. (8-2) which we derive next) there of 2.32 Å. Measurements by Erbil, Weber, Cargill, and Boehme (1986) gave a relaxed bond length of 2.41±0.02Å, longer than the silicon-silicon distances rather than shorter. Scheffler (1987) in a much more accurate total energy calculation obtained also a distance for Si-As shorter than that for silicon, but shorter by only half as much. Harrison and Kraut (1988) nonetheless used the natural bond lengths predicted by tight-binding theory in the studies used in the next section. For the present analysis we use bulk bond lengths as natural bond lengths where they exist - isovalent solutions - which are the cases for which the misfit energy is important. We may then follow a procedure given by Shih, Spicer, Harrison, and Sher (1985) for estimating the distortions and energy.

In Fig. 8-1 we show a dissolved atom, indium, in the center, with its four arsenic neighbors and twelve gallium second neighbors, all initially separated by the bond-length $d_H = 2.45$ Å of the host gallium arsenide. If we neglect the small displacements of the second neighbors, the arsenic atoms will be displaced outward by u as shown, compressing the gallium arsenic springs. For small displacements the change in bond length for these

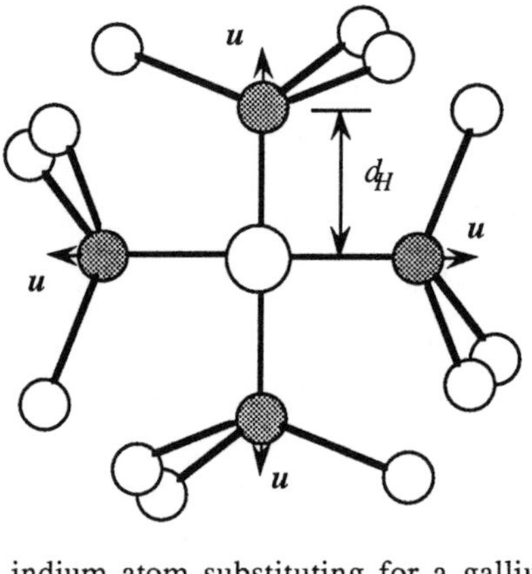

Fig. 8-2. An indium atom substituting for a gallium atom in gallium arsenide. The four neighboring arsenic atoms relax outward by u from their position at the host bond length d_H and we neglect the relaxation of the second neighbors.

springs is $-u\cos\theta = -u/3$ with θ the angle between $\boldsymbol{u}$ and the bond axis, and the component of force along $\boldsymbol{u}$ is $-\kappa_0 u \cos^2\theta$. We neglect any difference in the κ_0 for the different bonds. Thus we add up the force from these three bonds and add it to the force from the central bond, which has a natural length d_0 exceeding the host length d_H. That force is $\kappa_0 (d_0 - d_H - u)$ At equilibrium the total force is zero, or $-{}^3/_9\, \kappa_0 u + \kappa_0(d_0 - d_H - u) = 0$, and

$$u = \frac{3}{4}(d_0 - d_H). \tag{8-2}$$

The central bonds have relaxed three quarters of the way to their natural bond lengths. The strain energy of the four bonds before relaxation was $4 \times {}^1/_2\, \kappa_0(d_0 - d_H)^2$ and after relaxation the total strain energy of the sixteen bonds, with the outer twelve decreased in length by $(d_0 - d_H)/4$, is $4 \times {}^1/_2 \kappa_0(d_0 - d_H)^2/16 + 12 \times {}^1/_2 \kappa_0(d_0 - d_H)^2/16 = {}^1/_2\, \kappa_0 (d_0 - d_H)^2$, one quarter as large. Using the experimental $\kappa_0\ d_0^2 = 47.2\ eV$ for GaAs from Table 2-5 we obtain 0.10 eV for the energy increase for each indium atom in gallium arsenide, equal then to the predicted heat of solution per molecule of InAs in GaAs. As small as this misfit energy is, it is ordinarily the dominant term, or comparable to other terms, in the case of solutions of semiconductors of the same valence. Harrison and Kraut (1988) obtained an additional contribution for this case of InAs and GaAs, from changes in metallization energy, making a total of 0.16 eV (Harrison and Kraut (1988) Table VI). Experimental values from Stringfellow (1972, 1973) cover a range of values, 0.14, 0.26, and 0.17 eV, all similar to our prediction.

Eq. (2) also explains the structure of alloys as found experimentally by Mikkelsen and Boyce (1983). It has long been known that the lattice spacing of substitutional alloys, as measured by X-ray diffraction, interpolates approximately linearly between that of the pure compounds, called Vegard's law (Vegard (1921)). Thus the average spacing of the alloy $GaAs_{1-x}InAs_x$ will be approximately $(1-x)2.45 + x2.61$ Å. Most workers were misled by Vegard's Law to believe that the alloy lattice was close to the ideal average structure with all bond lengths equal. However, if we could construct such an "average alloy structure" with all bond lengths equal to the Vegard's-law value, and then modify the natural bond lengths surrounding each gallium atom to 2.45 Å and surrounding each indium atom to 2.61 Å, we would find that the arsenic atoms surrounding each would relax 75% of the way to those natural bond length (each with additional shifts from other neighbors which average to zero). Thus the structure is basically an alloy of metal atoms on a nearly undistorted face-centered-cubic lattice with the arsenic atoms shifted out of position to accommodate to the variation of bond-length from site to site. This is just what Mikkelsen and Boyce (1983) found using Extended X-ray Absorption Fine Structure (EXAFS) which measured the average distance to the neighbors from a gallium atom and independently the average distance to the neighbors from an indium atom. The results of such measurements are illustrated in Fig. 8-3. The relaxations in other structures, such as the rock-salt structure are analogous, as illustrated in Problem 8-2.

Tersoff (1995) has noted that when energies of distortion around impurities are large, they will be reduced near surfaces because the stress can be relaxed by relaxation at the surface. Thus the solubility of such impurities will be enhanced near the surface. For similar reasons, the activation energies for motion will be reduced so that diffusion is enhanced near the surface.

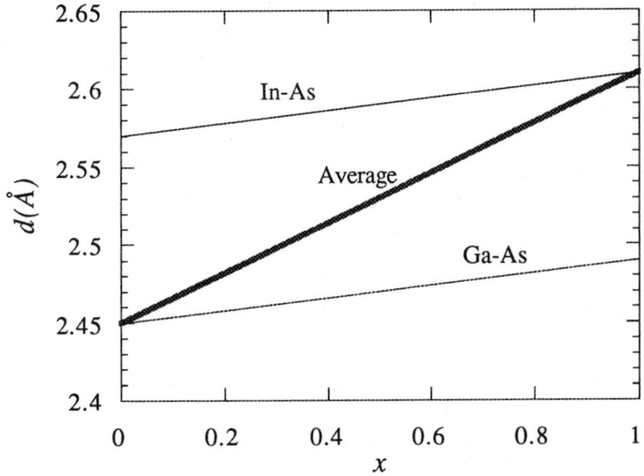

Fit. 8-3. The average bond length in an alloy of x molecular fraction In As in GaAs is shown as the heavy line (Vegard's Law). The average distances between Ga and As in the same alloy, and between In and As in the same alloy, as found by Mikkelsen and Boyce (1983) are shown as light lines.

C. Doping and Energies of Substitution

In our discussions of the solution of gallium arsenide in germanium as in Fig. 8-1, we dissolved equal numbers of gallium and arsenic atoms. A neutral gallium has one less electron than the germanium it replaced and a neutral arsenic atom has one more. There was no net addition of electrons to the germanium and no doping. Another problem arises in the more interesting case when only arsenic atoms are added to germanium (or more arsenic atoms than gallium atoms). If we simply transfer protons from nucleus to nucleus (a transfer of a proton from an arsenic nucleus to a germanium nucleus is equivalent to interchanging the two atoms) we would end up with net positive charge on the germanium side and net negative charge on the gallium arsenide side, still with two electrons in every bond. This would be a reasonable assumption if only a few arsenic atoms were transferred, but once we had transferred a finite percentage of the bulk atoms, the fields and electrostatic energy would be so great that electrons and holes would be generated to neutralize the charge. The germanium would have been doped n-type and the gallium arsenide would have been doped p-type. In this situation there are additional terms in the heat of solution. Each time we move an arsenic atom to the germanium we must also create an electron-hole pair, costing an energy equal to the band gap, placing the electron on the germanium side and a hole on the gallium arsenide side. This is an additional term in the energy needed to make the change. Similarly, if we were to now move a gallium atom to the germanium, we would reduce the number of electrons in the germanium by one and reduce the number of holes on the gallium arsenide side by one(*compensating* the doping on both sides) . We would gain an additional energy equal to the band gap. Thus with any heteropolar substitution, the energy required will depend upon whether an excess electron is added at the conduction-band energy or the valence-band energy. For the calculation of enthalpies of solution, electrons are added or subtracted at the thermodynamic Fermi energy, which is near the conduction-band edge or the valence band edge depending upon whether the system is n-type or p-type. These contributions to the enthalpy of solution are comparable to the enthalpies of solution predicted from the bonding terms and make it energetically even more expensive to dope semiconductors. It involves putting electrons in antibonding bands, or removing them from bonding bands.

This more complicated set of circumstances is better organized with the introduction of *energies of substitution* . At the same time, we shall consider the contribution of metallization to the energies of alloys. The energy of substitution $E_{AB}(X_A)$ is defined as the energy required to remove an atom of type A from a compound AB, to leave it as a free neutral atom in the ground state and to take a free neutral atom C and place it in the vacant A site of the compound. If this changes the number of electrons in the compound, the excess electrons are inserted at, or the deficit of electrons obtained from, the valence band maximum. (This is a part of the definition. When the system is such that electrons are added at the conduction-band edge, the corresponding correction is made explicitly.)

The tight-binding theory of this energy has been attempted by Kraut and Harrison (1984, 1985) and Harrison and Kraut (1988), with tables of predicted values for a large number of cases, and references to many other papers on this subject. In the end, it appears that the approximations made were not adequate to give a very useful account of these energies of substitution, though likely tight-binding theory itself without the approximations would be. We therefore restrict ourselves here to the principal lessons from the study without reproducing the analysis which is a direct extension of what we gave for enthalpies of solution in Section 8-1-A.

One additional feature which was included was a correction to the valence-band maximum, E_V, which did not arise before, and which comes from the nonorthogonality of neighboring orbitals. The effects of this nonorthogonality have previously been absorbed in the overlap repulsion V_0. It was noted by Enderlein and Harrison(1984) that V_0 includes approximately the effect of orthogonalization if all bond states are doubly occupied and all electrons are in bonds. If an electron is removed from the valence band, its contribution to the overlap repulsion should be subtracted or when it is added back, its contribution should be added. The average shift of a single valence electron, $V_0/2$, was used by Enderlein and Harrison, but Kraut and Harrison (1983) argued that an electron at the top of the valence band should be shifted by an amount smaller by a factor of $(1.28/3.22)^2$ based upon the ratio of the covalent shift of the top of the valence band to that of the average valence electron. (Both the S and the $V_{ss\sigma}$ in the $SV_{ss\sigma}$ discussed at the end of Section 1-2 are scaled.) This gave an additional +0.32 eV for example in the energy of substitution of a germanium for a gallium atom in GaAs.

Harrison and Kraut included also the change in overlap repulsion $V_0(d)$ and the change in the metallization energy due to the substitution. This was somewhat intricate since it involved a change in metallization energy not only for the four bonds which have been modified in the substitution, but for the twelve bonds which neighbor them. For this substitution energy of Ge for Ga in GaAs, $E_{GaAs}(Ge_{Ga})$, they obtained contributions of 3.15 eV for change in bond energies, -1.76 eV for promotion energy (including the change in E_v) , -1.52 eV for change in overlap repulsion, -2.86 eV for change in metallization, for a total of -2.99 eV. We note particularly that the change in metallization is almost as large as the entire contribution, as in our calculation of cohesion.

Values for these energies of substitution are surprisingly difficult to obtain. One would have thought that they could be directly obtained from solubilities. For example, the concentration of germanium in solution in gallium arsenide is proportional to $exp(-E_{GaAs}(Ge_{Ga})/ k_BT)$ if it goes predominantly on the gallium site. However, there are also temperature dependences to the entropy terms which are difficult to separate, making the determination of substitution energies from thermodynamic data such as that from Trumbore, Isenberg, and Porbansky (1959) difficult. (See, for example, Su and Brebrick (1985) and Tersoff (1990)). One careful attempt has been made by Su and Brebrick (1985) to obtain the energy of substitution for Zn, In, and Sn in Ge from such thermodynamic data. Their experimental values (in parentheses) are compared with the tight-binding estimates by Harrison and Kraut (1985) as $E_{GeGe}(Zn_{Ge})$ = 6.06 eV (3.59 eV), $E_{GeGe}(In_{Ge})$ = 3.91 eV (1.95 eV), and E_{GeGe} (Sn_{Ge})= 1.12 eV (0.83 eV). This is not very impressive agreement and we shall see that for elements to the right of column IV, cases such as $E_{GeGe}(As_{Ge})$ they are of different sign. It would be desirable to have a large set of experimental energies of substitution, such as Su and Brebrick have provided for these three cases, or more accurate theoretical values, which along with tables of cohesive energies would give direct predictions of a wide variety of properties.

We might for example obtain enthalpies of solution in terms of such energies of substitution. The enthalpy of solution, generalized from Section 8-1-A, is $H_{CD}(AB)$ equal to the change in energy per atom pair of AB dissolved in CD . We obtain it in steps as minus the cohesive energy (the cohesive energy is a positive quantity) of AB to remove an A atom and a B atom from the bulk, the energy $E_{CD}(A_c)$ to substitute the A atom for a C atom and $E_{CD}(B_D)$ to substitute the B atom for a D atom in CD and minus the cohesive energy of CD as these two atoms are returned to the bulk. This is

$$H_{CD}(AB) = E_{coh}(AB) + E_{CD}(A_c) + E_{CD}(B_D) - E_{coh}(CD) \qquad (8\text{-}3)$$

per atom pair. This would be for example the energy to dissolve a GaAs pair from the step in Fig. (8-1) into the bulk germanium . We list in Table 8-1 a short list of energies of substitution, containing most of the systems discussed here; they are from a very extensive table given by Harrison and Kraut

Table 8-1. Calculated energies of substitution $E_{CD}(A_C)$ and $E_{CD}(A_D)$ (in eV/ atom) for atom X substituted, respectively, for atom C or for atom D in compound CD (From Harrison and Kraut (1988)). The approximations used appear not to be sufficient to obtain very meaningful values.

$X =$	Ge	Sn	Ga	As	Zn	In
$E_{GeGe}(X_{Ge})$	-	1.12	3.11	-1.51	6.06	3.91
$E_{GaAs}(X_{Ga})$	-3.01	-2.35	-	-5.62	2.62	0.25
$E_{GaAs}(X_{As})$	3.21	5.21	8.68	-	14.23	10.34

(1988). Values from such a table would be used with the theoretical cohesive energies from Table 2-3 (each entry multiplied by four to obtain the energy per atom pair) in the evaluation of Eq. (8-3).

For the particular case of $H_{GeGe}(GaAs)$ we obtain 8.36 + 3.11 -1.51 - 9.20 = 0.76 eV. For the case $H_{GaAs}(GeGe)$ we obtain 9.20 - 3.01 + 3.21 - 8.36 = 1.04 eV. We estimated both of these values for heterovalent solutions from the change in bond energy alone in Section 8-1-A to be 2.04 eV. Again, the corrections are quite large but the experimental knowledge of the enthalpies of solution when they are large and positive is quite limited.

For homovalent solutions we may add the contribution from Eq. (8-3) to what we obtained for misfit energy alone. The contributions from Eq. (8-3) are also quite small and comparable to the estimates from lattice distortion. They are compared with experiment for a number of systems by Harrison and Kraut (1988)

Of particular interest are cases in which doping occurs, as we indicated at the beginning of this subsection. This can occur as we indicated there by mutual doping of two semiconductors separated by an interface, or in the more usual case of doping from a pure element. We may consider for example n-type doping of germanium by arsenic. We may again proceed as for Eq. (8-3), but now inserting an individual arsenic atom. $E_{coh}(AB)$ is replaced by the cohesive energy per atom for elemental arsenic for which we must use an experimental value, obtained from Table 1-3 as 2.96 eV per atom. We add the energy to insert the arsenic into germanium, $E_{GeGe}(As_{Ge})$ = -1.51 eV and transfer the extra electron from the valence-band maximum to the conduction band for +0.76 eV. Finally, we subtract the energy 4.60 eV gained from returning the removed Ge atom to the germanium for a net energy change of -2.39 eV. [I made numerical errors at the conclusion of Harrison and Kraut (1988) which incorrectly gave a positive result.] This negative value seems to be incorrect and it must be our calculated energy of substitution which is responsible; we could have used experimental values for each step except $E_{GeGe(As_{Ge})}$. We do not have a more meaningful value for this energy, but a theoretical estimate has been made by Berding (1995) for $E_{SiSi(As_{Si})}$ equal to + 2.3 eV, within a few tenths of an eV, based on full local-density calculations. Such a value seems more reasonable.

D. Alloy Energy Bands and Scattering

We should note at the start the comprehensive text on this subject, *Semiconductor Alloys* by Chen and Sher (1995). This book includes extensive tabulations of the band structures of semiconductors and their alloys as well as more complete analyses of the alloy properties than can be included here.

Our construction of energy bands in Chapter 5 depended directly on the translational symmetry of the crystal and in some sense the bands (energy as a function of wavenumber in the Brillouin Zone) do not exist in alloys where the periodicity is destroyed by the presence of different kinds of atoms, randomly arranged. On the other hand, the bands were only really defined for a perfect crystal and for periodic boundary conditions, neither of which we have in real systems. Similarly we may *define* bands in the alloy.

In tight-binding terms we do this by defining a *virtual crystal* in which the s-state term values are all taken as the weighted average for the constituents, $(1\text{-}x)\varepsilon_s(A) + x\varepsilon_s(B)$ for an alloy of x atomic fraction of B in A, and the corresponding average for p-states. We then place all the atoms on a perfect lattice, called the *average lattice* , with the weighted average of lattice distance for the two constituents, and therefore $V_{ll'm}$ which are the same for every set of nearest neighbors. This system is them mathematically equivalent to a perfect lattice and the energy bands are defined and calculated exactly as in Chapter 5. It is interesting that although the coefficients of the tight-binding states, $u_j = u_0 e^{i\mathbf{q}\cdot\mathbf{r}_j}$, are the same as for the perfect lattice, the states themselves are not: at each site with an A-atom, a combination of A-like atomic states appears, which may even have different numbers of nodes than those at a site with a B-atom. Similarly average pseudopotentials can be used with an average lattice to define energy bands though the corresponding real crystalline states have quite different character on different sites.

Obviously these alloy energy bands for the virtual crystal are an average of the energy bands for the constituents, defined as those obtained from the average parameters. In particular, the energy gap E_0 between the valence band maximum and the conduction-band Γ_1 state will be approximately a linear interpolation of that for the two bands. Experimentally, one finds a very slight nonlinearity as a function of concentration, or "bowing", of the gap which will come partly from the average we have defined for the virtual crystal and partly from other effects included in tight-binding theory and going beyond it. We are not concerned with these very small effects.

We may next correct each term value by it's appropriate value relative to the average and allow the atoms to move to their real positions relative to the average lattice. As we may deduce from the nature of the distortions discussed in Section 8-1-B, this will approximately shift each coupling three-quarters of the way to its pure-component value. We ordinarily regard these corrections, based upon a random arrangement of impurities, as a perturbation which causes scattering of the carriers between states in the virtual-crystal band structure. We may do this again using the Golden Rule, Eq. (1-32) for the transition rate. It is then not difficult to see from the dependence of the corrections on concentration x that the scattering will be proportional to $x(1-x)$, going to zero for either pure component.

This may be illustrated for a fictitious case of the scattering of electrons near a conduction-band minimum at $\mathbf{k} = 0$ and an alloy of x atomic per cent of element B in element A in which we may neglect the effects of lattice distortion. Then in the virtual crystal (for this homopolar case) the states near the conduction-band minimum at $\mathbf{k} = 0$ are approximately $1/\sqrt{N}\,\Sigma_j e^{i\mathbf{k}\cdot\mathbf{r}_j}|s_j\rangle$ with N the number of atoms in the crystal. To evaluate the scattering rate using the Golden Rule we need the matrix element of the perturbation in the alloy. Let the difference between the s-state energies on the two constituents be $\varepsilon_s(B) - \varepsilon_s(A) = \delta\varepsilon_s$. Then the perturbation on a site with an A atom is $\delta H_j = \varepsilon_s(A) - [(1-x)\varepsilon_s(A) + x\varepsilon_s(B)] = -x\,\delta\varepsilon_s$ and on a site with an atom B, $\delta H_j = (1-x)\delta\varepsilon_s$. The matrix element between a state of wavenumber $\mathbf{k}$ and one of wavenumber $\mathbf{k}'$ becomes

$$\langle\mathbf{k}'|H|\mathbf{k}\rangle = \frac{1}{N}\sum_j e^{i(\mathbf{k}-\mathbf{k}')\cdot\mathbf{r}_j}\,\delta H_j\,. \tag{8-4}$$

Now the expression we need is

$$|\langle\mathbf{k}'|H|\mathbf{k}\rangle|^2 = \frac{1}{N^2}\sum_{j,i} e^{i(\mathbf{k}-\mathbf{k}')\cdot(\mathbf{r}_j-\mathbf{r}_i)}\delta H_j\,\delta H_i\,. \tag{8-5}$$

We may see that only diagonal terms, $j = i$, in the sum contribute because for any fixed $\mathbf{d} = \mathbf{r}_j - \mathbf{r}_i$ the sum over j on sites for atom A will vanish because the sites i will be A or B in proportion to their concentration. Thus the perturbation for j as a B atom, $(1-x)\delta\varepsilon_s$ occurs with probability x and the perturbation for an A atom, $-x\delta\varepsilon_s$ occurs with probability $1-x$. They cancel when we sum over all sites j , holding $\boldsymbol{d}$ fixed, and only the diagonal terms $\boldsymbol{d} = 0$ contribute. There are $(1-x)N$ sites of type A, each having $\delta H_j^2 = x^2\delta\varepsilon_s^2$, and xN sites of type B, each having $\delta H_j^2(1-x)^2\delta\varepsilon_s^2$, leading to the final form.

$$|\langle \boldsymbol{k}'|H|\boldsymbol{k}\rangle|^2 = \frac{1}{N}x(1-x)\delta\varepsilon_s^2 \cdot \tag{8-6}$$

This is a constant in a sum over final state of the transition probability from Eq. (1-32),

$$P_{tot} = \Sigma_j P_{ij} = \frac{2\pi}{\hbar N}x(1-x)\delta\varepsilon_s^2\ \Sigma_{k'}\ \delta(\varepsilon_{k'} - \varepsilon_k)\,. \tag{8-7}$$

The sum over wavenumbers of the final state is obtained again using Eq. (5-3) as $\Sigma_{k'} = \Omega/(2\pi)^3\int 4\pi k'^2 dk' = \Omega/(2\pi)^3\int 4\pi k'^2(m_c/(\hbar^2 k'))d\varepsilon_{k'}$. In this last step we used the effective-mass approximation described in Section 6-3 to write $\varepsilon_k \approx E_c + \hbar^2k^2/(2m_c)$. The integral is immediate, and

$$P_{tot} = \frac{m_c\, k\, \Omega_0 \delta\varepsilon_s^2\, x(1-x)}{\pi\ \hbar^3} \tag{8-8}$$

with Ω_0 the atomic volume. It has the concentration dependence we anticipated, is proportional to the square of the difference in s-state energy between the two atom types and of course has the units of reciprocal time. As we noted in the treatment of scattering by acoustical phonons, the scattering rate relevant to transport theory has an additional factor $1-cos\theta$ under the sum in Eq. (8-7), but in this case the average over the $cos\theta$ would vanish (the scattering is isotropic) and the same result would be obtained.

Note that at small x this rate is proportional to concentration. We may define a cross-section σ_X for the scattering by each impurity. The total scattering rate is the velocity of the carrier $\hbar k/m_c$ times this cross-section times the total number of impurities divided by the volume of the system. Solving for the cross-section we obtain

$$\sigma_X = \pi\left(\frac{m_c\Omega_0\delta\varepsilon_s}{\pi\,\hbar^2}\right)^2 . \tag{8-9}$$

For this case, the cross-section is independent of the velocity of the carrier, but depends upon the square of the carrier mass.

2. Impurity States

In the preceding subsection we have treated the effects of different atoms as a perturbation on the electron states. In particular, in obtaining Eq. (8-9) we treated the effect of a single impurity as a perturbation. If the shift $\delta\varepsilon_s$ were negative, and too large, an electronic state could be pulled out of the band to form a state in the gap, just as an attractive potential may form a bound state of otherwise free electrons, with that state having energy lower than the free-electron states.

In fact, if the impurity atom is larger in valence by one than the atom it replaces, it will produce an attractive potential $-e^2/(\varepsilon_s r)$ solely due to this extra charge. Consider a case in which its extra electron lies in a simple conduction band of effective mass m_c. As we have seen in Section 6-3 and 4 this electron will have the dynamics of a free electron of effective mass m_c and will have a hydrogen like state with energy given by Eq. (1-2) relative to the conduction-band minimum, but with the mass replaced by the effective mass and the couplings reduced by the dielectric constant,

$$\varepsilon_D = E_c - \frac{e^4 m_c}{2\varepsilon_s{}^2\hbar^2}, \qquad (8\text{-}4)$$

with ε_s the static dielectric constant, Eq. (4-12) for homopolar semiconductors, with corrections, as in Section 4-3-C for polar semiconductors. This, and the series of states with factors $1/n^2$, are called *donor states*, signified by the subscript D, with each impurity atom donating an additional electron to the system. The energy below the conduction-band minimum is reduced from the hydrogen value of one Rydberg, equal to -13.6 eV, by a factor $m_c/(m\varepsilon_s{}^2)$, which for GaAs (which has such a conduction-band minimum) is 0.00055, so the binding is only 0.007 eV. This is much less than k_BT at room temperature so the electrons ordinarily leave their atom of origin and are truly donated to the conduction band. Further, the radius of this donor state is so large, corresponding to the effective Bohr radius $\hbar^2\varepsilon/(m_c e^2) = 88.\ \mathring{A}$, is so large that the use of the effective-mass approximation is justified and the probability density on the impurity atom itself is so small that the additional shift due to a different atomic energy e_s is negligible. Thus Eq. (8-4) is a good estimate for the energy of such a shallow donor state. This weak effect of the impurity atom is consistent with our treatment of scattering in the preceding section.

The corresponding shallow acceptor states arise from the negative charge of an acceptor atom, e. g., Al in Si, which will attract the hole which it produces in the valence band. There will in fact be acceptor states represented in analogy with Eq. (8-4) by $\varepsilon_A = E_v + e^4 m_h/(2\varepsilon_s{}^2 \hbar^2)$. There are acceptor states associated with the light hole, with m_h being the light-hole mass and acceptor states associated with the heavy hole, with m_h being the heavy-hole mass. The latter are of course deeper and the holes will tend to be bound in the deeper state, if they are bound at all.

The shallow donor states are much more complicated in indirect semiconductors such as silicon where there are six conduction-band minima in the six [100] directions. We could imagine constructing a hydrogen-like state for each minimum, based upon an anisotropic mass, heavy in the longitudinal [100] direction but light in the lateral two directions. It would be a pancake-like state, extending far in the lateral directions. Six such states could be constructed, one from each of the six minima, and these states are coupled. (The coupling may be thought of as arising from the nonparabolicity of the bands.) The lowest-lying state will be a symmetric combination of the six pancakes, all centered at the donor atom. There will also be combinations of other symmetries at slightly higher energies. We shall not explore these further.

If the impurity atom had valence larger by *two* , the estimate from Eq. (8-4) would be deeper by a factor of four, the radius would be reduced by a factor of two, and the effect of the central atom would be enhanced. This becomes a *deep level* and the effective mass estimate is no longer reliable. We must return to a description of the states in terms of individual atomic orbitals. In terms of bond orbitals, a substituted atom introduces four shifted bond levels of identical energy. However, we can proceed with more accuracy by expanding the region of the solid for which we solve. In this case, we shall choose a *bonding unit* consisting of the impurity atom and the four hybrids directed towards it from the host material. We then solve for the states in this bonding unit exactly, as we solved for the states in the single bond exactly. Expanding the bonding unit always should increase the accuracy, even up to solving exactly for the states of the entire crystal. However, we only expand as far as will still allow us to solve analytically for the states. Once we have obtained them we may again include the metallization of the bonding unit due to coupling with the material outside the bonding unit. We will have other occasions to use such enlarged bonding units, in particular when we discuss silicon dioxide.

In this analysis of impurity states we are following Dow and coworkers (Hjalmarson, Vogl, Wolford, and Dow (1980)) who clarified the principal trends in deep levels using a tight-binding theory. In particular, they elucidated the insensitivity of the deep levels to the specific impurity.

Our bonding unit containing the four host hybrids surrounding the impurity has the full symmetry of a regular tetrahedron. We may use that symmetry to construct the states as illustrated in Fig. 8-4, just as we did in constructing the states of the nitrogen molecule in Section 1-4-B. In particular, the system has reflection symmetry through any (110) plane through the impurity and states can be chosen to be even or odd with respect to these reflection. If it is even under all reflections, the normalized combination of the four hybrids which enters, $1/\sqrt{4}\ \Sigma_j\ |h_j\rangle$, will have the coefficients on each hybrid the same, a symmetry which is denoted by A_1 . Further, no p-states from the impurity will enter this fully-symmetric state so the impurity state will be a linear combination of this even set of hybrids and the s-state. The coupling between the two is $2V_{sh\sigma} = V_{ss\sigma} - \sqrt{3}\ V_{sp\sigma} = -3.78\ \hbar^2/md^2$. The two resulting states have the full symmetry of the defect and are referred to as states of A_1 symmetry. We find their energies as

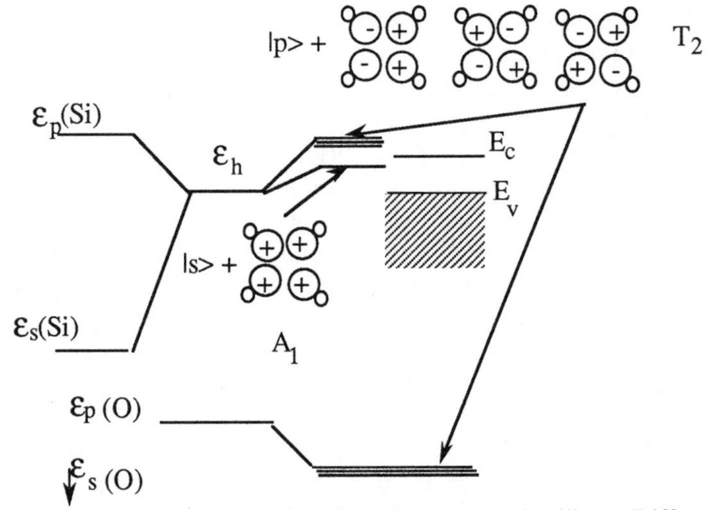

Fig. 8-4. The formation of a deep level from oxygen in silicon. Different combinations, shown, of the four silicon hybrids directed at the oxygen couple with the oxygen s-state (energy level out of the range of the figure) or oxygen p-state, forming the levels shown. Only the upper A_1 state lies in the gap. The lower A_1 state and the three lower T_2 states are doubly occupied, leaving two electrons for the upper A_1 state, a deep donor level.

$$\varepsilon_{A1} = \frac{\varepsilon_s + \varepsilon_h}{2} \pm \sqrt{\left(\frac{\varepsilon_s - \varepsilon_h}{2}\right)^2 + \left(\frac{-3.78\hbar^2}{md^2}\right)^2} \quad . \tag{8-5}$$

They are illustrated for an oxygen impurity in silicon in the energy-level diagram, Fig. 8-4.

There are also states which are odd under some reflections. There are three orthogonal combinations of hybrids which have the symmetry of p-states on the impurity atom. They can be chosen for various axes, but perhaps most convenient is for oxygen p-states oriented along the cube axes of the crystal and combinations of hybrids $1/\sqrt{4}\ \Sigma_j \pm |h_j\rangle$ each with two pluses and two minuses. These states are also illustrated in Fig. 8-4. The impurity p-state makes an angle with cosine of $1/\sqrt{3}$ with the axes of the hybrids so the coupling between the p-state and the appropriate combination of hybrids is $2V_{ph\sigma}/\sqrt{3} = V_{sp\sigma}/\sqrt{3} - V_{pp\sigma} = -3.04\ \hbar^2/md^2$. These are states of T_2 symmetry given by

$$\varepsilon_{T2} = \frac{\varepsilon_p + \varepsilon_h}{2} \pm \sqrt{\left(\frac{\varepsilon_p - \varepsilon_h}{2}\right)^2 + \left(\frac{-3.04\hbar^2}{md^2}\right)^2} \quad , \tag{8-6}$$

and are shown in Fig. 8-4. Also shown to the right are the energies of the valence-band maximum E_v and conduction-band minimum E_c . Only the upper A_1 state lies in the gap for the case of oxygen in silicon.

The levels with minus signs in Eqs. (8-5) and (8-6) lie below the valence-band maximum and are all occupied. With six valence electrons contributed by the oxygen atom and four from the surrounding silicon atoms, these levels are all filled, leaving two electrons to be occupied by the upper A_1 level in the neutral impurity. These electrons could be excited thermally into the conduction band so it is a donor level, but in this case it is a deep donor. This is a qualitatively valid description of the formation of the deep donor level by an impurity from column VI in silicon. The energy level itself which we find at 0.09 eV below the conduction band minimum is in reality much deeper for all of the column-VI impurities (see Sze (1969)).

For any meaningful quantitative results we must expand the bonding unit and treat the levels more completely, as did Hjalmarson, et al. (1980). Dow has indicated in private communication that about half the total probability density for these deep levels lies outside our bonding unit and we may qualitatively confirm this. An impurity atom with four nearest-neighbors and twelve second neighbors is shown in Fig. 8-5. For the combination $1/\sqrt{4}\ \Sigma_j \pm$

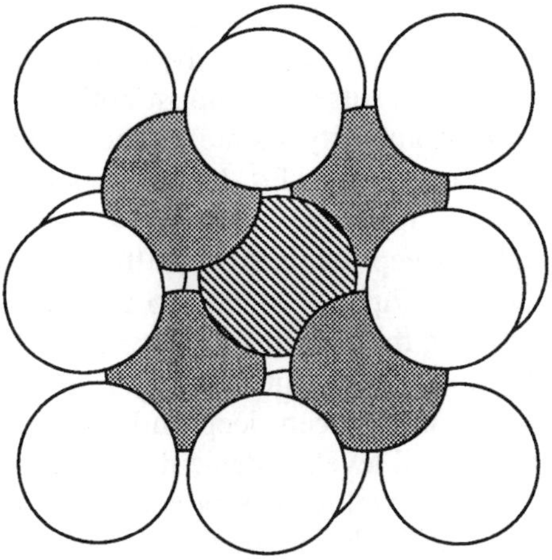

Fig. 8-5. An impurity atom shown with stripes surrounded by four nearest neighbors, shaded, and twelve second neighbors, white.

$|h_j\rangle$ on the nearest neighbors and oriented toward the impurity, for example, each of these hybrids is coupled to a bond, or antibond, between this nearest neighbor and a second-nearest neighbor. This bond or antibond lies outside the unit and is coupled by $V_1/\sqrt{2}$ in a homopolar semiconductor (since that bond or antibond is only half on the shared atom). If we estimate in perturbation theory the probability density on external bonds and antibonds as the squared coupling, divided by the squared energy difference V_2^2, summed over the twelve such bonds and twelve such antibonds we obtain $24V_1^2/(8V_2^2)$, equal to 0.49 for silicon. This is large enough that we must go beyond the simplest bonding unit, for example by expanding the impurity state in four terms, the central atomic state, the symmetric combination of four neighboring hybrids, the symmetric combination of twelve second-neighbor bonds and the symmetric combination of twelve second-neighbor antibonds. (In these latter combinations the three bonds, or antibonds, sharing one atom have equal coefficients, whether the central atomic state is an s- or p-state.) If we wish to use perturbation theory, we may test its validity in terms of solutions of the four-by-four problem. It may also be necessary to include Coulomb shifts due to charge redistributions, as done by Hjalmarson, et al. (1980). The energies of such deep levels are very well known experimentally. Values have been tabulated, for example, by Sze (1969).

The insight which Hjalmarson, Vogl, Wolford, and Dow (1980) derived from such an analysis is quite important. The deep level in the gap is derived from a combination of silicon hybrids and an s-state so low in energy that it is not seen in Fig. 8-4. The level is only slightly higher than the silicon hybrid energies and the state itself is almost entirely on these hybrids with only a tiny admixture of impurity s-state. [The occupied level at very low energy, derived from the minus in Eq. (8-5), lies of course almost entirely on the impurity s-state.] The level in the gap is also very insensitive to the energy of that s-state. Deep levels from sulfur and oxygen had generally been associated with the impurity s-state so that it had been puzzling that they were so insensitive to the electronegativity of the impurity atom.

There is an additional important insight based upon Hjalmarson, et al. concerning the distinction between deep and shallow states. If our tight-binding calculation gives all levels outside the gap, we would say there is no deep level and the shallow level is to be calculated as in Eq. (8-4). This is the case for the A_1 level far to the left in the diagram in Fig. 8-6. Such a case is a selenium atom substituted for an arsenic atom in an AlAs-GaAs alloy. As one shifts the alloy concentration, the conduction-band minimum shifts and the shallow level directly follows the conduction-band minimum according to Eq. (8-4). For a deep level, as for an arsenic atom on a gallium site (the *arsenic antisite defect*) in this same alloy, a deep level arises and remains quite fixed relative to the valence-band maximum given in terms of atomic term values as the concentration is shifted, shifting the conduction-band minimum.

In a similar way, deep acceptor levels may arise from atoms in Column I or II. These levels will be T_2 levels, estimated in Eq. (8-6), with the plus sign. They are slightly above the hybrid energy for silicon and will be composed almost entirely of the p-like combination of hybrid states. In Fig. 8-6 we show schematically the variation of these levels, as well as of the deep donors, as a function of the atomic energy of the impurity state, s-state in the case of the A_1 level and p-state in the case of the T_2 level. Note that for illustration we have taken the hybrid level near midgap, though our parameters put it near the valence-band maximum for silicon.

The structure of the impurity system can also be much different from the ideal substitution which we have considered here. In particular, studies of the antisite defect in gallium arsenide (an As atom sitting on a Ga site) by Dabrowski and Scheffler (1988,1989) and by Chadi and Chang (1988) (See also Chadi (1992)) have shown that although the stable geometry is that with tetrahedral symmetry as we have assumed here there is a metastable state in which the arsenic antisite atom takes a position which is reflected through one of the triangular planes making up the tetrahedron of its nearest

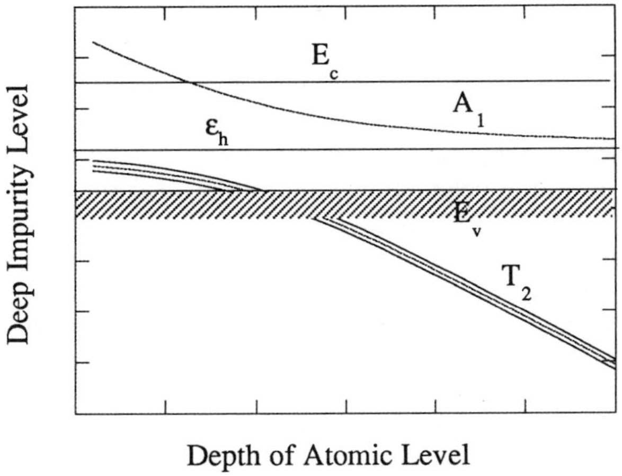

Fig. 8-6. A schematic representation of deep impurity levels as a function of the depth of the impurity atomic level, the valence s-state in the case of the A_1 level and the valence p-state in the case of the T_2 level. Both are dominated by the host hybrid states, with energies indicated by ε_h .

neighbors, leaving the initial antisite vacant. This feature seems to be essential to the ubiquitous EL2 optical center of gallium arsenide. (See the references to Dabrowski, Scheffler, Chadi and Chang just given.) In the ideal tetrahedral site there are two electrons in a donor state drawn from the conduction band, corresponding to two antibonding electrons. In the metastable state one arsenic-arsenic bond is broken, placing two electrons in each the two dangling hybrids so generated.

3. Vacancies and Interstitials

We begin with a discussion of the electronic states of an ideal vacancy, a missing atom with the remainder of the crystal unchanged. We discuss these in the bBond-Orbital Approximation and with corrections to that picture. We then turn to the distortions which spontaneously occur when such a vacancy is formed. We turn finally to the same subjects for an interstitial atom.

A. The Ideal Vacancy

We may again construct electronic states for the crystal by forming sp^3-hybrids on every atom, and then forming bonds except for the four hybrids extending into the vacancy. These "dangling hybrids" (frequently called, less appropriately, "dangling bonds") remain at the hybrid energy since the one orbital to which they were strongly coupled has been removed. We can understand these as the limit of an impurity state as the atomic levels in the impurity atom, as shown in Fig. 8-6, are taken to arbitrarily great depth, to the right in the figure. We only showed the upper A_1 state there, but there is also an upper T_2 state which approaches the hybrid energy from above. In the Bond Orbital Approximation, in which only nearest-neighbor coupling is included this yields four states at exactly the hybrid energy. If we include the metallic energy coupling each dangling hybrid to its neighboring bonds and neighboring antibonds, there arises an indirect coupling between the four hybrids through the lattice. Between any two of these hybrids there are in fact two paths through four bonds each, and five metallic-energy couplings, which couple the pair. It seems no major job to obtain the resulting states, but we do not do that here since the principal qualitative results are understandable without the numbers. The four levels are split, as with the impurity states, into an A_1 combination at lowered energy and a T_2 at raised energy.

In silicon the sp^3-hybrid at -9.39 eV is essentially at the valence-band maximum, -9.35 eV from Table 6-1. Thus the A_1 level will lie within the valence band and the T_2 level will lie in the gap. For the neutral vacancy, there are four electrons available to occupy these levels, two filling the A_1 level (really a resonance rather than a local state) in the valence band and two in the T_2 degenerate states which could accommodate six. In such a case of orbital degeneracy in the ground state, occupied and empty orbitals at the same energy, we may expect a spontaneous lattice distortion.

B. Jahn-Teller Distortion

This is exactly the situation considered in general by Jahn and Teller (1937) and applied to this case by Watkins and Messmer (1973). The effect is easily understandable. The three T_2 combinations might contain a combination of hybrids on the four atoms neighboring a vacancy with ± coefficients as illustrated in Fig. 8-7, with pluses for the atoms in the positive z-direction and minuses for those in the minus-z direction. The other two combinations are oriented along the x- and the y-direction. We imagine a

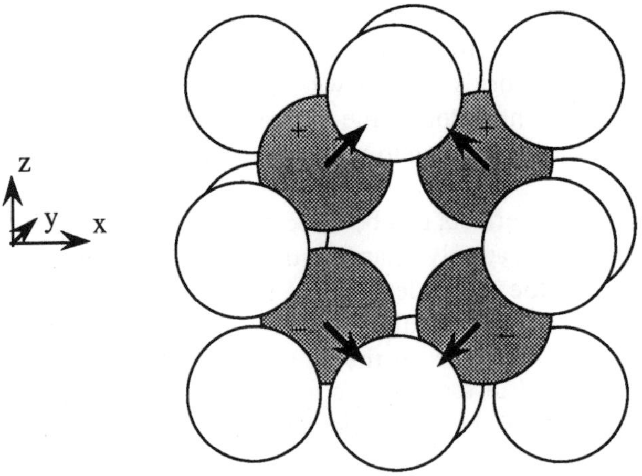

Fig. 8-7. Displacement of four neighbors to a vacancy which will split the levels of T_2 symmetry, to first-order in the displacement, and can therefore be a Jahn-Teller distortion.

small displacement of the nearest neighbors to a vacancy in the directions indicated in Fig. 8-7, which would shift the T_2-level shown in proportion to the displacement, possibly lowering it. This would be true if the displacement increased the effective coupling between the upper two hybrids while decreasing the effective coupling between the two hybrids to the right and between the two to the left. Symmetry will guarantee that such displacements will not change the average energy linearly in the distortion (only quadratically) so the corresponding x- and y-levels would then each rise half as much. Further, there would be energy changes, including elastic energy, increasing the energy in proportion to the square of the displacement from the equilibrium position. However, if we put the two electrons in the z-oriented state and leave the x- and y-oriented states empty, the energy will always drop linearly and the distortion will proceed until the quadratic terms prevent further displacement. If this distortion happened to raise the energy of the z-oriented T_2 state, we could add this z-oriented distortion to an equal y-oriented distortion and see that the x-oriented state drops in energy and the y- and z-oriented states rise half as much. In either case, energy is gained in the distortion and it is guaranteed to occur. This is the Jahn-Teller distortion arising from the splitting of orbitally degenerate ground states.

Indeed this is the distortion which is observed for the neutral vacancy (Watkins and Messmer (1973)). We could readily estimate the distortions in tight-binding theory by calculating the energy levels and including the needed terms in the total energy, though we could not expect high accuracy from the calculation. [In a full calculation the second-neighbor coupling between dangling hybrids, which we neglected, might be larger than the small indirect coupling through the lattice.] The net effect is to produce a doubly-occupied level in the gap with two degenerate empty levels above it in the gap.

These qualitative effects can be important to understand. We note that if the vacancy were negatively charged, the additional electron would reside initially in one of the upper levels, but then the same argument as above would indicate that there should be a Jahn-Teller distortion splitting those upper levels. We may turn to the original problem. Let us say that the energy level oriented in the z-direction varies as $\varepsilon_z = -\lambda(u_z - u_x/2 - u_y/2)$ with a distortion u_z of the same orientation (as in Fig. 8-7) and u_x and u_y in the other two orientations. The energies of the other two states, ε_x and ε_y are obtained by rotating indices. The dependence of the energy upon distortion is obtained by adding the energy of the occupied states to $^1/_2\kappa(u_x^2 + u_y^2 + u_z^2)$. We may readily find that with two electrons in the state ε_z the energy is minimized at $u_z = 2\lambda/\kappa$ and $u_x = u_y = -\lambda/\kappa$ for an energy of $-3\lambda^2/\kappa$. With one electron also in ε_y we find the energy minimized at $u_z = -u_x = -3\lambda/(2\kappa)$ and $u_y = 0$ for an energy $-^9/_4\ \lambda^2/\kappa$. The properties of such systems in which the equilibrium atomic positions depend upon the electronic occupation of states are quite interesting, particularly if we treat the atomic displacements as well as the electrons quantum-mechanically. We have seen here only how the basic Hamiltonian for such systems can be constructed. They can be used in computational approaches such as molecular dynamics (Seong and Lewis (1995), Wang, Ho and Chen (1994), Molteni, Colombo, and Miglio (1953)).

It is interesting that tight-binding theory correctly requires Jahn-Teller distortions in these cases of orbital degeneracy but that they may not occur in the more complete first-principles calculations. The reason is that such calculations almost universally are based upon supercells in which the defect is periodically repeated in the crystal in order to use the computational methods which were developed for periodic lattices. For a finite supercell there will be interactions between neighboring defects, broadening the levels into bands. In such a case, with partly filled *bands* , there may no longer be spontaneous distortions. The methods would be correct for sufficiently large supercells, but perhaps not for cells which are currently used.

C. The Tetrahedral and Shared-Site Interstitials

An interstitial atom is an extra atom inserted in the lattice. The natural site to assume for the interstitial is easily seen in a three-dimensional model of the structure as the largest open space in the crystal. We may understand the site in the cubic view of the structure given in Fig. 2-2. The face-centered cube was divided into eight "cubies" and four of those were filled with an additional atom to form the perfect crystal. We could insert an interstitial at the center of one of the empty cubies and it would have also four nearest neighbors arranged as a regular tetrahedron at the equilibrium distance, 2.35 Å in silicon. Since each of its neighbors already has bonds with four neighbors, we cannot form independent bonds with the interstitial; one will expect that the nearest neighbors might be pushed outward to accommodate this interstitial atom. The cube center in that figure is also such a tetrahedral site, but in different sublattice; if this were an interstitial in gallium arsenide, in one case it would be surrounded by arsenic atoms and in the other by gallium atoms. A cube edge is also such a tetrahedral site of the same type as the cube center.

We could begin a discussion of the electronic states of such an ideal self-interstitial in silicon by adding the s-orbital and p-orbitals for the added atom. The energy of the s-orbital is deep in the valence band and will be occupied by two electrons. The p-level, at -7.59 eV, is well above the conduction-band edge. The surrounding orbitals have all formed bonds and antibonds, and the interstitial p-level interacts with them. These interactions push the p-like state still further into the conduction band. The two electrons from the p-states are dumped into the conduction band. The remaining positive charge will form donor levels in the gap, but in this case we expect them to be of the nature of shallow levels, discussed in Section 2, though because of the double charge too deep to accurately treat in effective-mass theory. For silicon the occupied levels should be from the fully symmetric combination of pancake-like states discussed in Section 8-2.

The ground state of the neutral ideal interstitial is *not* unstable against a Jahn-Teller distortion because there is only a single fully symmetric orbital, doubly occupied. However, having these two electrons in states derived from essentially antibonding (conduction-band) states is energetically unfavorable. We might expect an alternate geometry to be favored.

Bar-Yam and Joannopoulos (1984) have found from careful total energy calculations that in the lowest-energy configuration the neutral interstitial moves out of the tetrahedral site as one of the atoms from the original lattice moves off its site so that there are finally two atoms sharing a site initially occupied by one atom. Blöchl, Smargiassi, Car, Laks, Andreoni, and Pantelides (1993) have confirmed this result. We have attempted to show this shared-site interstitial in Fig. 8-8. Consider the starting site at the center of this group, surrounded by four shaded atoms as a regular tetrahedron. If we construct two points, each obtained from the central site by reflection through one of the faces of that tetrahedron, and place atoms at those points with none at the central site, we would have two interstitials and a vacancy. If we then move those two interstitials back toward the central site until they are separated by 95% of a bulk nearest-neighbor distance, we obtain this shared-site interstitial. The two extra atoms are separated from each other in a [110] direction.

Another way of thinking of this shared-site interstitial is to sketch a close-packed zig-zag chain of silicon atoms lying along a [110] direction, as in Fig. 8-9. One atom is replaced by a pair of atoms in the same plane, as shown. A continuing motion of the first atom into the original site, with the second atom moving into a neighboring tetrahedral site completes a diffusion, by replacement, from one tetrahedral site to the next.

Chi-Lung Chou (1994) made a tight-binding analysis of this split interstitial in silicon, finding that there were enough states below the valence-band maximum to accommodate all of the eight electrons from the interstitial. This of course means that there are no electrons in antibonding bands, consistent with the stability of the structure.

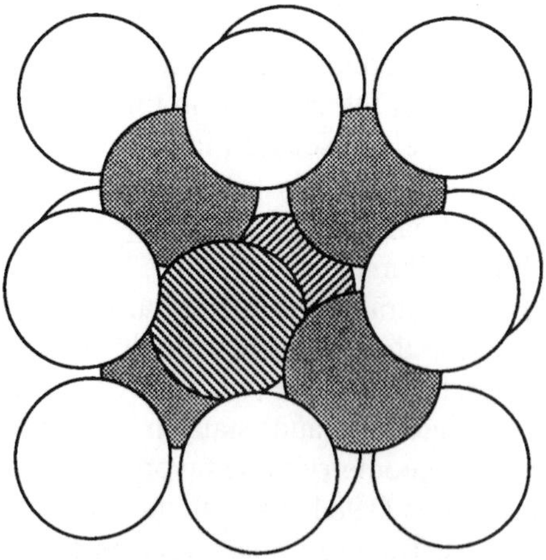

Fig. 8-8. The two striped atoms form a shared-site interstitial centered at the central position initially occupied by one atom, as described in the text.

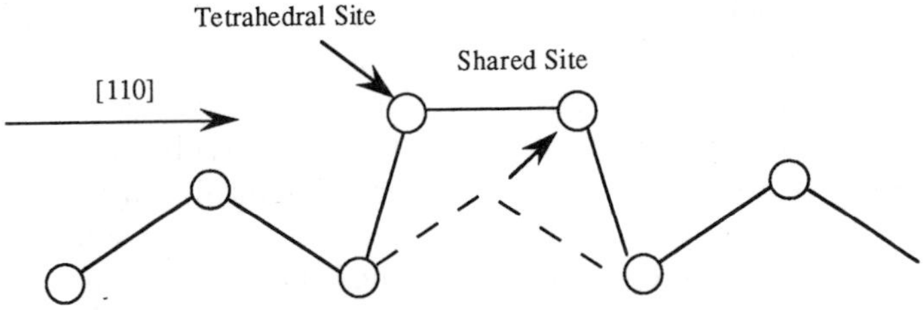

Fig. 8-9. An interstitial atom in a tetrahedral site can move into a shared-site position with the original atom moving to a symmetric position, equivalent to that shown in Fig. 8-8. A further motion of the first atom into the bulk site, with the original atom moving into an adjacent tetrahedral site, completes a diffusion step.

It also has the interesting consequence that if the interstitialcy moves, presumably with an intermediate geometry with a single atom in a tetrahedral site, electron states will rise from the valence band up into the conduction band and return. This has the possibility of carrying holes from the valence band to the conduction band and electrons from the conduction band to the valence band, accomplishing electron-hole recombination without radiation.

Harrison and Wills (1996) have carefully explored this possibility. Wills carried out full-potential, local-density-approximation calculations of the total energy of the two geometries, confirming the results of Bar-Yam and Joannopoulos. He also found that for the tetrahedral geometry there was a resonance in the conduction band, 1.22 eV above the valence-band maximum. (The local-density calculation gave the conduction-band edge only 0.49 eV above the valence-band edge, but the gap enhancement shifts discussed in Section 5-5-A raise it by 0.68 eV to the observed gap of 1.13 eV. The resonance, representing states occupied in the atom, is not shifted upward, but is still 0.09 eV above the gap.) He also found the occupied resonance very close to the valence-band maximum for the shared-site interstitial, with the valence-band also full for the neutral system. These confirmed the tight-binding expectations for the levels and the possibility of a recombination center.

It was possible to calculate the rate at which electrons enter and leave these resonances by interpreting Wills' calculations. They were performed on a "supercell", a crystal structure based upon a silicon lattice which initially was a perfect lattice, but with a cell of 32 (and in a separate calculation 16) atoms defined as a unit cell. Then one interstitial was introduced in each cell, forming a body-centered-cubic array of interstitials, with nearest-neighbor spacing $4d$. (The 16 atom cell is face-centered-cubic with nearest-neighbor distance $\sqrt{8/3}\, d$.)

Such supercell calculations cannot be taken at face value since there are many effects, such as the two electrons per interstitial filling the conduction band more than half an electron volt deep for the tetrahedral interstitial in silicon, which are artifacts of the small supercell. It was a sizable undertaking (Harrison (1997)) to interpret the results in terms of a model from which one could then estimate the corresponding results for an infinite supercell representing correctly the isolated interstitial. The calculation by Wills yielded a band of states from the interstitial resonance. For the tetrahedral site these are interpreted as p-like states coupled through the conduction-band, just as we shall treat the coupling between d-states in transition metals in Chapter 15. Thus the density of interstitial states is fit to individual resonances at the 1.22 eV above the valence-band as given above, coupled to teach other by a $V_{pp\sigma}$ and a $V_{pp\pi} = -\,{}^{1}/_{2}\, V_{pp\sigma}$ as in Section 17-1. The resulting $V_{pp\sigma}$ was fit in turn to the form $V_{pp\sigma} = (1/\pi)\,\hbar^2 r_p/md^3$ from Section 17-1 giving an $r_p = -1.4$ Å which is seen there to correspond to a coupling between free-electron states and local p-states of

$$\langle k|H|n\ell m,\mathbf{R}\rangle = -\sqrt{\frac{4\pi r_p^3}{3\Omega}\frac{\hbar^2k^2}{m^*}\frac{1}{kr_p}}\; e^{-i\mathbf{k}\cdot\mathbf{R}}\, Y_{l\,m}(\mathbf{k})\,. \tag{8-7}$$

We have introduced an effective mass m^* chosen to represent the many-valley conduction band. The proportionality of $V_{pp\sigma}$ to $1/R^3$ was approximately confirmed by comparison of the 32-atom supercell with the 16-atom supercell. These matrix elements could be used with the Golden Rule to obtain a rate of transition out of an occupied resonance of

$$\frac{1}{\tau(\varepsilon)} = \frac{2\pi}{\hbar}\sum_f |\langle k|H|nlm,\mathbf{R}\rangle|^2\,\delta(\varepsilon_k - \varepsilon_0) = \frac{4\hbar k^3|r_p|}{3m^*}\,, \tag{8-8}$$

with $\varepsilon = \hbar^2k^2/(2m^*)$ the energy of the resonance relative to the conduction-band minimum.

For capture of electrons by such a resonance, Harrison and Wills defined a cross-section σ such that the rate of capture was $N_e\langle v\rangle\sigma$, with N_e the density of electrons and $\langle v\rangle$ their average speed (${}^{1}/_{2}m^*\langle v\rangle^2 = k_BT$). Then for a system in equilibrium, with an equilibrium density of carriers, and a statistical distribution of interstitial configurations, they could equate the emission and capture rates for the resonance to obtain a capture cross-section. The same analysis was carried out for the resonance in the valence band, which was also found to correspond to p-like symmetry around the midpoint of the shared-site pair.

Finally, they allowed much larger concentrations of electrons and holes, which scaled up the capture rates. The steady-state condition that electrons and holes be captured at the same rate led to comparable fractions of interstitials in the tetrahedral and shared-site configurations and a recombination rate for electrons and holes which did not include any energy barrier to recombination, and thus no e^{-E/k_BT} factors. It was possible to define effective cross-sections for electrons and for holes, each of the order of the square of the corresponding wavelength for a thermal carrier, and obtain the capture rate in terms of these. This seemed a likely candidate for the dominant mechanism for radiationless recombination (Shockley-Read recombination) as long as the carrier concentration is not so high that recombination by Auger processes, where the energy is given to a third carrier, dominates, or the temperature is not so high that some activated process dominates.

Harrison and Wills (1996) went on to consider a similar mechanism for gallium arsenide. It was based upon an arsenic interstitial, which has a triply-occupied p-state at -8.98 eV without coupling, in the gap as seen from Table 6-1. As with the silicon tetrahedral interstitial, this p-level is coupled to the bonds and antibonds on the neighboring four atoms, the former pushing the level up and the latter pushing it down. However, since the bonds lie predominantly on the arsenic and the antibonds predominantly on the gallium, the interstitial surrounded by arsenics will have its level raised considerably and that surrounded by gallium will have it lowered significantly.

It is not difficult to estimate the shifted levels. A p-state oriented along a [100] direction will have a squared coupling with the bonds or antibonds given by $4/3(V_{sp\sigma}^2 + V_{pp\sigma}^2 + 2V_{pp\pi}^2)(1 \pm \alpha_p)/2$. These are divided by the energy denominator, $\varepsilon_p - \varepsilon_b$ or $\varepsilon_p - \varepsilon_a$, and added for each case to obtain a resonance at -10.92 eV for the gallium tetrahedron and -7.55 eV for the arsenic tetrahedron, both out of the gap from -9.64 eV to -8.12 eV. A subsequent calculation by Wills (private communication) confirmed this result with a full local-density calculation, suggesting essentially the same mechanism for recombination as in silicon.

D. The Role in Diffusion

The diffusion of the interstitial in silicon was studied in detail in Harrison (1998). In the ground state all interstitials are either in shared-site, or tetrahedral sites in this model. For intrinsic silicon, all interstitials are neutral and therefore in shared sites. In such a case the principal diffusion arises when thermal holes are captured by the interstitial and in the resulting singly charged positive state the interstitial moves easily through the lattice. This provides not only diffusion of the interstitial, but diffusion of other silicon atoms and impurities with which it interchanges as one of the atoms in a shared site moves to an adjacent site, sharing with a different atom.

If the silicon is doped p-type, there will also be many interstitials doubly-positively-charged, in the tetrahedral site, but again few singly-charged. The interstitial is what is called a *negative-U* center: the energy is much lower with a neutral and a doubly-positive interstitial, than with two singly-charged interstitials, as if the Coulomb repulsion between the electrons were negative, attracting the two electrons to the same interstitial. The principal mechanism for diffusion is again the capture of carriers to form singly-charged species which move through the lattice.

When there are excess carriers, and therefore carrier recombination through the interstitials, there are many more singly-charged interstitials present and therefore greatly enhanced diffusion. In addition, the energy from the recombination ends up in the interstitials, speeding them up, which will further enhance the diffusion.

In the course of studying the rate equations it was found that for recombination to occur, one must go beyond the capture of electrons from the conduction band by doubly-charged interstitials, the emission of electrons into the conduction band by singly-charged interstitials, and the corresponding capture and emission of holes in the valence band. One must include also the less frequent capture of electrons from the conduction band by singly-charged interstitials and the emission from neutral interstitials, and the corresponding processes in the valence band. Otherwise, doubly-charged interstitials are eliminated only by removing electrons from the conduction band and created only by emission into the conduction band and in steady state there is no net recombination. This was overlooked in Harrison (1998) and the recapture formulae could only be correct if there were other mechanisms of exchanging charge between interstitials. This oversight was indicated in an erratum to the article. It is easy to correct in the rate equations, and numerical calculations proceed easily. However, it complicates the derivation of general analytic formulae for recombination and diffusion.

Problem 8-1. Heats of Solution

Confirm that the heat of solution of germanium in gallium arsenide, steps *d* and *e* in Fig. 8-1 is the same as that for gallium arsenide in germanium, in the approximations which led to Eq. (8-1).

What is the difference in the heat of solution if the two germanium atoms are placed on adjacent sites in gallium arsenide? (This would be the binding energy for a germanium pair in gallium arsenide.)

Problem 8-2. Relaxation around impurities

Redo the derivation of atomic displacements for a potassium atom substituted for a sodium atom in NaCl . The physics is the same as that leading to Eq. (8-2) for the tetrahedral structure, but the structure is different from that illustrated in Fig. 8-2 and the result is different. Structures and lattice parameters are available in Chapter 9. You may again restrict relaxation to the nearest neighbors, though it would not be difficult to improve on that.

CHAPTER 9

Bonding in Ionic Crystals

Our tight-binding analysis of semiconductors has been based upon only those atomic states occupied, or partially occupied, in the free atom; atomic term values have been used for those atoms and nearest-neighbor coupling, of the form $V_{ll'm} = \eta_{ll'm}\hbar^2/md^2$, has been used. We continue with this basis and these couplings, as we turn to ionic crystals, by which we mean insulating compounds in which we may think of the nonmetallic atoms (from columns V, VI or VII of the Periodic Table) as receiving enough electrons to fill their valence p-shells from the metallic atoms, with all state higher in energy being empty. These are called "closed-shell systems" since we may think of only closed shells on each atom.

The atoms in such compounds ordinarily have more than four nearest neighbors. Use of the same coupling when there are more than four neighbors is questionable, and in fact Polatoglou, Theodorou, and Economou (1986) suggest that use of the free-electron formulae for systems with six neighbors may overestimate the matrix elements by a factor of as much as two. We can not have great quantitative confidence in results which depend strongly on these couplings.

If we were simply diagonalizing the tight-binding Hamiltonian by computer, there would be no important change from increasing the number of neighbors. Because we seek analytic solutions and elementary calculations of the properties, the difference is important. For crystal structures with more than four neighbors the transformation from one s- and three p-states to four hybrids directed at the nearest neighbors is not possible. The independent bonds which were the conceptual basis of our understanding of semiconductors are no longer a possibility. Fortunately, in ionic solids the occupied states of the nonmetallic atoms can serve the function for our analysis of ionic crystals which the bond orbitals served for semiconductors.

A typical case is rock salt, in which we imagine the electron from the sodium 3s-state being transferred to a chlorine 3p-state. Then the filled chlorine 3p-states are so low in energy relative to the empty sodium 3s-states that we may neglect the coupling between these states in comparison to the energy differences and think of the negatively-charged chlorine *ion* as an independent unit, interacting electrostaticly with the positively-charged sodium ions. We can later add the coupling between these sodium and chlorine states, as we added the metallization of the independent bonds in the semiconductor, and will regard this as a covalent contribution to the bonding. The resulting theory is nevertheless even simpler than that of semiconductors.

We shall actually construct a series of approximations, of increasing accuracy. We first consider the "independent-ion model" in which the coupling between electronic states on neighboring atoms is neglected altogether, allowing us to treat a number of properties. We then add this coupling in second-order perturbation theory, providing an extremely simple theory of the volume-dependent energy and related properties in Section 9-2-D. In Chapter 10 we shall use a more accurate form of this covalent energy arising from the coupling, which still allows in Section 10-3 analytic expressions for the properties. Finally, in Section 10-5 we shall incorporate Coulomb effects, which eliminates some of the largest discrepancies with experiment, but requires numerical solution for each property.

The concept of solids consisting of independent ions is very old and its analysis using universal-parameter tight-binding theory was included in the text, *Electronic Structure and the Properties of Solids* (Harrison, 1980). However, the parameters are now updated and there have been several simplifications and significant improvements which are included here (Harrison (1981a), (1985a), and (1990)).

1. The Independent-Ion Model

We begin the analysis of the properties of ionic crystals neglecting any effects of the coupling between electronic states on individual ions. We do, however, retain an overlap repulsion which for the present we regard as abrupt, but shall develop in detail later. At the outset we shall illustrate each calculation for potassium chloride, consisting of the atoms of atomic number one higher and one lower than that of the inert gas argon, as we may see from Table 1-1. Later we shall make a number of tabulations of properties for a wide range of ionic compounds.

The basic independent-ion model is very simple, as we indicated in Chapter 1. The cohesion of the potassium chloride is calculated by transferring an electron from its s-state on the potassium to a p-state on the chlorine, with a change in energy $\varepsilon_p(Cl) - \varepsilon_s(K)$, the magnitude being our first estimate of the cohesion. Then the lowest-energy excitation is returning one such electron in the crystal, costing an energy $\varepsilon_s(K) - \varepsilon_p(Cl)$, which becomes our estimate of the band gap. Such estimates will be listed in this chapter, based upon the term values in Table 1-1, and will turn out to be quite good. Our discussion in this section shows why this works and how corrections can be made. It requires addressing Coulomb effects in a way which was not necessary for semiconductors.

A. Coulomb Effects and Closed-Shell Ions

We noted in Section 5-5-A that there are Coulomb enhancements of the energy gaps in semiconductors. We viewed the system as consisting of independent neutral bonds, with an excitation energy $\varepsilon_a - \varepsilon_b$ required to lift an electron to the antibonding state in the same bond site. There is an additional energy U if the electron is carried away to another bond site, with the bonding state doubly occupied, but that additional energy is reduced by the dielectric relaxation to U/ε . We shall discuss these enhancements here also, but there is another correction for a system consisting of charged entities such as ions. In potassium chloride, with an additional electron on each chlorine, the term value ε_p is raised by such a Coulomb interaction, $U(Cl)$. If such transfers are made for all atoms in the crystal, this term value is also shifted by the charges on its neighboring ions. These tend to cancel each other and our simplest estimates assume exact cancellations. We explore also the effects of a lack of complete cancellation.

There is an additional subtle complication concerning the meaning of the energy levels. When we obtained bond energies and antibonding energies in Section 1-4, we did it by minimizing the total energy with respect to the occupation u_i^2 of the i-th state and the eigenvalues, or energy levels, obtained were the derivative of the total energy with respect to occupation of the state. There is a separate argument, called *Koopmans' Theorem* (Koopmans(1933), discussed for example in Harrison (1970), p. 76) in Hartree-Fock theory, which identifies these eigenvalues also with the removal energy for a complete electron from that state if the system is large enough. This theorem applies also approximately to the difference in the energy in transferring an electron from the bond to the antibonding state in the same site. However, for transferring an electron from the sodium to its neighboring chlorine there is a problem. Ignoring for the moment the Coulomb interaction between the two sites, the energy level on the chlorine is raised by $U(Cl)$ as indicated above and also the energy level on the potassium is lowered by $U(K)$ so the difference in energy levels is $\varepsilon_s(K) - \varepsilon_p(Cl) - U(Cl) - U(K)$. On the other hand, the energy gained in transferring a complete electron is the removal energy $\varepsilon(K)$ from potassium minus the energy at which it was added to chlorine, $\varepsilon_p(Cl) + U(Cl)$, differing from the difference in *energy levels* by $U(K)$. This discrepancy arises when we transfer simultaneously an electron from every potassium to a neighboring chlorine and is quite distinct from the enhancement of the gap when we transfer a single electron back to a site remote from the hole. Fortunately, much of the $U(K)$ discrepancy is canceled by interactions with the neighbors so it does not affect the simplest theory, but it becomes essential when we consider refinements of the model. We shall distinguish the two estimates of the transfer energy as "energy-level difference" and "transfer energy" in this section, and list both in our tabulations of the band gaps. The latter is obtained in a natural way using Born's model for the cohesion. We shall in fact conclude in the end, by comparison with full calculations, that contemporary band calculations give band gaps closer to the transfer energy than to the energy-level difference.

Inert gas systems (to be discussed in Section 20-2) are of course also closed-shell systems, as are ionic crystals. It is in fact useful to construct potassium chloride conceptually from the inert-gas atom as illustrated in Figure 9-1. The inert-gas atom argon has a filled 3p-shell, too deep to take part in bonding under ordinary circumstances; thus the main essential property of the shell is its overlap repulsion. Removing a proton from the argon nucleus yields a chlorine nucleus and a less tightly bound full 3p-shell, the electronic configuration of the chlorine ion (one electron added to the chlorine atom). It is nonetheless a full shell and electronically inert. Similarly adding a proton to an argon nucleus yields a potassium nucleus and a more tightly bound full 3p-shell, the electronic configuration of the potassium ion (one electron removed from the potassium atom). The ions now have charge so that there are Coulomb interactions between them and the overlap repulsion remains, but presumably modified in strength. This is

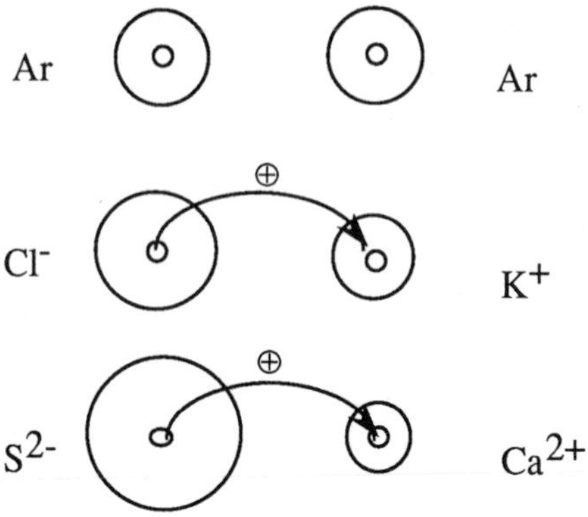

Fig. 9-1. Closed-shell ions have essentially the same electronic structure as inert-gas atoms which neighbor them in the periodic table. They differ, however, in nuclear charge. Since S, Cl, Ar, K, and Ca appear in sequence in the periodic table a K ion and a Cl ion may be constructed from two Ar atoms by transferring a single proton between nuclei. This is accompanied by a modification of the ion size as illustrated. The transfer of an additional proton produces a Ca and an S ion.

the traditional picture of ionic solids, charged balls with a rather abrupt repulsive interaction.

We may similarly construct the doubly-charged ions by transferring an additional proton, as indicated in Fig. 9-1, to obtain a sulfur ion and a calcium ion, now doubly charged ions with closed electronic shells, and an overlap repulsion. We might even try to go a step further to obtain scandium phosphide, but this compound apparently does not form. At each step the s-state energy on the metallic atom has become lower and the p-state energy on the nonmetallic atom has become higher until they are comparable and the independent-ion picture becomes less relevant. By ScP the levels have become comparable and the system apparently finds more stable arrangements.

This "theoretical alchemy" by which we convert Ar to KCl and CdS fits well with the overall organization of the periodic table which we discussed in Section 1-4-E, and distinguishes the compounds made by combining the nonmetals with metals on the right as ionic compounds, as opposed to the covalent crystals obtained by combining with metals on the left.

B. Crystal Structures and Electrostatic Energy

We may expect these charged closed-shell ions to arrange themselves in a crystal structure of low electrostatic energy with a spacing set by an abrupt overlap interaction. Indeed the ionic crystals with equal numbers of positive and negative ions usually form in one of the two structures, illustrated in Fig. 9-2, which have the lowest (and almost identical) electrostatic energies for fixed nearest-neighbor distance. Most, in fact form in the rock-salt structure of Fig. 9-2a in which the ions are arranged in a simple cubic lattice, alternately positive and negative (the cube having one half the edge shown as a in Fig. 9-2a). In the cesium chloride structure of Fig. 9-2b, one ion type forms a simple cubic lattice and the other an identical simple cubic lattice with each ion in the center of a cube of the first lattice. Nearest-neighbor distances are given for the monovalent and divalent compounds in Table 9-1.

From the point of view of the independent-ion model, the reason these are the stable structures is their low electrostatic energy. This energy is called the *Madelung energy*, calculated first by Madelung (1909). For N_c primitive cells, or pairs of point charges (ions), of charge $\pm Ze$ each, the Madelung energy *per ion pair* is defined by

$$E_{Mad.} = \frac{1}{2N_c} {\sum}'_{i,j} \frac{\pm Z^2 e^2}{|\boldsymbol{r}_i - \boldsymbol{r}_j|} \equiv -\frac{\alpha Z^2 e^2}{d} \tag{9-1}$$

with the plus when the ions at $\boldsymbol{r}_i$ and at $\boldsymbol{r}_j$ are of the same sign and minus when they are of opposite sign. (The prime of course indicates the omission

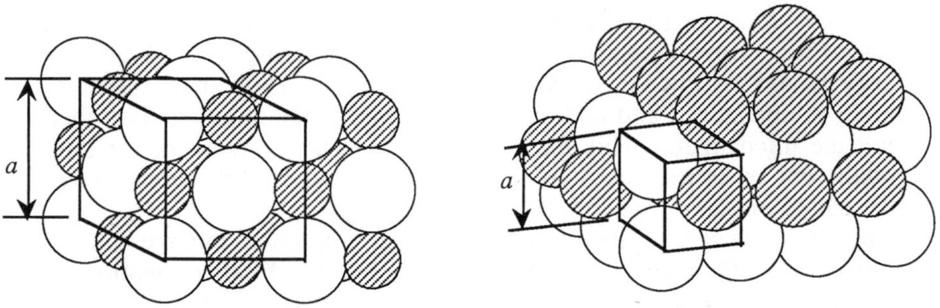

a. Rocksalt Structure　　b. Cesium-Chloride Structure

Fig. 9-2. In the rock-salt structure of part a large spheres (Cl^-) lie at the corners of a cube and in the center of each face. Small spheres (Na^+ or K^+) lie at the center of each edge and at the center of the cube. In the periodic structure there are four ions of each type per cube of volume a^3. The nearest neighbor distance is $d = a/2$; it will be listed for a number of compounds in Table 9-1.

In the cesium chloride structure of part b large spheres (Cl^-) lie at each cube corner and a small shaded sphere (Cs^+) at each cube center. In the periodic structure there is one ion of each type per cube of volume a^3. The nearest-neighbor distance is $\sqrt{3}a/2$. In both cases an equivalent structure can be obtained by interchanging positive and negative ions.

Table 9-1. Equilibrium spacings d in Angstroms for the alkali halides and their divalent counterparts, taken from the molecular weight and density listed in the CRC Handbook (Weast (1975)) where possible. All are for the rock-salt structure except the cesium compounds.

	F	Cl	Br	I
Li	2.01	2.57	2.75	3.00
Na	2.32	2.82	2.99	3.24
K	2.67	3.15	3.30	3.53
Rb	2.80	3.29	3.42	3.64
Cs	-	3.57*	3.71*	3.95*

	O	S	Se	Te
Mg	2.10	2.60	2.73	-
Ca	2.41	2.85	2.96	3.18
Sr	2.58	3.10	3.12	3.33
Ba	2.76	3.19	3.30	3.49

*CsCl structure.

of the term $i = j$.) In the final form d is the nearest-neighbor distance and α is called the *Madelung Constant* for that structure. If the point charges were replaced by nonoverlapping spherically symmetric charges, the Madelung energy would be unchanged since the potential outside these spherical charges is unchanged and therefore the energy change in assembling them would be unchanged. This feature makes the energy relevant for ionic crystals, but of course real ions overlap each other slightly which would make corrections.

In the independent-ion model the charges on each ion are integral numbers Z. When we introduce coupling between orbitals in Section 9-2 we will find it convenient to discuss transfer of charge between ions and effective charges Z^* which are quite different from integral, near 0.5 for alkali metal halides. Our discussions in this section based upon integral Z's remain meaningful. This is most clearly seen in the Born Model (Section 9-1-E) in which cohesion is calculated by producing separated ions of integral charge, bringing them together to the solid and gaining the Madelung energy. If at the end, when the ions contact, there is a redistribution of charge, the corresponding energy change will be small and we have still gained the full Madelung energy with integral charges in bringing the atoms together from widely separated positions.

Table 9-2. Madelung constants for simple structures

Compound	and Structure	α
NaCl	Rock Salt	1.75
CsCl	CsCl	1.76
ZnS (cubic)	Zincblende	1.64
ZnS (hexagonal)	Sphalerite	1.64

The calculation of α is straightforward, but tricky since the Coulomb interaction drops off so slowly. Values are given in Table 9-2 for a few important structures. They can be generalized to more complicated structures, such as $CaCl_2$ and values for many structures have been tabulated by Johnson and Templeton (1961) and reproduced in Harrison (1980), p. 305.

We see as we indicated that the Madelung energy is lower for the rock-salt and cesium-chloride structures, and very close for those two. The less compact tetrahedral structures are higher in energy and are presumably stabilized by the two-electron bonds discussed in preceding chapters.

We may also see from the first form in Eq. (9-1) that the Madelung energy $-\alpha Z^2e^2/d$ is also the electrostatic potential at one ion site due to the presence of all other ions. (There are $2N_c$ terms in the sum over i.) Then we think of the electrostatic energy per atom pair as the potential at each ion, times its charge, summed over ions, divided by the number of pairs N_c, and divided by two since each interaction was counted twice. This local potential is essential to our discussions. We noted in Part A that the energy level on a chlorine ion is raised by $U(Cl)$ due to the extra electron on it, but in the complete potassium chloride crystal it is also lowered by the Madelung potential at that site, $-\alpha e^2/d$, dominated by the six neighboring potassium ions. The residual shift is an effective Coulomb shift of

$$\delta\varepsilon(X) = U(X) - \frac{\alpha e^2}{d} \tag{9-2}$$

for the halogen level in a monovalent compound MX, where in our illustration the M is potassium and the X is chlorine. The corresponding $\delta\varepsilon(M) = -U(M) + \alpha e^2/d$ is defined also for the metallic atom; it enters with a minus sign since the electron occupation is *reduced* by one relative to the neutral atom.

For KCl, with spacing $d = 3.15$ Å (Table 9- 1) $\alpha e^2/d$ is 8.00 eV . Thus taking U from Table 1-1 we have $\delta\varepsilon(Cl) =$ 2.30 eV and $\delta\varepsilon(K)$ = 2.44 eV. Both shifts are small compared to the individual U and with an electron transferred from potassium to chlorine both levels are shifted upward by about the same amount. It makes perfectly good sense to ignore the shifts altogether in estimating the band gap and use free-atom term values directly when we estimate band gaps as energy-level differences.

We might note that the U-values which we use from Table 1-1 were obtained systematically from the differences in first and second ionization potentials, where really the difference between ionization energy and electron affinity is relevant here for the chlorine. The results are much the same if we use the experimental differences between ionization energy and electron affinity for F, Cl, Br, and I , which are 13.97 eV, 9.40 eV, 8.48 eV, and 7.39 eV, to be compared with the 15.75 eV, 10.30 eV, 9.78 eV, and 8.58 eV given in Table 1-1. The former were not used since they are only available for a limited number of elements.

The cancellation within each of the shifts is no accident since for the internuclear distances which occur in ionic crystals, the space occupied by the electronic state which was emptied (potassium 4s-state) and that of the state where the electron goes (chlorine 3p) are a similar distance from the chlorine nucleus or from the potassium nucleus. Thus what we think of as a transfer of electrons to form ions constitutes a rather subtle rearrangement of the charge so that atomic term values remain meaningful for the electronic structure. It is also this cancellation which allowed us to obtain a positive $\delta\varepsilon$ for potassium corresponding to a transfer of the electron from potassium bringing it closer to the potassium nucleus. The same cancellation occurs in the tetrahedral solids as we discussed in Section 2-3-E, where as in potassium the Madelung terms are typically larger than the U values.

C. The Band Gap.

Ignoring such $\delta\varepsilon$ shifts, we obtain immediately the difference between the highest-energy occupied level and the lowest empty level in an alkali halide as $\varepsilon_s(M)$ - $\varepsilon_p(X)$. Using Table 1-1 we subtract directly to obtain a band gap for KCl of -4.01 - (-13.78) = 9.77 eV, in comparison to the observed 8.4eV. This, and the corresponding prediction for the other alkali halides, are listed as the first entry in Table 9-3, and compared with experimental gaps. We shall see that the coupling $V_{sp\sigma}$ does not modify the band gap in the nearest-neighbor-coupling band calculation so this prediction does correspond to a crude band calculation. As such an exceedingly simple first-principles calculation, the agreement is extraordinary. The general magnitudes and most trends are well given. As in our analyses of semiconductors, the largest errors arise for systems from the first row of the periodic table (Li, Be, B, C, N, O, and F).

Making corrections $\delta\varepsilon$ to these simplest tight-binding estimates corrects the energy-level differences in a way consistent with our treatment of semiconductors. We may immediately use Eq. (9-2), and the counterpart for the metallic ion, to estimate the shift $\delta\varepsilon$ for each term value taking U from Table 1-1 and d from Table 9-1. Note that for KCl the shift of the Cl level was $\delta\varepsilon(Cl) = +2.30$ eV and that for the K was $\delta\varepsilon(K) = +2.44$ eV. Both are shifted upward so our predicted change in the gap is quite small in KCl. It is also small in the other systems as seen in Table 9-3, where the second entry includes these corrections. It is also seen that the agreement with experiment is not generally improved.

Table 9-3. Band gaps for the alkali halides (MX) in eV. The first entry was obtained as $\varepsilon_s(M) - \varepsilon_p(X)$ from Table 1-1. For the second, corresponding to a tight-binding band calculation, $\delta\varepsilon(M) + \delta\varepsilon(X)$ was added to give the appropriate energy-level difference. In the third entry, only $-\delta\varepsilon(X)$ was added, corresponding to the transfer energies discussed in the text. Experimental values for the gap (Poole, et al. (1975a, b)) are in parentheses. (No spacing d was known for CsF.)

	F	Cl	Br	I
	14.5	8.4	7.1	5.6
Li	15.7	9.5	7.5	5.6
	11.3	7.9	6.5	5.4
	(13.6)	(9.4)	(7.6)	-
	14.9	8.8	7.5	6.0
Na	16.7	10.3	8.4	6.8
	11.0	7.5	6.1	5.2
	(11.6)	(8.5)	(7.5)	-
	15.9	9.8	8.4	7.0
K	13.5	9.9	8.4	7.0
	9.6	7.5	6.3	5.6
	(10.7)	(8.4)	(7.4)	(6.0)
	16.1	10.0	8.7	7.2
Rb	13.2	10.0	8.5	7.3
	9.3	7.4	6.2	5.5
	(10.3)	(8.2)	(7.4)	(6.1)
	16.5	10.4	7.1	7.6
Cs	-	9.2	7.8	6.7
		7.2	6.1	5.4
	(9.9)	(8.3)	(7.3)	(6.2)

This draws attention to the two types of Coulomb corrections discussed in Part 9-1-A. We noted in particular that from the point of view of transferring full electrons, the $\delta\varepsilon(M)$ did not enter the energy transfer. Indeed we shall see this in detail for the Born model discussed in Part 9-1-E, so a better estimate of the gap - an improvement over the band calculation - includes only the shift $\delta\varepsilon(X)$, and no $\delta\varepsilon(M)$. This estimate is included as the third entry in Table 9-3.

Dropping the $\delta\varepsilon(M)$ has generally changed our overestimates of the gap to underestimates, which is in fact the right answer. We noted in Part A of this Section that there are enhancements of the gap for the transfer of a single electron back from the valence-band to the conduction band. We shall discuss these in Part F, finding that our estimates are comparable to, but generally larger than, the difference between our best estimate in the third entry and the experimental values for the gaps. This probably indicates the level of error in our table, not in the enhancement estimates.

The same theory is applicable to divalent compounds such as CaS, where the predicted band gap from free-atom term-value differences is -5.32 - (-11.60) = 6.28 eV. This value, and the corresponding estimates for the other divalent compounds, are listed as the first entry in Table 9-4. The agreement is comparable to that for the monovalent compounds and, indeed the systematics of the differences are very similar.

The corrected energy-level differences, arise from the transfer of two electrons and are therefore $\delta E_g = 4\alpha e^2/d - 2U(M) - 2U(X)$ for a divalent compound, listed as the second entry in Table 9-4. Again the correction generally makes the values too large, even without the Coulomb enhancements discussed in Part F of this Section.

Again a better prediction should come from consideration of removal energies for electrons. Initially ignoring Madelung potentials, the first electron is removed at the free-atom $\varepsilon_s(M)$ and added to the chalcogenide at $\varepsilon_p(X) + U(X)$. The second electron transferred from the metallic atom has removal energy $\varepsilon_s(M) - U(M)$ and is added to the chalcogenide in a state with removal energy $\varepsilon_p(X) + 2U(X)$. Again each of these U shifts is reduced by an $\alpha e^2/d$, as we shall see in detail in Section E, so the correction $3\alpha e^2/d - U(M) - 2U(X)$ is to be added to the gap $\varepsilon_s(M) - \varepsilon_p(X)$ to obtain the transfer energy, listed as the third entry in Table 9-4. This rectifies much of the overestimate, but still gives values too large considering the anticipated enhancement of the gap, discussed in Section F. The size of the discrepancies may not be at all surprising in view of the fact that the $3\alpha e^2/d$ term is typically 25 eV so there is much cancellation of large terms.

In trivalent compounds, where three electrons are transferred, one

Table 9-4 Band gaps for divalent ionic compounds, in eV, the first entry predicted as $\varepsilon_s(M) - \varepsilon_p(X)$. For the second value $4\alpha e^2/d - 2U(M) - 2U(X)$ was added to obtain the energy-level difference. For the third value $3\alpha e^2/d - U(M) - 2U(X)$ was added to obtain the transfer energy. Experimental values are in parentheses (Strehlow and Cook (1973)).

	O	S	Se	Te
Mg	9.8 14.3 9.6 (7.8)	4.7 10.0 7.6 -	3.8 8.0 6.0 (5.6)	2.7 - - (4.7)
Ca	11.4 11.5 7.4 (7.7)	6.3 10.0 7.6 (5.8)	5.4 8.5 6.4 (4.9)	4.2 7.0 5.6 (4.1)
Sr	11.9 10.6 6.5 (5.8)	6.8 9.0 6.6 (4.8)	5.8 8.5 6.1 (4.4)	4.7 8.4 6.3 (3.7)
Ba	12.4 8.6 5.2 (5.1)	7.3 8.6 6.4 (3.9)	6.4 7.4 5.5 (3.6)	5.3 6.7 5.2 (3.4)

electron must come from a state other than a metallic s-state. In ScN, the third electron comes from a d-state at -9.35 eV (Fischer (1972)) which is lower in energy than the s-state. Thus the predicted band gap is 13.84 - 9.35 = 4.49 eV. The compound does form but apparently has a vanishing gap (Monnier, Rhyner, and Rice (1985)), perhaps from Coulomb corrections due to charge transfer. ScP is predicted to have a gap of 0.2 eV, well within error of zero and the compound does not form.

There is no reason for the independent ion model to be less reliable for similar compounds with more complicated formulae. In Sc_2O_3, where one imagines transferring the three electron each from the two scandium atoms to the two empty p-states each on the three oxygen atoms, one predicts a band gap of 16.72 - 9.35 = 7.37 eV in comparison to the observed 6 eV; for Sc_2S_3 one predicts 2.3 eV in comparison to the observed 2 eV.

We conclude that for all of the simple ionic compounds, for which the electronic structure is understandable in terms of electrons transferred from higher-energy atomic states to fill low-energy atomic shells, it is reasonable to estimate the band gap by simply subtracting the appropriate term values listed in Table 1-1. There are significant electrostatic shifts in the levels, which can be obtained by adding a $\delta\varepsilon$ given by Eq. (9-2), or its polyvalent counterpart, but these may not improve the predictions. Estimates based upon removal energies are more successful but not always better than the simplest estimate. There are also the Coulomb enhancements of the gaps to be discussed in Part 9-1-F.

The electrostatic corrections become much more serious if one moves beyond scandium into the transition-metal compounds. The energy of the partly filled d-states in these systems depends rather sensitively upon the number of d-electrons present, becoming shallower with increased occupation for each element. The prediction of the electronic structure requires the self-consistent determination of occupation of the d-states and cannot meaningfully be done within the independent-ion model, but is readily carried out when one goes beyond this model, using the method of Froyen (1980). We discuss these transition-metal systems in Chapter 17.

D. The Cohesive Energy

The calculation of the energy gain in forming the solid from free atoms in the independent-ion model is also extraordinarily simple if we do not ask for Coulomb corrections; we simply transfer an electron from the alkali-metal atom to the halogen atom in an alkali halide, gaining an energy per ion pair equal to the band gap ($\varepsilon_s(M)$ - $\varepsilon_p(X)$ for the simplest estimate) . In the polyvalent ionic compound we transfer more electrons but the calculation is immediate. These estimates of the cohesive energy per ion pair for the alkali halides, which of course equal the estimates of the band gaps from the first entry in Table 9-3, are compared with experiment in Table 9-5. [Experimental values were obtained from the heats of formation of the compound (energy per mole to separate the crystal into elements in the standard state at 25°C from Weast (1975)) plus the cohesive energy of the elements from Table 1-3. Tables for the alkali halides in Kittel (1976) and in Harrison (1980) were for separation into *ions*. Corrections for temperature are of order k_BT and neglected.]

These simplest estimates are in reasonable accord with experiment, although less so than the estimates for the band gaps. As for the band gaps, the errors are worse for the fluorides, which contain an element from the first row of the periodic table. Values for the first-row element lithium do not appear worse than for the heavier alkali metals.

We might expect that the correction $\delta\varepsilon$ to the halogen p-state, which is occupied in the crystal, applies also to the cohesion but that the correction to the alkali-metal state does not. We shall see in Section 9-1-E that an analysis of the total energy confirms this. The cohesion corrected for $\delta\varepsilon(X)$

Table 9-5. Cohesive energy the alkali halides (MX) in eV per ion pair. The first entry was obtained as $\varepsilon_s(M) - \varepsilon_p(X)$ as was the first entry in Table 9-3. The second is corrected by subtracting $\delta\varepsilon(X)$. Experimental values (in parentheses) were calculated from heats of formation of the compound (Weast (1975)) and cohesive energies of the elements (Kittel(1976))

	F	Cl	Br	I
Li	14.5 11.3 (8.9)	8.4 7.9 (7.3)	7.1 6.5 (6.5)	5.6 5.4 (5.6)
Na	14.9 11.0 (7.9)	8.8 7.5 (6.8)	7.5 6.1 (6.1)	6.0 5.2 (5.2)
K	15.9 9.6 (7.6)	9.8 7.5 (6.9)	8.4 6.3 (6.1)	7.0 5.6 (5.4)
Rb	16.1 9.3 (7.4)	10.0 7.4 (6.7)	8.7 6.2 (6.1)	7.2 5.5 (5.4)
Cs	16.5 - (7.2)	10.4 7.2 (6.7)	9.1 6.1 (6.1)	7.6 5.4 (5.4)

appears as the second entry in Table 9-5, which is the same as the third entry in Table 9-3 for the gaps. Agreement is greatly improved by inclusion of this correction $\delta\varepsilon(X)$, more so than for the band gaps. To a large extent this may be because the Coulomb enhancements of the gap discussed in Section 9-1-F do not apply to the cohesion.

On the other hand, we shall obtain an attractive interaction between atoms arising from the coupling $V_{sp\sigma}$ in Section 9-2, half of which will be canceled by the overlap interaction, but which will have the net effect of increasing the estimated cohesion. We shall see that this coupling does not increase the band gap. The overall agreement for the cohesive energy is not modified greatly by including these corrections but they are needed to understand other properties of the compounds.

For the divalent compounds two electrons are transferred so the first estimate of the cohesion is $2(\varepsilon_s(M) - \varepsilon_p(X))$, which is twice the band gap estimates from the first entry of Table 9-4. These estimates for the cohesion

Table 9-6. Cohesive energy for divalent ionic compounds, in eV per ion pair, predicted as $2[\varepsilon_s(M) - \varepsilon_p(X)]$ in the first entry. For the second entry $U(M) + 3U(X) - 4\alpha e^2/d$ was subtracted. Experimental values, in parentheses, were obtained as in Table 9-5.

	O	S	Se	Te
Mg	19.8 17.1 (10.4)	9.4 12.6 (8.0)	7.6 10.0 -	5.3 - -
Ca	22.9 14.9 (11.0)	12.6 13.2 (9.7)	10.7 11.2 (7.3)	8.4 9.7 -
Sr	23.8 13.8 (10.4)	13.5 11.9 (9.3)	11.6 11.0 -	9.4 10.8 -
Ba	25.0 12.4 (10.3)	14.6 12.2 (9.4)	12.8 12.2 (10.3)	10.1 9.7 -

are listed as the first entry in Table 9-6. The agreement is similar to that for the alkali halides and quite good.

We may correct this estimate for electrostatic shifts by noting that for the first electron transferred from the M to the X the energy change is $\varepsilon_p(X) + U(X) - \alpha e^2/d - \varepsilon_s(M)$ and for the second is $\varepsilon_p(X) + 2U(X) - 2\alpha e^2/d - [\varepsilon_s(M) - U(M) + \alpha e^2/d\,]$. This corresponds to a correction to the energy change on formation of the compound of

$$\delta E_{tot} = U(M) + 3U(X) - \frac{4\alpha e^2}{d} \tag{9-3}$$

for the divalent compounds. We shall see from another point of view that this is correct in Section 9-1-F. This has been subtracted from the first entry in Table 9-6 to obtain the second. The agreement is again similar to that for the alkali halides. The second entry is generally better than the first but the correction is certainly not reliable when it is small because it is the small difference between large terms, as was the correction to the band gap.

The only trivalent compound for which we found an experimental value for the cohesion was LaN. For that case the predicted $2\varepsilon_s(La) + \varepsilon_d(La) - 3\varepsilon_p(N)$ is approximately twice the observed 12.52 eV per pair, as would be anticipated from comparison of the first entry in Tables 9-5 and 9-6 for CsF and BaO with which it is isoelectronic. It is anticipated that similar agreement would be obtained with other ionic compounds of more complex formulae, though this has not been explored. It would also be straightforward to make the appropriate electrostatic corrections to these cohesive energies.

E. The Born Electrostatic Model.

The traditional theory of the cohesion of ionic crystals, given by Born (1931), is an empirical approach which differs in concept from our discussion in the preceding section but we shall see that it is quite close to the method giving the third entry in Tables 9-5 and 9-6 . In the Born model, illustrated in Fig. 9-3, one begins with isolated atoms, for example potassium and chlorine atoms. One ionizes the potassium atom requiring experimentally an energy 4.3 eV. This electron is than added to a chlorine atom gaining an energy equal to the electron affinity, given experimentally by 3.6 eV and obtainable from values given in Part B. We then assemble these ions together to form a crystal, gaining an electrostatic energy equal to the Madelung energy of Eq. (9-1). It has a magnitude of $1.75\ e^2/d$ = 8.00 eV for potassium chloride. Together these give us the cohesive energy of 7.3 eV (compared to the experimental 6.9 eV from Table 9-5), if we neglect again the small repulsive overlap interaction between atoms as we did in the last section, but which Born estimated.

This is simply a different path between the atomic and crystalline end points, as illustrated in Fig. 9-3. Our simplest independent-ion calculation of

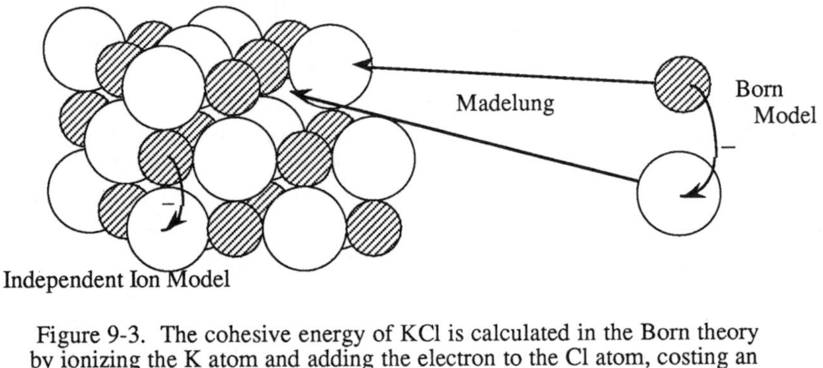

Figure 9-3. The cohesive energy of KCl is calculated in the Born theory by ionizing the K atom and adding the electron to the Cl atom, costing an energy equal to the ionization energy of K but gaining an energy equal to the electron affinity of the Cl. These are then brought together gaining the Madelung energy. In the independent-ion model the neutral atoms are brought together directly and the electrons then transferred, for a simpler and more general (though less accurate)theory.

the last section brought the neutral atoms together, without ion formation and without electrostatic energy, and formed the ionic solid at the last moment gaining the energy $\varepsilon_s(M)$ - $\varepsilon_p(X)$. The independent-ion model corresponding to the first entries in Tables 9-5 and 9-6 is not so accurate, but it is very much simpler, requiring only the free-atom term values, is applicable to divalent crystals such as CaS (although the S^{2-} ion is not stable in free space so the Born model is not directly applicable), and is a part of the same representation of the electronic structure which will give us almost all of the other properties of the ionic solid.

Once we make the electrostatic corrections to the term values,. our treatment of the cohesive energy becomes equivalent to the Born model, except for our use of atomic term values and U's from Table 1-1, as opposed to experimental values. Born's model in fact provides us a good way of understanding the electrostatic corrections to the cohesion. We may carry out the construction of the crystal following his path, but using our parameters. We first ionize the potassium atom, requiring an energy which we take to be the magnitude of the atomic term value $\varepsilon_s(K)$ (-4.0 eV from Table 1-1, rather than the experimental -4.3 eV). We add the electron to the chlorine which gains us an energy $\varepsilon_p(Cl) + U(Cl)$ (13.78 + 10.30 = -3.48 eV rather than the experimental -3.6 eV). We then gain the Madelung energy $-\alpha e^2/d$ and neglect the overlap repulsion. Thus the change in energy per atom pair in forming the solid becomes $\varepsilon_p(Cl) + \delta\varepsilon - \varepsilon_s(K)$, where we have used Eq. (9-2) for the correction $\delta U = U - \alpha e^2/d$ to the chlorine p-state energy for the band gap. It was this correction which brought the cohesion into good agreement with experiment in Table 9-5.

We may make the same analysis for the divalent ionic crystals. The energy change in the transfer the first electron from, for example, calcium to sulfur is $\varepsilon_p(S) + U(S) - \varepsilon_s(Ca)$. The change for the transfer of the second electron is $\varepsilon_p(S) + 2U(S) - (\varepsilon_s(Ca) - U(Ca))$ and the Madelung energy per pair is $-4\alpha e^2/d$ from Eq. (9-1). This gives just the correction $3U(S) + U(Ca) - 4\alpha e^2/d$ which we used for the second entry in Table 9-6. Again, the Born model and the independent-ion model become equivalent if we include the same effects.

It is interesting that these two different outlooks can have the same result. It has been traditional to say that ionic crystals are held together by electrostatic forces, while we would say it is from the transfer of electrons to the lower-energy states of the anion. Our view has the advantage that it provides an immediate estimate of the cohesive energy by simply subtracting term values. and that it gives us many other properties, such as the band gap discussed before. The Born view makes it clear that the charges which enter the Coulomb shifts U and the Madelung energies are the full integral charges, not some effective charge arising from partial transfer between atoms in the solid.

Perhaps the most remarkable aspect of these predictions is that the cohesive energy for a compound of valence Z is simply Z times the band gap. We are not aware that this approximate relation between the two experimental quantities had been noted previously. We see that the corrections to the simple theory are different for cohesion and for the band gap, but these corrections are not so large in any case.

F. Coulomb Enhancements of the Gap

We return to the gap enhancements which are the counterpart for ionic crystals of the difference in ionization energy and electron affinity in simple atoms. They also correspond to the enhancements we discussed for semiconductors in Section 5-5-A. It may be useful to think through these Coulombic enhancements explicitly for the ionic crystals using the outlook from the Born Model. We should emphasize again that these are fundamentally different from the electrostatic shifts in the band gaps discussed in Section 9-1-C, which correspond to self-consistency in a band calculation. Those involved shifts in the energy when an electron is transferred from *every* atom M to every atom X.

The enhancements we discuss now involve transferring a single electron in an entire crystal. They are enhancements relative to our estimates and also to the full band calculation. In the silicon atom, the Hartree-Fock calculation gives us the same eigenvalue ε_p for all three p-states in the atom. Though it is the correct removal energy for the occupied states, the energy at which an additional electron is added must be corrected as $\varepsilon_p + U$. We used this same U in both calculations.

To think through these enhancements using the Born model, we simply assemble the solid with a single electron left on an M atom, and one distant X atom left neutral. We again add up the changes in the total energy in bringing the system together.

We begin again with isolated atoms, and transfer one electron from each chlorine atom to each potassium atom except for one pair of atoms, saving us one transfer energy $\varepsilon_p(Cl) + U(Cl) - \varepsilon_s(K)$. (This is a negative number representing the energy gain.) When we then bring the atoms together gaining the Madelung energy, two of the sites are occupied by neutral atoms so we fail to gain the missing charge times the Madelung potential, $-\alpha e^2/d$, at each of those points, costing $2\alpha e^2/d$. Up till now there has been no dielectric polarization of the medium, which for independent ions would be associated with polarization of the ions. Since each ion has had an environment with cubic symmetry, there was no polarization. However, the removal of the charge at the two sites produces fields which now polarize the ions making up the dielectric medium around each. We use the formula, Eq. (5-34), for the associated energy of relaxation for each, $-{}^{1}/_{2}[U(K) + U(Cl)][1 - 1/\varepsilon]$. Again we neglect the overlap repulsion.

The total energy of this crystalline system of N_c pairs of atoms due to placing one electron in the alkali s-state, relative to the ground state of the system, is

$$\delta E = -[\varepsilon_p(Cl) + U(Cl) - \varepsilon_s(K)] + 2\frac{\alpha e^2}{d} - \frac{1}{2}[U(K) + U(Cl)][1 - \frac{1}{\varepsilon}]$$

(9-4)

$$= \varepsilon_s(K) + \delta\varepsilon(K) - [\varepsilon_p(Cl) + \delta\varepsilon(Cl)] + \frac{1}{2}[U(K) - U(Cl)] + \frac{\bar{U}}{\varepsilon}$$

with $\bar{U} = {}^{1}/_{2}[U(K) + U(Cl)]$, the average U. Terms have been arranged in the second form such that it begins with the quantity we identified with our tight-binding band calculation, $\varepsilon_s(K) - \delta\varepsilon(K) - [\varepsilon_p(Cl) + \delta\varepsilon(Cl)]$, the second entry in Table 9-3. Then there are two contributions to the correction. The first is ${}^{1}/_{2}[U(K) - U(Cl)]$ associated with transferring an entire electron, rather than an infinitesimal fraction, between each pair of atoms. The second is the enhancement we discussed in Section 5-5-A. These two are compared in the first three rows of Table 9-7 for the alkali halides for which we had experimental enhancements from Carlsson(1985). The comparison indicates that the $\bar{U}/\varepsilon$ is in much better accord than the combined terms, or that the first term is already included in some sense in the band calculations upon which Carlsson based his comparison. We believe that to be the case, but confirming it would require looking in detail at the treatment of electron-electron interactions in those calculations. That is not essential to our study here. If we group terms in Eq. (9-4) as corrections to the third entry in Table 9-3, then the first correction becomes $\alpha e^2/d - \bar{U}$, which is always very near zero, and the second correction is $\bar{U}/\varepsilon$ which we take to be the Coulomb enhancement of the band-theory gap. That gap from band theory corresponds to the third entry in Table 9-3, which we take to be the gap from a self-consistent tight-binding band calculation.

Similarly for divalent compounds, if we construct the counterpart of Eq. (9-4) we obtain the same two corrections to the second entry in Table 9-4, indicating again that the correction ${}^{1}/_{2}[U(M) - U(X]$ has already been made in the band calculation and that $\bar{U}/\varepsilon$ corresponds to the Coulomb enhancement of the gap which would be obtained from self-consistent band theory.

Table 9-7 Predicted enhancements of the band gaps relative to one-electron theory. The first entry is the correction for transferring a full electron, rather than an infinitesimal fraction of an electron. The second is the counterpart of the difference between ionization energy and electron affinity for the free atom The final column is the difference between one-electron theory and experiment as determined by Carlsson (1985).

Compound	$[U(M)-U(X)]/2$	$\bar{U}/\varepsilon$	Experiment
LiF	-3.8	6.4	4.7
NaCl	-2.1	3.7	3.7
CsCl	-2.6	2.8	3.3
MgO	-3.6	3.6	2.5
CaS	-1.53	1.8	2.8
BaS	-1.87	1.6	2.1

We could evaluate for all compounds the discrepancy between experimental band gaps and our band-gap estimates in the third entry in Tables 9-3 and 9-4 and compare this discrepancy with our estimate $\bar{U}/\varepsilon$ of the gap enhancement, but it seems not informative. The estimate $\bar{U}/\varepsilon$ is generally somewhat larger, presumably from inaccuracy of the band gap estimates, since the validity of the enhancement estimate seems confirmed by Table 9-7.

The band gaps given as the third entry in Table 9-3 and 9-4 remain appropriate for calculations such as of the dielectric constant for which the changes in electron states correspond to transfer of charges among nearest neighbors, and in fact to no net change in the charge of ions in the bulk if the fields are uniform, as we shall see. However, the corrections listed in Table 9-7, and readily calculated as $\bar{U}/\varepsilon$ for other systems, should apply to real excitations across the gap.

2. Coupled Ions

The coupling between adjacent ions lowers the energy of occupied states on the nonmetallic ion (though not affecting the gap at $\boldsymbol{k} = 0$, as we shall see). This may be thought of as a covalent contribution to the bonding of the ionic crystal since it arises from an interatomic coupling as did the covalent energy V_2 in semiconductors. In most ionic insulators the coupling is small enough that it becomes appropriate to treat it as a perturbation, as in expanding $(V_2^2 + V_3^2)^{1/2} = \pm (V_3 + V_2^2/(2V_3) + \ ...)$. This case is particularly simple and we shall treat it first. When this approximation is valid, the contributions to the cohesion are sufficiently small that we cannot expect improvement on the rather good estimates of cohesion obtained in the last section, but the resulting attraction between atoms will be important in other properties. When the coupling (V_2) is not small compared to the energy difference ($2V_3$) this approximation becomes inaccurate and we present a more accurate theory in terms of resonant bonding.

A. Covalent Corrections to the Energy

Since only the coupling between empty and full states affects the total energy, and since the contribution decreases with increasing energy difference, the dominant terms will be those coupling the metallic s-state $\varepsilon_s(M)$ and the nonmetallic p-state $\varepsilon_p(X)$, the states which also determined the band gap and the cohesion of the ionic crystal. We shall thus drop the coupling between the metallic s-state and the nonmetallic s-state for clarity. There is no difficulty in adding these terms but their effect appears to be smaller than other errors in the theory so its inclusion does not add much but complexity to the theory.

The only coupling, then, which we include is from Eq. (1-21),

$$V_{sp\sigma} = 1.42 \frac{\hbar^2}{md^2} \quad . \tag{9-5}$$

For KCl, with d = 3.15 Å, this is 1.09 eV. Its contribution to the energy is estimated by adding each term $V_{sp\sigma}{}^2/(\varepsilon_p(X) - \varepsilon_s(M))$ from perturbation theory, Eq. (1-30). The rock-salt structure is redrawn in Fig. 9-4, showing each p-state coupled to the s-states on the two ions lying along its axis; the coupling with the s-states on the other four neighboring ions is zero. The ion contains two electrons in each of the three p-states giving a total shift of

$$E_{cov} \approx \frac{12\, V_{sp\sigma}{}^2}{\varepsilon_p(X) - \varepsilon_s(M)} \tag{9-6}$$

per ion pair. For KCl this is -1.46 eV. As for semiconductors some half of this would be canceled by the overlap interaction and the remainder is of the order of the error in our estimates of the cohesive energy in Table 9-5; it increases the cohesion and slightly worsens the agreement with experiment in almost all cases.

We might at first think that even the interatomic force from this small term was not important. Noting that $V_{sp\sigma}$ varies as $1/d^2$, we see that the force arising from this covalent energy between a pair of nearest neighbors is $(4/d)\; 2V_{sp\sigma}{}^2/(\varepsilon_p(X) - \varepsilon_s(M))$, equal to -0.31 eV/Å for KCl, while the Coulomb force, Z^2e^2/d^2 , is -1.45 eV/Å for KCl. However, near the

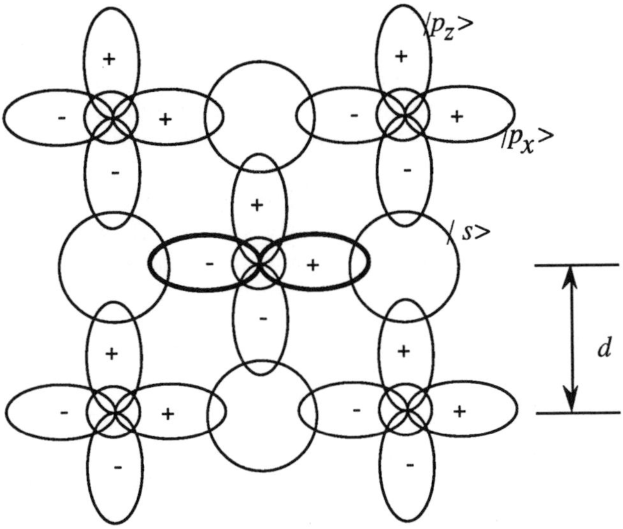

Figure 9-4 The central figure is a halogen atom in a the rock-salt structure. Its electronic p-states, $|p_x>$ (shown bold) and $|p_z>$ overlap with, and are coupled to, s-states on the four neighboring alkali atoms shown. Four other halogen atoms are also shown. A third p-state, $|p_y>$, on the central atom, is oriented perpendicular to the first two and is coupled to alkali s-states on atoms above and below the plane of the figure.

equilibrium spacing the same coupling which produced the covalent bonding energy, Eq. (9-6), also shifts charge back to the alkali, reducing the *effective charge* Z^*. Furthermore, Z^* is changing with spacing so it is necessary to calculate the covalent contributions and electrostatic contributions together, self-consistently. This was done by Straub and Harrison (1989) and we shall return to this in Chapter 10 after discussing the energy bands. It turns out that the final effect of the Coulomb terms is small enough that a meaningful treatment of the volume-dependent properties of at least the alkali halides could be made using the covalent attraction, from Eq. (9-6), alone, with an overlap interaction, as we shall see at the end of this chapter. Although this perturbation expression, Eq. (9-6), is adequate for alkali halides, it is not accurate for divalent compounds. Our consideration of the energy bands will give us a method for improving upon it.

B. The Energy Bands

The coupling, $V_{sp\sigma}$, which we treated as a perturbation above, in fact broadens the s-like and p-like atomic levels out into energy bands, as the metallic energy V_1 broadened the bond-like levels out into energy bands as discussed in Chapter 2. An accurate estimate of the total energy requires an averaging of the energy over all of the band and this average is only very approximately obtained by the perturbation formula entering Eq. (9-6). In order to understand how we can proceed more accurately we must consider the bands themselves.

We follow the procedure used for semiconductors in Section 5-3, but in this case the analysis is particularly simple. If there were no coupling between the metal s-states and the nonmetal p-states we could write a crystal state of wavenumber $\boldsymbol{k}$ in terms of all of the s-states.

$$/\, s, \boldsymbol{k} > = N_c^{-1/2} \sum_j e^{i\boldsymbol{k} \cdot \boldsymbol{r}_j} /\, s_j> \quad , \tag{9-7}$$

called again a Bloch sum, where $/\, s_j>$ is the s-state centered upon the metallic ion at $\boldsymbol{r}_j$. The number of ion pairs, or primitive cells, is again written N_c. Similarly Bloch sums of wavenumber $\boldsymbol{k}$ could be written in terms of the $/p_{xj}>$ states, and Bloch sums in terms of the $/p_{yj}>$ and $/p_{zj}>$ (or mixtures of these three p-like states of the same wavenumber). If we now allow coupling, $V_{sp\sigma}$ between neighboring s- and p-states, and take the particularly simple case of a wavenumber in the x-direction, it is not difficult to see that only the $/\, s,\boldsymbol{k}>$ and the $/p_x,\boldsymbol{k}>$ Bloch sums are coupled. We may see this, for example, by consideration of Fig. 9-4. The $/p_y>$ state at the center of the figure is coupled to s-states above and below it but because the lobes of the p-state are of opposite sign, and the phase of the s-state ($e^{ik_xx_j}$) is the same for both, the two terms in the coupling cancel. Thus the $/p_y,k_x>$ and $/p_z,k_x>$ Bloch sums remain uncoupled and have energy equal to ε_p, independent of k_x. These two bands are flat, as were two for semiconductors in Section 5-2-A. The two s-states to which the $/p_x>$ state at the center of Fig. 9-4 is coupled on the other hand have phases differing by e^{ik_xd} and e^{-ik_xd} and therefore cancel only at $k_x = 0$. The coupling between the two Bloch sums is in fact given by

$$V_{sp\sigma}(e^{-ikxd} - e^{ikxd}) = -2iV_{sp\sigma} \sin k_x d. \tag{9-8}$$

The energy of the two coupled states, for each wavenumber, is found immediately as we found the bond and antibonding energy From Eq. (1-10) in Chapter 1. The result is

$$\varepsilon_{kx} = \frac{\varepsilon_s(M)+\varepsilon_p(X)}{2} \pm \sqrt{\left(\frac{\varepsilon_s(M)-\varepsilon_p(X)}{2}\right)^2 + 4V_{sp\sigma}^2 \sin^2 k_x d} \; . \tag{9-9}$$

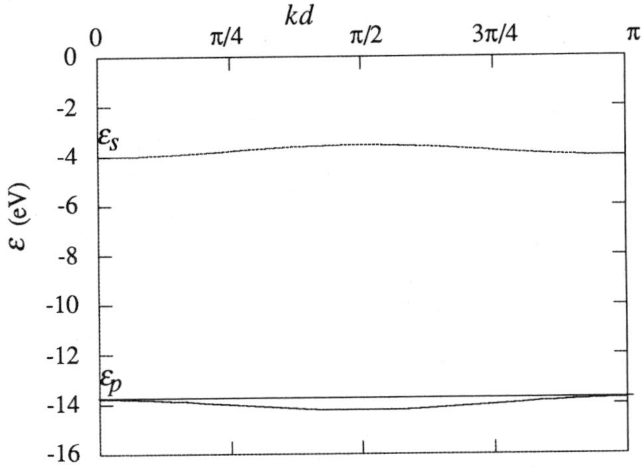

Figure 9-5 The energy bands (for $\boldsymbol{k}$ parallel to a cube direction) for KCl based upon potassium s-states and chlorine p-states with only nearest-neighbor $V_{sp\sigma}$ interactions, from Eq. (9-9).

These bands are plotted in Fig. 9-5. As energy bands they are very crude and inaccurate but illustrate some important features. First, the minimum energy difference between the valence and conduction bands remains $E_g = \varepsilon_s(M) - \varepsilon_p(X)$ so there is no correction to our estimates of the band gaps given is Section 9-1-C. This remains true when the wavenumbers are allowed to lie in other directions than that shown in Fig. 9-5. Second, we see that two bands remain completely independent of wavenumber, and this also remains true at all directions of wavenumbers.

This second point will be particularly relevant when we consider resonant bonding. Note that at each wavenumber we are forming bonding and antibonding combinations between p-states of a particular orientation and s-states; the p-states of the other two orientations remain unbonded. For wavenumbers in an x-direction, illustrated in Fig. 9-5, the particular p-state involved in the bonding is $|p_x\rangle$ on every ion. For wavenumbers in a y-direction it is $|p_y\rangle$, and for arbitrary directions of $\boldsymbol{k}$ it will be a p-state oriented in a different direction. In some sense we can think of the sp-bond as *resonating* in wavenumber space between various orientations of the p-state. With six neighbors and only four essential atomic orbitals per ion, we cannot make independent bonds as we could for four neighbors in the covalent solid. The concept of a resonant bond (Pauling (1960)) remains a good one in the present context and in fact takes on an even richer meaning as a resonance in wavenumber space.

The true energy bands for the alkali halides differ considerably from the elementary version given in Eq. (9-9). They have been discussed, for example, by Pantelides (1975a) and a review of published band calculations, and comparison with experiment, for all alkali halides has been given by Poole, Liesegang, Leckey, and Jenkin (1975a,b). Both the true valence and true conduction bands are considerably broader than shown in Fig. 9-5 and the bands at $k = \pi/d$ are separated much more than at $k = 0$, both effects presumably coming largely from the coupling between second-neighbor ions. Such coupling will also tend to reduce the estimated band gap and would not improve the predictions made in Section 9-1-C. We shall not discuss them in detail here, but make use of the crude bands from Eq.(9-9) for discussing other properties.

C. Special Points

To obtain the total energy, and therefore the bonding, in the crystal we need to average the energy bands over all wavenumbers. Because of the square root in Eq. (9-9) this becomes awkward even in this simplest approximation. There is, however, an ingenious method for obtaining an approximate average called the "special points method" (Baldereschi (1973) and Chadi and Cohen (1973) discussed also in Harrison (1980) 181ff.). In that method a single (or a few) special wavenumbers are selected which are expected to be representative. For example, in a one-dimensional chain, such as we discussed in Section 1-3, with a band we obtained as $\varepsilon_k = \varepsilon_s + 2V_{ss\sigma} \cos kd$, the average can be obtained by using $k^* = \pi/2d$. If coupling to more distant neighbors were included, there would be additional terms, $2V_{ss\sigma}^{(2)}\cos 2kd + 2V_{ss\sigma}^{(3)}\cos 3\pi kd + ...$, but we expect the coefficients to be increasingly small and a reasonably accurate estimate to be obtained by evaluating ε_{k^*} for these more accurate bands. The special point was chosen so that the contribution of the leading components of a Fourier expansion of the band, $\cos kd$ in this one-dimensional case, would be zero. In two dimensions one could pick a line in wavenumber space where $\cos k_xd + \cos k_yd$ vanished, or the intersection of this line with that for which $\cos (k_x + k_y)d + \cos (k_x - k_y)d$ vanishes, so that the leading *two* sets of Fourier components vanished. In three dimensions the selection of the special point is more complicated but it is possible in some structures to select a point where three sets of Fourier components vanish. (For face-centered-cubic symmetry, the two sets based upon $cos[(k_x \pm k_y)a/2]$, etc., and $cos[k_xa]$, etc., can be eliminated but the condition for the third set cannot simultaneously be satisfied.) We use a simpler approach here where we shall be able to obtain our result without knowing what wavenumber $\boldsymbol{k}$ * is selected.

The important feature of the calculation of the bands show in Fig. 9-5 for arbitrary wavenumber is that it gives two bands at energy $\varepsilon_p(X)$, as we indicated, and two of the form of Eq. (9-9) but with $4sin^2k_xd$ replaced by some more complicated function $f(\boldsymbol{k})$. If we knew the function and the special point $\boldsymbol{k}^*$ we could estimate the average energy for each band from

$$\varepsilon_{\boldsymbol{k}^*} = \frac{\varepsilon_s(M)+\varepsilon_p(X)}{2} \pm \sqrt{\left(\frac{\varepsilon_s(M)-\varepsilon_p(X)}{2}\right)^2 + f(\boldsymbol{k}^*)V_{sp\sigma}^2} \quad . \qquad (9\text{-}10)$$

We next note that this same form should be correct (with the same $f(\boldsymbol{k}^*)$) also when $V_{sp\sigma}$ is small enough to be treated as a perturbation and when Eq. (9-10) becomes, for the lower state near $\varepsilon_p(X)$,

$$\varepsilon_{\boldsymbol{k}^*} \approx \varepsilon_p(X)) - \frac{f(\boldsymbol{k}^*)V_{sp\sigma}^2}{\varepsilon_s(M)-\varepsilon_p(X)} + \ldots \qquad (9\text{-}11)$$

But the perturbation theory, which earlier led to Eq. (9- 6), gives the sum of corrections for the n nearest neighbors as $- nV_{sp\sigma}^2/(\varepsilon_s(M) - \varepsilon_p(X))$, with $n = 6$ for the rock-salt structure. (We use n for the coordination here, rather than X , to avoid confusion with the X used to indicate the nonmetallic, or halogen, ion.) Thus for the appropriate choice of $\boldsymbol{k}^*$ we may expect that $f(\boldsymbol{k}^*)$ is equal to n. Remarkably enough, if one makes a direct evaluation for $f(\boldsymbol{k}^*)$ using the $\boldsymbol{k}^*$ selected by Baldereschi (1973) for the face-centered-cubic structure, one confirms this value for both the rock-salt (where $n = 6$) and zincblende (where $n = 4$) structures, which have this fcc translational symmetry (Harrison (1981a)). We may multiply the electron energy by two for spin and subtract the energy for the two electrons without coupling to obtain an expression for the covalent contribution to the bonding,

$$E_{cov} = - \sqrt{(\varepsilon_s(M) - \varepsilon_p(X))^2 + 4nV_{sps}^2} + \varepsilon_s(M) - \varepsilon_p(X) \qquad (9\text{-}12)$$

per ion pair, which reduces to Eq. (9-6) as $V_{sp\sigma}$ becomes very small. This additional bonding energy is appropriate independent of the valence difference between metallic and nonmetallic atoms. The difference between Eq. (9-6) and (9-12) is significant for a number of the alkali halides and is quite important for the divalent compounds (alkaline-earth chalcogenides) and we shall use the more accurate form, Eq. (9-12) in Chapter 10, where we shall interpret Eq. (9-12) as describing the effects of a resonant bond.

D. Volume Dependent Properties in Second Order

Before beginning that full analysis, we should note the very simple theory of volume-dependent properties of the alkali halides which the approximate perturbation-theory form, Eq. (9-6), provides us.

In order to discuss volume dependence, we need an overlap repulsion, which we shall take as proportional to $1/d^8$, for reasons to be discussed in Chapter 10. We shall also see there that we need repulsion between second-neighbor halogens (or chalcogens), as well as nearest-neighbor ions. However, the total overlap-repulsion energy still scales as C/d^8 and in this section we simply fit the constant C to the equilibrium spacing and evaluate the higher derivatives of the energy per ion pair, $E_{cov}(d) + C/d^8$, with $E_{cov} \approx 12\, V_{sp\sigma}{}^2/(\varepsilon_p(X) - \varepsilon_s(M))$ from Eq. (9-6), and proportional to $1/d^4$. The analysis is exactly as in Section 2-4 for homopolar covalent solids except that the covalent energy goes as $1/d^4$ rather than $1/d^2$ and the repulsion goes as $1/d^8$ rather than $1/d^4$.

Setting the first derivative equal to zero at the equilibrium spacing d_o gives $-4E_{cov}(d_0)/d_0 = 8C/d_0{}^9$, or

$$\frac{C}{d_0{}^8} = -\frac{E_{cov}(d_0)}{2}\,. \tag{9-13}$$

The repulsion cancels half the bonding as we found in semiconductors. The second derivative is $5\times4\times E_{cov}/d^2 + 9\times8\times C/d^{10}$ and the third derivative is $-6\times5\times4\times E_{cov}/d^3 - 10\times9\times8\times C/d^{11}$, or, at $d=d_0$,

$$d^2\frac{\partial^2 E_{cov}}{\partial d^2} = -16\, E_{cov} \tag{9-14}$$

and

$$d^3\frac{\partial^3 E_{cov}}{\partial d^3} = 240\, E_{cov}\,. \tag{9-15}$$

We obtained $E_{cov} = -1.46\ eV$ for KCl following Eq. (9-6). As we indicated there, adding the $E_{cov}/2$ increases our estimate of the cohesion from the independent-ion model, given as the first entry in Table 9-5, slightly worsening the agreement with experiment.

The change in energy per pair for a fractional change in volume of $3\delta d/d$, with a starting volume per pair of $2d^3$, is related to the bulk modulus B by $^1/_2(2d^3)B(3\delta d/d)^2$. This is equal to the energy change per pair $^1/_2\partial^2E_{tot}/\partial d^2\ \delta d^2$ so the bulk modulus is given by

$$B = \frac{1}{18d^3}\,d^2\frac{\partial E_{tot}}{\partial d^2} = -\frac{8}{9}\frac{E_{cov}}{d^3}\ . \tag{9-16}$$

This is 0.0415 eV/Å^3 = 0.67×10^{11} ergs/cm^3 for KCl, compared with the experimental value of 0.109 eV/Å^3 = 1.7×10^{11}ergs/cm^3. This is essentially equal to the value we shall give in Chapter 10, along with values for other compounds, based upon the more accurate form for the covalent contribution to the energy, Eq. (9-12). Generally these predictions are too small by a factor of around two, but showed reasonable agreement with trends from material to material.

We may evaluate the Grüneisen constant γ defined in Eq. (2-40) as

$$\gamma = -\frac{1}{6}\frac{d^3\partial^3E_{tot}/\partial d^3}{d^2\partial^2E_{tot}/\partial d^2} = \frac{5}{2} \tag{9-17}$$

It is related to the volume derivative of the bulk modulus by Eq. (2-42). We postpone comparisons with experiment until Section 10-3-D, where we provide a more accurate expression.

These explicit formulae allow easy estimates of important properties of the alkali halides in terms of the basic parameters of the system, $V_{sp\sigma}$ and the term values of Table 1-1. The perturbation theory of Eq. (9-6) can be generalized directly to other systems and other crystal structures, though not usually as accurately. The more accurate form, Eq. (9-12), does not complicate the theory greatly, nor for the alkali halides does it improve the results appreciably. We shall see that incorporation of Coulomb effects does improve the accuracy considerably, but then requires numerical solutions rather than simple formulae for the properties.

Problem 9-1. Independent ions.

CaF_2 forms in the fluorite structure in which Ca forms a face-centered cubic structure, with cube edge 5.48 Å. [Obtained from the structure, density in g/cm^3 and molecular weight from Weast (1975).] If the cube is divided into eight cubies, a fluorine ion is at the center of each cubie, with four Ca neighbors each at $d = \sqrt{3}\, a/4 = 2.37$ Å. In contrast, arsenic atoms filled only half of the cubies in the gallium face-centered-cubic structure in Fig. 2-2. CaF_2 is an ionic crystal, as is NaCl, and the same approximations are appropriate. Use the simplest estimate for each (include a numerical value):

a) Estimate the band gap.
b) Estimate the cohesive energy per molecule.

Problem 9-2. Coupled Ions

For fluorite, described in Problem 9-1,

a) Find the additional covalent contribution to the cohesion per formula (CaF_2) unit in analogy with Eq. (9-6).

b) Taking a) as the principal attraction between neighbors, and an overlap repulsion C/d^8, what is the radial force constant κ_0 ?

c) Estimate the bulk modulus for fluorite using this value.

CHAPTER 10

Elastic Properties of Ionic Crystals

We saw in Chapter 9 that some properties of ionic solids can be described *without* introducing the coupling between neighboring atomic states which produces the energy bands. That coupling is however real, becomes increasingly important as one moves from alkali halides to compounds with divalent or trivalent ions, and will be essential to the understanding of transition-metal compounds which we discuss in Chapters 17 and 18. Once we go beyond the simple second-order theory of Eq. (9-6), we no longer can obtain simple formulae for properties such as Eqs. (9-16) and (9-17). However, it is quite simple to use the more accurate Eq. (9-12) to obtain numerical results and we do that here. The analysis is based upon Harrison (1986). We then proceed to the somewhat more intricate effects of Coulomb interactions studied by Straub and Harrison (1989) and find significant improvement in the results. First we must take a closer look at the overlap repulsion which we discussed only briefly in Chapter 9.

1. The Overlap Repulsion

The covalent energy and the electrostatic energy both continue to grow as the ions are brought closer together and, as in the covalent semiconductors, it is the overlap interaction which prevents the system from collapsing. In the covalent solids we utilized an overlap repulsion proportional to $1/d^n$ with $n = 4$, twice the value from the covalent interaction V_2. [n is again an exponent, not a number of nearest neighbors.] In Section 2-2-E we discussed four different justifications for this form. All of them, if reargued for ionic solids, would suggest higher exponents than four.

In particular, the first argument was based upon the virial theorem, which states that the potential energy should have twice the magnitude of the kinetic energy. Then with the covalent energy associated with potential energy, and varying as $1/d^2$, the overlap repulsion associated with kinetic energy should vary as the square, $1/d^4$, so that the equilibrium condition leads to a magnitude of overlap energy half that of the covalent energy. This is just as the spring energy ($^1/_2\kappa x^2$) cancels half of the gravitational energy gain ($-Mgx$) of a weight of mass M suspended from it. For ionic solids, with a covalent energy varying as $1/d^4$ (from Eq. (9-6), for $V_{sp\sigma}$ varying as $1/d^2$) we would expect that the overlap repulsion would vary as $1/d^8$. The same argument would suggest that for the inert gases, where the principal attraction is the Van-der-Waal's interaction, described in Section 20-2-D, varying as $1/d^6$, the repulsion should vary as $1/d^{12}$, which is in fact the value traditionally used.

A closer look at the origin of the overlap interaction, the third justification mentioned in Section 2-2-E, gives a different understanding of why this higher power arises in such closed-shell systems. We make that analysis here.

A. Based upon Local-Density Theory

Local-density theory is the basic starting point for contemporary first-principles total-energy calculations. It is based upon the proof by Hohenberg and Kohn (1964) that the total energy of a system of atoms in the ground state can be written as a functional of the electron density $n(\mathbf{r})$ (meaning that one could in principle determine the total energy knowing only $n(\mathbf{r})$). Kohn and Sham (1965) then noted that this energy could be approximated by writing the functional of $n(\mathbf{r})$ as a simple *function* of $n(\mathbf{r})$, called the *local-density approximation*, or *LDA*. (They actually calculated the electronic kinetic energy by a more accurate procedure but, in the subsequent evaluation of the overlap interaction which we shall discuss, the free electron formula is used also for kinetic energy.)

Nikulin (1971) and Gordon and Kim (1972, 1976) approximated the electron density of a collection of closed-shell atoms or ions by the superposition of free-atom densities, which were obtained from Hartree-Fock calculations, and used existing approximations for the energy density for a free-electron gas. An examination of their results (Harrison (1981a)) showed that the dominant term in the repulsion between such closed-shell systems is the excess kinetic energy of the overlapping charge densities, as we suggested above. Thus our general description of the overlap repulsion as kinetic energy is supported by the more complete calculations.

This approach was further simplified (Harrison (1981a)) by using the asymptotic form at large r for the free-atom wavefunction of the shallowest occupied state to estimate that density. That asymptotic form is $r^{\nu} e^{-\mu r}$ with μ related to the energy of the atomic state(in our case a p-state) by $-\hbar^2\mu^2/2m = \varepsilon_p$, and ν near zero (and equal to zero if the energy is near one Rydberg in energy, which our states mostly are). We take $\nu = 0$ and if we use exactly that form to obtain the normalization, we obtain a density for each of the six p-electrons of $(\mu^3/(8\pi))e^{-2\mu r}$. An evaluation of the excess kinetic energy for μ the same on both atoms gave (Harrison (1981a))

$$V_0(d) = 70.8\, \mu d\, |\varepsilon_p|\, e^{-5\mu d/3} . \tag{10-1}$$

The origin of the exponential $e^{-5\mu d/3}$ was the dominance of the contribution at the midpoint between atoms, $r = d/2$, where each density is proportional to $e^{-\mu d}$. The kinetic energy density is proportional to the five-thirds power of the electron density leading to this factor. The remaining factors and the coefficient came from consideration of the details.

This form of $V_0(d)$ seems to be quite correct, but the approximate normalization leading to a value of 70.8 for the leading factor is not. Larger and larger reductions in the density near the nucleus occur for heavier atoms and the appropriate value of the leading constant factor increases with atomic number. Brock (1981) studied the values of this leading factor needed to fit the observed spacing in the simple monovalent and divalent ionic compounds in the rock-salt structure and found that the value depended significantly only upon the row (in the periodic table) of the nonmetallic ion. The resulting values were near 70 or 80 for the compounds with nonmetal atoms in the 2p- and 3-p rows, chlorides, bromides, sulfides and selenides, but larger values for the iodides and tellurides, and values half as large for the fluorides and oxides. This degree of agreement for the fit, and the agreement for predicted bulk moduli for these compounds (Harrison

(1981a)) confirms the exponential form of the repulsion and the general picture. It also provides a perfectly good way to proceed. However, we subsequently found (Harrison (1986)) that the $1/d^n$ form works about as well and is more convenient. For that reason we shall use it here.

B. Fitting the Power-Law Form

We may use the exponential form, Eq. (10-1), to learn what values of exponent will be most appropriate in the algebraic form. We note that for an interaction $V_0 \propto 1/d^n$, the exponent n is given by

$$n = -\frac{d}{V_0}\frac{\partial V_0}{\partial d} \tag{10-2}$$

Thus if one wishes to model the repulsion by a form proportional to $1/d^n$ near the equilibrium spacing, the exponent n can be obtained by differentiating Eq. (10-1) to obtain from Eq. (10-2)

$$n = \frac{5\,\mu d}{3} - 1 \quad , \tag{10-3}$$

evaluated at the equilibrium spacing.

Returning to the semiconductor silicon, with ε_p = -7.58 eV and d = 2.35 Å, we obtain n = 4.52, rather close to the 4 which we used. For argon, with ε_p = -16.08 eV and d = 3.76Å, we obtain n = 11.87, close to the traditional Lennard-Jones form (discussed in Section 20-2-E) of $1/d^{12}$, and very far from the $1/d^4$ for semiconductors. For ionic solids we may use an average of the μ obtained for the highest occupied states on the two ions, and typically find values near $n = 8$, again very much more abrupt (rapidly varying) than the form for semiconductors. For the evaluations carried out in the following sections we shall proceed with the algebraic form, proportional to $1/d^8$. It seems remarkable that this detailed derivation, based upon a series of approximations to local-density theory, should fit so well with the intuitive arguments for the exponents 4, 8, and 12 based upon the virial theorem.

We may learn more about the power-law form of the overlap repulsion from the exponential form, Eq. (10-1), suggested by local-density theory. We note that it is proportional to the energy ε_p of the states dominating the repulsion, so we retain such a factor. Further, the remaining factors, $\mu d\, e^{-5\mu d/3}$, are functions only of μd which suggests a form

$$V_0(d) = \eta_0{}' \frac{|\varepsilon_p|}{(\mu d)^8} \qquad (10\text{-}4)$$

for the $1/d^8$ dependence, with again

$$\frac{\hbar^2\mu^2}{2m} = |\varepsilon_p| \,. \qquad (10\text{-}5)$$

A prime was put on $\eta_0{}'$ because we shall absorb other constants later and define a new dimensionless coefficient η_0.

When the states dominating the repulsion are of different energy on the two atoms, we would anticipate an intermediate value of ε_p being used. The construction of ions from inert-gas atoms, illustrated in Fig. 9-1, would suggest that for potassium chloride we use the energy ε_p for the argon atom, which lies between potassium and chlorine in the periodic table. This would also suggest that we use the same inert-gas ε_p for calcium sulfide, for which the atoms are two columns to the right and left of argon. Thus we follow the earlier treatment (Harrison (1986)) in using the $|\varepsilon_{ig}|$, the p-state energy from Table 1-1 for the inert gas atom for compounds arising from elements in the same row in the periodic table, and taking the geometric mean of the two p-state energies when the constituent elements are from two different rows. It is more convenient, and perhaps more appropriate, to use these inert-gas-atom energies than an average of the valence p-state energy for the halogen or chalcogen and the *core* p-state energy for the alkali or alkaline earth. We have not needed these core-state energies before, though they are listed in Table 1-1. We may see there that the difference in the two choices for ε_p is not so large. We use the same ε_p for the $|\varepsilon_p|$ factor in Eq. (10-4), and for determining the appropriate μ from Eq. (10-5).

Finally, we noted that the coefficient $\eta_o{}'$ cannot really be taken as universal, for the same reasons that the 70.8 of Eq. (10-1) could not be. A good procedure would be to adjust it to give the correct spacing for each compound, and use the same value to obtain other properties. We also wish to see here how well the theory accounts for the equilibrium spacings themselves, and for this we need more general values of $\eta_0{}'$. The experience of adjusting the numerical coefficient in Eq. (10-1) to give the correct equilibrium spacings (Harrison (1981a), discussed after Eq. (10-1)) suggests that it depends principally on the row of the nonmetallic atom. Thus we fit $\eta_0{}'$ for a single compound for each row, the corresponding potassium halide, and use that same value to predict equilibrium spacings for all halides and chalcogenides from that same row. This was the procedure used by Harrison (1986) and by Straub and Harrison (1989) and we carry it out here in Section 3 of this chapter.

C. Second-Neighbor Repulsions

We initially assumed that we would include only overlap repulsions between nearest-neighbor atoms, as we did in the covalent solids. However, we may also evaluate the overlap repulsion, either from Eq. (10-1) or Eq. (10-4), for the interaction between two chlorine ions in KCl, which are second-neighbor ions, using the same rules for parameters. The μ based upon two chlorine ions is enough smaller than that based upon an average of a chlorine and a potassium ion (or in our estimate, an argon atom), that these are far from negligible. The second-neighbor interactions between *metallic* ions remain negligible since they are based upon the much deeper core levels of the metallic ion and have larger μ values.

One would immediately suggest that if we are to include the overlap repulsion between second-neighbors, we should not neglect the couplings V_{ppm} between the states on second-neighbor nonmetallic ions. That may well be true, but since the coupled states in question are all occupied, and the coupling does not change the average energy of two coupled states, it does not contribute to the bonding of the system. Its principal effect is to give curvature and breadth to the valence (and conduction) energy bands as we indicated at the end of Section 9-2-C, and we neglect any effects upon the total energy other than in the repulsion.

We are therefore to include for the first time in this text some effect of second-neighbor interactions. We clearly are to use the valence ε_p values for the halogen or chalcogen in Eqs. (10-4) and (10-5) for second-neighbor interactions, and we select the η_0' appropriate to the nonmetallic atom, and therefore the same value of η_0' for both nearest-neighbor and second-neighbor interactions. All parameters are therefore chosen. For the treatment of resonant bonds in the following section we use this simple form of the repulsion, Eq. (10-4), proportional to $1/d^8$.

2. Resonant Bonds

Eq. (9-12) for the covalent contribution to the bonding, which we rewrite by taking a factor of 4 from under the square root,

$$E_{cov} = -2\sqrt{\left(\frac{\varepsilon_s(M) - \varepsilon_p(X)}{2}\right)^2 + nV_{sps}^2} + \varepsilon_s(M) - \varepsilon_p(X) , \qquad (10\text{-}6)$$

gives us a description of the bonding, which we have indicated can be thought of as a resonating of the sp-bond over wavenumbers in the Brillouin Zone. The factor n under the square root is again the number of nearest neighbors. E_{cov} is a correction to the independent-ion term in the energy, $\varepsilon_p(X) - \varepsilon_s(M)$ for alkali halides (or twice that for alkaline earth chalcogenides). It may be desirable to make that identification more explicit.

The first term in Eq. (10-6) gives the lowering in energy, relative to the average orbital energies corresponding to the $-\sqrt{V_2^2 + V_3^2}$ per electron for the covalent solids. We think of the factor 2 in the first term as representing the two electrons per ion pair in the resonant bond, corresponding to the one valence band which is dispersed by the coupling in Fig. 9-5. It is the same for monovalent and divalent compounds, which differ only in the independent-ion terms. The coupling enters through the factor $nV_{sp\sigma}^2$ which we write as a squared covalent energy V_2^2 with

$$V_2 = \sqrt{n}V_{sp\sigma} . \tag{10-7}$$

We think of the $\sqrt{n}$ as giving the enhancement of the bond due to its resonating among n neighbors. $n = 6$ for the rock-salt structure.

This enhancement is a very important characteristic of resonant bonding. This same enhancement of π-bonding in resonant systems such as benzene (an enhancement factor $\sqrt{2}$) and graphite (an enhancement factor $\sqrt{3}$) was derived in Harrison (1983c) and discussed in Section 2-5. Such an enhancement was noted from experiment many years ago by Pauling (1960) who simply defined the extra energy as a "resonance energy" but did not have a formula directly relating it to the bond energy and to the number of sites among which the resonance occurs. A consideration of the experimental (or theoretically predicted) bond lengths in complex molecules such as anthracene indicates that this enhancement only applies to the energy averaged over the molecule, not to the relative bond strengths within the molecule when different atoms have different coordinations.

As in the semiconductors, there is also a polar energy,

$$V_3 = \frac{\varepsilon_s(M) - \varepsilon_p((X))}{2} , \tag{10-8}$$

and the bonding energy (including the independent-ion term, which canceled the final two terms in Eq. (10-6)) takes the form $- 2\sqrt{V_2^2 + V_3^2}$ for the alkali halides. For the divalent compounds there is an additional $-2V_3$ for the second electron, transferred from the metallic ion to the nonbonding (flat) valence band.

Finally we may write the contribution of the overlap repulsion, for which we take the form given in Eq. (10-4), proportional to $1/d^8$. For nearest neighbors, Eq. (10-4) is conveniently written in terms of the covalent energy of Eq. (10-7) as

$$V_0(d) = \eta_0{}' \frac{|\varepsilon_p|}{(\mu d)^8} = \left(\frac{\eta_0{}'}{16n^2\eta_{sp\sigma}{}^4}\right)\frac{V_2{}^4}{|\varepsilon_p|^3} = \eta_0 \frac{V_2{}^4}{|\varepsilon_p|^3} \ . \qquad (10\text{-}9)$$

This gives the definition of the dimensionless parameter η_0 in terms of the original one, $\eta_0{}'$. As we indicated at the end of Part 10-1-B, we shall adjust it to give the correct equilibrium spacing for KF, KCl, KBr, and KI and use those values for all compounds with their nonmetallic atom from the same row as the halogen in the corresponding potassium halide.

For the six nearest-neighbor interactions per ion pair, the ε_p which enters is ε_{ig} , the geometric means of the p-state energies of the inert gas atoms in the same row as the two ions making up that compound. For the twelve second neighbors it is the ε_p for the nonmetallic ion. Half the contribution is associated with each of the coupled ions. Also, in the rock-salt structure, the second-neighbor distance is larger by a factor $\sqrt{2}$ than the nearest-neighbor distance, so the $1/d^8$ coupling is reduced by a factor 1/16. This leads to a total overlap interaction per pair in the rock-salt structure of

$$E_{overlap} = 6\eta_0 V_2{}^4\left(\frac{1}{|\varepsilon_{ig}|^3} + \frac{1}{16|\varepsilon_p|^3}\right) . \qquad (10\text{-}10)$$

In MgS, for example, with magnesium two columns to the right of neon in the periodic table and sulfur two columns to the left of argon in the periodic table (with chlorine between), we use an η_0 determined for KCl. For ε_{ig} we use the geometric mean of the neon and argon p-state energies from Table 1-1, and for ε_p we use the sulfur value from Table 1-1.

When we treat the cesium compounds, all in the cesium-chloride structure of Fig. 9-2b, there are eight nearest-neighbor interactions so that $V_2 = \sqrt{8}\, V_{sp\sigma}$, and with the modified neighbor distances Eq. (10-10) is replaced by (Harrison (1986))

$$E_{overlap} = \frac{9}{2}\eta_0 V_2{}^4\left(\frac{1}{|\varepsilon_{ig}|^3} + \frac{243}{2048|\varepsilon_p|^3}\right) \text{(CsCl structure)} \qquad (10\text{-}11)$$

We first use these forms directly to estimate volume-dependent properties. We then incorporate Coulomb effects and reevaluate these same properties and then turn to the elastic shear constants and lattice vibrational frequencies.

3. Volume-Dependent Properties

A. Equilibrium Spacing

In terms of these covalent (different for the rock-salt and cesium-chloride structures) and polar energies, the total energy per ion pair, relative to that of free atoms, is given for alkali halides by

$$E_{tot} = -2\sqrt{V_2^2 + V_3^2} + E_{overlap} . \qquad (10\text{-}12)$$

For divalent compounds there is an additional term $-2V_3$. Because the only dependence upon position in $E_{overlap}$ is from the factor V_2^4 , we may set the derivative of E_{tot} with respect to V_2 equal to zero as the equilibrium condition. Applying this first to the potassium halides, we adjust η_0 to give the correct spacing and obtain (Harrison (1986)) η_0 equal to 4.69 for KF, 7.93 for KCl, 9.02 for KBr, and 11.48 for KI . The increasing values for

Table 10-1. Equilibrium spacings d in Angstroms for the alkali halides. The first entry is the prediction from minimization of Eq. (10-12), with η_0 fit to the corresponding potassium halide (Harrison (1986)). The second entry includes Coulomb effects described in Section 10-5 (Straub and Harrison (1989)). The third entry, in parentheses, is the experimental value taken from the molecular weight and density listed in the CRC Handbook (Weast (1975)).

	F	Cl	Br	I
Na	2.36 2.46 (2.32)	2.79 2.87 (2.82)	2.93 3.03 (2.99)	3.15 3.23 (3.24)
K	fit (2.67)	fit (3.15)	fit (3.30)	fit (3.53)
Rb	2.79 2.71 (2.80)	3.29 3.20 (3.29)	3.44 3.38 (3.42)	3.70 3.62 (3.64)
Cs	2.97 - -	3.52 - (3.57)	3.70 - (3.71)	3.98 - (3.95)

lower rows arises from the larger core and corresponding scaling up of the electron density at large distances. We may then use these values of η_0 to predict the spacing for compounds with nonmetallic atoms in the same row as the corresponding halogen atom. This gives the spacings listed in Tables 10-1 and 10-2. (From Harrison (1986)). (Lithium compounds were not included in that study.)

The agreement with experiment is quite good, suggesting the formulae contain the principle origins of trends from material to material. The addition of Coulomb effects, to be discussed, gives the second entry, and does not generally improve the predictions. Perhaps of more interest is the direct application, with the same η_0 values obtained from the potassium halide, to the divalent compounds, given in Table 10-2. Comparison of these values with experiment suggests that we also have incorporated the essential dependence upon valence in the theory, though we overestimate the spacing on average by some 5%. We also see that Coulomb effects, to be added, lead to an *underestimate* by a comparable amount.

In Harrison (1981a) such estimates were also made for the equilibrium

Table 10-2. Equilibrium spacings d in Angstroms for the divalent compounds. The first entry is the prediction from minimization of Eq. (10-12), with η_0 as for Table 10-1 (Harrison (1986)). The second entry includes Coulomb effects described in Section 10-5 (Straub and Harrison (1989)). The third entry, in parentheses, is the experimental value as in Table 10-1.

	O	S	Se	Te
Mg	2.27 2.19 (2.10)	2.69 2.53 (2.60)	2.82 2.66 (2.73)	3.04 2.83 (2.99)*
Ca	2.55 2.33 (2.41)	3.00 2.79 (2.85)	3.15 2.86 (2.96)	3.37 3.04 (3.18)
Sr	2.66 2.37 (2.58)	3.12 2.76 (3.10)	3.28 2.92 (3.12)	3.51 3.10 (3.33)
Ba	2.80 2.42 (2.76)	3.28 2.83 (3.19)	3.44 2.99 (3.30)	3.70 3.19 (3.49)

*Interpolated from neighboring compounds. MgTe is hexagonal, not rock-salt structure.

spacing of the trivalent compounds ScN, YN, and LaN, which also form in the rock-salt structure, with spacings d of 2.20 Å, 2.44 Å, and 2.65 Å respectively. The cohesion in these cases includes also the transfer of a d-electron from the metallic atom to the p-state of the nitrogen, but the same formulae apply to the determination of equilibrium spacing. Its application again led to predictions some 5% larger than experiment. Presumably inclusion of Coulomb terms would convert this to an underestimate. In these cases we might also expect additional bonding terms from the coupling between the nitrogen p-states and the metallic d-states, to which we return in Chapter 17 on transition-metal compounds.

As we turn to other properties it will be convenient to use the equilibrium condition to write $E_{overlap}$ in terms of the bonding terms, as we did for the second-order theory in Section 9-2-D. We are in fact simply repeating the calculation with the more accurate Eq. (10-12). Since we evaluate properties at the observed spacing this is equivalent to readjusting η_0 to give the correct spacing for each individual compound.

B. Cohesive Energy

$E_{overlap}$ in Eq. (10-12) is a constant times V_2^4 and the equilibrium condition (that $\partial E_{tot}/\partial V_2 = 0$) gives the constant as $1/[2\sqrt{V_2^2 + V_3}]$, evaluated at equilibrium. This gives a cohesive energy per ion pair (equal to the magnitude of the energy per ion pair gained in the formation of the solid) of

$$E_{coh} = 2\sqrt{V_2^2 + V_3} - \frac{V_2^2}{2\sqrt{V_2^2 + V_3}} \tag{10-13}$$

for alkali halides, with an additional $+2V_3$ for the divalent compounds, and a further addition of $\varepsilon_d(M) - \varepsilon_p(N)$ for the nitrides. As we indicated in Chapter 9, the inclusion of the small covalent energy V_2 makes only a small correction to the values without it, and slightly worsens agreement with experiment. We did not include these corrections in the cohesive energies of Table 9-5.

C. Bulk Modulus

We can immediately evaluate the second derivative of the total energy with respect to d , again use the equilibrium condition to eliminate the coefficient in the overlap repulsion, and thereby directly predict the bulk modulus, related to the elastic constants by $B = (c_{11} + 2c_{12})/3$. We obtain

Table 10-3 Bulk modulus, in eV/Å^3, for rock-salt-structure compounds. (Values in ergs/cm^3 are obtained by multiplying these values by 1.60×10^{12}). The first value is from Eq. (10-14), the second includes Coulomb effects from Section 10-5. Experimental values, in parentheses, were compiled by Straub and Harrison (1989).

	Alkali Halides			
	F	Cl	Br	I
Na	0.223	0.094	0.073	0.051
	0.410	0.200	0.157	0.117
	(0.290)	(0.150)	(0.124)	(0.094)
K	0.080	0.041	0.034	0.025
	0.290	0.137	0.107	0.080
	(0.190)	(0.109)	(0.092)	(0.079)
Rb	0.057	0.029	0.026	0.020
	0.262	0.123	0.096	0.071
	(0.164)	(0.097)	(0.081)	(0.066)

	Divalent Compounds			
	O	S	Se	Te
Mg	0.579	0.233	0.191	0.173
	1.551	0.918	0.697	0.534
	(1.03)	(0.71)	-	-
Ca	0.216	0.114	0.099	0.074
	1.243	0.722	0.555	0.431
	(0.71)	(0.28)	(0.32)	(0.26)
Sr	0.132	0.063	0.067	0.052
	1.191	0.677	0.527	0.409
	(0.55)	-	(0.29)	(0.25)
Ba	0.080	0.048	0.043	0.035
	1.100	0.615	0.477	0.369
	(0.38)	(0.56)	(0.25)	(0.18)

Nitrides		
ScN	YN	LaN
0.486	0.249	0.134

$$B = \frac{4}{9}\frac{V_2}{d^3}\frac{V_2(3V_2^3 + 2V_3^2)}{(V_2^2 + V_3^2)^{3/2}} . \quad (10\text{-}14)$$

Predictions from this formula are tabulated and compared with experiment in Table 10-3. These values are dominated by the overlap interactions and are only qualitatively in agreement with experiment. We see from the second entry that the largest discrepancies are eliminated if one includes Coulomb effects, which will be done in Section 10-5.

D. Grüneisen Constants

We may also take the third derivative of the energy with respect to d and thus obtain the volume dependence of the elastic constants, related to the Grüneisen constant as discussed for covalent solids in Section 2-4-C. After some algebra we obtain (corrected from Harrison (1986), where a different γ was defined and the V_2^4 was incorrectly written V_3^4) for the Grüneisen constant of Eq. (2-40)

$$\gamma = -\frac{d}{6}\frac{\partial^3 E_{tot}/\partial d^3}{\partial^2 E_{tot}/\partial d^2} = \frac{5}{2} - \frac{V_2^4}{(V_2^2 + V_3^2)(3V_2^2 + 2V_3^2)} , \quad (10\text{-}15)$$

dimensionless, and slightly less than $^5/_2$. In Chapter 9 we kept only terms to lowest order in V_2/V_3 and the second term did not appear in Eq. (9-17).

It will be convenient to make comparison of these predictions with the experimental derivative of the bulk modulus with respect to pressure, $\partial B/\partial P = -(\Omega/B)\,\partial B/\partial\Omega$. We indicated in Eq. (2-42) that this is given by $2\gamma + {}^1/_3$, or

Table 10-4 Logarithmic volume derivative of the bulk modulus, $-(\Omega/B)(\partial B/\partial\Omega)$, dimensionless, is also the pressure derivative of the bulk modulus, dB/dP. The first value is from Eq. (10-16) , the second includes Coulomb effects from Section 10-5. Experimental values, in parentheses, compiled by Straub and Harrison (1989), were from Simmons and Wang (1971) and from Bartles and Schule (1965).

NaCl	KCl	KBr	KI
5.22	5.29	5.28	5.27
4.55	4.43	4.48	4.51
(5.13)	(5.34)	(5.39)	(6.28)

$$-\frac{\Omega}{B}\frac{\partial B}{\partial \Omega} = 5 + \frac{1}{3} - \frac{2V_2^4}{(V_2^2 + V_3^2)(3V_2^2 + 2V_3^2)} \cdot \qquad (10\text{-}16)$$

Values given as the first entry in Table 10-4 are compared with the available experimental values in parentheses. Note that keeping only the leading term in V_2/V_3 , as in Section 9-2-D, gives simply 5.33 for all systems. That prediction is impressive, depending only upon the exponents 4 and 8 . As is often the case, the use of a more accurate form of the energy, Eq. (10-12), does not improved the results.

4. Elastic Constants

All of the above analysis has assumed an undistorted cubic structure and some generalization is required in order to treat deformed structures. The overlap interaction was written in terms of two-body central-force interactions and generalizes directly. The second-neighbor interactions which were included in Eq. (10-10) are quite essential in stabilizing the structure since the rock-salt structure is not stable with respect to an e_4 shear under nearest-neighbor central-force interactions alone. If we used the approximate form for the covalent contribution to the bonding, Eq. (9-6), it is also a two-body central force interaction, but involving only nearest neighbors. Then the calculation of the elastic shear constants is immediate.

It is preferable, however, to generalize the more accurate covalent contribution, Eq. (10-6), and this generalization is not unique nor has it been fully explored. The generalization used in Harrison (1981a) is to note that the coupling between neighbors enters Eq. (10-6) through $nV_{sp\sigma}^2$, which we may regard as a sum of $V_{sp\sigma}^2(d_j)$ over the six neighbors to each halogen ion, and to simply modify the values of d_j for the individual neighbors as appropriate to the distortion of the lattice.

$$E_{cov} = -2\sqrt{\left(\frac{\varepsilon_s(M) - \varepsilon_p(X)}{2}\right)^2 + \Sigma_j V_{sps}(d_j)^2} + \varepsilon_s(M) - \varepsilon_p(X) \ , \qquad (10\text{-}17)$$

This was done in Harrison (1981a) for a strain e_1 to obtain the c_{11} given in Table 10-5. In that analysis, the exponential form of the overlap interaction, Eq. (10-1), was used, rather than the algebraic form, Eq. (10-4). These were combined with the values of the bulk modulus (the counterpart of Table 10-3, but using exponential repulsions) to obtain c_{12} , also given in Table 10-5. The bulk modulus predicted in that study was larger by a factor of two or more from that which we list in Table 10-5, presumably because of the different form of overlap interaction used, but we have not redone those calculations. Trends are expected to be similar, but values cannot be compared. (Note also the units are different.) Only a sampling of the alkali halides were included; the compounds of K, Cl, and Br should vary smoothly between these limiting systems both theoretically and experimentally. MgO, for which experimental values were found, was also included. We shall give a much more complete tabulation of elastic constants in Table 10-8, when Coulomb effects are also included.

Table 10-5 Elastic constants, predicted using the exponential form of the repulsion, Eq. (10-1), and the covalent energy of Eq. (10-17). Experimental values are in parentheses. For c_{44} , the covalent correction, Eq. (10-19), was added to the predicted c_{12} . Values are in 10^{11} ergs/cm^3. (from Harrison, 1981a).

	c_{11}	c_{12}	c_{44}
NaF	14.39 (9.7)	3.31 (2.44)	3.57 (2.81)
NaI	4.21 (3.03)	1.65 (0.89)	1.69 (0.73)
RbF	4.93 (5.52)	1.94 (1.40)	1.95 (0.95)
RbI	1.97 (2.56)	0.22 (0.36)	0.23 (0.28)
Mg0	17.7 (29.2)	11.25 (9.1)	12.32 (15.4)

A. The Shear Constant, $(c_{11} - c_{12})/2$

The predicted values for the shear constant $(c_{11} - c_{12})/2$ from this earlier study were found to be slightly less accurate than the bulk modulus but not significantly so. Furthermore, the accuracy was found to be similar to that for values obtained from the Born theory, based upon the Madelung term alone. The electrostatic values for c_{12} (also in units of 10^{11} ergs/cm^3) for the five compounds listed in Table 10-5 are 2.77, 1.73, 1.31, 0.46, and 16.5, respectively. It is remarkable that the agreement with experiment for the elasticity, obtained assuming two such different physical origins for the rigidity, is so similar.

If the total energy calculation were equivalent to a sum of two-body central-force interactions in equilibrium (as it would be, had we used the approximate form for the covalent contribution to the bonding, Eq. (9-6)) then the Cauchy relations (Born and Huang (1954), to be discussed in detail for simple metals in Section 14-3-B) would guarantee that c_{44} would have the same value as c_{12}. Strictly speaking we have not made that approximation for Table 10-4 since we used the square-root form, Eq. (10-17), for the covalent energy and there are cross terms between different d_j^2 in the expansion of the square root. However, with nearest-neighbor interactions only it is possible to see that for these two constants the calculation is equivalent to a central-force nearest-neighbor interaction and c_{12} and c_{44} would be found to have the same value. We seek corrections to this model in the following subsection.

B. The Shear Constant, c_{44}, and the Chemical Grip

The perturbation theory which we carried only to second order in Eq. (9-6) did give a two-body central-force interaction. If carried to fourth order, it would yield also cross terms in different d_j^2 as from an expansion of the square root in Eq. (10-17). However, it also leads to terms which depend explicitly upon the angles between neighbors. Such terms were evaluated in Harrison (1980), and called a "chemical grip".

Such fourth-order terms decrease the bonding energy, so one shouldn't call them bonding terms, but they give the leading term in the angular forces. We will write them for general angular-momentum quantum numbers for the central orbital in Section 18-3-C, and will apply them to d-states. Here we write just the simpler form for p-states appropriate here. The corresponding contribution to the energy is (Harrison (1980), p. 462)

$$E_{grip} = \frac{4}{[\varepsilon_s(M) - \varepsilon_p(X)]^3} \Sigma_{j>i} V_{sp\sigma}(d_i)^2 V_{sp\sigma}(d_j)^2 \cos^2\theta_{ij} , \qquad (10\text{-}18)$$

where θ_{ij} in the angle between $\mathbf{d}_i$ and $\mathbf{d}_j$. (Eq. (19-29) in Harrison, 1980 was in error by a factor of two in the first printing, and took the $V_{sp\sigma}^2$ out from under the sum, which would be allowed for the calculation of c_{44} but not in general.) This very interesting term in the energy has not been widely considered. It contains a variation with angle which increases the shear constant

$$\delta c_{44} = \frac{16}{d^3} \frac{V_{sp\sigma}^4}{[\varepsilon_s(M) - \varepsilon_p(X)]^3} \cdot \qquad (10\text{-}19)$$

There is no such angular change in the distortions e_1 and e_2 so this δc_{44} was directly added to the predicted c_{12} to obtain the values listed in Table

10-5. (A different approximation for the denominator, which we shall use in the following subsection on vibrational frequencies, was made in Harrison (1981a) so the numbers were somewhat different.) The contribution is of the same sign as the experimental difference in MgO and NaF, where the effects are largest due to the $1/d^{11}$ dependence of Eq. (10-19), but is of the wrong sign in the other cases. We regard this as a correct contribution, but not accurately given by Eq. (10-19) and lost among larger effects, such as Coulomb effects to be discussed in Section 10-5.

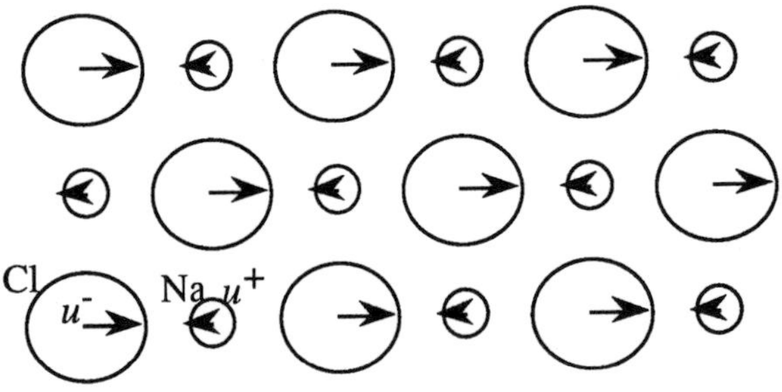

Fig. 10-1. Displacements of the ions in a $q = 0$ optical mode in NaCl.

C. Vibrational Frequencies

In calculating the shear constant $(c_{11} - c_{12})/2$ we used the form, Eq. (10-17) for the covalent energy, with a sum of couplings over the six individual bonds surrounding a given nonmetallic atom. We may use exactly that approximation in calculating the frequency of the $q = 0$ optical mode here, as for covalent solids in Section 3-3-B. That is the mode in which the metallic and nonmetallic ions move exactly out of phase. It is illustrated in Fig. 10-1. At the end of the section we shall discuss briefly the extension to other wavenumbers.

A convenient way to make this calculation is to equate the average kinetic energy and the average potential energy per ion pair, both to second order in the displacements. This is equivalent to equating force to mass-times-acceleration as we did in Section 3-3. We obtain the covalent energy directly from Eq. (10-17), without needing to calculate forces from the derivatives in this three-dimensional system. We take displacements to be along a cube direction [100], labeled u_+ for sodium and u_- for chlorine in opposite directions so that the relative displacement is $u = u_+ + u_-$. With the center of mass fixed we may note that $M_+u_+ = M_-u_-$. Then, with a vibrational frequency ω_{TO} (in radians per second), the kinetic energy per ion pair becomes $^1/_2(M_+ u_+^2 + M_- u_-^2)\omega_{TO}^2$. We have associated this frequency with the transverse optical mode (the subscript "TO") since, as we saw for polar covalent solids in Section 3-3-C, there are additional Coulomb forces for longitudinal optical modes, in which the amplitudes are modulated in a direction parallel to the displacements. We write the reduced mass $M_r = M_+M_-/(M_+ + M_-)$, and obtain an average kinetic energy per ion pair of

$$K.E. = {}^1\!/_2\, M_r \langle u^2 \rangle\, \omega_{TO}^2 \,, \tag{10-20}$$

in terms of the time average $\langle u^2 \rangle$ of u^2 .

We can also directly obtain the potential energy associated with a single nonmetallic ion due to interaction with its six nearest neighbor metallic ions. By inspection of Fig. 10-1 we see that one neighbor distance becomes $d + u$, one becomes $d - u$ and four become $\sqrt{d^2 + u^2}$. Thus we may substitute these six modified d_j values in Eq. (10-17) for the covalent energy and expand to second order in u to obtain

$$\delta E_{cov} = -\frac{1}{2}\,\frac{4V_2^2}{\sqrt{V_2^2 + V_3^2}}\left(\frac{u}{d}\right)^2, \tag{10-21}$$

where we have used the parameters defined in Eqs. (10-7) and (10-8). We may also sum the six nearest-neighbor overlap repulsions, each of which we write $\eta_0 V_2/|\varepsilon_{ig}|^3$, according to Eq. (10-9), C/d^8 , with the modified spacings given above, to obtain

$$\delta E_{overlap} = \frac{1}{2}\,112\,\frac{\eta_0 V_2}{|\varepsilon_{ig}|^3}\left(\frac{u}{d}\right)^2 . \tag{10-22}$$

Note that there is no relative displacement of second neighbors since they move together. However, the second-neighbor interactions did enter the equilibrium condition which fixed η_0 . From Eq. (10-10) and (10-12) that equilibrium condition is

$$\left(1 + \frac{1}{16}\left(\frac{\varepsilon_{ig}}{\varepsilon_p}\right)^3\right)\frac{6\eta_0 V_2{}^4}{|\varepsilon_{ig}|^3} = \frac{V_2^2}{2\sqrt{V_2^2 + V_3^2}}. \tag{10-23}$$

Setting the kinetic energy equal to the sum of changes in covalent and overlap energy gives an equation for the frequency,

$$M_r \omega_{TO}^2 d^2 = \left(\frac{28}{3(1 + (\varepsilon_{ig}/\varepsilon_p)^3/16)} - 4\right)\frac{V_2^2}{\sqrt{V_2^2 + V_3^2}}. \tag{10-24}$$

Table 10-6. Predicted (and experimental in parentheses) transverse optical mode frequencies (in units of 10^{13} radians/sec) for the alkali halides in the rock-salt structure. (Source for experimental values, Kittel (1966), p. 156.)

	F	Cl	Br	I
Li	9.51 (5.8)	5.15 (3.6)	4.23 (3.0)	3.43 -
Na	4.57 (4.5)	2.62 (3.1)	2.05 (2.5)	1.65
K	2.83 (3.6)	1.73 (2.7)	1.34	1.10 (1.9)
Rb	2.51 (2.9)	1.32	0.96 (1.4)	0.077

The results are compared with experiment for the alkali halides in the rock-salt structure in Table 10-6. It is gratifying that the values and trends are as well given as they are. The only parameter taken from the solid itself is the internuclear distance. The factor $1 + (\varepsilon_{ig}/\varepsilon_p)^3/16$ was near one, the largest value being 1.23 for LiI, so the expression in brackets could be taken as a universal 16/3 without significant loss of accuracy.

Eq. (10-24) is equally applicable to divalent compounds, leading to the values shown in Table 10-7. The only experimental value available is 7.5 $\times 10^{13}$/sec. for MgO (Kittel (1966)) in comparable agreement with experiment to that for the alkali halides.

In the divalent compounds, the factor $1 + (\varepsilon_{ig}/\varepsilon_p)^3/16$ becomes more significant, reaching 1.35 for MgTe. This reduces the factor in brackets in

Table 10-7. Predicted transverse optical mode frequencies (in units of 10^{13} radians per second) for divalent compounds.

	O	S	Se	Te
Mg	6.49	3.35	2.61	2.09*
Ca	4.31	2.50	1.91	1.54
Sr	3.31	1.76	1.35	1.06
Ba	2.66	1.55	1.06	0.83

*Assuming an extrapolated $d = 2.95$ Å .

Eq. (10-24) from 5.3 to 2.9. If this factor in brackets were to become negative, the system would become unstable and must form in a different structure. Even before that, the system may have lower energy in a different structure and, in fact, MgTe forms in a hexagonal structure rather than in the rock-salt structure.

Such a change in crystalline structure would be called a "soft-mode" transition. The frequency of the vibrational mode in question, the optical mode, drops until the crystal becomes unstable against a deformation in which the two ion types displace with respect to each other, as illustrated in Fig..10-1. Then it will seek a more stable arrangement of the ions.

The physical origin of the soft-mode tendency here is quite apparent. For the heavy chalcogenides ε_p becomes shallower, and the second term in $1 + (\varepsilon_{ig}/\varepsilon_p)^3/16$, which represents the relative effect of second neighbor repulsions, becomes larger. In more physical terms the chalcogen becomes so large that the lattice is held apart principally by the repulsion between them and the small magnesium "rattles around" in the interstices. Its covalent attraction to the neighbors pulls it out of rock-salt-structure site.

We have calculated the frequency of a mode of infinite wavelength but we could imagine modulating it in a direction perpendicular to the displacements and the frequency would be only slightly changed. This why we referred to this as the *transverse* optical frequency. The shift in the longitudinal optical modes to higher frequencies, discussed for polar semiconductors in Section 3-3-C, will be calculated in Chapter 11.

Acoustical vibrations of long wavelength (compared to the lattice spacing) can of course be calculated from the elastic constants. And, of course, it is possible to calculate the entire vibration spectrum using the forces arising from the total energy, Eq. (10-17) plus the overlap repulsion, just as we did for covalent solids in Section 3-3. This can be done either with the full expression for the energy or for force-constant models. The understanding of the role of electronic structure is perhaps more clearly seen from the limiting cases such as very long wavelengths, or possibly modes with special wavenumber such as $2\pi/a$ [100] at the Brillouin Zone face for rock-salt or zincblende structures. The interpolation between these limits will generally be quite smooth.

5. Coulomb Contributions

It was reasonable to ignore shifts in the term values due to charge redistribution in much of our analysis because the shift $U\delta Z$ in a term value from adding δZ electrons to the same atom was approximately equal and opposite to the Madelung shift $-\delta Z\alpha e^2/d$ from the charges on the neighboring atoms, as we indicated in Chapter 9. However, when d is changed, or the lattice is deformed, the changes are different and we must look into the effects. Majewski and Vogl (1987) did systematically include such Coulomb effects, as well as introducing a more intricate tight-binding representation, and found that it eliminated the general errors of a factor of two which we have found, as well as describing the relative stability of structures. It was not clear immediately whether the improvement came principally from the inclusion of Coulomb effects or the additional parameters which were introduced. Straub and Harrison (1989) then redid the simple theory we have described here with only the inclusion of Coulomb effects to find that they alone corrected most of the error. We give here a brief account of that work.

The $-2\sqrt{V_2^2 + V_3^2}$ in our covalent energy came from Eq. (9-10), which may be derived by minimizing the energy $u_s^2\varepsilon_s(M) - 2u_su_pV_2 + u_p^2\varepsilon_p(X)$ of the occupied state (at the special point $\boldsymbol{k}^*$) with respect to the coefficients u_s and u_p of the Bloch sums based upon s-states and upon p-states, assuming $\varepsilon_s(M)$ and $\varepsilon_p(X)$ were fixed. This is following the procedure used in Sections 1-4 and 2-2-C. We now need to add contributions to the energy from the Coulomb interactions. We first follow Straub and Harrison (1989) in introducing a single variable S^* for the state, such that $u_s^2 = (1 - S^*)/2$ and $u_p^2 = (1 + S^*)/2$ which makes the normalization explicit. After normalization S^* will take a value analogous to polarity in covalent solids. Note that the occupation of the s-state, from electrons of both spins, is $1 - S^*$, so the effective charge of the alkali ion in an alkali halide is $Z^* = S^*$, equal to one if the electrons were entirely on the halogen. In terms of S^* the energy in the absence of Coulomb effects which was minimized to obtain Eq. (9-9), and then (9-10), is an electronic energy per ion pair, relative to one electron on each atom, given by

$$E_{elec.} = -(\varepsilon_s(M) - \varepsilon_p(X))S^* - 2V_2\sqrt{1 - S^{*2}} . \tag{10-25}$$

Here we have made use of the fact that $2u_su_pV_2$ is equal to $V_2\sqrt{1 - S^{*2}}$.

We now add Coulomb energies again following Straub and Harrison(1989). For an atom with Z^* equal to the number of electrons on the atom minus the number on the neutral atom, we write the electronic energy of the atom as a Taylor expansion,

$$E(Z) = E(0) + E'Z^* + {}^1\!/_2\, E''Z^{*\,2} + \dots \tag{10-26}$$

For free atoms Z^* will be an integer, but it will be continuous in the solid, equal to S^* for an alkali ion and $-S^*$ for a halogen ion in alkali halides. We wish to identify the parameters E' and E'' with free-atom parameters, dropping terms in Eq. (10-26) of higher order than $Z^{*\,2}$. We identify the free-atom term value as the energy $\varepsilon = E(0) - E(-1) = E' - E''/2$ (the negative of the energy required to take an electron from the neutral atom and carry it to zero energy at infinite distance). The Coulomb energy U is defined (and listed in Table 1-1) as the difference between the ionization energy and the electron affinity, $U = -\varepsilon - E(0) + E(1)$. Solving for E' and E'' and substituting in Eq. (10-26) we obtain

$$E(Z) - E(0) = (\varepsilon + {}^1\!/_2 U)\, Z^* + \frac{UZ^{*\,2}}{2}\,. \qquad (10\text{-}27)$$

This gives the energy of an atom, relative to the neutral atom. Since we have dropped higher-order terms, we would obtain the second ionization energy as $E(-2) - E(-1) = -\varepsilon + U$, larger than the first ionization energy by U, in reasonable accord with the experimental values obtained from Weast (1975). Higher-order terms would be required to fit a larger set of ionization energies for any one atom

A. Volume-Dependent Properties

Straub and Harrison added the Coulomb energy, Eq. (10-27), for each atom in the compound to the electronic energy, Eq. (10-25). They then added the Madelung energy, $-\alpha S^{*2}e^2/d$ per ion pair for the alkali halides, with $\alpha = 1.75$. Finally, they added the overlap repulsion given in Eq. (10-10) to obtain the energy per ion pair, relative to neutral atoms, for alkali halides as

$$E_{sp\ bond} = -\,S^*(\varepsilon_s - \varepsilon_p) - 2V_2\sqrt{1-S^{*2}} + {}^1\!/_2 S^*(U^p - U^s) + {}^1\!/_2 S^{*2}(U^p + U^s) - \alpha e^2 S^{*2}/d + E_{overlap}\,. \qquad (10\text{-}28)$$

[Note that for neutral atoms this is $-2V_2 + E_{overlap}$, with the first term arising because the neutrality is accomplished by filling two nonpolar bond states, but this is not the state of minimum energy unless $\varepsilon_s = \varepsilon_p$.] The indices s and p indicate the alkali and the halide atoms, respectively. For divalent crystals they obtained

$$E_{sp\ bond} = -\,(S^*+1)(\varepsilon_s - \varepsilon_p) - 2V_2\sqrt{1-S^{*2}} + {}^1\!/_2(S^*+1)(U^p - U^s) + {}^1\!/_2(S^*+1)^2(U^p + U^s) - \alpha e^2(S^*+1)^2/d + E_{overlap}\,. \qquad (10\text{-}29)$$

Straub and Harrison used essentially the same parameters as given in Table 1-1, and minimized these equations numerically with respect to S*, to predict equilibrium spacings, cohesion, bulk modulus, and the Grüneisen constant. In general they found that where the predictions without Coulomb corrections were reasonably accurate, the addition of corrections left the predictions with comparable accuracy, and for the bulk modulus the discrepancies of something like a factor of two were largely eliminated. We listed the results from Straub and Harrison for the spacings in Tables 10-1 and 10-2. Comparison in those tables indicates that frequently the earlier error in the estimate of the spacing was reversed by Coulomb corrections. Results for Grüneisen constants in Table 10-4 indicate that the surprising accuracy without Coulomb corrections was lost with them. The results obtained by Straub and Harrison for cohesion were rather close to the values which we listed in Tables 9-5 and 9-6 with a correction δU. The improvement in the results from inclusion of these terms was significant, though the results without them were already surprisingly good. The real improvement came with the bulk modulus, and we listed the values from Straub and Harrison in Tables 10-3. We consider them in the context of the full elastic constants, which Straub and Harrison treated in detail.

B. Elastic Constants

For the shear constants Straub and Harrison (1989) used the individual sums over neighbor as we did in Eq. (10-17) for the covalent contribution. The Coulomb energies for the individual atoms were obtained as in Eq. (10-27) and the interatomic Coulomb energy, which is the Madelung energy given by $-\alpha Z^{*2}e^2/d$ for the undistorted crystal, now depends upon shear, as we indicated in Section 10-4-A. It is possible to write electrostatic contributions to each elastic constant as $\beta Z^{*2}e^2/d^4$ and Straub and Harrison listed values β for the corresponding constants, obtained from Wallace (1972). Then numerical calculation of all of the elastic constants, as well as the pressure derivatives were direct. The results for the elastic constants are given in Tables 10-8 and 10-9.

We see that indeed the that the predictions are considerably improved, as they were for the bulk modulus, by the inclusion of Coulomb corrections. We still overestimate the constants for the compounds of fluorine and oxygen, which Straub and Harrison ascribed to the overlap repulsion proportional to $1/d^8$ being too steep. Use of a $1/d^6$ repulsion for these systems removed much of the discrepancy. It is interesting to compare again c_{12} to c_{44}, which the Cauchy relations give as equal. We uniformly predict c_{44} to be less than c_{12}, while the experiment has them more nearly equal. We saw that the angular force is predicted by Eq. (10-19) to increase c_{44} relative to c_{12}, so it corrects some of this error. The experimental constants agree as well with the Cauchy relations as they do partly because

Table 10-8 Elastic constants for the alkali halides in the rock-salt structure, obtained by Straub and Harrison (1989) including Coulomb corrections, and experimental values in parentheses also compiled by Straub and Harrison. Values are in eV/A^3 . Multiplying by 16.02 gives values in units of 10^{11} ergs/cm^3.

	c_{11}	c_{12}	c_{44}
NaF	0.821 (0.605)	0.204 (0.152	0.115 (0.175)
NaCl	0.366 (0.303)	0.117 (0.078)	0.079 (0.079)
NaBr	0.280 (0.248)	0.096 (0.066)	0.064 (0.062)
NaI	0.198 (0.189)	0.076 (0.056)	0.053 (0.046)
KF	0.598 (0.410)	0.136 (0.091)	0.079 (0.078)
KCl	0.267 (0.253)	0.072 (0.041)	0.049 (0.051)
KBr	0.207 (0.216)	0.058 (0.035)	0.039 (0.032)
KI	0.150 (0.172)	0.045 (0.028)	0.032 (0.023)
RbF	0.542 (0.345)	0.122 (0.087)	0.072 (0.058)
RbCl	0.242 (0.222)	0.063 (0.037)	0.044 (0.029)
RbBr	0.187 (0.196)	0.050 (0.030)	0.035 (0.024)
RbI	0.136 (0.160)	0.039 (0.022)	0.028 (0.017)

of the two canceling corrections.

We listed in Table 10-4, for a few compounds, the volume dependence of the bulk modulus, calculated without Coulomb corrections and the results from Straub and Harrison with Coulomb corrections. For this case, the

Table 10-9 Elastic constants for the divalent compounds in the rock-salt structure, obtained by Straub and Harrison (1989) including Coulomb corrections, and experimental values in parentheses also compiled by Straub and Harrison. Values are in ev/Å^3. Multiplying by 16.02 gives values in units of 10^{11} ergs/cm^3.

	c_{11}	c_{12}	c_{44}
MgO	2.843 (1.91)	0.905 (0.59)	0.824 (0.98)
MgS	1.436	0.659	0.621
MgSe	1.075	0.507	0.496
MgTe	0.776	0.412	0.412
CaO	2.393 (1.39)	0.667 (0.37)	0.606 (0.51)
CaS	1.263	0.452	0.421
CaSe	0.969	0.348	0.334
CaTe	0.730	0.282	0.274
SrO	2.303 (1.08)	0.636 (0.28)	0.574 (0.35)
SrS	1.206	0.413	0.383
SrSe	0.937	0.322	0.305
SrTe	0.709	0.259	0.248
BaO	2.148 (0.70)	0.576 (0.22)	0.520 (0.21)
BaS	1.118	0.363	0.337
BaSe	0.868	0.281	0.266
BaTe	0.658	0.224	0.214

corrections worsened the agreement with experiment, and the same was found by Straub and Harrison for the volume dependence of the other elastic constants. It has turned out to be principally for the elastic constants themselves that the Coulomb corrections make major improvements.

Problem 10-1. Optical vibrational modes

The structure of fluorite, CaF_2, was described in Problem 9-1. Consider a vibrational mode in which the calcium ions are displaced u_+ and both fluorine ions in each cell are displaced by u_- in the opposite direction. Following the analysis of Section 10-4-C,

a) rewrite the elastic energy per molecular unit (3 ions) in terms of the reduced mass, $M_r = 2M_+M_- /(M_+ + 2M_-)$.

b) Similarly, calculate the change in covalent energy for the eight bonds around one calcium ion, in analogy with Eq. (10-21) and the change in overlap energy in analogy with Eq. (10-22).

c) Finally, obtain a formula for the frequency ω_{TO} in analogy with Eq. (10-24), and evaluate it numerically for CaF_2.

CHAPTER 11

Dielectric Properties of Ionic Crystals

In discussing the bonding properties of ionic solids we have focused upon the total energy of the crystal and how that total energy depends upon the positions of the constituent ions. In discussing the dielectric properties we must consider how the energy varies under the influence of applied electric fields, as we did for semiconductors in Chapter 4. This brings up the question of effective charges and effective dipoles since a dipole can be defined as the shift in energy of a system due to an applied field, divided by that applied field. We have addressed these same questions for semiconductors in terms of polarizable bonds. We now address them in terms of a crystal of ions. In this case we find that the approach we have been using becomes quite inaccurate for compounds of the heavier metals and see how we can rectify this difficulty.

1. The Optical Dielectric Susceptibility

At optical frequencies, well above the highest vibrational frequencies of a crystal, the displacement of the ions under the influence of the optical electric field is negligible. The response of the system is entirely in the variation of the electronic states, as it was in the semiconductor. However, in this case the occupied electronic states are represented to a first approximation as the p-states on the nonmetallic (halogen or chalcogen) ion, weakly coupled to the s-states on the metallic (alkali or divalent metal) atom. In this context, the electronic states are affected by an electric field only in the change in admixture of the metallic states. This is in contrast to the traditional view of the dielectric properties of ionic crystals which, quite naturally, viewed the effect of an electric field as in causing a polarization of the individual constituent ions. Such polarization would be described quantum-mechanically as an admixture of the higher-energy, empty (peripheral s-states in the language of Section 5-3-B) on the nonmetallic ion, or as a distortion of the core states on the metallic ion. Neither of these are included in our tight-binding description based upon valence states alone.

This does not mean that either of the descriptions of dielectric properties is necessarily incorrect, any more than either of the two alternative descriptions of cohesion of ionic crystals - the tight-binding description or the Born Electrostatic Model of Section 9-1-E - was wrong. The empty states on the nonmetallic ion occupy much the same space as empty states on the metallic ion and it is possible that the shift in charge can be described reasonably well in terms of either. The advantage of the tight-binding approach to the dielectric properties, like the advantage of this approach to bonding properties, is its simplicity and its generality. On the other hand, we shall see in Table 11-1 that when the halogen-ion spacings become large our approach becomes very inaccurate, while the polarizable-ion approach remains valid.

The polarizable ion model was thought to have been confirmed experimentally by fitting the observed dielectric constants of the sixteen compounds Li, Na, K, and Rb with F, Cl, Br, and I with adjusted polarizabilities of the eight ions. This worked to a few percent because there are systematic variations, as we shall see, and eight parameters are enough to fit the sixteen experiments to a few percents. Similarly the divalent comments could be well fit. The real test came in applying these ionic polarizabilities to a mixed compound such as $CaCl_2$ for which the model failed badly.

A simple way to obtain the susceptibility in our tight-binding context is to compute the shift in energy of the electronic states due to an applied electric field and then equate the corresponding change in energy density to $-\frac{1}{2}\chi^{(1)}E^2$, where $\chi^{(1)}$ is the susceptibility and E is the electric field, (The sign arises because a positive $\chi^{(1)}$ corresponds to a polarization parallel to the field, a lowering in energy.) We focus in particular upon the central halogen ion in Fig. 9-4, redrawn as Fig. 11-1. We add a term to the Hamiltonian due to the electric field, $-(-e)Ex$, where x is measured relative the nucleus of that ion, this being the effect of a field E in the x-direction. Measuring x from a

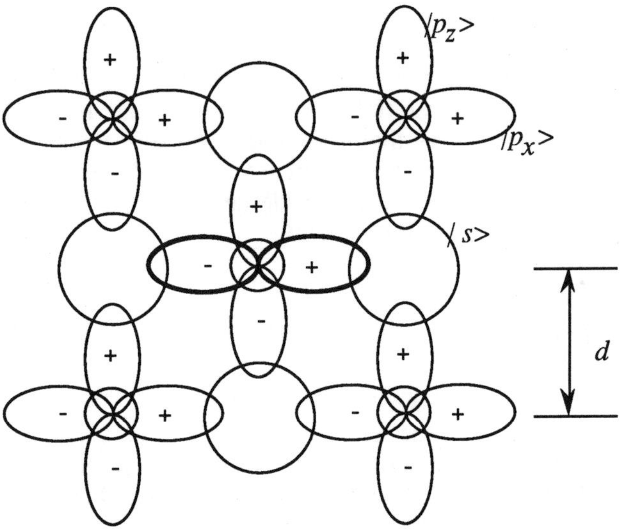

this

Figure 11-1. Redrawn from Fig. 9-4. The central figure is a halogen ion in a rock-salt structure. Its electronic p-states, $/p_x>$ (shown bold) and $/p_z>$ overlap with, and are coupled to, s-states on the four neighboring alkali atoms shown. Four other halogen atoms are also shown. A third p-state, $/p_y>$, on the central atom, is oriented perpendicular to the first two and is coupled to alkali s-states on atoms above and below the plane of the figure.

different point corresponds to adding a constant potential to the system, which has no effect in the energy of the full neutral crystal. If the central halogen were an isolated ion, such a field would lower the energy on the left of the ion and raise it on the right and (because x is measured from the nucleus) the average shift would be zero. However, the alkali s-state on the right is raised in energy by eEd relative to the central halogen p-states and the s-state on the left is lowered by $-eEd$ and this affects the covalent contribution to the energy which we found in second-order perturbation theory as $V_{sp\sigma}^2/(\varepsilon_p(X)-\varepsilon_s(M))$ in Eq. (9-6). For the shift of the central halogen p-state due to interaction with the s-state to the right, the energy denominator becomes $\varepsilon_p(X)-\varepsilon_s(M)-eEd$. The shift can be evaluated by expanded for small E to become

$$\delta\varepsilon_p(X) = \frac{V_{sp\sigma}^2}{\varepsilon_p(X)-\varepsilon_s(M)} + \frac{V_{sp\sigma}^2 eEd}{(\varepsilon_p(X)-\varepsilon_s(M))^2} + \frac{V_{sp\sigma}^2 (eEd)^2}{(\varepsilon_p(X)-\varepsilon_s(M))^3} + \dots \tag{11-1}$$

The corresponding shift due to coupling with the alkali s-state to the left is of the same form but with d replaced by $-d$. The first-order terms (in E) cancel and the second-order terms add. Only these two neighbors have a relative shift in energy due to the field and only this p-state is coupled to the corresponding alkali states. We therefore obtain the shift in energy per unit volume by multiplying the third term in Eq. (11-1) by two for the two neighbors, by two for spin, and by the halide density, $4/(2d)^3$. Setting this shift equal to $-\,^1/_2\,\chi^{(1)}E^2$ gives immediately

$$\chi^{(1)} = \frac{4e^2V_{sp\sigma}^2}{(\varepsilon_s(M) - \varepsilon_p(X))^3 d}\,. \tag{11-2}$$

This is the susceptibility which gives directly the optical dielectric constant $\varepsilon = 1 + 4\pi\chi^{(1)}$, and the refractive index $n = \sqrt{\varepsilon}$, and is therefore readily measured. We could go on to higher-order susceptibilities as we did for covalent solids, and as was done by Ching, Gan, and Huang (1995), but here we consider only the linear susceptibility.

A. The Alkali Halides

Equation (11-2) may be immediately evaluated using the experimental internuclear distances from Table 10-1 and the atomic term values from Table 1-l. The results are compared with experiment in Table 11-1. The prediction has given the correct general magnitudes, at least for the lithium and sodium halides and the principal trends, but very much overestimates the decrease in susceptibility for the heavier alkali halides. The trend with internuclear distance is in fact more nearly given by a proportionality to the density of halogen ions (or alkali-halide pairs) with a proportionality constant depending only upon the particular halogen ions, as would be appropriate to the picture of polarizable individual ions with that of the halogen being very much larger. This yields a susceptibility of the MX alkali halide of

$$\chi^{(1)}_{MX} = (d_{LiX}/d_{MX})^3\,\chi^{(1)}_{LiX}\,, \tag{11-3}$$

shown as the second entry in Table 1-1. The results are in much better accord with experiment.

Perhaps this suggests that the intra-ionic and inter-ionic viewpoints are both applicable if the alkali metal is light (Li or Na), but not for the halides of the heavier alkalis. Then using the inter-ionic viewpoint allows us to estimate the susceptibility of the lithium halides, knowing only the atom spacing, without an empirical halogen polarizability. We then represent the

Fig. 11-1. Optical dielectric susceptibilities of the alkali halides based upon Eq. (11-2). The second entry is the value from Eq. (11-2) for the corresponding lithium halide, scaled by ion density, $1/d^3$, as in Eq. (11-3). The value in parentheses is experimental, from Wemple (1977).

	F	Cl	Br	I
Li	0.07 0.07 (0.07)	0.10 0.10 (0.14)	0.12 0.12 (0.18)	0.16 0.16 (0.22)
Na	0.01 0.04 (0.06)	0.06 0.08 (0.10)	0.07 0.09 (0.13)	0.09 0.13 (0.17)
K	0.01 0.03 (0.06)	0.02 0.05 (0.10)	0.03 0.07 (0.11)	0.04 0.10 (0.14)
Rb	0.01 0.03 (0.07)	0.01 0.05 (0.10)	0.02 0.06 (0.11)	0.02 0.09 (0.14)

result as giving us the halogen polarizability and use the intra-ionic viewpoint to obtain the susceptibility of the halides of the heavier alkalis using Eq. (11-2), exactly as we have done for the second entry in Table 11-1.

We must nevertheless regard this as a failure of the approximation we are using which omits higher atomic s-states on the nonmetallic atom. (As we have indicated, these are the principal states needed to describe a polarizable ion.) These excited s-states were also needed in semiconductors to give the true shape of the conduction band in Section 5-3-B. The excited s-states were not important in the bonding properties of ionic crystals since without an applied field they are coupled only to the states on the neighboring metallic ions, which are also empty. For the same reason they will not be so important in the effective charges we estimate in the next section. However, in the susceptibility where the electric field couples them directly to the filled p-states on the halogen ion, they become significant and the effect is not swamped by a large interatomic-bond polarizability as in the semiconductors.

In a similar way we may expect the susceptibility under pressure to vary as the ion density, proportional to $1/d^3$, rather than the $1/d^5$ from Eq. (11-2). In fact experimental dependences range between $1/d^2$ and $1/d^3$ according to Bendow, Gianino, Tsay, and Mitra (1974).

B. Noble Metal Halides

Some of the noble metal halides also form in the rock-salt structure (they mostly form tetrahedral semiconductors). The electronic structure contains again fully occupied halogen p-states and empty s-states on the metal. The difference is that the noble-metal atoms have full d-states which are at similar energy to the halogen p-states so that the valence bands consist of both. These d-states, however, do not directly add to the susceptibility in our theory since they would be coupled only to halogen states which are also fully occupied so there are no additional covalent terms. Further, the applied field does not couple the occupied d-states to the empty s-states on the same atom as it did the occupied p-states to the empty s*-states on the halogen. The only coupling which enters the susceptibility is that between a full and an empty state. Thus Eq. (11-2) is directly applicable to the noble-metal halides in the rock-salt structure using again the term values from Table 1-1. In Table 11-2 we have evaluations for the two cases for which we had values for the experimental susceptibility. We see that the predicted susceptibilities are comparable to the lithium halides, as are the experimental values (from Wemple (1977)) and the agreement is rather good, as it was for the lithium halides.

Table 11-2. Optical dielectric susceptibilities $\chi^{(1)}$ of the noble metal halides in the rock-salt structure from Eq. 11-2.

Compound	$d(\text{Å})$	Theoretical	Experimental Wemple (1977)
AgCl	2.77	0.14	0.24
AgBr	2.89	0.22	0.29

C. Divalent Compounds

Eq. (11-2) for the susceptibility is directly applicable also to divalent compounds in the rock-salt structure. In these compounds we have seen that the band gaps are predicted, and found, to be smaller and the internuclear distances are decreased. Both effects increase the susceptibility and Eq. (11-2) predicts a large increase in going for example from LiF to MgO, which is isoelectronic with it. Such an increase is seen in the experimental values for the divalent compounds in Table 11-3, and the prediction is reasonably good for MgO. (A recent study of MgO has been made by Schönberger and Aryasetiawan (1995).) However, Eq. (11-2) overestimates the decrease with increasing size of the metallic ion, as it did for the alkali halides. Again the susceptibilities lower in the table are more accurately described as free-ion polarizabilities, using the counterpart of Eq. (11-3), now $\chi^{(1)}{}_{MX} = (d_{MgX}/d_{MX})^3\, \chi^{(1)}{}_{MgX}$, which gives the values shown as the second entry in Table 11-3.

D. Ten-Electron Compounds

Essentially the same theory can be written down also for compounds in which the metallic atom retains two electrons in its valence s-shell. This occurs mostly in compounds of the very heavy metals, presumably because the relativistic increase in the mass of an electron when it is near the nucleus of these atoms of high atomic number. This lowers the energy of the valence s-electrons, which do penetrate the core, and the states remain occupied in the compound like core states. Lead telluride is such a compound in which the lead gives its two valence p-electrons to the tellurium to fill the tellurium valence p-shell. Lead is said to be divalent in this compound, and the total number of valence electrons present is ten. These are called IV-VI compounds, referring to the columns in the periodic table from which the constituents come. Tin telluride and lead sulfide are also compounds of this type. They all occur in the rock-salt structure.

Table 11-3. Optical susceptibility of the divalent ionic compounds in the rock-salt structure. The first entry is from Eq. (11-2), the second is based upon a generalization of Eq. (11-3) which scales the predictions for the magnesium compounds by $1/d^3$. The values in parentheses are experimental, from Weast (1975) and Wemple (1977).

	0	S	Se	Te
Mg	0.17 0.17 (0.16)	0.54 0.54 (0.33)	0.83 0.83 (0.39)	1.50 1.50 -
Ca	0.06 0.11 (0.18)	0.14 0.41 (0.28)	0.19 0.65 (0.33)	0.28 1.25 (0.42)
Sr	0.04 0.09 (0.20)	0.09 0.32 (0.27)	0.12 0.56 (0.31)	0.18 1.18 (0.38)
Ba	0.02 0.07 (0.23)	0.05 0.29 (0.29)	0.07 0.47 (0.33)	0.09 0.94 (.39)

The electronic structure in such compounds consists of filled s- and p-shells on the nonmetallic ion and filled s-shells on the metallic ion. Of the set of orbitals we include in the theory, only the metallic p-states are empty and the deformation of the electronic system will arise from transfer of electrons from the nonmetallic p-states to the metallic p-states. This leads to exactly Eq. (11-2), with $V_{sp\sigma}^2$ replaced by $V_{pp\sigma}^2 + 2V_{pp\pi}^2$ and $\varepsilon_s(M) - \varepsilon_p(X)$ replaced by $\varepsilon_p(M) - \varepsilon_p(X)$. The replacement of $V_{sp\sigma}^2$ by the appropriate combination of p-state couplings scales Eq. (11-2) up by a factor of 2.83. Then taking the energies $\varepsilon_p(M)$ and $\varepsilon_p(X)$ for the energy denominator from Table 1-1 leads to the values given in Table 11-4. These values are much larger than for the other ionic compounds, largely because of the much smaller energy denominator. In fact the gaps in these materials are small enough that they can be made to conduct and may be regarded as ionic semiconductors. Indeed, the agreement of the susceptibility with experiment is similar to that which we obtained for the tetrahedral semiconductors. Because these predictions are large, we expect that the intra-atomic coupling of the occupied s-state on the metallic ion to the empty p-state on the metallic ion to be relatively unimportant.

Another class of ten-electron compounds are the thallous halides, III-VII compounds in which the thallium behaves as monovalent and retains two electrons in its s-shell. These compounds occur in the cesium chloride structure of Fig. 9-2b. It is necessary to rederive the susceptibility for this structure. For cesium chloride itself we obtain exactly the formula Eq. (11-

Table 11-4. Optical susceptibility of ten-electron compounds, based upon Eq. (11-2) with $V_{sp\sigma}^2$ replaced by $V_{pp\sigma}^2 + 2V_{pp\pi}^2$ for those in the rock-salt structure and with an additional factor of $\sqrt{3}$ for those in the cesium chloride structure. In all cases the energy denominator is replaced by the difference in p-state energies. Experimental values are from Wemple (1976).

Compound	$d(\text{Å})$	$\chi^{(1)}$(Theoretical)	$\chi^{(1)}$(Experimental)
	In the rock-salt structure		
PbS	2.98	0.62	1.14
PbTe	3.23	2.00	2.45
SnTe	3.16	2.83	3.66
	In the cesium chloride structure		
TlCl	3.32	0.13	0.27
TlBr	3.44	0.18	0.33
TlI	3.64	0.27	0.41

2) multiplied by $\sqrt{3}$. For the thallous halides we obtain the same formula as for PbTe, but with this additional factor of $\sqrt{3}$. It is evaluated using term values from Table 1-1 to give susceptibilities which are compared with experiment in Table 11-4. The predictions are very much smaller than those for the lead compounds, as are the experimental values, and the agreement is again comparable to that for the tetrahedral semiconductors.

E. The Implications of the Errors

We see that the success of the simple theory of the susceptibility based upon only the valence orbitals and transfer of charge between atoms is comparable to that in the tetrahedral semiconductors when the calculated contribution is large enough to make the susceptibility similar in magnitude to that of a semiconductor. However, when the spacing between ions becomes large enough that the coupling becomes small, the polarizable-ion picture retains its validity (as it must when the ions become sufficiently widely spaced) and the interatomic theory greatly underestimates the susceptibility. This has clearly become a serious problem in the heavier alkali halides and alkaline earth chalcogenides. A way out, which avoided introducing empirical polarizabilities, was to apply the theory to the lithium compounds, to reinterpret the result as arising from polarizable halogen or chalcogen ions, and to utilize these for estimating the susceptibilities for the compounds of the heavier metals.

This difficulty with the susceptibility will not disrupt the theory of other properties such as the effective charges. It is a difficulty which arises most directly from our neglect of the coupling between states on the same atom from the applied electric field.

F. Local-Field Effects

In all of our discussion of susceptibilities we have ignored any effects of the local charge redistribution on the fields seen by an individual ion. Such "local-field" effects are easier to understand in terms of a derivation based upon dipole formation than they are in terms of the derivation based upon energy which led to Eq. (11-2). We therefore make that derivation here.

In Section 2-2-C we found the states resulting from two levels, differing in energy by $2V_3$ and coupled by V_2. For the lower-energy (bond) state, the squared coefficient of the higher level was found to be (Eq. (2-11))

$$u_{+}^{b2} = \frac{1 - \alpha_p}{2} = \frac{\sqrt{V_2^2 + V_3^2} - V_3}{2\sqrt{V_2^2 + V_3^2}} \approx \frac{V_2^2}{(2V_3)^2} \,. \tag{11-4}$$

In the second step we substituted from Eq. (2-12) $\alpha_p = V_3 / \sqrt{V_2^2 + V_3^2}$ and in the third step we expanded for small coupling V_2. This gives the probability density transferred to the second state by the coupling V_2 and is the counterpart for probability densities of the second-order perturbation theory for the energy, Eq. (1-30). As applied to the shift of charge to a neighboring alkali ion we have $V_2 = V_{sp\sigma}$ and $V_3 = (\varepsilon_s(M) - \varepsilon_p(X))/2$ and multiplying by two for spin we have the charge transferred from the halogen ion to each alkali nearest neighbor given by

$$\delta Z^* = \frac{2V_{sp\sigma}^2}{(\varepsilon_s(M) - \varepsilon_p(X))^2} \ . \tag{11-5}$$

This result will be useful when we discuss effective charges in the following section.

Now if the field E_x raises the energy of the alkali s-state to the right by eE_xd, it causes a change in the charge transferred of $-e\Delta\delta Z^* = 4e^2Ed/(\varepsilon_s(M) - \varepsilon_p(X))^3$, of first order in the field. We may imagine that charge as transferred from the metal ion to the left, a distance $2d$, to give a dipole of $8e^2Ed^2/(\varepsilon_s(M) - \varepsilon_p(X))^3$. We may multiply by the pair density, $1/(2d^3)$, and divide by field to obtain Eq. (11-2) for the susceptibility. This did not change the charge on the central halogen ion, and as much charge is transferred onto each metal ion by the electric field as is transferred off so the metal ions retain their same charge. Thus the dielectric susceptibility in this picture arises from the transfer of charge through the crystal with no change in the local charge distribution. The charge of the last ion in the crystal, at the surface, does change. This gives rise to a surface charge density equal to $\chi^{(1)}E_x$ as one discusses in the theory of continuous dielectric media. This reduces the field within the material by a factor of the dielectric constant and it is that reduced, but constant, field which we represented by E_x in the calculation of the susceptibility. In a similar way, when we think of the bond polarizability in the semiconductor as the simple transfer of charge from one atom to another, there were no local-field effects. Furthermore, a metal may be thought of as having an infinite static susceptibility. When the metal is placed in a field, charge transfers through it, producing surface charges which reduce the field within the material to zero.

However, in the halides of the heavy alkali metals, where we saw that the crystal is behaving more like independent polarizable ions, each ion produces a dipole field which is seen by each other ion. It is correct to add these fields (and even higher multipole fields) in the evaluation of the susceptibility. The effects can be very significant, particularly in the halides of the heavy alkali metals. They depend upon the size of the ions and the geometry of the crystal. We have not attempted to add these effects here.

G. Application to Other Ionic Solids

The same theory has been applied also in Harrison (1980) to compounds which do not consist of equal numbers of cations and anions with agreement with experiment consistent with that obtained here. Such a calculation for fluorite, CaF_2, is outlined in Problem 11-2. A more recent study of antifluorite compounds has been made by Corkill and Cohen (1993). The theory is directly applicable for any ionic compound for which the crystal structure is known and, if the constituent atoms are not transition metals, the needed parameters appear in Table 1-1. Our feeling is that this interatomic theory will ordinarily be more accurate than an estimate based upon polarizabilities of individual ions, fit to rock-salt-structure compounds, but that depends upon the compound in question and probably will not be true when very heavy ions are involved and the predicted susceptibilities are so small.

In such a case one may return to the polarizable-ion approach, and the rather complete set of susceptibilities in the rock-salt structure in Tables 11-1 and 11-3 provides an experimental basis for such a study.

2. Effective Ionic Charges

We saw in Section 11-1-F that the coupling $V_{sp\sigma}$ transfers a charge, given by Eq. (11-5), from each halogen ion to each neighboring alkali. This transfer modifies the charge on the individual ions from the integral values corresponding to the independent-ion model. We shall first calculate the resulting "static effective charge" Z^* but should recognize that this static charge has only limited physical significance and cannot be directly measured. As we indicated in Sections 2-3-E and 4-1, the transfer of charge between ions is really a subtle rearrangement of charge and that, for example, it makes only a small shift in the energies of the electronic states. We shall then discuss the transverse charge, which can be measured. A discussion of the corresponding effects in tetrahedral semiconductors was given in Section 4-3-A.

A. The Static Effective Charge

We saw in Section 11-1-F how the transfer of probability density from an occupied halogen p-state to an empty alkali s-state can be calculated in perturbation theory; that is, in the approximation that the coupling is very small compared to the energy difference. Without coupling between ions the charge on an alkali atom is $+e$ and would be $+Z\,e$ with $Z = 2$ for the alkaline earths. This is reduced by the transfer δZ^*, Eq. (11-5), for the σ-oriented p-state from each of the six neighboring chlorine ions, giving a total *static effective charge* of

$$Z^* = Z - \frac{12V_{sp\sigma}^2}{(\varepsilon_s(M) - \varepsilon_p(X))^2} \tag{11-6}$$

in the rock-salt structure. For the cesium chloride structure with eight neighbors the result is $Z^* = Z - 16V_{sp\sigma}^2/(\varepsilon_s(M) - \varepsilon_p(X))^2$.

For the IV-VI ten-electron compounds such as PbS, the charge in the absence of any coupling is two. The metallic s-state on the lead is occupied so coupling it with the occupied p-states on the sulfur does not transfer charge. The only charge transfer comes from the coupling between p-states on the two atoms. (We again neglect the s-state on the nonmetallic atom.) Consider again the geometry of Fig. 11-1, imagining a lead atom in the center, but with the s-states on the neighbors replaced by sulfur p-states. The lead p_x-state receives $2V_{pp\sigma}^2/(\varepsilon_p(M) - \varepsilon_p(X))^2$ of an electron (including both spins) from the p_x-state on the right, an equal contribution from the left and four contributions of $2V_{pp\pi}^2/(\varepsilon_p(M) - \varepsilon_p(X))^2$ from each of the four sulfur atoms oriented laterally (above, below, and above and below the plane of the figure). We need multiply by three for the three orientations of lead p-state to obtain the contribution to the static ionic charge on the lead,

$$Z^* = 2 - \frac{12(V_{pp\sigma}^2 + 2V_{pp\pi}^2)}{(\varepsilon_p(M) - \varepsilon_p(X))^2} \tag{11-7}$$

applicable to the IV-VI compounds in the rock-salt structure. When we evaluate this expression we shall find that the second term is as large or larger in magnitude than the first, indicating that the perturbation theory leading to this equation is inadequate. A more appropriate approach can be based upon the special-points method, as indicated in Problem 11-1, but we shall proceed with the simpler form, Eq. (11-7), here. We shall see that this overshoot in the transfer term does not occur for the transverse effective charge. The thallous halides occur in the cesium-chloride structure and the corresponding analysis gives $Z^* = 1 - 16(V_{pps}^2 + 2V_{pp\pi}^2)/(\varepsilon_p(M) - \varepsilon_p(X))^2$.

These have been evaluated for a large number of ionic crystals in Table 11-5. We note that Z^* is very considerably reduced from the value Z corresponding to the independent-ion model. One might at first think that this should reduce the electrostatic Madelung energy entering the Born

Table 11-5. Effective ionic charges and transverse charges for ionic crystals. Only the transverse charges e_T*, obtained from Eqs. (11-9) through (11-12), are measurable. Experimental values from the splitting of the longitudinal and transverse optical modes are from Lucovsky, Martin, and Burstein (1971).

	Z	Z*	e_T*	e_T* (exp)
		Rock-salt Structure		
Alkali Halide				
LiF	1	0.59	1.14	0.85
LiCl	1	0.55	1.15	1.23
LiBr	1	0.51	1.16	1.28
NaF	1	0.78	1.07	1.03
NaCl	1	0.72	1.09	1.11
NaBr	1	0.69	1.10	1.13
NaI	1	0.65	1.12	1.25
KF	1	0.89	1.04	1.17
KCl	1	0.85	1.05	1.13
KBr	1	0.83	1.06	1.13
KI	1	0.81	1.06	1.17
RbF	1	0.91	1.03	1.28
RbCl	1	0.88	0.04	1.16
RbBr	1	0.87	1.04	1.15
RbI	1	0.85	1.05	1.17
Noble-Metal Halide				
AgCl	1	0.47	1.18	1.37
AgBr	1	0.30	1.23	1.50
Monoxide				
MgO	2	1.25	3.35	1.77
CaO	2	1.68	2.11	1.96
SrO	2	1.77	2.08	2.11
Ten-electron compound				
PbS	2	0.03	2.66	4.8
PbTe	2	-2.04	3.35	6.5
SnTe	2	-3.17	3.72	8.1
		Cesium-Chloride Structure		
Alkali Halide				
CsCl	1	0.89	1.04	1.31
CsBr	1	0.88	1.04	1.30
CsI	1	0.87	1.04	1.31
Ten-electron compound				
TlCl	1	0.58	1.20	1.96
TlBr	1	0.48	1.24	2.06

theory of cohesion. However, as we indicated, the ions were produced at very wide spacing and then brought together to form the crystal. During most of this process the ions retain their full integral charge. It is only at the last moment when the ions come into contact that the charge redistributes leading to the charge Z^*. Almost all of the energy gain corresponds to work done with forces for which Z is appropriate and that is the charge which should enter the cohesion if one is proceeding in that mode.

The effective charge Z^* on the other hand, may be appropriate if one wishes to estimate potentials in the crystal. For example, the electrostatic corrections to the band gap which we made in Section 9-1-C could reasonably be scaled down by a factor of Z^*/Z. Those corrections were not found to be quantitatively very interesting in any case.

These values of Z^* may also be useful for thinking about these systems, as they were for the semiconductors, but they can be different in different formulations of the problem, and there is therefore no unique means of measuring them. One might imagine obtaining the true electron density distribution in the crystal and then dividing it into individual atomic cells to obtain effective charges, but of course the result depends upon where one constructs the cell boundaries; the same ambiguity arises in any other way of assigning effective charges. We may note in passing that the effective charges Z^* decrease with pressure in all cases because of the d -dependence of the interatomic-coupling energies. Of more interest are the measurable charges such as the transverse charge e_T^* which we discussed for polar semiconductors in Section 4-3-A.

B. The Transverse Charge

The transverse charge is defined in terms of the electric dipole arising from the displacement of one ion or atom. If one evaluates that dipole, either theoretically or experimentally, for a certain displacement $\boldsymbol{u}$, then the transverse charge e_T^* is defined by setting that dipole equal to $ee_T^*\boldsymbol{u}$, just as it was in Section 4-3-A. In systems with cubic symmetry such as rock salt the magnitude of e_T^* does not depend upon the direction of displacement and a single value obtains for the metallic ions, with the nonmetallic ion having a transverse charge of the same magnitude and opposite sign (since displacement of the entire lattice cannot give any dipole). In systems such as fluorite, CaF_2, the magnitude of the e_T^* for the fluorine must be half that of the calcium We noted also in Section 4-3-A that this charge could be directly measured in terms of the differences in the longitudinal and transverse optical-mode vibrational frequencies.

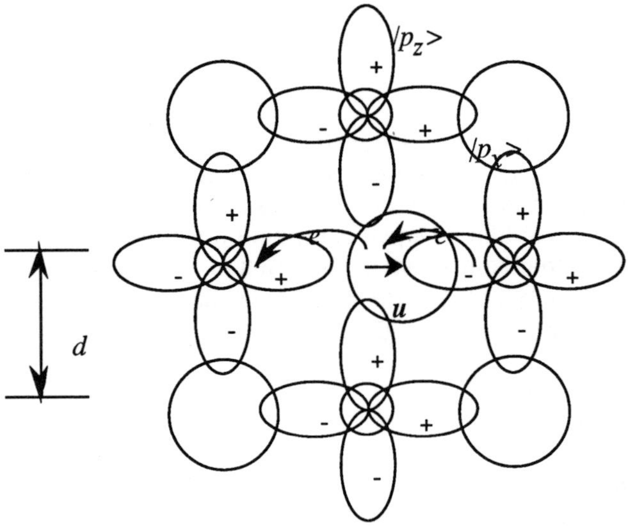

Fig. 11-2. When an alkali ion, in the center of the figure, is displaced it produces an electric dipole due to its static effective charge Z^*. In addition, the increased coupling to the halogen ion on the right increases charge transferred from it to the central ion, indicated by the curved arrow. Similarly, the decreased coupling to the halogen ion on the left transfers an equal charge to that ion. Thus the displacement causes a transfer between the two neighboring halogen ions which enhances the dipole, contributing to the transverse charge e_T^* an amount equal to $2d$ times the transfer.

If the charges Z^*e which we discussed in the preceding section moved rigidly with the ion as it was displaced, the transverse charge would simply equal Z^*. However, displacing an alkali ion, as indicated in Fig. 11-2, shortens the distance to the halogen ion to its right, thereby increasing the coupling, and increases the charge transfer δZ^* of Eq. (11-5). This additional charge transfer adds to the dipole and must be included in the calculation. It is sometimes called a "dynamic" contribution to the transverse charge but of course it is present also in static displacements so we prefer to call it a "transfer" contribution.

The transfer contribution is particularly simple for eight-electron compounds, such as the alkali halides, in the rock-salt structure. For a displacement $\boldsymbol{u}$ to the right, as in Fig. 11-2, the coupling between the central s-state and the p-state to the right changes by $-u\partial V_{sp\sigma}/\partial d = 2uV_{sp\sigma}/d$. This changes the fraction of an electron transferred to the alkali due to this coupling by

$$-u\frac{\partial \delta Z^*}{\partial d} = \frac{8V_{sp\sigma}^2}{(\varepsilon_s(M) - \varepsilon_p(X))^2 d}\, u\,. \tag{11-8}$$

The dipole induced by this transfer is obtained by multiplying by the distance transferred, $-d$, and the electronic charge $-e$. The electronic charge has been transferred to the left so this gives a dipole to the right, adding to the direct term. We add an equal contribution for the effect of transfer to the ion to the left to obtain a total contribution of $16eV_{sp\sigma}^2u/(\varepsilon_s(M) - \varepsilon_p(X))^2$ to the dipole. This is added to the direct contribution, Z^*eu using Eq. (11-6), to obtain the dipole ee_T^*u with

$$e_T^* = Z + \frac{4V_{sp\sigma}^2}{(\varepsilon_s(M) - \varepsilon_p(X))^2} \tag{11-9}$$

for alkali halides in the rock-salt structure. Note that it is always greater than the independent-ion charge Z, in contrast to the static effective charge Z^*. This is in agreement with the experimental observations, as seen in Table 11-5 where values from Eq. (11-9) are listed for the alkali halides along with experimental values. The same formula applies for the divalent compounds, with $Z = 2$ in Eq. (11-9). We list values for the monoxides in the rock-salt structure, for which we had experimental values, also in Table 11-5.

In the cesium-chloride structure, all eight nearest neighbors enter, rather than only two, but the component of displacement along the internuclear distance is reduced by a factor $1/\sqrt{3}$, as is the component of the dipole along the displacement. Thus the transfer contribution is increased by a factor 4/3 and the static charge, scaling with the number of neighbors, is increased by a factor 8/6. The resulting transverse charge is

$$e_T^* = Z + \frac{16V_{sp\sigma}^2}{3(\varepsilon_s(M) - \varepsilon_p(X))^2} \tag{11-10}$$

for cesium-chloride-structure compounds.

As for the static effective charge, the results for the ten-electron compounds are modified by the replacement of $V_{sp\sigma}^2$ by $V_{pp\sigma}^2 + 2V_{pp\pi}^2$ and the replacement of $\varepsilon_s(M) - \varepsilon_p(X)$ by $\varepsilon_p(M) - \varepsilon_p(X)$. For the IV-VI compounds (in the rock-salt structure) this gives

$$e_T^* = 2 + \frac{4(V_{pp\sigma}^2 + 2V_{pp\pi}^2)}{(\varepsilon_s(M) - \varepsilon_p(X))^2}\,. \tag{11-11}$$

For the III-VII compounds (in the cesium-chloride structure) it gives

$$e_T^* = 1 + \frac{16(V_{pp\sigma}^2 + 2V_{pp\pi}^2)}{3(\varepsilon_p(M) - \varepsilon_p(X).)^2} \quad . \tag{11-12}$$

These also have been evaluated and compared with experiments in Table 11-5.

The predictions for e_T^* are all qualitatively correct in the sense that they give values in excess of the independent-ion charge Z and the trends from system to system are mostly correctly given. In the IV-VI compounds, where the predictions (and experimental values) are largest, the perturbation theory which we have used becomes inaccurate. We noted that for these Z^* is predicted to be near zero or negative, in the latter case indicating that more than two electrons have been transferred to the lead. We should treat the coupled levels more accurately, as we did in the case of the tetrahedral semiconductors and in the treatment of resonant bonding leading to Eqs. (9-12) and (10-6) and as outlined for Z^* in Problem 11-1. This would not necessarily improve the agreement with experiment and we may be gratified that the perturbation-theory predictions of e_T^* give as adequate a description of the transverse charges as they do.

We note one peculiarity in the experimental values, 0.85 for LiF, less than the independent-ion value of unity. A similar anomaly arises in the fluorites, for which the fluorine effective charges are obtained in Problem 11-2. They are always greater than one, while Samara (1976) found values experimentally of 0.86, 0.88, and 0.92 for CaF_2, SrF_2 and BaF_2, respectively.

This may not be as surprising as one might first think. The very large reduction in static effective charge depends on $V_{sp\sigma}$ but the large additive transfer term depends upon $\partial V_{sp\sigma}/\partial d$. Thus if, for example, $V_{sp\sigma}$ varied as $1/d^n$, the transfer term would be scaled by $n/2$ relative to what we derived. The experimental value of $e_T^* = 0.85$ for LiF would correspond to a value of $n = 1.87$, which is not so far from our value two. Such a failure for the fluorides did not arise in the theory of the susceptibility leading to Table 11-1, since the results did not depend upon $\partial V_{sp\sigma}/\partial d$. The modification of the two contributions to e_T^* through a more accurate inclusion of coupling, as in Problem 11-1 for Z^*, would appear to be very similar for both contributions so that by itself it would be unlikely to lead to an e_T^* less than Z .

C. The Static Dielectric Constant

In Section 11-1-A we calculated the electric polarization arising from an electric field due to the redistribution of electronic charge. The corresponding susceptibility is relevant at optical frequencies, much greater than the vibrational frequencies of the lattice, where any displacement of the ions themselves can be neglected. In static fields there is an additional contribution due to the displacement of the ions. The force on an individual ion due to an electric field E is given by $ee_T{}^*E$, so if the elastic force on that ion due to a displacement u relative to its neighbors is $-\kappa_{eff}u$, then the displacement in equilibrium will be $ee_T{}^*E/\kappa_{eff}$ and the dipole per ion pair is $e^2e_T{}^{*2}E/\kappa_{eff}$. It may be convenient to write this in terms of the transverse optical mode frequency, $\omega_{TO}{}^2 = \kappa_{eff}/M_r$, with

$$M_r = \frac{M_+M_-}{M_+ + M_-} \qquad (11\text{-}13)$$

the reduced ion mass. [A reduced mass twice this size was inadvertently introduced in Harrison (1983b).]Then the dipole may be multiplied by the density of ion pairs (or primitive cells), $N_c = 1/(2d^3)$ for the rock-salt structure, divided by the electric field, and added to the optical susceptibility $\chi^{(1)}$ of Eq. (11-2) to obtain a static susceptibility of

$$\chi_s = \chi^{(1)} + \frac{N_c e^2 e_T{}^{*2}}{M_r \omega_{TO}{}^2}\,. \qquad (11\text{-}14)$$

We may immediately evaluate the displacive component, $N_c e^2 e_T{}^2/(M_r\omega_{TO}{}^2)$, for the monovalent and divalent compounds in the rock-salt structure using the predicted transverse frequencies from Table 10-6 and 10-7 and predicted transverse charges from Table 11-5. The only *experimental* parameter which enters is the internuclear distance d. The results are compared with experiment for the alkali halides in Table 11-6.

We see that these displacive contributions are considerably larger than the electronic contributions, $\chi^{(1)}$ given in Table 11-1. They are predicted to be very nearly the same for each alkali halide, as is observed, and the magnitude is quite well given, even better than the predicted individual parameters, ω_{TO} from Table 10-6 and $e_T{}^*$ from Table 11-5, entering Eq. (11-14).

The predicted values are very much larger for the divalent compounds, also listed in Table 11-6. The only experimental value given by Kittel

Table 11-6. Lattice-displacive contribution $\chi_S - \chi^{(1)}$ to the static dielectric susceptibility. Experimental values (in parentheses) for the alkali halides are from Kittel (1976).

		Alkali Halides		
	F	Cl	Br	I
Na	0.33 (0.27)	0.31 (0.29)	0.34 (0.30)	0.42 -
K	0.31 (0.32)	0.30 (0.20)	0.30 (0.30)	0.31 (0.19)
Rb	0.34 (0.37)	0.32 -	0.31 -	0.38 (0.23)
		Divalent Compounds		
	O	S	Se	Te
Mg	1.65	6.47	27.08	-
Ca	1.12	1.34	1.33	1.76
Sr	1.13	1.19	1.24	1.37
Ba	1.19	1.17	1.19	1.21

(1976) was that for MgO; it was 0.55, a factor of three smaller than that predicted. It is interesting that the predicted values, as for the alkali halides, are rather constant over the set of compounds except for the heavy chalcogenides of magnesium for which very high values are predicted. These high values arise from small values of ω_{TO} entering Eq. (11-14). These were discussed in Section 10-4-C, where they were seen to be associated with small magnesium ions moving easily in the large interstices of the chalcogenide lattice. We expect large static susceptibilities to occur in these systems, but have at present no experimental confirmation.

Problem 11-1. Effective charges.

Return to the second form in Eq. (11-4) in order to obtain a more accurate form than Eq. (11-7) for the static charge for IV-VI compounds. $u_b{}^{+2}$ gives the fraction of each special-point state which lies on the metallic atom. V_3 becomes $(\varepsilon_p(M) - \varepsilon_p(X))/2$ and $V_2{}^2$ becomes the sum of squared couplings for each orbital, $2(V_{pp\sigma}{}^2 + 2V_{pp\pi}{}^2)$. There are two spins and three bands occupied. Obtain the formula for Z^* and confirm that it leads to Eq (11-7) in the limit that $V_2 << V_3$.

Problem 11-2. Dielectric properties of fluorite.

The structure of fluorite, CaF_2, was given in Problem 9-1. For that system

a) What is the effective charge Z^* on the fluorine (just to second order in $V_{sp\sigma}$) ?
b) What is the dielectric susceptibility $\chi^{(1)}$?
c) What is the transverse charge $e_T{}^*$ on the fluorine?

CHAPTER 12

Covalent Insulators, SiO_2

The most important insulating material for the semiconductor industry is silicon dioxide, which does not form in the ionic structures discussed in Chapters 9, 10, and 11. The silicon-dioxide structure is characterized by silicon atoms each surrounded by four oxygen atoms, forming a regular tetrahedron, and each oxygen atom is bonded to two silicon atoms, shown schematically in Fig. 12-1.

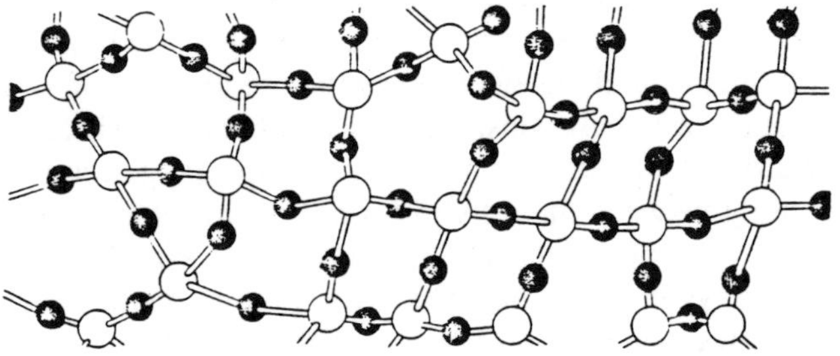

Figure 12-1. A schematic diagram of silicon dioxide. The white circles are silicon atoms, each connected to four neighboring silicon atoms by a bent oxygen-atom bridge (smaller dark circles). Because of the bend it is not difficult to complete the network without crystalline order; that is, as an amorphous glass as shown here. (Taken from Birchall and Kelly (1983)).

The open structure, with no more than four nearest neighbors to each atom, would suggest a description in terms of two-electron bonds between hybrid states and a covalent description analogous to that which we have made for semiconductors. Indeed we shall see that the band gap arises principally from interatomic couplings (covalent energies) rather than from differences in atomic energies on neighboring sites (polar energies), so this speculation is confirmed. The resulting gap is 9 volts, much larger than needed to qualify as an insulator, as discussed in Section 1-5-A.

1. The Structure and Bonding Unit

In the silicon-dioxide structure, which was illustrated in Fig. 12-1, the oxygen atom is equidistant from its two silicon neighbors and the directions to the two neighbors make an angle of about 144°; the oxygen forms a bent bond, or bridge, between the two silicons. The simplest crystal lattice of SiO_2 formed in this way is the beta-cristobalite structure in which the silicon atoms are arranged as in the elemental silicon structure but with the silicon-silicon bonds replaced by oxygen bridges. It is often assumed that the oxygen lies on a straight line between the silicons in this structure but that is apparently not true. Because of the bend at the oxygen it is not difficult to form noncrystalline structures; that is a glass, in the case of SiO_2 called *vitreous silica.* The oxygen bridges can be rotated to allow the bond lengths and angles to remain near their ideal values. Such an amorphous structure was illustrated in Fig. 12-1. There is even theoretical indication (Sarnthein, Pasquarello, and Car (1994)) that some residual structure of this type remains in liquid SiO_2. In our tight-binding description we focus mostly upon the nearest neighbors to each atom and the long-range order is of little consequence. We treat glasses on the same footing as we treat solids.

In order to understand the electronic structure in terms of weakly coupled two-center bonds, we shall construct sp-hybrids on the silicon atoms, each directed toward one of the four nearest-neighbor oxygen atoms. We shall treat the coupling between these hybrids on the same atom, a metallic energy equal to the V_1 of elemental silicon, as weak. We shall also construct hybrids on the oxygen atom, directed toward the two silicon atoms. In Fig. 12-2 we show an oxygen atom along with the two silicon hybrids directed toward it. The angle $\theta \approx 18°$ characterizing the bend in the oxygen bridge is half the difference between the bond angle, 144°, and 180°.

In Pantelides and Harrison (1976) and Harrison (1980) we regarded the p-states on this oxygen atom plus the two silicon hybrids as a bonding unit, analogous to the bonding unit we constructed for semiconductor impurities in

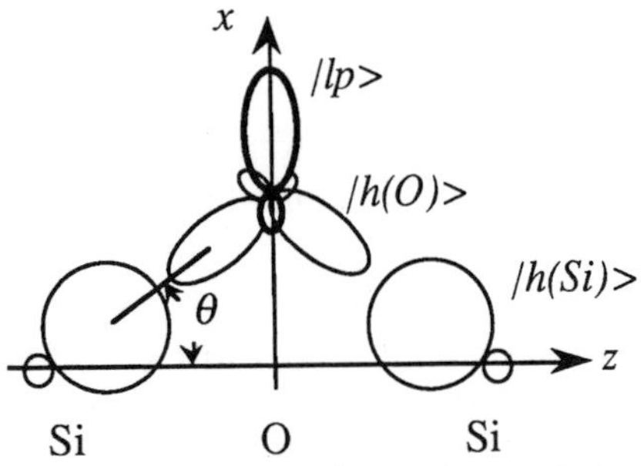

Figure 12-2. The silicon-dioxide bonding unit, consisting of an oxygen atom and the two silicon sp^3-hybrids. In order to treat this as two individual bonds, sp^n-hybrids on the oxygen are constructed which are orthogonal to each other but are directed at the neighboring silicon atoms. The remaining orthogonal hybrid, constructed from the s-state and the p_x-state, points upward in the figure and is called the lone-pair hybrid. The remaining orbital on the oxygen is a p-state oriented perpendicular to the plane of the figure. It is the π-state and is not shown.

Section 8-2. A most important feature for a bonding unit for a solid is that it be *stoichiometric* ; the unit must have the same ratio of constituent atoms as the compound. Then it is possible to construct the crystal out of such units. The oxygen atom in Fig. 12-2, plus one quarter of the hybrids from the two silicon atoms, meets this criterion. A silicon atom with four oxygen neighbors is not allowed as a bonding unit. Such units with components "shared" with neighboring units can be very deceiving. On the other hand, in some structures we might choose more than one set of bonding units, as we did for impurities in semiconductors (Section 8-2), as long as the entire crystal can be constructed out such bonding units.

In the earlier studies we obtained electronic states for the bonding unit in terms of all of the orbitals in Figure 12-2, except the oxygen s-state. The resulting states of the bonding unit are like molecular orbitals, even or odd with respect to a vertical line through the oxygen atom in Fig. 12-2, and each such orbital has the same weighting (coefficient squared) on each of the two silicon atoms. Any effect of the oxygen s-state was neglected.

Here we divide the unit in Fig. 12-2 into smaller bonding units by constructing individual silicon-oxygen bond orbitals on the two sides (including the oxygen s-state in the analysis) and then treating the coupling between these bond orbitals as a perturbation. We shall in fact confirm that this coupling is sufficiently small that the new procedure is justified; it is simpler and may be more accurate than the larger bonding unit because of the inclusion of the oxygen s-state. In Section 12-10 we shall return briefly to the larger bonding unit in constructing the energy bands.

2. The Parameters

The first step in obtaining the electronic structure is to take atomic term values from Table 1-1 for the constituent atoms; these are indicated graphically in Fig. 12-3. We next form sp^3-hybrids on the silicon of the form $|h(Si)\rangle = {}^1/_2\,(|s\rangle + \sqrt{3}\,|p\rangle)$ which have energies $(\varepsilon_s + 3\varepsilon_p)/4$ as indicated in Fig. 12-3. These values are given in Table 12-1 where we shall collect the various parameters that characterize the electronic structure of the system. The coupling between two hybrids on the same silicon atom, $-V_{1+}$, as in Section 2-2-D is $(\varepsilon_s - \varepsilon_p)/4$ based upon silicon term values.

The next step is the construction of hybrids on the oxygen. If they are to be directed at the neighboring silicon atoms, the oxygen p-state entering the hybrid pointing to the left must be given by $-cos\theta|p_z\rangle - sin\theta\,|p_x\rangle$, in terms of the coordinate system and angle θ given in Fig. 12-2 (note that the x-axis is vertical). That for the oxygen hybrid pointing to the right will be the same

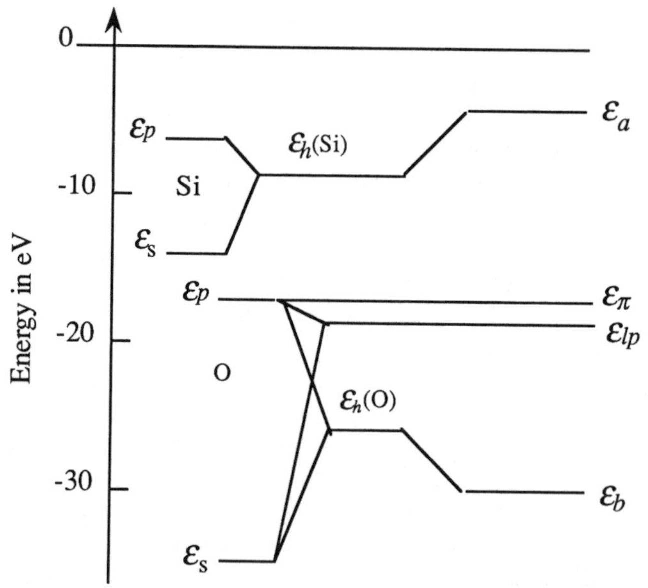

Figure 12-3. Energy-level diagram depicting the formation of Si-0 bonds in SiO_2. Silicon sp^3-hybrids are formed above, while two oxygen hybrids are formed below, along with a lone-pair hybrid and the remaining π-state. Then bonding and antibonding orbitals are formed from the silicon and oxygen hybrids. The bond, lone-pair, and π-states form the valence bands and there are enough electrons to fill them; the conduction bands are formed from the antibonding orbitals.

Table 12-1 Parameters for the electronic structure of SiO_2. In quartz the volume per molecule is Ω_{mol} = 38.0 Å^3 ; in vitreous silica it is 45.7Å^3.

d = 1.61 Å	$V_0 = \frac{49.8 \text{ eV-Å}^4}{d^4}$
θ = 18°	V_{1+} = 1.80 eV
$\varepsilon_h(Si)$ = -9.38 eV	V_{1-} = 7.74 eV
$\varepsilon_h(O)$ = -24.46 eV	V_{1lp} = 3.76 eV
$\varepsilon_{lp}(O)$ = -18.55 eV	V_2 = 9.47 eV
$\varepsilon_\pi(O)$ = -16.77 eV	V_3 = 7.54 eV
α_p = 0.623	η = 48.03°

with the coefficient of the state $|p_z\rangle$ changed in sign. The relative contribution of the s- and the p-states is obtained by requiring that the two hybrids be orthogonal to each other. If we write the coefficients of the s-and p-states as $\cos\eta$ and $\sin\eta$, respectively, we shall guarantee that each hybrid is also normalized and η is simply a parameter of the system depending upon the angle θ. For the special case of sp^3-hybrids it would be 60° giving the form of the hybrid $|h(Si)\rangle$ written above.

Thus the oxygen hybrids to the right and left become

$$|h(O)_l\rangle = \cos\eta\,|s\rangle + \sin\eta\,\left(-\cos\theta\,|p_z\rangle - \sin\theta\,|p_x\rangle\right),$$
$$|h(O)_r\rangle = \cos\eta\,|s\rangle + \sin\eta\,\left(\cos\theta\,|p_z\rangle - \sin\theta\,|p_x\rangle\right). \qquad (12\text{-}1)$$

The requirement that the two be orthogonal, $\langle h(O)_r\,|\,h(O)_l\rangle = 0$ becomes immediately

$$\cos^2\eta + \sin^2\eta\,(\sin^2\theta - \cos^2\theta) = 0 \qquad (12\text{-}2)$$

We may solve for $\tan^2\eta$. A convenient form for the result is

$$\cos^2\eta = \frac{1}{2}(1 - \tan^2\theta),$$
$$\sin^2\eta = \frac{1}{2}(1 + \tan^2\theta). \qquad (12\text{-}3)$$

For the θ = 18° of SiO_2 this gives η = 48.03° and $\cos^2\eta$ = 0.45, corresponding to 45% s-state and 55% p-state, an $sp^{1.24}$hybrid. The hybrid energy is

$$\varepsilon_h(O) = \cos^2\eta\ \varepsilon_s(O) + \sin^2\eta\ \varepsilon_p(O)$$
$$= \frac{1}{2}(1 - \tan^2\theta)\ \varepsilon_s(O) + \frac{1}{2}(1 + \tan^2\theta)\ \varepsilon_p(O)\,, \qquad (12\text{-}4)$$

or -24.46 eV for $\theta = 18^o$ and the oxygen atomic energies from Table 1-1. This is listed also in Table 12-1.

Similarly we may obtain the coupling between the two oxygen hybrids, as we obtained it for tetrahedral hybrids in Section 2-2-D. It is $-V_{1-}$, with the metallic energy given in this case by

$$V_{1-} = (\varepsilon_p(O) - \varepsilon_s(O))\cos^2\eta\,, \qquad (12\text{-}5)$$

equal to 7.74 eV at $\theta = 18^o$. We may also obtain the coupling between the silicon and the oxygen hybrids in the same bond $\langle h(Si)|H|h(O)\rangle = -V_2$. This covalent energy is given by

$$V_2 = -\frac{1}{2}\cos\eta\ V_{ss\sigma} + \left(\frac{1}{2}\sin\eta + \frac{\sqrt{3}}{2}\cos\eta\right)V_{sp\sigma} + \frac{\sqrt{3}}{2}\sin\eta\ V_{pp\sigma}, \qquad (12\text{-}6)$$

which accidentally turns out to be $3.22\,\hbar^2/md^2$ for $\theta = 18^o$ ($\eta = 48.03^o$), equal to the covalent energy for tetrahedral semiconductors; that would not be true at a different angle θ. The value for SiO_2, with d = 1.61 Å is given in Table 12-1.

It will be interesting to note at the same time that the coupling between, for example, the left silicon hybrid and the oxygen hybrid pointing to the right is very small as would be suggested by the Fig. 12-2. Like V_1 this matrix element couples bonds and antibonds in neighboring sites and it was given the symbol V_1^x by Pantelides and Harrison (1976). It is the negative of the V_2 as written in Eq. (12-6), but with $\sin\eta$ replaced by $-\sin\eta\cos 2\theta$. We obtain

$$V_1^x = \frac{1}{2}\cos\eta\ V_{ss\sigma} - \left(-\frac{1}{2}\sin\eta\cos 2\theta + \frac{\sqrt{3}}{2}\cos\eta\right)V_{sp\sigma} + \frac{\sqrt{3}}{2}\sin\eta\cos 2\theta\, V_{pp\sigma}$$
$$= -0.32\frac{\hbar^2}{md^2}\,. \qquad (12\text{-}7)$$

This is smaller than V_2 by a factor of ten and enters squared in perturbation theory so we are safe in neglecting its effect. Pantelides and Harrison defined two additional interatomic matrix elements which had tiny effects and which we drop here.

We may evaluate the polar energy,

$$V_3 = \frac{1}{2}(\varepsilon_h(Si) - \varepsilon_h(O)) . \tag{12-8}$$

It depends upon angle through $\varepsilon_h(O)$ with the value for $\theta = 18^o$ given in Table 12-1. In terms of these parameters we have the antibonding and bonding energies,

$$\varepsilon_{a,b} = \frac{1}{2}(\varepsilon_h(Si) + \varepsilon_h(O)) \pm \sqrt{V_2^2 + V_3^2} , \tag{12-9}$$

which are -4.81 eV and -29.03 eV, respectively.

We may also evaluate the repulsive overlap interaction V_0 which we discussed for semiconductors in Section 2-2-E and for ionic crystals in Section 10-1. It would seem appropriate in this covalent material to use the $V_0 = C/d^4$ which we used for the semiconductors and as did Pantelides and Harrison (1976). Fitting it at the observed angle in Section 12-4 will give a value of $C = 49.8$ eV-Å^4, which is listed in Table 12-1 and which we shall use throughout. It is interesting that if we were instead to estimate C from Eq. (2-19), using one factor of core radius r_c from silicon and one from oxygen (from the Solid-State Table at the end of the text) we would obtain $C = 92.3$ eV-Å^4, nearly a factor of two larger. Using this to estimate the equilibrium spacing would yield a value $(92.3/49.8)^{1/4}\times 1.61$ Å = 1.88 Å, enough too large to confirm that Eq. (2-19) is not so useful for predicting spacings for different compounds. We shall see, following Eq. (12-20), that adding metallization would reduce this error by approximately 0.11 Å, to a prediction of some 1.76 Å, but we still regard Eq. (2-19) as a poor predictor of bond length.

We may also obtain parameters for the oxygen lone-pair orbital, which we write

$$|h_{lp}\rangle = \cos\eta_{lp}|s\rangle + \sin\eta_{lp}|p_x\rangle. \tag{12-10}$$

The orbital $|h_{lp}>$ could be obtained by requiring orthogonality to the other hybrids, but in fact the square of the sums of the coefficients of the s-state must equal one so we have immediately $cos^2\eta_{lp} = 1 - 2cos^2\eta$, or, using Eq. (12-3),

$$\begin{aligned} cos^2\eta_{lp} &= tan^2\,\theta, \\ sin^2\eta_{lp} &= 1 - tan^2\theta\,. \end{aligned} \qquad (12\text{-}11)$$

We may immediately obtain the lone-pair energy as

$$\varepsilon_{lp}(O) = tan^2\theta\;\varepsilon_s(O) + (1 - tan^2\theta)\;\varepsilon_p(O) \qquad (12\text{-}12)$$

and the coupling $-V_{1lp}$ between the lone-pair orbital and the adjacent oxygen hybrids as

$$-V_{1lp} = cos\eta\;cos\eta_{lp}\;(\varepsilon_s(O) - \varepsilon_p(O)\,)\,. \qquad (12\text{-}13)$$

The coupling between the lone pair and the silicon hybrids is very small and, like $V_1{}^x$, is neglected here.

These parameters, given by the formulae above and evaluated in Table 12-1 at the observed geometry, are all that are needed to directly estimate all of the bonding and dielectric properties of SiO_2, which we do in the following sections. The corresponding calculations for GeO_2 were carried out by Pantelides and Harrison (1976) and in Harrison (1980), using the full bonding unit of Fig. 12-2. Problem 12-1 here makes the same extension of our smaller bonding unit to GeO_2.

These parameters will also be used to obtain the energy bands themselves, as we did for the tetrahedral solids in Chapter 5. The highest-lying occupied states are the π-states on the oxygen, and these give rise to the highest valence-band states. These bands are quite narrow, in agreement with more complete calculations by Chelikowsky and Schlüter(1977) and there are a series of lower bands arising from the lone-pair states and the bonding states. The antibonding states give rise to the conduction bands, unoccupied in SiO_2.

3. Cohesive Energy

The first property we calculate is the energy per bond gained in forming the crystal, the cohesive energy, or the heat of atomization. We may do this step by step, as for the semiconductors in Section 2-3 and for ionic crystals in Chapter 9. The first step is the promotion energy in placing the electrons on each atom into appropriate hybrid states. For the silicon, the energy per atom required to place the electrons in sp^3-hybrids is $\varepsilon_p(Si) - \varepsilon_s(Si) = 4V_{1+}$ as in the semiconductor. There are four silicon-oxygen bonds per silicon atom so the promotion energy per bond is V_{1+}. For each oxygen atom the six electrons at a total energy of $2\varepsilon_s(O) + 4\varepsilon_p(O)$ are raised to $2\varepsilon_h(O) + 2\varepsilon_{lp}(O) + 2\varepsilon_\pi(O)$. Using Eqs. (12-4) and (12-12) we may rewrite the change in energy per oxygen as $2(\varepsilon_s(O) - \varepsilon_p(O)) \cos^2\eta$, and since there are two bonds per oxygen atom, this becomes, according to Eq. (12-5), V_{1-} per bond. Thus the promotion energy per bond takes the very simple form for this case,

$$E_{pro} = V_{1+} + V_{1-}. \qquad (12\text{-}14)$$

The energy gained by the two electrons forming each bond is, of course, $2\sqrt{V_2^2 + V_3^2}$. In the Bond-Orbital Approximation, in which we neglect any coupling between bonds and antibonds, we also neglect coupling of the lone-pair and π-states with the neighboring silicon states. Thus we need only add the overlap interaction $V_0 = C/d^4$ to obtain the bond energy (negative for an energy gain),

$$E_{bond} = V_0 + V_{1+} + V_{1-} - 2\sqrt{V_2^2 + V_3^2} \qquad (12\text{-}15)$$

in the Bond Orbital Approximation. Substituting values from Table 12-1 we obtain -7.25 eV per bond. This is 40% larger than the observed -4.69 eV. [The experimental value was obtained by adding the heat of formation of SiO_2 from Weast (1975) of 8.91 eV per molecule to the heat of atomization of silicon and twice the heat of atomization of oxygen from Table 1-3 and dividing by four, for the number of bonds per molecule.] This error is as we might expect; we found in Chapter 2 that the first-row solid, diamond, was overestimated by a factor of two and the second-row system, silicon, was close to correct. A 40% overestimate for a compound of silicon and oxygen is not out of line.

A. Metallization

The Bond Orbital Approximation is the neglect of coupling between full and empty states. We consider now corrections to the Bond-Orbital Approximation and will conclude from the analysis in this subsection, and from that in Subsection 12-3-B, that the corrections are indeed small and we can generally proceed without them.

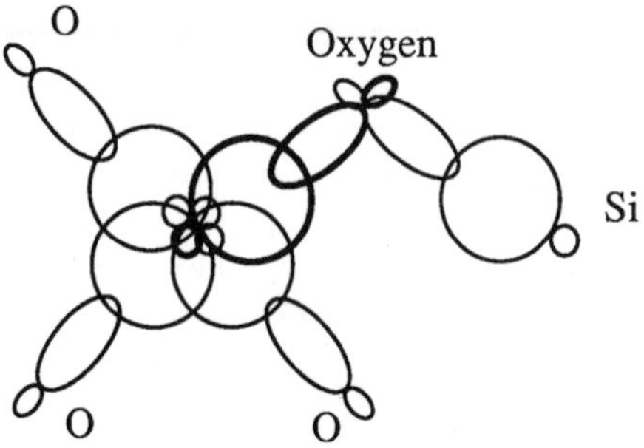

Figure 12-4. The metallized antibond in SiO_2. The antibond, represented by the pair of heavy-lined hybrids, is coupled to a nearest-neighbor bond to the right through the oxygen at the center, with coupling $-V_{1-}$, to the nonbonding lone-pair state (not shown), with coupling $-V_{1lp}$, and to three nearest-neighbor bonds to the left through a silicon atom, with coupling V_{1+}. There is also coupling, via inter-atomic metallization between some of these and orbitals beyond this group.

In this case all states within our description are occupied except the silicon-oxygen antibonding states, so only coupling with those states enters. It is easier to compute the metallization of the antibond, equal and opposite to the metallization of all of the occupied states. We noted in Chapter 2 that the principal coupling in semiconductors is the intra-atomic coupling arising from the metallic energy for two hybrids sharing the same atom. The corresponding intra-atomic metallization for each antibond in SiO_2 can be evaluated directly and is equal to the metallization energy per bond for the crystal.

The calculation is illustrated in Fig. 12-4. There is coupling of the central antibond to the neighboring bond sharing an oxygen and to the three neighboring bonds sharing a silicon. The coefficient of the oxygen hybrid in the antibond is $\sqrt{1/2(1-\alpha_p)}$ and the coefficient of the oxygen hybrid in the bond is $\sqrt{1/2(1+\alpha_p)}$. Since the coupling between the two hybrids sharing the oxygen is $-V_{1-}$, the coupling between bond and antibond is $-\sqrt{1/4(1-\alpha_p^2)}\,V_{1-}$, to be squared, divided by the energy difference between the two states, $2\sqrt{V_2^2+V_3^2}$, and multiplied by two for spin to obtain the magnitude of the metallization contribution per bond. The same analysis leads to the same form, but with V_{1-} replaced by V_{1+}, for metallization of the antibond with each of the three bonds sharing the silicon atom.

Similarly there is a coupling of $-\sqrt{1/2(1-\alpha_p)}V_{1lp}$ between the antibond and the lone pair sharing the oxygen atom, and an energy difference of $\varepsilon_a - \varepsilon_{lp}$. The π-state is not coupled to the neighboring antibonds so the total intra-atomic metallization energy for the two electrons per bond becomes

$$E_{met} = -\frac{1-\alpha_p^2}{4}\frac{V_{1-}^2 + 3V_{1+}^2}{\sqrt{V_2^2 + V_3^2}} - \frac{V_{1lp}^2(1-\alpha_p)}{\varepsilon_a - \varepsilon_{lp}} , \qquad (12\text{-}16)$$

which is -0.88 - 0.39 = -1.27 eV. Adding this to the value from the Bond-Orbital Approximation worsens the agreement with the experimental value of the cohesion, but not in an important way. An important feature of this intra-atomic metallization is that, as pointed out for quite polar semiconductors in Section 2-3-B, the metallization energy rises with increasing spacing, causing additional tension which tends to reduce the spacing. In detail this arises from the $1 - \alpha_p^2 = V_2^2/(V_2^2 + V_3^2)$ in the first term in Eq. (12-16). This factor reduces the metallization coupling when the system becomes more polar . Also, of course, the interatomic metallization which we consider next also tends to decrease the spacing, in this case because the V_{1+} and V_{1-} are replaced by interatomic couplings which drop as $1/d^2$.

A particularly important feature of these results is that the terms in V_{1-}^2 and $V_{1lp}{}^2$ are the principal terms omitted here by treating the system as individual two-center bonds rather than a bonding unit containing two silicon hybrids and all of the states on the intervening oxygen. These terms are indeed small in comparison to the terms from the σ-bonding and their treatment in perturbation theory should be adequate.

B. Interatomic Metallization

We may also consider the effects of interatomic metallization, discussed for tetrahedral semiconductors in Section 3-2-C. This is the effect of coupling between bonds and neighboring antibonds through interatomic coupling, such as the $V_1{}^x$ which we obtained in Eq. (12-7). This is a coupling between a silicon hybrid and the hybrid on the oxygen toward which it points, but the oxygen hybrid which points into the opposite bond. It would enter the coupling between bonds and antibonds which share an oxygen atom. We saw that it was very small, $-0.32\ \hbar^2/md^2$. The corresponding coupling with the lone-pair hybrid is even weaker and we neglect both.

The largest term in the interatomic metallization is that from coupling between , for example, the π-state on the oxygen near the center of Fig. 12-4 and the three antibonding silicon-oxygen states shown to the left in Fig. 12-4. The coupling in each case is $V_{pp\pi}$ times the coefficient of the silicon hybrid, $\sqrt{1/2(1+\alpha_p)}$ times the fractional occupation of the p-state for that hybrid,

$\sqrt{3}/2$, times the cosine of the angle that p-state makes with the orientation of the oxygen π-state. In fact, when this coupling is squared and summed over the four antibonds on the silicon, the total is $(1 + \alpha_p)V_{pp\pi}^2/2$. The difference in energy between the two coupled states is $\varepsilon_a - \varepsilon_p(O)$ and two electrons occupy the π-state so the resulting interatomic metallization becomes

$$E_{met\text{-}\pi} = -(1 + \alpha_p)\frac{V_{pp\pi}^2}{\varepsilon_a - \varepsilon_p(O)} \tag{12-17}$$

The value at the observed angle and bond length is -0.46 eV.

The corresponding metallization of the lone pair, arising from $V_{pp\pi}$, is similar but the coupling is reduced by the fraction $sin^2\eta_{lp}$ of the lone pair which is p-like and the component $cos\theta$ of that p-state which is π-like, and of course the energy denominator is modified. Thus the counterpart of Eq. (12-17) becomes, for the lone pair,

$$E_{met\text{-}lp} = -(1 + \alpha_p)(1 - tan^2\theta)\, cos^2\theta \frac{V_{pp\pi}^2}{\varepsilon_a - \varepsilon_{lp}} \tag{12-18}$$

or -0.33 eV per bond. There is a similar contribution through $V_{pp\pi}$ from the coupling of, for example, the oxygen hybrid in the bond orbital to the right of the central oxygen in Fig. 12-4 and the silicon p-states in the three antibonding states to the left in Fig. 12-4. It is $-{}^1/_2(1 + \alpha_p)^2\, sin^2 2\theta\, sin^2\eta\, V_{pp\pi}^2/(\varepsilon_a - \varepsilon_b)$,which is only 0.04eV and we neglect it as we neglected the contributions from V_1^x .

These metallization contributions to the cohesion also slightly worsen the agreement with experiment; Eqs. (12-17) and (12-18) would raise our estimate to -7.43 eV per bond. This is not significantly worse than before and the more important point is that they are small compared to other contributions. Proceeding with the Bond Orbital Approximation for other properties is quite reasonable. Also these two formulae for the largest metallization contributions could be used to estimate corrections to the Bond-Orbital Approximation if we wished.

4. Radial Force Constant

The cohesive energy per bond without metallization, Eq. (12-15), was evaluated at the observed equilibrium spacing. It is also of interest to evaluate that expression as a function of internuclear distance and as a function of the bond angle θ in order to obtain the overlap repulsion by fitting the bond length, and then to predict the radial force-constant and bond angle at oxygen.

We consider first the evaluation as a function of d , at the observed angle of $\theta = 18^o$, without metallization. We minimize the E_{bond} of Eq. (12-15), which includes the overlap repulsion C/d^4, with respect to d , and solve for C to obtain

$$C = \frac{V_2{}^2 d^4}{\sqrt{V_2{}^2 + V_3{}^2}} = 49.8 \ eV\text{-}Å^4 \ . \tag{12-19}$$

We have taken d , V_2 , and V_3 from Table 12-1, and this is the overlap repulsion we listed in that table.

We next evaluate the force constant, $\kappa_0 = \partial^2 E_{bond} / \partial d^2$. Using Eq. (12-15) we obtain

$$\kappa_0 = \frac{20C}{d^6} - \frac{4V_2{}^2(3V_2{}^2 + 5V_3{}^2)}{d^2(V_2{}^2 + V_3{}^2)^{3/2}} = \frac{8V_2{}^4}{d^2(V_2{}^2 + V_3{}^2)^{3/2}} \cdot \tag{12-20}$$

Of course it had to come out the same as that for the tetrahedral solids, Eq. (2-38), (with modified covalent and polar energies) because of the same simple forms for the bond energy and overlap repulsion. Using values from Table 12-1 we obtain κ_0 = 14.0 eV-Å^2. We shall use this value when we predict vibrational frequencies in Section 12-7.

We may also use the force constant in order to estimate the effect of metallization on bond length. We obtained the value for C in the overlap repulsion by setting the repulsion $4C/d^5$ equal to the bond tension without metallization, $T = 4V_2{}^2/(d\sqrt{V_2{}^2 + V_3{}^2}\,)$ = 18.4 eV/Å, to obtain Eq. (12-19). We may obtain the additional tension due to metallization from Eqs. (12-16), (12-17), and (12-18). These values (obtained numerically) are 0.41, 0.66, and 0.47 eV/Å , respectively. The total of δT = 1.54 eV/Å is small enough that we may obtain the shift in bond length as $\delta d = -\delta T/\kappa_0$ = -0.11 Å. This is large enough that if we are to seriously include metallization in our estimates of properties, we should also include it in the determination of the C entering the overlap repulsion. Here we shall principally use the Bond-Orbital Approximation and only obtain estimates of the general magnitude of the metallization corrections.

For example, metallization also contributes to the force constant. The contributions may be readily obtained numerically from the second derivatives of Eqs. (12-16), (12-17), and (12-18). We obtain 2.30, 0.94, and

0.66 eV/Å^2 for a total force constant of κ_0 = 17.9 eV/Å^2 , a rather small correction to the value from the Bond Orbital Approximation, 14.0 eV/Å^2

5. The Angle at Oxygen

We may also evaluate the various contributions to the bond energy as a function of angle θ in order to attempt a prediction of the bond angle. We proceed carefully since the earlier attempts (Pantelides and Harrison (1976) and Harrison (1980)) gave predictions near $\theta = 45^o$ rather than the observed 18°. We shall again find values too large, but shall see the likely cause of the discrepancy.

With d fixed, the overlap interaction V_0 entering Eq. (12-15) becomes a constant. The covalent energy V_2 , given in Eq. (12-6), depends upon angle θ through η , as does the hybrid energy $\varepsilon_h(O)$ and therefore the polar energy V_3 . So also does the metallic energy V_1.. Thus the bond energy

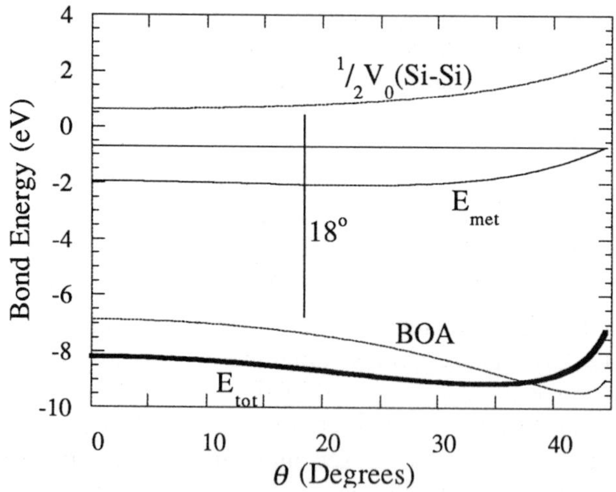

Figure 12-5. Contributions to the crystal energy, per bond, as a function of the angle θ describing the bend in the oxygen bridge, shown in Fig. 12-2. The curve BOA corresponds to the Bond-Orbital Approximation with only the promotion energy, the bond formation energy, and the nearest-neighbor overlap interaction included. The minimum energy occurs at 43°, far from the observed 18°. Adding the principal terms in the metallization (E_{met}) and an estimate of a share of the silicon-silicon repulsion ($^1/_2V_0(Si\text{-}Si)$) to give E_{tot}, shown as the heavy line, shifts the minimum to 34°.

from Eq. (12-15) may directly be obtained as a function of angle. It is plotted in Fig. 12-5 as BOA. Note that at the observed angle it shows the value -7.25 eV given after Eq. (12-15). As in the earlier treatments, it shows a deep minimum at an angle near 45°, corresponding to a 90° angle between Si-O bonds at the oxygen.

We may think of the origin of this effect as a dehybridization energy. The sum of the two oxygen hybrid energies and the lone-pair energy is independent of angle but each individual one is not. The lone pair is doubly occupied but the other two hybrids share in bonds, with bond energy less sensitive to the hybrid energy. Thus energy is gained by increasing θ, making the lone pair more s-like and lower in energy. In systems where there are no electrons in the lone-pair state, we do not expect a bend in the bond. Such a case is Ag_2O which is in an SiO_2-like structure, with the silver atoms two-fold coordinated between two oxygen atoms, but with empty π- and lone-pair states, and the silver lies on a straight-line between the oxygens (Gillespie (1972)).

As to the observed angle being much greater than 90° in SiO_2, we might imagine that this energy could be gained by modifying the hybrids without changing the bond angle but the earlier calculations mentioned above made no assumption of directed hybrids such as we have here and still obtained the minimum energy near 45°. It would appear that the discrepancy with experiment must be understood in another way.

We may similarly evaluate the various contributions to the metallization, Eqs. (12-16), (12-17), and (12-18), as a function of angle θ. The sum of these contributions is shown as E_{met} in Fig. 12-5. They do in fact favor a smaller θ, but their effect is not nearly large enough to dominate the dehybridization energy of the BOA curve. It is large enough to be a significant contribution to the cohesive energy but is quite independent of angle near the observed angle and therefore is not important in other properties. Again it appears quite reasonable for many properties to proceed with the Bond-Orbital Approximation, the neglect of metallization.

A likely candidate for the remaining error in the prediction of the angle might be the overlap interaction between the two silicon atoms. These are second neighbors but at the angle we predict, $\theta \approx 43°$, their separation would almost exactly equal the equilibrium spacing of elemental silicon for which we have used an overlap repulsion of the form C/d^4. It is reasonable to use this same form, with the same constant equal to $V_2(d_0)d_0^4$ with $d_0 = 2.35$Å and V_2 given by $3.22\hbar^2/(md_0^2)$ at least when the silicon-silicon separation is similar (near $\theta = 45°$). There is one such repulsion arising for each oxygen, and therefore half this energy per bond. The resulting $^1/_2V_2(d_O)(d_O/d)^4$ is the $V_O(Si\text{-}Si)$ shown in Fig. 12-5, and the sum of all three contributions is shown as E_{tot}, which has a minimum at 33°, still much larger than the observed 18°.

We might suggest that the estimate of the silicon-silicon interactions is not very accurate but, in fact, it is presumably an overestimate since the occupation of the silicon orbitals is depleted in SiO_2. Furthermore, even if we were to arbitrarily scale up this overlap repulsion to bring the angle into agreement with experiment it would exert large radial forces in the bonds and expand the bond lengths far beyond what is observed. In the context of our analysis it does not seem appropriate to experiment further. It may be that even an exact calculation of the total energy in terms of our tight-binding parameters would reproduce this discrepancy. We will find an analogous error in our discussion of the silicon (111) surface in Chapter 19. Tight-binding will suggest a similar dehybridization should cause alternate silicon atoms to move outward and inward, with two electrons in the nonbonding state of the outward atom and none on the inward atom. Coulomb effects which we omit there, and have not discussed here, in fact prevent that and it is possible that such effects are important here. We must concede that we do not now have a suitable theory of the observed angle at the oxygen, nor presumably of the corresponding angular force constant which would be the second derivative of the energy as a function of θ (as in Fig. 12-5) at the θ for minimum energy.

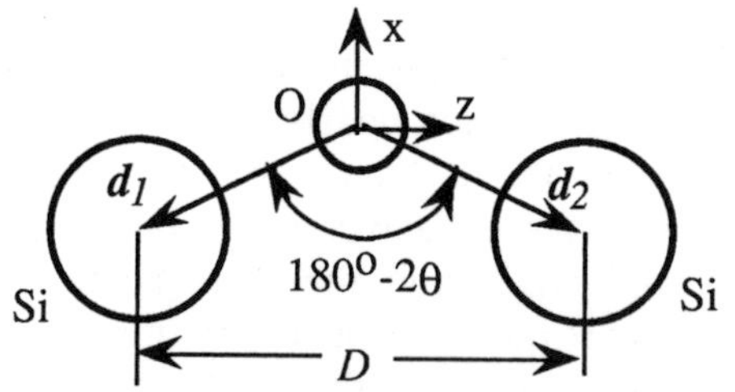

Fig. 12-6. Notation for the bonding unit containing an oxygen atom and its bonds to two neighboring silicon atoms. $D = (d_1^2 + 2d_1d_2\cos 2\theta + d_2^2)^{1/2}$.

6. Elasticity and the Bonding Unit

In order to treat the elastic properties of the solid we again focus upon the bonding unit consisting of an oxygen atom and the two silicon sp-hybrids directed at it. We assume that it can be oriented in the solid such that the four hybrids on each silicon are tetrahedrally oriented so the silicon hybrids can be exactly oriented along the bond directions as in Fig. 12-2. We shall return later to misorientations of these hybrids. Then the geometry of the bonding unit is specified by giving the two bond lengths d_1 and d_2 (equal in the equilibrium configuration) and the angle between them which we write $180^\circ - 2\theta$ (with θ equal to 18° in the equilibrium configuration), as illustrated in Fig. 12-6. We saw in the preceding sections that the metallization energies were small and rather slowly varying with the parameters of the bonding unit so we proceed with the energy in the Bond Orbital Approximation, Eq. (12-15). We combine the energy for the two bonds in the unit to write the total energy

$$E_{bu} = V_0(d_1) + V_0(d_2) + 2V_{1+} + 2V_{1-} - 2\sqrt{V_2(d_1,\theta)^2 + V_3(\theta)^2} - 2\sqrt{V_2(d_2,\theta)^2 + V_3(\theta)^2}\,, \tag{12-21}$$

where we have specified the parameters of the bonding unit upon which each term depends.

The energy E_{bu} should in principal be minimum at the observed geometry so the first derivative of E_{bu} with respect to the three parameters d_1, d_2, and θ should be zero. Then the elastic energy is defined in terms of the second partial derivatives of E_{bu} with respect to the these parameters.

$$E_{elast} = \frac{1}{2}\frac{\partial^2 E_{bu}}{\partial d^2}(\delta d_1^2 + \delta d_2^2) + \frac{1}{2}\frac{\partial^2 E_{bu}}{\partial \theta^2}\delta\theta^2 + \frac{\partial^2 E_{bu}}{\partial d\partial\theta}\delta\theta(\delta d_1 + \delta d_2)\,. \tag{12-22}$$

Each derivative is to be taken at the equilibrium geometry so we did not need to distinguish the derivatives with respect to d_1 and d_2. The first two terms are simply $^1/_2\kappa_0(\delta d_1^2 + \delta d_2^2)$ with the κ_0 given in Eq. (12-20). The second term could have been extracted from one of the curves in Fig. 12-5 had that prediction been successful, but when the minimum in the curves is so far from the observed angle, it is difficult to attach meaning to the second derivative at either angle. We must instead define a $\kappa_\theta = \partial^2 E_{bu}/\partial\theta^2$ and fit it to an appropriate experiment. The third term may be expected to be quite small. Note that only the final two terms in Eq. (12-21) depend upon both d and θ and V_2 is quite insensitive to θ. There is uncertainty in this term also because of the poor prediction of the equilibrium angle so it seems best to drop it altogether. When there is a successful theory of the bond angle it should be checked to see if this cross term is in fact negligible as we assume here. Thus Eq. (12-22) is reduced to

$$E_{elast} = \frac{1}{2}\ \kappa_0 \,(\delta d_1{}^2 + \delta d_2{}^2) + \frac{1}{2}\ \kappa_\theta\, \delta\theta^2\ . \tag{12-23}$$

We may now directly calculate the bulk modulus, assuming again that the coordination of atoms around each silicon remains a regular tetrahedron; then the compression occurs entirely from the compression of the individual bonding units. A dilatation Δ (equal to the change in volume divided by the volume) will change the inter-silicon distance, D in Fig. 12-6, as $\delta D/D = \Delta/3$. Given this δD we may minimize the energy with respect to the oxygen position to find $\delta d_1 = \delta d_2 = \delta d$ and $\delta\theta = -\,(\kappa_0/\kappa_\theta)\ tan\theta\ d\ \delta d$. Then after some algebra we find that the change in energy per bonding unit is given by

$$E_{elast} = \frac{\kappa_\theta\ \kappa_0\ d^2}{\kappa_\theta + 2\kappa_0\ d^2\ tan^2\theta}\left(\frac{\delta D}{D}\right)^2 . \tag{12-24}$$

It is easy to check the limiting cases. When κ_θ becomes extremely large, θ will not change under the distortion and each bond length changes as $\delta d/d = \delta D/D$, so the elastic energy becomes $\kappa_0\ \delta d^2 = \kappa_0\ d^2\ (\delta D/D)^2$, in accord with Eq. (12-23) which approaches this limit for large κ_0 . When κ_0 becomes extremely large, d does not change and with $D = 2d\ cos\theta$, one finds $\delta\theta = -\delta D/(D\ tan\theta)$. The elastic energy becomes $^1/_2\ \kappa_\theta\ \delta\theta^2 = {}^1/_2\ \kappa_\theta\ (\delta D/Dtan\theta)^2$, in accord with this limit for Eq. (12-23).

The bulk modulus B gives the change in energy per unit volume as $^1/_2 B\Delta^2$. We may equate this to twice the energy Eq. (12-24) (which is for one oxygen atom) for each SiO_2, divided by the volume per molecule Ω_{mol} to obtain

$$B = \frac{4\kappa_\theta\ \kappa_0\ d^2}{9\Omega_{mol}(\kappa_\theta + 2\kappa_0\ d^2\ tan^2\theta)}\ . \tag{12-25}$$

Given our estimate of 17.9 eV/Å^2 for κ_0 and the experimental parameters for vitreous silica from Weast (1975) ($\Omega_{mol} = 45.7$ Å^3 and $B = 3.65\times10^{11}$ ergs/cm^3 = 0.227 eV/Å^3) we obtain our estimate of the angular force constant, $\kappa_\theta = 10.00$ eV . Note that it is small in comparison to the radial force constant of the same units, $\kappa_0 d^2 = 46.4$ eV, as expected.

It is not clear that we can estimate the shear elastic constants without specifying the structural geometry in more detail since it would seem that the structure may partially accommodate to a shear by appropriate rotation of the bonding units. Given a specific geometry, the calculation of the rigidity would appear to be straightforward but complicated; we have not attempted it.

7. The Vibrational Frequencies

As in other compounds, there are two categories of vibrational modes: the acoustic modes which are describable in terms of the elastic constants, and the optical modes which correspond to molecular-like vibrations within the individual molecular cells. In the case of SiO_2 the latter may be described approximately as local oxygen modes (Kleinman and Spitzer (1962), Harrison (1980)) in which the light oxygen atom vibrates within a stationary lattice. We consider only these oxygen local modes.

There are three such modes for each oxygen atom which may be specified in terms of Fig. 12-6. In the highest-frequency mode (highest because of the high radial force constant) the oxygen is displaced in the z-direction, to the right and left in Fig. 12-6. Its frequency is written as ω_z, given in radians per second. The next-highest frequency is ω_x with displacements vertical in Fig. 12-6. The third is perpendicular to both, ω_y. The first two are easily calculated in terms of the force constants we obtained in the last section. With z-displacements θ changes only to second-order in the displacement so only the force-constant κ_0 enters. Only the component of displacement along the bond enters and only the component of the radial force along the displacement enters so the force from one bond, for a displacement u, is $-\kappa_0 \cos^2\theta\, u$. Both changes in d_1 and d_2 occur so we obtain (from setting mass times acceleration equal to the force)

$$\omega_z^2 = \frac{2\,\kappa_0 \cos^2\theta}{M_O}, \tag{12-26}$$

where M_O is the mass of oxygen (it should be a slightly reduced mass to account for the finite mass of the silicon atoms, which also move). Using the parameters from Section 12-6 we obtain $\omega_z = 1.39\times10^{14}$ rad/sec., in fair agreement with the observed (Kleinman and Spitzer (1962)) 2×10^{14} rad/sec.. Note that no adjusted parameters entered this estimate. Similarly the second mode frequency is obtained as

$$\omega_x^2 = \frac{2\,\kappa_0 \sin^2\theta + (\kappa_\theta / d^2)\cos^2\theta}{M_O} \tag{12-27}$$

or $\omega_x = 0.64\times10^{14}$ rad/sec., compared to the experimental 1.47×10^{14} rad/sec..

Vibrations in the y-direction correspond to a rocking motion in which the bonding unit rotates without deforming. This misaligns the silicon and oxygen hybrids, assuming that the silicon hybrids do not rotate during the vibration, and the rigidity is calculated as was the shear constant for the twist distortion in Section 3-2-A. A displacement u of the oxygen atom misaligns the hybrids by an angle $\psi = u/d$. This changes the covalent energy of Eq. (12-6) by replacing $V_{sp\sigma}$ by $V_{sp\sigma} \cos\psi$ and $V_{pp\sigma}$ by $V_{pp\sigma} \cos^2\psi + V_{pp\pi} \sin^2\psi$. The change in V_2 for small ψ becomes $\delta V_2 = -2.51\hbar^2\psi^2/(md^2)$. The change in energy of the four electrons in the two bonds is then $-4V_2\delta V_2/\sqrt{V_2^2 + V_3^2}$,which has the value $\frac{1}{2}(17.82\ eV/\text{Å}^2)u^2$. The corresponding local mode frequency is given by

$$\omega_y^2 = \frac{8V_2}{M_O\sqrt{V_2^2 + V_3^2}} \frac{2.51\hbar^2}{md^4} , \tag{12-28}$$

or $\omega_y = 1.03\times10^{14}$ rad/sec, compared to the observed 0.90×10^{14} rad/sec.

The predictions of ω_z and ω_y were entirely in terms of the parameters of the electronic structure and the agreement is not so bad, although the ω_z error corresponds to an error of a factor of two in the radial force constant. The error is largest for ω_x which depended upon the force constant κ_θ which we sought to evaluate in terms of the bulk modulus. It seems likely that the error lies there. If we were to adjust the force constants to fit the two frequencies we would obtain $\kappa_0 = 36.64$ eV/Å^2 and $\kappa_\theta = 82.20$ eV (corresponding to a curvature similar to that near the minimum in E_{tot} in Fig. 12-5). Substituting these in Eq. (12-25) gives a bulk modulus 3.6 times the experimental value. The error may well be in the approximations leading to Eq. (12-25).

8. The Refractive Index

We turn next to the theory of the dielectric properties, proceeding in the Bond Orbital Approximation since the corrections to the susceptibility due to metallization for elements in the oxygen and silicon rows were rather small, only 10% and 40% for diamond and silicon, respectively. The electronic contribution to the dielectric susceptibility comes from the polarization of the silicon-oxygen bonds. We saw in Eq. (4-9) that for a field in the x-direction the dipole from a single bond arising in the x-direction is given by

$$\delta p_x = \frac{e^2 d_x^2 (1 - \alpha_p^2) E_x}{2\sqrt{V_2^2 + V_3^2}} . \tag{12-29}$$

We may again sum the contributions for the four bonds to a single silicon (again the average d_x^2 is $d^3/3$ for the four bonds, whether or not they are in [111] directions) and divide by the volume per SiO_2 molecule to obtain the polarization density, and thus a susceptibility of

$$\chi^{(1)} = \frac{2 e^2 d^2 (1 - \alpha_p^2)}{3 \Omega_{mol} \sqrt{V_2^2 + V_3^2}} . \tag{12-30}$$

Using the molecular volume for quartz, we obtain $\chi^{(1)} = 0.033$. This is considerably smaller than the experimental 0.11. We might expect metallization to bring it up near half the experimental value, as in the other systems we have treated.

This susceptibility arising from bond polarizability again is the optical susceptibility, determining the optical dielectric constant $\varepsilon = 1 + 4\pi\chi^{(1)}$ and the refractive index, $n = \sqrt{\varepsilon}$, a property of particular importance in SiO_2. We obtain 1.19 in comparison to an experimental value near 1.54. Experimentally it depends upon temperature, partly through thermal expansion, and it depends upon wavelength, partly from the effects such as those discussed in Section 4-3-G

9. Effective Transverse Charge

We consider first the effective charge Z^* on the oxygen. In the Bond-Orbital Approximation it contains six nuclear valence charges, two electrons in the π-state, two in the lone-pair state, and twice $1+\alpha_p$ from the two bonds, or a total of $Z^* = -2\alpha_p = -1.25$.

To obtain the transverse charge we may make a displacement of the oxygen, say a displacement u_z in the z-direction in Fig. 12-6. This gives a dipole eZ^*u_z due to the displacement of the oxygen charge, but also changes the polarity of the forward bond by

$$\delta\alpha_p = -\frac{V_3 V_2\, \delta V_2}{(V_2^2 + V_3^2)^{3/2}} = \frac{2 V_3 V_2^2 u_z \cos\theta}{(V_2^2 + V_3^2)^{3/2} d} . \tag{12-31}$$

The change in the polarity of the bond to the left is the negative of this, giving a net transfer of $|\delta\alpha_p|$ electrons from the silicon on the left to the silicon on the right and an additional dipole of $-e\,/\delta\alpha_p/\ 2d\ \cos\theta$ and a transverse charge with a magnitude of

$$e_z^* = 2\alpha_p + 4\alpha_p\,(1 - \alpha_p^2)\cos^2\theta = 2.63\ . \tag{12-32}$$

The corresponding calculation for a displacement in the x-direction includes a similar transfer term and the Z^* contribution, but contains also a term due to a change in θ linear in u_x. The result is

$$e_x^* = 2\alpha_p + 4\alpha_p\,(1 - \alpha_p^2)\sin^2\theta + 2\frac{\partial\alpha_p}{\partial\theta}\sin\theta\cos\theta = 1.30. \tag{12-33}$$

For displacements in the y-direction there is no change in the polarity linear in displacement and we obtain simply $Z^* = 2\alpha_p$,

$$e_y^* = 2\alpha_p = 1.25\ . \tag{12-34}$$

The term in $\partial\alpha_p/\partial\theta$ entering e_x^* is equal to -0.09 but in the earlier treatments (Pantelides and Harrison (1976) and Harrison (1980)) was larger and led to a value of e_x^* less than e_y^* ; here it reduces e_x^* to a value near ε_y^* . We expect that inclusion of metallization would bring our results closer to the earlier calculations and reduce our estimate of e_x^* below ε_y^* . The fit to the infrared intensities made by Kleinman and Spitzer (1962) led to magnitudes of the transverse charges of e_x^* , ε_y^*, and e_z^* equal to 0.93, 1.50, and 3.93, respectively, compared to our predictions of 1.30, 1.25, and 2.63.

The agreement is comparable to that for the tetrahedral solids shown in Table 4-5. The considerable variation between directions in SiO_2, which arises entirely from the charge-transfer terms, is qualitatively reproduced. It is interesting that a phenomenological two-parameter fit to the experimental magnitudes by Kleinman and Spitzer (1962) gave similar magnitudes to ours, but a reversed sign for e_z^* . Only the magnitudes entered the intensities they used. This indicates that a correct physical description of the electronic structure is needed to properly interpret the experiments, and an *ad hoc* model can lead to a qualitatively incorrect physical description.

The same analysis which we used in Section 4-3-C and 11-2-C may be used to write the static dielectric constant in terms of the transverse charges and optical mode frequencies. For this case the result, Eq. (4-36) or (11-14), is replaced by

$$\chi_s - \chi^{(1)} = \frac{2e^2}{3\Omega_{mol}M_0}\left(\frac{e_x{}^{*2}}{\omega_x{}^2} + \frac{e_y{}^{*2}}{\omega_y{}^2} + \frac{e_z{}^{*2}}{\omega_z{}^2}\right), \tag{12-35}$$

with the factor two coming from the two oxygen atoms and the 1/3 from an angular average for each mode.

Pantelides and Harrison (1976) also considered the piezoelectric effect in quartz. A knowledge of the internal displacements under elastic strains is necessary and they took it from earlier analyses based upon force constants and the quartz structure. These had given a poor account of the piezoelectricity and correcting these for the transfer effects which were included above did not improve the situation greatly.

10. Energy Bands

The fact that we have not discussed the energy bands to this point emphasizes the fact that most properties are most simply calculated without their use. However, the bands are important to other properties and are obtainable approximately in the tight-binding context. This was in fact done by Pantelides and Harrison (1976) and by Harrison (1980) and we make a similar analysis here.

There is a difference, however, in that we have chosen the two-center bonds, rather than the Si-O-Si bonding unit, as a basis for our study of the electronic structure. We now include the various metallic energies which broaden these levels into bands. The largest metallic energies are seen immediately from Table 12-1 to be the V_{1-} = 7.74 eV coupling two bonds through the oxygen atom, more than twice any other matrix elements we have neglected. We may now include these directly, coupling the left- and right-hand bonds in Fig. 12-2, which we write $|b_1>$ and $|b_2>$, respectively. The two resulting states are exactly $(|b_1> \pm |b_2>)/\sqrt{2}$, with energies $\varepsilon_b \mp {}^1/_2(1 + \alpha_p)V_{1-}$, including the factors of the coefficients of the oxygen hybrid in each bond. These are illustrated in the energy-level diagram to the left in Fig. 12-7. The odd combination is exactly the B_z bonding-unit orbital introduced by Pantelides and Harrison (1976), (the oxygen s-state which they dropped does not enter this combination) and the even combination is approximately the oxygen s-state which they dropped. It has the same symmetry as the lone-pair orbital which becomes the counterpart of their bonding B_x orbital. We may also take even and odd combinations of the antibonding orbitals, seen also in Fig. 12-7, which are high in energy, $\varepsilon_a \mp {}^1/_2(1 - \alpha_p)V_{1-}$ and will form the conduction bands in the crystal.

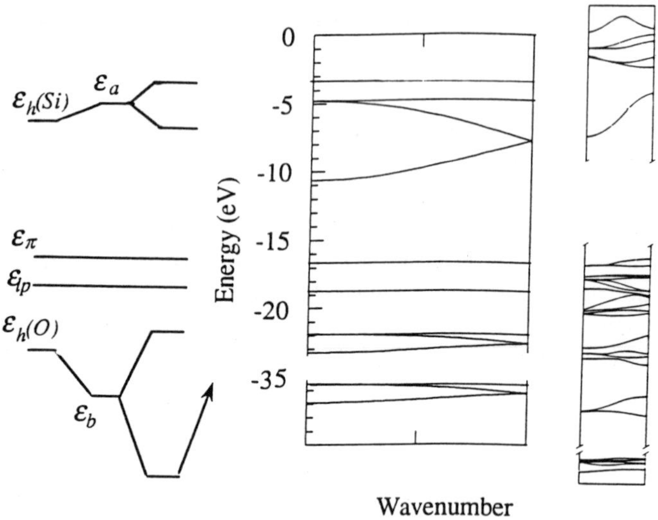

Fig. 12-7. The formation of bands in SiO_2. To the left the two-center bonds at energies ε_b and ε_a are formed, and even and odd combinations made of each, splitting them by $-(1 \pm \alpha_p)/2$ times the oxygen metallic energy V_{1-}. Each of these levels is then coupled through the silicon, forming the bands shown corresponding to the [100] direction in a beta-cristobalite structure. Coupling involving the lone-pair and π-state levels is neglected so their bands are shown flat. To the right are bands for the alpha-quartz structure, calculated by Chelikowsky and Schlüter (1977).

We may next incorporate the coupling between adjacent bonding units through the silicon metallic energy V_{1+} to calculate bonding bands for the beta-cristobalite structure. We include only the coupling $-V_{1+}(1 \pm \alpha_p)/4$, with the plus applying to the conduction bands. Thus the conduction bands are broadened more than the valence bands because the antibonds lie more heavily on the silicon than do the bonds. The additional factor of 1/2 is from combining the right- and left-hand bond, or antibond, orbitals. In addition we include only coupling between orbitals of the same type; e. g., B_x orbitals. Thus the resulting bands are exactly of the form of the simplest silicon bands constructed in Chapter 5 (Fig. 5-4) , and are shown in Fig. 12-7 for the even and odd combinations of bond orbitals and the even combination of antibonding orbitals. The odd combination of antibonding orbitals is so high in energy its bands are not meaningful and are not drawn.

Since we neglect the V_{1lp} coupling between the lone-pair orbital and the bonds and antibonds, it produces a flat band at ε_{lp} . We also neglect the coupling between the oxygen π-state and the neighboring silicon states, so it also becomes a flat band, but at energy $\varepsilon_\pi = \varepsilon_p(O)$. Both are shown in Fig. 12-7, and they form the highest-energy occupied bands in SiO_2.

The lowest-lying empty states are the antibonding states, predominantly on the silicon, and these form the conduction bands. The lowest state among these contains a sum with equal coefficients of the antibonds at each silicon, and therefore has the symmetry of a silicon s-state. It is a state of wavenumber $\boldsymbol{k} = 0$ and has also s-like symmetry around the oxygen. Chelikowsky and Schlüter (1977) in fact chose to view this state as derived from the excited (3s-) oxygen s-state. This is another representation, and perhaps the easiest way to understand the rather free-electron form with minimum energy at Γ, the center of the Brillouin Zone, as seen in Fig. 12-8.

To the right are shown bands for alpha-quartz as obtained by Chelikowsky and Schlüter (1977). Comparison is somewhat schematic since they calculated for a different structure with more molecular units per primitive cell so there are more bands. They took their origin of energy at the valence-band maximum, so we have shifted it to match our ε_π.

Our calculated gap between valence and conduction band is 6.1 eV. We expect a Coulomb enhancement of the gap, as discussed in Section 5-5-A for semiconductors and 9-1-F for ionic crystals. If we take the average of the U values for silicon and oxygen from Table 1-1, and the dielectric constant of 2.4, we obtain an enhancement of $U/\varepsilon = 4.6$ eV, or a predicted gap of 10.7 eV, not so far from the observed 9 eV. There are other Coulomb corrections, discussed for ionic crystals in Section 9-1-F, which complicate the issue, and even the experimental gap is not unambiguous in such large-gap systems.

Such a Coulomb enhancement should apply also to the local-density calculation by Chelikowsky and Schlüter (1971), though they obtained a gap of 9 eV, close to the observed value. This presumably came from some empirical adjustment since an LDA calculation without adjustment should have given an unenhanced gap, near 4.6 eV. We have argued against such adjustment and prefer to make Coulomb corrections to the bands for particular properties. They found the lowest bands at about -41 eV on our scale, lower than our -35 eV, but close enough. These we associate primarily with oxygen s-states. The remaining bands we calculate fall at appropriate energies and with appropriate widths in comparison to the more complete calculation of Chelikowsky and Schlüter.

A discussion of the experimental electronic spectra, and their relation to the energy bands, was given by Pantelides and Harrison (1976) and Harrison (1980). The spectra are complicated by an exciton peak, from absorption into a state in which the electron in the conduction band is bound to the hole in the valence band. This strong peak occurs very near the conduction-band minimum. It occurs also in ionic crystals and is noticeable also in wide-gap semiconductors. Such excitonic effects are discussed for example in Harrison (1980), but will not be discussed for other systems here.

11. Summary of the SiO_2 Study

We have attempted the prediction of a considerable range of bonding and dielectric properties of SiO_2, using exactly the same parameters as those for the semiconductors and ionic insulators. The formulation followed closely that of the semiconductors, and was in fact based upon the two-electron bond orbital; thus SiO_2 is appropriately regarded as a covalent solid although its large gap, as for diamond, makes it insulating. Also, with a V_2 which represents the bond larger than the V_3, the covalent energy dominates the bonding-antibonding splitting. Because the analysis has this local bond basis, the calculated results depended upon the local geometry and applied as well to the amorphous form of silica as to crystalline quartz.

Within this electronic structure there were electrons in lone-pair and π-states as well as in the bond states. If we make the Bond-Orbital Approximation, by neglecting the coupling between the antibonding states and these various occupied orbitals, the theory becomes almost as simple as that for tetrahedral semiconductors. We obtained in this way reasonable estimates of the bond length, cohesive energy, radial force constants, and predicted a bend (although too large) in the oxygen bridge between neighboring silicons. Based upon this formulation of the total energy we could represent the elasticity in terms of force constants, but the angular force constant for the oxygen bridge was not meaningfully given. This aspect was not improved by including metallization, the principal correction to the Bond-Orbital Approximation. The other aspects of the bonding properties were given about as well as the corresponding theory of semiconductors and ionic insulators. We also treated the optical dielectric susceptibility and the transverse charge which characterizes the coupling of the vibrational modes with infrared and in terms of which the static dielectric constant could be written. Finally we constructed approximate energy bands for the beta-cristobalite structure of SiO_2, based upon our tight-binding parameters, and compared them with the results of more complete calculations on crystalline quartz.

Altogether this constitutes a theory of the electronic structure and properties of SiO_2 which is reasonably complete but not terribly accurate. The theory provides a framework in which the entire range of properties can be described if one wishes to make empirical adjustments of the coefficients.

12. Other Covalent Insulators

Almost all of our discussion of semiconductors and ionic compounds was directed at simple structures with no more than two atoms in the repeating primitive cell. SiO_2 has a rather complex structure but we were able to treat the electronic structure in terms of simple covalent bonding units, in fact, simple two-electron bonds in this treatment. In the earlier analyses a larger bonding unit, consisting of an oxygen atom and two silicon hybrids, was used. An extension of that approach has proven most useful in more complex systems. We shall not analyze any such systems in detail but mention a number of representative examples. In all cases it appears that the minimal basis we have been using does describe the principal features of the electronic structure, the parameters which we introduced in Chapter 1 are sufficient to allow estimates of essentially all properties, and the problem can be reduced to elementary hand calculations through choice of stoichiometric bonding units.

A. GeO_2

We begin with direct generalizations of the SiO_2 system. First is GeO_2, which is in a two-four-coordinated structure (oxygen having two neighbors and germanium having four) similar to that of SiO_2, with a bond length of 1.71 Å rather than 1.61 Å and an angle $\theta = 25^o$ rather than 18°. The slightly longer bond is as expected, and it may not be surprising that θ would be larger and more in accord with the simple theory; the H-O-H angle in water is greater than the 90° we would expect, but decreases for the heavier molecules H_2S and H_2Se. In general the simplest theory seems to work best in the silicon row, and the lower rows, and to be most in error in when elements from the carbon row - in this case oxygen - are involved. That does not mean however that the reason is clear. The analysis of the electronic structure for GeO_2 and calculation of properties is identical to that for SiO_2, and was carried out in Pantelides and Harrison (1976), in Harrison (1980), and here in Problem 12-1.

B. Oxyanions, Sulfates, etc.

Pantelides and Harrison (1976) also discussed an interesting set of compounds which are a generalization of SiO_2, analogous to the replacement of alternate silicon atoms in pure silicon by aluminum and phosphorus atoms, to produce aluminum phosphide. The compound $AlPO_4$, aluminum phosphate, is obtained in just this way from SiO_2. Similarly, transferring another proton from the nucleus of each aluminum to that of a neighboring phosphorus produces $MgSO_4$, magnesium sulfate. Transferring another produces $NaClO_4$, sodium perchlorate. The change in terminology, to *per*chlorate, is chemical terminology, based upon the normal products from solution of oxides in water, rather than upon structure - or electronic structure- of the resulting complex. Such a choice was unfortunate from the point of view of electronic structure. However, the ClO_4^- ion is the "perchlorate" ion, and the term "chlorate" applies to ClO_3^-. We shall return later to a discussion of such complexes with three oxygens. The four compounds we listed are polar counterpart of SiO_2 and their analysis is quite straight-forward. It seems remarkable that such familiar chemical groups arise in solids in such a natural way. We look a little closer, following Harrison (1980).

In $AlPO_4$ the electron density shifts to the phosphorus atom and we may think of the PO_4 group as a molecular ion (called an oxyanion) of charge minus-three, forming an ionic compound with aluminum ions of charge plus-three, the customary chemical view. If this is derived from the beta-cristobalite structure, the aluminum ions form a face-centered-cubic lattice. The phosphate ions also form a face-centered cubic lattice combining with the aluminum in a zincblende structure. This choice of structure presumably comes from the inherent tetrahedral symmetry of the phosphate group and the covalent contribution to the energy which favors the oxygen atoms being close to the aluminum ions, though some might argue that it was electrostatic forces.

We wish to consider the general tetrahedral complex, AO_4 , with phosphorus or some other atom as A in the center, surrounded by four oxygen atoms, as illustrated in Fig. 12-8. The electronic structure of the AO_4 is most simply understood by forming sp^3-hybrids on the central A-atom, making two-center bonds with the sigma-oriented p-state on each oxygen atom (or an sp^n oxygen hybrid such as we constructed for SiO_2). Filling all but the antibonding states provides the stability of the structure. In all cases, including PO_4^{3-}, the complex contains a total of 32 valence electrons, corresponding to two electrons in each of the four two-center bonds, four π-electrons on each oxygen and a doubly-occupied s-state on each oxygen. (If we form a hybrid on each oxygen for the phosphorus-oxygen bond, then the remaining orbital is a lone-pair orbital rather than an s-state, orthogonal to the bonded oxygen hybrid, as for the lone pair we discussed for SiO_2.). Only the four antibonding states are empty. With the twenty-four valence charges of the oxygen atoms and the five phosphorus valence charges (for PO_4^{3-}) this

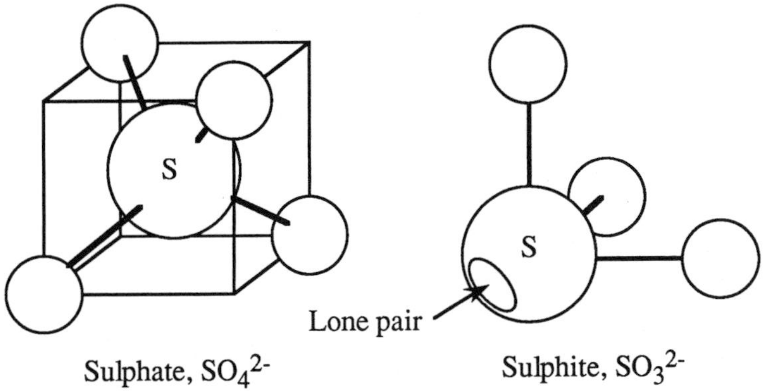

Fig. 12-8. Examples of a 32-electron AO_4 oxyanion and a 26-electron AO_3 oxyanion. The position of the lone pair is shown for the latter. Its shift toward the energy ε_s drives the complex to the pyramidal shape, with angles to the oxygens near 90°.

leaves a net charge of 3 electrons. If the phosphorus atom is replaced by a sulfur, the additional valence charge for this sulfur reduces the charge of the sulfate ion to -2. Replacing the sulfur atom by a chlorine atom gives a net charge of -1.

We may compare the stability of the various oxyanions relative to separating the complex into four neutral oxygen atoms, and the A-atom with the full charge which the complex had, formally carrying eight electrons. The energy gain in forming the complex is that in taking these eight electrons on atom A (in atomic states, or in sp^3-hybrid states, with the same total energy) to the bonding states of the complex. The four two-center bonds are characterized by a covalent energy $V_2 = -V_{hp\sigma} = {}^1/_2(V_{sp\sigma} + \sqrt{3}V_{pp\sigma}) = 2.63\,\hbar^2/md^2$, and a polar energy $V_3 = (\varepsilon_h(A) - \varepsilon_p(O))/2$. Thus the energy gain is $E_{tot} = -8(V_3 + \sqrt{V_2^2 + V_3^2}\,)$, shown in Fig. 12-9 as the energy of the complex AO_4. We see that it becomes more stable as the polarity - and the net charge - increases. At the same time, it is quite unrealistic to imagine a complex or an atom stable as an isolated entity with a charge exceeding one. Further a more relevant comparison is with the four oxygen atoms forming molecules rather than isolated atoms. To be consistent with the simplest description of the complex, we should say that the energy gained in then forming the two oxygen molecules is $4V_{pp\sigma}(O)$ but $4V_2$ is close, so that energy is also shown in Fig. 12-9. The neutral molecule, ArO_4, is not stable, though Fig. 12-9 suggests a weak stability. The molecule XeO_4 , which would lie to the right of ArO_4 in Fig. 12-9, is weakly stable.

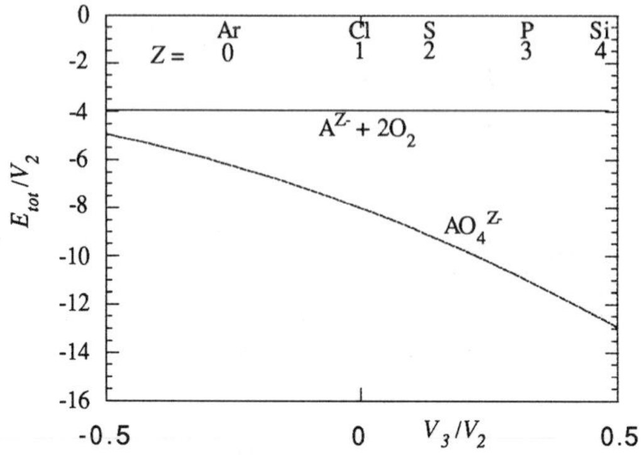

Fig. 12-9. The energy of a 32-electron complex AO_4, where A can be any atom, relative to isolated atoms, with the atom A carrying the net charge $-Z$ from the complex. Also shown is the energy of the competing arrangement (relative to the same basis) for two oxygen molecules and an isolated A with charge $-Z$. Approximate positions for atoms A taken from the silicon row of the table, and the corresponding Z, are shown above.

Perhaps a more meaningful comparison is between a complex such as OS_4^{2-} as opposed to SO_4^{2-}. This simply changes the sign of V_3 and from Fig. 12-7 we see that it is considerably less stable. It helps us to understand why the sulfates are so common, but the inverse arrangement is not. Sulfur *can* replace a single oxygen in the sulfate, to produce the thiosulfate ion, $S_2O_3^{2-}$, which is stable, but less so than the SO_4^{2-} ion.

We may return to the oxyanions with only three oxygen atoms. Again all levels except the antibonds are filled, 26 electrons. This then includes a lone pair on the central atom, which would be a p-state if the group were planar, but drops to a pure s-state if the central angles to the oxygens go to 90°. This of course forces the system to be pyramidal, just as the lone pair on the oxygen in SiO_2 forced the bend in the bond at oxygen. In the case of AO_3 ions, the angle approaches ninety degrees as shown in Fig. 12-7. Again, it tends to be closer to 90° for the heavier compounds, but this is not uniformly true. ClO_3^- has an angle of 107°, BrO_3^- has an angle of 112°, and IO_3^- has an angle of 97° (Gillespie (1972)). Given the angle, the formulation of the electronic structure is immediate, the simplest treatment containing the three bonds and the lone pair, plus full π-states and s-states on the oxygens.

In such AO_3 complexes in which the lone pair is *not* occupied, we expect the lowest energy when all four atoms lie in the same plane, with 120°

between the oxygen atoms. Thus the neutral SO_3 molecule is planar, as is the ion NO_3^- (Gillespie (1972)). When these groups occur in solids or as part of larger molecules, there can be small distortions from the planar structure because the p-like lone-pair state will be coupled to neighboring atoms. Then there can be shifts in the total energy proportional to displacement out of the plane, admixing s-states into the p-like lone-pair state. Perhaps more usually in molecules the neighboring ions will lie in the same plane, as in HNO_3, and then this distortion does not occur. (Gillespie (1972))

A slight improvement on the simplest two-center-bond description of these electronic structures would use also a hybrid on each oxygen atom to form the bond, and then instead of a filled oxygen s-state we would have a filled oxygen lone-pair orbital, but it is makes no qualitative difference. Neither does the full solution for AO_4 units with the A-atom s- and p-states and the four oxygen atom hybrids. This yields states with s-like and p-like symmetry relative to the A-atom, as we found for substitutional impurities in Section 8-2 and drew in Fig. 8-4. This incorporates the largest missing coupling from the simplest picture, the V_1 on the A-atom, exactly. It differs from the simplest two-center representation by correctly splitting the four bonding levels to a three-fold degenerate level, and a single level at lower energy. Similarly the antibonds are spit into a three-fold level and a lower single level.

C. Thallium Arsenic Selenide

A compound Tl_3AsSe_3, turns out to be analogous to the oxyanion systems. It has been treated by essentially the same approach as given here. (Ewbank, Kowalczyk, Kraut, and Harrison (1981)). As a preliminary step one may obtain the term values for the constituent elements and count the electrons to find that there are enough electrons to fill all but the thallium and arsenic p-states (leaving the s-states on both filled as in the ten-electron compounds discussed in Section 11-1-D), with a gap of a little more than a volt between full and empty levels. We correctly expect semiconducting behavior. We may then look at the structure and local coordination to find a closely-bound group consisting of an equilateral triangle of selenium atoms with the arsenic atom displaced out of the plane of the triangle at its center. This then may be treated as a bonding unit, or as a molecular ion. Ewbank, et al., find that three antibonding levels arise from this group and filling the remaining levels gives an $AsSe_3^{3-}$ covalently-bonded unit. It is in fact a 26-electron group analogous to the phosphite PO_3^{3-} ion and forms an ionic solid with three Tl^+ ions, for which the approximations we used for ionic solids become appropriate. The fact that it is a 26-electron AB_3 group, as to the right in Fig. 12-7, explains why it is pyramidal, though the angles are far from 90°.

D. Other Mixed Covalent-Ionic Solids

Beta-eucryptite, with formula $LiAlSi_3O_8$, has a continuous two-four-coordinated structure such as SiO_2, with one quarter of the silicon atoms replaced by aluminum. This can then be considered a covalent insulator, with two-electron bonds coupling each nearest-neighbor pair, and with filled lone-pair and π-states on each oxygen. The extra electron is provided by the lithium which then appears as an ion embedded in the charged covalent framework. Thus it is a mixed covalent-ionic solid, as are those with oxyanions, or as thallium-arsenic selenide, or oxynitrides (Xu and Ching (1995)).

We might at first think that Al_2O_3 , aluminum oxide, was a covalent insulator in the same sense, but the structure indicates not. Although aluminum has only three nearest-neighbor oxygens at 1.89 Å, there are three more oxygen neighbors at 1.93 Å so it is in a sense six-fold coordinated. (Wyckoff (1963) is an excellent source of structural information for any such compounds.) Correspondingly, with a nearest-neighbor distance of 1.89 Å the covalent energy is smaller than in SiO_2 and with a larger polar energy, it is the polar energy which dominates the bonding-antibonding splitting. The free-atom s-state energy for aluminum is higher by six electron volts than the oxygen p-state so description as an ionic insulator makes good sense. We have not treated its properties, but would assume that free ions would be the most suitable starting point. It is interesting to note in passing that the ancients considered the two systems, sapphire and ruby, as two extremes of solid-state properties, though in the end both turned out to be Al_2O_3, with traces of iron and cobalt, respectively. [This, and many other interesting historical facts concerning solids and crystals, appear in Queisser (1990).]

E. A Complex Semiconductor, SiP_2

A compound which clearly should be considered covalent, but which requires a novel choice of bonding unit, is silicon phosphide, SiP_2, in the pyrite structure. In this structure each silicon atom is surrounded by six phosphorus atoms, forming an octahedron (as with the six neighbors in the rock-salt structure). We cannot form six independent bonds with the four silicon valence orbitals and yet it cannot be analyzed as a simple ionic solid since placing the electrons in the lowest atomic levels leaves a partly occupied phosphorus p-shell. We may, however, form a bonding unit in analogy with the Si-O-Si unit in SiO_2. In the pyrite structure each phosphorus is surrounded tetrahedrally by four neighboring atoms (three silicons and one other phosphorus). We quite naturally form sp^3-hybrids on the phosphorus atoms and treat the phosphorus metallic energy which couples them as a weak perturbation. Then we may form one phosphorus-phosphorus bond, and consider the orbitals in the bonding unit consisting of a single silicon atom with six phosphorus hybrids directed at it; these are then all weakly coupled between bonding units and to the phosphorus-phosphorus bond.

The calculation of the states in the bonding unit is quite straightforward, in analogy with impurity states in Section 8-2. It can be simplified by using symmetry considerations but that is not necessary. The results are illustrated in Fig. 12-10. The phosphorus-phosphorus bond shown is just as in a tetrahedral crystal. The combination of four phosphorus hybrids shown as the nonbond is not coupled to the s- nor to the p-states on the central silicon;

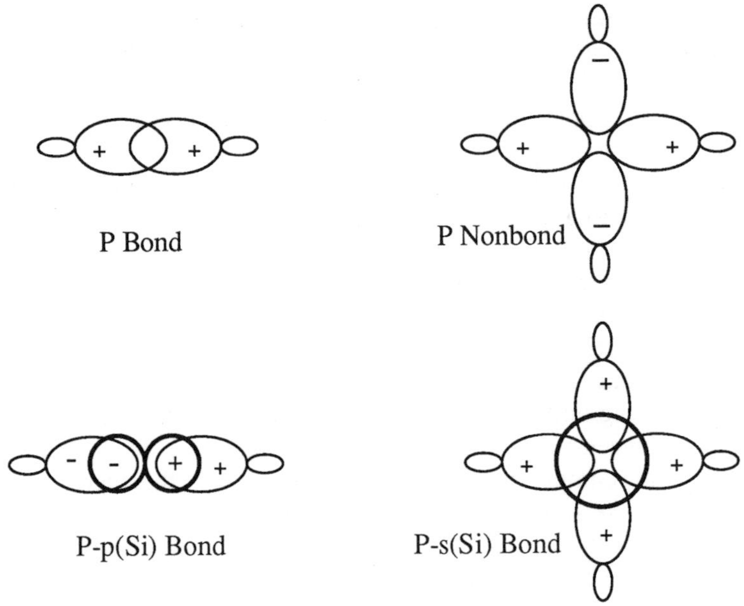

Figure 12-10. A schematic representation of bond orbitals for SiP_2 in the pyrite structure. The P-P-bond consists of sp^3-hybrids on neighboring phosphorus atoms. The Si-P bonds are multi-center orbitals, each with the silicon atom at the center (and its orbitals drawn with heavy lines), with combinations of the σ-oriented hybrids from the six surrounding phosphorus atoms.

the coupling from the individual hybrids cancels out in all four cases. This is a nonbonding orbital with energy equal to the phosphorus hybrid energy. This, and another combination at the same energy (containing equal coefficients on the four hybrids shown, and coefficients twice as large and of opposite sign on the hybrids perpendicular to the plane of the figure), are doubly occupied. There are three P-p(Si) bonds, each doubly occupied, from the three silicon p-states. The corresponding antibonding states are empty. The P-s(Si) bond shown is doubly occupied and its antibond empty. Calculation of all of the bonding and dielectric properties of such a system proceeds directly in terms of these orbitals and their metallization. It is complicated by the number of orbitals, in comparison to SiO_2, but all parameters are known and the procedure is the same as was carried out for SiO_2 here.

An analogous kind of geometry was present in CaF_2 in the fluorite structure, which was treated as an ionic crystal in Problem 9-1. There each fluorine is tetrahedrally surrounded by four calcium atoms, but each calcium has eight fluorine neighbors arranged as the corners of a cube. It would be possible to construct hybrids on the fluorine and make a bonding unit consisting of a calcium ion and eight fluorine hybrids, much as we did with the SiP_2 for the silicon and its six phosphorus neighbors. In this case, however, the ionic approximation is simpler, and probably accurate enough.

In each of these complex structures it has been necessary to consider the local geometry of the structure, which suggested a starting approximation or bonding unit in terms of which the calculation could be carried out. The choice is not necessarily unique - in SiO_2 one could use either a two-electron bond basis or a three-atom bonding unit. Ordinarily, increasing the size of the bonding unit increases the accuracy of the approximation, but makes the calculation more complicated. Inclusion of metallization should make the results for different formulations more nearly equivalent. In each case the choice of a bonding unit can be tested by calculation of the correction terms, the coupling between levels from different bonding units. There is no reason to think that the complexity of these systems makes the theory any less accurate than for the simple systems. The accuracy has varied from property to property and with position of the constituent in the periodic table, and these trends may be expected to be the same in the more complex solids.

Problem 12-1 Germania, GeO_2

For GeO_2 the structure is similar to that of SiO_2, but $d = 1.71$ Å, and $\theta = 25^o$. Obtain all of the other parameters of Table 12-1 for GeO_2. These could be used in the formulae subsequently derived to predict all of the properties considered for SiO_2.

Problem 12-2 CaF_2 in the SiO_2 Structure.

If we put CaF_2 in the 2-4-coordinated SiO_2 structure, rather than the 4-8-coordinated structure, we expect the bond length to decrease by a factor $(1/2)^{1/4}$, suggested by Eq. (2-58). Assume the simple SiO_2 structure with $\theta = 45^o$, a ninety-degree angle at the fluorine. Then construct two-center bonds as we did for SiO_2

a) What are the sp^3-hybrid energies on the Ca? Assume a valence p-state energy $\varepsilon_p = {}^1/_2\, \varepsilon_s$. It was a core energy given in parentheses in Table 1-1.

b) What enters as the F hybrid energy for the bond, and for the lone-pair hybrid? (These would be different if we didn't pick $\theta = 45^o$.)

c) What would be V_2 , V_3 , and the polarity of the bond?

d) What would be the Z^* on the fluorine? It is quite different from that obtained for the real fluorite structure.

CHAPTER 13

Simple Metals, Electronic Structure

The roots of our understanding of the electronic structure of metals are very old. Only a few years after the discovery of the electron by J. J. Thomson (1897) Drude (1900) suggested that the electrons in a metal form a gas of free electrons. This accounted for the conductivity of the metal, but gave the incorrect prediction that the electrons contribute $^3/_2 k_B T$ per electron to the heat capacity of the metal. It was only with quantum-mechanics and the Pauli principle, that this error was rectified in the work of Sommerfeld (1928). Also with quantum mechanics came the demonstration by Bloch (1928) that electrons would not be scattered by a perfectly periodic crystalline array of atoms, explaining why the mean free path between scattering events could be many times the inter-atomic distance. Somewhat later, Landau's Fermi-liquid theory (Landau (1957)) clarified why electron-electron collisions did not preclude the view that electronic "quasiparticles" only weakly interacted with each other.

These last two contributions made the free-electron view understandable, but did not necessitate a free-electron dispersion of the electrons, $E = \hbar^2 k^2/2m$. That aspect of free-electron theory was generally viewed as very naïve until the time of the Fermi Surface Conference of 1960 (Harrison and Webb (1960)), when it became apparent that the dispersion in real metals was indeed close to that for free electrons. This was learned both from experimental studies of Fermi surfaces and from detailed calculations of energy bands for metals. It was made understandable in terms of pseudopotential theory (Harrison (1963), (1966)). This pseudopotential theory also could be applied to liquid metals, yielding a conductivity only slightly higher than that of the crystal, indicating that it was really the weakness of the pseudopotential - not the periodicity of the lattice - which made metallic mean-free paths so long. Considerably later (Wills and Harrison (1984)) it was realized that the opposite view, tight-binding theory, also provided a meaningful description of metals. We made that comparison in Section 1-3 and we shall make use of it here.

1. Crystal Structures

To the extent that the electronic structure is that of free electrons, the crystal structure becomes unimportant. Conversely, the electronic structure does not require special atomic arrangements, as the two-electron bonds did for the covalent solids. The atoms arrange themselves in very simple structures.

The *face-centered-cubic structure,* or "fcc" structure, common to many metals, is also as dense a packing as can be achieved with a fixed minimum spacing. That is, in terms of the packing of hard spheres of diameter d , it is a packing of maximum density. Or, thinking in terms of softer spheres, it minimizes the elastic distortion of the sphere given an overall density chosen to minimize the free-electron energy.

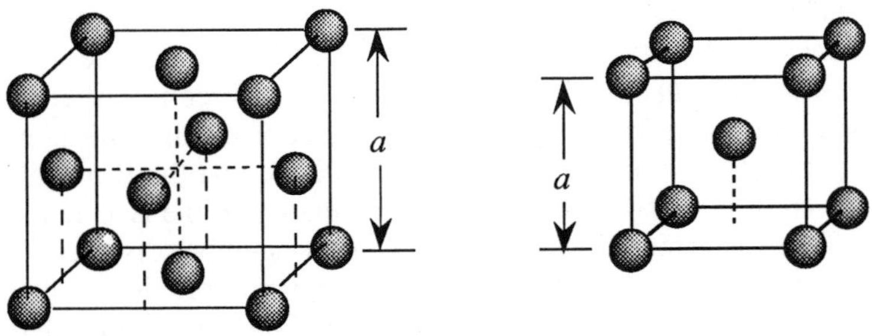

a) fcc b) bcc

Fig. 13-1. In part a), a unit cube of the face-centered-cubic structure with eight cube-corner atoms and six face-center atoms. When the cube is repeated as a three-dimensional array, sharing the corner and face atoms, the two sites are seen to be equivalent, just depending upon where the cube is drawn. The dashed lines may help to indicate the positions of the atoms at the face centers. The density of atoms is $4/a^3$. In part b), a unit cube of the body-centered-cubic structure, with atom density $2/a^3$.

The fcc structure is illustrated in Fig. 13-1. We noted that this was the arrangement of the zinc atoms in the zincblende structure, Fig. 2-2, and that the sulfur atoms form an identical face-centered-cubic structure, displaced by $[111]a/4$ from the zinc atoms, with a the cube edge, also indicated in Fig. 13-1. There are four atoms per unit cube; the eight atoms at the cube corners are each shared by eight cubes and the six face-center atoms are each shared by two cubes, for $1 + 3$ atoms per cube. The volume per atom, Ω_0, is thus $a^3/4$, often equated to an *atomic sphere of radius* r_0 with

$$\Omega_0 = \frac{4\pi r_0^3}{3} = \frac{a^3}{4}. \tag{13-1}$$

To understand that this is a close-packed structure we may construct the structure by stacking close-packed planes, as illustrated in Fig. 13-2. The first close-packed plane, with atomic positions designated by A, has each atom - or sphere - in contact with six neighboring atoms in the plane, like pennies on a table. A second close-packed plane can be placed above the first, with each atom nestled between three neighboring atoms from the A plane. There are two distinguishable types of sites for doing this, indicated by B and C in Fig. 13-2. There we chose the site B and placed the lightly shaded spheres there. If we then place the third layer at site C and repeat the ordering ABCABCABC, we obtain the face-centered-cubic structure. A single cube is picked out of that structure to the upper right, and the atoms reshaded to distinguish the cube corners from the cube faces. The cube is also extracted from the crystal and redrawn to the left below.

Any other ordering of layers (as long as each successive layer differs) is of equal packing density. The ordering ABABAB is called the *hexagonal close-packed structure*, or "hcp" structure. The direction of stacking is called the c-axis and is inequivalent to other directions. Also, the atoms from adjacent planes have a different environment. Note that the B-neighbors to an A-atom are rotated 60° around the c-axis, relative to the A-neighbors to a B-atom. Thus displacing the crystal by a nearest-neighbor distance $\boldsymbol{d}$ in one of the stacked planes takes every atom into a position previously having an atom, but the smallest displacement out of the plane which does that is $\boldsymbol{c}$, parallel to the c-axis and equal to the distance between two A-planes (or two B-planes). As in the silicon (or diamond) structure, there are two atoms per primitive cell, an A and a B-atom.

There is no reason for this distance between second-neighbor planes, which is $\sqrt{8/3}\, d$ for the fcc structure, to have exactly this relation to the

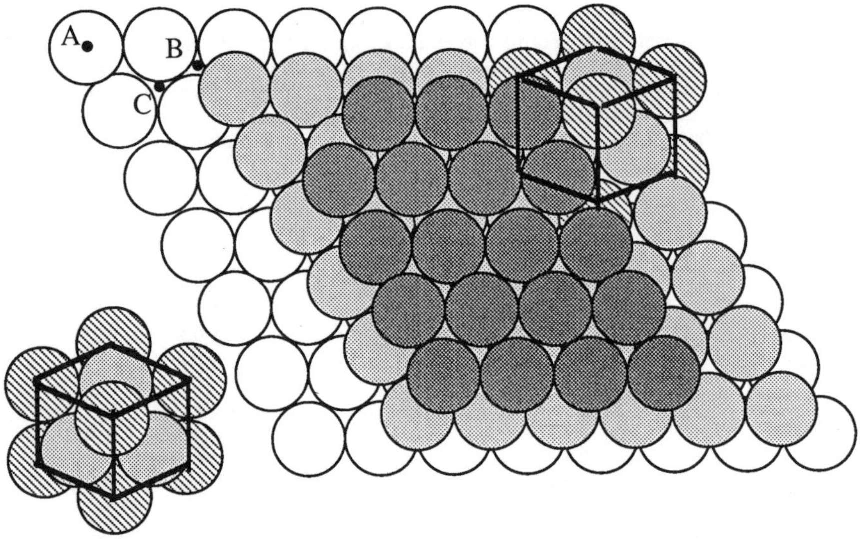

Fig. 13-2. A stacking of close-packed planes of spheres to obtain the face-centered-cubic structure. The empty circles represent atoms in a layer with centers A designated to the upper left. The lightly shaded circles are at positions B and the heavily shaded circles are at positions C. One unit cube is picked out at the upper right and redrawn at the lower left, with the circles reshaded.

spacing within the planes in the hexagonal-close-packed structure. Indeed that c/a ratio is always different, close to $\sqrt{8/3} = 1.633$ in magnesium, but considerably larger in zinc and cadmium, all of which have the hcp structure. Values for these ratios will be given for hcp structures in Table 13-1.

The third common structure is the *body-centered cubic*, or "bcc", structure, shown in Fig. 13-1b. There are two atoms per cube, and the nearest-neighbor distance for this case is $d = \sqrt{3}\, a/2$. The atomic volume, for the same spacing d , is 9% larger than for the fcc structure. Some metals occur in different structures, such as Li and Na which are bcc at room temperature but hcp at low temperatures. A few, Ga, In, Sn, and Hg, have individual structures of lower symmetry, but each is rather closely packed in terms of stacking of hard spheres, and the structure is usually named after the element in question. In Table 13-1 we indicate the structure which each simple metal takes at room-temperature. The systematics for the simple metals are better represented with beryllium and magnesium placed to the

Table 13-1. The simple metals and their structure, if it is fcc, bcc or hcp. The first number is the Fermi wavenumber k_F (Å^{-1}) for Z electrons per atom at the observed volume (compilation of Hafner (1987), p. 68 was used).The second number is the radius r_0 (Å) of a sphere of volume equal to the volume per atom. Nearest-neighbor distances (Å) are given as the third number for each element in a close-packed (fcc) structure, except for those listed as bcc, in which case the nearest-neighbor distance for bcc is given.

Z = 1	2	3	4	1	2
k_F(Å) r_0(Å) d(Å)$_{cubic}$				Li(bcc) 1.11 1.72 3.03	Be(hcp*) 1.94 1.25 2.26
		Al(fcc) 1.75 1.58 2.86		Na(bcc) 0.92 2.08 3.66	Mg(hcp) 1.37 1.77 3.20
Cu(fcc) 1.36 1.41 2.56	Zn(hcp) 1.58 1.53 2.77	Ga 1.66 1.67 3.02		K(bcc) 0.75 2.57 4.52	Ca(fcc) 1.11 2.18 3.94
Ag(fcc) 1.20 1.60 2.89	Cd(hcp) 1.40 1.73 3.12	In 1.51 1.84 3.33	Sn 1.64 1.86 3.36	Rb(bcc) 0.70 2.75 4.84	Sr(fcc) 1.08 2.25 4.06
Au(fcc) 1.21 1.59 2.88	Hg 1.36 1.77 3.21	Tl(fcc) 1.46 1.89 3.42	Pb(fcc) 1.58 1.93 3.49	Cs(bcc) 0.65 2.97 5.23	Ba(bcc) 0.98 2.47 4.34

* c/a ratios are 1.57 for Be, 1.63 for Mg, 1.86 for Zn, and 1.89 for Cd.

right over calcium as in Table 13-1, rather than to the left over zinc as for semiconductors.

A convenient measure of the lattice spacing in the cubic structures is the interatomic distance d or the cube edge a for the cubic structures. The values of d are listed for the cubic metals as the third entry for each element in Table 13-1. For the elements in noncubic structures we list the nearest-neighbor distance which these elements would have if they were in the fcc structure at the same density. This is some average of the in-plane and out-of-plane distances for the hcp-structures, and would equal the distance of all twelve neighbors if the axial ratio were the ideal $\sqrt{8/3}$. For these noncubic structures it be better to give the atomic-sphere radius r_0 , the radius of a sphere with volume equal to the volume per atom. Such a value is given for each metal as the second entry in Table 13-1. We shall return to the first entry shortly.

2. Free-Electron States

For constructing free-electron states the organization of wavenumbers for a crystal lattice, given in Section 5-1, is unnecessary. We may apply periodic boundary conditions (ψ and $\nabla\psi$ each equal on opposite faces) on a large rectangular box of dimensions L_x, L_y, and L_z to obtain wavenumbers of components $k_x = 2\pi n_x/L_x$, , etc., with integers n_x , n_y, and n_z . Then the volume per state, in wavenumber space, is $(2\pi)^3/L_xL_yL_z$. To accommodate Z electrons for each of the $L_xL_yL_z/\Omega_0$ atoms in wavenumbers in a *Fermi sphere* of volume $4\pi k_F{}^3/3$ in wavenumber space, at two electrons per state, we require $(2L_xL_yL_z/(2\pi)^3)\,4\pi k_F{}^3/3 = ZL_xL_yL_z/\Omega_0$, or

$$k_F{}^3 = \frac{3\pi^2 Z}{\Omega_0} = \frac{9\pi Z}{4r_0{}^3} \; . \tag{13-2}$$

The corresponding values of the Fermi wavenumber are listed for each element as the first entry in Table 13-1. It is the wavenumber to which the free-electron levels are filled in the ground state. The corresponding energy $E_F = \hbar^2 k_F{}^2/2m$ is several electron volts in all cases, much greater than the thermal energy k_BT , 0.025 eV at room temperature. The Fermi velocity $v_F = \hbar k_F/m$ is of order 10^8 cm/sec., a percent or so of the speed of light.

We saw in Section 5-1 that the Brillouin Zone in wavenumber space could accommodate two electrons per primitive cell, two electrons per atom for the face-centered-cubic and body-centered-cubic structures which have a single atom per cell. Thus this Fermi sphere has half the volume of the Brillouin Zone for monovalent metals in these structures, and has a volume equal to that of the Brillouin Zone for the divalent metals, etc.

An additional quantity which will be needed for calculating a number of properties is the density of states $n(E)$, the number of states per unit volume per unit energy. That is, there are $n(E)\Omega\, dE$ states, including a factor of two for spin, in the energy range dE in a system of volume Ω . Since there are $2\Omega 4\pi k^2 dk/(2\pi)^3$ states in the wavenumber range dk , that density of states is

$$n(E) = \frac{k^2}{\pi^2 \partial E/\partial k} \quad \left(= \frac{3Z}{2\Omega_0 E_F} \text{ at the Fermi energy} \right). \tag{13-3}$$

3. Tight-Binding Theory

We now combine the free-electron picture with the tight-binding picture as we did for the chain of lithium atoms in Section 1-3. It is the combination which we also used to determine our tight-binding parameters for covalent systems. We noted already in Section 1-3 , where we matched the free-electron and tight-binding bands for a chain of atoms, that this gave an estimate of the cohesion, $(\pi^2/24)\hbar^2/(md^2)$ for the chain of alkali-metal atoms. We follow the generalization to three-dimensional metals by Wills and Harrison (1984).

The generalization of the one-dimensional bands of Eq. (1-20) to the face-centered-cubic structure, and wavenumbers in a [100] direction, is

$$\varepsilon_k = \varepsilon_s + 4V_{ss\sigma}(1 + 2\cos k_x a/2)\,. \tag{13-4}$$

(Only the four neighbors $a/2$ to the right and the four to the left have different phase.) Froyen (1980) noted that the minimum in the conduction band is $\varepsilon_s + 12V_{ss\sigma}$ and the band energy at the Zone face, $2\pi/a$ in the [100] direction, is $\varepsilon_s - 4V_{ss\sigma}$. The energy difference can be set equal to the free-electron difference of $(\hbar^2(2\pi/a)^2/2m)$ giving the conduction-band minimum at

$$\varepsilon_0 = \varepsilon_s - \frac{3\pi^2\hbar^2}{4md^2}\,. \tag{13-5}$$

Froyen found that exactly the same formula in terms of the nearest-neighbor distance applied for body-centered cubic and hexagonal close-packed (at ideal c/a ratio) structures.

A. Cohesive Energy

The average energy of the occupied states is $^3/_5\hbar^2 k_F{}^2/2m$ above this. Thus if we estimate the change energy in the formation of the solid as the change in the energies of the occupied states we obtain, as did Wills and Harrison (1984), an energy per atom for monovalent or divalent metals of

$$E_{atom} = -Z\left(\frac{3\pi^2\hbar^2}{4md^2} - \frac{3}{5}\frac{\hbar^2k_F{}^2}{2m}\right) \equiv -A(Z)\frac{\hbar^2k_F{}^2}{m}. \qquad (13\text{-}6)$$

The last form is for use in the following section on the bulk modulus. $A(Z)$ can be evaluated for any structure using Eq. (13-2) which relates k_F to the atomic volume and the relation between the atomic volume and spacing for each structure ($\Omega_0 = d^3/\sqrt{2}$ for fcc and $4d^3/(3\sqrt{3})$ for bcc). For the fcc structure (or the hcp structure with the ideal axial ratio of $c/a = \sqrt{8/3} = 1.633$ so that all nearest-neighbor distances are the same) $A(Z) = (3\pi^2Z/128)^{1/3} - 3Z/10$, equal to 0.314 for Z = 1 and 0.173 for Z= 2.

Again we should add the overlap repulsion which prevents the crystal from collapsing. If it is taken to vary as $1/d^4$, or equivalently as $k_F{}^4$, it will cancel half of the net energy gain (which varies as $1/d^2$). Thus the cohesive energy, the energy gain per atom in forming the metal from free atoms is predicted to be

$$E_{coh} = \frac{Z}{2}\left(\frac{3\pi^2\hbar^2}{4md^2} - \frac{3\hbar^2k_F{}^2}{5m}\right) = \frac{A(Z)}{2}\frac{\hbar^2k_F{}^2}{m} \qquad (13\text{-}7)$$

for Z = 1, 2. Again $A(1) =$ 0.314 and $A(2) =$ 0.173 for fcc structures.

There are important differences for trivalent and tetravalent metals. First, one of the electrons in the atom is in a p-state so its initial energy was ε_p though in the metal it enters the same band with minimum related to ε_s by Eq. (13-5). Second, $A(Z)$ evaluated from Eq. (13-6) is negative, -0.015 for Z = 3. We still obtain an energy decrease in the formation of the metal because of the energy of that one electron $\varepsilon_p - \varepsilon_s$ higher in the atom. However, we no longer have a term in the energy decreasing with decreasing spacing. We no longer have an equilibrium condition. We nonetheless make estimates using the negative $A(Z)$ from Eq. (13-6) and with no added repulsion. Thus, for Z = 3 and 4 we have

$$E_{coh} = (Z-2)(\varepsilon_p - \varepsilon_s) + A(Z)\frac{\hbar^2k_F{}^2}{m} \qquad (13\text{-}8)$$

for Z = 3 and 4. $A(3) =$ -0.015, $A(4) =$ -0.226 for fcc structures.

In Table 13-2 we compare such predictions (using everywhere the $A(Z)$ for fcc structures) with experimental values for the cohesive energy, using the Fermi wavenumber from Table 13-1. We chose here to make all evaluations relating d to the Fermi wavenumber as for an fcc structure. (The corresponding test given by Wills and Harrison did utilize the observed structures but did not include the reduction by the overlap repulsion.) It is slightly simpler to treat all metals the same and probably is just as meaningful since the cohesion would be similar even for bcc metals and the atomic volume would change little. Use of the relation for bcc structure increases the prediction slightly, improving the agreement for most alkali metals and barium, but we do not regard the difference as significant.

The general agreement is already remarkable in view of the extreme simplicity of the model and the absence of any experimental parameters but the equilibrium spacing . In particular, the light alkali metals (Li, Na) and beryllium and magnesium are in good agreement. For the heavier alkalis we have underestimated the cohesion, perhaps partly because in the heavier metals the repulsion proportional to $1/d^n$ will correspond to a larger n , as we indicated for the semiconductors in Section 2-5-B, and therefore the repulsive energy will be smaller than our estimate.

Even for the metals of higher valence, the trends are correctly predicted. Our underestimate for the noble metals is significant and indicates the role of the d-electrons, which also are responsible for the much reduced atomic volume for these metals in comparison to the alkalis. They must be treated as transition metals, which we shall do in Chapter 15.

Table 13-2. The cohesive energy of metals (eV per atom) in the fcc structure, obtained from Eq. (13-7) and (13-8) using the experimental k_F from Table 13-1 and atomic term values from Table 1-1. Experimental values (Kittel (1976), also in Table 1-3 here) are given in parentheses.

				Li 1.47 (1.63)	Be 2.49 (3.32)
		Al 4.66 (3.39)		Na 1.01 (1.11)	Mg 1.24 (1.51)
Cu 2.21 (3.49)	Zn 1.65 (1.35)	Ga 5.57 (2.81)		K 0.67 (0.93)	Ca 0.81 (1.84)
Ag 1.72 (2.95)	Cd 1.30 (1.16)	In 4.52 (2.52)	Sn 7.94 (3.14)	Rb 0.59 (0.85)	Sr 0.77 (1.72)
Au 1.75 (3.81)	Hg 1.22 (0.67)	Tl 4.35 (1.88)	Pb 7.63 (2.03)	Cs 0.51 (0.80)	Ba 0.63 (1.90)

B. The Bulk Modulus

We may press this simple tight-binding view a step further in evaluating the bulk modulus for the monovalent and divalent metals. Eq. (13-5) came from two terms which may be combined and written $-A(Z)\hbar^2 k_F{}^2/m$ plus a term in $k_F{}^4$ with a coefficient adjusted to give the minimum energy at $k_F = k_F{}^0$, the value at the observed equilibrium spacing. That condition yields a repulsion, $(A(Z)/2k_F{}^{02})\hbar^2 k_F{}^4/m$, yielding Eq. (13-5), $-{}^1/_2A(Z)\,\hbar^2 k_F{}^2/m$, at $k_F = k_F{}^0$. To obtain the bulk modulus we evaluate the second derivative of the total energy with respect to k_F, which is $-2A(Z)\hbar^2/m + 6A(Z)\,\hbar^2/m$. The bulk modulus is $(1/\Omega_0)\Omega^2\partial^2 E_{tot}/\partial\Omega^2$ with E_{tot} the total energy per atom, and Ω_0 the atomic volume. This is ${}^1/_9(1/\Omega_0)k_F{}^2\partial^2 E_{tot}/\partial k_F{}^2 = {}^4/_9A(Z)\,\hbar^2 k_F{}^{02}/(m\Omega_0)$ or

$$B = \frac{8}{9\Omega_0}\,E_{coh}\,, \tag{13-9}$$

in terms of the cohesive energy per atom given in Eq. (13-7). These are compared with experiment in Table 13-3 for the monovalent and divalent

Table 13-3 Bulk modulus (10^{12} ergs/cm^3) predicted as $B = 8E_{coh}/(9\Omega_0)$ using the predicted E_{coh} from Table 13-2. Experimental values, in parentheses, were compiled by Kittel (1976).

		Li 0.096 (0.12)	Be 0.437 (1.00)
		Na 0.038 (0.068)	Mg 0.077 (0.35)
Cu 0.268 (1.37)	Zn 0.156 (0.60)	K 0.014 (0.032)	Ca 0.027 (0.15)
Ag 0.143 (1.01)	Cd 0.085 (0.47)	Rb 0.010 (0.031)	Sr 0.023 (0.12)
Au 0.149 (1.73)	Hg 0.074 (0.38)	Cs 0.007 (0.020)	Ba 0.014 (0.10)

metals. It works reasonably well only for the alkali metals, and would generally have done slightly better had we used experimental cohesions. Thus Eq. (13-9) might be taken as an empirical relation, with or without the factor $^8/_9$, for the alkali metals, like our empirical relation that the band gap in alkali halides is approximately equal to the cohesive energy per atom pair, Section 9-1-D. For the heavy alkalis and the divalent metals the underestimates become more serious and the direct extension to $Z = 3$ and 4, as we did for Table 13-2, gives negative values. Perhaps we could not expect more from such a simple description of these metals.

We made efforts to apply this tight-binding approach to an understanding of alloys, their equilibrium volumes and heats of solution. This did not lead to a simple understanding such as it provided for the pure materials. Indeed it is a complicated problem, treated in detail, and from many points of view by Hafner (1987). We return to this problem in Section 14-7.

Tight-binding theory is indeed a crude description of metals. We turn to a description based more directly upon the free-electron theory. Where we obtained tight-binding parameters from the free-electron bands before, we now obtain pseudopotential parameters from the known electronic structure of the atoms.

4. Pseudopotentials

It is indeed remarkable that the bands in solids could be so much like free-electron bands when the potentials arising from the atoms are intrinsically so strong. If we imagine beginning with a free-electron gas, which we have used to represent a lithium metal, and slowly introducing the atomic potentials for each lithium atom, these potentials will produce a bound state (the core 1s-state) at each lithium site. For elements in lower rows of the periodic table, additional core states will be produced. L. D. Landau is quoted to have said, for this reason, that the free-electron Fermi surface which had been proposed could not be correct, that "there is no small parameter". Indeed, that is precisely the correct question, but the answer is that there *is* a small parameter, the pseudopotential. The idea was apparently first introduced by Fermi in discussing the scattering between neutrons and protons. The interaction potential is strong enough to produce a bound state, but Fermi introduced an effective potential which described the scattering and could be treated as a perturbation. Phillips and Kleinman (1959) used the concept to reduce the number of parameters in band calculations for semiconductors.

Once it was realized that simple-metal bands were indeed free-electron-like, we utilized the pseudopotential concept for metals (Harrison (1963)), not requiring the principal approximations used by Phillips and Kleinman, but treating the pseudopotential in perturbation theory. This led to a theory of the entire range of metallic properties.

The early formulation for metals was quite rigorous in the sense that it would lead to the same energy bands as full first-principles numerical calculations if the perturbation theory were not used. One such first-principles method expanded the states in *orthogonalized plane waves*, or *OPW's,* plane waves with terms added to make them orthogonal to the core states, as the true band states must be. The matrix elements of the pseudopotential $<\mathbf{k} + \mathbf{q}|W|\mathbf{k}>$ were defined to be the matrix elements of the Hamiltonian between these OPW's. This is simply a definition, but its purpose is to suggest useful approximations. We shall in fact outline this formulation in Chapter 15, as a prelude to the construction of transition-metal pseudopotentials. We use a simpler formulation in this chapter. The development of different forms for pseudopotentials continues, as in the recent work by Fiolhais, Perdew, Armster, MacLaren, and Brajczewska (1995), but they will not be discussed here.

An OPW band calculation is the diagonalization of the corresponding Hamiltonian matrix, and its success depended on the fact that a limited number of OPW's was adequate, whereas a vast number of plane waves would be required for a plane-wave expansion. The approximation made by Phillips and Kleinman (1959) to simplify the semiconductor band calculation was to assume, as would be true for a simple potential, that $<\mathbf{k} + \mathbf{q}|W|\mathbf{k}>$ was independent of $\mathbf{k}$. They also assumed that only matrix elements for the three smallest q-values entering the band calculation are nonzero. This allowed a three-parameter fit to the energy bands, and those three parameters could be adjusted to fit known bands, or experimental properties. We shall return to the bands of semiconductors from this outlook in Section 14-6.

The use of pseudopotentials in perturbation theory for the simple metals did not require either of these approximations. However, once the calculations of properties were carried out (Harrison (1966)) it was noted that very similar results could be obtained for a "local approximation" in which the dependence of the matrix elements upon $\mathbf{k}$ was neglected. That simpler view is what we use here, and in particular we use the approximate "empty-core" pseudopotential proposed first by Ashcroft (1966).

A. Empty-Core Pseudopotentials

The idea behind pseudopotentials, and the empty-core pseudopotential in particular, is easily understandable in terms of atomic states, and even the 2s-state of hydrogen. For the hydrogen atom, the potential is $v(r) = -e^2/r$ (we use lower case for the potentials for atoms, reserving capitals for the total potential of the metal). The spherically symmetric 1s-state is $\psi_{1s}(r) = \sqrt{\mu^3/\pi}\, e^{-\mu r}$ and the spherically symmetric 2s-state is $\psi_{2s}(r) = \sqrt{\mu^3/8\pi}\,(1-\mu r/2)\, e^{-\mu r/2}$. Both are solutions of the radial Schroedinger Equation,

$$-\frac{\hbar^2}{2m}\frac{1}{r^2}\frac{\partial}{\partial r}r^2\frac{\partial\psi(r)}{\partial r} + v(r)\,\psi(r) + \frac{\hbar^2\,\ell(\ell+1)}{2mr^2}\psi(r) = \varepsilon\,\psi(r) \qquad (13\text{-}10)$$

with angular-momentum quantum number ℓ taken equal to zero for s-states. It is usually simplified by writing $\chi(r) = r\psi(r)$. Then the equation for $\chi(r)$ is readily found to be

$$-\frac{\hbar^2}{2m}\frac{\partial^2\chi(r)}{\partial r^2} + v(r)\,\chi(r) + \frac{\hbar^2\,\ell(\ell+1)}{2mr^2}\chi(r) = \varepsilon\,\chi(r)\,. \qquad (13\text{-}11)$$

For the hydrogen states they are found analytically, but for general $v(r)$ they can be obtained numerically by noting that χ_ℓ approaches a constant times $r^{\ell+1}$ as $r \to 0$. Then Eq. (13-11) may be integrated from zero to large r , adjusting the energy ε until $\chi(r)$ approaches zero at large r . (See Problem 13-2.)

The 2s-state $\psi_{2s}(r)$ has a single node, and is plotted in Fig. 13-3, along with the potential $v(r) = -e^2/r$. There is an *empty-core pseudopotential*, (defined earlier in Chapter 1, and shown also in Fig. 1-11) given by

$$w(r) = \begin{cases} \dfrac{-Ze^2}{r} & \text{for } r > r_c \\[2ex] 0 & \text{for } r < r_c\,, \end{cases} \qquad (13\text{-}12)$$

($Z = 1$ for hydrogen) which will yield an s-state with no nodes, and with the energy ε equal to the 2s-state energy of $-e^4m/(8\hbar^2)$. It is sketched as $\phi(r)$, the dotted line in Fig. 13-3, joining the $\psi(r)$ for $r > r_c$. ($\phi(r)$ is not normalized if $\psi(r)$ is, and is only exactly proportional to $\psi(r)$ because we have chosen $w(r)$ to be exactly equal to $v(r)$ for $r > r_c$.) r_c has been

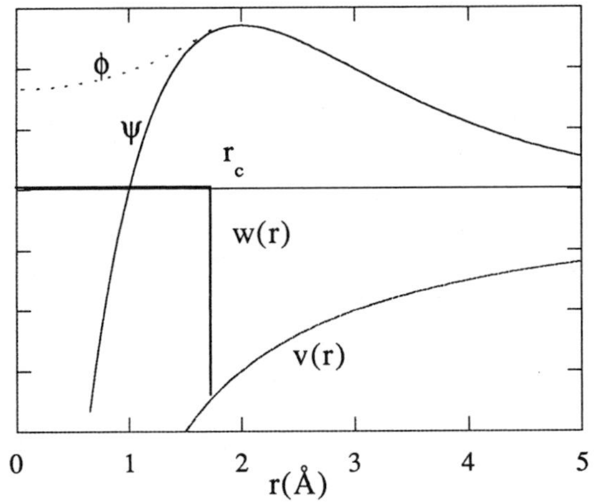

Fig. 13-3. The potential, $v(r) = -e^2/r$ for atomic hydrogen and the 2s-state $\psi(r)$ for that potential. Also shown is the empty-core pseudopotential, $w(r)$, which has an eigenstate $\phi(r)$ of the same energy.

adjusted to give the correct 2s-state energy and the corresponding $w(r)$ is also shown in Fig. 13-3. (See Problem 13-2.)

This pseudopotential $w(r)$ is sufficiently weak that it does not allow the lower-energy 1s-core state, but its lowest-energy eigenvalue, with a nodeless pseudowavefunction, has the hydrogen 2s-state energy. The core radius could similarly be adjusted to give a nodeless pseudowavefunction $\phi(r)$ with the valence 2s-state energy of -5.34 eV for lithium, or the valence 3s-state energy of -4.96 eV for sodium, or the valence s-state energy for any other atom. These empty-core pseudopotentials are very similar for all elements in one column of the periodic table, since the valence s-state energies are quite similar. They thus manifestly show the similarity of the properties of different rows in the periodic table. It turns out (See Problem 13-3) that these empty-core pseudopotentials also give a lowest p-state energy (with a node in $\phi(r)$ only at $r = 0$) approximately equal to the valence p-state energy, about half the valence s-state energy. Thus the true atomic potential can as an approximation be replaced by the weak empty-core pseudopotential if we are interested only in the valence states.

Similarly, when we go to a solid, where the potential is approximately equal to a superposition of atomic potentials $v(\mathbf{r} - \mathbf{R}_j)$ centered at the nuclear

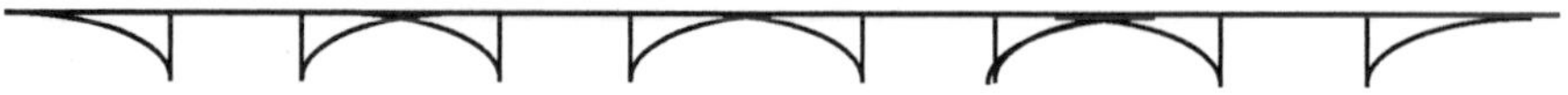

Figure 13-4. A schematic superposition of atomic pseudopotentials along a line of atoms.

positions $\boldsymbol{R}_j$, we may use a pseudopotential for the solid which is a superposition of atomic pseudopotentials $w(\boldsymbol{r} - \boldsymbol{R}_j)$, as illustrated in Fig. 13-4. It is not difficult to imagine that these superimposed pseudopotentials provide a rather weak perturbation of the electrons in a metal.

There are other formulations of pseudopotential operators which give the atomic states, and energy bands, exactly the same as the full potentials (the Orthogonalized-Plane-Wave method given in Section 15-1-A is such a way), but they are more complicated than the empty-core form.

There are various ways to obtain the empty-core radii r_c for each element. One is to fit the atomic term values, as in Problem 13-2. Another is to tune them to fit more complete band calculations, basically the way the standard values listed as the first entry in Table 13-5, and in the Solid-State Table at the end of the book, were obtained. Another is to fit some experimental property, the way the second and third entries in Table 13-5 were obtained.

We shall screen the pseudopotential in Section 13-6, which means we shall include the effects of electron redistribution by the potential in the potential itself. We do this in a way appropriate to the metal, but use the simple empty-core pseudopotential, Eq. (13-12) as the starting point. It would be preferable, particularly for polyvalent metals, to incorporate a contribution from the valence electrons in that starting point. Since the screening calculation is a correction from the potential arising from a uniform electron gas, it would seem appropriate to include in the potential of Eq. (13-12) the electrostatic potential of a uniform distribution of Z electrons over a sphere of volume equal to the atomic volume of the metal, and then the screening potential added to that. The r_c of the resulting empty core potential (with the $w(r)$ taken as zero for $r < r_c$ either before or after adding these terms) could be adjusted to fit the atomic term values. We shall in fact obtain core radii by various other approaches, and find them similar to each other. Values for all of the simple metals will be given in Table 13-5 in the following section.

Given the core radius for any metal, we can directly estimate a wide range of properties. We begin with the total energy of the solid, retaining only terms of first order in the pseudopotential.

B Madelung and Total Free-Electron Energy

If we neglect any distortion of the states by the weak pseudo-potentials, so that the states remain free-electron-like, $|k\rangle = (1/\sqrt{\Omega})e^{i k \cdot r}$ we may calculate the total energy simply in terms of

$$\sum_{k<k_F} \varepsilon_k = \sum_{k<k_F} \langle k| \frac{-\hbar^2 \nabla^2}{2m} + W(r)|k\rangle. \tag{13-13}$$

The sum over the kinetic energy gives simply $^3/_5 Z \hbar^2 k_F^2/(2m)$ per ion, as we have seen in Part 13-3-A. . The potential $W(r)$ contains the sum over the N_a empty-core pseudopotentials, $\Sigma_j w^0(r - r_j)$. It is convenient to separate out the effect of the empty core as $\langle k|w^0(r)|k\rangle = \langle k|-Ze^2/r|k\rangle + (1/\Omega)\int_{0,r_c} (Ze^2/r)4\pi r^2 dr$, with the second term equal to $2\pi Ze^2 r_c^2/\Omega$, summing to $2\pi Z^2 e^2 r_c^2/\Omega_0$ per ion with Ω_0 the volume per ion. The sum over the first term is simply the electrostatic energy of interaction between a set of point positive charges, Ze , at the atomic positions and a uniform compensating background. $W(r)$ also contains the potential arising from the other electrons, a uniform charge density of Z electrons per ion (times $(ZN_a-1)/(ZN_a)$ if we actually subtracted off the self interaction, negligible here). As we noted in Section 2-2-F, to obtain the total energy we must recognize that in summing over individual electron energies we count the interaction between each pair of electrons twice, once in the energy for each of them. Thus Eq. (13-13) contains twice the self-Coulomb energy of a uniform charge density $-ZN_a e/\Omega$ with itself and we should subtract a contribution equal to one times that self-Coulomb energy. We must also add, as noted in Section 2-2-F, the interaction between the positively-charged point ions (that is not affected by the empty core since they do not overlap) at the atomic positions. These Coulomb terms are combined as E_{es} , the electrostatic energy per ion of point positive ions at the atomic positions, imbedded in a uniform compensating charged background. Then the total energy per ion is

$$E_{ion} = \frac{3Z}{5}\frac{\hbar^2 k_F^2}{2m} + \frac{2\pi Z^2 e^2 r_c^2}{\Omega_0} + E_{es} . \tag{13-14}$$

The individual contributions to the electrostatic energy diverge for an infinite system, but the combined interaction for the neutral system is well-defined. Its calculation is difficult, as was the electrostatic energy of ionic crystals discussed in Section 9-1-B, because of the long-range nature of the Coulomb interaction. However, as for the ionic crystals, there are methods for carrying it out and the values are known. (The methods are discussed in most solid-state texts and in some detail in Harrison (1966).) This electrostatic energy depends upon the crystal structure and is written

$$E_{es} = -\frac{1}{2}\frac{Z^2e^2}{r_0}\alpha \,, \tag{13-15}$$

showing the correct dependence upon charge Z and upon atomic-sphere radius r_0 defined by $4\pi r_0^3/3 = \Omega_0$. The values of α for the common metallic structures, and two others, are given in Table 13-4. They are of course known with mathematical precision, and differ slightly among the first three structures, but that difference is never significant in real systems so it is not given here. The coefficients are significantly smaller, corresponding to weaker bonding, for simple-cubic and diamond structures, and this is generally believed to be the reason that these structures never occur in simple metals for which the theory we are describing is appropriate.

These two contributions, Eqs. (13-14) and (13-15), are certainly not adequate for a meaningful description of metals. Aside from the effects of the nonuniformities due to the ions, to which we return, real electrons avoid each other to lower their energy. The simplest such correlation of the electrons arises from the fact that real electron states are antisymmetric with respect to the interchange of electrons (giving rise to the Pauli principle which we have used throughout the text in occupying each state with no more than one electron of each spin). The simplest description including this antisymmetry is Hartree-Fock theory. In that theory the wavefunctions for two electrons of the same spin in one-electron states $\psi_1(\mathbf{r})$ and $\psi_2(\mathbf{r})$ would have a state $(\psi_1(\mathbf{r})\,\psi_2(\mathbf{r}') - \psi_2(\mathbf{r})\,\psi_1(\mathbf{r}'))/\sqrt{2}$ with $\mathbf{r}$ being the position of one and $\mathbf{r}'$ being the position of the other. This explicitly causes a change in

Table 13-4 The Madelung constants for the electrostatic energy of several metallic structures.

Face-centered cubic	1.79
Body-centered cubic	1.79
Hexagonal-close-packed (ideal close packing)	1.79
Simple cubic	1.76
Diamond	1.67

sign if the coordinates of the two electron are interchanged. This state goes to zero obviously when $\mathbf{r} = \mathbf{r}'$ so that the expectation value of the electron-electron interaction $e^2/|\mathbf{r}'-\mathbf{r}|$ is reduced. This reduction is called the *exchange* energy and is readily evaluated for a free-electron gas by direct integration over $\mathbf{r}'-\mathbf{r}$, and over the wavenumbers $\mathbf{k}$ and $\mathbf{k}'$ of the plane-wave states. The result depends only upon the density of electrons, or equivalently upon the Fermi wavenumber. The exchange energy per electron thus will be expected to be a universal constant times e^2k_F. It is, and we multiply by the number Z of electrons per ion to obtain the exchange energy per ion in our model metal of

$$E_{ex} = -\frac{3Ze^2k_F}{4\pi} . \tag{13-16}$$

There are other corrections for a uniform electron gas. These have been calculated in a number of ways, summarized by Hedin and Lundqvist (1969), principally for high electron densities, as an expansion in the average electron spacing, or in our terms in $1/k_F$. These terms are called *correlation energy* and in our context the correlation energy for every Z electrons in a uniform electron gas, given by Hedin and Lundqvist, becomes

$$E_{cor} = -Z(0.42\ ln\left(\frac{k_F}{3.63Å^{-1}}\right) + 1.564)\ \frac{eV}{ion} . \tag{13-17}$$

This correlation energy is frequently included in the theory of metals, but it has so little effect that we do not consider it worth the conceptual complication and we drop it here. We shall note its effect at one or two places where we have evaluated it.

We may now collect terms to obtain our expression for the energy of the metal, to first order in the pseudopotential and therefore treating the electrons as a uniform gas. We rewrite the empty-core contribution, noting that $1/\Omega_0 = k_F^3/(3\pi^2Z)$, and similarly write the r_0 of the electrostatic energy in terms of k_F to obtain the total energy entirely in terms of Z, k_F, and r_c,

$$\underset{\text{total}}{\frac{E_{eg}}{ion}} = \underset{\text{kinetic}}{\frac{3Z\hbar^2k_F^2}{10m}} + \underset{\text{pseudopotential}}{\frac{2Ze^2r_c^2k_F^3}{3\pi}} - \underset{\text{exchange}}{\frac{3Ze^2k_F}{4\pi}} - \underset{\text{Madelung}}{\frac{Z^{5/3}e^2k_F}{(18\pi)^{1/3}}}\alpha . \tag{13-18}$$

We note first that this total energy does not provide a prediction of the cohesive energy, the energy change in going from free atoms to the solid. That would require the corresponding evaluation of the total energy for the atom, presumably using the same pseudopotential. Probably the errors in this atomic calculation are more serious than those in the solid in which we are interested, so that would not be a very fruitful exercise.

It may also be true in the more sophisticated calculations of electronic structure that the errors are larger in the atom. In fact, Chelikowsky (1981) used a fact noted by Teller (1962), that a complete calculation of cohesion in terms of the Fermi-Thomas approximation gives a value near zero. Since the Fermi-Thomas approximation is believed to be a reasonably good description of the metal, it would follow that the cohesive energy of the solid is equal to the error in the Fermi-Thomas treatment of the atom. Chelikowsky used the difference between Fermi-Thomas energies for the atoms, and more complete local-density calculations for the atom, to confirm that this was semiquantitatively true. The analysis based upon tight-binding theory, carried out in Section 13-3-A is probably more relevant to the cohesion than Eq. (13-18).

C. Equilibrium Spacing and Bulk Modulus

The energy in Eq. (13-18) can however be used to discuss the equilibrium spacing. It depends upon the relative positions only through the volume and through the Madelung constant α given in Table 13-3. We shall add higher-order terms shortly but in the mean time it is of some interest to see the consequences of the approximation containing the pseudopotential only to first order.

We may directly predict the equilibrium spacing for the close-packed structures by inserting $\alpha = 1.79$, the valence, and r_c and minimizing with respect to k_F. An equivalent exercise was made by Hafner (1987) in fitting the r_c which will yield the experimental k_F as the value at which E_{eg} is a minimum. We have redone that fit using Eq. (13-18) with results shown as the second entry in Table 13-5. The values obtained from fits to the pseudopotential form factors (see Section 13-6-B, and Fig. 13-6 in particular) from more complete calculations, taken from Harrison (1980), are listed as the first entry and comparison of the two is a test of Eq. (13-18) as a prediction of the spacing based upon these standard values. Our fitted values differ by a few percent from those obtained by Hafner because he included the correlation energy of Eq. (13-17). (His values, incidentally, were in atomic units rather than Angstroms.)

Values obtained using Eq. (13-18) are of some interest in their own right, since we may use these fitted r_c to discuss other properties. We also can

Table 13-5. Values of the empty-core pseudopotential radius (Å). The first entry is taken from Harrison (1980) and is considered the standard value. It appears also in the Solid-State Table at the end of the book. The second is the value which will yield a minimum of E_{eg} of Eq. (13-18) (energy to first order in the pseudopotential) at the observed spacing. (These r_c values are close to those given by Hafner (1987), p. 68.) The third entry yields (Eq. (14-14)) the observed bulk modulus in a calculation based upon interatomic interactions from Eq. (14-9).

				Li	Be
				0.92	0.58
				0.65	0.55
				0.38	*
		Al		Na	Mg
		0.61		0.96	0.74
		1.01		0.89	0.96
		0.23		0.76	0.49
Cu	Zn	Ga		K	Ca
-	0.59	0.59		1.20	0.90
0.49	0.77	1.08		1.17	1.26
0.80	0.30	0.31		1.12	0.75
Ag	Cd	In	Sn	Rb	Sr
-	0.65	0.63	0.59	1.38	1.14
0.60	0.92	1.24	1.46	1.26	1.40
0.99	0.53	0.46	0.68	1.33	0.90
Au	Hg	Tl	Pb	Cs	Ba
-	0.66	0.60	0.57	1.55	1.60
0.59	0.96	1.28	1.51	1.40	1.47
1.15	0.55	0.48	0.45	1.42	0.96

*The observed bulk modulus cannot be fit with a positive r_c for Be.

obtain values for the noble metals in the same way, though they were not included in the earlier estimates.

Comparison of the first two entries for each element in Table 13-5 certainly shows that the general magnitudes are correct, and many general trends are reproduced, but indicates that Eq. (13-18) is not very useful otherwise as a predictive theory. The alternative comparison, made in Harrison (1980), of using original r_c values to predict the Fermi wavenumber (as a measure of the equilibrium spacing) appeared similar though the fractional errors were not as large; e. g., the predicted Fermi wavenumbers for tin and lead were too large by 17% and 25%, respectively.

The second derivative of the energy of Eq. (13-18) can also be evaluated at the equilibrium spacing, using the r_c fit to the equilibrium spacing, and the bulk modulus derived. This was done using Eq. (13-18) in Harrison (1980) (where it appeared as Eq. (15-16)) to obtain an analytic formula for the bulk modulus in terms of k_F and r_c. It gave values which were typically a factor of two too large. It was interesting to note that the predicted bulk modulus contained a term independent of r_c and varying as $1/d^5$ at constant Z, as we found for semiconductors. There was an additional term, proportional to r_c^2, which varied as $1/d^6$. We shall not repeat those results here, but shall fit r_c to the observed bulk modulus using a formula based upon interatomic interactions, rather than free-electron energy, in Chapter 14, leading to the values listed as the third entry in Table 13-5.

Generally the free-electron-gas energy alone does not provide a very good set of predictions for spacing or bulk modulus, nor does it allow predictions of a wide variety of properties. We shall incorporate terms of second order in the pseudopotential in Chapter 14 and see in that this greatly extends the range of properties which can be treated, and in many cases predicted with better accuracy. It is not that these corrections to Eq. (13-18) improve the predictions we have discussed here, but that it allows treatment of new properties. It seems frequently to be true in pseudopotential theory that adding higher-order terms for the estimate of any one property does not improve the agreement with experiment. The bulk modulus of the simple metals will turn out in Section 14-2 to be one exception.

The semiquantitative agreement of the three entries in Table 13-5 lends support to the treatment of the pseudopotential as a perturbation and to the empty-core form, but it also warns against expecting accurate results from the resulting estimates of properties.

D. Higher-Order Terms

If we are to go beyond first order in the pseudopotential, we incorporate the matrix elements of the pseudopotential between states of different wavenumbers, $\langle \mathbf{k} + \mathbf{q}|W|\mathbf{k}\rangle$, with $\mathbf{q} \neq 0$. These enter the second-order perturbation theory discussed in Section 1-5-B and we shall calculate the corresponding contributions to the total energy for the bonding properties of metals in the following chapter. Here we look more carefully at these matrix elements and how they enter other properties.

5. Structure Factors and Form Factors

These matrix elements between plane-wave pseudowavefunctions are given by

$$
\begin{aligned}
\langle \boldsymbol{k} + \boldsymbol{q}|W|\boldsymbol{k}\rangle &= \int d^3r \frac{e^{-i(\boldsymbol{k}+\boldsymbol{q})\cdot \boldsymbol{r}}}{\sqrt{\Omega}} \Sigma_j\, w(\boldsymbol{r} - \boldsymbol{r}_j) \frac{e^{i\boldsymbol{k}\cdot \boldsymbol{r}}}{\sqrt{\Omega}} \\
&= \frac{1}{\Omega} \Sigma_j \int d^3r\, e^{-i\boldsymbol{q}\cdot \boldsymbol{r}} w(\boldsymbol{r} - \boldsymbol{r}_j) \\
&= \frac{1}{N} \Sigma_j\, e^{-i\boldsymbol{q}\cdot \boldsymbol{r}_j} \frac{1}{\Omega_0} \int d^3r e^{-i\boldsymbol{q}\cdot (\boldsymbol{r} - \boldsymbol{r}_j)} w(\boldsymbol{r} - \boldsymbol{r}_j) \\
&\equiv S(\boldsymbol{q})\, w_{\boldsymbol{q}} \ .
\end{aligned}
\qquad (13\text{-}19)
$$

In the first step we wrote the pseudopotential as a superposition of atomic pseudopotentials. In the final step we were able to factor the matrix element into a *structure factor*,

$$
S(\boldsymbol{q}) = \frac{1}{N} \Sigma_j\, e^{-i\boldsymbol{q}\cdot \boldsymbol{r}_j} \ , \qquad (13\text{-}20)
$$

which depends only upon the positions of the ions, but is independent of the form of the identical individual atomic pseudopotentials, and a *form factor* ,

$$
w_{\boldsymbol{q}} = \frac{1}{\Omega_0} \int d^3r\, e^{-i\boldsymbol{q}\cdot \boldsymbol{r}} w(r) , \qquad (13\text{-}21)
$$

which depends only upon the individual atomic pseudopotentials, and is independent of the atomic positions, except through Ω_0 . For Eq. (13-21) we changed the variable of integration from $\boldsymbol{r} - \boldsymbol{r}_j$ to $\boldsymbol{r}$. This separation into factors is very familiar in diffraction theory, which is in fact what we are considering here, the diffraction of electrons by the pseudopotential.

The factorization is an extraordinary simplification. For the simple pseudopotential given in Eq. (13-12), for example, we note that $w(r)$ is spherically symmetric so we may take the angular average, $\sin qr/qr$, of $e^{-i\boldsymbol{q}\cdot \boldsymbol{r}}$ in Eq. (13-21). With $w(r) = -Ze^2/r$ at large r we need to insert a convergence factor $e^{-\kappa r}$, taking $\kappa = 0$ in the end. We obtain the form factor as

$$w_q = \frac{1}{\Omega_0}\int dr\, 4\pi r^2 \frac{\sin qr}{qr} w(r) = -\frac{4\pi Ze^2}{\Omega_0 q}\int_{r_c,\infty} dr\; e^{-\kappa r}\sin qr$$

$$= -\frac{4\pi Ze^2 e^{-\kappa r_c}\cos qr_c}{\Omega_0(q^2+\kappa^2)} - \frac{4\pi Z\kappa e^2 e^{-\kappa r_c}\sin qr_c}{\Omega_0 q(q^2+\kappa^2)} \rightarrow -\frac{4\pi Ze^2\cos qr_c}{\Omega_0 q^2}. \tag{13-22}$$

We shall find that screening of the pseudopotential will have the effect of introducing a finite κ , which further justifies our use of this convergence factor.

Even with a more complicated pseudopotential, we may obtain w_q as a function of q and use that same function with the structure factors for *any* arrangement of atoms of interest to calculate properties of the corresponding metal. The exact final form in Eq. (13-22) will ordinarily not be of interest because of the effects of screening to which we turn next.

6. Screening of Pseudopotentials

An important effect of the pseudopotentials is to modify the electronic states. This in turn modifies their contribution to the charge distribution and therefore introduces additional potentials, called *screening* of the pseudopotential. The straight-forward way to do this is first-order perturbation theory, which corrects the plane-wave pseudowavefunction as

$$|k\rangle \rightarrow |k\rangle + \sum_q |k+q\rangle \frac{\langle k+q|W|k\rangle}{\varepsilon_k - \varepsilon_{k+q}} \tag{13-23}$$

for the presence of the pseudopotential W . This must be done self-consistently in the sense that initially we know only the *unscreened* pseudopotential and we much calculate the charge density from the corrected wavefunctions in order to determine the screened pseudopotential which enters Eq. (13-23). At the end of this part we shall indicate the result of carrying out that calculation. However, it will be adequate, and extraordinarily simpler, to proceed approximately using the linearized Fermi-Thomas approximation. We may think of these results as coming from a full self-consistent calculation, using the perturbation theory of Eq. (13-23), and then taking the limit as the wavenumber q in Eq. (13-23) becomes small compared to the Fermi wavenumber k_F . That is the way that these results were derived in Harrison and Wills (1982).

A. Fermi-Thomas Theory

We describe first the general method, and will then carry it out in detail for small perturbations. The Fermi-Thomas approximation is basically a local evaluation of the kinetic energy of a collection of electrons as the kinetic energy of a uniform electron gas of the same density. That is, the Fermi wavenumber for a uniform electron gas of density ρ electrons per cubic centimeter is given by $k_F{}^3 = 3\pi^2\rho$, as we saw in Eq. (13-2). Then the kinetic energy density is given by $^3/_5\hbar^2k_F{}^2\rho/(2m)$. If we evaluate the kinetic energy locally, the total kinetic energy of the system becomes

$$E_{ke}(total) = \int d^3r\ \frac{3\hbar^2k_F{}^2\rho(\mathbf{r})}{10m} = \frac{3(3\pi^2)^{2/3}\hbar^2}{10m}\int d^3r\ \rho(\mathbf{r})^{5/3}\ . \qquad (13\text{-}24)$$

This could, for example, be divided by the number of ions and substituted for the first term in Eq. (13-18), giving the same result for the uniform electron gas.

It can also be used, as an approximation to the Schroedinger Equation, to treat nonuniform systems entirely in terms of $\rho(\mathbf{r})$ without use of wavefunctions. The total potential energy is $\int d^3rV(\mathbf{r})\rho(\mathbf{r})$ and $V(\mathbf{r})$ consists of the pseudopotential arising from the metallic atoms $W_0(\mathbf{r})$ plus the potential $V_{sc}(\mathbf{r})$ due to the electrons obtained directly from $\rho(\mathbf{r})$ using Poisson's Equation. Then the ground state of the system is obtained by varying the electron density (subject to the total electron count remaining constant) to minimize the energy so calculated. This is the Fermi-Thomas approximation. The basic approximation is treating the electron gas as locally uniform and it becomes a good approximation when $\rho(\mathbf{r})$ is slowly varying over an electron wavelength. This approximation is frequently very good in the simple metals we treat here.

The Fermi-Thomas approximation has also been used to treat individual atoms and we may see that this results in serious difficulties, because the density variation in atoms is large over an electron wavelength. If we imagine a neutral atom, the potential must of course be zero far from the atom. Then as we approach the atom, the potential must remain zero until some electron density is present. This cannot however occur in the Fermi-Thomas approximation until the potential drops below the Fermi energy, which means that the Fermi energy must occur *at the vacuum energy level.*

The removal energy of the highest-occupied state, the ionization energy of the atom, is found to be zero. Even with this defect, other aspects of the Fermi-Thomas theory of large atoms were meaningful enough to be useful. We mentioned a second related defect of the Fermi-Thomas approximation in Section 13-4-B. It was the fact noted by Teller (1962) that the cohesive energy of a metal is negative and near zero in the Fermi-Thomas approximation.

These are serious difficulties with the Fermi-Thomas approximation for the atom, but it remains an excellent approach for bulk metals for which the electron density is quite uniform. In that case we may in fact treat the deviations from uniformity as small, calculating the energy to first order in those deviations, the linearized Fermi-Thomas approximation. It is then very simple. We begin with a uniform electron gas (with a compensating positive background charge which we initially take to be uniform) and allow a small deviation $V(\mathbf{r})$ of the potential from a constant. The Fermi energy of the system remains constant and adjusted so that the total number of electrons calculated from $\int d^3r\ \rho(\mathbf{r})$ is the correct total number of electrons. This is like a swimming pool, as illustrated in Fig. 13-5. An added potential $V(\mathbf{r})$ shifts the energy of a stationary (zero-kinetic-energy) electron at that point, but the Fermi energy remains the same so the depth in energy to which states are occupied is changed by $-V(x)$. If this change is small, and if $n(\varepsilon_F)$ is the density of states per unit volume and per unit energy, (defined in Section 13-2) at the Fermi energy, the change in the electron density at that point will be

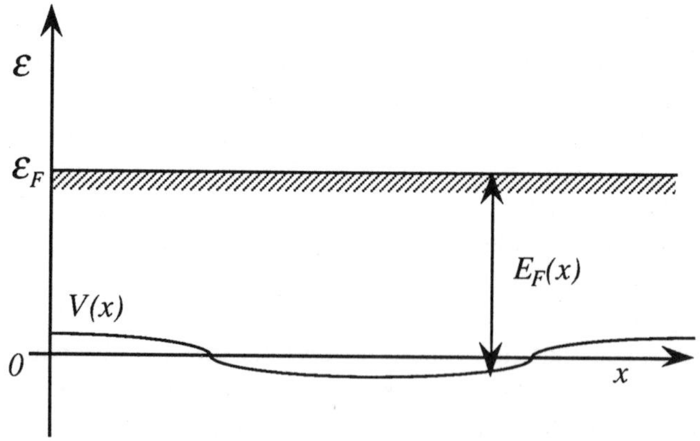

Fig. 13-5. The linearized Fermi-Thomas approximation. The Fermi level ε_F is a constant of the system, adjusted to give the correct number of electrons. If a small potential $V(x)$ is introduced, the total width $E_F(x)$ at each point x is shifted by $-V(x)$, changing the electron density by approximately $-n(\varepsilon_F)V(x)$.

$-V(\mathbf{r})n(\varepsilon_F)$, with $n(E_F)$ obtainable from Eq. (13-3).

The potential $V(\mathbf{r})$ of course includes any applied potential (such as the pseudopotential from the ions in the metal) $W_0(\mathbf{r})$ and the potential which arises from the electron density, which we write $V_{sc}(\mathbf{r})$. This screening potential is related to the shifted electron density $-V(\mathbf{r})n(\varepsilon_F)$ through Poisson's Equation by

$$-\nabla^2 V_{sc}(\mathbf{r}) = -4\pi e^2 n(\varepsilon_F)\, V(\mathbf{r}) \equiv -\kappa^2 V(\mathbf{r})\,, \tag{13-25}$$

where we have defined the *Fermi-Thomas screening parameter* κ by

$$\kappa^2 = 4\pi e^2 n(\varepsilon_F) = \frac{6\pi Z e^2}{\varepsilon_F \Omega_0} = \frac{4e^2 k_F m}{\pi\hbar^2}\ . \tag{13-26}$$

B. Screened Form Factors

We have noted in Section 13-5 that the pseudopotential enters our calculations through the form factor $w_q = \int d^3r w(r) e^{-i\mathbf{q}\cdot\mathbf{r}}$ and since we obtain the screening only to first order in the potential, we may directly sum the screening for each term in the unscreened pseudopotential $W_0 = \Sigma_j w_0(\mathbf{r} - \mathbf{r}_j)$, and in fact simply screen each Fourier component of the atomic pseudopotential; that is, the form factor. The screened final potential $V(\mathbf{r})$ becomes simply the sum of screened atomic pseudopotentials $w(\mathbf{r})$. For each wavenumber $\mathbf{q}$ the Laplacian ∇^2 in Eq. (13-25) is simply $-q^2$ and Eq. (13-25) becomes $q^2 V_{sc} = -\kappa^2 V$. We may add $q^2 W_0$ to each side, noting that $V = W_0 + V_{sc}$, and solve for $V = q^2 W_0 /(q^2 + \kappa^2)$. This is more usually written in terms of a *dielectric function*

$$\varepsilon(q) = 1 + \frac{\kappa^2}{q^2} \tag{13-27}$$

as $w_q = w_{0q}/\varepsilon(q)$ in analogy with the reduction of uniform electric fields or potential gradients by a dielectric constant.

Obtaining the dielectric function constitutes a solution of the screening problem. In this case we have obtained it in the Fermi-Thomas approximation. Then the pseudopotential form factor which we found in Eq. (13-22) as $-4\pi Z e^2 \cos qr_c/(\Omega_0 q^2)$ is divided by the dielectric function to become the screened pseudopotential form factor,

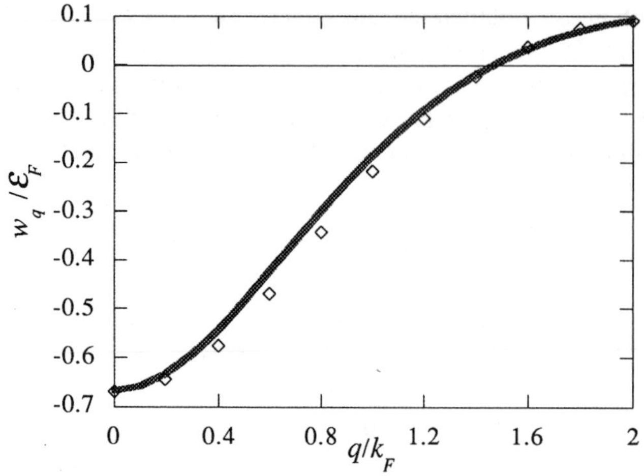

Fig. 13-6. The pseudopotential form factor in aluminum, from Eq. (13-28) using the k_F and r_c which we have given. The points are the calculated form factor for coupling states on the Fermi surface from Animalu and Heine (1965).

$$w_q = -\frac{4\pi Ze^2 \cos qr_c}{\Omega_0(q^2+\kappa^2)} = -\frac{2\varepsilon_F}{3}\frac{\cos qr_c}{1+\frac{q^2}{\kappa^2}} . \tag{13-28}$$

The last form was obtained by substituting for Z/Ω_0 from Eq. (13-26).

This pseudopotential form factor is plotted for aluminum in Fig. 13-6 as a function of q/k_F. We see that it approaches $-{}^2/_3\varepsilon_F$ as q goes to zero, as follows from Eq. (13-28) and follows for all metals. This is a property of the screening, and remains true even with the more accurate dielectric function we shall give in Eq. (13-31). At higher q it becomes positive. We have shown also in Fig. 13-6 the pseudopotential form factor, obtained for the model potential of Heine and Abarenkov (1964) from the detailed calculations by Animalu and Heine (1965). These calculations included the dependence upon $\boldsymbol{k}$ but were evaluated for both $\boldsymbol{k}$ and $\boldsymbol{k+q}$ lying on the Fermi surface so they become a function of the magnitude of q alone and are restricted to $0 < q < 2k_F$. It is no coincidence that they cross zero at the same value of q/k_F since the r_c values taken from Harrison (1980) were obtained by making that fit. The crossing will be at slightly different q/k_F for other, equally valid, pseudopotential calculations.

We may proceed just as simply as without screening. We saw that the effects of the pseudopotential arise directly through these form factors, multiplied by the appropriate structure factor. Similarly if other potentials are applied to the metal, they may be Fourier transformed, divided by the dielectric function and reexpanded in real space to obtain the screened potential.

C. The Real-Space Pseudopotential, $w(r)$

One might have thought that the form factor of Eq. (13-28) was too complicated to reexpand in real space analytically, but it is actually quite simple (Harrison and Wills (1982)). We may see by dividing both sides of Eq. (13-21) by the number of atoms N that w_q/N is the Fourier transform of the screened pseudopotential from a single atom. Thus we may re-expand in real space, converting the sum over wavenumbers to an integral over wavenumber space, noting as we did in Section 13-2 that the density of states in wavenumber space is $\Omega/(2\pi)^3$ and that w_q, given in Eq. (13-28) is spherically symmetric,

$$w(\mathbf{r}) = \sum_q \frac{w_q}{N} e^{i\mathbf{q}\cdot\mathbf{r}} = \frac{\Omega}{(2\pi)^3}\int dq\, 4\pi q^2 \frac{w_q}{N}\frac{\sin qr}{qr}$$
$$= -\frac{2Ze^2}{\pi r}\int dq \frac{q\cos qr_c \sin qr}{q^2+\kappa^2} \tag{13-29}$$

The integral in Eq. (13-29) extends from $q=0$ to ∞ but we may equivalently replace $\sin qr$ by $e^{iqr}/2$ and integrate from $-R$ to R with R very large. We then close a contour in the upper half plane. The integral around the upper portion of the contour goes to zero at large R because of the factor e^{iqr}, as long as $r > r_c$. Thus our real-space result applies only for $r > r_c$. The only pole in the upper half plane is at $q = i\kappa$ (arising from $1/(q^2 + \kappa^2)$). The integral is then $i\pi$ times the residue at that pole, $i\kappa \cosh \kappa r_c\, e^{-\kappa r}/(2\kappa i)$. We have

$$w(\mathbf{r}) = -\frac{Ze^2\cosh \kappa r_c\, e^{-\kappa r}}{r} \quad \text{for } r > r_c. \tag{13-30}$$

This is an extraordinarily simple result. One limit is quite familiar, $r_c = 0$. The screening of a Coulomb potential $-Ze^2/r$ in Fermi-Thomas theory simply reduces the potential by a factor $e^{-\kappa r}$. It can be obtained quite directly from Eq. (13-25) by writing $V(r) = V_{sc}(r) - Ze^2/r$ in the final form and solving the differential equation. The general solution with spherical symmetry is $Ae^{-\kappa r}/r + Be^{+\kappa r}/r$. The second is not allowed and we have a charge $-Ze$ at $r = 0$ requiring $A = -Ze^2$. For the empty-core pseudopotential the only other effect outside the core radius is the scaling up of the potential by a factor $\cosh\kappa r_c$, arising because the empty-core pseudopotential attracts less electronic charge to the atom locally than the full Coulomb potential. This form will prove very useful in the following chapter where we use it to obtain the form of the interatomic interactions.

D. More Accurate Screening

A more accurate solution of the screening can be obtained by using the perturbation theory of Eq. (13-23) and summing over occupied states to obtain the $\rho(\mathbf{r})$ which in the Fermi-Thomas approximation we obtained as $-n(\varepsilon_F)V(\mathbf{r})$. The solution for the screened potential is then carried out, just as we did for the Fermi-Thomas approximation, to obtain a quantum-mechanical dielectric function called the *Lindhard dielectric function* or the *Hartree dielectric function* , derived for example in Harrison (1970), 290 ff. It is

$$\varepsilon(q) = 1 + \frac{\kappa^2}{2q^2}\left(\frac{1 - q^2/(2k_F)^2}{q/k_F}\ln\left|\frac{2k_F + q}{2k_F - q}\right| + 1\right), \tag{13-31}$$

with κ^2 given again by Eq. (13-26). It is easy to confirm that as q/k_F becomes small, this more accurate and more complicated form approaches the Fermi-Thomas dielectric function. At large q this form also approaches $\varepsilon(\infty) = 1$, but more rapidly with increasing q as $1 + 4\kappa^2 k_F^2/(3q^4)$. At $q = 2k_F$ the behavior is singular leading to Friedel oscillations which we discuss next. They may be the most important consequence of the full quantum calculation.

The transformation back to real space is more complicated if we use the Lindhard dielectric function rather than the Fermi-Thomas dielectric function. The screened potential $V(r)$ arising from an unscreened e^2/r was calculated numerically for a typical metallic electron density using Eq. (13-31) and is shown as the "quantum" form in Fig. 13-7a. It is also compared with the unscreened e^2/r and the Fermi-Thomas $e^2e^{-\kappa r}/r$. Indeed the Fermi-Thomas theory gives a good account of the potential, and its reduction from the unscreened values. Proportionately the deviations are significant at large r and in fact if we expand the scale as in Fig. 13-7b we see that there are oscillations in the quantum form, the so-called Friedel oscillations. It is not difficult to obtain the large-r form [see, for example, Harrison (1970)] as $V(r) \approx me^4/4\pi\hbar^2\varepsilon(2k_F)^2)\, cos(2k_Fr)/(k_Fr)^3$. They arise, as we indicated before, from the singularity of the Lindhard dielectric function at $q = 2k_F$. It is only for special properties that these oscillations are significant and we proceed with the much simpler Fermi-Thomas theory.

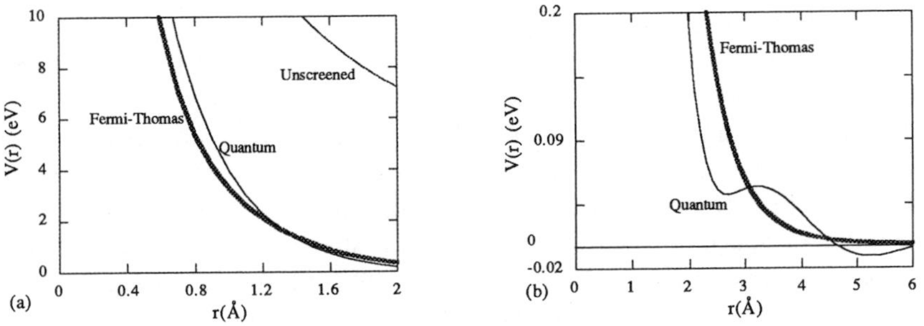

Fig. 13-7. In part a is shown the unscreened potential $V(r) = e^2/r$, its screened form $e^2e^{-\kappa r}/r$ from the Fermi-Thomas approximation, and the result with quantum screening. Parameters are for sodium, $k_F = 0.91$ Å^{-1}, $\kappa = 1.48$ Å^{-1}. In Part b the scale is expanded to show the Friedel oscillations.

7. Diffraction and Fermi Surfaces

We saw in Section 6-4-A that the dynamics of carriers in energy bands was described by a velocity

$$v = \frac{1}{\hbar}\frac{\partial\varepsilon_k}{\partial k} \qquad (13\text{-}32)$$

and an acceleration corresponding to the wavenumber change

$$\hbar\frac{dk}{dt} = \boldsymbol{F} = -e(\boldsymbol{E} + \frac{1}{c}\, v\times\boldsymbol{H}\,)\,. \qquad (13\text{-}33)$$

A. Motion in a Uniform Magnetic Field

The dynamics of electrons in metals is perhaps best understood in terms of behavior in a uniform magnetic field. Then we learn two things from the second equation, Eq. (13-33), depending upon how we express $\boldsymbol{v}$. First, noting Eq. (13-32) for $\boldsymbol{v}$ we see that the changes in $\boldsymbol{k}$ with time are perpendicular to the gradient of energy with respect to wavenumber, so the energy does not change with time. We also note that the change in $\boldsymbol{k}$ with time is perpendicular to $\boldsymbol{H}$, so the component of $\boldsymbol{k}$ along the field does not change with time. Together this means that the motion in wavenumber space is the intersection of a surface of constant energy and a plane perpendicular to the field, as illustrated in Fig. 13-8.

By writing $\boldsymbol{v}$ as $d\boldsymbol{r}/dt$ in Eq. (13-33) we see that the shape of the orbit in real space (perpendicular to the magnetic field) is exactly the same as the shape in wavenumber space, scaled by a factor $-eH/\hbar c$ and rotated 90° . This applies for arbitrary energy bands and shapes of constant-energy surfaces. For free-electron bands it corresponds to circular orbits in wavenumber space and helical orbits in real space, circular when projected on a plane perpendicular to $\boldsymbol{H}$ with possible uniform motion along the field.

With a pseudopotential present, there may also be diffraction by that pseudopotential. As we indicated in Part A, our separation of the matrix elements into a structure factor and a form factor was following traditional diffraction theory. We may in fact say that the regular arrangements of atoms produces a periodic pseudopotential which diffracts the electrons.

We may organize that diffraction theory for a perfect lattice, which will lead us again to a description of energy bands, which we came to earlier

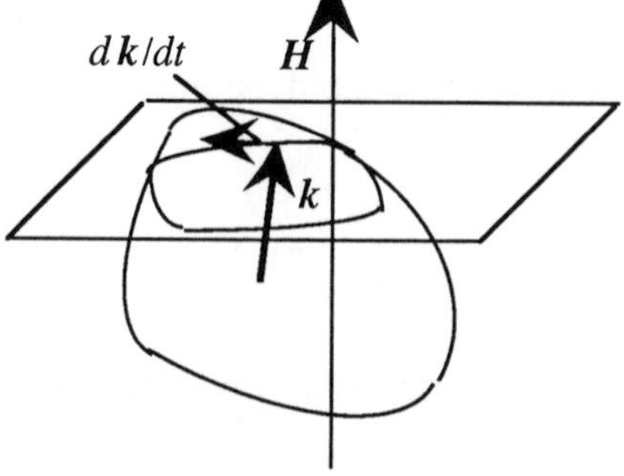

Fig. 13-8. The change of wavenumber with time $d\boldsymbol{k}/dt$ in the presence of a uniform magnetic field $\boldsymbol{H}$ is along the intersection of a surface of constant energy and any plane perpendicular to the magnetic field.

through the tight-binding description. Here we first evaluate the structure factor. For a linear chain it would be simply $S(q) = \Sigma_j e^{-iqdj}/N$ with the sum running from 0 to $N-1$. This is simply a geometric series ($\Sigma_n a^n$) with a sum $(1 - e^{-iqdN})/(1 - e^{-iqd})$. The numerator is always zero because of periodic boundary conditions giving wavenumbers such that $qdN = 2\pi n$, with n an integer. Thus the structure factor is zero unless the denominator is zero; that is, if qd is 2π times an integer. Then every term in the sum is unity and the structure factor is $N/N = 1$. We obtain nonzero structure factors only if q is some integral multiple of $2\pi/d$. This generalizes to three dimensions, as we saw in Section 5-1, where we introduced primitive lattice wavenumbers $\boldsymbol{k}_1$, $\boldsymbol{k}_2$, and $\boldsymbol{k}_3$, each corresponding to the $2\pi/d$ in the one-dimensional chain. The sum over N_c cell positions, $n_1\tau_1 + n_2\tau_2 + n_3\tau_3$, becomes three sums over $n_i\, \boldsymbol{k}_i \cdot \tau_i$ and is zero unless $\boldsymbol{q} = m_1\boldsymbol{k}_1 + m_2\boldsymbol{k}_2 + m_3\boldsymbol{k}_3$ with integral m_1, m_2, and m_3; in that case it is one. These wavenumbers are called the *wavenumber lattice* and matrix elements vanish except for $\boldsymbol{q}$ a lattice wavenumber. If there are two atoms per primitive cell, as for the silicon lattice discussed in Chapter 5-1, there is an additional factor from the sum within each cell of $(1 + e^{-i\boldsymbol{q} \cdot \tau})/2$, with τ the separation between the two atoms in the cell. (In metals there are two atoms per primitive cell in the hexagonal-close-packed structure, so such a factor occurs. It causes structure factors to vanish on some of the lattice wavenumbers. We shall not consider such cases here.)

Diffraction will occur only if the two coupled states, $\boldsymbol{k}$ and $\boldsymbol{k} + \boldsymbol{q}$ have the same energy. That, for free electrons, means $|\boldsymbol{k} + \boldsymbol{q}|^2 = k^2 + 2\boldsymbol{k} \cdot \boldsymbol{q} + q^2 = k^2$, or

$$\boldsymbol{k} \cdot \boldsymbol{q} = -\frac{q^2}{2}, \tag{13-34}$$

the *Bragg condition*. This means that diffraction can occur only if the wavenumber $\boldsymbol{k}$ lies on a plane bisecting a lattice wavenumber, call *Bragg reflection planes*. Then the electron will change its wavenumber by the negative of that lattice wavenumber to a state of equal energy.

This is simplest to understand for a one-dimensional system, with $\varepsilon_k = \hbar^2k^2/2m$, and a lattice spacing d, so that the lattice wavenumbers are $2\pi n/d$, with n any integer. This is illustrated in Fig. 13-9. Then an electron will only be diffracted if its wavenumber is some $\pi n/d$; two such diffractions are indicated by the arrows. If an electron is accelerated by an electric field such that its wavenumber moves steadily to the right, when it reaches a value

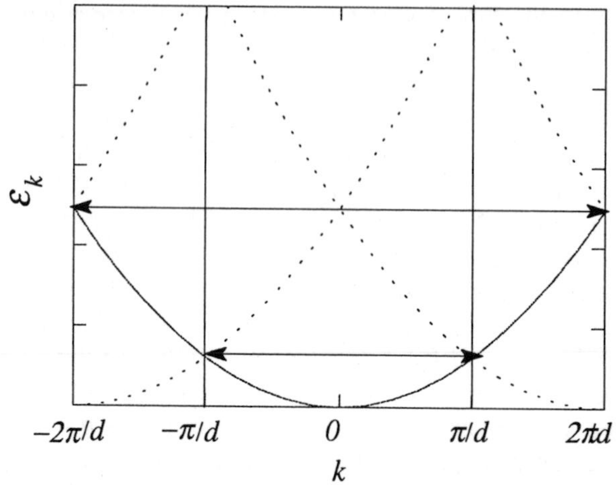

Fig. 13-9. The free electron energy as a function of wavenumber, the solid line, for an electron moving in a one-dimensional lattice of spacing d. Diffraction conditions are indicated by the arrows. The dotted lines are sections of the parabola, translated by a lattice wavenumber, which provide the corresponding energy band within the Brillouin Zone.

$\pi n/d$, its wavenumber will be diffracted to the left and again commence moving to the right until it again matches that condition.

In our tight-binding description, and in most representations of energy bands, wavenumbers which differ by a lattice wavenumber are regarded as equivalent and all bands are plotted as a function of the equivalent wavenumber which lies within the Brillouin Zone, $-\pi/d < k < \pi/d$ in this case. We have constructed these bands in Fig. 13-9 by plotting a free-electron parabola originating from each of the lattice wavenumbers, as dotted lines. Then within the Brillouin Zone we have the entire free-electron band replotted. Then each diffraction corresponds to a change in wavenumber to an equivalent value at the opposite Brillouin Zone face.

We may illustrate the generalization of this plot to three dimensions by considering the face-centered-cubic lattice of many metals. The lattice wavenumbers are obtained from Eq. (5-1) $\mathbf{k}_1 = 2\pi\, \tau_2 \times \tau_3 / \tau_1 \cdot \tau_2 \times \tau_3 = 2\pi/a\ [-1,1,1]$, $\mathbf{k}_2 = 2\pi/a\ [1,-1,1]$, and $\mathbf{k}_3 = = 2\pi/a\ [1,1,-1]$. The corresponding wavenumber lattice is a body-centered-cubic lattice as we noted in Section 5-1. This is shown in Fig. 13-10, which views the

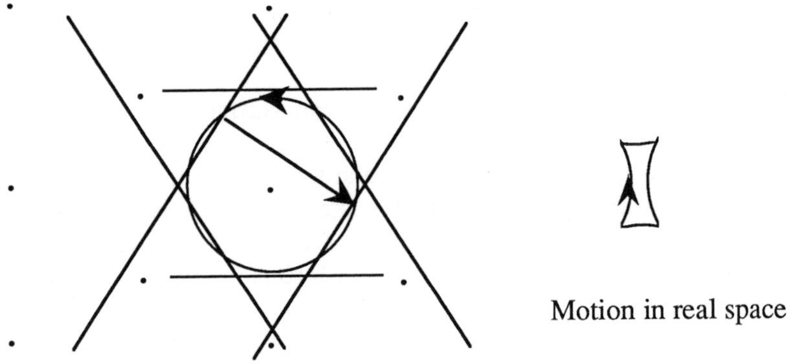

Fig. 13-10. The wavenumber lattice for a face-centered-cubic system, viewed along a [110] direction with the z-axis vertical. The figure cuts through a diagonal plane of the body-centered-cubic wavenumber lattice containing four corners and the central wavenumber of each cube. The intersections of the nearest Bragg planes and the plane of the figure are shown, each perpendicular to the plane of the figure and making up a part of the Brillouin Zone. If a magnetic field is applied perpendicular to the plane of the figure, electrons will move around the constant-energy surface as shown. When they reach the Bragg plane the wavenumber will be diffracted to the opposite plane as indicated. The motion of the electron in real (as opposed to momentum) space is shown to the right.

wavenumber lattice from a [110] direction, so that there are four primitive lattice wavenumbers, $\pm \boldsymbol{k}_1$ and $\pm \boldsymbol{k}_2$, in the plane of the figure. We also in Section 5-1 defined the Brillouin Zone as the region of wavenumber space closer to $\boldsymbol{k} = 0$ than to any other lattice wavenumber, as in the one-dimensional case. This is of course bounded by the bisectors of lattice wavenumbers and therefore, by Eq. (13-34) , by Bragg reflection planes. The nearest Bragg planes, which make up the Brillouin Zone are also shown in Fig. 13-10.

In Fig. 13-10 we also show a constant-energy surface which intersects the nearest Bragg planes. If a magnetic field is applied perpendicular to the plane of the figure, states on that circle will move along the circle, the direction, in this case counter-clockwise, depending upon whether the field is into or out of the figure. When the wavenumber reaches a Bragg plane, it is diffracted to the opposite Bragg plane, as shown and as in the one-dimensional case, with the electron changing its direction of motion, but of course not its position. It then continues to move in a counterclockwise direction until it meets another Bragg plane. Thus the motion in real space is as depicted to the right, continuing through four Bragg reflections.

B. Construction of Fermi Surfaces

In metals, the important electrons are at the Fermi surface, the constant-energy surface separating full and empty states, because it is only these states which change occupation in the presence of small electric fields. Thus we may imagine that the sphere drawn in Fig. 13-10 is the Fermi surface for the free-electron metal.

We should now relate this to our finding, in the discussion of Fig. 13-8, that the orbit in real space is the same as that in wavenumber space, but rotated by 90°. As in the one-dimensional case, all states could be represented by wavenumbers within the Brillouin Zone. In the tight-binding case, that again was because adding any lattice wavenumber, such as $\boldsymbol{k_1}$, to some wavenumber $\boldsymbol{k}$ in the Brillouin Zone produces coefficients $e^{i(\boldsymbol{k}+\boldsymbol{k_1})\cdot \boldsymbol{r}_j}$ which differ only by a factor $e^{2\pi i n} = 1$ and do not change the state. These wavenumbers within the Brillouin Zone describe the properties of the state under these translations of the crystalline lattice. Thus the segment of the Fermi surface which lies to the upper right of Fig. 13-10, and lies just outside the Zone, can be translated back by a lattice wavenumber and replotted just inside the Zone to the lower left in the figure. Such a translation of the Fermi sphere into the Brillouin Zone was made many years ago by A. V. Gold (1958) in the study of the Fermi surface of lead. There is of course an easier way, as we saw in Fig. 13-9, since each such translated spherical segment becomes a segment of sphere centered not at $\boldsymbol{k} = 0$, but at some lattice wavenumber. Thus we may construct spheres around each lattice wavenumber and this will include any Fermi surface which could be translated into the Brillouin Zone, just as the additional parabolas in Fig. 13-9 gave all of the one-dimensional bands in the Brillouin Zone. This construction of Fermi surfaces is illustrated for a simple-cubic lattice in Fig. 13-11.

This corresponds to a Fermi energy just above the lowest arrow in Fig. 13-9, which would then represent bands plotted along the k_x direction in Fig. 13-11. We see that the first sphere-crossing along k_x is at the Fermi surface in the second band, the band of second-lowest energy at that wavenumber. We could in fact deduce that directly from Fig. 13-11 by noting that this surface divides the region within two-spheres from the region within one sphere. Thus there is also Fermi surface in the first band near the corner of

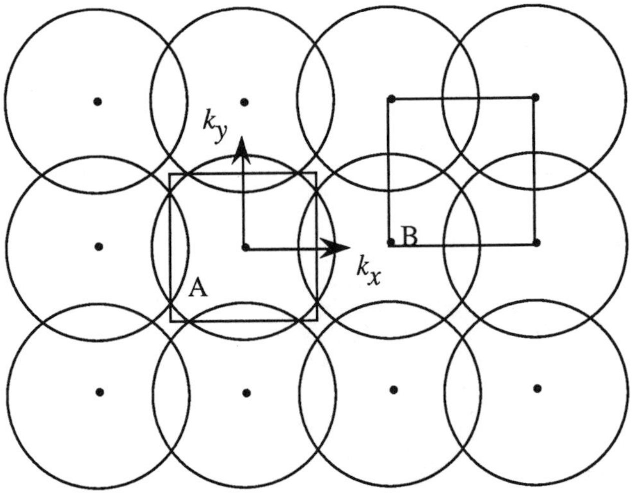

Fig. 13-11. The points represent lattice wavenumbers, relative to $\boldsymbol{k} = 0$ in the center of the Brillouin Zone A. Fermi spheres were constructed around every lattice wavenumber, producing all of the Fermi surface to be found within that Brillouin Zone. Since the construction is periodic it could equally well be represented by a Brillouin Zone constructed at some other point, such as B.

the section of Brillouin Zone shown, dividing a region within one sphere and a region within no spheres.

In this periodic representation of the Fermi surface one clearly could have chosen the Brillouin Zone, which is a primitive cell of this wavenumber lattice, centered wherever one liked. For example, if one chose it as indicated by B in Fig. 13-11, one would see this Fermi surface in the first band as square section, rather than as four lines at the corners of the square. This has the advantage that the section of Fermi surface then has a shape identical to the orbit of an electron in real space. Our construction has plotted every state which differs by a lattice wavenumber - and which can therefore be connected by a diffraction - at the same point so there need be no jumps in wavenumbers. Since the spheres represented states of the same energy, their intersection satisfies the second condition, energy conservation, for Bragg reflection.

This construction generalizes directly to any three-dimensional crystal structure. In the face-centered-cubic structure, a monovalent metal such as copper will have a Fermi sphere with half the volume of the Brillouin Zone, and that sphere does not intersect the Brillouin Zone. (It does come quite

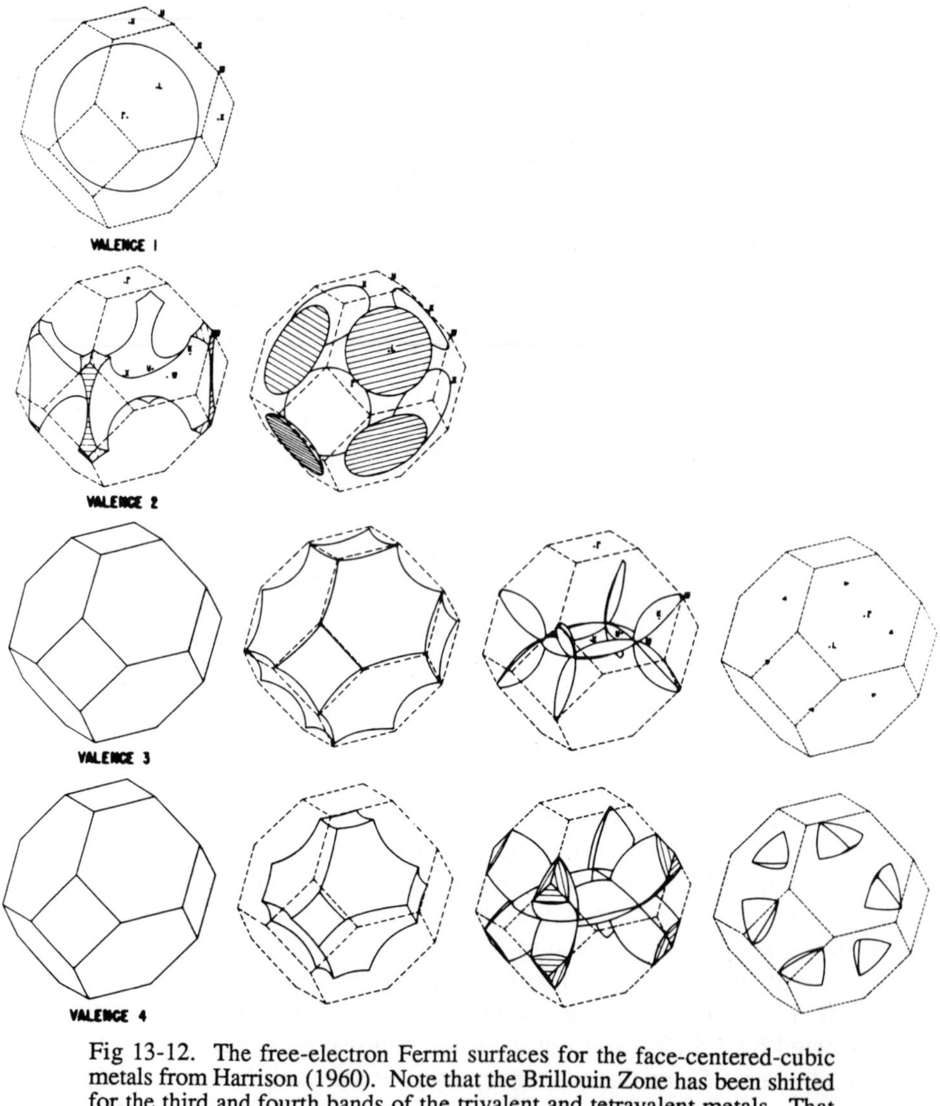

Fig 13-12. The free-electron Fermi surfaces for the face-centered-cubic metals from Harrison (1960). Note that the Brillouin Zone has been shifted for the third and fourth bands of the trivalent and tetravalent metals. That for the tetravalent metal had been made earlier by Gold (1958).

close and in fact in the noble metals there are distortions of the sphere sufficient for the real Fermi surface to reach the Brillouin Zone.) A divalent fcc metal such as calcium has a Fermi sphere of volume equal to that of the Brillouin Zone and its surface is analogous to that in Fig. 13-11. These constructions, as well as those for a trivalent metal (aluminum) and a tetravalent metal (lead) are shown in Fig. 13-12.

Experimental confirmation of these Fermi surfaces played a very important role in the 1960's in suggesting that a weak pseudopotential might describe the properties of the simple metals, and more fundamentally, in showing that the one-electron approximation was a viable approach for the understanding of solids. The reduction of the Fermi surfaces to the Brillouin Zone, as in Fig. 13-12, helped one to visualize the electron orbits in wavenumbers space, and therefore in real space. Then various experiments, de Haas-van Alphen effect, ultrasonic attenuation in a magnetic field, and cyclotron resonance, made direct measurements of these orbits. Having confirmed this understanding, it is ordinarily better to return to the basis of free-electron states and the large Fermi sphere when one wishes to take advantage of the weak pseudopotential to estimate properties.

C. Fermi-Surface Distortions

It is also of interest to move the first step beyond regarding the effect of the pseudopotential as diffraction. Near the diffraction condition, the two coupled states are of nearly the same energy and the coupling cannot be treated in simple perturbation theory. One must solve the coupled system more accurately. Thus if we consider a wavenumber in the [100] direction in a simple-cubic crystal, $|k\rangle$ with k near π/d, it will be coupled by a matrix element (equal to the form factor since $S(q) = 1$) to a state $|k - 2\pi/d\rangle$ of approximately the same energy. We may write the state arising from this coupling as a linear combination of the two states, as we first did in Eq. (1-11),

$$|\psi\rangle = u|k\rangle + v|k-2\pi/d\rangle \; . \tag{13-35}$$

The variational conditions for obtaining the state of lowest energy, subject to the condition that $u^2 + v^2 = 1$, are

$$\begin{aligned} \frac{\hbar^2k^2}{2m}u + \quad w_{2\pi/d}v \quad &= \varepsilon u \\ w_{2\pi/d}u + \frac{\hbar^2(k-2\pi/d)^2}{2m}v &= \varepsilon v \, , \end{aligned} \tag{13-36}$$

and the energy obtained from solution of the quadratic secular equation is

$$\varepsilon = \frac{1}{2}\left(\frac{\hbar^2k^2}{2m} + \frac{\hbar^2(k-2\pi/d)^2}{2m}\right) \pm \sqrt{\frac{1}{4}\left(\frac{\hbar^2k^2}{2m} - \frac{\hbar^2(k-2\pi/d)^2}{2m}\right)^2 + w_{2\pi/d}{}^2} \,. \tag{13-37}$$

The effects of coupling to other plane waves, states of much different energy, are small and could be included in perturbation theory. We plot the corresponding energies, from Eq. (13-37), with a small pseudopotential in Fig. 13-13, which can be compared with the result of the diffraction approximation of Fig. 13-9.

We see that the successive bands within the Brillouin Zone are separated from each other by a small gap. This makes the diffraction more understandable in that a wavenumber changing with time due to an applied field follows its individual band and when it reaches the Zone face we could think of it proceeding on the continuation of that band outside the Zone. We could alternatively note that the same state may be represented by a wavenumber at the opposite face and we could think of it continuing from that point within the Brillouin Zone. If we focus on the lower bands within the Zone, and the upper bands outside the Zone, we have bands which follow

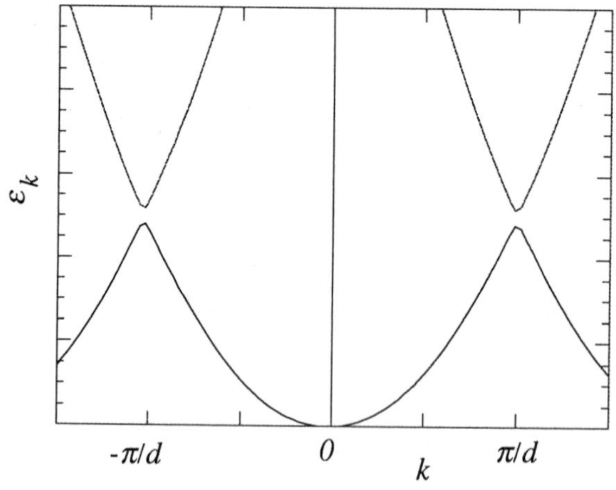

Fig. 13-13. Nearly-free-electron bands, with a small pseudopotential introduced. The bands which most closely follow the free-electron parabola are the *extended-zone* bands. Those within the Brillouin Zone, $-\pi/d < k \leq \pi/d$, are the *reduced-zone* bands.

closely the original free-electron bands, but the pseudopotential has opened up small gaps at the Bragg-reflection planes. That "extended-zone" picture is often a useful view to take.

This same opening of a gap in the bands at the Bragg planes ordinarily rounds off the sharp edges of the Fermi surfaces shown in Fig. 13-12, and in some cases, such as in the fourth-band pockets in the trivalent and tetravalent metals, it raises the entire band above the Fermi energy and eliminates the Fermi surface in that band altogether.

This rounding off is seen in the experimental studies of the Fermi surfaces, and gives an experimental measure of the magnitude of the pseudopotential form factors. The points on the surfaces sometimes remain sharp, as in the case of the second-band surfaces of the trivalent and tetravalent face-centered-cubic metals. In these cases a pair of degenerate bands crosses a nondegenerate band, rather than the crossing of two nondegenerate bands as in Fig. 13-13. Then the odd combination of the degenerate bands is not coupled to the nondegenerate band; the two bands are not separated as they were in Fig. 13-13, and correspondingly the point on the Fermi surface remains sharp. This is called a "contact" between the two bands. The lower band is still called "first band" and the upper the "second band". The contact occurs along a line which threads the point of the Fermi surface. Such details may be interesting, but rarely have important implications for the properties of the metal.

D. Magnetic Breakdown

It *is* interesting to imagine the transition from the band picture, with diffraction at the Bragg planes, to the free-electron picture without diffraction, as the pseudopotential is reduced toward zero. Of course that transition must occur, and at the intermediate situation we may imagine the wavenumber following the lower band in Fig. 13-13 through a Bragg plane. As it crosses, some probability density will arise in the upper band, and some probability remains that the electron will continue on the upper band. This occurs experimentally in strong magnetic fields, and is called *magnetic breakdown.* It can have dramatic effects on the transport properties in a magnetic field if an increasing field produces a change from *open* orbits, in which an electron successively diffracts so that it continues to move through the crystal (until it scatters from a defect), to *closed* orbits such as occur for a free electron or for orbits on any closed Fermi surface.

8. Scattering by Defects

Pseudopotential theory is in no sense restricted to perfect crystals such as we have been discussing. The ionic positions enter only through the structure factors of Eq. (13-20),

$$S(\boldsymbol{q}) = \frac{1}{N} \Sigma_j e^{-i\boldsymbol{q} \cdot \boldsymbol{r}_j} . \tag{13-38}$$

and we can consider any arrangement of ions which we choose. We saw in Section 13-5 that for a perfect crystal, the structure factors were zero except at the lattice wavenumbers, multiples of $\boldsymbol{k}_1$, $\boldsymbol{k}_2$, and $\boldsymbol{k}_3$. This is a sparse set of wavenumbers relative to those allowed by periodic boundary conditions, which are more numerous by a factor equal to the number of primitive cells - or atoms in simple structures - in the crystal.

Let us now imagine a crystal with a single atom removed from the site at $\boldsymbol{r}_0$. We again evaluate the structure factor given in Eq. (13-38). For $\boldsymbol{q}$ equal to a lattice wavenumber of the original crystal we have one less term in the sum, but N is the number of atoms in the crystal and is also reduced by one. The structure factors, and the matrix element leading to the energy bands as in Fig. 13-13 are unchanged. We may say that the energy bands are unaffected. However, for wavenumbers different from a lattice wavenumber, for which the sum in the complete crystal was zero, the sum is now the negative of the missing term,

$$S(\boldsymbol{q}) = -\frac{e^{-i\boldsymbol{q} \cdot \boldsymbol{r}_0}}{N} . \tag{13-39}$$

These tiny ($1/N$) structure factors arise at *every* wavenumber. Their effect is described as electron scattering, as we discussed for impurities in semiconductors in Section 8-1-D. Just as for x-rays, where a perfect crystal provides diffraction spots on photographic film, defects produce a general fogging of the film. In Section 1-5-B we wrote Fermi's Golden Rule for such transition rates in Eq. (1-32) as

$$P_{ij} = \frac{2\pi}{\hbar} |\langle j|H_1|i\rangle|^2 \delta(E_i - E_j) . \tag{13-40}$$

Pseudopotential theory has given us a way of estimating the matrix elements between a state of wavenumber $\boldsymbol{k}$ and one of wavenumber $\boldsymbol{k}+\boldsymbol{q}$ as the product of the structure factor of Eq. (13-39) and the form factor,

$$<\boldsymbol{k}+\boldsymbol{q}\,/\,W/\boldsymbol{k}> = -\frac{e^{-i\boldsymbol{q}\cdot\boldsymbol{r}_0}}{N}w_q \tag{13-41}$$

in place of the $\delta\varepsilon_s/N$ which entered Eq. (8-6) in the context of tight-binding theory. There is a significant difference here in that the matrix element now depends upon $\boldsymbol{q}$, in contrast to the simple $\delta\varepsilon_s/N$, so we redo the calculation of scattering rate.

We can obtain a total scattering rate by summing Eq. (13-40) over all final states $/j> = /\boldsymbol{k}+\boldsymbol{q}>$. Better still we can obtain a relaxation rate $1/\tau$ for momentum by weighting each event by the fraction of momentum lost, $1 - \cos\theta$, with θ the angle between initial and final states. From Eq. (13-40),

$$\frac{1}{\tau} = \frac{2\pi}{\hbar}\Sigma_{k'}/<\boldsymbol{k}+\boldsymbol{q}/W/\boldsymbol{k}>/^2\,(1-\cos\theta)\,\delta(\varepsilon_{\boldsymbol{k}+\boldsymbol{q}}-\varepsilon_{\boldsymbol{k}})\,. \tag{13-42}$$

We again replace the sum over final states by an integral, $\Sigma_{k'} = \Omega/(2\pi)^3\int d^3k' = \Omega/(2\pi)^3\int 2\pi\sin\theta\,d\theta\int k'^2dk'$. In the integral over k' we write $\int dk' = \int d\varepsilon_{k'}/(\partial\varepsilon_{k'}/\partial k')$ and use the delta function to obtain

$$\frac{1}{\tau} = \frac{\Omega k^2}{2\pi\hbar\,\partial\varepsilon_k/\partial k}\int d\theta\,\sin\theta\,(1-\cos\theta)\ /<\boldsymbol{k}+\boldsymbol{q}/W/\boldsymbol{k}>/^2\,. \tag{13-43}$$

With $\boldsymbol{k}+\boldsymbol{q}$ and $\boldsymbol{k}$ having the same length, the law of cosines gives $q^2 = 2k^2(1-\cos\theta)$ so $\sin\theta\,d\theta = qdq/k^2$ and we may write the integral as an integral over the magnitude of q from 0 to $2k$. We also write $\partial\varepsilon_k/\partial k = \hbar^2k/m$ and insert the matrix element to obtain

$$\frac{1}{\tau} = \frac{\Omega m}{4\pi\hbar^3k^3N^2}\int dq\,q^3w_q^2\,. \tag{13-44}$$

This requires a simple integral over a form factor, such as the empty-core form, Eq. (13-28), and then yields the counterpart of Eq. (8-8) for the scattering in tight-binding theory. [In Eq. (8-8) we gave a total scattering rate, but since the matrix element did not depend upon θ the integral is not affected by the $(1-\cos\theta)$ factor and $P_{tot} = 1/\tau$.] The result here is proportional to $1/N$ (since $\Omega/N = \Omega_0$ is independent of the size of the system) as it should be. For a larger crystal, the rate of colliding with a single defect is reduced. In Eq. (8-8) for an alloy the scattering rate depended upon concentration of defects x as $x(1-x)$, which goes to $1/N$ for a single defect among a large number N of atoms.

This vacancy is the simplest defect to treat. An interstitial atom, in the absence of any displacement of the neighbors, would give the negative of the vacancy matrix element, Eq. (13-41), and therefore the same relaxation time. The displacement of a single bulk atom from a position $\boldsymbol{r_0}$ to $\boldsymbol{r_0} + \delta$ would give a structure factor

$$S(\boldsymbol{q}) = \frac{e^{-i\boldsymbol{q}\cdot\boldsymbol{r_0}}}{N}(e^{-i\boldsymbol{q}\cdot\delta} - 1) \ . \tag{13-45}$$

The replacement of an atom of type a by an atom of type b , at position $\boldsymbol{r_0}$, would give a matrix element of $\langle \boldsymbol{k} + \boldsymbol{q}|W|\boldsymbol{k}\rangle = (e^{-i\boldsymbol{q}\cdot\boldsymbol{r_0}}/N)(w_q{}^b - w_q{}^a)$.

Any combination of these changes could be added together *in the matrix element* . This is an important distinction. There is interference between different terms. Thus, insertion of an interstitial gives one term in the matrix element, and the displacement of the neighboring atoms away from that interstitial gives a correction which ordinarily tends to reduce the matrix element and the scattering. In that sense, the relaxation partially heals the disruption of the lattice, rather than giving additional scattering.

9. Electron-Phonon Interaction

Since the positions of the atoms in the metal enter the matrix elements only through the structure factor, if we displace the atoms as in a lattice vibration, we need only evaluate the modification of the structure factors and we can evaluate all matrix elements between electronic plane-wave states arising from the presence of that lattice vibration. These matrix elements provide for metals the counterpart of the electron-phonon interaction matrix elements which we obtained for semiconductors in Section 7-3.

A. The Structure Factor

In Eq. (3-28) we wrote the displacements of individual atoms, numbered by j due to lattice vibration amplitudes $\boldsymbol{u_q}^i$. We rewrite that here, using $\boldsymbol{Q}$ rather than $\boldsymbol{q}$ for the phonon wavenumber since we reserve $\boldsymbol{q}$ for the wavenumber in the structure factor. We also take the case of a single atom per cell (as in bcc or fcc structures, but not in hcp where there are two atoms per cell) for simplicity, so we replace the number of cells N_C by the number of atoms N and drop the super i which specified the atom within the cell. Then Eq. (3-28) becomes

$$\delta r_j = \Sigma_Q \frac{u_Q}{\sqrt{N}} e^{i Q \cdot r_j} . \tag{13-46}$$

The sum over Q should be taken to include a sum also over the three modes at each wavenumber in the Brillouin Zone. u_Q is the amplitude of the mode, which we take to be small. It contains the time dependence which can be written $e^{-i\omega t}$ for a mode of frequency ω. We must require $u_{-Q} = u_Q{}^*$ in order that the displacements be real. Since u_Q contains a real and an imaginary part , this leaves one degree of freedom per wavenumber.

We may now evaluate the structure factor. We take r_j to be the positions in the undisturbed crystal, so that the positions in the crystal with the vibrations present are $r_j + \delta r_j$ and the structure factor may be written, and expanded for small u_Q as

$$\begin{aligned} S(q) &= \frac{1}{N} \Sigma_j e^{-iq \cdot (r_j + \delta r_j)} \\ &= \frac{1}{N} \Sigma_j e^{-iq \cdot r_j} (1 - i\, q \cdot \delta r_j - {}^1/_2 (q \cdot \delta r_j)^2 + \dots) . \end{aligned} \tag{13-47}$$

The first term gives the structure factors for the undistorted crystal, nonzero only at lattice wavenumbers. The corresponding matrix elements produce the band structure for the crystal as we have discussed. The second term gives the electron-phonon interaction and may be written

$$\begin{aligned} \delta S(q) &= \frac{1}{N} \Sigma_j e^{-iq \cdot r_j} \Sigma_Q \frac{-iq \cdot u_Q}{\sqrt{N}} e^{iQ \cdot r_j} \\ &= \Sigma_Q \frac{-iq \cdot u_Q}{\sqrt{N}} \frac{1}{N} \Sigma_j e^{-iq \cdot r_j} e^{iQ \cdot r_j} . \end{aligned} \tag{13-48}$$

The final sum is exactly of the form of the sum for the perfect crystal and is nonzero only if the wavenumber which enters, $q + Q$, is a lattice wavenumber; that is, q differs from a lattice wavenumber by-Q . That structure factor is

$$\delta S(\boldsymbol{q}) = \frac{-i\boldsymbol{q}\cdot \boldsymbol{u_Q}}{\sqrt{N}} . \tag{13-49}$$

Similarly for each **Q** there is a structure factor of $-i\boldsymbol{q}\cdot \boldsymbol{u_Q}^*/\sqrt{N}$ at a wavenumber $\boldsymbol{q}$ differing from the lattice wavenumber by $+\boldsymbol{Q}$. Thus we see that there are satellite structure factors to each lattice wavenumber produced by the phonon of wavenumber $\boldsymbol{Q}$, as illustrated in Fig. 13-14. These structure factors are proportional to $\boldsymbol{u_Q}$ and will enter squared in the scattering of electrons, described in Section 13-8, correctly giving scattering rates quadratic in the phonon amplitudes.

There are also matrix elements from the terms in $(\boldsymbol{q}\cdot \delta\boldsymbol{r}_j)^2$ in Eq. (13-47), but they contribute to scattering only to order $\boldsymbol{u_Q}^4$, and are negligible for small amplitudes. These terms however can be important for $\boldsymbol{q}$ a lattice wavenumber, where the zero-order structure factor was one. Then the $S^*(\boldsymbol{q})S(\boldsymbol{q})$ which enters for example the total energy calculation also contains terms from $(\boldsymbol{q}\cdot \delta\boldsymbol{r}_j)^2$ which are of order $\boldsymbol{u_Q}^2$ which are not negligible. These corrections reduce the structure factor at the lattice wavenumbers by the *Debye-Waller Factor* . With lattice vibrations present, the $S^*(\boldsymbol{q})S(\boldsymbol{q})$ at the lattice wavenumbers is reduced approximately enough to compensate for the new $S^*(\boldsymbol{q})S(\boldsymbol{q})$ which appears in the satellites around that same lattice

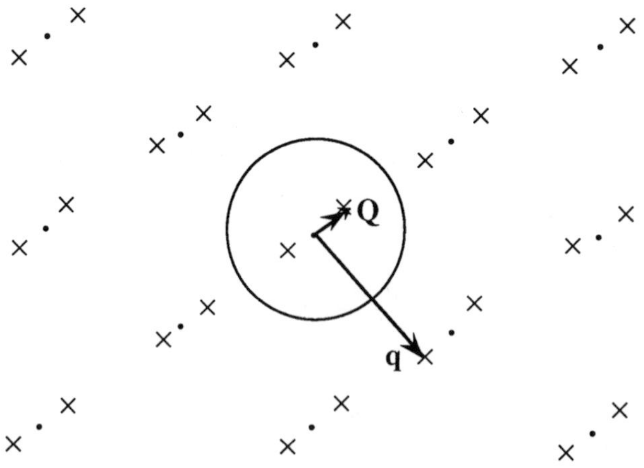

Fig. 13-14. Structure factors $-i\boldsymbol{q}\cdot \boldsymbol{u_Q}/\sqrt{N}$ or $-i\boldsymbol{q}\cdot \boldsymbol{u_Q}^*/\sqrt{N}$ introduced at satellites (×) to the lattice wavenumbers (•) by a phonon of wavenumber $\boldsymbol{Q}$ and amplitude $\boldsymbol{u_Q}$. Note that the matrix element for the $\boldsymbol{q}$ indicated can couple states on the Fermi surface, indicated by a circle.

wavenumber. Thus the fogging of the photographic plate by lattice vibrations in x-ray diffraction, is compensated by a reduction in the intensity of the diffraction spots which were also there without phonons present. We will note the effect of these second-order terms when we discuss the total energy in Chapter 14, but will not carry out the detailed evaluation.

B. Electron Scattering

We treated electron scattering by lattice vibrations in semiconductors in Section 7-3-H. We saw that the phonon energy $\hbar\omega$ is ordinarily negligible in comparison to the electron energies. That is particularly true in metals with electrons at the Fermi energy of several volts rather than thermal energies of a few millivolts. Then the scattering can be treated as if the displacements $\delta\boldsymbol{r}_j$, and the resulting matrix elements, were independent of time and we may proceed exactly as we did for defects in Section 13-8. The matrix elements are given by

$$\langle \boldsymbol{k}+\boldsymbol{q}|W|\boldsymbol{k}\rangle = w_q\,\delta S(\boldsymbol{q}) = \frac{-i\boldsymbol{q}\cdot\boldsymbol{u}_Q\, w_q}{\sqrt{N}} . \tag{13-50}$$

For a given initial $\boldsymbol{k}$ we are to add the scattering to any states $\boldsymbol{k}+\boldsymbol{q}$ which satisfy energy conservation.

It is interesting first to consider vibrations with $\boldsymbol{Q}$ small enough that we distinguish the lattice wavenumber with which each satellite is associated, as in Fig. 13-14. Processes arising from satellites to the lattice wavenumber $\boldsymbol{q}_0 = 0$ are called *normal scattering*. For these, $\boldsymbol{q} = \boldsymbol{Q}$ so the matrix element contains $\boldsymbol{Q}\cdot\boldsymbol{u}_Q$. To the extent that modes are longitudinal and transverse (as for elastically isotropic systems), the dot product is zero for transverse waves and equal to Qu_Q for longitudinal modes, and normal scattering arises only from longitudinal modes, as we found true for electron scattering in semiconductors in Section 7-3-H. For displacements $\boldsymbol{u}_Q\, e^{i\boldsymbol{Q}\cdot\boldsymbol{r}}/\sqrt{N}$ the dilatation is $i\boldsymbol{Q}\cdot\boldsymbol{u}_Q\, e^{i\boldsymbol{Q}\cdot\boldsymbol{r}}/\sqrt{N}$ so the matrix element of Eq. (13-50) corresponds to a deformation-potential constant (compare with Eq. (7-20)) of $D = -w_Q$. At small Q the deformation-potential is $^2/_3E_F$ according to Eq. (13-28). In metals, however, we are frequently interested in larger Q, comparable to the Fermi wavenumber, and the deformation-potential "constant" depends upon Q.

Processes arising from satellites to other lattice wavenumbers, as for the $\boldsymbol{q}$ shown in Fig. 13-14, are called *Umklapp* processes. The wavenumber of the electron is "flipped over" as if a diffraction of the lattice had occurred at the same time a phonon is absorbed or emitted. $\boldsymbol{q}$ is no longer parallel to $\boldsymbol{Q}$ so both longitudinal and transverse modes give scattering. The matrix elements with the $\boldsymbol{q}$ shown in Fig. 13-14 can take an electron very nearly across the Fermi surface even when the phonon wavenumber, and therefore frequency, is quite small. These low-frequency modes have large amplitudes for thermal vibrations and Umklapp can dominate the resistivity since the normal processes from these same small-wavenumber modes only slightly deflect the electrons at the Fermi surface and disrupt the current flow very little.

The distinction between normal and Umklapp scattering is no longer sharp when the wavenumbers $\boldsymbol{Q}$ become large. If we follow the scattering as we increase $\boldsymbol{Q}$ and follow $\boldsymbol{k} + \boldsymbol{Q}$ along the Fermi surface, at some point $\boldsymbol{Q}$ will move out of the Brillouin Zone, but the mode itself and its frequency will continue to vary smoothly without regard to our choice of a Brillouin Zone to represent all of our states. The scattering has then become Umklapp by our definitions but the behavior varied continuously from one regime to the other.

Obtaining the matrix elements, as in Eq. (13-50), is the main task in the scattering calculation. It becomes quite easy to use them with the Golden Rule to obtain scattering rates when the geometry is simple, as for normal scattering. If we seek more complete solutions, as for the wavenumber $\boldsymbol{q}$ shown in Fig. 13-14, it remains simple in principal, but the geometry for the calculation may become complicated.

One interesting calculation is the total scattering rate for electrons due to lattice vibrations at high temperatures. Such a calculation led to the total scattering rate for semiconductors in Eq. (7-41). The high-temperature assumption allowed the matrix elements to be written in terms of the temperature so that the result was proportional to k_BT. This allows us here to define the *dimensionless electron-phonon coupling constant* λ. We retain the $2\pi/\hbar$ from the Golden-Rule formula, and the k_BT and the remaining factor in the total scattering rate is that constant,

$$P_{tot} = \frac{2\pi k_BT}{\hbar}\lambda \ . \tag{13-51}$$

λ for metals becomes an integral over the Fermi surface, analogous to Eq. (7-41). For normal scattering the screened deformation-potential constant D_c is replaced by the pseudopotential form factor w_q and we obtain a coupling constant

$$\lambda = \frac{m\Omega_0}{4\pi^2\hbar^2 k_F M v_s^2}\int w_q^2 q\, dq\,. \tag{13-52}$$

For spherical Fermi surfaces of radius k_F the integral extends from $q = 0$ to $q = 2k_F$. This formula is stretching the assumption of vibrational frequencies proportional to the speed of sound, $\omega = v_s q$.

The focus upon this constant arose when its role in superconductivity became clear (see for example McMillan (1968) or Chakraborty, Pickett, and Allen (1976)). It not only gives the high-temperature total scattering, Eq. (13-51), but the reduction of the electron velocity at the Fermi surface (by a factor $1/(1+\lambda)$) due to a wake of phonons which accompanies a moving electron. This decrease in velocity gives also the corresponding increase in the density of states and electronic specific heat. It also relates the superconducting critical temperature T_c to the Debye temperature θ_D as $T_c = \theta_D e^{-1/(\lambda - \mu^*)}$, with μ^* a dimensionless electron-electron coupling constant. These were discussed in Harrison (1980), 397ff, and the three effects treated in detail in Harrison (1970), 414 ff, with additional references but without the definition of the specific parameter λ. Our focus here is upon the determination of the parameter, rather than the physical phenomena which it enters.

The relaxation rate $1/\tau$ for conductivity is obtained by a combination of Eqs. (13-51) and (13-52), but with the integrand $w_q^2 q$ replaced by $w_q^2 q^3/(2k^2)$ as we indicated following Eq. (7-41).

Problem 13-1. Packing Fractions.

In order to get accustomed to the metallic crystal structures we may calculate the "packing fractions", the fraction of the total volume which lies within spheres packed together, and touching, in a particular structure. Obtain these for the face-centered-cubic structure, the body-centered-cubic structure and the simple-cubic structure.

Problem 13-2. Atomic s- and p-states

An empty-core pseudopotential radius, r_c, can be obtained for an alkali metal atom by finding the value which will give the correct s-state energy by solution of the Schroedinger Equation

$$-\frac{\hbar^2}{2m}\frac{1}{r^2}\frac{\partial}{\partial r}r^2\frac{\partial}{\partial r}R + w(r)R = \varepsilon_s R$$

with

$$w(r) = \begin{cases} 0 & \text{for } r<r_c \\ -e^2/r & \text{for } r>r_c. \end{cases} \qquad (1)$$

a) Obtain r_c for lithium (ε_s = -5.34 eV) and sodium (ε_s = -4.96 eV), integrating the Schroedinger Equation numerically . Use a step in r of 0.01 Å and adjust r_c to within 0.01 Å so $R(8\text{Å})=0$. Compare these results with the values in Table 13-5, obtained in different ways.

Suggestions: It is easier to work with $\chi(r) = r\,R(r)$ for which

$$-\frac{\hbar^2}{2m}\frac{\partial^2}{\partial r^2}\chi + w(r)\chi = \varepsilon_s\chi \ . \qquad (2)$$

It is helpful again to use $\hbar^2/m$ = 7.62 eV-Å^2 and e^2 = 14.4 eV-Å. Then we can seek a solution of Eq. (2) by numerical integration. At small r, you may take $\chi = r$ (not normalized) to set $\chi(0)$, $\chi'(0)$, and $\chi''(0)$, and proceed using Eq. (2) for χ'' at larger r .

The value of r_c which gives a nodeless wavefunction that does not diverge at large r is correct. Using $R(8\text{Å})$ rather than $R(\infty)$ is good to around 0.01 Å. You probably need to print out $\chi(r)$ values to see that there is no node between r= 0 and r= 8 , but a tenth of the values of r is plenty.

Problem 13-3. p-state from the pseudopotential.

Use the sodium r_c from Problem 13-2 and find ε_p , the lowest p-state energy (within about 0.01 eV). It is interesting to see how close this is to the rule of thumb, $\varepsilon_p = {}^1/_2\,\varepsilon_s$, from Table 1-1.

Suggestions: You need to add $\hbar^2\ell(\ell+1)/(2mr^2)$ to $w(r)$ in Eq. (2), fix r_c and adjust the energy ε_p to get a nodeless solution with $\chi(8\text{Å})$ = 0. At small r, you may take $\chi = r^2$ (not normalized).

The full free-atom calculation is a direct generalization of this procedure with w(r) replaced by a more complete potential.

CHAPTER 14

Simple Metals, Bonding Properties

Just as we calculated the scattering of states in perturbation theory, we may calculate the shift in the energy of each state to second order in the pseudopotential using perturbation theory, as in Eq. (1-30). This yields an additional term in the energy to be added directly to the energy given in Eq. (13-13), which was the sum of the one-electron energies to first order in the pseudopotential. After correcting these second-order terms for self-interactions counted twice, the additional term is called *band-structure energy* since it arises from the deviations of the bands from free-electron parabolas. In principle this gives a more accurate estimate of the total energy. More importantly, we shall see that it is possible to rewrite this band-structure energy as a contribution to the effective interaction between ions, which will prove to be more generally useful as an approach for calculating properties.

1. The Band-Structure Energy

The shift of the energy of an occupied state of wavenumber $\boldsymbol{k}$ due to interaction with a state of wavenumber $\boldsymbol{k} + \boldsymbol{q}$ due to a pseudopotential W is given, according to Eq. (1-30), by

$$\delta\varepsilon_k = \frac{<\boldsymbol{k}/W/\boldsymbol{k+q}> <\boldsymbol{k+q}/W/\boldsymbol{k}>}{\varepsilon_k - \varepsilon_{k+q}} \qquad (14\text{-}1)$$

with $\varepsilon_k = \hbar^2 k^2/2m$. We sum this over all states $/\boldsymbol{k+q}>$ though if both states are occupied one is raised in energy and the other lowered by the same amount. To obtain the total energy shift, we must sum over occupied states, $k < k_F$. It is not obvious that we can ignore the distortion of the Fermi surface by the pseudopotential, but careful consideration of the difference shows it to be of third order in the pseudopotential (Harrison (1966) p. 98). Then the total shift in energy to second order is given by

$$\delta E_{bs} = \sum_{k<k_F} \sum_q \frac{<\boldsymbol{k}/W/\boldsymbol{k+q}> <\boldsymbol{k+q}/W/\boldsymbol{k}>}{\varepsilon_k - \varepsilon_{k+q}}$$
$$= \sum_q S^*(\boldsymbol{q})S(\boldsymbol{q}) \sum_{k<k_F} \frac{w_q{}^2}{\varepsilon_k - \varepsilon_{k+q}} . \qquad (14\text{-}2)$$

The last step is an important one. We have written the matrix elements out as $<\boldsymbol{k+q}/W/\boldsymbol{k}> = S(\boldsymbol{q})w_q$ as in Eq. (13-19). We have also interchanged the two sums and taken the structure factors out since they do not depend upon $\boldsymbol{k}$. The final sum depends upon $\boldsymbol{q}$ but for a given magnitude of $\boldsymbol{q}$ it depends on the positions of the ions only through the electron density which determines k_F and the screening of the pseudopotential (e. g., κ^2 in Eq. (13-28)).

At this point we should correct for the electron-electron interactions which are counted twice in the calculation, the interaction between the screening charge and the screening potential. It is plausible, and can be verified (Harrison (1966), 289ff), that this is accomplished by using one factor of the pseudopotential as w_q and the other as the unscreened w_{0q}. The final sum over $\boldsymbol{k}$ is divided by the number of atoms in the system, N , to obtain the *energy-wavenumber characteristic* ,

$$F(q) = \frac{1}{N} \sum_{k<k_F} \frac{w_q w_{0q}}{\varepsilon_k - \varepsilon_{k+q}} \approx - \frac{\Omega_0 q^2 w_{0q}{}^2}{8\pi e^2 \varepsilon(q)} (\varepsilon(q) - 1) . \qquad (14\text{-}3)$$

The *first* form can be evaluated using the full dependence of the matrix elements of the pseudopotential operator $<\boldsymbol{k+q}/w/\boldsymbol{k}>$ upon $\boldsymbol{k}$, and that was in fact done in the first treatments of the band-structure energy (Harrison (1963), (1964)). The calculation of properties in terms of the resulting $F(q)$, given for example for sodium, magnesium and aluminum in Harrison (1966), p. 40, is just as simple as with the additional approximations made here.

To obtain the second form we noted that the sum over $\boldsymbol{k}$ in the first form, ignoring any $\boldsymbol{k}$- dependence of the form factors, is identical to the sum performed in evaluating the Lindhard dielectric function of Eq. (13-31). (Harrison (1980), p. 385,6). It entails an additional approximation if we use the Fermi-Thomas form, Eq. (13-27), for the dielectric function rather than the full quantum form, Eq. (13-31). The $F(q)$ obtained using the empty-core pseudopotential with the Fermi-Thomas dielectric function and with the Lindhard quantum dielectric function are compared for aluminum in Fig. 14-1. The logarithmic singularity at $q/k_f = 2$ in the quantum form, which we discussed for Eq. (13-31), is scarcely noticeable. However, the difference in the two curves probably represents the most serious approximation in the evaluation of the second-order energy.

In either case we have a very simple form for the band-structure energy, the terms second-order in the pseudopotential. That band-structure energy per ion from Eq. (14-2) is

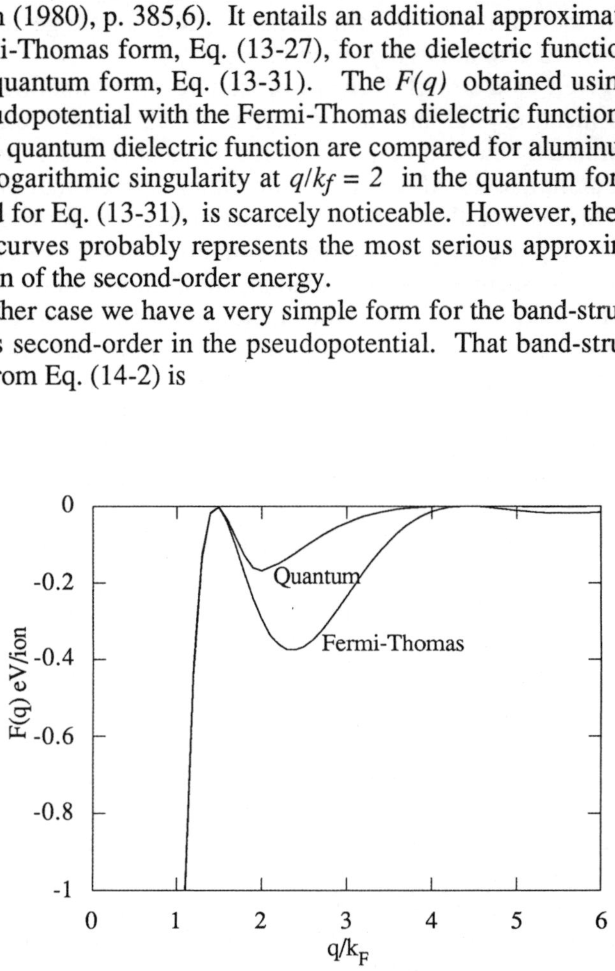

Fig. 14-1. The energy-wavenumber characteristic for aluminum ($k_F = 1.75$ $Å^{-1}$, $r_c = 0.61$ Å) from the final form in Eq. (14-3), based upon the quantum dielectric function (Eq. (13-31)) and based upon the Fermi-Thomas dielectric function (Eq. (13-27)).

$$\frac{E_{bs}}{ion} = \sum_q S(\boldsymbol{q})^* S(\boldsymbol{q})\, F(q)\,. \tag{14-4}$$

At fixed volume the only terms in the energy which depend upon the atomic arrangement are this energy and the electrostatic energy. For any arrangement of ions, the structure factor $S(\boldsymbol{q}) = {}^1/_N \sum_j e^{-i\boldsymbol{q}\cdot\boldsymbol{r}_j}$ can be evaluated and the sum over $\boldsymbol{q}$ converges rapidly as suggested by the approach of $F(q)$ to zero at large q, seen in Fig. 14-1.

This has given us a straightforward procedure for estimating the total energy of the metal crystal and the properties which derive from it. The accuracy is limited by the fact that we include terms only to second order in the pseudopotential, even if we make no other approximations in the calculation. This has not always been so apparent because, for example, we could select an empty-core-pseudopotential radius which gives the correct bulk modulus in the second-order calculation, and then it would give quite a good vibration spectrum. On the other hand, if we fit the core radius to obtain the correct atomic s-state energy, or to fit some other property, the vibration spectrum might be twenty or thirty percent in error. There is even a range of rigorous pseudopotential operators, all of which are expected to give accurate results if carried to all orders, (see, for example, Harrison (1970), p. 100) but which give significantly different results for the vibration spectrum when calculated only to second order. One cannot test the validity of a calculation simply from its agreement with experiment. For this reason, we consider only a few properties in terms of the band-structure energy, and electrostatic energy, and then proceed to use the Fermi-Thomas approximation and obtain a simpler, and more easily applicable, theory in terms of interatomic interactions. There is not a great loss in total accuracy in going to this simpler theory.

A. Vibrational Frequencies and the Kohn Effect

We can calculate the vibrational frequency at constant total volume of the crystal, so all we need is the band-structure energy of Eq. (14-4) and the electrostatic energy of the distorted crystal. The complete procedure is given in Harrison (1966), Chapter 7. For the band-structure energy we need the structure factors, which we found for a crystal with a vibrational distortion in Chapter 13 and showed in Fig. 13-14. A lattice vibration produces structure factor at two satellites to each lattice wavenumber. If we write the lattice wavenumber $\boldsymbol{k}_j$ the new $S(\boldsymbol{q})^*S(\boldsymbol{q})$ at these two points is $S(\boldsymbol{k}_j \pm \boldsymbol{Q})^*S(\boldsymbol{k}_j \pm \boldsymbol{Q}) = [(\boldsymbol{k}_j \pm \boldsymbol{Q})\cdot \boldsymbol{u}_{\boldsymbol{Q}}]^2/N$. We also noted that $S(\boldsymbol{q})^*S(\boldsymbol{q})$ was reduced by a term proportional to the square of the amplitude at each lattice wavenumber. The change is in fact given by $\delta(S(\mathbf{k}_j)^*S(\mathbf{k}_j)) = -2(\mathbf{k}_j \cdot \mathbf{u}_Q)^2/N$. Thus we must insert these three contributions into Eq. (14-4) (each with the $F(q)$ appropriate to the wavenumber, $\mathbf{k}_j \pm \mathbf{Q}$ or $\mathbf{k}_j$) for each lattice wavenumber $\mathbf{k}_j$, and sum over j . This sum converges rather well, but in any case once a program has been written we can easily sum over as many lattice wavenumbers as are needed. This sum gives us an elastic energy proportional to the square of the amplitude $\mathbf{u}_Q$, as a function of the wavenumber $\mathbf{Q}$ of the vibration. It is not difficult to see (Harrison (1966), p. 247)) that the electrostatic energy can also be written in the form of Eq. (14-4) with $F(q)$ replaced by $2\pi Z^2e^2/(\Omega_0 q^2)$ with a convergence factor. Thus it is obtained by the same kind of sum over the wavenumber lattice. For a vibration with $\mathbf{Q}$ lying along a symmetry direction, $\mathbf{u}_Q$ will correspond to longitudinal or transverse waves so its direction is known. One may then equate the elastic energy to the kinetic energy of the lattice, and solve for ω. For arbitrary wavenumbers the band-structure and electrostatic energies yield a three-by-three matrix in the three components of $\mathbf{u}_Q$ and one must solve for the three normal modes by diagonalizing that matrix. For two atoms per primitive cell we must diagonalize a six-by-six matrix.

We shall complete such a calculation of the spectrum in Section 14-4 using interatomic interactions, but here only note some aspects of the solution. We saw in Eq. (14-3) that $F(q)$ was proportional to $1 - 1/\varepsilon(q)$ and from Eq. (13-31) that the quantum dielectric function $\varepsilon(q)$ contains a term proportional to $(1 - (q/2k_F)^2)\, ln|(2k_F - q)/(2k_F + q)|$. This has a logarithmic singularity at $x = q - 2k_F$ of the form $x\, ln\, x$. This function is continuous at $x = 0$ but its derivative with respect to x is negatively infinite at that point. This is a very weak singularity, as we noted in Fig. 14-1, in some sense invisible at all magnifications (Harrison (1970), p. 298). However, if we imagine calculating frequencies as we change the wavenumber, including contributions from each $\times$ in Fig. 13-14, every time an $\times$ crosses the circle $q = 2k_F$ (a circle of radius twice that of the Fermi surface shown in the figure) from inside to outside, the frequency will rise with infinite $\partial\omega/\partial Q$. This is called an upward Kohn anomaly. Similarly when an $\times$ crosses from outside to inside, the frequency will drop with negatively infinite $\partial\omega/\partial Q$, called a downward Kohn anomaly. This effect was first predicted by Kohn (1959). It is of course included in any calculation of the vibration spectrum which includes quantum screening, and it is observed in the spectra of many metals.

The wavenumbers at which they occur can be predicted by making a construction analogous to the Fermi-surface construction of Fig. 13-11, only with spheres of radius $2k_F$. Whenever the phonon wavenumber in the Brillouin Zone crosses such a sphere, an anomaly is predicted. In the alkali metals the corresponding form factors are so small that these anomolies are not observable, but they are seen in polyvalent metals. Because the pseudopotentials associated with the lattice wavenumbers are not infinitesimal, the real Fermi surface is distorted and not spherical, so the wavenumber at which the anomaly occurs is not exactly where it is predicted for a spherical Fermi surface, but depends upon some dimension of the real distorted Fermi surface.

B. Other Properties

Calculating the vibration spectrum using the band-structure energy allows a prediction of the elastic constants (from the speed of long-wavelength sound frequencies) and a number of related properties which we shall discuss in Section 2 of this chapter.

The band-structure energy does not directly allow us to estimate the cohesion, the difference between the energy of the crystalline metal and free atoms, because it is not suitable for calculating the energy of free atoms. However, we were able to describe the cohesion from a different point of view in Section 13-3-A.

The next problem which comes to mind is the comparison of the energy for different crystal structures. For the perfect crystal we have seen that $S(\boldsymbol{q}) = 1$ at lattice wavenumbers and zero otherwise. Thus for each structure Eq. (14-4) becomes a rapidly convergent sum over the wavenumber lattice. Further, we saw in Section 13-4-B and Table 13-4 that the Madelung energies are known for each structure and these two are all that are needed to compare energies at fixed volume. The first use of pseudopotential perturbation theory to predict structures (Harrison (1964)) used not only the full quantum-mechanical screening, but also a much more accurate pseudopotential operator, and successfully predicted the stable structures for sodium, magnesium, and aluminum. However, a simple model such as we use here also gave correct structures for these three, but incorrect structures for most others (Harrison (1966) p. 194). We have not thought that second-order theory is adequate for these small differences, though an extensive review by Heine and Weaire (1970) concluded that it was. A more recent review of the question by Hafner ((1987) p. 77) is sympathetic to their view and indicates that full pseudopotential calculations are indeed reliable for the energy differences. This same approach and the same considerations apply to the determination of the axial (c/a) ratio for hexagonal-close-packed structures and for the elastic shear constants. All of these are obtained from the combination of Madelung energy and the sum given in Eq. (14-4). We remain skeptical about perturbation theory for the question of structure determination. It does appear that full density-function theory of the relative energies of different structures, such as those by Moriarty (1995) and Moriarty and Althoff (1995), can reliably predict structures, but the validity of the approximate approaches remains in question.

The band-structure energy provides a suitable approach for a number of properties, some of which were treated in Harrison (1966). One is the energy of formation of a point defect, such as a vacancy (Harrison (1966), p. 205), or the interaction between defects. Another is the energy of a stacking fault (an error in the stacking shown in Fig. 13-2, such as in the sixth plane in the series ABCABABCABC) (Harrison (1966), p. 207). The theory seemed to give reasonable estimates, though of limited accuracy, for all of these cases.

We shall not carry the application of this approach further, but turn instead to the transformation to interatomic interactions which allows quite simple estimates for a number of properties.

2. Interatomic Interactions

The band structure energy of Eq. (14-4) can be rewritten as a contribution to an effective interaction between ions, as suggested first by Cohen (1962) and carried out in Harrison (1966). This has the advantage that it can be combined with the electrostatic Madelung energy, providing a screening of the Coulomb interactions and giving results in terms of short-range interactions. Thus it avoids the intricacies of the Madelung calculation and the less-intuitive sum over wavenumbers for the band-structure energy. We will find this a major simplification for the treatment of these terms, but an essential feature which we shall discuss further is that there are additional terms in the energy, not included in this interaction.

A. The Effective Interaction

We may obtain the band-structure contribution to the interatomic interaction by writing out the structure factors in Eq. (14-4) to obtain a total energy of

$$N\frac{E_{bs}}{ion} = \frac{N}{N^2}\sum_{i,j}\sum_q e^{i\boldsymbol{q}\cdot(\boldsymbol{r}_i-\boldsymbol{r}_j)}F(q)\,. \qquad (14\text{-}5)$$

We have also interchanged the sums over ions and the sum over wavenumber. Then for fixed $\mathbf{r} = \mathbf{r}_i - \mathbf{r}_j$ we may define an indirect interaction

$$V_{ind}(r) = \frac{2}{N}\sum_q e^{i\mathbf{q}\cdot\mathbf{r}}F(q) = \frac{2\Omega_0}{(2\pi)^3}\int 4\pi q^2 \frac{\sin qr}{qr}F(q)dq \qquad (14\text{-}6)$$

and the total band-structure energy becomes

$$E_{bs} = \frac{1}{2}\sum_{i,j} V_{ind}(\mathbf{r}_i - \mathbf{r}_j) \ . \qquad (14\text{-}7)$$

In the last step of Eq. (14-6) we replaced the sum Σ_q by the integral $(\Omega/(2\pi)^3)\int d^3q$ and carried out the angular integral. In Eq. (14-7) we see that the total band-structure energy can indeed be written as a sum of two-body, central-force interactions. Eq. (14-7) also contains diagonal terms with $j = i$,

$$E_{bs}(diagonal) = \frac{1}{2}\sum_i V_{ind}(0) = \frac{N\Omega_0}{2\pi^2}\int F(q)q^2 dq \ . \qquad (14\text{-}8)$$

The interaction in Eq. (14-6) can be evaluated numerically using the full $F(q)$ with quantum screening, yielding a radial interaction with Friedel oscillations such as appeared in Fig. 13-7b. We will give the form of that interaction in connection with transition-metal interactions in Section 15-5-C.

It was not realized until relatively recently (Harrison and Wills (1982)) that with the Fermi-Thomas form for $\varepsilon(q)$, Eq. (13-27), in Eq. (14-3), the integral in Eq. (14-6) can be evaluated analytically. We replace $\sin qr$ by $e^{iqr}/2$ and extend the integral from $-\infty$ to ∞ as in the evaluation of Eq. (13-29). We again close a contour in the upper half plane and again obtain a contribution from the pole at $q = i\kappa$, but there is now also a pole at $q = 0$ which gives a contribution $-Z^2e^2/r$. This, however, just cancels the direct interaction between ions in the electrostatic energy so that the two may be combined to obtain the total effective interaction between ions

$$V(r) = \frac{Z^2e^2\cosh^2\kappa r_c\, e^{-\kappa r}}{r} \ . \qquad (14\text{-}9)$$

This simple result is now almost obvious. If the core radius is zero, we obtain simply the interaction of a charge Ze and the screened potential $Zee^{-\kappa r}/r$ from a second such charge, as obtained from Eq. (13-30). We also saw from Eq. (13-30) that the effect of a nonzero core radius for this second ion simply scales up the potential by a factor of $\cosh \kappa r_c$, increasing the energy of the first charge by that factor. Clearly a second such factor must arise for the nonzero core radius of the first ion. We saw in detail how each of these factors arises in the contour integration as $\cos(i\kappa r_c)$. A final interesting feature of the integration can be noted. If it were true that $r < 2r_c$ so that the empty cores overlapped each other, the derivation would be seen to fail because one of the terms e^{-2iqr_c} in $\cos^2 qr_c$ becomes larger than the factor e^{iqr} which allowed us to discard the integral over the closing contour. That will not occur in ordinary circumstances.

B. The Total Energy

This combination of terms and screening has eliminated the long-range Coulomb interaction between ions which made the Madelung calculation tricky. However, it raises immediately a complication. We have extracted the divergent interaction between the point ions of the Madelung energy, but are left with the divergent self-interaction of the uniform negative compensating background in the Madelung energy. This is a purely volume-dependent term, which must be added to obtain the correct volume dependence of the energy, so we evaluate it here.

We do this by replacing the uniform background by a fine-meshed grid of N_g point charges of Z/N_g in each atomic cell, and will let N_g approach infinity in the end. Then the interaction between grid points and the ions, and between the grid points and other grid points is screened in just the way the interaction, Eq. (14-9), was but with $r_c = 0$ for the grid points. In this context we redo the entire electrostatic calculation including the corresponding band-structure terms which led to Eq. (14-9).

The interaction between an ion and the neighboring grid points came from the first-order term $\Sigma_k \langle k|W|k\rangle$ and we already included the empty-core contribution as the second term in Eq. (13-14), so we are working now with the full Coulomb potential $-Ze^2/r$. Summing the interaction derived from Eq. (14-9) over grid points gives

$$\delta E_{es}\,(ion\text{-}grid) = -\int \frac{4\pi r^2 dr}{\Omega_0/N_g}\, Z\frac{Z}{N_g}\frac{e^2 e^{-\kappa r}}{r} = -\frac{4\pi Z^2 e^2}{\kappa^2 \Omega_0} = -\frac{4Ze^2 k_F{}^3}{3\pi\kappa^2} \quad (14\text{-}10)$$

for each ion. Similarly for each grid point there is just such a term, but positive and reduced by a factor $1/N_g$. Summing over all the grid points in one atomic cell, and dividing by two since each pair of grid points has been counted twice, the contribution per ion cancels half of that given in Eq. (14-10). This combination of ion-grid and grid-grid terms was not discussed in Harrison and Wills (1982) and would not have affected the properties they considered, all at constant volume, and which we discuss in Part 5 of this chapter. The interaction of a grid point with itself (made finite by replacing it by a small but finite uniform charge) of course contains a factor $1/N_g^2$ and adding the N_g grid points per atom still leaves a negligible contribution.

Thus in our total energy we are to eliminate the Madelung final term in Eq. (13-18) for the electron-gas contribution, but add $^1/_2\, N_a\, \delta E_{es}\,(ion\text{-}grid)$ from Eq. (14-10) and the sum over interatomic interactions using Eq. (14-9). Finally, we add the diagonal term from Eq. (14-8), which can be evaluated for the empty-core pseudopotential as (Harrison and Wills (1982))

$$
\begin{aligned}
E_{bs}(diagonal) &= -\frac{N_a Z^2 e^2}{\pi} \int \frac{dq\, \cos^2 q r_c}{q^2 + \kappa^2} \\
&= -{}^1/_2 N_a Z^2 e^2 \kappa \cosh(\kappa r_c) e^{-\kappa r_c} \, .
\end{aligned}
\tag{14-11}
$$

The total energy of the system becomes

$$
\begin{aligned}
E_{tot} = {} & N_a Z \left(\frac{3\,\hbar^2 k_F^2}{10m} - \frac{3e^2 k_F}{4\pi} + \frac{2e^2 k_F^3}{3\pi\kappa^2}(\kappa^2 r_c^2 - 1) \right) \\
& - \frac{Z^2 e^2 \kappa}{2} \cosh(\kappa r_c) e^{-\kappa r_c} + \frac{1}{2} \sum_{i \neq j} V(\mathbf{r}_i - \mathbf{r}_j) \, .
\end{aligned}
\tag{14-12}
$$

This includes the Madelung terms, but also screening terms which make them convergent. If we eliminate screening by letting κ go to zero, the sum over $V(\mathbf{r}_i - \mathbf{r}_j)$ diverges but one term in the first line diverges as $1/\kappa^2$ and the two terms together converge. These two terms constitute one solution of the Madelung problem discussed in Section 13-4-B.

C. Volume Dependence of the Total Energy

Of most interest will be the calculations at constant volume using only the sum over $V(\mathbf{r}_i - \mathbf{r}_j)$, but we look briefly at the volume dependence of the full expression. Eq. (14-12) is basically the free-electron total energy of Eq. (13-18) in Section 13-3-B, but with screening and second-order terms added. As we indicated in that section, adding higher-order terms does not generally improve the estimates.

We can first take the derivative of that total energy from Eq. (14-12) with respect to volume, and set it equal to zero, to predict the equilibrium volume in terms of our given core radius r_c for each element. As in our discussion of the free-electron energy in Chapter 13, we prefer to see what core radius will yield the correct equilibrium volume, and compare that with our standard set of core radii given in Table 13-5. Such a comparison of the r_c values was made in that table for the free-electron energy, and it is made in Fig. 14-2 for Eq. (14-12). Again these do not correspond to quantitative predictions of spacing, nor even a uniform improvement on the earlier estimate, but again it gives a reasonable account of the trends from metal to metal, and of the general magnitude of r_c. A plot of the E_{tot} from Eq. (14-12) obtained using these values of r_c fit to the equilibrium spacing for lithium and potassium is given for illustration in Fig. 14-3.

Such curves can of course be used to estimate other properties. We may for example evaluate the second derivative of this energy with respect to volume to obtain the bulk modulus,

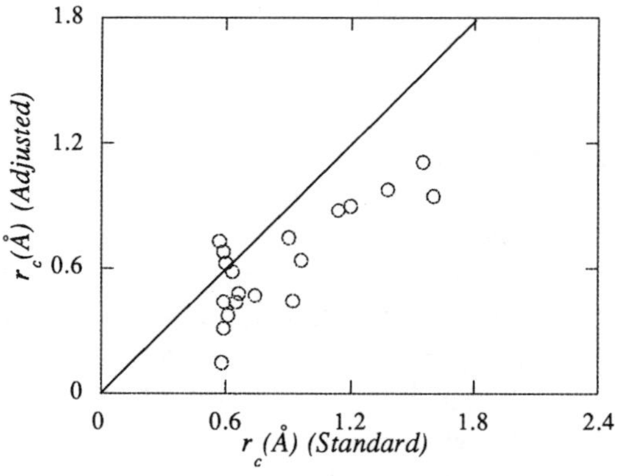

Fig. 14-2. A plot of the core radii fit to yield the correct equilibrium spacing based upon Eq. (14-12), using the body-centered-cubic structure for the evaluation of the final term. They are given as a function of the standard values obtained from the first entry in Table 13-5.

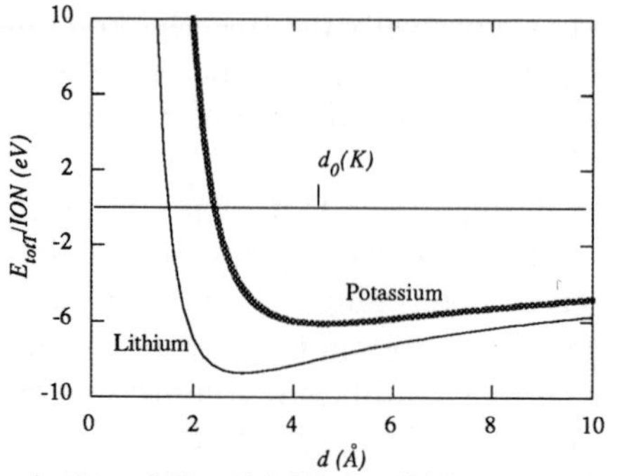

Fig. 14-3. A plot of Eq. (14-12) for lithium and potassium in the body-centered-cubic structure using r_c values chosen so that the minimum energy comes at the observed spacing.

$$B = \Omega\frac{\partial^2 E_{tot}}{\partial\Omega^2} = \frac{k_F{}^2}{9\Omega}\frac{\partial^2 E_{tot}}{\partial k_F{}^2} + \frac{4k_F}{9\Omega}\frac{\partial E_{tot}}{\partial k_F}. \quad (14\text{-}13)$$

Here E_{tot} is the total energy of the system and Ω is the total volume. Of course the final term will be zero since r_c has been chosen such that the energy is minimum at that k_F . As we indicated at the close of Section 13-4, such an evaluation using the free-electron energy gave estimates generally a factor of two too large. The corresponding prediction of the bulk modulus based upon Eq. (14-12) is given as the first entry in Table 14-1 and compared with experiment. Except for the noble metals, this would seem indeed to provide an excellent description of the bulk modulus, and in this case an improvement on the free-electron formula, with the only experimental input being the equilibrium density. For the noble metals, the d-shells play an important role in the elasticity, as we shall see in Chapter 15, and their treatment as a simple metals is not so appropriate.

We might also carry out an evaluation of the Grüneisen constant in terms of the third derivative of E_{tot} with respect to k_F , but we choose to make that study in terms of the two-body interactions $V(r)$ alone. Finally, we might think of using the value of E_{tot} for a calculation of the cohesive energy, requiring a treatment of the free atom also in the Fermi-Thomas approximation. We noted however in Section 13-4-B and 13-6-A that Teller has pointed out that Fermi-Thomas theory should give us a vanishing cohesive energy. We stay with our tight-binding estimates from Section 13-3-A.

Table 14-1. Predicted bulk modulus (10^{12} ergs/cm^3) as $(k_F{}^2/(9\Omega))\partial^2 E_{tot}/\partial k_F{}^2$, Eq. (14-13), using the r_c adjusted to give the correct equilibrium spacing. Experimental values, in parentheses were compiled in Kittel (1976).

				Li	Be
				0.084	0.63
				(0.12)	(1.00)
		Al		Na	Mg
		1.00		0.053	0.39
		(0.72)		(0.068)	(0.35)
Cu	Zn	Ga		K	Ca
0.16	0.47	0.38		0.022	0.15
(1.37)	(0.60)	(0.57)		(0.032)	(0.15)
Ag	Cd	In	Sn	Rb	Sr
0.14	0.13	0.46	1.31	0.018	0.14
(1.01)	(0.47)	(0.41)	(1.11)	(0.031)	(0.12)
Au	Hg	Tl	Pb	Cs	Ba
0.12	0.27	0.61	0.90	0.014	0.13
(1.73)	(0.38)	(0.36)	(0.43)	(0.020)	(0.10)

3. Properties from Interatomic Interactions

The idea of representing the total energy of a solid by interactions between a collection of atoms dates from the old molecular theory (see Sommerfeld (1950)) of Navier, Cauchy, and Poisson from the early 1800's, even before it was established that matter was in fact composed of atoms. In modern terms it is associated with the work of Lennard-Jones (1924, 1925) , who postulated an interatomic interaction of the form $-A/r^6 + B/r^{12}$ for inert-gas atoms. The essential feature of this approach, and subsequent modifications, was the adjustment of the parameters to fit measured properties, such as the equilibrium spacing and the cohesive energy, and then using the potential to predict all of the remaining properties of the system.

Our finding that the principal terms in the energy could be written in terms of an effective interaction given in Eq. (14-9) is in some ways a justification of that view. There is however the important distinction that this is not the *only* contribution to the energy. There are the additional terms which in pseudopotential theory are volume-dependent energies. Because of this the system of atoms is not in equilibrium under the interatomic interactions alone. In fact, the interaction written in Eq. (14-9) is purely repulsive whereas all phenomenological potentials contained an attractive interatomic term which held the atoms near a minimum in the interatomic interaction. Thus this new view is quite counter-intuitive and therefore in contrast to the traditional view of solids, and metals in particular. It was really known from the beginning that the traditional view of atoms in equilibrium under the influence of two-body interactions was not tenable. Cauchy (see Born and Huang (1954) or Sommerfeld (1950)) had shown that in such a system, with cubic symmetry, (and one atom per cell, as in fcc and bcc structures) a relation between the elastic constants, $c_{12} = c_{44}$, must apply, as noted also in Section 10-4-A. It was known experimentally to be far from the truth. Nonetheless, the traditional and intuitively plausible view was widely accepted and deviations from the Cauchy relations were associated with angular, multi-atom forces.

The view of atoms with purely repulsive, two-body interactions, held together by a somewhat structureless volume-dependent energy (ultimately a multi-atom interaction) was suggested in Harrison and Wills (1982). Surprisingly almost the entire range of bonding properties are obtainable by calculating at constant volume, using only that repulsion, $^1/_2 \ \Sigma_{i \neq j} V(\mathbf{r}_i - \mathbf{r}_j)$ from Eq. (14-12). Clearly one can calculate the vibration spectrum holding the total volume fixed, from which one can obtain the elastic constants. From the elastic constants one can obtain the bulk modulus, the Poisson ratio (the reduction in cross-section of an elongated sample) and, by calculating the vibration spectra at two neighboring volumes, the thermal expansion coefficient. These are all properties which we associate with changes in volume, but all can be obtained from calculations at constant volume. One can also, of course, predict distortions around defects, which we shall do in Problem 14-2. Here we outline the results obtained by Harrison and Wills. We first check the validity of our empty-core radii by looking at the bulk modulus, and then consider a number of ratios of elastic properties, for which the magnitude of the core radius cancels out and which therefore test the form of the theory.

A. The Bulk Modulus

We do not actually need to calculate the vibration spectrum in order to obtain the bulk modulus in terms of these interactions which are a function of internuclear spacings d_j . We may write the total energy per atom as $^1/_2 \Sigma_j V(d_j)$ with the sum over the distances to all neighbors, each of which scales as $\Omega^{1/3}$. The result is analogous to Eq. (14-13).

$$B = \Omega\frac{\partial^2 E_{tot}}{\partial\Omega^2} = \frac{1}{2\Omega_0}\Sigma_j\left(\frac{d_j^2}{9}\frac{\partial^2 V(d_j)}{\partial d_j^2} - \frac{2d_j}{9}\frac{\partial V(d_j)}{\partial d_j}\right). \qquad (14\text{-}14)$$

$V(d_j)$ is given by Eq. (14-9). We find each term in brackets above to be given by $1/9 Z^2e^2 \cosh^2\kappa r_c\, e^{-\kappa d_j}(\kappa^2 d_j + 4\kappa + 4/d_j)$ and Ω_0 is the atomic volume. Again we check the validity by fitting r_c for each metal to obtain the observed bulk modulus, with results (taken from Harrison and Wills (1982)) shown as the third entry in Table 13-5, where they can be compared with the standard values for r_c. The fit seems to be somewhat better than that found by adjusting r_c in the free-electron energy to obtain the observed equilibrium spacing. It generally supports this description of the electronic structure, and provides r_c values which may be better for estimates of atomic properties than the standard values. It sets the scale for the vibration spectrum and therefore for all of the elastic properties.

The other elastic properties considered by Harrison and Wills can also be obtained directly without carrying out the calculation of the full vibration spectrum. The results are mathematically equivalent. We discuss these next, and then return to the full vibration spectrum.

B. Deviations from the Cauchy Relations

We noted at the beginning of this section that if the system were in equilibrium under interatomic interactions alone, the Cauchy relation between the elastic constants, $c_{12} = c_{44}$, would follow, and that the experimental values deviate significantly from this. We note further that each of these constants is determined entirely by the interatomic interaction, given by $Z^2e^2\cosh^2\kappa r_c\, e^{-\kappa r}/r$, and that the empty-core radius enters only through the leading factor. Thus the *Cauchy ratio,* c_{12}/c_{44}, does not depend upon the scaling of the interactions by the core radius. For any given structure the predicted ratio is a function only of κr_0. That function is nearly the same for body-centered-cubic and face-centered-cubic structures. The comparison of the predicted and observed ratio from Harrison and Wills (1982) is given in Table 14-2. Each was calculated in the observed crystal structure. We see that the theory correctly predicts values greater than one, and in fact overestimates the deviations considerably. The trend, as a function of κr_0, turned out to be no better than the values themselves. Qualitatively, the deviations from Cauchy's relation are explained, but the magnitudes, with this Fermi-Thomas screening, are not well given.

Table. 14-2. The Cauchy ratio c_{12}/c_{44} , predicted for the cubic metals and compared with observed values. It would be 1.00 for two-body, central-force interactions alone.

	Predicted	Observed*
Cu	4.07	1.61
Ag	3.78	2.03
Au	3.78	3.88
Al	3.10	2.15
Pb	2.68	2.84
K	2.72	1.67
Na	3.05	1.48
Li	3.43	1.30

*Huntington (1964)

C. Poisson's Ratio

Poisson's ratio is the ratio of the decrease in lateral dimension to the increase in longitudinal dimension when a bar of the material is stretched at the ends and the lateral surfaces are free. It can be directly calculated from the elastic constants, so it is, interestingly, obtainable from our constant-volume calculation of the elastic constants. It is given by

$$\sigma = \frac{c_{12}}{c_{11} + c_{12}} \ . \qquad (14\text{-}15)$$

Table. 14-3. Poisson ratio σ from Eq. (14-15) predicted for the cubic metals and compared with observed values.

	Predicted	Observed*
Cu	0.48	0.36
Ag	0.48	0.37
Au	0.48	0.36
Mg	0.47	0.35
Zn	0.47	0.25
Al	0.47	0.33
Tl	0.47	0.35
Sn	0.46	0.33
Pb	0.46	0.40-0.45
Na	0.49	0.43
K	0.49	0.44

*Baumeister (1967)

Predictions resulting from our interatomic couplings were given by Harrison and Wills (1982) and are compared with experiment in Table 14-3. As for the Cauchy ratio, we have a ratio of elastic constants and the effect of the core radius has canceled out. Predictions and experiment are quite independent of material, though in both cases slightly larger for bcc structures (Na and K). We overestimate values, which could be interpreted, as for the Cauchy ratio, as a relative overestimate of c_{12} .

There is no general prediction for Poisson's ratio, such as a Cauchy relation, but for a simple nearest-neighbor radial spring constant κ_0 we note that the fcc structure is stable (though the bcc structure needs further interactions to make it rigid). Then in the fcc structure all elastic constants can be written in terms of this κ_0. We easily obtain $c_{44} = c_{12} = \kappa_0/a$ and $c_{11} = 2\kappa_0/a$ with a of course the unit cube edge, $\sqrt{2}\, d$. Thus from Eq. (14-15) we obtain $\sigma = {}^1/_3$, seen in Table 14-3 to often be very close to experiment.

D. The Grüneisen Constant

We defined the Grüneisen constant in Section 2-4-C in terms of the volume dependence of the vibrational frequencies. With the simple method of calculating vibrational frequencies by equating elastic and kinetic energies, described in Section 14-1-A we could readily estimate a Grüneisen constant for each mode. We instead obtain a characteristic value in terms of the volume-dependence of the bulk modulus

$$\gamma = -\frac{1}{6} - \frac{1}{2}\frac{\Omega \partial B}{B \partial \Omega} = -\frac{1}{6}\left(\frac{d}{B}\frac{\partial B}{\partial d} + 1\right) \qquad (14\text{-}16)$$

as we did in Eq. (2-42). The values have a weak dependence upon r_c through a derivative of κ in $\cosh \kappa r_c$ with respect to spacing. Harrison and Wills (1982) predicted values for the Grüneisen constant, using the core radii fit to the bulk modulus (the third entry in our Table 13-5). Use of the standard radii gave predictions 2% to 9% higher. They actually chose a $\gamma = -{}^1/_6\, (d/B)\, \partial B/\partial d$ and we corrected their values by subtracting 1/6 from each, predicted and observed. The results for Eq. (14-16) based upon the values given by Harrison and Wills (1982) are listed in Table 14-4. We noted in Section 2-4-C that if the Grüneisen constant is the same for all vibrational modes, the thermal expansion coefficient is directly related to that Grüneisen constant. The experimental values listed in Table 14-4 are based upon that assumption.

Table. 14-4. The Grüneisen constant γ of Eq. (14-16) predicted for the cubic metals and compared with experimental values from thermal expansion data. It would vanish for the nearest-neighbor spring-constant model.

	Predicted	Observed*
Cu	1.02	1.79
Ag	1.09	2.23
Au	1.14	2.77
Mg	0.99	1.31
Zn	0.91	2.02
Cd	1.00	2.01
Al	0.93	2.00
Tl	1.03	2.04
Pb	1.06	2.53
Na	1.02	1.08
K	1.14	1.17
Ca	1.10	0.90
Sn	1.14	4.17

*Pearson (1958)

The predictions are rather good for the low-electron-density metals, Na, K, Ca, and Mg, but are underestimated by a factor of about two for the high-electron-density metals, which experimentally have values similar to those in the semiconductors listed in Table 2-6.

For the nearest-neighbor spring-constant model we discussed for Poisson's ratio, with $B = (c_{11} + 2c_{12})/3 = 4\kappa_0/(3a)$, the Grüneisen constant of Eq. (14-16) is zero. For the Lennard-Jones potential (see Section 20-2-E) it is large (we obtained $23/6 \approx 4$) because of the large powers of $1/d$.

4. Calculation of Vibration Spectra

The calculation of the vibration spectrum in terms of the interatomic interactions of Eq. (14-9) is straight-forward but a little intricate. This is partly because there are interatomic forces even before the vibrational mode is introduced and a displacement rotates these forces as well as changing their magnitude. The forces sum to zero in the undistorted crystal, but not when the mode is introduced. For a single atom per primitive cell, as in the fcc and bcc structures, the displacements of each atom for a vibrational mode of wavenumber $\mathbf{Q}$ and frequency ω can be written as the real part of

$$\delta r_j = \frac{u_Q}{\sqrt{N}} e^{i(Q \cdot r_j - \omega t)} \qquad (14\text{-}17)$$

as in Eq. (13-46). (We have taken a single term and made the time-dependence explicit.) Using periodic boundary conditions, the volume of the system remains unchanged by the displacements so we use only the $V(r)$. Then the forces on each atom are immediately obtained in terms of the relative displacements of neighbors. Force is set equal to mass times acceleration for a single atom, giving three simultaneous equations in the three components of $\boldsymbol{u}_Q$. Solving these gives the frequencies of the three vibrational modes for that wavenumber in the Brillouin Zone.

We may illustrate this for the simple case of $\boldsymbol{Q}$ along a [100] direction in an fcc lattice. Calculation for the longitudinal mode is illustrated in Fig. 14-4. The displacement $\delta\boldsymbol{r}_j$ of an atom initially at the position $\boldsymbol{r}_j$, and of its twelve nearest neighbors, are shown in Fig. 14-4, all in an x-direction. We first calculate the force on the central atom due to interaction with the neighbor to the upper right. The position of that neighbor, relative to the central atom, is $\boldsymbol{r} = \boldsymbol{d} + \delta\boldsymbol{d}$ with $\delta\boldsymbol{d} = \delta\boldsymbol{r}_{j+1} - \delta\boldsymbol{r}_j = \delta\boldsymbol{r}_j(e^{iQa/2} - 1)$. The force $\boldsymbol{F}_j = (\boldsymbol{r}/r)F_j$ on the central atom arising from $V(r)$ between these two

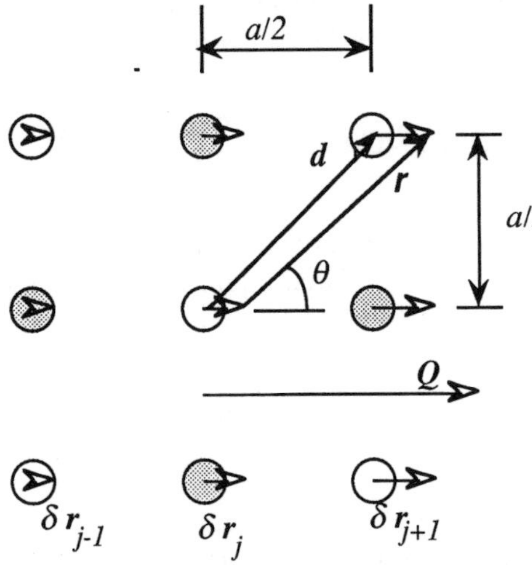

Fig. 14-4 Calculation of the vibrational frequency of a mode with wavenumber $\boldsymbol{Q}$ in a [100] direction in a face-centered cubic crystal. Open circles are atoms in the plane of the figure, shaded circles are in a plane $\pm a/2$ above or below that plane. Displacements $\delta\boldsymbol{r}_j = Re\, \boldsymbol{u}_Q e^{i(\boldsymbol{Q}\cdot\boldsymbol{r}_j - \omega t)}$ are shown for a longitudinal mode and are the same for all atoms in the same plane perpendicular to $\boldsymbol{Q}$.

corresponds to $F_j = \partial V(r)/\partial r = \partial V(d)/\partial d + \partial^2 V(d)/\partial d^2 \delta r + ...$, where the derivatives with respect to d indicate an evaluation at the equilibrium spacing and δr is the change in the magnitude of $\boldsymbol{r}$. The force is directed along the internuclear distance, which makes an angle θ with the x-axis, with $\cos\theta = (a/2 + \delta d)/r = (d/\sqrt{2} + \delta d)/(d + \delta d/\sqrt{2}) = 1/\sqrt{2} + \delta d/(2d) + ...$ We need only the component of the force along the x-direction, since other components from the various neighbors will cancel. We also need the force only to first-order in δd and $\delta r = r - d = \delta d \cos\theta$. Thus the x-component of the force becomes

$$F_j \cos\theta = \left(\frac{\partial V(d)}{\partial d} + \frac{\partial^2 V(d)}{\partial d^2}\delta r\right)\left(\frac{1}{\sqrt{2}} + \frac{\delta d}{2d}\right)$$
$$= \frac{1}{\sqrt{2}}\frac{\partial V(d)}{\partial d} + \frac{\partial V(d)}{\partial d}\frac{\delta d}{2d} + \frac{1}{2}\frac{\partial^2 V(d)}{\partial d^2}\delta d \tag{14-18}$$

plus terms of higher order in δd. The zero-order term cancels when we sum over neighbors. The first-order terms in $\delta d = \delta r_j (e^{iQa/2} - 1)$ add for the four neighbors to the right with the $\delta r_j(e^{-iQa/2} - 1)$ for the four neighbors to the left to give $8(\cos(Qa/2) - 1)\delta r_j = -16\sin^2(Qa/4)\delta r_j$. There is no relative motion for the four neighbors in the central plane so the net force is simply

$$F_j \cos\theta = -8\,\delta r_j \sin^2(Qa/4)\left(\frac{1}{d}\frac{\partial V(d)}{\partial d} + \frac{\partial^2 V(d)}{\partial d^2}\right) \tag{14-19}$$

We may similarly add the forces arising from second and more-distant neighbors and set the total force equal to the mass of the atom times its acceleration $-M\omega^2\delta r_j$ and solve for ω. For nearest neighbors only it is

$$\omega^2 \approx \frac{8}{M}\left(\frac{1}{d}\frac{\partial V(d)}{\partial d} + \frac{\partial^2 V(d)}{\partial d^2}\right)\sin^2(Qa/4)\,. \tag{14-20}$$

Without the term linear in $\partial V/\partial d$ it is a familiar result for longitudinal modes in a [100] direction in fcc crystals with a nearest-neighbor spring constant $\kappa_0 = \partial^2 V/\partial d^2$, of $\omega = \sqrt{8\kappa_0/m}\,\sin(Qa/4)$. Here we have an additional term for nearest neighbors and terms for more distant neighbors with different argument for the $\sin^2$ factor, leading to a modified curve of $\omega(Q)$.

Wills carried out this calculation (Harrison and Wills (1982)) along symmetry lines to obtain a more complete spectrum. He plotted the ratio of the calculated frequency to the maximum frequency ω_m , with the result shown in Fig. 14-5. All values of ω^2 are linear in $V(d)$ and therefore all frequencies proportional to $\cosh\kappa r_c$, with no other dependence upon the empty-core radius. The validity of various choices of core radius was tested from the other elastic properties so for the spectrum he chose r_c to give the correct maximum vibrational frequency, giving a value close to that obtained by fitting the bulk modulus.

It is interesting that it is possible to duplicate the spectrum this closely using a single adjustable parameter, r_c . An analogous result could be obtained using a single nearest-neighbor radial spring constant, $\kappa_0 = \partial^2 V/\partial d^2$, and omitting the terms in $\partial V/\partial d$, again giving all frequencies in terms of a single scale parameter. The relative accuracy of the two models could be checked by comparing the two predicted spectra, but we chose instead to make the comparisons in terms of the Cauchy ratio, the Poisson ratio, and the Grüneisen constant, given for a range of materials in Tables 14-3, 14-4, and 14-5.

In Part b of Fig. 14-5 we show the density of vibrational frequencies, also obtained by Wills (Wills and Harrison (1982)) and the comparison with experiment. The shape of the curves agrees about as one would expect by

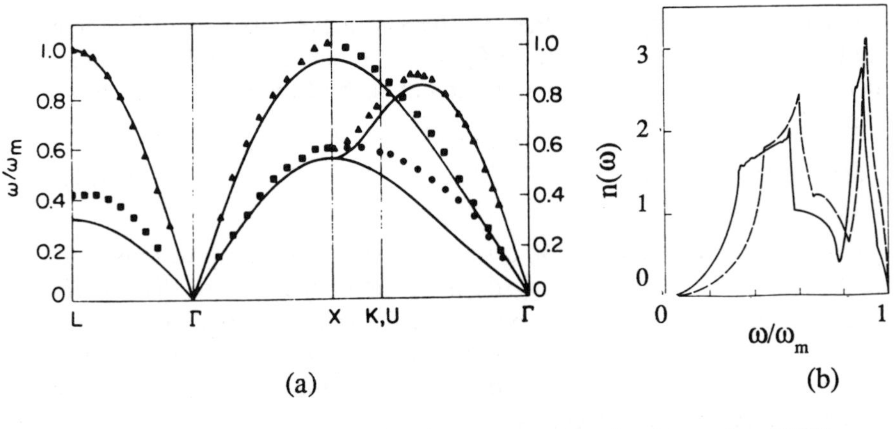

Fig. 14-5. In Part a is the vibration spectrum for aluminum along [100], [111], and [110] directions obtained from the interaction in Eq. (14-9), with each frequency divided by the calculated maximum frequency. Experimental points are from Stedman and Nilsson (1966). Part b, also from Harrison and Wills (1982), shows the total density of vibrational states per unit frequency range from the same calculation. The dashed curve is experimental, from Stedman and Nilsson (1967).

looking at Part a, and since both curves are normalized to give the same number of total modes, the vertical scales must match.

As we had hoped, the simple theory gives us a way of predicting the vibration spectrum, but it may not be such an impressive accomplishment in this case.

5. Liquid Metals

Most of the study of condensed matter systems has been directed at crystalline solids, for which the translation symmetry of the lattice could be used to reduce the problem of calculating the electronic states, or the vibrational states, to the treatment of a single primitive cell. It has also been possible to treat defects in an otherwise perfect crystal, as we have done. The idea of treating a liquid might seem formidable, but with pseudopotentials treated as a perturbation we have a clear way to proceed. Again any effects of the atoms enter through the matrix elements of the pseudopotential, which we have factored as $S(\mathbf{q})w_q$. The form factors are the same in the solid and the liquid and the atomic positions enter only through the structure factors.

A. Liquid Structure Factors

In our perturbation theory, the $\langle \mathbf{k}|H|\mathbf{k}\rangle$ which enters first-order theory is independent of the structure ($S(0) = 1$) and the second-order terms depend upon $S(\mathbf{q})^*S(\mathbf{q})$. We may write out each of the structure factors to obtain

$$S(\mathbf{q})^*S(\mathbf{q}) = \frac{1}{N^2} \sum_{i,j} e^{-i\mathbf{q}\cdot(\mathbf{r}_j - \mathbf{r}_i)} . \tag{14-21}$$

For each atom, at position $\mathbf{r}_j$, we are to sum over the positions of the neighbors at $\mathbf{r}_i$, which we may do in terms of the *radial distribution function* $P(r)$, the probability per unit volume of finding a neighbor at distance r . We must also include the term $i = j$. Since the result is averaged over atoms j this should give exactly the right answer, and the sum over j gives simply a factor N .

$$S(\mathbf{q})^*S(\mathbf{q}) = \frac{1}{N}\left(1 + \int 4\pi r^2 dr\, P(r)\, e^{-i\mathbf{q}\cdot\mathbf{r}}\right) = \frac{1}{N} + \frac{4\pi}{Nq}\int r\, dr\, P(r)\, \sin qr . \tag{14-22}$$

$P(r)$ can be directly measured experimentally by neutron diffraction, or x-ray diffraction, exactly because the diffraction intensity is calculated in terms of the same matrix elements, except that the form factor which enters is the delta-function interaction with the nucleus for neutrons, or a charge-density form factor for x-rays. Thus we have exactly the structural information needed for any second-order calculation, such as the scattering of electrons to predict the resistivity of the liquid metal.

Actually, the structure factor is sufficiently insensitive to the details of the theory that theoretical models for the structure are often adequate. A view that is *too* simple is that the positions of the neighbors are completely random as in an ideal gas. Then we would return to Eq. (14-21) and note that an average over each term $e^{-i\mathbf{q}\cdot(\mathbf{r}_j-\mathbf{r}_i)}$ for $i \neq j$ gives zero and only the diagonal terms, giving $1/N$, survive. This remains constant as q goes to zero, whereas it can be verified that the correct $S^*(q)S(q)$ goes to a very small value; the density is quite uniform on the large scale, with deviations calculable in terms of thermally excited longitudinal vibrations. [$S(q)^*S(q)$ at small q will arise from a vibrational mode, Eq. (14-17) with $Q = q$, which gives a local change in atomic density of $-\nabla\cdot\delta\mathbf{r}_j$ and an $S(q)^*S(q)$ of $q^2u_q{}^2/N$. Setting the energy of the mode at the thermal value of k_BT gives a $NS(q)^*S(q) = k_BT/(Mv_s{}^2)$, with v_s the speed of sound. The denominator is of the order of electron volts.]

The next simplest view would simply exclude neighbors within an atomic diameter of each atom j. This corresponds to subtracting the $P(r) = 1/\Omega_0$ of the random distribution for $r < d$. The result, from Eq. (14-22), is

$$S(\mathbf{q})^*S(\mathbf{q}) = \frac{1}{N}\left(1 - \frac{4\pi}{\Omega_0 q^3}(\sin qd - qd\cos qd)\right). \tag{14-23}$$

We may again look at the small-q limit, where the expression in brackets approaches $1 - 4\pi d^3/(3\Omega_0)$. If it is to be small at small q then we must choose $4\pi d^3/3 = \Omega_0$, smaller than reasonable values for d. For example, for hard spheres in the body-centered-cubic structure $4\pi d^3/3 = \pi\sqrt{3}\,\Omega_0$. The resulting structure factor is shown as the $n_0 = 0$ curve in Fig. 14. We could make one more improvement in assuming that not only are neighbors excluded for a distance d, but there are extra atoms in contact, adding a delta-function in $P(r)$ at $r = d$. If we add a term $n_0\delta(r - d)/(4\pi d^2)$ to $P(r)$ it will correspond to n_0 atoms in contact and will add a term $n_0 \sin qr / qr$ to $NS(q)^*S(q)$. If we now require that the relation between d and Ω_0 be that for a bcc crystal and that $S(q)^*S(q)$ go to zero at q

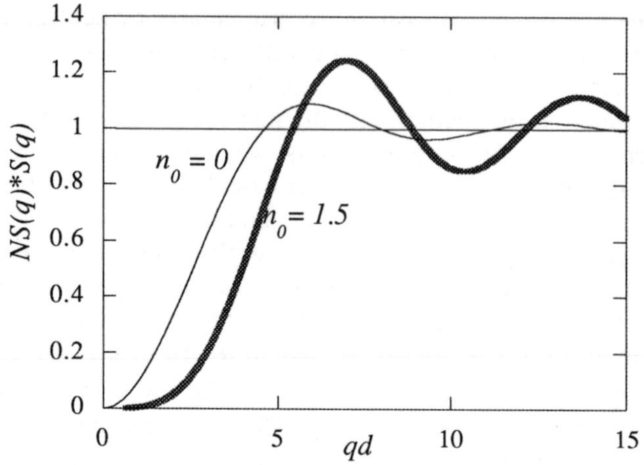

Fig. 14-6. Approximate structure factors for a liquid. The $n_0 = 0$ curve is for a radial distribution function $P(r) = 1/\Omega_0$ for $r > d$ and zero for $r < d$, with d adjusted to give $NS(0)S(0) = 0$. The $n_0 = 1.5$ curve has a delta function added to $P(r)$ at the cut-off $r = d$.

$= 0$, we find $n_0 = \pi\sqrt{3} - 1 = 4.4$ but the resultant $S(q)*S(q)$ is negative at small q . The structure factor with the smaller value of $n_0 = 1.5$ is shown in Fig. 14-6. This structure factor is qualitatively correct. A more accurate form, depending upon an assumed hard-sphere diameter and packing fraction, has been given by Percus and Yevick (1958) and Percus(1962). It is also possible in any calculation to use an experimentally determined structure factor.

These structure factors, with the pseudopotential form factors we have given, provide the basis for discussion of the electronic structure of the liquid. Without translational symmetry we do not have a Brillouin Zone, nor the energy bands except to zero - or first - order in the pseudopotential, which are free-electron bands. Some properties we have treated, such as the bulk modulus, are insensitive to structure and our estimates for the crystal could be applied to the liquid. The total energy itself, relative to the energy in the crystal, appears to be much too sensitive to structure for use of these approximate, or even experimental, structure factors. The Madelung energy, for example, can be obtained from Eq. (14-4), with $F(q)$ replaced by

$2\pi Z^2e^2/(\Omega_0 q^2)$ and converted to an integral. This leads to an energy per ion of $(2Z^2e^2/\pi)\int NS(q)^*S(q)\, dq$. To obtain it to the needed millivolt accuracy would require the area under the curve of Fig. 14-6, minus the area under $NS(q)^*S(q) = 1$ to a part in a thousand. This could better be used as a way of adjusting the model than a way of estimating the change in energy under melting. One property which has been successfully treated is the resistivity of the liquid metal.

B. Resistivity

The resistivity of the liquid metal was first calculated using pseudopotentials in Harrison (1963). It used calculated pseudopotential form factors for zinc and experimental structure factors. It followed a method given by Ziman (1961). The procedure is exactly what we used in Section 13-3-B, leading to a relaxation rate for electrons due to a vacant site, given in Eq. (13-44). We now use the structure factor for the liquid, rather than $S(q) = -e^{-i\mathbf{q}\cdot\mathbf{r}_0}/N$ for the vacancy, to obtain

$$\frac{1}{\tau} = \frac{\Omega m}{4\pi\hbar^3 k^3}\int dq\, q^3 w_q{}^2 S(q)^*S(q) \; . \qquad (14\text{-}24)$$

The factor $1/N$ in $S(q)^*S(q)$ cancels against the Ω to give a result of order $w_q{}^2/(\hbar E_F)$, which is quite small on the scale of 1 eV/$\hbar$ since the form factor is going through zero in just the wavenumber range where the factors q^3 and $S(q)^*S(q)$ are becoming large. Eq. (14-24) gives scattering rates near those obtained from the scattering by lattice vibrations in the solid at temperatures near the melting temperature. Thus the change in resistivity upon melting is not large. This is an important conceptual point. The long electronic mean-free paths in metals do not arise so much from the translational periodicity, as traditionally thought, but more from the weakness of the atomic pseudopotential, as we indicated in the introduction to this chapter.

Ashcroft and Lekner (1966) carried out extensive calculations using exactly this approach and the structure factors of Percus and Yevick (1958). They found in general rather good accord with the experimental resistivities.

6. Pseudopotentials and Covalent Bonding

We return to the application of the pseudopotential method to covalent solids, and in particular to the insight it provides into the bonding in these systems, given in more detail in Chapter 18 of Harrison (1980). We may again begin with free-electron states and slowly introduce the pseudopotential arising from the lattice. As we saw in Fig. 13-13, gaps are introduced in the energy bands at the Bragg planes, and small pieces of Fermi surface are eliminated. What clearly happens in the case of the semiconductors is that finally all of the Fermi surface disappears into Bragg planes. The volume in wavenumber space of occupied states remains constant in the process. It was pointed out many years ago (Mott and Jones (1936)) that the Bragg planes perpendicular to the lattice wavenumbers $[220]2\pi/a$ form a dodecahedron, called the *Jones Zone*, which has just the volume needed to accommodate the four electrons per atom. These Jones-Zone planes, then, are regarded as the essential planes for the formation of an insulating covalent crystal, and other Bragg planes which may intersect the initial free-electron sphere are regarded as inessential.

A. Coupling Through the $[100]2\pi/a$ States

A closer look at this description raises serious problems which we believe are central to the understanding of semiconductors. First, $q/k_F = 2\sqrt{2}\,(\pi/12)^{1/3} = 1.81$ for the [220] lattice wavenumber, just past the crossing for the aluminum pseudopotential shown in Fig. 13-6, and for the similar pseudopotential for silicon. Thus the form factor for this diffraction is not strong. A clue to the resolution was given by Heine and Jones (1969), who noted that a combined diffraction by $[111]\,2\pi/a$ and $[11\text{-}1]\,2\pi/a$ would also take an electron between two states differing in wavenumber by $[220]2\pi/a$ and the much large form factor for the corresponding "second-order diffraction" with $q/k_F = \sqrt{3}\,(\pi/12)^{1/3} = 1.11$ may be the important one. This turns out to be correct, but the outlook, expanding energies for small pseudopotentials, seems to be inappropriate.

We should consider the four coupled states, $\boldsymbol{k}_1 = [110]2\pi/a$, $\boldsymbol{k}_2 = [\text{-}1\text{-}10]\,2\pi/a$, $\boldsymbol{k}_3 = [001]2\pi/a$, $\boldsymbol{k}_4 = [00\text{-}1]2\pi/a$, as we considered the two coupled states at a Zone face in Eq. (13-36). The first two have energy $\varepsilon_2 = 2(\hbar^2/2m)(2\pi/a)^2$ and the second two have energy $\varepsilon_1 = (\hbar^2/2m)(2\pi/a)^2$. This yields four simultaneous equations, rather than two, and a Hamiltonian matrix which we may write

$$H = \begin{pmatrix} \varepsilon_2 & W_{220} & W_{11\text{-}1} & W_{111} \\ W_{\text{-}2\text{-}20} & \varepsilon_2 & W_{\text{-}1\text{-}1\text{-}1} & W_{\text{-}1\text{-}11} \\ W_{\text{-}1\text{-}11} & W_{111} & \varepsilon_1 & W_{002} \\ W_{\text{-}1\text{-}1\text{-}1} & W_{11\text{-}1} & W_{00\text{-}2} & \varepsilon_1 \end{pmatrix}. \qquad (14\text{-}25)$$

Each matrix element is given with a subscript corresponding to the wavenumber difference, in units of $2\pi/a$. There is a complication with these matrix elements which did not come up for metals, with one atom per primitive cell. Then the structure factor of Eq. (13-20), $S(\boldsymbol{q}) = \Sigma_j e^{-i\boldsymbol{q}\cdot\boldsymbol{r}_j}/N$, contained a sum over translationally equivalent atoms and was one for lattice wavenumbers and zero for other wavenumbers. Now at the lattice wavenumbers there is a second term, differing by a factor $e^{-i\boldsymbol{q}\cdot\delta}$ with $\delta = [111]a/4$, for the second atom in the primitive cell, as described at the beginning of Chapter 5. Thus the structure factors entering these matrix elements at the lattice wavenumbers $\boldsymbol{q}$ are

$$S(\boldsymbol{q}) = \frac{1 + e^{-i\boldsymbol{q}\cdot\delta}}{2}. \qquad (14\text{-}26)$$

The structure factor for W_{220}, with $\boldsymbol{q}\cdot\delta = [220]\cdot[111](2\pi/a)a/4 = 2\pi$, is one, as it is for $W_{2\text{-}20}$. The structure factors for W_{002} and $W_{00\text{-}2}$ are zero (there is no diffraction at these wavenumbers). The structure factors for W_{111} and $W_{\text{-}1\text{-}11}$ are $(1 + i)/2$ and those for $W_{\text{-}1\text{-}1\text{-}1}$ and $W_{11\text{-}1}$ are $(1 - i)/2$. These matrix elements can be substituted in Eq. (14-25) and the matrix diagonalized analytically (Harrison (1976a)). Of more interest is the result of keeping only the four matrix elements of the W_{111} type which would seem to be the dominant couplings, neglecting W_{200} and $\varepsilon_2 - \varepsilon_1$ in comparison. Then measuring the energies from $\varepsilon_1 \approx \varepsilon_2$ the Hamiltonian matrix becomes

$$H \approx \begin{pmatrix} 0 & 0 & W_{111}^* & W_{111} \\ 0 & 0 & W_{111}^* & W_{111} \\ W_{111} & W_{111} & 0 & 0 \\ W_{111}^* & W_{111}^* & 0 & 0 \end{pmatrix}. \qquad (14\text{-}27)$$

[This has the complex conjugates differently placed than in Harrison (1980), p. 414, but the eigenvalues are the same.] This is readily solved to obtain two eigenvalues of zero. [A vector $\boldsymbol{u}$ such that $H\boldsymbol{u} = 0$ is $[1, -1, 0, 0]/\sqrt{2}$. Another is $[0, 0, W_{111}, -W_{111}^*]/\sqrt{2W_{111}^*W_{111}}$.] The other two eigenvalues are $\pm 2\sqrt{W_{111}^*W_{111}}$.

We may directly identify these eigenvalues with the true energy bands of Fig. 5-6. In Fig. 14-7 we have reproduced the bands for silicon from that figure. The center of the Bragg plane bisecting $[220]2\pi/a$ is outside the Brillouin Zone, but equivalent to the points X in the [001] direction (at the wavenumber $[001]2\pi/a$ which we included in our expansion of the states above). Thus we identify the band values labeled X_4 with the zero eigenvalues of Eq. (14-27) and the band values labeled X_1 with the eigenvalues $\pm 2\sqrt{W_{111}^*W_{111}}$. (There is a second band at each of these X_1 points arising from the sets with $\pm[1\text{-}10]2\pi/a$.) The band gap in the semiconductor is $2\sqrt{W_{111}^*W_{111}}$ between the middle and upper eigenvalues, as if there were a coupling of magnitude $\sqrt{W_{111}^*W_{111}}$ between the states at opposite faces. If we change the wavenumber along Δ towards Γ that gap remains about the same, but the bands rise as do the corresponding free-electron bands. It is interesting that the gap arises predominantly from the W_{111} matrix element and is predicted therefore to be the same as the gaps at L , which is at the center of the Brillouin Zone face in the [111] direction. That is seen to be approximately true not only

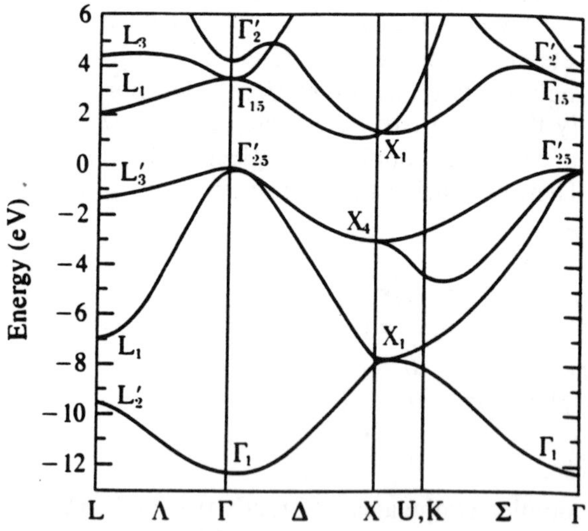

Fig. 14-7. The energy bands for silicon from Chelikowsky and Cohen (1976), which were given also in Fig. 5-6.

for the gap between the conduction and valence band, but for the lower gap within the valence band.

The most significant point from this is that the general understanding of the bands is based upon the W_{111} matrix element, neglecting any effects not only of W_{220}, but also the energy difference $\varepsilon_2 - \varepsilon_1 = (\hbar^2/2m)(2\pi/a)^2$. This is exactly the opposite outlook from that for simple metals where the kinetic energies were regarded as the large quantities and we expanded for small pseudopotentials. Then the parallel with the tight-binding outlook becomes very clear. There we regarded the interatomic matrix elements - the covalent energy V_2 - as the large quantity, producing the bonding and antibonding states. The sp-splitting, which was characterized by the metallic energy V_1, and broadened the bands as kinetic energy, was treated as the small quantity. From both tight-binding and pseudopotential outlooks, we are to take just the opposite expansion for metals which we take for covalent solids.

This is closely related to the negative gaps discussed in Section 6-5-B, and to the evolution of the bands as atoms are brought together. As silicon atoms, at large spacing but in the geometry of the diamond lattice, are brought together, initially the two s-states per cell broaden into two bands, as illustrated schematically in Fig. 14-8. The six p-states per cell broaden into

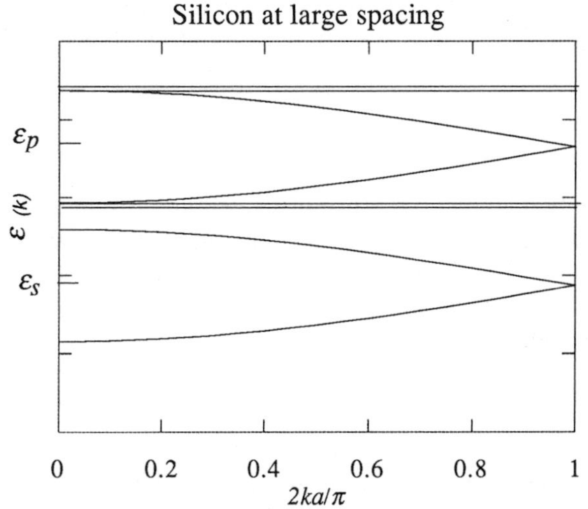

Fig. 14-8. A schematic representation of the bands for silicon when the states on neighboring atoms have just begun to overlap and the gap is between the s- and p-bands, rather than between bonding and antibonding bands as in Fig. 5-6.

six bands, also shown. The eight electrons per primitive cell fill the s-bands and partially fill the p-bands, leaving the system metallic. The bands are schematic, but essentially correct and correspond to the negative gap illustrated in Part c) of Fig. 6-3. They are also strikingly similar to the tight-binding bands of Fig. 5-4, but here the lower of the two flat double bands is attached to the p-like complex. The upper two sets of bands at $k = 0$ correspond to pure p-like states each in bonding or antibonding relation to its neighbor. Similarly the two lower sets at $k = 0$ correspond to pure s-states in bonding or antibonding relation to their neighbors. As we bring the silicon atoms closer, eventually the bonding p-states drop below the antibonding s-states at $k = 0$ and the double band is attached to the lower complex, as in the bonding bands of Fig. 5-4. A new gap has opened up, now with four bands below, all filled and now a semiconductor. The system has become a semiconductor with the dominant coupling being the bonding-antibonding splitting, a positive gap, rather than the sp-splitting.

The analysis of semiconductor properties in terms of pseudopotentials is carried much further in Chapter 18 of Harrison (1980), identifying the matrix element W_{111} qualitatively and quantitatively with the covalent energy V_2 (though V_2 is of order twice W_{111}). Estimates are then made in the pseudopotential context of the dielectric susceptibility and angular rigidity. This is not really as useful a formulation for such properties as the tight-binding formulation, though it may enrich one's understanding of the electronic structure. Here we shall just outline the extension to polar systems.

B. Polar Systems

When the two atoms in the primitive cell are different, the form factors of the pseudopotentials on the two atoms add, as the two identical terms added in Eq. (14-26), to give a [111] matrix element $W_{111} = (w_{111+} + iw_{111-})/2$. Thus the magnitude of W_{111} , which plays the role of half the bonding-antibonding splitting can be written

$$\sqrt{W_{111}{}^*W_{111}} = \sqrt{{}^1/_8(w_{111+} - w_{111-})^2 + {}^1/_8(w_{111+} + w_{111-})^2} \,. \quad (14\text{-}28)$$

For the homopolar case this goes to $w_{111}/\sqrt{2}$. The second term under the square root becomes the squared covalent energy, based upon the average form factor. The first term becomes the counterpart of the squared polar energy, with a polar energy given by $(w_{111+} - w_{111-})/\sqrt{8}$. The pseudopotential for the polar crystal is also fitting beautifully with the tight-binding view. That would not have been the case if the dominant coupling across the Jones Zone had been from W_{220} since that is given by $(w_{220+}+ w_{220-})2$ and depends only upon the average. This would have given qualitatively different dependence of properties on the polarity of the system.

Finally, we should mention the results for ionic crystals, discussed in detail in Harrison (1980), Chapter 18. The essential difference in the ionic crystals is the different crystal structure, in which there are ordinarily six neighbors and no possibility to form independent bonds with each neighbor. In both cases the atoms of one element form a face-centered-cubic structure, and the atom of the other element is displaced from the first by a δ , equal in the case of rock-salt structures to $[100]a/2$. The lattice wavenumbers are the same and if we think in terms of the same Jones Zone, we focus on the matrix element $W_{111} = (w_{111+} - w_{111-})/2$ for this structure (with $\boldsymbol{q} \cdot \delta = \pi$) . Thus to lowest order the band gap and the cohesion arise only from the polar energy, as we found from our tight-binding analysis of ionic crystals. Once again the two descriptions are remarkably consistent, but only if we focus on the splitting $2\sqrt{W_{111}{}^*W_{111}}$ rather than $2W_{220} = w_{111+} + w_{111-}$ for the rock-salt structure, as we did for the tetrahedral structure.

7. Alloys

Hafner (1987) has given a full account of the pseudopotential theory of alloys of the simple metals. It appears that when the full theory is carried out accurately and completely for well-specified systems, the predicted changes in energy upon alloying are quantitatively accurate. This requires relaxation of the neighboring atoms and self-consistent screening, and has not been done in many cases. Hafner also discusses many simple theories and compares them with experiment, though experimental numbers are much more limited than one might expect. There are similarities with our findings for semiconductor alloys in that the energies of solution of homovalent alloys tend to be dominated by size effects, while heterovalent alloys are dominated by chemical effects. However, the metal alloys are very much more intricate and complicated.

We sought to apply the view developed here to give a simple, understandable theory of metallic heats of solution. In the end we did not feel that the attempt was successful in the sense that it was for semiconductor alloys where the results could be summarized in terms of misfit energy for homopolar solutions and positive shifts of average bond energies from different polarities of bonding in heterovalent mixtures.

Nonetheless it seems appropriate to give here a brief account of these considerations. The approach is direct, simple, and useful for understanding the diverse number of questions which arise in metallic alloy systems. It also suggests a number of aspects which determine the relative solubilities.

In addition it provides a basis for judging semiempirical schemes such as that due to Miedema, de Chatel, and de Boer (1980). Their scheme has two adjustable parameters per element which are enough not only to fit the limited data available but to give the appearance of the parameters having some physical basis in themselves. Pettifor (1987) has critiqued this approach and we similarly do not find that basis persuasive. However, the scheme does systematize the data sufficiently to guess results for unknown systems.

In principle the treatment of alloys in pseudopotential theory is direct. We simply insert the appropriate atomic pseudopotential at each atomic site in the alloy and the screening is based upon the total number of electrons present in the system. This is how we proceeded when we calculated scattering by impurity atoms. It will also be appropriate for an ordered alloy of, for example, an equal number of two atom types, or for a disordered alloy where the ions are rather uniformly distributed and the screening electron density is based upon the total number of electrons present in the total volume.

There are problems in using Eq. (14-12) for the total energy in other systems. It is clear that use of this average electron density would give poor results for a segregated alloy, a system with crystals of two different metals separated by an interface. Then certainly for the total energy we should use electron densities appropriate to the individual metals. Use of the same average density on both sides would give a different, and an incorrect, result. This is connected with the fact, which we noted, that a volume-dependent term is really a many-atom-interaction term of limited range which is only approximated by a volume-dependent energy. This same difficulty faces us if we seek the heat of solution of one metal in another. We consider that problem first.

A. Heats of Solution

We seek $\delta E_{sol}(A\ in\ B)$, the heat of solution of metal A in metal B, which we define to be the energy change when an A atom is taken from bulk metallic A, at zero temperature, and inserted substitutionally in bulk metal B, increasing the volume of that metal by one atomic volume; if it is negative heat is evolved in the solution. As a first step we do not deform the structure of B when the atom is substituted but increase the number of atomic cells in that metal by one. The change in energy is of order an electron volt, while the total energy of the system, Eq. (14-12), is of order N_a electron volts, with N_a the total number of atoms present, so we must be careful to keep all terms of order one electron volt. Thus there are interaction terms between the transferred atom in which $V_{BB}(\mathbf{r}_i - \mathbf{r}_j)$ of Eq. (14-12) is replaced by $V_{AB}(\mathbf{r}_i - \mathbf{r}_j)$, given by

$$V_{AB}(r) = \frac{Z_A Z_B e^2 \cosh(\kappa r_{cA}) \cosh(\kappa r_{cB})\, e^{-\kappa r}}{r}, \tag{14-29}$$

but there are also very small (smaller by a factor of order $1/N_a$ if we change the number of electrons in the solvent by one) changes in $V_{BB}(\mathbf{r}_i - \mathbf{r}_j)$ which apply to *all* B atoms and make therefore comparable contributions.

We can avoid many of these terms by imagining that the two metals are initially in contact, which means that the Fermi energies are the same in both metals and that if the A atom we transfer to metal B has a different number of electrons than the host B , there will be no energy required to transfer the extra electron between metals. Equality of Fermi energy also means that there is no voltage difference between the two metals and no contribution arises which would be involved if we were to move a proton across a voltage drop between the two metals. We shall write down at the end of the chapter (Eq. (14-34)) the change in electron energy which would arise from Eq. (14-12) due to the transfer of electrons between metals due to the voltage difference (or work-function difference) between the two metals which is implied by Eq. (14-12), but do not believe it affects the heat of solution since extra nuclear charges equal to the number of electrons are also transferred.

Thus we consider two pure metals in contact, N_A atoms of type A of valence Z_A and Fermi wavenumber k_{FA} and empty-core radius r_{cA} and N_B atoms of type B with Z_B , k_{FB} and r_{cB} . We will remove one A atom, leaving a perfect A crystal with one less atom and substitute it for one B atom, increasing the number of sites in that crystal by one. We look first at the first two terms, kinetic and exchange energy, in the total energy, Eq. (14-12). The electron density remains the same in metal A but has Z_A fewer electrons. If the valence Z_A and Z_B are the same, this similarly increases the total kinetic and exchange energy in the metal B by the values for those energies per atom. If Z_A is different from Z_B we may initially transfer the A nucleus and Z_B electrons, giving the same result, and then transfer the Z_A -

Z_B electrons across at no cost in energy. The corresponding change in kinetic and exchange energy is given as the first line in

$$\delta E_{sol}(A\ in\ B) = Z_B\left(\frac{3}{10}\frac{\hbar^2 k_{FB}^2}{m} - \frac{3e^2 k_{FB}}{4\pi}\right) - Z_A\left(\frac{3}{10}\frac{\hbar^2 k_{FA}^2}{m} - \frac{3e^2 k_{FA}}{4\pi}\right)$$

$$+ \frac{2e^2(k_{FB}^3 - k_{FA}^3)Z_A r_{cA}^2}{3\pi} - \frac{2e^2 Z_A}{3\pi}\left(\frac{k_{FB}^3}{\kappa_B^2} - \frac{k_{FA}^3}{\kappa_A^2}\right) \qquad (14\text{-}30)$$

$$+ \sum\nolimits_{j(B)} (V_{AB}(\mathbf{r}_j) - {}^1\!/_2 V_{BB}(\mathbf{r}_j)) - {}^1\!/_2 \sum\nolimits_{j(A)} V_{AA}(\mathbf{r}_j)$$

$$- \frac{Z_A^2 e^2 \kappa_B}{2} cosh(\kappa_B r_{cA})e^{-\kappa_B r_{cA}} + \frac{Z_A^2 e^2 \kappa_A}{2} cosh(\kappa_A r_{cA})e^{-\kappa_A r_{cA}} .$$

The first term in the second line is the change in the energy from moving the empty-core pseudopotential of the dissolved atom from metal A to metal B. The second term in the second line is the change in the screened interaction between the charge $Z_A e$ of the impurity atom and the uniform background (reduced by the screened interaction of the background with itself) due to the shift of the impurity atom to metal B.

For the third line it is noted that the interaction energy, per atom, in a pure material A is ${}^1\!/_2\Sigma_j V_{AA}(\mathbf{r}_j)$ with V_{AA} given by Eq. (14-29) (with all parameters evaluated for metal A) and with a sum over neighbors to one atom. Thus such an energy is subtracted for the removed impurity atom, and added (with all parameters evaluated for metal B) for the additional cell in B. In addition the sum of neighbors to the impurity site is changed from V_{BB} to V_{AB} , so the two in combination give the sum over $j(B)$ indicated. Note that in the first sum, the atoms have the spacing for the B crystal but in the second it is for the A crystal. The fourth line gives the change in the diagonal interaction term for the impurity in moving it to metal B.

One might question our holding the total volume of the metal B constant at $N_B + 1$ host-atom cells when an A atom is substituted. However, our empty-core radii were chosen so the energy was minimum at the observed spacing so a uniform change in volume of order one cell makes a change in energy of order $1/N_B^2$ and is negligible. This aspect of the calculation is correct.

We might also allow local distortions such that the impurity-cell neighbors are moved while the total volume is held fixed. In this case the changes come only from the third line. We estimated the changes for

Table 14-5. Heats of solution (eV per atom) for one atom from a metal listed at the top of each column dissolved in a host, listed at the left of each row They are obtained from Eq. (14-30) using r_c (second entry, Table 13-5) and k_F (Table 13-1) values fit to the pure-metal spacing (the only numerical input other than Z and fundamental constants). Positive values imply that the energy increases when the solution is formed.

-	Li	Na	K	Rb	Cs	Be	Mg	Ca
Li	0	0.29	1.58	2.17	3.34	9.84	2.12	1.42
Na	0.27	0	0.28	0.49	0.93	13.7	4.06	1.73
K	0.80	0.23	0	0.01	0.10	17.4	6.21	2.97
Rb	0.96	0.33	0.02	0	0.03	18.3	6.76	3.35
Cs	1.22	0.52	0.10	0.04	0	19.7	7.65	4.00
Be	3.85	8.19	17.0	20.5	27.0	0	2.45	14.0
Mg	-0.63	0.35	2.86	3.90	5.91	5.37	0	1.51
Ca	-1.21	-1.07	-0.11	0.35	1.26	9.03	1.04	0
Sr	-1.21	-1.30	-0.72	-0.40	0.24	10.5	1.69	-0.01
Ba	-1.18	-1.35	-0.92	-0.67	-0.13	11.2	2.03	0.07
Cu	0.22	1.38	4.31	5.53	7.90	6.22	0.93	2.69
Ag	0.01	0.54	2.25	3.01	4.49	8.69	1.67	1.62
Au	0.02	0.58	2.36	3.14	4.67	8.53	1.61	1.66
Zn	0.55	2.55	6.96	8.73	12.0	2.89	0.14	4.71
Cd	-0.47	0.68	3.49	4.65	6.87	4.88	-0.05	1.95
Hg	-0.67	0.27	2.71	3.73	5.68	5.50	0.02	1.41
Al	2.60	5.41	11.3	13.6	18.0	2.27	1.61	8.81
Ga	1.47	3.76	8.69	10.6	14.3	2.70	0.79	6.36
In	-0.16	1.32	4.75	6.13	8.76	3.75	-0.19	2.91
Tl	-0.50	0.81	3.90	5.16	7.56	4.07	-0.33	2.22
Sn	1.53	3.62	8.18	9.99	13.4	3.28	0.95	5.97
Pb	0.81	2.64	6.73	8.35	11.4	3.40	0.43	4.66

(continued)

reasonable empty-core radii differences and found that the total interaction-energy was generally less than one electron volt and changes were of order one or two percent of that small value and seemed negligible. This may not be true in all circumstances. It is always possible to correct this aspect by minimizing the energy with respect to the positions of the neighboring atoms.

It is quite straightforward to evaluate this expression, Eq. (14-30), using the second entry from Table 13-5 for r_c (fit to the observed spacing for each element) and the observed Fermi wavenumber which we listed in Table 13-1. Then both of these parameters are functions only of Z and the observed

Table 14-5 (continued). Heats of solution (eV/atom).

-	Sr	Ba	Cu	Ag	Au	Zn	Cd	Hg
Li	2.04	2.56	0.26	0.03	0.04	4.34	2.45	2.05
Na	1.40	1.37	0.86	0.40	0.43	7.13	4.56	3.93
K	2.13	1.83	1.60	0.99	1.04	9.89	6.83	6.06
Rb	2.44	2.08	1.80	1.16	1.21	10.5	7.41	6.60
Cs	2.98	2.56	2.12	1.44	1.50	11.6	8.33	7.48
Be	22.3	27.3	1.99	3.21	3.06	-0.45	1.59	2.69
Mg	3.48	4.78	-0.77	-0.72	-0.74	1.15	0.09	-0.01
Ca	0.40	0.79	-0.88	-1.15	-1.14	3.39	1.39	0.96
Sr	0	0.17	-0.75	-1.12	-1.09	4.39	2.12	1.59
Ba	-0.08	0	-0.66	-1.07	-1.04	4.88	2.49	1.92
Cu	4.86	6.28	0	0.10	0.07	2.01	1.00	0.92
Ag	2.65	3.41	0.14	0	0.00	3.57	1.93	1.61
Au	2.75	3.55	0.13	-0.00	0	3.46	1.86	1.56
Zn	8.60	11.0	-0.12	0.29	0.23	0	-0.06	0.21
Cd	4.23	5.71	-0.69	-0.58	-0.61	0.89	0	-0.05
Hg	3.31	4.56	-0.79	-0.76	-0.77	1.22	0.12	0
Al	14.3	17.7	1.49	2.20	2.11	0.40	1.15	1.74
Ga	10.8	13.6	0.65	1.16	1.09	0.22	0.49	0.88
In	5.87	7.74	-0.56	-0.33	-0.37	0.27	-0.25	-0.16
Tl	4.84	6.51	-0.81	-0.64	-0.67	0.36	-0.34	-0.32
Sn	10.1	12.7	0.80	1.25	1.19	0.61	0.70	1.03
Pb	8.32	10.6	0.21	0.57	0.05	0.41	0.25	0.48

(continued)

spacing of the pure metal. We have taken a face-centered cubic lattice for the evaluation of the sums of interatomic interactions and included only nearest neighbors, the others being a small correction. The results are listed in Table 14-5. We shall see that many trends agree with experiment so that such a complete table could be used as a basis for interpolation. It contains much more information than the heat of formation of 50% alloys which we shall see may be estimated as $1/4(\delta E_{sol}(A\ in\ B) + \delta E_{sol}(B\ in\ A))$ per atom pair.

In order to make sense of these results we consider the analytical form, Eq. (14-30). We note again that since we have evaluated the core radii to obtain the minimum energy at the observed spacing, these r_c values depend only upon the atomic volume of the pure material and the valence Z, which also enters the pseudopotential. Thus, given the valence, and the Fermi wavenumber, we may determine r_c as we did for Table 13-5 and of course the Fermi-Thomas screening parameter of Eq. (13-26) depends only upon the Fermi wavenumber so we have all the parameters entering Eq. (14-30). For a given solute, the heat of solution for solvents of each valence

Table 14-5 (continued). Heats of solution (eV per atom).

-	Al	Ga	In	Tl	Sn	Pb
Li	9.91	7.52	4.04	3.39	4.99	3.52
Na	15.5	12.5	7.86	6.86	11.2	9.19
K	21.0	17.5	12.0	10.8	18.2	15.8
Rb	22.3	18.8	13.1	11.9	20.0	17.5
Cs	24.4	20.7	14.8	13.5	22.8	20.2
Be	-2.06	-1.07	3.72	5.71	4.74	8.35
Mg	4.07	2.41	0.55	0.39	-0.15	-0.70
Ca	9.18	6.71	3.04	2.33	4.12	2.54
Sr	11.3	8.60	4.43	3.57	6.45	4.61
Ba	12.3	9.50	5.13	4.21	7.62	5.68
Cu	5.07	3.42	1.65	1.51	1.10	0.59
Ag	8.33	6.15	3.14	2.63	3.51	2.29
Au	8.11	5.96	3.02	2.53	3.32	2.14
Zn	0.88	0.07	0.29	0.78	-0.75	-0.04
Cd	3.41	1.89	0.38	0.32	-0.48	-0.83
Hg	4.24	2.54	0.61	0.42	-0.06	-0.65
Al	0	-0.08	1.91	2.98	1.32	3.15
Ga	0.51	0	0.92	1.64	-0.03	1.11
In	1.93	0.74	0	0.18	-1.08	-0.96
Tl	2.38	1.04	-6.80	0	-1.09	-1.19
Sn	1.13	0.47	1.04	1.65	0	0.92
Pb	1.35	0.48	0.51	0.95	-0.59	0

depends only upon k_F for the solvent. A plot of the heat of solution for aluminum in materials of the four valences is given in Fig. 14-9 as a function of the Fermi wavenumber for each solvent. This set of curves includes all of the entries in the column below aluminum in Table 14-5 and a similar set of curves can be constructed for each column. Note that for such comparisons the alkali and noble metals fall on the same $Z=1$ curve and the Mg to Ba results are on the same $Z=2$ curve with the Zn to Hg results. In discussing these results, we should keep in mind that we are talking about results for this simple calculation, not experiment.

a. Homovalent solutions. We note first the heavy curve, corresponding to metals of valence 3, as is aluminum. It must go to zero at the Fermi wavenumber for aluminum since the heat of solution of aluminum in itself is zero. In fact it is very near the minimum of the curve, which rises to both sides. We may note the individual numerical contributions from Eq. (14-30) to see that the rise in heat of solution for large solvent wavenumbers

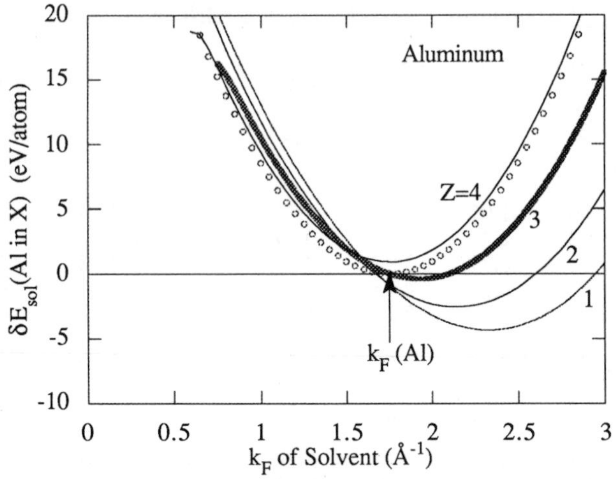

Fig. 14-9. The heat of solution of aluminum in metals of each valence Z from Eq. (14-30) as a function of the wavenumber of the solvent, the only parameter which enters the evaluation. The dots are a plot of $2\hbar^2(k_{FX} - k_{FAl})^2/m$.

comes from the first line in Eq. (14-30), the energy increase coming from compressing the electron gas from the transferred electrons. It may be thought of as elastic energy of inserting a large atom in a small atomic cell. The rise to the left, at small solvent k_F , comes largely from the fourth line of Eq. (14-30), which we may associate with the relaxation of the electron gas around the transferred atom. (It is of second-order in the pseudopotential and associated with a single site.) For small k_F for the solvent, κ is small and little energy is gained in comparison to the undissolved state with the atom in its native aluminum.

One can note that the curves for $Z = 3$ and $Z = 4$ are beginning to turn over at small k_F . This is because of the third row of Eq. (14-30), the interatomic interactions, and in particular because of the reduced number of solvent-solvent interactions which would be very large if these unrealistic wavenumber-valence combinations occurred. They do not, so this drop is not of interest and the curves were stopped there.

In spite of the different mechanisms dominating different portions of the curve, they are quite parabolic as indicated by the dotted parabola

$$\delta E_{sol(Homopolar)} \approx \frac{\gamma \hbar^2}{m} (k_{FA} - k_{FB})^2 \tag{14-31}$$

also shown in Fig. 14-9 with $\gamma = 2$. This is simply an empirical rule, but a similar form can be fit to the other homopolar combinations in Table 14-5. For magnesium (with Z=2 solvents) the coefficient is perhaps $\gamma = 1.5$, and for sodium (with Z=1 solvents) the coefficient is $\gamma = 1$. There are significant deviations from this simple form, Eq. (14-31), which is an approximation to Table 14-5. It would correspond to $\delta E_{sol}\,(B\ in\ A) = \delta E_{sol}\,(A\ in\ B)$ if the coefficients were the same for all solutes of the same Z, and this can be seen to be only approximately true in Table 14-5. Deviations come both from the fact that the calculated curve (Z=3 in Fig. 14-9) is not exactly parabolic and does not have its minimum exactly at $k_{FB} = k_{FA}$ (though it crosses that point) and the curves are not identical for each element of the same valence.

The general result, Eq. (14-31), would indicate that homovalent metals would only be soluble in each other if their wavenumbers were very close to each other so the heat of solution is small (or negative in some cases). Indeed there is such is a well-established Hume-Rothery (1931) empirical rule, discussed by Hafner (1987, p. 171), and justified by a number of outlooks and calculations. He states it in terms of the ratio of atomic radii, which is exactly the reciprocal of the ratio of Fermi wavenumbers for homovalent systems. Hafner states that "a continuous series of solid solutions is formed in the systems (the number in parentheses gives the ratio of atomic radii) K-Rb (1.07), K-Cs (1.15), Rb-Cs (1.08), Ca-Sr (1.032), Ca-Ba (1.13), Sr-Ba (1.10), Mg-Cd (1.03), Mg-Hg (1.01), Cd-Hg (1.03) - for all other systems the radius ratio exceeds the critical value of 1.15 and the solid solubility is virtually zero." At this critical ratio, Eq. (14-31) with $\gamma = 1$ and $k_F = 0.92\ \text{Å}^{-1}$ for sodium, the heat of solution is 0.15 eV. Our qualitative prediction seems to be correct, but its interpretation is not so simple in view of the complexity of the form, Eq. (14-30), upon which it is based.

b. Heterovalent solutions. We see from Fig. 14-9 that very similar curves are obtained for heterovalent solutions, but the curves are shifted relative to the homovalent curve. The shift is in fact along a straight line and linear in solvent valence Z . This seems to be true in other systems also; there is such a linear shift for magnesium and sodium, but in different directions. Where the shift with Z for aluminum is up and to the left, for magnesium it is down and to the left and for sodium it is straight down. For barium also it is straight down. Thus incorporating these corrections would seem to require specifying the direction for each element and a proliferation of parameters. We have not explored these further.

One effect of the shift in these parabola is that it will no longer be even approximately true that $\delta E_{sol}(B\ in\ A)$ is equal to $\delta E_{sol}(A\ in\ B)$. Indeed that is the case and, as suggested by the variation in directions of shifts indicated in the preceding paragraph, the systematics are not simple. We might summarize this aspect of Table 14-5 by saying that it almost always predicts that a monovalent atom (alkali or noble-metal) would have lower heat of solution in the polyvalent metal than the polyvalent atom dissolved in the monovalent solvent. On the other hand, when both elements are polyvalent, usually the *higher-valent* atom has the lower heat of solution in the lower-valent solvent.

Another Hume-Rothery Rule (Hume-Rothery (1931), discussed by Hafner (1987), p. 170) suggests generally the rule we find for the polyvalent atoms and polyvalent solvents, that the higher-valent solute has the lower heat of solution in the lower valent solvent. Hafner indicates that this rule is violated for the Mg-Al system, and indeed our prediction for that particular case indicates a violation of both the Hume-Rothery rule and our general trends. For the two cases involving both monovalent and polyvalent atoms which Hafner discussed, Li-Al satisfies our ordering of heats, violating the Hume-Rothery rule, and Li-Mg satisfies the Hume-Rothery rule, but conflicts with our prediction. For these three cases, two predictions from Table 14-5 appear to agree with this aspect of experiment and one not.

In spite of the fact that the theoretical form, Eq. (14-30) depends only upon valence and wavenumber (since the core radius was fit to those two parameters), the predicted dependence of the heats of solutions on those parameters seems intricate. We did not succeed in obtaining a point of view which ordered the predictions, beyond the general matching of atomic radii corresponding to Eq. (14-31). There is the separate question of the validity of the predictions themselves. We learn a little more about that by going to 50-50 alloys, for which there is more experimental information. We shall find that in many cases our predictions based upon Table 14-5 are much too high.

B. Alloy Formation Energies

The formation energy for higher concentrations has a number of experimental complications. First, when the heats of solution for the two limits are of opposite sign, we shall see that they cancel against each other for the 50% alloy, making even the sign of the result less certain. Second, there are frequently specific concentrations at which ordered alloys form with much lower energy - not really predictable from the heats of solution alone, but arising presumably from the level of filling of the electron energy-bands for the ordered alloy.

Also for the theory additional problems arise when we discuss the properties of alloys which are not extremely dilute. One is the interpolation of the atomic volume for the alloy between that for the two pure elements, which we saw that we could ignore in the dilute limit if the pure elements were at their equilibrium volumes. There are two commonly used rules for this interpolation: *Vegard's law* which linearly interpolates the lattice distance and *Zen's law* which linearly interpolates the volume. Hafner (1987) points out that neither rule is very good and that use of pseudopotential theory to predict spacings gives sizable corrections to either rule, and corrections in good accord with experiment. Even use of the electron-gas model, which incorporates only the first two lines in Eq. (14-30), gives a rule different from either law and corrections which are rather good. We shall not follow that further, but use a simple interpolation similar to Vegard's and Zen's.

We first seek to relate the $\delta E_{sol}(A\ in\ B)$ to the energy of formation of the random 50% AB alloy. This may best be done by thinking in terms of the interatomic interactions given in the third line in Eq. (14-30). In the pure element A the sum of the interatomic interactions might be called E_{AA} per atom and those in the pure element B as E_{BB}. Let us now imagine an alloy consisting of $c_A N_a$ atoms of type A and $c_B N_a$ atoms of type B, with concentrations $c_A + c_B = 1$. In the random alloy a fraction c_A of the neighbors to an A atom are of type A and a fraction c_B are of type B so the total interactions are $N_a\,[c_A\,(c_A\,E_{AA} + c_B E_{AB}) + c_B(c_A\,E_{AB} + c_B E_{BB})]$ and the change in interaction energy, relative to the separated alloys is $N_a\,[\,c_A\,(c_A\,E_{AA} + c_B E_{AB}) + c_B(c_A\,E_{AB} + c_B E_{BB}) - c_A\,E_{AA} - c_B E_{BB}\,]$, for a total change in energy

$$E_{form} = N_a(2E_{AB} - E_{AA} - E_{BB})\,c_A(1 - c_A)\,. \tag{14-32}$$

This could also be written with $c_A(1 - c_A) = c_A c_B$.

We can apply this form to the dilute case (as c_A approaches zero) to see that the heat of solution $\delta E_{sol}(A\ in\ B)$ is equal to the value of $2E_{AB} - E_{AA} - E_{BB}$ (when c_A approaches zero if the value depends upon concentration). Similarly, $\delta E_{sol}(B\ in\ A)$ is equal to the value of $2E_{AB} - E_{AA} - E_{BB}$ when c_B approaches zero. We have found in Table 14-5 different values for these two heats of solution so the theory indicates that $2E_{AB} - E_{AA} - E_{BB}$ does depend upon alloy concentration. If we assume it varies linearly with concentration, (this is not Vegard's law, Zen's law, or even a calculated interpolation of lattice distance) taking the average value at a 50% concentration we obtain a heat of formation for the 50% concentration of

$$E_{form}(AB) = N_a(\ \delta E_{sol}(A\ in\ B)\ +\ \delta E_{sol}(B\ in\ A\ )\ )\ /8\ . \qquad (14\text{-}33)$$

This will only be approximately true even for the theory because of our approximate interpolation, which could easily be improved. However, it provides us a simple form and an approximate way to compare with other calculations and experiment.

We look first at alloys of alkali metals with each other. Analyses of these have been extensively reviewed by Hafner (1987). In Table 14-6 we list predicted heats of formation which come directly from our Table 14-5, using Eq. (14-33), for the systems for which he gave values for random solid solutions. Our results are very close to those obtained from the most complete pseudopotential calculations as we would hope. The main additional approximation which we have made is use of the empty-core pseudopotential adjusted to the pure-metal atomic volume, the use of the Fermi-Thomas approximation rather than the full quantum screening, and the approximate interpolation of the interactions for the alloy. This comparison supports all three approximations.

Hafner points out that these heats of formation are also roughly consistent with the Hume-Rothery rule for heats of solution based on atomic radii, which we discussed at the end of Part 14-7-Aa. The idea there was that solution is inhibited by mechanical misfit. This was redone in terms of pseudopotential core radii by Girifalco (1976) who found that combinations of alkalis for which the ratio of empty-core pseudopotential radii exceeded

Table 14-6. Energy of formation of alkali-metal alloys (eV per atom) obtained from Eq. (14-33) using Table 14-5. The second column is from Hafner's second-order first-principles pseudopotential calculation. The third omits second-order terms, and the fourth is experiment (Hafner (1987)).

Alloy	Eq. (14-33)	Hafner(1987)	Electron Gas	Experiment
NaK	0.064	0.050	0.055	
NaRb	0.102	0.137	0.088	
NaCs	0.182	0.223	0.157	
$K_{0.3}Rb_{0.7}$	0.003*	0.006	-0.004*	0.002
KCs	0.025	0.030	0.022	0.007
RbCs	0.009	0.005	0.008	0.001

$*(0.7\delta E_{sol}(K\ in\ Rb)+0.3\,\delta E_{sol}(Rb\ in\ K))\times 0.3\times 0.7)$

1.15 were insoluble in each other as liquids, but that those with a ratio closer to one were soluble. Similar rules can be made using ratios of atomic-sphere radii or pseudopotential empty-core radii since for a single valence the empty-core radii will be monotonic with the atomic spacing. We would obtain Girifalco's rule with our empty-core radii and a ratio of 1.25. We see that a mechanical view is not inconsistent with our purely electronic view.

These comparisons are all with experimental heats of *solution* and Hafner indicated that there were no experimental values for the heats of *formation* of the solid-state alloys. He made estimates based upon the experimental heats of solution of the liquids and the heats of fusion. These also are listed in Table 14-6. They are of a similar size but considerably smaller than all of the theoretically estimated heats of solution.

One model discussed by Hafner (1987) for the alkali metals is the electron-gas model, the pseudopotential picture including only the terms of first order in the pseudopotential. This corresponds to including only the first two lines in Eq. (14-30). That result is given as the third column in Table 14-6. Indeed it is rather close to the full estimate. Values differ from those given by Hafner largely because he used a different interpolation for the atomic volume of the alloy.

The situation is very much the same with alloys of the divalent metals with each other. Heats of formation of their alloys from Eq. (14-33) and Table 14-5 are given in Table 14-7. Hafner notes that only the alloys Ca-Sr, Ca-Ba, Sr-Ba, Mg-Cd, Mg-Hg, and Cd-Hg have appreciable solubility in each other and indeed these are the ones which are found in Table 14-7 to have heats of solution less the 0.1 eV (except for Ca-Ba at 0.107). Hafner notes also that these are just the ones with empty-core radius ratios (his values) less that 1.15.

Table 14-7. Heat of formation (eV per atom) of 50% alloys of the divalent metals from Eq. (14-33) and values from Table 14-5.

-	Be	Mg	Ca	Sr	Ba	Zn	Cd	Hg
Be	0							
Mg	0.98	0						
Ca	0. 29	0.32	0					
Sr	4.11	0.65	0.05	0				
Ba	4.82	0.85	0.11	0.01	0			
Zn	0.31	0.16	1.01	1.62	1.98	0		
Cd	0.81	0.01	0.42	0.80	1.02	0.10	0	
Hg	1.02	0.00	0.30	0.61	0 .81	0.18	9.5	0

We may note a distinction between the divalent and monovalent cases in that the heats of solution $\delta E_{sol}(A\ in\ B)$ are frequently of opposite sign from $\delta E_{sol}(B\ in\ A)$ in the divalent case, where they are not for monovalent solutions. Then the heat of formation for the alloy from Eq. (14-33) may have strong cancellation and less reliability. This is even more so in heterovalent alloys, those between metals of different valence.

In Fig. 14-10 we plot the predicted heats of formation of the 50% alloys containing sodium. The little experimental information quoted by Hafner (1987) indicates that the predicted values for higher valence are too high and that the estimates for each valence should be lowered sufficiently that the heaviest elements had small negative heats of formation for each series. This same discrepancy applied to earlier calculations by Schlüter and Varma (1981), though their estimates for columns II, III, and IV were lower than ours, and therefore in better accord with experiment. The trends within each series seem in any case to be correctly given.

We make a similar plot for aluminum alloys in Fig. 14-11. There is limited data with which to compare but we may expect successes and discrepancies comparable to those for the sodium alloys.

Except for a few cases our estimates for the heterovalent alloys give

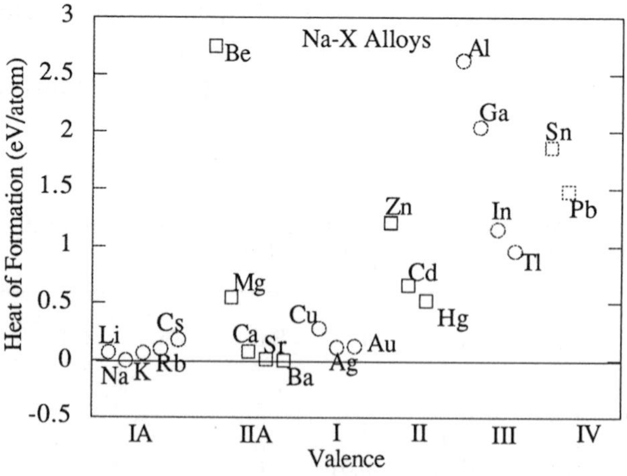

Fig. 14-10. The calculated heats of formation of 50% alloys with sodium, based upon Eq. (14-33) and Table 14-5.

positive heats of formation, as in Figures 14-10 and 14-11 while experimentally there are a number of cases of negative heat of formation. It seems likely that the principal source of that discrepancy, and in the values in Fig. 14-10 being too large, is that we have not optimized the volume nor the local distortions nor local order in the alloy, all of which will reduce the value of the heat of formation since the pure-metal energies correspond to such optimized systems.

We noted that in the heterovalent compounds the $\delta E_{sol}(A\ in\ B)$ are frequently of opposite sign from $\delta E_{sol}(B\ in\ A)$. This is true experimentally in the two heterovalent cases discussed by Hafner (1987), Li-Mg and Al-Mg. In both cases the higher-valent metal dissolving in the lower valent material had the positive heat of solution. We find this also for Li-Mg and we find a much larger heat of solution for Al in Mg than for Mg in Al, though not the negative value for Mg in Al as in the experiment.

The various full pseudopotential treatments of alloys discussed by Hafner (1987), obtaining heats of formation, stable crystal structures, and ordering, seem to give quite good results. The few comparisons made here suggest that the more approximate treatment based upon Eq. (14-30) may give almost as reliable results. Even the use of the electron-gas model - the first two lines of Eq. (14-30) - may be useful.

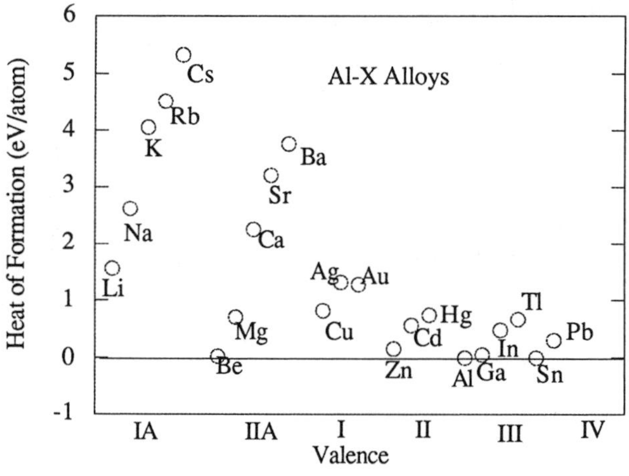

Fig. 14-11. The calculated heats of formation of 50% alloys with aluminum, based upon Eq. (14-33) and Table 14-5.

The very large heats of solution in Table 14-5 would seem not to be reasonable. We suspect that they may come from the inadequacy of linear screening, as well as the inadequate treatment of atomic relaxation which we discussed. We obtain these high values when the two pure-metal Fermi wavenumbers or pseudopotentials are very different (Ba in Be for example). A full screening calculation and the full relaxation of the neighboring atoms - particularly in a liquid - should result in an environment for each atom much more like that in the pure element, and a much lower heat of solution. Linear screening seems not to accomplish this. Again, this defect would not be so difficult to remedy. In the end it is not so serious since the corresponding heats of solution are in fact large enough that the metals are essentially insoluble.

Finally, it may be desirable to give also the additional terms which would be obtained by using Eq. (14-12) directly to obtain the heat of solution. These additional terms give the shift related to each of the terms in Eq. (14-30) but are the contributions arising from the shift in the terms in the bulk-metal B due to the change in Fermi energy from the transfer of an atom A. We have argued that they should not be included because they are canceled by the transfer of an equal number of electrons and protons. However, they might be useful in other contexts. They are

$$\delta E_{trans} = (Z_A - Z_B)\Big\{ \frac{\hbar^2 k_{FB}^2}{2m} - \frac{e^2 k_{FB}}{\pi} + \frac{2e^2 k_{FB}^3 r_{cA}^2}{3\pi} - \frac{4e^2 k_{FB}^3}{9\pi\kappa_B^2}$$

$$+ \frac{Z_B e^2 \kappa_B}{12} \sum_{j(B)} [\kappa_B r_{cB} \sinh(2\kappa_B r_{cB})\frac{e^{-\kappa_B r_j}}{r_j} - \cosh^2(\kappa_B r_{cB})] \tag{14-34}$$

$$- \frac{Z_B e^2 \kappa_B}{12}(\cosh(\kappa_B r_{cB})e^{-\kappa_B r_{cB}} - \kappa_B r_{cB}\, e^{-2\kappa_B r_{cB}})\Big\} .$$

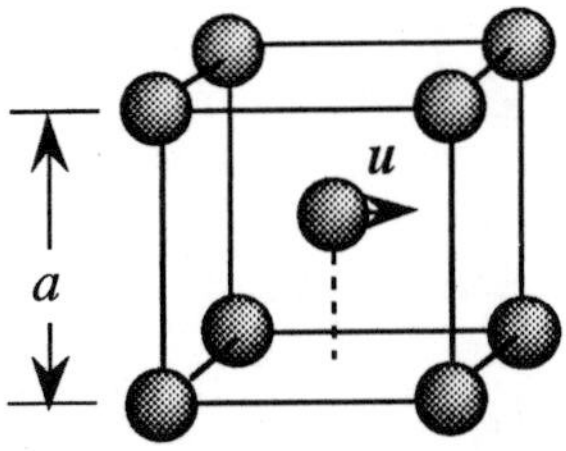

Problem 14-1. Vibrations in potassium.

a) Calculate the effective force constant (κ_{eff} in eV/Å^2) due to nearest-neighbor interactions for displacement of a single potassium atom in body-centered-cubic potassium. The displacement $\boldsymbol{u}$ is illustrated above and the force in the opposite direction is to be calculated.

Hint: It may be best to proceed numerically, using Eq. (14-9), the nearest-neighbor distance from Table 13-1, empty-core radii from Table 13-5 (use the first entry) and the screening parameter from Eq. (13-26). You can add the eight nearest-neighbor interactions for $\boldsymbol{u} = 0$ and for $\boldsymbol{u} = \pm 0.01$ Å and fit the results to an energy equal to $^1/_2 \kappa_{eff} u^2$.

b) Use this κ_{eff} (units need to be changed) to calculate the highest vibrational frequency for such a bcc crystal, in which all cube-center atoms move together and all cube-corner atoms move together in the opposite direction.

Note that second-neighbor interactions do not affect this frequency, so omitting them in Part a) was not a problem.

Problem 14-2. Distortions near an impurity

Calculate the displacement to the nearest neighbors to an aluminum atom substituted for a potassium atom in the bcc potassium treated in Problem 14-1. For this problem it is good enough to calculate the change in force on the nearest neighbors due to the substitution, and use the κ_{eff} calculated there to estimate the displacement.

CHAPTER 15

Transition Metals

We saw in Section 1-1 that transition-metal atoms were characterized by partially filled d-shells, atomic states with $\ell = 2$ units of angular momentum. In Section 1-7 we noted that these states, of similar energy to the valence s-states, are inevitably much more strongly localized around the nucleus, as illustrated for classical orbits in Fig. 15-1. Thus any attempt to expand them in plane waves, as we did for s- and p-states in the discussion of pseudopotentials for simple metals, may require so many terms as to be quite inappropriate.

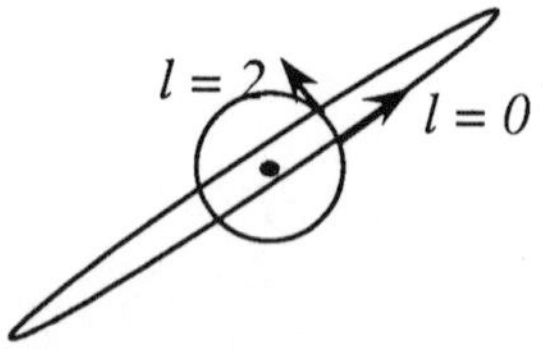

Fig. 15-1. An orbit of high angular momentum ($\ell = 2$), with the same energy as one of low angular momentum ($\ell = 0$) will have the same speed where the orbits cross, but that with low angular momentum will be directed outward or inward, leading to a larger orbit.

In an early study (Harrison (1969)) the pseudopotential theory, which we have outlined in the last two chapters, was extended to transition metals by extending the basis states for expansion to include the atomic d-states themselves as well as plane waves, or OPW's as discussed at the beginning of Section 13-4, plane waves orthogonalized to the core states. This led to a modified pseudopotential W which coupled plane waves to each other and to a hybridization term $<d|H|\boldsymbol{k}>$ which coupled the atomic d-states to the plane waves. Moriarty (1975) subsequently showed that tight-binding matrix elements between two d-states on neighboring atoms in the metal could be written in terms of this hybridization, clarifying the relation between these interatomic matrix elements and an intra-atomic resonance discussed earlier by Pettifor (1969).

In an alternative analysis of transition metals, Andersen (1973) constructed a muffin-tin potential, spherically symmetric within a sphere constructed around each atom, but constant between. He noted that since the flat portion occupied so little volume in the metal, the choice of the value of the potential in that region made little difference. He therefore made the convenient choice as equal to the energy of the d-state. This choice of "muffin-tin zero" is precisely what was needed to simplify the transition-metal pseudopotential theory.

Harrison and Froyen (1980) carried out the reformulation of the pseudopotential theory, using Andersen's choice of muffin-tin zero, in the manner we shall follow here. This led to hybridization and interatomic matrix elements written in terms of a "d-state radius", r_d , which characterized each transition metal. Values for this parameter could be obtained by fitting to more complete calculations, but it was subsequently found by Straub and Harrison (1985) that they could be calculated directly from an Atomic-Surface Method, thus making this elementary theory of d-state solids self-contained. The analysis by Harrison and Froyen (1980) led to the result, already obtained by Andersen, Klose, and Nohl (1978), of interatomic matrix elements, $V_{dd\sigma}$, $V_{dd\pi}$, and $V_{dd\delta}$ varying with spacing as $1/d^5$ and in the proportion 6:4:1.

This theory was extended to f-shell metals (Harrison (1983)) and to the interatomic forces in transition metals by Wills and Harrison (1983). It is this final complete, though approximate, analytic approach to these systems which we shall describe. The earlier aspects of it were described rather completely in Harrison (1980), from a number of different points of view. There may be advantage here is staying with a single outlook.

1. Transition-Metal Pseudopotentials

For the formulation of transition-metal pseudopotentials, we should utilize a formulation of pseudopotentials which was based upon orthogonalized plane waves (as in, for example, Harrison (1966)). We will not need that formulation once we begin the calculation of properties, but it clarifies the approximations made at the outset. It also provides a deeper understanding of the pseudopotentials which we used in understanding simple metals.

A. Pseudopotentials and OPW's

In this theory the wavefunctions for the valence states of a solid are expanded in plane waves $|\mathbf{k}\rangle = e^{i\mathbf{k}\cdot\mathbf{r}}/\sqrt{\Omega}$ which have been made orthogonal to the core states $|\alpha\rangle$ on each atom by subtracting those core states from the plane wave, with a coefficient $\langle\alpha|\mathbf{k}\rangle = \int\psi_\alpha(\mathbf{r})^* e^{i\mathbf{k}\cdot\mathbf{r}}\, d^3r/\sqrt{\Omega}$. Thus the *orthogonalized plane wave* , or *OPW*, $|\mathbf{k}\rangle\rangle$ distinguished by a double ">>" on the right, is written

$$|\mathbf{k}\rangle\rangle = |\mathbf{k}\rangle - \Sigma_\alpha |\alpha\rangle \langle\alpha|\mathbf{k}\rangle , \tag{15-1}$$

with the sum over all core states on every atom in the system. There are several essential features of these OPW's:

a) They are indeed orthogonal to every core state since multiplying on the left by any core-state $\langle\beta|$ and integrating gives $\langle\beta|\mathbf{k}\rangle$ for the first term and vanishes for every term in the sum except for the term $\alpha = \beta$, for which it gives - $\langle\beta|\mathbf{k}\rangle$.

b) Since the valence states *are* orthogonal to all core states, and the plane waves are a complete set, these OPW's are a complete set for expansion of the valence states.

c) The motivation for using this expansion ,

$$|\psi_v\rangle = \Sigma_k u_k |\mathbf{k}\rangle\rangle , \tag{15-2}$$

for a valence state $|\psi_v\rangle$ was that a relatively small number of OPW's would suffice for a good description of the valence states of simple metals, or of semiconductors. Correspondingly the *pseudowavefunction* $|\phi_v\rangle = \Sigma_k u_k |\mathbf{k}\rangle$

with the same coefficients, but with plane waves, will be smooth. The nodes and structure of the real wavefunction near each atomic nucleus are supplied by the orthogonalization, $|\psi_v\rangle = (1 - \Sigma_\alpha|\alpha\rangle\langle\alpha|)\,|\phi_v\rangle$.

d) This last form for $|\psi_v\rangle$ can be substituted into the Schroedinger Equation to obtain an equation for $|\phi_v\rangle$ which looks like the Schroedinger Equation, but the potential V is replaced by a pseudopotential operator

$$W = V + \Sigma_\alpha(\varepsilon - \varepsilon_\alpha)\,|\alpha\rangle\langle\alpha| \tag{15-3}$$

for a valence state of energy ε. When we use the pseudopotential, we obtain the pseudowavefunction, expanded in plane waves $|\boldsymbol{k}\rangle$, so we seek matrix elements between plane waves, $\langle\boldsymbol{k}+\boldsymbol{q}|W|\boldsymbol{k}\rangle$, which correspond to matrix elements of the Hamiltonian between orthogonalized plane waves, $\langle\langle\boldsymbol{k}+\boldsymbol{q}|H|\boldsymbol{k}\rangle\rangle$.

Note that in Chapter 13 we wrote the pseudopotential as a sum of individual atomic pseudopotentials, $\Sigma_j\, w(\boldsymbol{r} - \boldsymbol{r}_j)$, and we may do that here, including the core state from a particular atom with its individual pseudopotential. Then a matrix element of the pseudopotential, $\langle\boldsymbol{k}+\boldsymbol{q}|W|\boldsymbol{k}\rangle$ could be written as a structure factor $S(\boldsymbol{q})$ times a form factor $w_q = \langle\boldsymbol{k}+\boldsymbol{q}|w|\boldsymbol{k}\rangle$ based upon the individual atomic pseudopotentials $w(\boldsymbol{r})$,

$$\langle\boldsymbol{k}+\boldsymbol{q}|W|\boldsymbol{k}\rangle = \frac{1}{\Omega}\int e^{-i(\boldsymbol{k}+\boldsymbol{q})\cdot\boldsymbol{r}}\,\Sigma_j w(\boldsymbol{r} - \boldsymbol{r}_j)\,e^{i\boldsymbol{k}\cdot\boldsymbol{r}}\,d^3r$$

$$= \frac{1}{N}\Sigma_j\, e^{-i\boldsymbol{q}\cdot\boldsymbol{r}_j}\;\frac{1}{\Omega_0}\int e^{-\boldsymbol{q}\cdot\boldsymbol{r}}\,w(r)\,d^3r \;\equiv S(\boldsymbol{q})\;w_q\,. \tag{15-4}$$

The first term from Eq. (15-3) can be written exactly this way but the second term is an operator, leading to a form factor of

$$\langle\boldsymbol{k}+\boldsymbol{q}|w|\boldsymbol{k}\rangle = v_q + \Sigma_{\alpha\,(atom)}(\varepsilon - \varepsilon_\alpha)\,\langle\boldsymbol{k}+\boldsymbol{q}|\alpha\rangle\langle\alpha|\,\boldsymbol{k}\rangle\,, \tag{15-5}$$

with the sum over the core states on the atom in question and now $\langle\alpha|\boldsymbol{k}\rangle = (1/\Omega_0)\int\psi_\alpha^*(\boldsymbol{r})e^{i\boldsymbol{k}\cdot\boldsymbol{r}}\,d^3r$.

The presence of the operator $\Sigma_\alpha\,(\varepsilon - \varepsilon_\alpha)|\alpha\rangle\langle\alpha|$ makes accurate treatment of this term somewhat complicated, but it can be carried through in detail (Harrison (1966)). More often a *local* approximation is made, as in the empty-core pseudopotential of Chapter 13, in which this term is replaced by a simple spherical potential, or equivalently $\langle\boldsymbol{k} + \boldsymbol{q}|w|\boldsymbol{k}\rangle$ is taken to be a function only of q as in Eq. (15-4). In Harrison (1966), one way this was done was to evaluate $<\boldsymbol{k}+\boldsymbol{q}|w|\boldsymbol{k}>$ for the pseudopotential operator with the magnitude of both $\boldsymbol{k}$ and $\boldsymbol{k}+\boldsymbol{q}$ equal to the Fermi wavenumber and ε equal to the Fermi energy. Then again the matrix element is a function only of the magnitude of q, running from 0 to $2k_F$. A more recent discussion, with references, of ways to do this is by Fiolhais, Perdew, Armster, MacLaren, and Brajczewska (1995).

The empty-core pseudopotential which we introduced for simple metals in Chapter 1 is a simpler such local form for the pseudopotential. It was chosen to have eigenstates for the atom of the same energy as the real atomic valence eigenstates, but led to smooth pseudowavefunctions. As with the empty-core pseudopotential, the eigenstates corresponding to atomic core states are eliminated (here because the wavefunctions are constructed to be orthogonal to them). The principal purpose, for simple metals, was to allow the pseudopotential representing the effect of the atoms to be treated in perturbation theory and keeping the operator nature was not a major complication.

B. Atomic d-States

Our discussion of atomic states to this point has been restricted almost entirely to s- and p-states. As we move to states of higher angular momentum, we must systematically introduce spherical harmonics which describe the angular properties of all states in spherically symmetric potentials. This will be familiar to most readers and the formulation is given in any quantum-mechanics text; we happen to follow Schiff (1968). It is a direct generalization of the treatment we have made for s- and p-states.

For a spherically-symmetric potential the electronic eigenstates can be factored into a radial function $\psi(r) = R_\ell(r)$, which is a solution of the radial Schroedinger Equation, Eq. (13-10), and a spherical harmonic $Y_\ell^m(\theta,\phi)$, where the angle θ is measured from the z-axis and ϕ is an azimuthal angle measured relative to the x-axis. The $Y_\ell^m(\theta,\phi)$ may be factored further into functions of θ and ϕ alone as $P_\ell^m(\theta)\, e^{im\phi}$. The s- and p-states are special cases for ℓ equal to zero and one, respectively, ℓ being the total-angular-momentum quantum number. The s-state is spherically symmetric, $Y_0^0 = 1/\sqrt{4\pi}$, chosen to be normalized under an integral over all solid angles, as are all the spherical harmonics,

$$\int Y_\ell^m(\theta,\phi)^* Y_\ell^m(\theta,\phi)\, \sin\theta\, d\theta\, d\phi = 1\ . \tag{15-6}$$

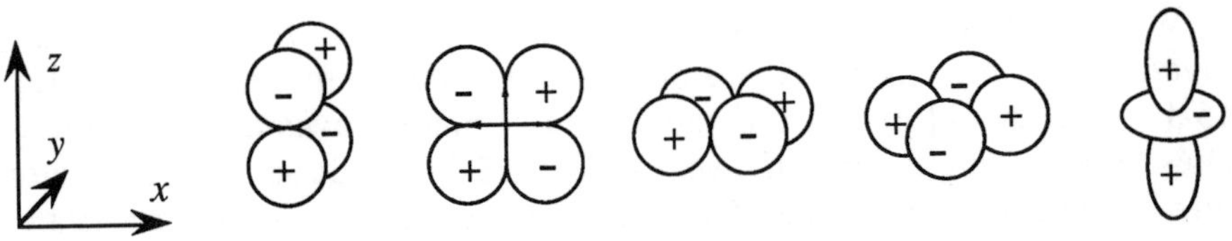

Symmetry yz zx xy $x^2 - y^2$ $3z^2 - r^2$

Fig. 15-2. Combinations of $\ell = 2$ spherical harmonics, analogous to the s- and p-states of Fig. 1-1 , which are useful for calculations.

The index m represents the component of angular momentum along the z-axis and is restricted to $-\ell \leq m \leq \ell$. There is only the $m = 0$ value for the s-state. There were three m values for the p-state. We took combinations of the three spherical harmonics which were proportional to $\sin\theta \cos\phi$, to $\sin\theta\sin\phi$, and to $\cos\theta$, or to x/r , y/r , and z/r . For $\ell = 2$, d-states, there are five and it will be convenient to proceed formally with the spherical harmonics, though for specific applications it will be convenient to take real combinations of them, called cubic harmonics, as we did for the p-states. Those we choose are illustrated in Fig. 15-2. It is readily verified that they are orthogonal to each other.

We rewrite the radial Schroedinger Equation, Eq. (13-10),

$$-\frac{\hbar^2}{2m}\frac{1}{r^2}\frac{\partial}{\partial r}r^2\frac{\partial}{\partial r}R_\ell(r) + \frac{\hbar^2\ell(\ell+1)}{2m\,r^2}R_\ell(r) + V(r)R_\ell(r) = \varepsilon_\ell\, R_\ell(r)\,, \qquad (15\text{-}7)$$

which gives the same $R_\ell(r)$ and ε_ℓ for all $2\ell + 1$ states of the same ℓ . This is the equation which is solved to obtain the atomic states, and we shall make use of it directly in our analysis.

C. Hybridization with d-states

As we noted at the beginning of this chapter, the simple expansion of solid-state valence wavefunctions in orthogonalized plane waves, Eq. (15-2), fails for transition metals where the valence d-states are so strongly localized around the nucleus that many terms in the expansion are required. For 3d-states the orthogonalization to core states is in fact irrelevant since there are no core states of that symmetry and the pseudowavefunction equals the true wavefunction.

We may however proceed by including also the free-atom d-state wavefunctions $|d\rangle$ in the expansion of states, along with the OPW's of Eq. (15-1). We should in fact orthogonalize our plane waves also to these d-states by including them in the sum in Eq. (15-1). If these atomic d-states were in fact eigenstates of the Hamiltonian in the solid, as the core states are, the pseudopotential theory of the remaining free-electron states would proceed exactly as in Chapters 13 and 14. However, the potential $V(\mathbf{r})$ in the metal differs from that $V_a(\mathbf{r})$ for the free atom so there is a coupling between the OPW's and the atomic d-states we have introduced. We call it a *hybridization* matrix element. Writing $V(\mathbf{r}) - V_a(\mathbf{r}) = \delta V$ it becomes

$$\langle\langle \mathbf{k}|H|d\rangle = \langle\langle k|\delta V|d\rangle = \langle k|\, \delta V - \langle d|\delta V|d\rangle\, |d\rangle, \tag{15-8}$$

where we noted that $\langle\langle \mathbf{k}|-\hbar^2\nabla^2/2m + V_a|d\rangle = \varepsilon_d\langle\langle k|d\rangle = 0$.

We now make the first application of Andersen's muffin-tin zero in taking the potential in the metal equal to the free-atom potential within a sphere and equal to the atomic d-state energy ε_d outside. The radius r_{mt} of the sphere is chosen such that $V_a(r_{mt}) = \varepsilon_d$, as illustrated in Fig. 15-3. The $\delta V(r)$ which determines the hybridization in Eq. (15-8) is seen to depend only upon the free atom and is thus independent of the density or structure of the metal, a very great simplification.

The other two factors in the integrand in the final form in Eq. (15-8) are the atomic d-state, well localized around the atom, and the plane wave. The plane wave can also be expanded in spherical harmonics since it is a solution of the Schroedinger Equation with a constant potential, therefore spherically symmetric. We take the z-axis along $\mathbf{k}$ and normalizing in a volume Ω it is (e. g., Schiff, (1968))

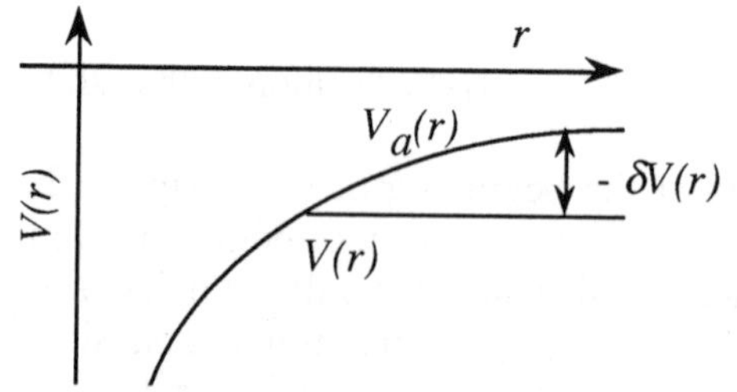

Fig. 15-3. The muffin-tin potential, equal to the free-atom potential $V_a(r)$ within the muffin-tin sphere but with Andersen's choice of a constant value equal to ε_d outside. The sphere radius is chosen such that the potential is continuous.

$$|k\rangle = \frac{1}{\sqrt{\Omega}} \Sigma_\ell (2\ell + 1) i^\ell j_\ell(kr) \left(\frac{4\pi}{2\ell + 1}\right)^{1/2} Y_\ell^0(\theta,\phi) . \tag{15-9}$$

The $j_\ell(kr)$ are solutions of Eq. (15-7) for $V(r) = 0$, called spherical Bessel functions. We return to their form shortly. In the integral, Eq. (15-8), with a d-state, only a spherical harmonic in Eq. (15-9) with $\ell = 2$ can contribute since $\delta V - \langle d|\delta V|d\rangle$ is spherically symmetric.

For arbitrary $\mathbf{k}$ we now have two coordinate systems if we retain the original axes for the d-state. We may relate the two in terms of the orientation, θ_k, ϕ_k of $\mathbf{k}$ relative to the axes in which we have written our d-states. Then with $\mathbf{r}_j$ the position of the nucleus for the d-state, the hybridization matrix element of Eq. (15-8) becomes (Harrison and Froyen (1980))

$$\langle\langle k|H|d\rangle = \frac{-4\pi}{\sqrt{\Omega}} Y_2^m(\theta_k,\phi_k) e^{-i\mathbf{k}\cdot\mathbf{r}_j} \int j_2(kr)(\delta V - \langle d|\delta V|d\rangle) R_2(r) r^2 dr. \tag{15-10}$$

For the special case in which $\theta_k = 0$ and $\mathbf{r}_j = 0$, this follows immediately from Eq. (15-9). Then only $m = 0$ enters and $Y_2^0(0,\phi_k) = \sqrt{5/(4\pi)}$ so there is only one coordinate system and the angular integral over the normalized $Y_2^0(\theta,\phi)^* Y_2^0(\theta,\phi)$ has been performed. The factors $Y_2^m(\theta_k,\phi_k)e^{-i\mathbf{k}\cdot\mathbf{r}_j}$ for the more general case are plausible.

For evaluation we let the wavelength be large enough that kr is small over the range where the well-localized d-state $R_2(r)$ is appreciable. Then we may use the expansion of $j_2(kr) \approx k^2r^2/15$ for small r. [Schiff (1968). The proportionality to k^2r^2 follows easily from Eq. (15-7). The coefficient follows from the definition of the j_2 .] This enables us to take a factor k^2 out of the integral, leaving an integral of $(\delta V - \langle d|\delta V|d\rangle)R_2(r)r^4$ from the muffin-tin radius to infinity, an integral which is characteristic of the atom only. It is convenient to absorb this integral in a constant *d-state radius* r_d for the element, keeping all the factors which depend upon $\mathbf{k}$ or the volume, to write Eq. (15-10) as

$$\langle\langle k|H|d\rangle = \langle k|\delta V - \langle d|\delta V|d\rangle|d\rangle$$

$$= \sqrt{\frac{4\pi r_d^3}{3\Omega}} \frac{\hbar^2 k^2}{m} Y_2^m(\theta_k,\phi_k) e^{-i\mathbf{k}\cdot\mathbf{r}_j} \tag{15-11}$$

with the d-state radius given by

$$r_d^{3/2} = -\frac{1}{3}\sqrt{\frac{4\pi}{3}\frac{m}{\hbar^2}} \int r^4(\delta V(r) - <d|\delta V|d>)\, R_2(r)\, dr\,. \qquad (15\text{-}12)$$

Eq. (15-11) is our principal result, giving the hybridization matrix element in terms of the atomic parameter r_d. There are various ways to obtain the r_d. The values we list in Table 15-1 as "MTO" are very close to those obtained by Harrison and Froyen (1980) by fitting, in a way we shall describe, energy bands calculated in the Muffin-Tin-Orbital approximation by Andersen and Jepsen (1977). Harrison and Froyen (1980) also obtained them directly from Eq. (15-12), noting that the integral extends from the muffin-tin radius, r_{mt} to infinity and that $\delta V(r) = \varepsilon_d - V(r)$ times the d-state is equal to the kinetic-energy operator times the d-state (a direct consequence of the Schroedinger Equation). Thus the integral could be evaluated directly from tabulated Hartree-Fock atomic wavefunctions, without use of the potential. This gave values of r_d approximately 1.5 times the MTO values listed in Table 15-1, apparently because of the approximate $k^2r^2/15$ expression for $j_2(kr)$.

Harrison and Froyen indicated that the MTO values are to be preferred over those obtained from Eq. (15-12). r_d values may also be calculated very directly by the Atomic-Surface Method, and these ASM values are also listed in Table 15-1. We shall describe this simple method in detail in Section 15-3 and give formulas, Eqs. (15-25) and (15-27), from which r_d can be calculated in terms of the tabulated Hartree-Fock wavefunctions. This Atomic-Surface Method is to be preferred over the Froyen-Harrison approach which depended upon the k^2 expansion.

This hybridization matrix element of Eq. (15-11) is the key to the electronic structure of transition metals. We shall use it first to obtain a correction to the simple-metal pseudopotentials which can be used at the beginning or the end of the transition-metal series where the d-states are well above, or well below, the Fermi energy of the free-electron states. We use it then to obtain the coupling between d-states on neighboring atoms, and finally in coupling these two types of Bloch states to form the energy bands.

D. Contribution to the Pseudopotential

Such a hybridization matrix element between plane waves and atomic d-states gives rise to an indirect coupling between plane-wave states, and

Table 15-1. Atomic sphere radius, r_0 , d-state radius, r_d , obtained with Eq. (15-33) from the band width W_d (also tabulated) from Muffin-Tin-Orbital theory (MTO) from Andersen and Jepsen (1977). The second r_d is from the Atomic- Surface Method (ASM) from Straub and Harrison (1985). All compiled by Straub and Harrison (1985). The MTO values are regarded as preferred for calculations. r_c was adjusted such that the derivative of energy in Eq. (15-47) with respect to r was zero at the observed r_0 in Section 15-4-F. (They are plotted in Fig. 15-12.) U_d was obtained by Straub from Hartree-Fock calculations as described in Section 15-4-B.

Z	r_0 (Å)	r_d (Å) MTO	r_d (Å) ASM	r_c (Å)	W_d (eV) MTO	U_d (eV) Straub
3 Sc	1.80	1.332	1.163	0.86	5.27	5.3
4 Ti	1.60	1.159	1.029	0.83	6.26	5.4
5 V	1.49	1.060	0.934	0.80	6.83	5.3
6 Cr	1.41	0.963	0.939	0.74	6.75	5.1
7 Mn	1.42	0.925	0.799	0.67	5.78	5.6
8 Fe	1.41	0.864	0.744	0.57	4.88	5.9
9 Co	1.39	0.814	0.696	0.48	4.38	6.3
10 Ni	1.38	0.767	0.652	0.39	3.80	6.5
11 Cu	1.41	0.721	0.688	0.37	2.83	6.9
3 Y	1.98	1.696	1.602	1.08	6.76	1.9
4 Zr	1.77	1.515	1.415	1.02	8.43	2.4
5 Nb	1.61	1.370	1.328	0.93	10.02	2.7
6 Mo	1.55	1.285	1.231	0.84	9.99	3.0
7 Tc	1.50	1.197	1.109	0.71	9.52	3.4
8 Ru	1.48	1.127	1.083	0.57	8.49	3.7
9 Rd	1.49	1.066	1.020	0.42	6.95	4.0
10 Pd	1.51	1.012	1.008	0.21	5.56	4.3
11 Ag	1.59	0.960	0.889	0.23	3.67	4.6
3 Lu	1.92	1.693	1.603	1.00	7.84	4.0
4 Hf	1.75	1.553	1.455	0.99	9.62	2.8
5 Ta	1.61	1.433	1.346	0.87	11.46	3.0
6 W	1.56	1.361	1.268	0.75	11.50	3.1
7 Re	1.51	1.284	1.201	0.57	11.36	3.2
8 Os	1.49	1.219	1.142	0.33	10.39	3.4
9 Ir	1.49	1.159	1.085	0.26i	8.93	3.5
10 Pt	1.52	1.116	1.069	0.46i	7.22	3.5
11 Au	1.58	1.081	1.007	0.54i	5.41	3.6

therefore a contribution to the pseudopotential (Eq. (15-3)) and its matrix elements ($S(\boldsymbol{q})$ times the form factor in Eq. (15-5)) coupling plane-wave states. This was the central feature of the original (Harrison (1969)) transition-metal pseudopotentials.

Such second-order matrix elements can be understood as arising from a first-order correction to a plane-wave pseudowavefunction, in this case $|\boldsymbol{k}\rangle$ +

$\Sigma_d\,|d\rangle\langle d|H|\boldsymbol{k}\rangle\rangle/(\varepsilon_{\boldsymbol{k}} - \varepsilon_d)$ from the form $|i\rangle + \Sigma'_j V_{ij}|j\rangle\,/\,(\varepsilon_i^0 - \varepsilon_j^0)$ given after Eq. (1-30) . This first-order state is then coupled to some other plane-wave pseudowavefunction $|\boldsymbol{k}+\boldsymbol{q}\rangle$ by

$$\langle \boldsymbol{k}+\boldsymbol{q}|W_{tm}|\boldsymbol{k}\rangle = \langle \boldsymbol{k}+\boldsymbol{q}|W|\boldsymbol{k}\rangle + \Sigma_d \frac{\langle\langle \boldsymbol{k}+\boldsymbol{q}|H|d\rangle\langle d||H|\boldsymbol{k}\rangle\rangle}{\varepsilon_{\boldsymbol{k}} - \varepsilon_d} . \tag{15-13}$$

Here W_{tm} is called the transition-metal pseudopotential, and W is the ordinary pseudopotential of Eq. (15-3). Note that the terms are again additive for each atom so the terms from d-states on a particular atom may be added directly to the pseudopotential form factor $\langle \boldsymbol{k} + \boldsymbol{q}\,|w|\boldsymbol{k}\rangle$. There is also a term from the first-order correction to $|\boldsymbol{k}+\boldsymbol{q}\rangle$ but it is canceled by the cross term between the two first-order corrections if one includes the normalization of the modified state. We do not go into this analysis, which is complicated by the appearance of $\varepsilon_{\boldsymbol{k}+\boldsymbol{q}}$, rather than $\varepsilon_{\boldsymbol{k}}$, in the denominator of these terms. We will be interested in matrix elements between states of nearly the same energy $\varepsilon_{\boldsymbol{k}}$ and we proceed with the plausible form, Eq. (15-13). Moriarty (1970) used a Green's-function method to improve the accuracy and to calculate a number of properties, much as we calculated properties of simple metals in the preceding two chapters.

We may use Eq. (15-11) for the matrix elements in the final term in Eq. (15-13), for a single atom, measuring energies from the bottom of the band so $\varepsilon_k = \hbar^2k^2/2m$, etc., and that term becomes

$$\Sigma_d \frac{\langle\langle \boldsymbol{k}+\boldsymbol{q}|H|d\rangle\langle d||H|\boldsymbol{k}\rangle\rangle}{\varepsilon_{\boldsymbol{k}} - \varepsilon_d} =$$

$$\frac{16\pi r_d^3}{3\Omega}\,\frac{\varepsilon_k\varepsilon_{k+q}}{\varepsilon_k - \varepsilon_d}\,e^{-i\boldsymbol{q}\cdot\boldsymbol{r}_j}\,\Sigma_m Y_2^m(\theta_{\boldsymbol{k}+\boldsymbol{q}}, \phi_{\boldsymbol{k}+\boldsymbol{q}})^*\,Y_2^m(\theta_{\boldsymbol{k}}, \phi_{\boldsymbol{k}}) . \tag{15-14}$$

As one might expect, the final sum over m depends only upon the angle θ between $\boldsymbol{k}$ and $\boldsymbol{k}+\boldsymbol{q}$ and can be written from the relation (e. g., Mathews and Walker (1964), p. 170)

$$\Sigma_m Y_\ell^m(\theta_{\boldsymbol{k}+\boldsymbol{q}}, \phi_{\boldsymbol{k}+\boldsymbol{q}})^*\,Y_\ell^m(\theta_{\boldsymbol{k}}, \phi_{\boldsymbol{k}}) = \frac{2\ell+1}{4\pi}P_\ell(\cos\theta) , \tag{15-15}$$

where $P_\ell(\cos\theta)$ are Legendre Polynomials, given by (Mathews and Walker (1964) p. 163) $P_2(\cos\theta) = {}^1/_2\ (3\cos^2\theta - 1)$ and $P_3(\cos\theta) = {}^1/_2\ (5\cos^2\theta - 3\cos\theta))$. Then for d-states, $\ell = 2$, this term becomes

$$\Sigma_d \frac{<<k+q|H|d><d||H|k>>}{\varepsilon_k - \varepsilon_d} = \frac{10\, r_d^3}{3\Omega} \frac{\varepsilon_k \varepsilon_{k+q}}{\varepsilon_k - \varepsilon_d} e^{-iq\cdot r_j} \ . \qquad (15\text{-}16)$$

We may now combine this term with the pseudopotential, as in Eq. (15-13) to obtain the transition-metal-pseudopotential form factor,

$$<k+q|w|k> = v_q + \Sigma_\alpha (\varepsilon - \varepsilon_\alpha)\ <k+q|\alpha><\alpha|\,k> + \frac{10\, r_d^3}{3\Omega_0} \frac{\varepsilon_k \varepsilon_{k+q}}{\varepsilon_k - \varepsilon_d} . (15\text{-}17)$$

The operator nature of the final term, which comes from the dependence upon the energies of the coupled states rather than the $\Sigma_d\ (\varepsilon - \varepsilon_\alpha)|\alpha><\alpha|$, makes this term also complicated. However, as we noted for the pseudopotential, it is possible to make a local approximation. In particular we could evaluate it for both states at the free-electron Fermi surface, as we did following Eq. (15-5), and the final factor becomes $\varepsilon_F{}^2/(\varepsilon_F - \varepsilon_d)$.

The perturbation theory which led to the final term is then only valid if the Fermi energy is well removed from the d-state energy, so that it is useful, as we indicated, at the beginning or the end of the transition-metal series. Its role in those cases is interesting. In particular, it can be used directly in such systems to treat the electron-phonon interaction exactly as we did for the simple metals in Section 13-9. A more general approach for the electron-phonon interaction in transition metals has been given by Savrasov, Savrasov, and Andersen (1994).

The first term in Eq. (15-17) is large and negative from the attractive atomic potential, and the second term large and positive, producing the weak pseudopotential. The third term is negative in alkali metals and alkaline earths where the d-state is above the Fermi energy and the bands are free-electron-like, but becomes positive in the noble-metals and their divalent neighbors where the d-state is below the Fermi energy. This may be partly responsible for the very different properties of these two sets of metals. Copper and zinc, like potassium and calcium, have 4s valence states but differ in whether the d-states lie above or below. This difference was *not* included when we fit our pseudopotentials to the atomic s-state since these d-like terms do not contribute to the s-state energy. It *is* included in some approximation when we fit pseudopotential form factors to more complete calculations, as in the points given in Fig. 13-6. A second difference between the two sets is that copper and zinc have an addition ten nuclear charges, and a ten-electron d-shell, producing a Coulomb potential which is largely responsible for the much-deeper s-state energies. The role of these effects in determining the crystal structures has been reviewed by Hafner (1987). It is intricate, without simple and clear trends, and we do not pursue it further here.

2. Coupling Between d-States

Just as the hybridization with d-states could give a coupling between plane-wave states, that hybridization can give coupling between the d-states themselves. We repeat the analysis by which Harrison and Froyen (1980) obtained it.

A. The dd-Coupling

An atomic d-state, since it is not an eigenstate of the Hamiltonian in the solid, contains a first-order correction, $|d\rangle + \Sigma_k |\boldsymbol{k}\rangle\rangle\langle\langle \boldsymbol{k}|H|d\rangle/(\varepsilon_d - \varepsilon_k)$, and exactly as we wrote the coupling between two plane waves in the final term in Eq. (15-13) we have a coupling between two d-states, $|d\rangle$ and $|d'\rangle$, on different atoms,

$$\langle d'|H|d\rangle = \sum_k \frac{\langle d'|H|\boldsymbol{k}\rangle\rangle\ \langle\langle \boldsymbol{k}|H|d\rangle}{\varepsilon_d - \varepsilon_k} . \qquad (15\text{-}18)$$

There is of course no coupling between d-states on the same atom since they appear with different spherical harmonics $Y_\ell^{\,m}(\theta,\phi)$ in the plane waves as well as the d-states. We choose our coordinate axes with the z-axis along the distance between the nuclei for the two d-states, $\boldsymbol{d} = \boldsymbol{r}_j(d) - \boldsymbol{r}_j(d')$. Then θ and ϕ become the angle for $\boldsymbol{k}$ relative to $\boldsymbol{d}$. The integration over ϕ eliminates all but d-states with the same value of m relative to this axis, so the coupling may be written V_{ddm} . We substitute the hybridization matrix elements, Eq. (15-11), and write the sum over $\boldsymbol{k}$ as $(\Omega/(2\pi)^3)\int d^3k$ and have

$$V_{ddm} = \frac{4\pi r_d^3}{3\Omega}\,\frac{\Omega}{(2\pi)^3}\int\left(\frac{\hbar^2k^2}{m}\right)^2 \frac{Y_\ell^{m*}(\theta,\phi)Y_\ell^{m}(\theta,\phi)\,e^{ikd\cos\theta}}{\varepsilon_d - \varepsilon_k}\,d^3k. \qquad (15\text{-}19)$$

We follow Harrison and Froyen ((1980) in using the zero of energy chosen by Andersen (1973). Just as choosing the zero of energy at the d-state for evaluating the hybridization matrix element, this second use of it replaces the denominator in Eq. (15-19) by simply $-\hbar^2k^2/2m$, tremendously simplifying the evaluation. This was not appropriate when we sought coupling with individual free-electron states, but turns out - for the reasons which motivated Andersen - to work well when we sum over all states; the results are not so sensitive to the choice.

The integration over ϕ gives a factor of 2π and for the integration over k we change variables to $u = kd$, allowing us to extract the dependence upon d as

$$V_{ddm} = -\frac{\hbar^2 r_d^3}{md^5}\frac{2}{3\pi}\int Y_\ell^{m*}(\theta,0)Y_\ell^m(\theta,0)\, e^{iu\cos\theta}\sin\theta\, d\theta\, u^4 du. \tag{15-20}$$

$$\equiv \eta_{ddm}\frac{\hbar^2 r_d^3}{md^5}\ .$$

If we were to think of this as a coupling $\int\psi_d(\mathbf{r}-\mathbf{d})H\psi_d(\mathbf{r})d^3r$ between d-states on atoms separated by $\mathbf{d}$, we would conclude that the atomic wavefunctions decreased with distance as $1/r^5$, or for general angular-momentum quantum-number ℓ , as $1/r^{2\ell+1}$. This is in fact exactly the form obtained in Muffin-Tin-Orbital theory (Andersen (1973)) as the form when the potential at large distance is chosen equal to the d-state energy.

To evaluate the numerical constant we may substitute for the spherical harmonics and integrate over θ , giving a series of terms in $u^ne^{iu} - (-u)^ne^{-iu}$. The remaining integral over u can be taken along the entire real axis, a contour closed in the upper plane and the only contribution comes from a pole from the term with $n = -1$. From this analysis Harrison and Froyen (1980) obtained

$$\eta_{dd\sigma} = -\frac{45}{\pi}$$

$$\eta_{dd\pi} = \frac{30}{\pi} \tag{15-21}$$

$$\eta_{dd\delta} = -\frac{15}{2\pi}\ .$$

for $m = 0, 1,$ and 2 respectively. The ratios 6:-4:1 are the same as obtained in Muffin-Tin-Orbital theory (Andersen, Klose, and Nohl (1978)), as is the proportionality to $1/d^5$.

Moriarty (1975) had obtained such interatomic couplings earlier, essentially from Eq. (15-19), without this simplifying choice of energy denominator. Then the contribution dominant at large d came from a pole arising from the energy denominator. Our view at this point is that, like the Friedel oscillations discussed in Section 13-6-D, these oscillatory terms are correct, but not the most important aspect of the problem. If we take the limit of Moriarty's form as ε_d approaches the band minimum, we again obtain the same ratio of matrix elements as found here and the $1/d^5$ dependence.

We take these matrix element from Eq. (15-20) to be the coupling between d-states, analogous to our matrix elements $\eta_{ll'm}\,\hbar^2/md^2$ for coupling between s- and p-states. The thought is that the initial d-states had negligible overlap, or were cut off artificially so that they had no overlap, and the only coupling came from the corrections to the states which we wrote at the beginning of this section. It may be interesting that these couplings between s- and p-states were viewed as empirical fits to accurate energy bands prior to the derivation of Eq. (15-20) for d-states. Once it was seen that such simple forms could be *derived* for the d-states, it seemed obvious that there must be a derivation of the form for s- and p-states, and the fit to free-electron bands came immediately to mind (Froyen and Harrison (1979)).

We shall return to the generalization of Eq. (15-20) to f-states in Section 16-2, and to the generalization to coupling between d- and p-states in Section 17-1. We proceed now to the use of these dd-matrix elements.

B. The Slater-Koster Tables

We chose in Section 15-1-B to take real combinations of the spherical harmonics to represent the d-states, as we chose p-states having symmetry of the functions x, y, and z. Just as in that case we can write the matrix elements between the different real d-states as illustrated in Fig. 15-3.

Once we had defined couplings $V_{pp\sigma}$ and $V_{pp\pi}$, we saw in Fig. 1-6 how to obtain the coupling between states of arbitrary orientations by splitting the states like components of vectors. That is not so simple here, but is quite straightforward. The results were given by Slater and Koster (1954), are necessary for many calculations, and we reproduce them here. They defined states in terms of a coordinate system as in Fig. 15-2. They wrote the direction cosines of the vector $\boldsymbol{d}$ leading from the first state as ℓ, m, and n;

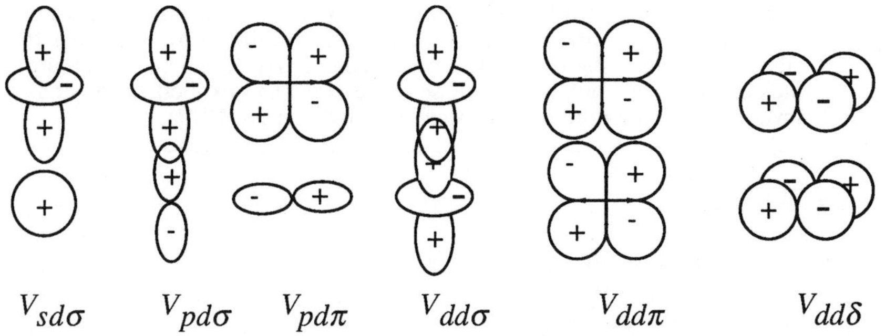

Fig. 15-3. The definition of matrix elements in terms of cubic harmonics, as in Fig. 1-6 for s- and p-states alone. All here are negative except $V_{pd\pi}$ and $V_{dd\pi}$.

that is, the vector $\boldsymbol{d} = [\ell,m,n]\, d$. Then, for example, the matrix element between a state of the form xy and a second state of the form yz at $\boldsymbol{d}$ was called $E_{xy,yz}$ and the matrix element written in terms of $V_{dd\sigma}$, $V_{dd\pi}$, $V_{dd\delta}$, ℓ, m, and n . Their results are given in Table 15-2. We include also the matrix elements between s- and p-states only for completeness and to help clarify the notation. Then, for example, the matrix element between a $3z^2$-r^2-state, and an s-state lying in the x-direction from it, is - $^1/_2V_{sd\sigma}$. In this chapter we will need only the dd-matrix elements and the hybridization since our states are constructed only of OPW's (or plane waves for the pseudowavefunctions) and d-states.

C. Energy Bands

The energy bands were written down explicitly by Harrison and Froyen (1980) for fcc and bcc structures and $\boldsymbol{k}$ along a [001] direction and for hexagonal-close-packed structures and $\boldsymbol{k}$ along the c-axis. For the cubic structures the symmetry is high enough that if we choose d-states of the form given in Fig. 15-2, and we take the wavenumber to lie along the z-direction, there is no coupling between d-states of different symmetry. (The matrix elements to d-states of different symmetry in one plane perpendicular to $\boldsymbol{k}$ sum to zero.). Further, for wavenumbers in the z-direction the bands for the zx- and yz-symmetry states are the same. Thus energy bands for the d-states could be written down immediately, making use of Table 15-2. For the face-centered-cubic structures, with nearest-neighbor interactions only, the five equations are

Table 15-2. The Slater-Koster (1954) tables of interatomic matrix elements as functions of the direction cosines, ℓ, m, and n, of the vector from the left state to the right state. Other matrix elements are found by permuting indices. General formulae for these expressions and explicit expressions involving f- and g-orbitals have been given by Sharma (1979).

$E_{s,s} =$	$V_{ss\sigma}$
$E_{s,x} =$	$\ell V_{sp\sigma}$
$E_{x,x} =$	$\ell^2 V_{pp\sigma} + (1-\ell^2)V_{pp\pi}$
$E_{x,y} =$	$\ell m V_{pp\sigma} - \ell m V_{pp\pi}$
$E_{x,z} =$	$\ell n V_{pp\sigma} - \ell n V_{pp\pi}$
$E_{s,xy} =$	$\sqrt{3}\,\ell m V_{sd\sigma}$
$E_{s,x^2-y^2} =$	$\tfrac{1}{2}\sqrt{3}\,(\ell^2 - m^2)V_{sd\sigma}$
$E_{s,3z^2-r^2} =$	$[n^2 - \tfrac{1}{2}(\ell^2 + m^2)]V_{sd\sigma}$
$E_{x,xy} =$	$\sqrt{3}\ell^2 m V_{pd\sigma} + m(1 - 2\ell^2)V_{pd\pi}$
$E_{x,yz} =$	$\sqrt{3}\ell mn V_{pd\sigma} - 2\ell mn V_{pd\pi}$
$E_{x,zx} =$	$\sqrt{3}\ell^2 n V_{pd\sigma} + n(1 - 2\ell^2)V_{pd\pi}$
$E_{x,x^2-y^2} =$	$\tfrac{1}{2}\sqrt{3}\ell(\ell^2 - m^2)V_{pd\sigma} + \ell(1 - \ell^2 + m^2)V_{pd\pi}$
$E_{y,x^2-y^2} =$	$\tfrac{1}{2}\sqrt{3}m(\ell^2 - m^2)V_{pd\sigma} - m(1 + \ell^2 - m^2)V_{pd\pi}$
$E_{z,x^2-y^2} =$	$\tfrac{1}{2}\sqrt{3}n(\ell^2 - m^2)V_{pd\sigma} - n(\ell^2 - m^2)V_{pd\pi}$
$E_{x,3z^2-r^2} =$	$\ell[n^2 - \tfrac{1}{2}(\ell^2 + m^2)]V_{pd\sigma} - \sqrt{3}\ell n^2 V_{pd\pi}$
$E_{y,3z^2-r^2} =$	$m[n^2 - \tfrac{1}{2}(\ell^2 + m^2)]V_{pd\sigma} - \sqrt{3}mn^2 V_{pd\pi}$
$E_{z,3z^2-r^2} =$	$n[n^2 - \tfrac{1}{2}(\ell^2 + m^2)]V_{pd\sigma} + \sqrt{3}n(\ell^2 + m^2)V_{pd\pi}$
$E_{xy,xy} =$	$3\ell^2 m^2 V_{dd\sigma} + (\ell^2 + m^2 - 4\ell^2 m^2)V_{dd\pi} + (n^2 + \ell^2 m^2)V_{dd\delta}$
$E_{xy,yz} =$	$3\ell m^2 n V_{dd\sigma} + \ell n(1 - 4m^2)V_{dd\pi} + \ell n(m^2 - 1)V_{dd\delta}$
$E_{xy,zx} =$	$3\ell^2 mn V_{dd\sigma} + mn(1 - 4\ell^2)V_{dd\pi} + mn(\ell^2 - 1)V_{dd\delta}$
$E_{xy,x^2-y^2} =$	$\tfrac{3}{2}\ell m(\ell^2 - m^2)V_{dd\sigma} + 2\ell m(m^2 - \ell^2)V_{dd\pi} + \tfrac{1}{2}\ell m(\ell^2 - m^2)V_{dd\delta}$
$E_{yz,x^2-y^2} =$	$\tfrac{3}{2}mn(\ell^2 - m^2)V_{dd\sigma} - mn[1 + 2(\ell^2 - m^2)]V_{dd\pi}$
	$+ mn[1 + \tfrac{1}{2}(\ell^2 - m^2)]V_{dd\delta}$
$E_{zx,x^2-y^2} =$	$\tfrac{3}{2}n\ell(\ell^2 - m^2)V_{dd\sigma} + n\ell[1 - 2(\ell^2 - m^2)]V_{dd\pi}$
	$- n\ell[1 - \tfrac{1}{2}(\ell^2 - m^2)]V_{dd\delta}$
$E_{xy,3z^2-r^2} =$	$\sqrt{3}\ell m[n^2 - \tfrac{1}{2}(\ell^2 + m^2)]V_{dd\sigma} - 2\sqrt{3}\ell mn^2 V_{dd\pi}$
	$+ \tfrac{1}{2}\sqrt{3}\ell m(1 + n^2)V_{dd\delta}$
$E_{yz,3z^2-r^2} =$	$\sqrt{3}mn[n^2 - \tfrac{1}{2}(\ell^2 + m^2)]V_{dd\sigma} + \sqrt{3}mn(\ell^2 + m^2 - n^2)V_{dd\pi}$
	$- \tfrac{1}{2}\sqrt{3}mn(\ell^2 + m^2)V_{dd\delta}$
$E_{zx,3z^2-r^2} =$	$\sqrt{3}\ell n[n^2 - \tfrac{1}{2}(\ell^2 + m^2)]V_{dd\sigma} + \sqrt{3}\ell n(\ell^2 + m^2 - n^2)V_{dd\pi}$
	$- \tfrac{1}{2}\sqrt{3}\ell n(\ell^2 + m^2)V_{dd\delta}$
$E_{x^2-y^2,x^2-y^2} =$	$\tfrac{3}{4}(\ell^2 - m^2)^2 V_{dd\sigma} + [\ell^2 + m^2 - (\ell^2 - m^2)^2]V_{dd\pi}$
	$+ [n^2 + \tfrac{1}{4}(\ell^2 - m^2)^2]V_{dd\delta}$
$E_{x^2-y^2,3z^2-r^2} =$	$\tfrac{1}{2}\sqrt{3}(\ell^2 - m^2)[n^2 - \tfrac{1}{2}(\ell^2 + m^2)]V_{dd\sigma} + \sqrt{3}n^2(m^2 - \ell^2)V_{dd\pi}$
	$+ \tfrac{1}{4}\sqrt{3}(1 + n^2)(\ell^2 - m^2)V_{dd\delta}$
$E_{3z^2-r^2,3z^2-r^2} =$	$[n^2 - \tfrac{1}{2}(\ell^2 + m^2)]^2 V_{dd\sigma} + 3n^2(\ell^2 + m^2)V_{dd\pi}$
	$+ \tfrac{3}{4}(\ell^2 + m^2)^2 V_{dd\delta}$

$$\varepsilon_k(xy) = 3V_{dd\sigma} + V_{dd\delta} + 4(V_{dd\pi} + V_{dd\sigma}) \cos ({}^1/_2 ka) ,$$

$$\varepsilon_k(yz) = \varepsilon_k(zx) = \varepsilon_d + 2V_{dd\pi} + 2V_{dd\delta} + (3V_{dd\sigma} + 2V_{dd\pi} + 3V_{dd\delta})\cos({}^1/_2 ka) ,$$

(15-22)

$$\varepsilon_k(x^2\text{-}y^2) = \varepsilon_d + 4V_{dd\pi} + ({}^3/_2 V_{dd\sigma} + 2V_{dd\pi} + {}^3/_2 V_{dd\delta})\cos({}^1/_2 ka) ,$$

$$\varepsilon_k(3z^2\text{-}r^2) = \varepsilon_d + V_{dd\sigma} + 3V_{dd\delta} + {}^1/_2(V_{dd\sigma} + 12V_{dd\pi} + 3V_{dd\delta})\cos({}^1/_2 ka),$$

with $a = \sqrt{2}d$ the cube edge. Only the states with $\varepsilon_k(3z^2\text{-}r^2)$ are coupled to the plane wave. The matrix element obtained from Eq. (15-11), using $Y_2^0(0,0) = \sqrt{5/(4\pi)}$, is $\sqrt{5r_d^3/3\Omega_0}\,\hbar^2k^2/m$. However, we shall see that for the bands the small-k expansion is inadequate and we shall include a correction factor $f(k)$. Here we shall use an approximate correction corresponding to $f(k) = 1 - ka/2\pi$ chosen to drop to zero at the Brillouin Zone face, for reasons we discuss after Eq. (15-24). We measure ε_d from the free-electron minimum, with that energy to be determined later, and so the fourth band in Eq. (15-22) combines with the free-electron bands to give

$$\varepsilon_k = \frac{1}{2}\left(\varepsilon_k(3z^2\text{-}r^2) + \frac{\hbar^2k^2}{2m}\right) \pm \sqrt{\frac{1}{4}\left(\varepsilon_k(3z^2\text{-}r^2) - \frac{\hbar^2k^2}{2m}\right) + \frac{5r_d^3}{3\Omega_0}\frac{\hbar^4k^4}{m^2}f(k)^2} .$$

(15-23)

For the body-centered-cubic structure they noted that second-neighbors were only 15% more distant than nearest neighbors so they included both, writing the second-neighbor matrix elements $V_{dd\sigma}^{(2)}$, etc. They obtained

$$\varepsilon_k(xy) = \varepsilon_d + ({}^8/_3 V_{dd\sigma} + {}^{16}/_9 V_{dd\pi} + {}^{32}/_9 V_{dd\delta})\cos ({}^1/_2 ka) + 4V_{dd\pi}^{(2)} + 2V_{dd\delta}^{(2)}\cos ka ,$$

$$\varepsilon_k(yz) = \varepsilon_k(zx) = \varepsilon_d + ({}^8/_3 V_{dd\sigma} + {}^{16}/_9 V_{dd\pi} + {}^{32}/_9 V_{dd\delta})\cos ({}^1/_2 ka) + 2V_{dd\pi}^{(2)}(1 + \cos ka) + 2V_{dd\delta}^{(2)},$$

(15-24)

$$\varepsilon_k(x^2\text{-}y^2) = \varepsilon_d + ({}^{16}/_3 V_{dd\pi} + {}^8/_3 V_{dd\delta})\cos ({}^1/_2 ka) + 3V_{dd\sigma}^{(2)} + V_{dd\delta}^{(2)}(1 + 2\cos ka),$$

$$\varepsilon_k(3z^2\text{-}r^2) = \varepsilon_d + ({}^{16}/_3 V_{dd\pi} + {}^8/_3 V_{dd\delta})\cos ({}^1/_2 ka) + V_{dd\sigma}^{(2)}(1 + 2\cos ka) + 3V_{dd\delta}^{(2)}.$$

The last of these, $\varepsilon_k(3z^2-r^2)$, is again coupled to the free-electron band, providing two bands given again by Eq. (15-23), again including a correction factor $f(k)$ which we discuss next. They also gave expressions for the bands for the hexagonal-close-packed structure.

Noting that Eq. (15-11) for $<<k|H|d>$ becomes inaccurate for large k Froyen (Harrison and Froyen (1980)) evaluated Eq. (15-10) by direct integration with $j_2(kr)$ instead of using the $j_2(kr) \approx k^2r^2/15$ approximation. He found that $<<k|H|d>$ rises as k^2 at small k, as it must, but turns over and crosses zero near a lattice wavenumber. This is in fact as it should be since at the point $\mathbf{k} = [001]2\pi/a$ in the bcc structure the d-states are coupled only to the d-like component in an expansion of the plane wave in spherical harmonics, as in Eq. (15-9). (We shall see this symmetry restriction when we discuss the d-band maximum and minimum at the end of this subsection.) The leading terms in that d-like component are then removed by the orthogonalization of the plane wave to the d-states. This crossing is accomplished by a choice $f(k) = 1 - ka/2\pi$ which they used for the bcc structure. We use the same form for fcc-structures and with the fcc cube-edge a this goes to zero at the fcc Brillouin Zone face.

Froyen took values of ε_d from Andersen and Jepsen (1977) and obtained the bands shown in Fig. 15-4. Correct structures were used, except for manganese which was taken as body-centered-cubic, technetium was assumed hexagonal, and all c/a ratios were taken as ideal. These bands are certainly crude in comparison to those of Moruzzi, Janak, and Williams (1978) who made the first comprehensive set of calculations of the electronic structure of transition metals. The resulting book is particularly useful for a uniform and complete analysis for all systems all in the same approximation. More recent studies have been given by Skriver (1985), Papaconstantopoulos (1986) , Watson, Fernando, Weinert, Wang, and Davenport (1991), and by Paxton, Methfessel and Polatoglou(1990). However, our bands are qualitatively correct and show the appropriate trends. It may be useful to have them in a single figure such as this one produced by Froyen.

We shall approximate this electronic structure for the purposes of calculating properties, and even for calculating the position ε_d of the d-bands which Froyen took from Andersen and Jepsen (1977). However, we shall first define a band width in terms of the band structure, which will be useful later in making approximations, and examine in more detail one example of the transition-metal band structure, that of body-centered-cubic chromium, in order to understand the Fermi surfaces and densities of states .

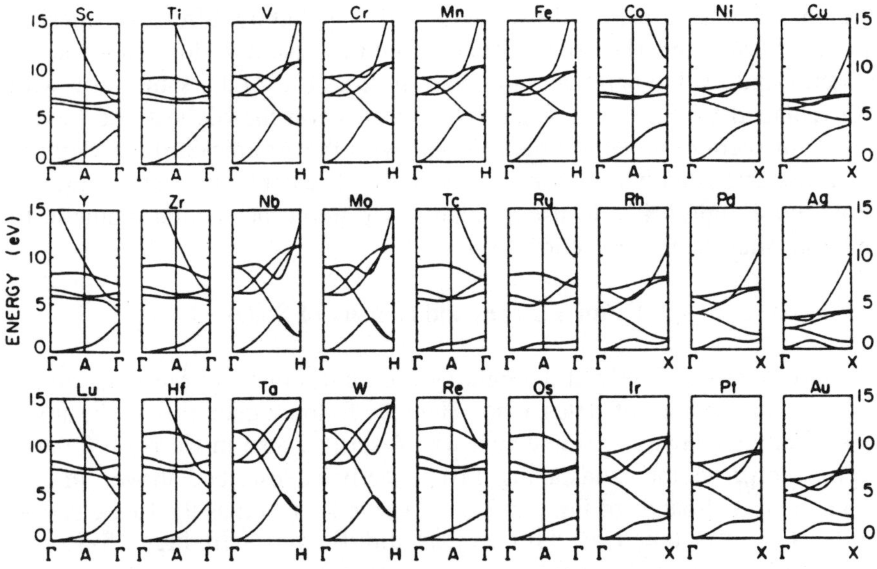

Fig. 15-4. Energy bands for the transition elements, obtained by Froyen, using the approximations described in the text, from Harrison and Froyen (1980). Bands for fcc structures extend from Γ to X, indicated below the figure. For bcc structures they extend from Γ to H. For hcp structures they extend from Γ to A to Γ in a double Brillouin Zone.

The band width is most simply defined for the body-centered-cubic structure at the point H (see for example vanadium in Fig. 15-4), $\boldsymbol{k} = [001]2\pi/a$. At this point, the bands are purely d-like, s-like, or p-like (as at Γ, $\boldsymbol{k} = 0$, since the states have the full rotation-reflection symmetry of the lattice). For d-like bands they reach their maximum and minimum values. This is also the point then at which the plane waves are very nearly uncoupled from the d-states, which is why we chose the correction factor, $f(k) = 1 - ka/2\pi$, equal to zero at this point. There are three states at high energy and two at low energy, differing in energy by the d-band width,

$$W_d\,(band) = -{}^8/_3 V_{dd\sigma} + {}^{32}/_9 V_{dd\pi} - {}^8/_9 V_{dd\delta} - 3V_{dd\sigma}^{(2)} + 4V_{dd\pi}^{(2)} - V_{dd\delta}^{(2)} \qquad (15\text{-}25)$$

$$= 114.97\,\frac{\hbar^2 r_d^3}{md^5} = 6.83\,\frac{\hbar^2 r_d^3}{m r_0^5}$$

obtained by subtracting the expressions in Eq. (15-24) at $k = 2\pi/a$. There are different ways to define the d-band width, giving slightly different numerical factors. Straub and Harrison (1985) defined the band width using second moments and twelve neighbors to obtain in Eq. (15-33) the same form with 6.83 replaced by 5.53. The value 6.83 based upon the highest and lowest states will seem more appropriate for use with the Atomic-Surface Method. We have designated it as a "band" value since for calculating total energies it seems more appropriate to use a value based upon the second moment, to which we shall return in Section 15-4.

D. Fermi Surfaces and Density of States

We look more closely at a single system, chromium, to see what Fermi surfaces and densities of states arise. Here we follow rather closely the study by Rath and Callaway (1973) which was described also in Harrison (1980). We first construct the bands along [001] exactly as described above. In Fig. 15-5a, the five d-bands of Eq. (15-24) are plotted, as well as the free-electron band. ε_d was taken as 8.00 eV, as did Froyen. In Fig. 15-5b the hybridization is included, with $f(k) = 1 - ka/2\pi$. Also the Fermi energy is indicated in approximately the correct position. This is essentially the same plot as that given for Cr in Fig. 15-4. Plots for the other elements are equally easily constructed. The Fermi surface for chromium is the two-dimensional intersection in three-dimensional wavenumber space of these bands with the Fermi energy. It is obviously complicated because of the

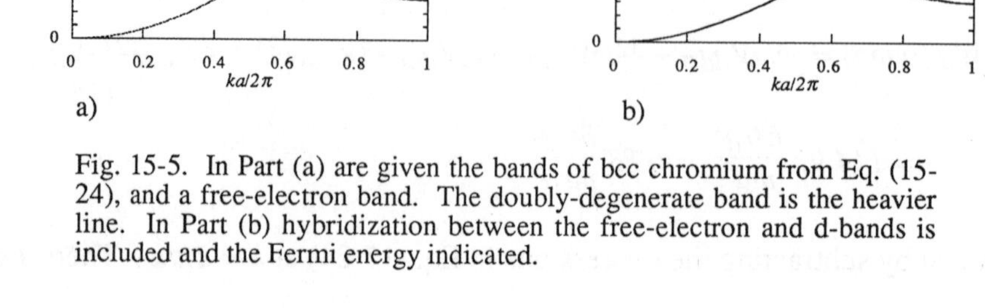

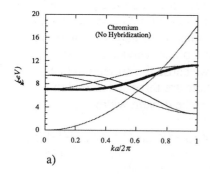

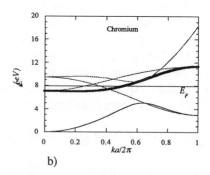

Fig. 15-5. In Part (a) are given the bands of bcc chromium from Eq. (15-24), and a free-electron band. The doubly-degenerate band is the heavier line. In Part (b) hybridization between the free-electron and d-bands is included and the Fermi energy indicated.

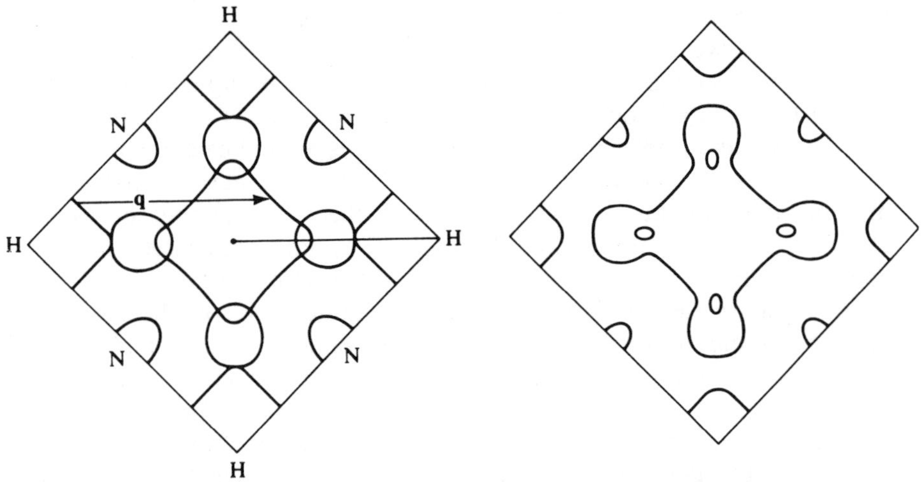

Fig. 15-6. (100) sections of the body-centered-cubic Brillouin Zone showing the Fermi surface of chromium, to the left for the central section of the Brillouin Zone, as determined by Rath and Callaway (1973). To the right are the intersections in a plane displaced slightly away from the center of the Brillouin Zone, which separates the crossings. The bands in Fig. 15-5 were for $\boldsymbol{k}$ along the horizontal axis from the center to H in the left figure. The surfaces separated by $\boldsymbol{q}$ are responsible for the antiferromagnetic order.

multiple intersections along this one line. The cross-section of the Fermi surfaces in a (100) plane are shown in Fig. 15-6.

These surfaces are understandable in terms of the bands given in Fig. 15-5b. At the center of the Zone, $k = 0$, there are four bands occupied. Between the first and second points where a band crosses the Fermi energy one less band is occupied and beyond the third crossing there are only two bands occupied. It may be easier to understand the geometry of the Fermi surface from the sketch to the right in Fig. 15-6 where the plane in wavenumber space is shifted away from the center of the Zone, the degeneracies between the bands are all eliminated and the Fermi surfaces separate. There is a large surface in the fourth band surrounding the central point, occupied in the first four bands as we can see in Fig. 15-5b. It is in the shape of an octahedron with a spherical bulge around each of the six apexes; the figure is called the "jack", as in the game of jacks. At the neck of each of these ball-shaped protrusions is a small pocket within which one of the four bands rises above the Fermi energy, as seen in Fig. 5-5b, leaving a region of holes in the fourth band. Thus both the jack and the pockets are fourth-band

Fermi surface. Just outside the jack only three bands are occupied and nearer the corner, H, we cross Fermi surface in the third band (further toward N in Fig. 15-6 than in Fig. 15-5); then only two bands are occupied. This surface around H , generally octahedral in shape, contains holes in the third band. Another pocket of holes in the third band occurs around each point N at the center of each face of the Brillouin Zone. The Fermi surfaces for molybdenum and tungsten, also bcc with the same number of valence electrons, are similar as one would expect. They have been carefully studied experimentally, as were the Fermi surfaces of the simple metals.

Calculating electronic properties of such systems is complicated by the geometry of the surface, but in principal is the same as for the simple metals. Friedel oscillations in screening arise from discontinuity in the occupation of levels at the Fermi surface, as discussed in Section 13-6-D for simple metals. They in fact become much stronger when two portions of the Fermi surface are flat and parallel (and occupied between both surfaces as for free-electron spheres, or empty between both surfaces), as for the two octahedron surfaces separated by $\mathbf{q}$ as indicated to the left in Fig. 15-6. Lomer (1962) first pointed out that this "nesting" of Fermi surfaces can give rise to an electron-spin density, fluctuating with distance with wavenumber $\mathbf{q}$, which is the antiferromagnetic state of metallic chromium. We shall discuss such a mechanism for antiferromagnetism in the cuprates in Section 18-4-B.

Given the energy bands, one may also calculate the density of states $n(\varepsilon)$ for the system. The number of states per atom in a region of wavenumber space d^3k is given by $[2\Omega_0/(2\pi)^3\,]\,d^3k$ so the number of states per atom in an energy range $\delta\varepsilon$ is given by $n(\varepsilon)\delta\varepsilon$ equal to $2\Omega_0/(2\pi)^3$ times the volume of wavenumber space between the constant energy surfaces for ε and $\varepsilon + \delta\varepsilon$. This is generally done by dividing the Brillouin Zone up into small tetrahedra, evaluating the band energies at their corners, interpolating the bands linearly with $\mathbf{k}$ within the tetrahedral, and adding analytically the contributions to $n(\varepsilon)$ from each tetrahedron. This was done for chromium by Rath and Callaway (1973) giving the density of states shown in Fig. 15-7. All of the structure, except for the fluctuations on the scale of a millimeter or so in the figure, is real and reproducible in different calculations. The Fermi energy, to which the states are filled, is also shown in that figure. Such a curve is of course the key to a calculation of the total energy, for which the principal contribution is $\int_{0,E_F} \varepsilon\, n(\varepsilon)d\varepsilon$. We will wish to simplify the form of the density of states for such calculations but it is helpful to know what the real density of states looks like.

There is a small density of states arising from the free-electron states, with an additional ten states per atom arising from the d-bands in the energy

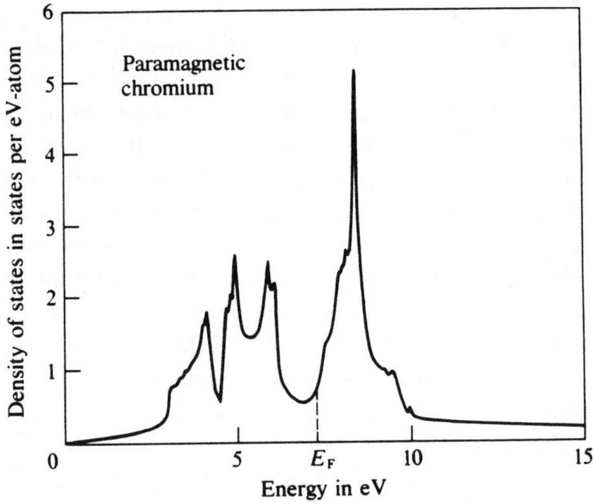

Fig. 15-7. The density of states for body-centered-cubic chromium, calculated by Rath and Callaway (1973), with the Fermi energy indicated.

range from 3 eV to around 10 eV. There are a total of six states per atom below the Fermi energy for chromium, near the deep minimum in the density of states characteristic of body-centered-cubic metals.

We should not miss a feature of this result, that the Fermi energy actually occurs in Fig. 15-7 slightly above the minimum in the density of states. If we consider the unhybridized bands, as in Fig. 15-5 (a), they would suggest that the density of states minimum occurs exactly at the center of the band if there are five d-electrons per atom so we appear to have slightly more than five d-electrons and slightly less than one free electron. Once we begin treating properties in Section 15-4 we will be working with hybridized bands and will define a number of free electrons Z_s which is the *weighted* occupation of s-states, the sum over all occupied states, each weighted by the squared coefficient of the wavefunction from the free-electron states. We will then find an occupation near 1.5, even in the noble metals, where the Fermi energy lies well above the d-like bands. These are two distinct numbers, like the occupation of all of the chlorine-like bands in rock salt, corresponding to a fully ionized sodium, and yet a weighted occupation of the sodium states such that $Z^* \approx 1/2$. In chromium, and presumably in other real transition metals, the position of the Fermi energy within the d-bands corresponds to a free-electron-band occupation of one or less, but the weighted occupation Z_s is near 1.5.

The density of states is analogous for the fcc and hcp structures, but does not show this deep minimum. We illustrate this with the density of states for hexagonal-closed-packed osmium in Fig. 15-8a, which we have taken from Harrison (1980). There the additional ten states per atom occur over the energy range from 4 eV to 14 eV. This is seen also from the integrated density of states which is shown in the same curve. In both of these cases the peaks in the density of states shift significantly in energy as we go from element to element in the same structure - as is apparent from the bands in Fig. 15-4 - but the principal features occur near the same number of electrons per atom in the same structure. Thus the prominent minimum for the bcc structure occurs near 6 electrons per atom and the sharp minimum in the hcp structure occurs at four electrons per atom. Because of this, many properties are quite systematic in the number of electrons per atom. This has led some to believe that the energy bands were fixed, the *rigid band model* , and simply filled as one increased the electron-to-atom ratio. That would be consistent with the observed systematics, but it is clearly not true. The bands shift, and the structure shifts with it.

As we shall see, the gross properties of transition metals are describable in terms of the crudest representation of this density of states, the Friedel model, as $n(\varepsilon) = 10/W_d$ over an energy range W_d , as is illustrated below in Fig. 15-8b for osmium. It seems a very crude representation of a curve with so much structure, but the integrated density of states is quite accurately described, so for many properties it is a very meaningful approach. We shall return to that after seeing how the parameters r_d may be obtained from the Atomic Surface Method.

3. The Atomic Sphere Approximation

One of the earliest methods for calculating electron states in solids was the cellular method, as carried out by Wigner and Seitz (1933). They sought the lowest-energy valence-band state in an alkali metal, a state which would have the full symmetry of the lattice. Thus if the crystal were divided into atomic cells, the wavefunction would have normal derivatives equal to zero at each cell surface, in order to have continuous gradients. They then replaced the cell by a sphere with radius r_0 chosen so that its volume would equal the cell volume. These are the r_0 values listed in Table 15-1. They took the potential spherically symmetric within the cell so that obtaining the state required the solution of the spherical Schroedinger Equation, Eq. (15-7), with the boundary condition $\partial R_0(r)/\partial r = 0$ at $r = r_0$. This gave a lower energy than that of the atomic state obtained with the same radial equation

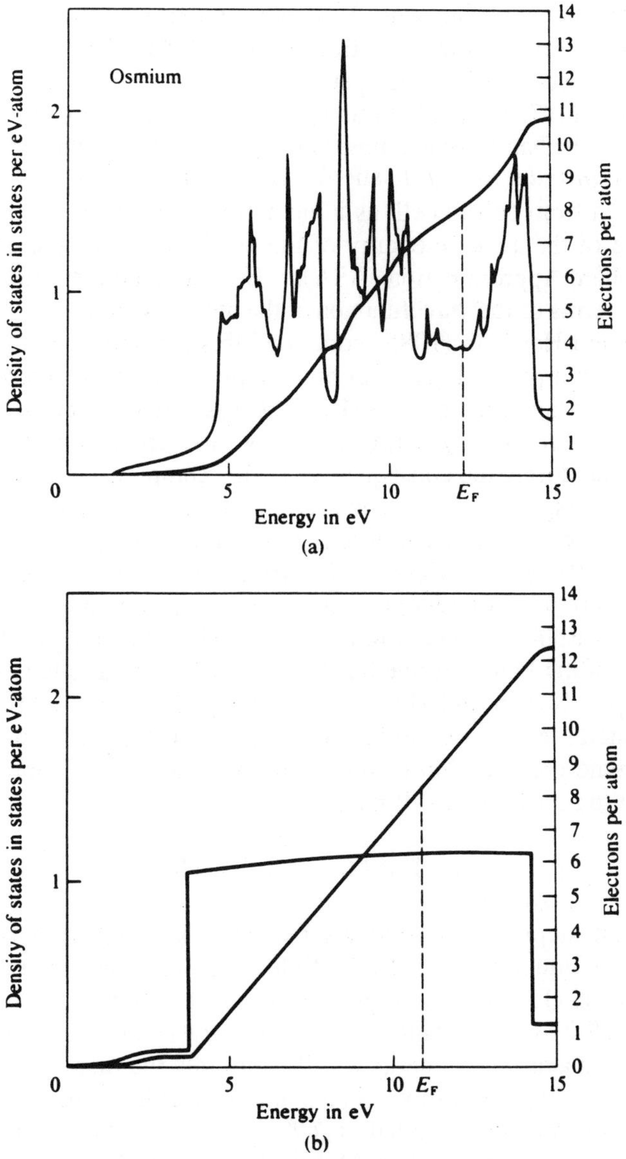

Fig. 15-8. The density of states in hexagonal-close-packed osmium, in Part (a) as obtained by Jepsen, Andersen, and Mackintosh (1975). The scale is to the left. The monotonically rising line is the integral over the density of states, with scale given to the right. Those in Part (b) are from the Friedel model with parameters defined in the text.

and the same potential but with the condition that R_0 go to zero at large r. The kinetic energy for a free-electron band was added, and the net change in energy gave the cohesive energy per atom.

Andersen (1973) extended this by noting that an estimate of the energy for the top of the band (for example, a d-band) could be obtained by requiring that the *magnitude* of $R_l(r)$ should vanish at r_0, so the wavefunction changed sign from cell to cell, as in an antibonding state. He also obtained an estimate of the average energy of the band and called the procedure the Atomic-Sphere Approximation (ASA). He also obtained estimates of these energies in terms of the wavefunction at the atomic sphere.

Straub and Harrison (1985) combined this method with a tight-binding view to obtain a particularly clear and simple form of this theory, calling it the Atomic Surface Method (ASM). This was actually a conscious effort to use *complementarity* by taking two very different views of a problem as correct and exploring the consequences of this complementary validity. The term is more specifically associated with the wave-particle duality of quantum theory. (For a recent discussion and references, see Beller (1998).) Harrison (1987b) outlines this approach, introducing the term "accommodation", not recognizing it as the complementarity approach discussed by Bohr and others in the 1930's. The matching of free-electron and tight-binding bands in Section 1-3-B is of course another case of using this outlook. Straub and Harrison (1985) applied the ASM to transition metals, f-shell metals, and the bands which develop from atomic core levels under extreme compression of the solid. We outline the approach and the application to transition metals next.

A. The Atomic Surface Method

The first steps in the analysis concern exact solution of the radial Schroedinger Equation, Eq. (15-7), following much the same procedure as used by Friedel (1952) in deriving his sum rule, to which we shall return. We write Eq. (15-7) with a solution R_ℓ of energy ε and again for a similar solution R_ℓ' at a different energy ε'. We multiply the first equation by r^2R_ℓ, the second by r^2R_ℓ', and subtract so that the terms in the potential and in the centrifugal potential proportional to $\ell(\ell+1)$ cancel. We then integrate from $r = 0$ to r_0, and perform a partial integration on the kinetic-energy term, to obtain the exact result

$$-\frac{\hbar^2}{2m}r_0^2\left(R_\ell'\frac{\partial R_\ell(r)}{\partial r} - R_\ell\frac{\partial R_\ell'(r)}{\partial r}\right)_{r=r_0} = (\varepsilon - \varepsilon')\int_{0,r_0} r^2 R_\ell' R_\ell(r)dr\,. \quad (15\text{-}26)$$

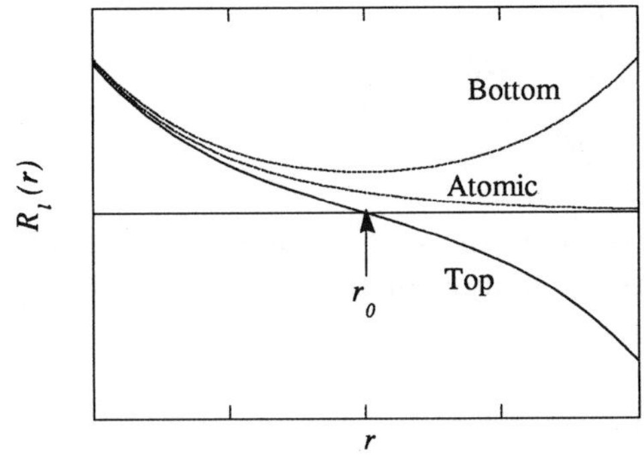

Fig. 15-9. The radial wavefunction $R_\ell(r)$ near the atomic sphere radius r_0. The curves labeled "Bottom" and "Top" correspond to the bottom and top of the corresponding energy band. That with "Atomic" is the free-atom solution.

We now let R_ℓ' be the solution for the top of the band $[R_\ell'(r_0) = 0]$ and R_ℓ be the solution for the bottom of the band $[\partial R_\ell/\partial r = 0$ at $r = r_0]$, as illustrated in Fig. 15-9. We obtain the width of the band, $W_\ell = \varepsilon' - \varepsilon$, as

$$W_\ell = \frac{-\frac{\hbar^2}{2m} r_0^2 R_\ell \frac{\partial R_\ell'}{\partial r}\Big|_{r=r_0}}{\int_{0,r_0} r^2 R_\ell' R_\ell dr} . \tag{15-27}$$

This is still exact within the Atomic-Sphere Approximation and all we require are the values and derivatives of the solutions at r_0.

At this point Straub and Harrison (1985) used the tight-binding view that the two solutions could be written as linear combinations of the atomic solution. Then, as is seen also in Fig. 15-9, the solution for the bottom of the band, where the atomic states add, will be approximately twice the value of each atomic solution on the atomic cell between the two atoms. Similarly, the solution for the top of the band will have a slope approximately twice that of either atomic solution. Finally, the atomic solution $R_\ell^a(r)$ should be close to the average, or geometric mean, of the top and bottom solutions so that $R_\ell^a(r)$ can be used in the denominator of Eq. (15-27) also, and Eq. (15-27) becomes

$$W_\ell \approx \frac{-\dfrac{\hbar^2}{2m}4r_0^2 R_\ell{}^a \dfrac{\partial R_\ell{}^a}{\partial r}\Big|_{r=r_0}}{\int_{0,r_0} r^2 R_\ell{}^{a\,2} dr} . \tag{15-28}$$

$R_\ell{}^{a\,2}$ is equal to $1/4\pi$ times the spherical average of the electron density ρ_ℓ associated with an electron of quantum number ℓ in the atom, so this can also be written in the form

$$W_\ell \approx \frac{-\dfrac{\hbar^2}{m}4\pi r_0^2 \dfrac{\partial \rho_\ell}{\partial r}\Big|_{r=r_0}}{\int_{0,r_0} 4\pi r^2 \rho_\ell \, dr} . \tag{15-29}$$

This is the essential result, giving the band width entirely in terms of the radial density for the atomic state. The denominator, if continued to large r, would be the normalization integral and it is near one, and the correction can readily be estimated in terms of the ρ_ℓ and its derivative at the surface so that the entire expression depends only upon values at the sphere. This simplification was used in obtaining values for Table 15-1. Taking the denominator as one, the width becomes simply $\hbar^2/m$ times the area A_0 of the atomic sphere times $\partial\rho_\ell/\partial r$ at the sphere, $W_\ell \approx -(\hbar^2 A_0/m)\partial\rho_\ell/\partial r$.

B. Values for r_d

Straub evaluated W_ℓ for the transition-metal d-states from Eq. (15-29), using the $R_2(r)$ from Hartree-Fock calculations by Mann (1967). He then used Eq. (15-25) (but with a coefficient 5.53, rather than 6.83) to obtain the d-state radii, r_d , which are listed in Table 15-1 as ASM values. This formulation gives exactly the Eq. (15-25) (except for the 5.53 in place of 6.83) which will relate r_d to the tabulated atomic wavefunctions. The smaller coefficient came from use of a second-moment representation of the band width which we shall obtain in Eq. (15-33). Use of Eq. (15-25) would have given values of r_d some 7% lower than those listed in Table 15-1. It may not be an important distinction numerically, though conceptually Eq. (15-25) based upon the highest and lowest band states seems more appropriate for use with the Atomic Surface Method. The contribution of additional neighbors, which we have neglected, to the electron density at the atomic cell surface is probably comparable to the differences in different expressions. We use the MTO values from Table 15-1 in our calculations and the utility of Eq. (15-29) is in providing such a clear and simple estimate and understanding of this essential parameter, r_d , for the electronic structure of transition metals.

C. Further Approximations

Straub and Harrison (1985) went on to use an approximate form, the hydrogenic $e^{-\mu r}$ with μ related to the atomic term value by $\varepsilon = -\hbar^2\mu^2/2m$, for the atomic wavefunctions to obtain expressions for r_d in terms of the d-state energy and r_0 alone. Their equation, Eq. (24), for the band width W_∞ needs to be multiplied by a factor two to give the correct expression $W_\infty = -16\varepsilon_\ell(\mu r_0)^2 e^{-2\mu r_0}/(1 - 2(\mu\rho_0)^2 e^{-2\mu r_0})$. This is the form they used to obtain the points in their Figs. 3, 4, and 5, but the displayed equation was in error.

They also accommodated the theory to the tight-binding shift in the term values due to nonorthogonality of the overlapping states, as we did first in Eq. (1-14), leading to the overlap repulsion of tight-binding theory. This provided the shift in the band center, relative to the atomic term value. We shall obtain this shift in another way when we evaluate the total energy in Section 15-4-D.

Straub finally demonstrated the power of this method, and Eq. (15-29) in particular, by using it to calculate the bands formed by the core levels of sodium when the pressure was so high that the cores overlapped. Then tabulated core-state wavefunctions could be used to construct the band widths, in essential agreement with full band calculations at comparable densities. The qualitative results were interesting: The core 2s-band width and the 2p-band width grew with decreasing spacing until the two bands contacted each other at an r_0 of 0.8 Å. At just this point both bands took free-electron widths and the sodium became again a free-electron metal, but with eleven valence electrons, rather than one, per atom.

4. The Friedel Model

For properties which arise from integrals over the states, such as the total energy, it becomes appropriate to simplify the electronic structure. This was seen for osmium in Fig. 15-8b where a replacement of the true density of states by a Friedel (1969) model of $n(\varepsilon) = 2(2\ell+1)/W_\ell$, over the width of the band, gave an integrated number of electrons as almost the same function of energy as the full calculation. We proceed with that simple form, but first obtain the band width from the second moment of the bands.

A. Band Width from the Second Moment

In setting the band width W_ℓ for the Friedel Model, it would seem that a rectangular density of states extending all the way to the minimum and maximum band energies would correspond to too large a range. It would seem preferable to take a value such that the second moment of the rectangular density of states was equal to that of the true density of states. We saw in Section 2-5-A, and in Eq. (2-54) in particular, that the second moment of a set of bands is given by

$$M_2 = \frac{1}{nN} \Sigma_{i,j} H_{ij}H_{ji} , \tag{15-30}$$

if there are n orbitals for each of the N atoms. For the transition metal with $2\ell + 1 = 5$ orbitals, and the same energy $\varepsilon_\ell = \varepsilon_d$ for each, we may sum the couplings for each of the X nearest neighbors to obtain the second moment measured from ε_d as

$$M_2 = \frac{X}{5}(V_{dd\sigma}{}^2 + 2V_{dd\pi}{}^2 + 2V_{dd\delta}{}^2). \tag{15-31}$$

This may be equated to the second moment of a square distribution of width W, which is $\int_{-W/2,W/2} \varepsilon^2 d\varepsilon / \int_{-W/2,W/2} d\varepsilon = W^2/12$, to obtain

$$W_d{}^2 = \frac{12\,X}{5}(V_{dd\sigma}^2 + 2V_{dd\pi}^2 + 2V_{dd\delta}^2) . \tag{15-32}$$

For twelve nearest neighbors, as in face-centered-cubic structures, this gives

$$W_d = 107.2\frac{\hbar^2 r_d^3}{md^5} = 5.53\,\frac{\hbar^2 r_d^3}{mr_0^5} , \tag{15-33}$$

the form used by Froyen in estimating the r_d for Table 15-1, and the values of band width from Andersen and Jepsen (1977) which he used are also listed there. The result for the body-centered-cubic structure, with the sum in Eq. (15-32) including second-neighbor couplings, replaces the 5.53 by 6.08. These are smaller band widths than obtained in Eq. (15-25), as expected, and we use the value from Eq. (15-33) with the MTO values for r_d in our calculations.

B. Coulomb Shifts and Values for ε_d

One can obtain values for ε_d and ε_s from an atomic structure calculation, as we obtained them for s- and p-states. However, these values depend strongly upon the assumed configuration, whether it is $d^{Z-2}s^2$ or $d^{Z-1}s$ for example. The difference in the energy for the d-state in these two configurations is thought of as the Coulomb repulsion U_d between a pair of electrons in the d-shell (relative to one in a d-shell and the other in the s-shell). This is our definition of U_d here, and we return to a more careful discussion in Section 16-2-E. It was evaluated by Straub as the shift in the Hartree-Fock term value due to such a transfer, by interpolation between Mann's Hartree-Fock term values (Mann (1967)) for $d^{Z-2}s^2$ and $d^{Z-1}s$ configurations and the values of U_d he obtained are listed in Table 15-1. These are large enough that they dominate the energy of the d-bands relative to the free-electron bands in the metal. Whatever the atomic values may have been, if they place the d-band low enough that electrons transfer into the d-shells, the bands shift up in energy by U_d times the transfer until the Fermi energy in the two sets of bands is the same.

Moriarty (1982) has found from self-consistent band calculations that if one decomposes each band state into s- and d-like parts, and sums over occupied states, the net occupation of all transition metals in the 3d-series is near $Z_s = 1.5$, including copper with $Z_s = 1.66$. Pettifor (1977) made such estimates for the 4d-series, finding uniformly Z_s near 1.6. Straub (unpublished (1997)) found that similar values could be obtained directly from an *atomic* calculation, without a band calculation, by allowing partial occupation of s- and d-states within the free atom in Mann's (1967) program and minimizing the energy.

It is interesting that these effective valences Z_s differ considerably from those suggested by Rose and Shore (1991, 1993) who propose a free-electron model for the cohesion in which the ions are represented by a ball of positive charge, each contributing a corresponding number of free electrons. They find valences in the neighborhood of three for the 3d-metals, and four for the 4d and 5d metals. A similar theory has been proposed by Perdew, Tran and Smith (1990). We prefer the view based upon the Friedel model as better describing the physics of these systems, but should note the apparent success of these alternatives. Perhaps an assertion that both are appropriate would lead to interesting consequences, as did the accommodation of tight-binding theory and free-electron theory in Section 5-3-D, and the accomodation of tight-binding theory and the Atomic Sphere Approximation which we have just discussed in Section 15-3-A

Here we use this $Z_s = 1.5$ directly to set ε_d relative to the free-electron bands. It is a simple approach, and probably more accurate than simple attempts to determine it self-consistently starting with free-atom values, as done by Froyen(1980), or with an improved renormalized-atom approach (Watson, Ehrenreich, and Hodges (1970)) by Straub and Harrison (1985). Such approaches would *predict* values for Z_s, presumably near 1.5 since the values obtained by more accurate methods, as we have indicated, are near 1.5.

We illustrate this choice for the determination of ε_d for chromium and will then compare the corresponding values for the d-state energy with those obtained from self-consistent band calculations. For chromium, with a total of $Z = 6$ valence electrons, we take $Z_d = Z - 1.5 = 4.5$ electrons in the d-band. Thus with the Friedel model, the Fermi energy in the d-band lies at $\varepsilon_d - W_d/2 + (4.5/10)W_d$ which must equal the Fermi energy $\hbar^2 k_F^2/2m + \varepsilon_0$ of the free-electron gas, with ε_0 the free-electron-band minimum and k_F determined from Eq. (13-2) to be

$$k_F = \left(\frac{9\pi \times 1.5}{4}\right)^{1/3} \frac{1}{r_0} = \frac{2.20}{r_0} \;;$$

$$\varepsilon_F\,(Z_s = 1.5) = \varepsilon_0 + 2.42\,\frac{\hbar^2}{m r_0^2}\,. \qquad (15\text{-}34)$$

This gives the d-state energy, relative to ε_0, as

$$\varepsilon_d - \varepsilon_0 = \frac{\hbar^2 k_F^2}{2m} - \frac{Z_d}{10} W_d + \frac{W_d}{2}\,, \qquad (15\text{-}35)$$

with of course k_F obtained from Eq. (15-34).

The corresponding values for chromium are $\varepsilon_F - \varepsilon_0 = 9.25$ eV, $W_d =$ 6.75 eV, and $\varepsilon_d = 9.58\ eV + \varepsilon_0$, higher than the 8.00 eV, relative to the free-electron minimum, we took for the band calculation in Fig. 15-5. We make the same comparison with values from band calculations for all of the transition metals in Fig. 15-10. The values for chromium appear for $Z = 4$ for the 3d-series. This complex set of curves indicates that the energies of the center of the d-bands from our simple calculation are roughly correct and they reproduce the trends in the accurate values. These "correct" values should indeed be different since they correspond to a different representation of the bands than our rectangular density of states. Values of $\varepsilon_d - \varepsilon_0$ from Froyen (1970) or Straub and Harrison (1985) are consistent with the same conclusion. Differences in values for ε_d have little effect on the results we shall be discussing since in order to obtain the correct total number of electrons, the Fermi energy needs to shift by about the same difference, and the properties - such as the bands we discussed in Section 15-2-C - come out nearly the same. It is however essential that we use an ε_d value consistent with our description of the density of states, and we proceed with the values from Eq. (15-35), consistent with the Friedel model.

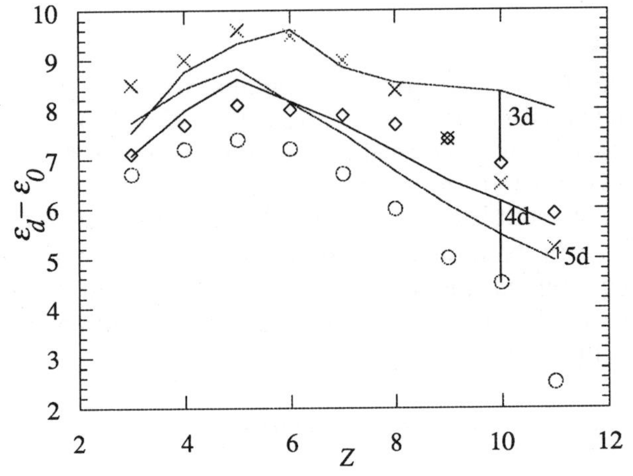

Fig. 15-10. d-band energy, relative to the free-electron band minimum for all transition metals. The lines are the values calculated here and the points are the values from Andersen and Jepsen (1977).

C. Hybridization Already Included

Our description of the Friedel model would suggest that we were thinking quite separately of the d-bands - with their rectangular density of states - and the free-electron bands, as in Fig. 15-5a, where there was no hybridization included. However, with our choice of $Z_s = 1.5$ s-electrons, which we noted was the expectation value for the s-orbital content, summed over *all* states, we are actually including the effects of hybridization. This remarkable fact was shown by Wills and Harrison(1983) and we repeat that argument, focusing first on the noble metals.

In the bands of copper, silver, and gold, seen in Fig. 15-4, the Fermi energy lies above the highest d-like band, so we might expect to regard the d-bands as full in the Friedel model, leaving a $Z_s = 1$. However, there are hybridization matrix elements $<<\mathbf{k}|H|d>$ coupling the free and d-like states. If the two coupled states are both occupied, or both empty, it is of little consequence but the coupling of the full d-states to the empty free states above the Fermi energy changes the orbital occupation. In perturbation theory it adds $<d|H|\mathbf{k}>> <<\mathbf{k}|H|d> /(\varepsilon_d - \varepsilon_k)^2$ to Z_s for every such coupling, leading to a

$$Z_s = 1 + \sum_d \sum_{k>k_F} \frac{<d|H|\mathbf{k}|>> <<\mathbf{k}|H|d>}{(\varepsilon_d - \varepsilon_k)^2} . \tag{15-36}$$

We cannot actually perform the integration with our simple hybridization matrix elements, proportional to k^2 , from Eq. (15-11) because the resulting integral, $\int 4\pi k^2\, dk$, over the final term would diverge at large k . However, with a more complete form for the hybridization matrix element the integral would converge and it is this final term which we associate with the extra $\delta Z_s \approx 0.5$ obtained in the more complete calculations we mentioned in Part B. This same coupling will lower the energy of the system, to second order in the hybridization, by the same sum as in Eq. (15-36) with only a single factor of $\varepsilon_d - \varepsilon_k$ in the denominator.

$$\delta E_{tot} = \sum_d \sum_{k>k_F} \frac{<d|H|\mathbf{k}|>> <<\mathbf{k}|H|d>}{\varepsilon_d - \varepsilon_k} . \tag{15-37}$$

We may expect the dominant terms in both cases to come for $\mathbf{k}$ near the Fermi surface, so the lowering in energy per atom will be approximately

$$\delta E_{tot} \approx \delta Z_s(\varepsilon_d - \varepsilon_F) . \tag{15-38}$$

In Part D we shall calculate the gain in energy due to the finite band width and the partial occupation of that band, Eq. (15-40). We may obtain the change in energy due to a change $\delta Z_d = -\delta Z_s$ by taking the differential of that expression, and find that it is given by $-W_d/2 + Z_d W_d/10$ times δZ_d . But this $-W_d/2$ is the energy of the bottom of the band relative to ε_d and $Z_d W_d/10$ is the Fermi energy relative to the bottom of the band, so this is exactly the energy gain given by Eq. (15-38). Thus our treatment of the noble metals as having $1 + \delta Z_s$ electrons in the conduction band and δZ_s holes in the d-band will give us the correct contribution to the total energy, though the true bands of predominantly d-character are completely full. Wills and Harrison added the hybridization shift in the energy of occupied s-states due to coupling with empty d-states, if the band was partly filled prior to introducing hybridization, to find that the same formula, Eq. (15-38), applies. Again the inclusion of this δZ_s shift in our calculation correctly includes the hybridization energy. They also noted that inclusion of this extra electron occupation δZ_s per atom is appropriately included in the free-electron energy which we discussed in Chapter 13.

D. The Total Energy

There is every indication that full local-density-theory calculations describe well the cohesive properties of the transition metals. This was seen in particular in the calculations by Moruzzi, Janak, and Williams (1978) for the light metals up through the 4d-series. Many of these properties of transition metals were understandable more simply (Andersen (1975), Mackintosh and Andersen (1979)) in the context of the Atomic Sphere Approximation which we have discussed. Pettifor (1977,1978) also used this approach to clarify the relative role of the s- and d-states in the properties, showing that many of the details of the electronic structure were inessential to the properties.

In Chapter 16 we shall look carefully at Coulomb correlation effects which lie beyond the local-density theory which has proven so successful in the theory of the transition metals. These become important in the f-shell metals and dominate the properties of the rare earths. From the formulae derived there we shall see that these same effects on properties such as the cohesion of the transition metals are far from negligible. We shall nevertheless postpone the detailed discussion of correlation effects till the next chapter and proceed here with the simpler theory, based upon the same local-density theory which has been used in the more complete calculations. However, when we make the detailed evaluation of the terms in the cohesion we shall directly add corrections for the correlated motion of the electrons which will be derived in Chapter 16. We shall find that they are significant, and appear in most cases to improve agreement with experiment.

For the present discussion we follow generally the analysis by Wills and Harrison (1983,1984) which is along the lines we have developed in this book and does not include the effects of correlation. The basic approach is to imagine bringing transition-metal atoms together, neglecting at first any hybridization terms, and therefore also any coupling between d-states on different atoms. The d-shell states, like the atomic cores, remain exactly as in the atom. Then the change in energy is precisely that upon forming a simple metal and we use exactly the theory we have given for simple metals. In our treatment, this energy will be computed for Z_s equal to two, or one, electrons depending upon the configuration of the atom.

The d-shell in the atom, partly occupied, will not be at its final correct energy $\varepsilon_d - \varepsilon_0$ relative to the conduction-band minimum because it will ordinarily have a configuration of $d^{Z-2}s^2$, and in some cases $d^{Z-1}s$. We should transfer charge between the d- and s-like levels until the d-band center is at the appropriate energy, at a cost in energy we call the "atomic preparation energy". Wills and Harrison (1984) explicitly neglected this term and computed the simple-metal energy using $Z_s = 1.5$ electrons in the simple-metal formula. In fact their value for the simple-metal energy is around twice (because they did not include repulsion) the average of what we obtain for $Z_s = 1$ and $Z_s = 2$. It is quite easy to estimate the atomic preparation energy if we work back from the final metallic state, viewed in the tight-binding context. Once the half an electron has been transferred from the doubly-occupied s-shell to the d-shell, the d-state energy will be at the energy $\varepsilon_d - \varepsilon_0$ which we calculated in Part B as Eq. (15-35). If we were to transfer that half an electron back, the energy in the d-shell would drop by $U_d/2$ (relative to the bottom of the band which is fixed by the s-state energy and the $V_{ss\sigma}$ coupling) and the Fermi energy would move up by $\varepsilon_F(Z_s = 2) - \varepsilon_F(Z_s = 1.5)$ due to the greater filling. Thus as we begin the transfer, Z_s is 2 so the d-state energy is lower by $U_d/2$ than the final d-state energy ε_d and the energy change per electron is $\varepsilon_d - U_d/2 - \varepsilon_F(Z_s = 2)$. At the end of the transfer ε_d is at its final value and the energy change per electron is $\varepsilon_d - \varepsilon_F(Z_s = 1.5)$. Taking the average, the atomic preparation energy per atom in transferring half an electron from the s- to the d-shell is

$$E_{prep.} = {}^1/_2(\,\varepsilon_d - \varepsilon_0 - U_d/4 - 1.106(\varepsilon_F - \varepsilon_0\,)) \qquad (\text{ for } d^{Z-2}s^2) \qquad (15\text{-}39)$$

$$= -0.128\frac{\hbar^2}{mr_0^2} - \frac{Z_d-5}{20}W_d - \frac{U_d}{8}\ .$$

The numerical coefficient in the first form is from $[\varepsilon_F(Z_s = 2) + \varepsilon_F(Z_s = 1.5)]/2\varepsilon_F(Z_s = 1.5) = (2^{2/3} + 1.5^{2/3})/2\times 1.5^{2/3} = 1.106$ (with each ε_F measured from ε_0). For the second form $\varepsilon_d - \varepsilon_0$ was taken from Eq. (15-35) with $\varepsilon_F - \varepsilon_0$ defined at $Z_s = 1.5$. When there is only a single electron in the s-shell in the free atom (the noble metals, Cr, Mo, and a number of others to be indicated in Fig. 15-12), half an electron is transferred from the d-state to the free electron Fermi energy. Initially the cost per electron is $\varepsilon_F(Z_s = 1) - (\varepsilon_d + U_d/2)$ and at the end it is $\varepsilon_F(Z_s = 1.5) - \varepsilon_d$. Thus the preparation energy becomes $E_{prep.} = -{}^1/_2(\varepsilon_d - \varepsilon_0 + U_d/4 - 0.881(\varepsilon_F - \varepsilon_0))$ with the $0.881 = (1 + 1.5^{2/3})/2\times 1.5^{2/3}$. Note the change in sign in front, because the transfer was in the opposite direction, as well as the different coefficient than in Eq. (15-39). This result for a single s-electron in the atom can be rewritten as $E_{prep.} = -0.143\hbar^2/(mr_0^2) + (Z_d-5)W_d/20 - U_d/8$, analogous to the final form in Eq. (15-39).

After preparing the atom, we now broaden out the d-band to its true width W_d. With its $Z_d = Z-1.5$ electrons, it will fill the band just to the Fermi energy for both bands, as illustrated in Fig. 15-11. This fills the band to an energy $Z_dW_d/10$ above the bottom of the band, so the average energy becomes ${}^1/_2(-W_d/2 + (-W_d/2 + Z_dW_d/10)$ relative to ε_d. This produces a gain in bonding energy per atom of

$$E_{bond} = -\frac{Z_d(1-Z_d/10)}{2}W_d, \tag{15-40}$$

relative to having all states at the energy ε_d at the center of the band. This is the important contribution to the bonding of the transition metals over that for the simple metals. We shall see in Eq. (16-6) how this contribution is corrected for the correlated motion of the electrons by replacing W_d by $\sqrt{W_d^2 + U_d^2} - U_d$. That form approaches W_d as U_d approaches zero, in which case the effects of correlation are negligible. It approaches zero when U_d is large and the correlation prevents the formation of bands. We use the corrected form, to be derived in Chapter 16, in our prediction of the energy and the volume-dependence of the energy.

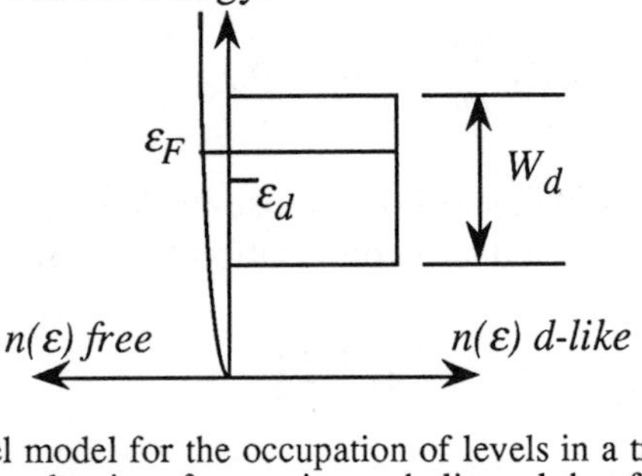

Fig. 15-11. The Friedel model for the occupation of levels in a transition metal. The free-electron density of states is parabolic and that for the d-bands is taken to be a constant $10/W_d$ over a range W_d centered at ε_d. Both are filled to the same Fermi level ε_F.

There is an additional term which was found to be essential in the earlier treatment of the f-shell metals (Harrison (1983d)) and Wills and Harrison (1983,84) included it also for transition metals. It is the repulsion between neighboring d-states arising from nonorthogonality, exactly as the overlap repulsion in tight-binding theory arises from nonorthogonality. We saw already in Eq. (1-14) that there is an upward shift of $- <2/1><1/H/2>$ in energy for two levels coupled by $<1/H/2>$. In this case, the d-state is corrected to first order in the hybridization matrix element, as at the beginning of Section 15-2-A, $/d> + \Sigma_k/k>><<k/H/d>/(\varepsilon_d - \varepsilon_k)$. Then Wills and Harrison (1983) found a nonorthogonality between d-states on neighboring sites of

$$S_{dd'm} = \sum_k \frac{<d/H/k>><<k/H/d'>}{(\varepsilon_d - \varepsilon_k)^2} = \sigma_{ddm} \frac{r_d^3}{d^3} , \tag{15-41}$$

where d is the distance between the atoms, and m again is the angular momentum quantum number with respect to that axis. The coefficient σ_{ddm} was obtained just as the coefficients η_{ddm} for V_{ddm} were obtained, with an integration over wavenumber taking ε_d at the bottom of the free-electron band. The result is $\sigma_{ddm} = (5/\pi, -5/\pi, 5/2\pi)$ for $m = (0,1,2) = (\sigma, \pi, \delta)$. Then for a crystalline solid they found a shift in the center of gravity of the d-bands $\Delta\varepsilon_d$ due to this nonorthogonality. The increase in energy per atom E_{dd} is obtained by multiplying by the number Z_d of electrons per atom in the d-shell.

$$E_{dd} = Z_d \Delta\varepsilon_d = - Z_d \frac{1}{5} \frac{\hbar^2 r_d^6}{m d^8} (\sigma_{dd\sigma}\eta_{dd\sigma} + 2\sigma_{dd\pi}\eta_{dd\pi} + 2\sigma_{dd\delta}\eta_{dd\delta})X$$

$$= 11.40\, X Z_d \frac{\hbar^2 r_d^6}{m d^8} = 1.19\, Z_d \frac{\hbar^2 r_d^6}{m r_0^8} \quad \text{(for } X = 12 \text{ , fcc).} \tag{15-42}$$

The final form obtains only for close-packed structures ($X = 12$) The result can also be interpreted as a contribution to the overlap repulsion of

$$V_0^{dd}(d) = 22.80(Z - 1.5) \frac{\hbar^2 r_d^6}{m d^8} . \tag{15-43}$$

All of this is closely parallel to the tight-binding results for the sp-bonded systems. The overlap repulsion is a term of higher order, but it has an appreciable effect on the cohesion and, because of its rapid variation with distance, it has a large effect on the elastic properties.

E. The Cohesive Energy

For the cohesive energy, we follow Wills and Harrison (1984) and utilize the tight-binding description of the simple metals which we gave in Section 13-3-A since it treats the atom and the metal in the same context. We gave in Eq. (13-7) the cohesive energy for a simple metal, the negative of the energy E_{sm} of the simple metal relative to that of the free atom, $E_{sm} = (Z/2)[3\pi^2\hbar^2/(4md^2) - 3\hbar^2k_F{}^2/(10m)]$. The contribution from the overlap repulsion, the leading factor of one-half, assumed that the metal was in equilibrium under the influence of the s-electrons alone. Now when we add the atomic preparation energy, Eq. (15-39), the d-bonding energy, Eq. (15-40), and the dd-repulsion, Eq. (15-42), we have additional terms dependent upon spacing. When we treat the volume-dependent properties in Part F, it will be important that the energy we use be minimum at the observed spacing and we shall adjust the pseudopotential core radius until that is the case. It is not important for the cohesion and we saw in Section 13-3-B that the volume-dependent energy was very poorly given for the simple metals by the tight-binding expression. It is much more appropriate for the cohesion to evaluate every contribution in terms of the parameters given in Table 15-1 and to evaluate the small simple-metal term exactly as before, with the factor of one half as if it were independently in equilibrium.

Thus we are to evaluate the simple-metal energy using a valence Z appropriate to the free atom (one or two), using Eq. (13-7), and then add the atomic preparation energy necessary to change the number of metallic electrons to 1.5 per atom. This brings the d-state energy to its final value and we use Eq. (15-40) to estimate the gain due to d-band formation, and add the dd-repulsion energy from Eq. (15-42). For the final evaluation we shall also correct Eq. (15-40) for correlation by replacing W_d by $\sqrt{W_d^2 + U_d^2} - U_d$, the form we obtain in Eq. (16-6).

To facilitate the evaluation we rewrite the simple-metal contribution E_{sm}, the negative of Eq. (13-7), in more convenient form, assuming a face-centered-cubic structure to relate d, r_0, and k_F,

$$E_{sm} = -\frac{Z}{2}\left(\frac{3\pi^2\hbar^2}{4md^2} - \frac{3\hbar^2k_F^2}{10m}\right) = -0.506\frac{\hbar^2}{mr_0^2} \qquad (15\text{-}44)$$

for $Z_s = 2$, the case for most transition metals. For the case in which we have only a single s-electron in the free atom, the coefficient in the final form is -0.578 rather than -0.506 . Similarly writing the other contributions in terms of r_0 we obtain the energy of the metal minus that of the free atom, the negative of the cohesive energy per atom,

$$-E_{coh} = -0.634\frac{\hbar^2}{mr_0^2} - \frac{Z_d - 5}{20}W_d - \frac{U_d}{8}$$

$$- \frac{Z_d(10-Z_d)}{20}\left(\sqrt{W_d{}^2 + U_d^2} - U_d\right) + 1.19Z_d\frac{\hbar^2 r_d^6}{mr_0^8} \qquad (15\text{-}45)$$

for systems with two s-electrons in the free atom. For a single s-electron in the free atom, the leading term in the first line has -0.634 replaced by -0.721, and the second term (the term in W_d) is changed in sign. The cohesive energies can then be directly evaluated for all the transition metals using the parameters from Table 15-1. The results are shown, along with experiment, in Fig. 15-12. The elements with a single s-electron in the atom are indicated with an asterisk, but Pt and Au were found to have negative cohesive energies using the forms for one electron per atom and were plotted as for two electrons per atom .

The predictions are lower than those from the calculation by Wills and Harrison (1984), largely from the inclusion of the correlation correction, the replacement of W_d by $\sqrt{W_d^2 + U_d^2} - U_d$ as will be seen in Eq. (16-5). This brings our predictions closer to experiment except for the 5d-series for which we underestimate the cohesion. We have also added the simple-metal repulsion reducing the simple-metal terms by a factor of two, as we did in Chapter 13, calculated the simple-metal energy depending upon the atomic configuration and added the atomic preparation energy. These differences in the simple-metal terms shift the cohesion by only an electron volt or so.

Inclusion of the d-state terms has eliminated most of the large discrepancy for the noble metals which we noted when we treated them as simple metals in Chapter 13 and describes well the increased cohesion at the center of each series, as follows from the $Z_d(10 - Z_d)$ form of Eq. (15-40), which is characteristic of the Friedel model. We seem to have reasonable estimates of the essential contributions, including the correlation corrections which we discuss in Chapter 16.

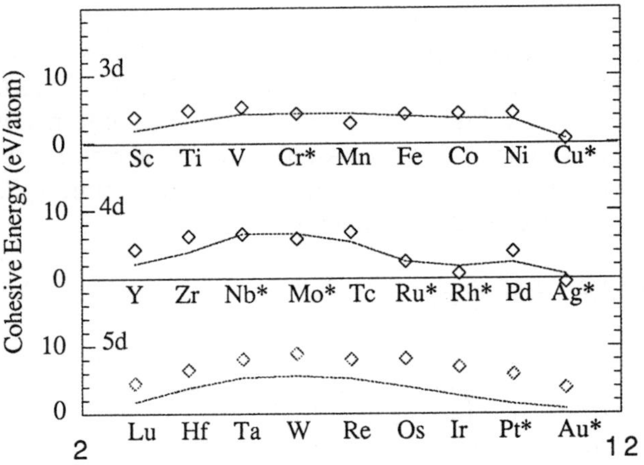

Fig. 15-12. Lines are the predicted cohesive energies, Eq. (15-45), of the transition metals. Points are experimental values from Kittel (1976). The elements indicated an asterisk * were assumed to be $d^{Z-1}s$ in the atom, the others $d^{Z-2}s^2$. (For Pt and Au, negative values were found for $d^{Z-1}s$, and $d^{Z-2}s^2$ was used.)

F. Volume-Dependent Properties

For the volume-dependence of the energy, we follow Wills and Harrison (1983) in taking the free-electron-based formulation for the s-like states, rather than the tight-binding formulation. It is more appropriate for the volume dependence of the energy of the solid, though the tight-binding view was preferable for the cohesion where we wished to compare the solid-state energy with the free-atom energy. We thus replace the simple-metal energy, Eq. (15-44), by the energy given in Eq. (13-18) which we used in Section 13-4 to study the volume-dependent properties of the simple metals. It was

$$\frac{E_{eg}}{ion} = \frac{3Z\hbar^2k_F^2}{10m} + \frac{2Ze^2r_c^2k_F^3}{3\pi} - \frac{3Ze^2k_F}{4\pi} - \frac{Z^{5/3}e^2k_F}{(18\pi)^{1/3}}\alpha \qquad (15\text{-}46)$$

and in the metal we are taking $Z = 1.5$. We also take the Madelung constant $\alpha = 1.8$ for the metallic structures. The four terms were the kinetic energy, the empty-core shift in the band, the exchange energy and the Madelung

Coulomb energy, respectively. There are no longer terms which depend upon the initial electronic structure of the atoms, as is appropriate for the volume-dependent properties. We write Eq. (15-46) in terms of the atomic sphere radius using Eq. (15-34) and add the d-state terms, which are all the terms in Eq. (15-45) minus the simple-metal energy, Eq. (15-44). For the volume dependence it is convenient to write the atomic-sphere radius as r , equal to r_0 at equilibrium, to obtain

$$E_{tot} = \frac{2.17\hbar^2}{mr^2} + \frac{3.37e^2r_c^2}{r^3} - \frac{2.81e^2}{r}$$

$$- \frac{Z_d(10-Z_d)}{20}\left(\sqrt{\left(5.53\ \frac{\hbar^2 r_d^3}{mr^5}\right)^2 + U_d^2} - U_d\right) + 1.19Z_d\frac{\hbar^2 r_d^6}{mr^8} . \quad (15\text{-}47)$$

We take parameters from Table 15-1 (choosing again the $r_d(MTO)$) and, following Wills and Harrison (1983), adjust r_c so that the minimum in this energy comes at the observed $r = r_0$. The resulting values are plotted as lines in Fig. 15-13 and are the values of r_c listed in Table 15-1 and 15-3. To the right in the 5d-series the required r_c values are imaginary. This simply means that a negative term $3.37e^2r_c^2/r^3$ is needed to produce the correct volume. Use of the slightly smaller Atomic-Surface-Method d-state radii from Table 15-1 yields real values for all but gold, and an r_d of 0.99 Å eliminates the problem for gold. This basically reflects a high sensitivity of our adjustment of a soft repulsion in the presence of strong, and perhaps inaccurate, forces from the d-states.

This sensitivity is illustrated for chromium in Fig. 15-14 where the total energy is plotted as a function of radius r . In this case the curvatures from the two contributions are nearly equal and opposite (chromium is the worst case in this regard) so that the position of the minimum is very sensitive to the adjusted core radius. The sensitivity would be resolved by even a small adjustment of the d-state radius, r_d , or as illustrated by the dotted line in Fig. 15-14 by eliminating the correlation correction (and readjusting the pseudopotential core radius from 0.74 Å to 0.84 Å to place the minimum again at r_0). We choose here to proceed with our $r_d(MTO)$, and our inclusion of the correlation correction which we believe appropriate, and r_c fit to give the minimum-energy r_0 at the observed value. Small adjustments could rectify the peculiarities, such as imaginary r_c , where they arise.

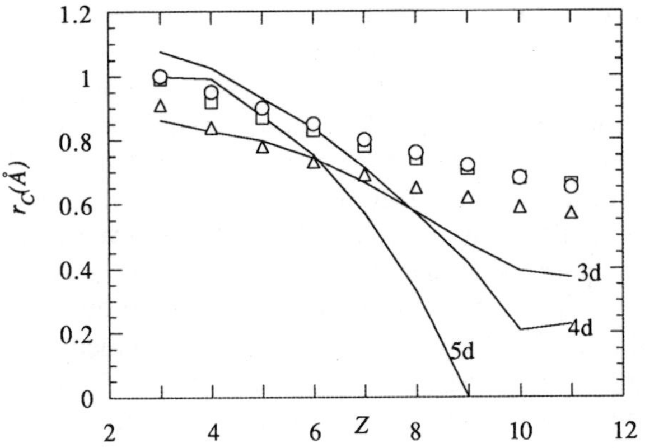

Fig. 15-13. The lines represent pseudopotential core radii adjusted so that Eq. (15-47) was minimum at the observe spacing. (Fit to r_0.) The triangles are core radii determined from the free-atom s-state energy for 3d-metals, the squares for the 4d-metals, and the circles for 5d metals.

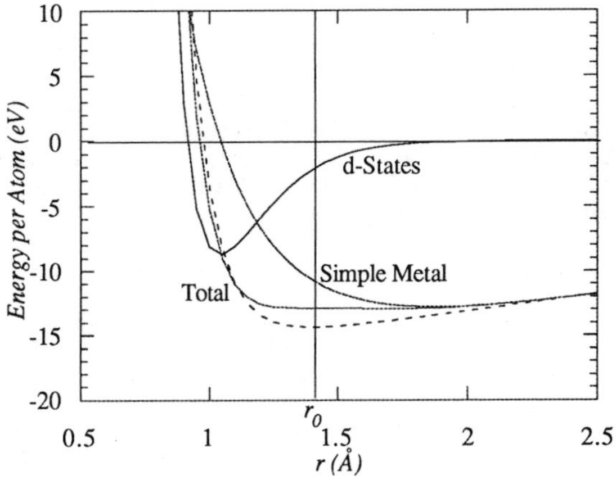

Fig. 15-14. The energy per atom for chromium, as a function of atomic sphere radius, from Eq. (15-47). The simple-metal terms (first line in Eq. (15-47)), the d-state terms, and their sum, are shown separately as solid lines. Setting U_d equal to zero and readjusting the core radius to obtain the correct minimum yields the dotted line.

In this regard it is also interesting to compare our fit values of r_c to predictions of the core radii from other sources. We obtain these, using the procedure and the curve from Harrison (1980), Problem 16-2, p. 360, which relates the core radius to the free-atom s-state energy. We used s-state energies from Mann (1976), some of which were for a $d^{Z-2}s^2$ configuration, the others for a $d^{Z-1}s$ configuration, as indicated in Fig. 15-12. These energies were different because of the difference in potentials, and we interpolated curves sketched for the two configurations to obtain values appropriate to $Z_s = 1.5$. These values are also plotted in Fig. 15-13, and listed in Table 15-3. They are remarkably in agreement with the fit values at the beginning of the series, and as we indicated in discussing the imaginary values to the right in the 5d-series, only small adjustments of the d-state radii, well within expected uncertainties, could bring these into accord. We take this as strong support for our description of the electronic structure. We proceed directly with Eq. (15-47) and with the parameters given in Table 15-1.

We may immediately take the second derivative of Eq. (15-47) with respect to r and obtain the bulk modulus, as did Wills and Harrison (1983),

$$B = \Omega_0 \frac{\partial^2 E_{tot}}{\partial \Omega_0^2} = \frac{1}{9\Omega_0} r^2 \frac{\partial^2 E_{tot}}{\partial r^2} . \tag{15-48}$$

Table 15-3. Pseudopotential core radii r_c (Å) adjusted so that $\partial E_{tot}/\partial r$ from Eq. (15-47) is zero at the observed volume, using other parameters (MTO values) from Table 15-1. Values with "i" are imaginary, negative r_c^2 values in Eq. (15-47). $r_c(atomic)$ values were fit to the free-atom s-state energies as described in the text.

	r_c	$r_c(atomic)$		r_c	$r_c(atomic)$		r_c	$r_c(atomic)$
Sc	0.86	0.91	Y	1.08	0.99	Lu	1.00	1.00
Ti	0.83	0.84	Zr	1.02	0.92	Hf	0.99	0.95
V	0.80	0.78	Nb	0.93	0.87	Ta	0.87	0.90
Cr	0.74	0.73	Mo	0.84	0.83	W	0.75	0.85
Mn	0.67	0.69	Tc	0.71	0.78	Re	0.57	0.80
Fe	0.57	0.65	Ru	0.57	0.74	Os	0.33	0.76
Co	0.48	0.62	Rh	0.42	0.71	Ir	0.26i	0.72
Ni	0.39	0.59	Pd	0.21	0.68	Pt	0.46i	0.68
Cu	0.37	0.57	Ag	0.23	0.66	Au	0.54i	0.65

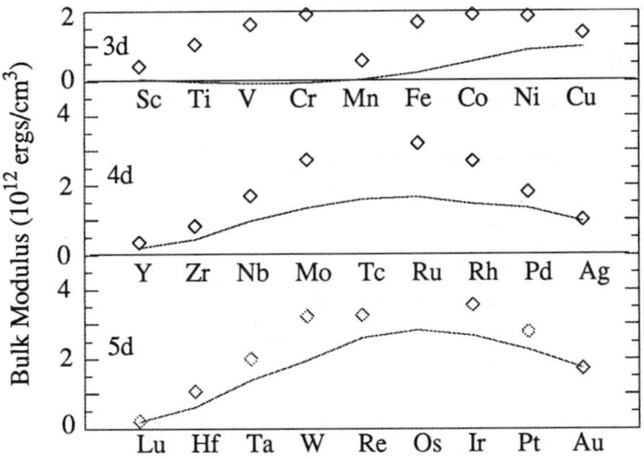

Fig. 15-15. The bulk modulus (10^{12} ergs/cm^3 = 1 megabar), predicted from Eq. (15-48) shown as the line, based on Eq. (15-47) with r_c fit to give the correct r_0. Experimental values (from Kittel (1976)) are shown as points.

Our result is compared with experiment in Fig. 15-15. The accidental softening in the 3d-series, which leads here even to negative bulk moduli, was discussed above and illustrated in Fig. 15-14. It is seen here to be a continuation of a trend from 5d- to 4d- to 3d-metals.

Wills and Harrison (1983) also calculated a modified Grüneisen constant, defined by

$$-\frac{\Omega_0}{2B}\frac{\partial B}{\partial \Omega_0} = -\frac{1}{2}\left(1+\Omega\frac{\partial^3 E_{tot}/\partial\Omega^3}{\partial^2 E_{tot}/\partial\Omega^2}\right) = -\frac{1}{2}\left(\frac{r}{3}\frac{\partial^3 E_{tot}/\partial r^3}{\partial^2 E_{tot}/\partial r^2} - 1\right) \quad (15\text{-}49)$$

and our values for this constant (larger by $1/6$ than the constant given by Eq. (14-16)) are compared with experiment in Fig. 15-16. The divergent region in the 3d-series, where our $r^2 \partial E_{tot}{}^2/\partial r^2$ goes to zero, was not included. The predicted values, with no experimental parameters except the equilibrium spacing, are not very impressive, but both figures suggest much in our description is essentially correct.

We have considered only the elements Sc, Y, Lu, and those to the right of them in the periodic table, as transition elements but it is interesting that even

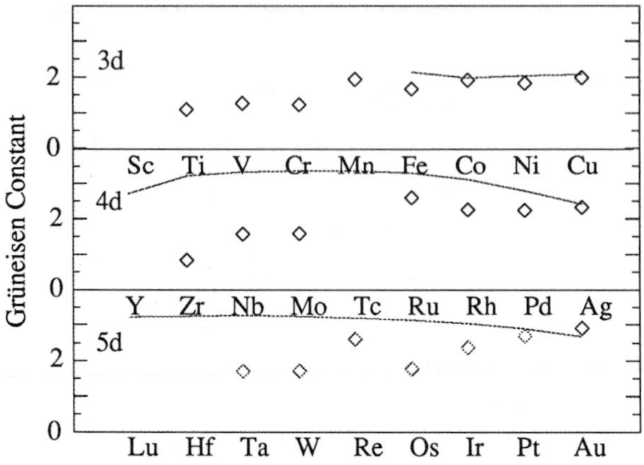

Fig. 15-16. A Grüneisen constant, $-{}^1/_2(\Omega/B)\partial B/\partial\Omega$, predicted from Eq. (15-49), shown as the line, and experimental values (calculated by Wills and Harrison (1983) from coefficients of thermal expansion from Pearson (1958)). $^1/_6$ should be subtracted from all values to obtain the Grüneisen constant γ defined in Eq. (14-16).

the alkali metals become transition metals under sufficient pressure. Band calculations (McMahan (1984) and references therein) indicate that under increasing pressure the d-state content increases, culminating in a volume collapse of 9% at high enough pressure, with face-centered cubic structure before and after the transition. There has recently been rather direct experimental confirmation by Abd-Elmiguid, Mohsen, Pattyn, and Bukshpan (1994) of this change in electronic character in cesium. The reduced volume can be understood in terms of the contraction of the lattice by the occupation of the lower part of a d-band, as in the rest of the transition-metal series, or as the appearance of a second minimum due to the d-states, analogous to the second minimum due to f-states in cerium which we shall see in Fig. 16-4.

G. Crystal Structure Determination

With the simplest form of the density of states, the Friedel Model, the energy is almost independent of crystal structure. There is some dependence through terms depending upon d rather than r_0 but those are the same for face-entered-cubic and hexagonal-close-packed structures and the relative size for body-centered-cubic depends upon whether the second neighbors, 15% more distant, are included. These terms seem to be inessential to the determination of structure. It became clear in the work of Pettifor (1970, 1972) that deviations of the d-band densities of states from the Friedel model were responsible for the observed crystal structures of the transition metals.

We saw in Section 15-2-D that the d-band density of states for the body-centered-cubic structure was split into two peaks, with a broad minimum between. Such a deviation from the constant density of states *is* important to the structure. For a given second moment M_2 for the band, it favors the bcc structure near half filling and fcc or hcp for nearly full or nearly empty bands. We may see this by using the Friedel model for $n(\varepsilon)$ for fcc and hcp structures but two separated peaks for $n(\varepsilon)$ for body-centered cubic. For the Friedel model the band extends an energy $\sqrt{3M_2}$ from the center, while if there were two symmetric sharp peaks (appropriate to bcc structure) they would lie $\sqrt{M_2}$ from the center. Thus for a small filling of the band, or an almost complete filling, the hcp or fcc structure is favored by the electrons at very low energy (or holes at very high energy in a symmetric band). For half filling, with an average lowering in energy of only $(\sqrt{3}/2)\sqrt{M_2}$ for the fcc and hcp structures, the bcc structure is favored.

We saw also in Section 15-2-D that the occupation of the d-like bands (as opposed to the weighted d-like occupation) was approximately $Z - 1$, so we should expect body-centered cubic structures for $Z = 6$, chromium, molybdenum, and tungsten, which are in fact all bcc. So also are the $Z = 5$ elements vanadium, niobium, and tantalum, though not the $Z = 7$ elements. If we were to take the minimum in the hcp density of states, seen in Fig. 15-8 to occur at $Z = 4$, as similarly favoring such structures, we would expect the $Z = 4$ elements, titanium, zirconium, and hafnium to be hcp, as they are, and as are the $Z = 3$ elements.

We shall return to the question of more complicated structure in the density of states, such as we saw for osmium in Fig. 15-8, when we discuss actinide metals in Section 16-4. We see there that it has nothing to do with the orbital geometry of the f-states, any more than the analysis here had to do with the orbital geometry of the d-states. Some studies have attempted to associate crystal-structure determination with lobes in the d- or f-states, but our study does not support such a view.

Pettifor (1970) made a quantitative study of just these effects, using a calculated density of states based only upon the parameters ε_d and W_d, and found that indeed it predicted a series of structures with increasing atomic number of hcp-bcc-hcp-fcc-bcc, in complete agreement with experiment except for predicting bcc structures for the noble metals. The addition of a hard-core repulsion, the counterpart of our Eq. (15-43), rectified this one defect. The actual energy differences predicted for different structures were considerably higher than the best experimental estimates, but the predicted structures were correct. More recently Ahuja, Eriksson, Wills, and Johansson (1995) studied high-pressure phases of calcium, in which the d-states also played a determining role.

Our qualitative discussion of the effect of these features of the density of states can be made more quantitative, without Pettifor's more complete numerical analysis, in terms of the higher moments of the density of states. The third moment describes something like a skewness in the density of states and the fourth moment describes the tendency to split into two peaks, as in the body-centered-cubic structure. One of the earliest such studies was by Cyrot-Lackmann, (1970), who pointed out that the high fourth moment in the bcc structure makes it favorable for the center of the transition-metal series. It is possible to made a direct expansion of the energy in second, third, and fourth moments, analogous to the expansion we made in order to discuss structural trends in semiconductors in Section 2-5. Recently Muller and Silcox (1995) (see references therein and also Muller, Singh, and Silcox, (1997)) have used this same expansion to interpret electron energy loss spectra (EELS) studies of alloys, which directly measure the density of states, projected onto the individual ℓ -components of the states. [This projection occurs because electrons are excited from core states and selection rules favor final states of particular symmetry.] This analysis then provides a direct link between the energetics of the alloys and the EELS spectra. We turn next to the general question of such alloys.

H. Transition-Metal Alloys

There has been a long history of studying the heats of formation of alloys involving transition metals. In particular, there has been an empirical scheme due to Miedema, de Chatel, and de Boer (1980), which we mentioned in connection with simple metals in Chapter 14, and which purported to provide the sign of the heats of formation for the entire range of alloys. Williams, Gelatt, and Moruzzi (1980) made an important point about this approach, in noting that with expected systematics among the thirty-six alloy combinations of the nine 4d-transition metals, only two parameters were required to separate those with positive, from those with negative, heats of formation. Miedema, et al., accomplished it with *eighteen* adjustable parameters. Further study, discussed for example by Hafner (1987) p. 165, has failed to substantiate the physical picture behind the Miedema scheme. The work by Pettifor (1979, 1995) seems to provide a much more convincing and useful description. It follows very much the lines we have described here.

For discussing the total energy of pure transition metals, in terms of the Friedel model, the d-levels shift in energy so that they contain Z - 1.5 electrons, with the remaining $Z_s = 1.5$ electrons as a free gas. We might expect this to be true also in alloys of the transition elements. To the extent that each atom behaved independently, there would be no change in energy upon mixing different elements, except that the free-electron density (Z_s/Ω_0) differs for different metals. The contribution to the heat of formation arising from this free-electron difference is the same as we have described for simple metals. It is quite a small contribution in comparison to the d-electron effects so Pettifor ignored it, and we do also.

The essence of Pettifor's (1979) approach is to construct a Friedel-Model d-band density of states common to the constituents in the alloy. This is very distinct from adding the density of states, weighted by concentration, for the pure constituents. This also dismisses any Coulomb effects associated with charged ions. This also neglects the effects of the dd-repulsion, to which we return afterward. The heat of formation then follows immediately from the description of the electronic structure we have given.

We imagine a random alloy of concentration x of atoms of type A and concentration 1-x of atoms of type B. We again obtain the band width from the second moment, so we imagine a large matrix, $5N$ by $5N$, for the five d-states on each of the N atoms and wish to evaluate $M_2 = (1/5N)\Sigma_{i,j}H_{ij}H_{ji}$. The diagonal elements for atom A, measured from the average energy $\bar{\varepsilon}$ = $x\varepsilon_{dA} + (1\text{-}x)\varepsilon_{dB}$, are $(1\text{-}x)(\varepsilon_{dA} - \varepsilon_{dB})$. These are squared and multiplied by the number, $5xN$, of such terms. We add the contribution for the atoms B and divide by $5N$ to obtain a contribution to M_2 of $x(1\text{-}x)(\varepsilon_{dA} - \varepsilon_{dB})^2$. We next add the coupling of each A atom to each of its neighbors. This could readily be written down for an ordered alloy, but we do it for a random alloy. If it were a pure material, we would use the M_2 which we used to obtain the band width $W_d = \sqrt{12M_2} = 5.53\ \hbar^2 r_d^3/(mr_0^5)$ in Eq. (15-33). Now a fraction x of these atoms are of type A and a fraction x of their neighbors are of type A and a fraction 1-x of the neighbors are of type B. Adding the fraction of the atoms which are of type B, and their two types of neighbors, we obtain a weighted r_d for the alloy of

$$r_d^3 = x^2 r_{dA}^3 + 2x(1\text{-}x)\sqrt{r_{dA}^3 r_{dB}^3} + (1\text{-}x)^2 r_{dB}^3 \qquad (15\text{-}50)$$

This leads to a band width for the composite band of

$$W_d = 12\sqrt{M_2} = \sqrt{12x(1-x)(\varepsilon_{dA} - \varepsilon_{dB})^2 + (5.53\hbar^2 r_d^3/mr_0^5)^2} \qquad (15\text{-}51)$$

for the alloy. We also need an r_0 for the alloy, for which we follow Pettifor (1979) in using Zen's law,

$$r_0^3 = x\, r_{0A}^3 + (1-x)\, r_{0B}^3 , \qquad (15\text{-}52)$$

though other choices could equally as easily be made.

This W_d for the alloy is then substituted into Eq. (15-40) , taking of course

$$Z_d = xZ_{dA} + (1-x)Z_{dB} , \qquad (15\text{-}53)$$

to obtain the bonding energy of the alloy. Then the heat of formation for the alloy is

$$E_{form} = -\frac{Z_d(1-Z_d/10)}{2} W_d + x\frac{Z_{dA}(1-Z_{dA}/10)}{2} W_{dA} + (1-x)\frac{Z_{dB}(1-Z_{dB}/10)}{2} W_{dB} \qquad (15\text{-}54)$$

per atom of the alloy.

Pettifor actually only addressed the case of the 50% alloy, for which Eq. (15-54) becomes

$$E_{form} = -\frac{Z_d(1-Z_d/10)}{2}\sqrt{3(\varepsilon_{dA} - \varepsilon_{dB})^2 + (5.53\hbar^2 r_d^3/mr_0^5)^2} + \frac{Z_{dA}(1-Z_{dA}/10)}{4} W_{dA} + \frac{Z_{dB}(1-Z_{dB}/10)}{4} W_{dB} . \qquad (15\text{-}55)$$

Pettifor expanded essentially this result for small differences in parameters for the two constituents, and discussed a "bond-order" part coming from the expressions which would be obtained with $\varepsilon_{dA} = \varepsilon_{dB}$, and a mismatch term from the difference. We do not make that expansion.

Values obtained from Eq. (15-55) and the data from Table 15-1 are shown in Figs. 15-17 and 18 for the systems for which data was collected by

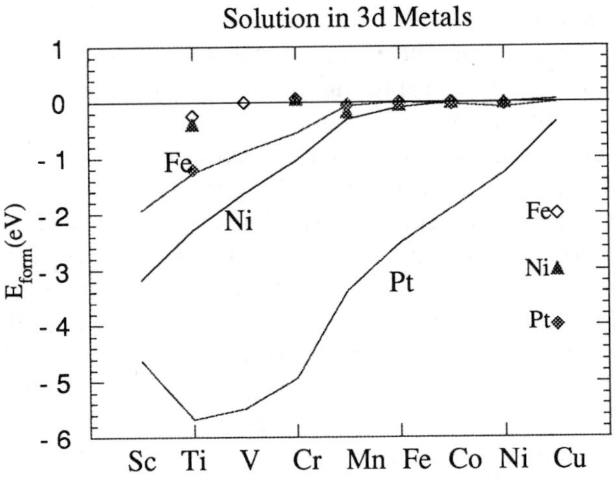

Fig. 15-17. Heats of Formation of 50% alloys, predicted by Eq. (15-55) and the parameters of Table 15-1. The points are experimental values collected by Watson and Bennett (1979) from references therein.

Watson and Bennett (1979). Surprising little data is available, and most theoretical results have been compared with Miedema's "semiempirical" values, which we choose not to do. We appear to have considerably overestimated the magnitudes of the heats of formation, but the trends with solvent (and with solute in the case of Fig. 15-17) are correct. Pettifor used different parameters, but our results are quite close to his, as expected, for iron in the 3d-metals. Watson and Bennett (1979), using a different

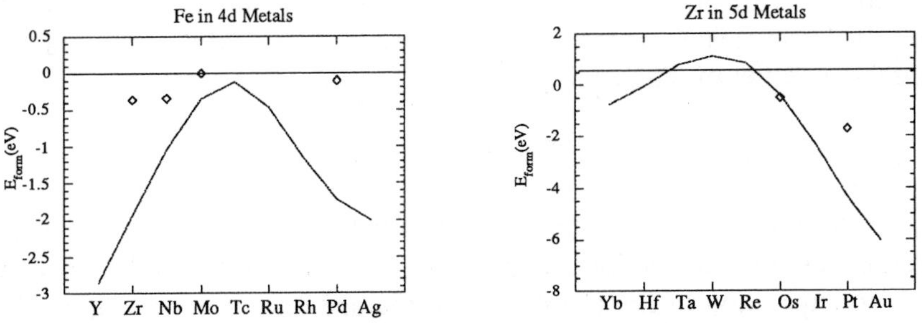

Fig. 15-18. Heats of Formation of 50% alloys, predicted by Eq. (15-55) and the parameters of Table 15-1. The points are experimental values collected by Watson and Bennett (1979) from references therein.

approach, predicted smaller values, by a factor of around two for alloys with iron, nickel and platinum, and therefore in better agreement with experiment. We regard Pettifor's theory as capturing the essential features of the alloying of transition metals, but not as very accurate.

We now return to the question of the dd-repulsion and can see that it is expected to add another negative term making the heat of formation larger and therefore in worse agreement with experiment. It is reasonable to expect that the dd-repulsion between an atom of type A and one of type B would be near the geometric mean of the AA and BB interactions, which we might write as A^2 and B^2, respectively, so that the AB interaction is AB. Then in the 50% alloy, one quarter of the interactions are A^2, on half AB and one quarter B^2, rather than one half A^2 and one half B^2 if the atoms are segregated. The change upon alloying is then $-\,{}^1/_4\,(A-B)^2$, and always negative. As for the cohesion, it will be a smaller effect than the bonding term.

We note that a similar theory could be constructed for alloys of transition metals (A) and simple metals (B). To the extent that the energy is dominated by the d-like terms, this energy is lost for each transition-metal atom which is introduced as a dilute solute in B. Similarly there is a loss in this bonding energy in A for every atom B substituted in the transition metal A. Thus we expect generally simple metals and transition metals to be quite insoluble in each other.

5. Interatomic Interactions and Elasticity

Wills and Harrison (1983) went on to rewrite the calculated total energy in terms of an effective two-body interaction which could be used to calculate the full elastic tensor, as well as properties of defects and the entire range of atomic properties of the metal.

A. The Effective Interaction

We have already in Chapter 13 written the simple-metal energy in terms of an interatomic interaction $-Z_s^2e^2\cosh^2\kappa r_c e^{-\kappa d}/d$, and the dd-repulsion as such an interaction in Eq. (15-43). It remained only to represent the bond term, Eq. (15-40), as an interatomic interaction. That bond energy is proportional to W_d, which from Eq. (15-32) is given by

$$W_d = \sqrt{\frac{12}{5}\frac{1}{N}\sum_{ij}(V_{dd\sigma}^2 + 2V_{dd\pi}^2 + 2V_{dd\delta}^2)}\,, \tag{15-56}$$

where $(1/N)\Sigma_i$ is equal to one for the undistorted crystal, and where we have replaced the coordination number X by a Σ_j , summing over nearest neighbors. The total bond energy for the N atoms in the crystal is the E_{bond} of Eq. (15-40) times N. The first variation of this, with respect to the change in a single nearest-neighbor distance, is equal to what it would be if the energy were represented by a nearest-neighbor interaction

$$V_{bond}(d) = -Z_d(1 - Z_d/10)\sqrt{12/X}\sqrt{{}^1/_5(V_{dd\sigma}^2 + 2V_{dd\pi}^2 + 2V_{dd\delta}^2)} \tag{15-57}$$

$$= -8.93\sqrt{12/X}\,Z_d(1 - Z_d/10)\frac{\hbar^2 r_d^3}{md^5}\,,$$

the form given by Wills and Harrison (1983).

This would seem to be the best representation of the total energy we can make in terms of interatomic interactions, but they noted that a straightforward expansion of the total energy in this way, to terms quadratic in the displacements, would give terms of the form $V_{bond}(d)^2/V_{bond}(d_0)$ not included in Eq. (15-57), and unphysical cross-terms between different displacements arbitrarily far from each other. The representation is inevitably approximate

It may be noted that this dependence upon coordination is as for a resonant bond, as we discussed in Section 10-2, where the total contribution of each electron is enhanced by a factor of the square-root of the coordination. Here the contribution to the total energy would be Eq. (15-57) times the coordination, leading to such a factor. The factor 12 which appears with it came from integration over a square density of states and is only accidentally equal to the number of neighbors X in a face-centered-cubic crystal.

The total effective interaction between atoms then consists of three terms,

$$V_{eff}(d) = Z_s^2 e^2 \cosh^2(\kappa r_c)\frac{e^{-\kappa d}}{d} + 22.80\,(Z - 1.5)\frac{\hbar^2 r_d^6}{md^8} \tag{15-58}$$

$$-8.93\sqrt{12/X}\,Z_d(1 - Z_d/10)\frac{\hbar^2 r_d^3}{md^5}\,.$$

There are also volume-dependent terms from the free-electron energy and from the d-states, but as with the simple metals most properties can be calculated without including them. Further, values for properties such as the bulk modulus will not ordinarily equal the values obtained before because of these volume-dependent terms and multibody interactions which are inherent in the full expression for the energy, but eliminated in Eq. (15-58).

Wills and Harrison (1983) used Eq. (15-58) to calculate the speed of longitudinal and transverse sound (at constant volume) and thereby deduce the elastic constants. They scaled the empty-core radius r_c in Eq. (15-58) to fit the observed bulk modulus $(c_{11} + 2c_{12})/3$ rather than to fit the equilibrium volume as we did for Table 15-1. However, they compared the r_c fit to the bulk modulus with values fit to the equilibrium volume and found them similar. The procedure was straightforward except for gold, for which parameters could not be fit without changing the Z_s to 2, similar to the problem we noted in Table 15-3, where the r_c^2 -repulsion needed to be made negative for gold in order to fit the equilibrium spacing.

B. Elastic Constants

The elastic constants which they obtained for the cubic metals with this r_c adjusted to give the correct bulk modulus are compared with experiment in Table 15-4. We would expect similar values for our choice of r_c. Furthest to the right is listed the Cauchy ratio, c_{12}/c_{44}. Had they made the corresponding fit without including the two final contributions in Eq. (15-58), they would have obtained values in excess of four for that ratio at the observed κr_0. The inclusion of the effects of the d-electrons brought the predicted values into good accord with experiment, indicating the importance of the d-electrons in the vibrational properties of the transition metals.

C. Multibody Forces

Of course representing the interaction between atoms in terms of two-body forces is always an approximation. Even in simple metals, going beyond the second order in the pseudopotential leads to multi-atom forces beyond the two-body interactions we obtained in Section 14-2-A. It may be interesting that it was even possible to obtain the asymptotic form of such interactions explicitly in the context of pseudopotentials. It was done (Harrison (1973a)) using the energy-wavenumber characteristic $F(q)$ with quantum screening, rather than the Fermi-Thomas screening which led to the simple exponential form of the interatomic interaction, Eq. (14-9). The

Table 15-4. Elastic constants of the cubic transition metals, in units of 10^{12} ergs/cm^3, (equal approximately to one megabar) calculated by Wills and Harrison (1983). Experimental values are mostly from Simmons and Wang (1971), and some from Kittel (1976).

	c_{11}		c_{12}		c_{44}		c_{12}/c_{44}	
	Theory	Expt.	Theory	Expt.	Theory	Expt.	Theory	Expt.
V	1.73	1.96	1.57	1.33	0.90	0.67	1.75	1.95
Cr	2.03		1.84		1.05		1.76	
Fe	1.78	2.28	1.64	1.33	0.85	1.11	1.93	1.20
Ni	2.09	2.44	1.76	1.58	0.68	1.02	2.57	1.55
Cu	1.51	1.70	1.30	1.15	0.46	0.61	2.83	1.88
Nb	1.79	2.46	1.66	1.39	1.24	0.29	1.34	4.73
Mo	2.84	4.75	2.67	1.67	1.81	1.08	1.34	1.55
Rh	3.14		2.49		1.28		1.95	
Pd	2.09	2.27	1.67	1.76	0.82	0.72	2.03	2.44
Ag	1.15	1.20	0.94	0.90	0.42	0.44	2.26	2.05
Ta	2.08	2.67	1.96	1.61	1.51	0.83	1.30	1.95
W	3.34	5.13	3.18	2.06	2.20	1.53	1.45	1.34
Ir	4.17	5.80	3.24	2.42	1.75	2.56	1.86	0.95
Pt	3.24	3.47	2.56	2.51	1.28	0.77	2.01	3.27
Au	2.01	1.94	1.59	1.61	0.78	0.47	2.03	3.40

n-atom interaction, at large distances, is found to be of the form

$$V_n \approx \frac{4}{\pi}\left(\frac{-3\pi Z}{4}\right)^n \sum_{paths} \frac{w_1 w_2 \dots w_n}{\varepsilon_F{}^{n-1}} \frac{\cos k_F(l_1 + l_2 + \dots l_n)}{(k_F l_1)(k_F l_2)\dots(k_F l_n) k_F(l_1 + l_2 + \dots l_n)} . \tag{15-59}$$

This is evaluated by constructing all sets of paths, each consisting of n straight-line segments of lengths l_j, connecting the n atoms. The w_j are the matrix elements of the pseudopotential connecting an electron state at the Fermi energy ε_F (measured from the band minimum) in the *j'th* segment to that in the $j+1st$ segment. There are $2(n-1)!$ paths, counting the two directions of traversing them.

The simplest case is of course the two-body interaction, with only the two segments $l_1 = l_2 = d$ which is

$$V_0(d) \approx \frac{9\pi Z^2}{8} \frac{w_2 k_F{}^2}{\varepsilon_F} \frac{\cos 2k_F d}{(k_F d)^3} , \tag{15-60}$$

which was given already in Harrison (1966). These show the familiar Friedel quantum oscillations which we saw in the screening field in Fig. 13-7.

The general multi-atom form, Eq. (15-59), has never proven very useful, nor even has the two-body form, Eq. (15-60). These oscillations are quite small and most properties are dominated by the short-distance contributions which we obtained for simple metals, Eq. (14-9), using Fermi-Thomas screening. On the other hand, it was noted in the earliest formulation (Harrison (1973a)) that the largest multi-atom correction is for three-body interactions when the three atoms are nearest neighbors lying on a line (so one of the w_j was - $^2/_3\varepsilon_F$ (for forward scattering). One might assume that this was the dominant term in structure determination, and that the important distinction between different elements was whether the hybridization contribution, the final term in Eq. (15-17), came from d-states above or below the Fermi energy. Then one is led to the correct assignment of hexagonal or cubic structures among the monovalent and divalent metals. This also suggested the correct assignment of high and low c/a ratios among the hexagonal elements. This suggestive finding has not been followed up. Our general feeling has been that these long-range limits of the interaction are not the important feature. They were eliminated for transition metals in taking Andersen's choice of muffin-tin zero in going from Eq. (15-19) to Eq. (15-20).

The form of multi-ion interactions in transition metals which arises without that choice of muffin-tin zero has been derived by Moriarty (1988) using a first-principles generalized pseudopotential theory. It leads to analytic three-body and four-body potentials (Moriarty (1990, 1994)). They were applied to molybdenum for treating cohesive, structural, elastic, vibrational, thermal, and melting properties (Moriarty (1994) and more recently (Xu and Moriarty (1997)) to the study of vacancy and interstitial formation and migration, and screw dislocations.

6. Ferromagnetism

One of the terms which entered the free-electron-energy in Eq. (15-46) was the exchange energy. Its origin is the antisymmetry of electrons under interchange, which we described in Section 13-4-B. This had the consequence that the wavefunction for electrons of the same spin is zero for the two electrons being at the same point, and the average Coulomb repulsion between them is thereby reduced. We gave in Eq. (13-16) this reduction, the exchange energy, for a free-electron gas.

Another consequence of this exchange energy is that if we have a partly-filled shell (orbitals of the same energy) in an atom or molecule, the energy will be lowered if states of the *same* spin are filled, called Hund's Rule. Thus in the oxygen molecule, with two electrons in the π^* state (See Fig. 1-7 and Problem 1-2), the ground state will have one electron in the π^*_y state and one in the π^*_z state, with parallel spin.

This same exchange energy would favor transferring electrons to the same spin in a metal, but this would also cost kinetic energy in shifting electrons from below the Fermi energy of one spin to above the Fermi energy of the other and the exchange energy is ordinarily small enough that the net energy change is positive. Both changes are proportional to the square of the number transferred. The exchange energy is proportional to e^2/r_0 and the kinetic energy to $\hbar^2/(mr_0^2)$ and the exchange energy would only win for smaller densities (larger r_0) than occur in any simple metal, slightly smaller than the electron density in cesium.

A. The Exchange Energy

In a transition metal, with its much higher d-band density of states, the change in band energy required to shift electrons between spins is much smaller and this exchange energy becomes competitive. For the transition metal we should take the atomic view of the d-states and the exchange energy U_x is the lowering in the energy of interaction between two d-states on the same atom when their spins are made parallel. It can be evaluated from an integral over the two states,

$$U_x = \iint d^3r\, d^3r'\, \psi_1(\mathbf{r})^* \psi_2(\mathbf{r}) \frac{e^2}{|\mathbf{r} - \mathbf{r}'|} \psi_2(\mathbf{r}')^* \psi_1\mathbf{r}') . \tag{15-61}$$

This is in fact exactly the form, with the $\psi_j(\mathbf{r})$ replaced by plane-wave states, which was evaluated to give the free-electron exchange given in Eq. (13-16). It is similar to the integral which gives the Coulomb interaction U_d between two atomic states which was listed in Table 15-1. However, U_x is much smaller since for U_d the wavefunctions entered as $\psi_1(\mathbf{r})^* \psi_1(\mathbf{r})$ $\psi_2(\mathbf{r}')^* \psi_2(\mathbf{r}')$ which is positive for all $\mathbf{r}$ and $\mathbf{r}'$. In Eq. (15-61) the integrand changes sign as a function of position and much smaller values are found. We shall take our U_x values from experiment but follow this argument a little more quantitatively in order to more clearly understand the value.

For the same reason, a fluctuating sign of the integrand, the exchange energy in a free-electron gas is much smaller than the direct interaction between an electron and the uniform distribution of electrons in that gas. That direct interaction can be estimated as the potential $^3/_2\, e^2/r_0$ at the center of a uniform sphere of radius r_0 if $4\pi r_0{}^3/3$ is the volume per electron (as in our discussion of the Madelung energy of Eq. (13-15)). This gives us an estimate of the ratio U_x/U for a free-electron gas and thus an estimate for the atom. The exchange energy of Eq. (13-16) (with $Z = 1$ so r_0 is the radius of the electron sphere) is combined with Eq. (13-2) for the Fermi wavenumber, and divided by the direct $^3/_2\, e^2/r_0$ to obtain

$$\frac{U_x}{U} = \frac{1}{2\pi}\left(\frac{9\pi}{4}\right)^{1/3} = 0.305 \qquad (15\text{-}62)$$

for the free-electron gas. Since the ratio does not depend upon density, one might expect it to apply for the atom, though the U_d defined from the atom is really not the integral for the total Coulomb energy, but the energy change of a d-orbital as an electron is moved from the s-orbital to another d-orbital. It is preferable to use U_x values obtained in Harrison (1980) from the atomic spectra (Moore (1949, 1952)) by comparing the energies of atoms with different numbers of d-state spins aligned. These are listed in Table 15-5. However, it is interesting to compare these with values obtained from Eq. (15-62) and the values of U_d from Table 15-1; these appear in parentheses in Table 15-5, and are very roughly in accord, though Eq. (15-62) generally overestimates U_x even though U_d is smaller than full U which appears in that equation.

B. The Ferromagnetic State

To determine whether ferromagnetism is expected, we consider the spin-up and spin-down bands separately, treating each with the Friedel model, following Harrison (1980). We fix the total number of d-electrons at $Z_d = Z - 1.5$ as before, allowing a fraction $1/2 + x$ to be of spin up and a fraction $^1/_2 - x$ to be of spin down. The band energy, or bonding energy, is recalculated separately as in Eq. (15-40) for each spin. (Now the fractional filling is $Z_{d\pm}/5$ rather than $Z_d/10$.) Of course the linear term cancels when we add the two to obtain

$$E_{bond} = -\frac{Z_d(1-Z_d/10)}{2}W_d + \frac{Z_d^2}{5}W_d x^2\,. \qquad (15\text{-}63)$$

Table 15-5. Values for U_x obtained (Harrison(1980)) from atomic spectra (Moore (1949, 1952)) in eV, values for U_x obtained from U_d of Table 15-1 and the free-electron ratio, and values of $W_d/5$ from Table 15-1. Systems with $U_x > W_d / 5$ are predicted to be ferromagnetic.

Z	4	5	6	7	8	9	10
3d Series	Ti	V	Cr	Mn	Fe	Co	Ni
U_x	0.90	0.68	0.64	0.78	0.76	1.02	1.60
Eq. (15-62)	(1.65)	(1.62)	(1.56)	(1.71)	(1.80)	(1.92)	(1.99)
$W_d/5$	1.25	1.37	1.35	1.16	0.98	0.88	0.76
4d Series	Zr	Nb	Mo	Tc	Ru	Rh	Pd
U_x	0.63	0.48	0.60	-	0.88	0.81	-
Eq. (15-62)	(0.73)	(0.73)	(0.92)	(1.04)	(1.13)	(1.22)	(1.31)
$W_d/5$	1.69	2.00	2.00	1.90	1.70	1.39	1.1
5d Series	Hf	Ta	W	Re	Os	Ir	Pt
U_x	0.70	0.60	0.39	0.36	0.61	0.86	-
Eq. (15-62)	(0.86)	(0.92)	(0.95)	(0.98)	(1.04)	(1.07)	(1.07)
$W_d/5$	1.93	2.29	2.30	2.27	2.08	1.79	1.44

We are neglecting for the moment any effects of correlation, which we return to at the end.

We may also compute the exchange energy, and of course it involves only electrons of the same spin. For n electrons of a given spin we obtain the exchange energy from $-\int_{0,n} nU_x dn$, For spin up there are $(1/2 + x)Z_d$ electrons for an exchange energy of $-(1/2 + x)^2 Z_d{}^2 U_x/2$. Adding for spin down gives an exchange energy of

$$E_{exch} = -Z_d^2 U_x\left(\frac{1}{4} + x^2\right). \tag{15-64}$$

If the coefficient of the x^2 term in Eq. (15-64) exceeds that in Eq. (15-63) the system becomes unstable against transferring spins, forming a ferromagnetic state. This state should be formed if

$$U_x > W_2/5 \tag{15-65}$$

Comparing W_d with the U_x values obtained from the spectroscopic data, we see that the criterion is only close to being satisfied to the right in the 3d-series, where in fact the only magnetic behavior is found experimentally among the transition metals. Strictly speaking we predict ferromagnetism only for cobalt and nickel, though iron is also of course ferromagnetic. The two neighbors to the left, manganese and chromium also show magnetic behavior, but they are antiferromagnetic (magnetic moments alternating from atom to atom). We saw how this arose from the particular Fermi surface in the case of chromium, and the criteria for an instability would be quite different in these two cases.

In each case x will of course continue to increase until the majority band is filled with 5 d-electrons. We saw in Section 15-2-D that the real occupation of the d-like bands (the relevant measure here), as opposed to the weighted average of d-orbital content, differs from Z by one or less, so we expect an imbalance of spin, or a magnetic moment per atom, of less than one in nickel, two in cobalt, and three in iron. The observed moment (in units of the magnetic moment of the electron) determined from the saturation magnetization, is 0.6 in Ni, 1.5 in Co, and some 2.2 in Fe. There are orbital contributions to the moment, arising from spin-orbit coupling, so this crude agreement is satisfactory. The measured moments in antiferromagnetic manganese and chromium are smaller which could be explained by a half-filling of the sub-bands due to a split density of states such as we have seen for bcc metals (suggested by Friedel (1969) and described in Harrison (1980), p. 525). As we look further at systems without ferromagnetic order, this suggestion seems more and more plausible.

It is well known that ferromagnets lose their magnetism above a critical temperature called the Curie temperature. One might at first think that this corresponded to a disappearance of unequal occupation of spin states on each atom, but in fact it is ordinarily to be associated with a disordering of the magnetic moments on different atoms, with the individual moments remaining. This can in fact be seen in the observed magnetic susceptibility above the Curie temperature, which indicates the existence of moments which can be partially ordered by an applied field.

Actually, the electronic structure of this disordered system is easier to understand in terms of our approximate description of electronic structure, than in the full band theory. The U_x which causes ferromagnetism is basically an intra-atomic parameter and the band width W_d which inhibits ferromagnetism was obtained from a sum of squared nearest-neighbor couplings, which does not depend upon order. Thus we can imagine Eq. (15-

65) as being applicable also to a disordered state. Then Friedel's explanation for the reduced magnetic moment in antiferromagnetic manganese and chromium could be understood in the same terms.

Finally, we should note that it would have been more appropriate in Eq. (15-63) to take into account the correlated motion of the electrons by replacing W_d by $\sqrt{W_d^2 + U_d^2} - U_d$, as we did in the analysis of the total energy. This is not obvious, but can be derived by carrying out the analysis we make in Section 16-1 separately for the spin-up and spin-down states, with interatomic coupling only between states of the same spin, and exchange energy associated with each set. This requires a generalization of Eq. (16-4) for the inclusion of repulsions U_f or U_d in which the $^{13}/_{14}$ for all f-levels is replaced by $^{6}/_{7}$ for those of one spin, or $^{9}/_{10}$ replaced by $^{4}/_{5}$ for d-states. Then the total energy of the system with equal numbers of each spin is seen to become a maximum, rather than a minimum, when $U_x > {}^{1}/_{7}(\sqrt{W_f^2 + U_f^2} - U_f)$ for f-states (this was simplified by replacing $U_f - U_x$ by U_f on the right side) and

$$U_x > \frac{1}{5}\left(\sqrt{W_d^2 + U_d^2} - U_d\right) \tag{15-66}$$

for d-states.

This weakens the condition, Eq. (15-65), for the formation of the ferromagnetic state, particularly as seen from Table 15-1 at the right end of the 3d-series. Such criteria based upon the Friedel model are sufficiently crude that we did not complicate the discussion with these correlations, but their inclusion is seen to favor ferromagnetism. It quantifies the intuitive view that the formation of the correlated state reduces the effect of the interatomic coupling which often prevents the same spin alignment in solids as occurs in atoms. We shall discuss this further in Section 16-5 when we consider the formation of a correlated state for an isolated f-shell-metal atom, and derive a condition which depends only upon the hybridization with free-electron states and upon the Coulomb energy U_f. In this case the *principal* driving force for the formation of the state is not exchange, U_x, but once the state is formed the spins will align because of U_x.

Problem 15-1. d-bands

a) Obtain the formulae for d-bands in a simple-cubic lattice with nearest-neighbor interactions only, and $\boldsymbol{k}$ in a [100] direction, in analogy with Eq. (15-22) for fcc structures.

b)Plot them using parameters for manganese, assuming the same r_0 as in Table 15-1, but a simple-cubic structure.

Problem 15-2. The Friedel Model

a) Obtain the band width W_d for the Friedel model from the second moment for the simple cubic manganese of Problem 15-1, to compare with the W_d given in Table 15-1.

b) Find the d-state energy from Eq. (15-35) for this system, to compare with the value plotted in Fig. 15-10.

Problem 15-2. Bands for Nb_3Sn.

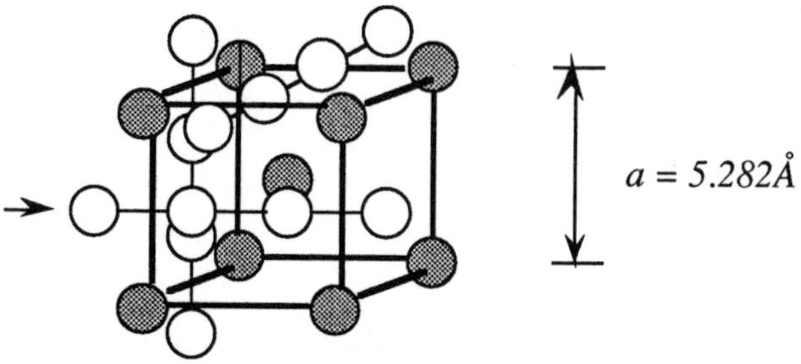

Nb_3Sn used to be the highest-T_c superconductor. It has the A15 structure shown above. The shaded atoms are tin, forming a body-centered-cubic lattice with cube edge a . Across the front face of this unit cube is a horizontal chain of niobium atoms (unshaded), spaced by $a/2$, indicated by an arrow. A parallel line on the back face is not shown. There is a vertical chain shown on the left face of the cube; the one on the right is not shown. A chain along the top face is shown, that on the bottom face is not. The niobium atoms in each chain are much more closely spaced than other atoms and the dd-coupling between them dominates the band structure.

a) Sketch the corresponding bands (relative to ε_d) along a [100] direction including only nearest-neighbor dd-couplings from Eqs. (15-20) and (15-21) with r_d from Table 15-1. Let $0 < k < 2\pi/a$, though really these should be folded back to π/a in the cubic Brillouin Zone.

b) Indicate the degeneracy of each of the bands, fifteen in total for the three orientations of chains. This is a very simple problem once you see what you are doing. The model is crude, but has proven conceptually useful. [The full bands have been calculated by Mattheiss (1975).]

CHAPTER 16

f-Shell Metals

The argument for small spatial extent of d-orbitals because of their two units of angular momentum, which we gave at the beginning of Chapter 15, applies even more strongly to the f-states with their three units of angular momentum. We shall see in Section 16-2 that all of the d-state parameters which we have introduced in Chapter 15 generalize directly to the f-states. We must note first, however, a qualitative change in the electronic behavior for f-shells caused be the Coulomb repulsion between them, which also arises from this extreme compression of the orbitals close to the nucleus: the electrons often behave as if they were localized and did not form bands at all.

1. Correlated States

This localized behavior is already understandable in terms of the simplest molecule, Li_2 , which we introduced in Section 1-2. The question arose in the first applications of quantum theory to molecules in which different workers utilized quite different forms of solutions. The molecular orbitals which we have been using were referred to as the Hund-Mulliken-Bloch form, but an alternative formulation was based upon electrons in individual atomic orbitals and is referred to as the Heitler-London scheme. It is this latter scheme which becomes newly relevant in many f-shell metals.

For Li_2 we obtained a bonding state which is occupied by an electron of spin up and one of spin down. With each equally shared by the two atoms there is a fifty percent chance of both electrons being on the same atom at the same time. We could retain that solution as the two atoms are taken far apart, so that $V_{ss\sigma}$ becomes very small, but then this becomes a high-energy solution for the two electrons, with an energy about $U/2$ higher than if they were each placed in an atomic state on a different atom. We can then be sure that in the real ground state the two electrons will localize on separate atoms, as suggested in the Heitler-London scheme. We now obtain an exact solution for this problem at intermediate spacing, following Harrison (1984), and see how we can use it in a way which will enable us to treat the f-shell metals.

A. An Exact Solution

In order to describe the interacting system, we must add the Coulomb repulsion U , defined now to be the extra energy to place both electrons on the same atom rather than on separate atoms, to our Hamiltonian. Because of this added term which couples the two electrons we can no longer write the state of the system as two independent (one-electron) solutions, nor the energy as the sum of the one-electron energies. We can, however, obtain an exact two-electron solution within this context of a single atomic orbital on each atom. In that context the electronic states can be written as linear combinations of six *two-electron* basis states. One basis state is with a spin-up electron on atom one and a spin-down electron on atom two, which we write $(1\uparrow,2\downarrow)$. Another is $(1\downarrow,2\uparrow)$. There are also two basis states with both electrons on the same atom, $(1\uparrow,1\downarrow)$ and $(2\uparrow,2\downarrow)$ and two basis states with parallel spin, $(1\uparrow,2\uparrow)$ and $(1\downarrow,2\downarrow)$. These six basis states have energies $2\varepsilon_s$, $2\varepsilon_s$, $2\varepsilon_s + U$, $2\varepsilon_s + U$, $2\varepsilon_s$, and $2\varepsilon_s$, respectively, before introducing the $V_{ss\sigma}$. In addition, the first basis state is coupled to the third and fourth by $V_{ss\sigma}$ (which of course couples individual electron states of the same spin on the two atoms) and so also is the second basis state couple to the third and fourth by $V_{ss\sigma}$. All other states are uncoupled. The corresponding six-by-six Hamiltonian matrix can be solved easily. The last two basis states, with parallel spin, are eigenstates with energy $2\varepsilon_s$, uncoupled to the other basis states or each other. We could think of these states as having one electron on each atom or one in a bonding state and the other in an antibonding state of the same spin. (When the state is written out, terms with both electrons on the same atom cancel as the Pauli Principle tells us they must.)

The remaining four eigenstates are even and odd combinations of the remaining four basis states. Of most interest is the ground state, which will be even. One even combination is: $[(1\uparrow,1\downarrow) + (2\uparrow,2\downarrow)]/\sqrt{2}$, with energy of $2\varepsilon_s + U$. The other is: $[((1\uparrow,2\downarrow) + (1\downarrow,2\uparrow)]/\sqrt{2}$ with energy $2\varepsilon_s$. We may verify that they are coupled by $2V_{ss\sigma}$ and solve the quadratic equation for the two even eigenstates as

$$\varepsilon_{\pm} = 2\varepsilon_s + \frac{U}{2} \pm \sqrt{\left(\frac{U}{2}\right)^2 + 4V_{ss\sigma}^2} \; . \qquad (16\text{-}1)$$

The minus sign gives the ground state. Note that if U is neglected it gives $2(\varepsilon_s + V_{ss\sigma})$, with $V_{ss\sigma}$ negative, the solution we obtained in Chapter 1. If on the other hand, $V_{ss\sigma}$ becomes very small, the energy approaches $2\varepsilon_s$. The electrons indeed separate onto different atoms. Furthermore, the energy and the state vary smoothly between the two limits as the atoms are separated from each other. Because of the smooth variation, over the entire range of $V_{ss\sigma}/U$, all states are correlated to some extent but the correlations only become important when U is of order or larger than $V_{ss\sigma}$. [In passing we note that in addition to the ground state there is a high-energy state obtained with the plus in Eq. (16-1). There is also one odd state with energy $2\varepsilon_s$ which, with the two parallel-spin states mentioned before, form a triplet, three states of the same energy. There is also an odd state with energy $2\varepsilon_s + U$ We are only interested here in the ground state.]

We shall generalize this ground state directly to transition metals, and f-shell metals, soon, and use it in our analysis. However, it may be useful first to make an approximate description in terms of the Unrestricted Hartree-Fock approximation. This is partly to clarify the nature of the correlated state, and partly to place the theory in the context of other analyses. We should not forget in the course of this that it is the exact solution (with the limited basis set) which we use in our treatment of the total energy and bonding properties.

B. The Unrestricted Hartree-Fock Approximation.

We consider now approximate one-electron states as we have for other systems, but let them break symmetry and each become more heavily weighted on a different atom. Thus we rewrite the molecular orbital of Eq. (1-4) for spin up with $u_1 = \cos\gamma$ and $u_2 = \sin\gamma$ (using the cosine and sine so the state is normalized). We obtain $\cos\gamma|s_1\uparrow\rangle + \sin\gamma|s_2\uparrow\rangle$. The state with spin down is then written $\sin\gamma\,|s_1\downarrow\rangle + \cos\gamma\,|s_2\downarrow\rangle$. The two are orthogonal because of opposite spin states. The expectation value of the energy, for our two-level model, for the corresponding two-electron state is

$$E(\gamma) = 2\varepsilon_s + 2U\sin^2\gamma\cos^2\gamma + 4V_{ss\sigma}\sin\gamma\cos\gamma. \qquad (16\text{-}2)$$

It is plotted in Fig 16-1(a) for two choices of parameters. The minimum energy of this one-electron solution is our best estimate of the ground-state energy for this form of state. We see that when $V_{ss\sigma}/U$ is large, there is one minimum for $0 < \gamma < \pi/2$. It is at $\gamma = \pi/4$ and is exactly the solution $u_1 = u_2 = 1/\sqrt{2}$, but has a higher energy by $U/2$ because of the repulsion, as we would expect. Once $2V_{ss\sigma}/U$ is less than one, this solution has maximum energy and there are new minima as seen in Fig. 16-1(a), at $\sin 2\gamma = -2V_{ss\sigma}/U$, corresponding to excess spin up to the right, the other to the left, or the other way around.

The resulting energy of the state is plotted in Fig. 16-1(b) as UHF (Unrestricted Hartree-Fock) as a function of $U/|V_{ss\sigma}|$. Since $|V_{ss\sigma}|$ decreases with increasing spacing, we may think of this as a plot as a function of the distance between atom . For $U/|V_{ss\sigma}|$ less than two the energy is equal to the symmetric Hartree-Fock solution, which it joins with continuous slope at $U/|V_{ss\sigma}| = 2$. Above $U/|V_{ss\sigma}| = 2$ the Hartree-Fock solution rises to $U/2$, but the unrestricted solution stays below zero, as does the exact solution, Eq.

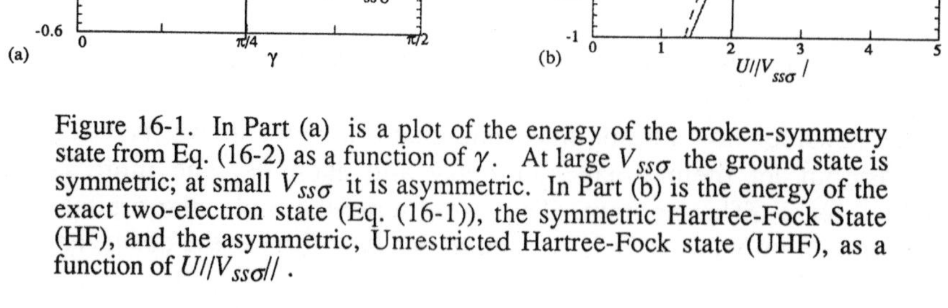

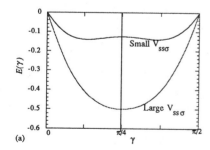

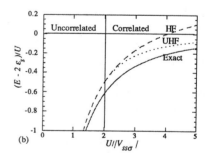

Figure 16-1. In Part (a) is a plot of the energy of the broken-symmetry state from Eq. (16-2) as a function of γ. At large $V_{ss\sigma}$ the ground state is symmetric; at small $V_{ss\sigma}$ it is asymmetric. In Part (b) is the energy of the exact two-electron state (Eq. (16-1)), the symmetric Hartree-Fock State (HF), and the asymmetric, Unrestricted Hartree-Fock state (UHF), as a function of $U/|V_{ss\sigma}|$.

(16-1), also shown in the figure. Thus the UHF solution rectifies the most serious error of the Hartree-Fock theory.

It is nevertheless only an approximate solution. At large spacing (large $U/|V_{ss\sigma}|$) the difference between E and $2\varepsilon_s$ is only half as large as the exact solution. At small spacing, it also lies above the exact solution because it neglects the correlated motion of the electrons. Furthermore, it divides systems sharply into uncorrelated and correlated states, indicated by the vertical line in Fig. 16-1(b), an artificial distinction.

Here we shall use the UHF approximation when we are far from the dividing line and in the discussion of isolated f-shell, or transition-metal, atoms and in the discussion of magnetism in terms of local moments. For the bonding properties we shall improve upon the Unrestricted Hartree-Fock approximation by generalizing the exact result to the transition-metal and f-shell-metal systems.

C. Self-Interactions

Before doing that we should put both analyses in perspective with other approaches. We note that the exchange energy which we discussed in the preceding section on ferromagnetism did not enter the calculation here. Here there were no electrons of the same spin in the ground state. That is physically correct, but the question becomes confused in Hartree-Fock theory in which it is customary to add an artificial direct interaction potential between each electron and its own charge distribution. At the same time one adds an artificial exchange interaction, Eq. (15-61) of each electron with itself. We may see from Eq. (15-61) that this exchange integral becomes exactly equal to the direct interaction when ψ_1 is equal to ψ_2 , so $-U_x$ cancels the direct term and no error is made. This procedure has mathematical advantage in that then the Schroedinger Equation for every electron is the same, and eigenstates are automatically orthogonal. It would not affect the result of the Unrestricted Hartree-Fock calculation of Eq. (16-2), but the direct potential would have been symmetric and the asymmetry would have arisen in the asymmetric exchange potential so that it would have appeared that exchange had caused the asymmetry.

The real problem arises when one goes to the local-density theory which is almost universally used in full electronic-structure calculations. Then the exchange (which is really an integral operator in the Schroedinger Equation), and correlation (all corrections to Hartree-Fock theory) are replaced by a simple potential, which is a function of electron density. This potential no longer exactly cancels the artificial direct self-interaction which was introduced. It would be possible to again allow, and obtain, asymmetric solutions but here also there is a difficulty when there are more electrons than in our simple Li_2 example. We have seen in Section 15-6 that the direct interaction U between two different orbitals is much larger than the exchange interaction U_x between those orbitals. But, for the specific case of self-interactions the exchange interaction should equal the direct interaction. Local-density theory lumps all of these exchange interactions together and thus dilutes the effect of self-exchange and underestimates the driving force for the formation of a correlated state. This is the root of much of the difficulty of contemporary calculations in describing highly correlated systems.

A considerable number of workers (Cowan (1967), Lindgren (1971), Bryant and Mahan (1978), Perdew (1979), Zunger, Perdew, and Oliver (1980), Perdew and Zunger (1981), Svane and Gunnarsson (1988)) sought to rectify this difficulty with a *Self-Interaction Correction* (SIC). The approach is basically carrying out a local-density calculation, allowing different spin densities as we have done, and then correcting for the effects of the self-interactions. In the course of this, the orthogonality of the states is required as a variational condition. (See Svane and Gunnarsson (1988).) This can be a correct approach, and is presumably on the path to successively more accurate solutions. However, it is not relevant to our approximate approach where we omit all self-interactions throughout. Questions such as nonorthogonality are dealt with separately.

D. Generalization to d- and f-Bands

When generalizing the two-level problem to f-shell and transition metals, there are two quite distinct questions. One concerns an isolated f-shell, or d-shell, atom in a simple metal, and the other is the pure f-shell or d-shell metal. We describe briefly the isolated atom first, using Unrestricted Hartree Fock. We then return in Section 16-5 to a generalization of the total energy from the exact Eq. (16-1) to the pure f-shell metal and explore the bonding properties.

a. Local Moments in the UHF Approximation. Anderson (1961) used the Unrestricted Hartree-Fock approximation of Part B to see how a local moment could arise on an isolated transition-metal atom in a metal. He imagined a local atomic state, which we call $|d\rangle$ though he took it to be a single state, rather than a five-fold degenerate level. The energy for a single electron in the level, ε_d , lies well below the Fermi energy of the simple metal so it is occupied, but the energy $\varepsilon_d + U$ for an additional electron lies well above the Fermi energy so the level is only *singly* occupied. With no hybridization between the d-level and the free-electron states, this is the ground state of the system and there is a local moment associated with the spin of the electron occupying the d-state. It could be reoriented by a magnetic field.

Anderson included hybridization between the d-state and the free-electron states, which replaced the local level by a "scattering resonance", as we shall describe in Section 16-5, where we also generalize the model to a degenerate set of levels appropriate to a real atomic d- or f-level. We combined this approach (Harrison (1984)) with the Unrestricted Hartree-Fock method, allowing some of the f-level resonances to lie below the Fermi energy and others to lie above, to describe the formation of local moments, or the "localization" of f-electrons. This also is described in Section 16-5. It allows the treatment of electron states in individual transition-metal, or f-shell-metal, atoms in a metal. It is essentially treating the set of f-levels as a two-level (occupied and unoccupied) system, and then applying the Unrestricted Hartree-Fock approximation of Part B of this Section.

In Section 16-5 we shall find a criterion, corresponding to the vertical line in Fig. 16-1(b), for the formation of a local moment in the Unrestricted Hartree-Fock approximation. To the left of that line, the state is uncorrelated as in our treatment of the d-bands in Chapter 15. To the right, the states will have separated into occupied and unoccupied orbitals. Once they are separated we could imagine one such moment forming on each atom in a pure f-shell metal and including the coupling between orbitals on adjacent atoms (which was done in perturbation theory in Harrison (1984)), providing the effects of energy-band formation in the correlated state. This would provide what we regard as a physically correct description of the electronic structure of f-shell metals, although the artificial discontinuity (in second derivative of energy with respect to coupling) is not real.

b. Generalizing the Exact Result. We note instead that the two sets of f-levels, occupied and unoccupied, their coupling, and the Coulomb repulsion U, are so closely analogous to the molecular two-level system of Part A, that we may generalize the *exact* solution for that problem, rather than the Unrestricted Hartree-Fock solution, to the f-shell metals. This is the way we shall proceed. This incorporates, and improves upon, the corrections to the Unrestricted Hartree-Fock approximation made in perturbation theory by Harrison (1984), and eliminates the artificial separation into two types of solutions, with all solutions correlated in varying degrees.

We look first for a generalization of Eq. (16-1) to the f-shell metals. When the Coulomb U is negligible, the ground-state energy of Eq. (16-1) becomes $2\varepsilon_s - 2|V_{ss\sigma}|$. In our treatment of d-levels in terms of the Friedel Model, neglecting Coulomb correlations, in Section 15-4, the d-band width W_d played the role of the coupling $V_{ss\sigma}$ of Part A. Where the energy gain per atom in Eq. (16-1) (when the Coulomb repulsion U is not playing an important role) was $-|V_{ss\sigma}|$, it is $-Z_d(10-Z_d)W_d/20$ for the d-bands from Eq. (15-40). For the f-bands (seven orbitals rather than five per atom) it will be

$$E_{bond} = -\frac{Z_f(14-Z_f)W_f}{28} \tag{16-3}$$

based upon f-shell parameters, Z_f and W_f analogous to Z_d and W_d, which we shall obtain in the following section.

When these coupling terms are very large, the effect of the Coulomb term under the square root in Eq. (16-1) goes to zero as $2U^2/(8V_{ss\sigma})$, but the leading Coulomb term remains. We now choose it to represent the effects of the Coulomb interaction - relative to that in the free atom - when the effects of the couplings $V_{\ell\ell m}$ are very large. In the atom the Coulomb energy is $Z_f(Z_f - 1)U_f/2$ for the interaction of the Z_f electrons with each other. The U_f is of course the Coulomb repulsion associated with the f-states, which will be different for the d-states on the same atom. In the energy bands of a solid all of the f-orbitals are equally occupied, as in the molecular state of Part A, and an electron in any of the fourteen orbitals sees the charge of $Z_f/14$ electrons in each of the other thirteen orbitals. (We again leave out self-interactions.) Again, counting each interaction only once we obtain a Coulomb energy for the metallic bands of $Z_f(13Z_f/14)U_f/2$. Thus the change in Coulomb energy in the formation of the bands is

$$E_{coul} = Z_f\frac{13Z_f}{14}\frac{U_f}{2} - Z_f(Z_f - 1)\frac{U_f}{2} = \frac{Z_f(14 - Z_f)\,U_f}{28}, \tag{16-4}$$

which replaces the leading $U/2$ in Eq. (16-1).

Now when the effects of the coupling become very small, the energy relative to that of the isolated atoms, with Z_f f-electrons each, must go to zero, so the $U/2$ under the square root in Eq. (16-1) must also go to Eq. (16-4). Thus the energy per f-shell atom, relative to the free atom, is generalized from the exact Eq. (16-1) as

$$E_{bond} = -\frac{Z_f(14 - Z_f)}{28}\left(\sqrt{W_f^2 + U_f^2} - U_f\right) \qquad (16\text{-}5)$$

This remarkably simple result, like Eq. (16-1), should be more accurate than the corresponding Unrestricted Hartree-Fock result of Harrison (1984) for all ratios of U_f to W_f. It was chosen to give the correct leading terms (in the Friedel Model) when W_f/U is small and when it is large. We have called it E_{bond} since it replaces, for f-states, the E_{bond} which we gave for the transition metals in Eq. (15-40). It did not come from a variational calculation, as our other formulae have, since the variational calculation requires a choice of a form for the wavefunction which leads to errors such as those of Unrestricted Hartree-Fock theory, illustrated in Fig. 16-1(b).

In our prediction of the cohesive energies and volume-dependent properties for the transition metal we similarly replaced Eq. (15-40) by the d-state counterpart of Eq. (16-5), which is clearly

$$E_{bond} = -\frac{Z_d(10 - Z_d)}{20}\left(\sqrt{W_d^2 + U_d^2} - U_d\right). \qquad (16\text{-}6)$$

Since we are using this more accurate form for the energy, we will not have occasion to use the counterpart of the Unrestricted Hartree-Fock criterion $U > 2|V_{ss\sigma}|$ for the formation of a correlated state, indicated in Fig. 16-1. However, we will wish to compare such a criterion for energy bands with that which we calculate for a single atomic resonance in Section 16-5, and also when we discuss transition metal compounds in Chapter 17. We may obtain such a criterion by comparing the "exact" energy for the bands from Eq. (16-5) or (16-6) with that for the diatomic molecule in Eq. (16-1). For the molecule, the Unrestricted Hartree-Fock correlated state occurred when the Coulomb term under the square root in Eq. (16-1) was one quarter of the coupling term $4V_{ss\sigma}^2$. Generalizing that to Eqs. (16-5) and (16-6) suggests that the correlated state arises in the band structure if the corresponding U exceeds one half of the corresponding band width, $U_f > {}^1/_2 W_f$ or $U_d > {}^1/_2 W_d$.

2. Parameters for f-Shell Systems

In order to proceed we will need the parameters describing the hybridization between |f> states and free-electron states, and coupling between f-levels, which are analogous to the corresponding parameters for d-levels. We first obtain the formulae for these parameters and then give values for both the rare earths and the actinides.

A. ff-Coupling

We are now extending our basis for the states to include atomic f-states, $2\ell + 1 = 7$ levels of each spin. We obtain their hybridization with plane-wave states by expanding the plane waves in spherical harmonics, Eq. (15-9), and obtaining the matrix element with the states of the same symmetry as in Eq. (15-10). Now, however, the expansion of the spherical Bessel function for small r is $j_3(kr) \approx k^3r^3/(7{\cdot}5{\cdot}3{\cdot}1)$. The extra factor of k shows up in the hybridization matrix element which is written, in analogy with Eq. (15-11), as

$$<<k|H|f>= <k|\delta V - <d|\delta V|f>|f> = \sqrt{\frac{4\pi r_f^3}{3\Omega}}\, kr_f \frac{\hbar^2k^2}{m} Y_3^m(\theta_k,\phi_k)e^{-i\mathbf{k}\cdot\mathbf{r}_j}. \tag{16-7}$$

An integral for the f-state radius r_f , analogous to Eq. (15-12), can also be written down, but it is preferable to obtain it as we did for d-states by fitting more complete calculations of the band width, or from the Atomic Surface Method.

We may next obtain the coupling between f-states on neighboring atoms as

$$<f'|H|f> = \sum_k \frac{<f'|H|k>> <<k|H|f>}{\varepsilon_f - \varepsilon_k} \tag{16-8}$$

in analogy with the calculation in Section 15-2-A, to obtain

$$V_{ffm} = \eta_{ffm} \frac{\hbar^2 r_f{}^5}{md^7}, \tag{16-9}$$

with (Harrison (1983d)), also obtainable from Eq. (16-17) here,

$$\eta_{ff\sigma} = 20\frac{525}{2\pi},\ \eta_{ff\pi} = -15\frac{525}{2\pi}$$
$$\eta_{ff\delta} = 6\frac{525}{2\pi},\ \eta_{ff\phi} = -\frac{525}{2\pi}. \tag{16-10}$$

with of course σ, π, δ, and ϕ signifying $m = 0, 1, 2,$ and 3.

The f-band width can be related to these r_f values using the second-moment formula for the band width, the generalization of Eq. (15-32),

$$W_f^2 = \frac{12X}{7}(V_{ff\sigma}^2 + 2V_{ff\pi}^2 + 2V_{ff\delta}^2 + 2V_{ff\phi}^2)\,. \tag{16-11}$$

In calculating the volume-dependent properties we will wish to have the result in terms of the atomic sphere radius r_0 which we do for face-centered-cubic structures for which $X = 12$ and $4\pi r_0^3/3 = d^3/\sqrt{2}$. Then

$$W_f = 181.\,\frac{\hbar^2 r_f^5}{m r_0^7}\,. \tag{16-12}$$

We have not obtained explicit f-bands for any structure. Because of the large effects of the Coulomb U_f they would be of limited interest. We could not therefore explore the use of the levels at the Brillouin-Zone corner, H, for the body-centered-cubic metals to define a band width. That was the method which led to Eq. (15-25) for the d-bands, and would presumably lead to a coefficient slightly larger than that in Eq. (16-12). The difference does not seem important.

Values for W_f were calculated by Straub (Straub and Harrison (1985)) with the Atomic Surface Method for all the f-shell metals. His values for the rare earths are listed in Table 16-1. (Values of W_f will be more directly useful than r_f values, which are directly related through Eq. (16-12), and the r_0 values listed also in Table 16-1.) The values obtained are considerably smaller (by a factor of three) than values obtained on lanthanum and cerium by Glötzel and Fritsche (1977) and we shall see a similar discrepancy for thorium among the actinides. Other parameters which we need for the rare earths are also listed in Table 16-1 and will be derived as we proceed.

f-band widths for the actinides were also obtained by Skriver (1984)*, using Linear Muffin-Tin-Oribtal Theory. (He did not give values for the rare earths.) Eq. (16-12) was used to obtain values for r_f from both sources of band widths. We compare Straub's Atomic-Surface-Method values with Skriver's LMTO values in Fig. 16-2.

LMTO values from Skriver (1984) appeared also in Straub and Harrison (1985) but differ from those (Skriver (1983)) given in (Harrison (1983d)).

*These can be extracted from Skriver (1984) using Eq. (4.11) from that book, and the values of $S\phi^2(-)$ from his Table 10.2 on page 266.

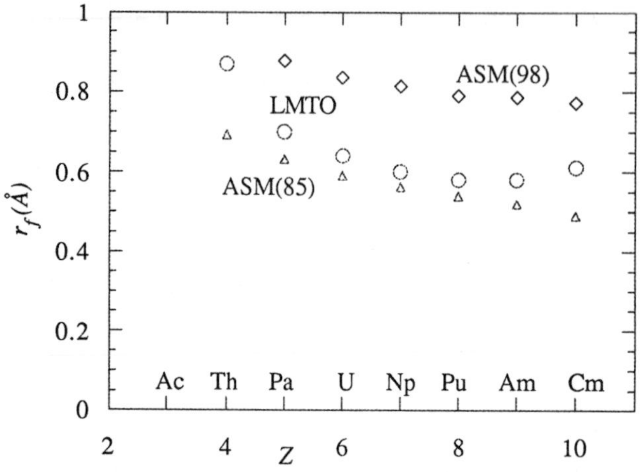

Fig. 16-2. r_f values obtained using Eq.(16-12) from f-band widths with the Atomic Surface Method (ASM) (Straub and Harrison (1985) and Straub (1998)) and with LMTO theory (Skriver (1984)) for the actinide metals.

ASM values from Straub and Harrison (1985) were based upon Mann's (1967) wavefunctions. Straub (1998) minimized the energy of the atom allowing nonintegral configurations, leading to the values ASM (1998). The variations shown could correspond to a factor of three difference in band width. We choose the values from LMTO theory for Table 16-1.

When f-bands are formed and partially occupied in the actinides, a bonding energy arises, the counterpart of the Friedel-Model transition-metal bonding energy of Eq. (15-40). When the effects of U_f are included it is given by Eq. (16-5) based upon the band width from Eq. (16-12).

If we are to include that bonding term, it is also important to obtain the ff-repulsion arising from the nonorthogonality of the hybridized f-states. This was found in Harrison (1983d), and generalized to transition-metal d-states (Wills and Harrison (1983)) as described in Section 15-4-D. The nonorthogonality between f-states on neighboring atoms is given by the counterpart of Eq. (15-41) as

$$S_{ff'm} = \sum_k \frac{\langle f|H|k\rangle\rangle\langle\langle k|H|f'\rangle}{(\varepsilon_f - \varepsilon_k)^2} = \sigma_{ff'm}\frac{r_f^{\,5}}{d^5}, \qquad (16\text{-}13)$$

Table 16-1. Parameters for the rare-earth metals. Z_m is the number of metallic electrons per atom. U_f values were obtained using Eq. (16-20) from Brewer's (1971) experimental values. We have listed $W_f = 181\ \hbar^2 r_f^5/(m r_0^7)$, from the r_f (e. g., $r_f = 0.461$ Å for Ce, from Straub and Harrison (1985)), since it is the relevant number to compare with U_f. The r_c were adjusted in Eqs. (16-25) to correspond to the correct equilibrium spacing.

	Z	Z_m	r_0 (Å)	r_c (Å)	W_f(eV)	U_f(eV)
Cs	1	1	3.00	1.39		
Ba	2	2	2.47	1.05		
La	3	3	2.08	0.84		
Ce	4	3	2.02	0.81	0.21	7.2
Pr	5	3	2.02	0.81	0.16	7.3
Nd	6	3	2.01	0.80	0.13	8.2
Pm	7	3			0.11	
Sm	8	3	1.99	0.79	0.09	3.3
Eu	9	2	2.27	0.95	0.06	
Gd	10	3	1.99	0.79	0.06	
Tb	11	3	1.95	0.77	0.06	
Dy	12	3	1.96	0.78	0.05	
Ho	13	3	1.95	0.77	0.05	
Er	14	3	1.94	0.77	0.04	
Tm	15	3	1.93	0.76	0.04	
Yb	16	2	1.99	0.80	0.07	
Lu	17	3	1.92	0.76	0.03	

Table 16-2. Parameters for the actinides. r_f values were based upon Skriver's (1984) LMTO calculations. U_f, based upon experimental values from Brewer (1971). The r_c were adjusted in Eq. (16-26) to correspond with the observed lattice distance. Spin-orbit coupling V_{SO} was obtained from Herbst, Watson and Lindgren (1976). $Z_f = Z - 3$ is the number of f-electrons per atom.

	Z	r_0 (Å)	r_c (Å)	V_{SO} (eV)	r_f (Å)	U_f (eV)
Ac	3	2.10	0.85		1.11	3.00
Th	4	1.99	0.88		0.87	3.20
Pa	5	1.80	0.79	0.17	0.70	3.35
U	6	1.69	0.74	0.20	0.64	4.09
Np	7	1.66	0.71	0.24	0.60	3.90
Pu	8	1.67	0.69	0.26	0.58	4.61
Am	9	1.91	0.76	0.33	0.58	4.96
Cm	10	2.03	0.82	0.36	0.61	5.10

with coefficients given by (Harrison (1983d)).

$$\sigma_{ff\sigma} = -8\,\frac{105}{2\pi}\,, \quad \sigma_{ff\pi} = 7\,\frac{105}{2\pi}\,,$$

$$\sigma_{ff\delta} = -4\,\frac{105}{2\pi}\,, \quad \sigma_{ff\phi} = \frac{105}{2\pi}\,. \qquad (16\text{-}14)$$

The shift $-S_{ffm}V_{ffm}$ from each of the X nearest neighbors for the Z_f occupied f-orbitals on each atom gives a repulsion per atom, as in Eq. (15-42), of

$$E_{ff} = -\,Z_f\,\frac{1}{7}\,\frac{\hbar^2 r_f{}^{10}}{md^{12}}\,(\sigma_{ff\sigma}\eta_{ff\sigma} + 2\sigma_{ff\pi}\eta_{ff\pi} + 2\sigma_{ff\delta}\eta_{ff\delta} + 2\sigma_{ff\phi}\eta_{ff\phi})\,X$$

$$= 83780X\,Z_f\,\frac{\hbar^2 r_f{}^{10}}{md^{12}} = 816\,Z_f\,\frac{\hbar^2 r_f{}^{10}}{mr_0{}^{12}} \quad (\text{for } X = 12\text{ , fcc}). \qquad (16\text{-}15)$$

B. General Formulae

It is clear that these formulae can be generalized to the coupling between atomic states of *any* angular-momentum quantum numbers, and Wills (Wills and Harrison (1983)) has provided the corresponding formulae,

$$V_{\ell'\ell m} = \eta_{\ell'\ell m}\,\frac{\hbar^2\sqrt{r_{\ell'}{}^{2\ell'-1}r_{\ell}{}^{2\ell-1}}}{md^{\ell'+\ell+1}} \qquad (16\text{-}16)$$

with

$$\eta_{\ell'\ell m} = \frac{(-1)^{\ell+m+1}}{6\pi}\,\frac{(\ell+\ell')!}{2^{\ell+\ell'}}\,\frac{(2\ell)!(2\ell')!}{\ell!\,\ell'!}\,\sqrt{\frac{(2\ell+1)(2\ell'+1)}{(\ell+m)!(\ell-m)!(\ell'+m)!(\ell'-m)!}}\,. \qquad (16\text{-}17)$$

Again, the r_ℓ are free-atom parameters and can be obtained as before from the individual band widths W_ℓ.

Wills also generalized the nonorthogonality which enters the overlap repulsion, given for d-states in Eq. (15-41) and f-states in Eq. (16-15) , as

$$S_{\ell'\ell m} = \sigma_{\ell'\ell m} \frac{\sqrt{r_{\ell'}^{2\ell'-1} r_{\ell}^{2\ell-1}}}{d^{\ell'+\ell-1}} \quad \text{for } \ell + \ell' > 1 \tag{16-18}$$

and

$$\sigma_{\ell'\ell m} = -\frac{1}{2(\ell+\ell')}\left(1 + \frac{4m^2-1}{(2\ell-1)(2\ell'-1)}\right)\eta_{\ell'\ell m} \tag{16-19}$$

with $\eta_{\ell'\ell m}$ given in Eq. (16-17). We shall make use of these formulae in the following chapter in the study of transition-metal compounds.

C. Coulomb U_f

As we have indicated in Section 16-1, the Coulomb repulsion between electrons in the f-shells is very central to the properties of the f-shell metals. Of particular interest is the change in energy as an electron is transferred from the f-shell on one atom to that on a neighboring atom, the counterpart of the U introduced for the pair of lithium atoms in Section 16-1-A. In the metal such a transfer is accompanied by a screening from the sd-electrons, which we can represent here, as we did for the transition metals, as the addition of an atomic s-electron on the atom where the f-electron was removed, and the removal of an s-electron from the atom where the f-electron was added. Thus we can obtain the Coulomb U_f from the corresponding free-atom total energies. We actually have only a limited choice for doing this since the atom can only have two, one, or no s-electrons, so we must start with one s-electron on each atom. Further we keep the number of d-electrons fixed in the process. Thus, since almost all of the f-shell metals are trivalent, the formula for the element with Z_f f-electrons must be

$$U_f = E(f^{Z_f+1}d^2) + E(f^{Z_f-1}d^2s^2) - 2E(f^{Z_f}d^2s)\ . \tag{16-20}$$

Here E is the total energy, or term value, for the atom with the configuration shown in parentheses.

These could be directly calculated with an atomic Hartree-Fock program, but they also can be obtained from experiment, as was done by Johansson (1975). A variety of such term values have been given for the f-shell metals by Brewer (1971). (He gives the energies in cm^{-1}, so we must multiply by 0.124 eV-cm to obtain values in eV.) In some cases the complete set of three values needed for Eq. (16-20) is not given, but another set with, for example, one less d-electron and one more f-electron in each configuration can be used. We have used such a procedure to obtain the estimates of U_f listed in Tables 16-1 and 16-2.

For cerium, $Z_f = 1$, using Brewer's (1971) values and Eq. (16-20) we obtain $7.22\ eV$. Brewer's data for this case also allows us to use a starting state fdsp, and the counterpart of Eq. (16-20) leads to $U_f = 6.10\ eV$. The difference is some measure of the uncertainty of the number. Brewer's data allows evaluations of Eq. (16-20) for Pr and Nd , giving 7.33 eV and 8.21 eV, respectively. The only other estimate for the rare earths allowed by the data was for Sm, but with the starting configuration $f^{Z_f+1}ds$. It leads to a low U_f of 3.33 eV but we listed that estimate. For all of the actinides we had data for the configurations in Eq. (16-20) and the corresponding values are listed in Table 16-2.

Johansson (1975) based his estimates of U_f for the actinides upon transfer of electrons from the d-state to the f-state, rather than the s-state to the f-state. This yields smaller values, as would be expected since conceptually they should correspond to our $U_f - U_d$, though we shall see in Part E that the difference is larger than the U_d we obtain there. Thus in place of Eq. (16-22) he uses $U_f = E(f^{Z_f+1}s^2) + E(f^{Z_f-1}d^2s^2) - 2E(f^{Z_f}ds^2)$. Using this expression we obtain 1.54, 1.66, 2.30. 2.54, 2.90, and 2.73 eV for Th, Pa, U, Np, Pu, and Am, respectively, close to his 1.5, 1.6, 2.3, 2.6, 3.5, and 5. eV for the first four, but considerably smaller than his for Pu and Am. This is an important distinction since, as we shall see, it is just at these elements that the correlation is becoming important. We see no reason for the larger values, and use the values from sf-transfer, listed in Table 16-2, in any case.

D. Spin-Orbit Coupling

We noted in Section 5-3-G that the effects of spin-orbit coupling become rapidly larger with increasing atomic number, so they are largest in the actinides. Even here, following Harrison (1983d) we shall find that their effects on the total energy are not large.

We saw in Eq. (5-28) that spin-orbit coupling is represented by a term in the Hamiltonian, $H_{SO} = \xi(r)\,\mathbf{L}\cdot\boldsymbol{\sigma}$, with $\mathbf{L}$ the angular-momentum operator and σ the spin operator. Here we are interested in the coupling between different f-states, $\ell = 3$. We noted in Section 5-3-G that the square of the total angular momentum, $\mathbf{J}^2 = (\mathbf{L} + \boldsymbol{\sigma})^2 = \mathbf{L}^2 + \boldsymbol{\sigma}^2 + 2\,\mathbf{L}\cdot\boldsymbol{\sigma}$, has eigenvalues $\hbar^2 J(J+1)$, with J equal here to $\ell \pm \sigma = 7/2$ or $5/2$, just as $\mathbf{L}^2$ has eigenvalues

$\hbar^2\ell(\ell+1) = 12\hbar^2$ and σ^2 has eigenvalues $\hbar^2\sigma(\sigma+1) = {}^3/_4\hbar^2$. Thus $\boldsymbol{L}\cdot\sigma$ has eigenvalues of ${}^1/_2(J(J+1) - \ell(\ell+1)- \sigma(\sigma+1)\,)$. In this case there are $2J+1$ = 8 eigenvalues of $({}^{63}/_4 - 12 - {}^3/_4)\ \hbar^2/2 = 3\hbar^2/2$ and six eigenvalues of $({}^{35}/_4 - 12 - {}^3/_4)\hbar^2/2 = -2\hbar^2$. The matrix elements of H_{SO} also contain an integral $\int\psi_f(r)^*\xi(r)\psi_f(r)\ d^3r$ called $V_{SO}/\hbar^2$ or sometimes simply ξ , with V_{SO} just twice the counterpart of the λ we introduced in Eq. (5-29). In the case of f-levels in the free atom, we see that the fourteen (including spin) f-levels are split into eight $J = {}^7/_2$ levels up by ${}^3/_2V_{SO}$ and six $J = {}^5/_2$ levels down by $2V_{SO}$.

When we add this spin-orbit coupling to the Hamiltonian matrix representing the solid, we add it in each fourteen-by-fourteen submatrix for the f-states on an individual atom. Its contribution to the second moment is unchanged if we transform that submatrix to eigenstates of $\boldsymbol{L}\cdot\sigma$, in which case it clearly adds directly to the second moment

$$M_2{}^{SO} = \frac{8({}^3/_2V_{SO})^2 + 6(2V_{SO})^2}{14} = 3V_{SO}{}^2 \tag{16-21}$$

to be added to the second moment M_2 which is related to the band width by $M_2 = W_f^2/12$. Thus if the band width of Eq. (16-12), without spin-orbit coupling , is written W_f^0 the full bandwidth with spin-orbit coupling is given by $W_f^2 = W_f{}^{0\,2} + 36V_{SO}{}^2$.

Values for V_{SO} were obtained by fitting band calculations by Herbst, Watson, and Lindgren (1976) for the actinides, tabulated in Harrison (1983d), and these values appear in Table 16-2. Very nearly the same values were given by Nugent, Baybarz, Burnett, and Ryan (1971). When squared, even when multiplied by 36, they make a small contribution to the second moment and the band width

All of this discussion of spin-orbit coupling has been in the context of the one-electron description which we have used throughout this text. The multi-electron effects complicate the spin-orbit coupling considerably, even for the free atom. Each electron spin interacts also with the orbital motion of the other electrons. If there were no spin-orbit interaction it would be possible to define a total orbital angular momentum, $\boldsymbol{L}$, which is quantized in units of $\hbar$, and a total spin angular momentum $\boldsymbol{S}$, even when the electrons interact with each other so that individual electrons did not have well-defined angular momentum. Thus if the spin-orbit interaction is weak it is reasonable to retain the $\boldsymbol{L}$ and $\boldsymbol{S}$ and couple them as we did for individual electrons, giving what is called $\boldsymbol{L}\cdot\boldsymbol{S}$, or Russel-Saunders coupling. Where in the one-electron case we had eigenvalues proportional to $^1/_2(J(J+1) - \ell(\ell+1) - \sigma(\sigma+1))$, we might expect the same proportionality constant times this expression with ℓ replaced by L and σ replaced by S. However, Nugent, Baybarz, Burnett, and Ryan (1971) give this expression divided by Z_f for $Z_f < 7$ (and divided by $Z_f - 14$ for $Z_f > 7$). We have not traced down the origin of this expression, but it is consistent with the fact that when spin-orbit coupling becomes quite strong, one generally treats the spin-orbit coupling for the individual electrons, associating a total-angular-momentum quantum number j with each and then treats the coupling, "jj coupling", between the electrons. In any case, it makes the estimates small as were our one-electron estimates.

There exists intricate theory of the various spin-orbit-split states, their quantum numbers, and selection rules for transitions between them. The only place where it could be important in our analysis is in the actinides and in the formation of correlated states. We choose to make all of our estimates without including spin-orbit coupling, and when we compare our predictions with experiment, which of course includes them, it will appear to support our decision. Spin-orbit splitting is a significant aspect of the electronic structure of the actinides, but all of these considerations suggest that they can be safely ignored for the properties we treat here.

E. The sd-Electrons

We have treated the trivalent metals at the beginning of each transition series as transition metals in Chapter 15. It would be natural to do this also for the non-f electrons in the f-shell series. Certainly the atomic d-orbitals contribute significantly to the band states in these metals. However, the 4f- and 5f-series metals seemed to be treated with some success considering these electrons as free in Harrison (1983d). We may look more carefully into that question by obtaining the appropriate d-state parameters for these metals.

In order to do this, we obtained d-band widths for these metals from Skriver (1984) (as we described for f-band widths in Section 16-2-A) and deduced the corresponding r_d values using Eq. (15-33). These are listed in Table 16-3.

For U_d we proceed exactly as we did for U_f in Section 16-2-C. We use the same starting configuration and the counterpart of Eq. (16-20) is

$$U_d = E(f^{Z_f}d^3) + E(f^{Z_f}ds^2) - 2E(f^{Z_f}d^2s)\,. \tag{16-22}$$

Table 16-3. Parameters describing the d-states in the rare earths and actinides, in Å and eV. W_d and r_d were based upon the LMTO calculations by Skriver (1984) and U_d values were from Brewer's (1971) data as described in the text. E_F would be the free-electron Fermi energy for three free electrons per atom.

	Z	r_0	r_d	U_d	W_d	E_F(3 el'ns)
				Rare Earths		
La	3	2.08	2.13	0.88	10.42	6.75
Ce	4	2.02	2.08	0.94	11.22	7.15
Pr	5	2.02	2.04	0.74	10.65	7.15
Nd	6	2.01	1.08	0.13	10.50	7.23
Pm	7				10.47	
Sm	8	1.99	1.97	0.68	10.40	7.37
Eu	9	2.27	2.31	1.00	8.64	5.66
Gd	10	1.99	1.96	1.20	10.23	7.37
Tb	11	1.95	1.92	1.35	10.52	7.68
Dy	12	1.96	1.93	1.66	10.57	7.60
Ho	13	1.95	1.92	1.87	10.62	7.67
Er	14	1.94	1.91	2.00	10.72	7.75
Tm	15	1.93	1.90	2.22	10.82	7.84
Yb	16	1.99	1.97	2.48	10.45	7.37
Lu	17	1.92	1.90	2.89	10.98	7.92
				Actinides		
Ac	3	2.11	2.39	0.69	14.1	6.62
Th	4	1.99	2.17	0.94	14.0	7.37
Pa	5	1.81	2.03	0.99	18.1	8.91
U	6	1.70	1.93	1.05	21.2	10.10
Np	7	1.66	1.91	0.62	23.3	10.59
Pu	8	1.68	1.91	0.43	21.9	10.34
Am	9	1.91	2.06	0.62	14.4	8.00
Cm	10	2.03	2.26	0.59	14.2	7.08

For cerium this leads to $U_d = 0.94\ eV$. If we use the starting configuration $f^{Z_f+1}ds$ we obtain almost the same $U_d = 0.92\ eV$. For the other rare earths we used Eq. (16-22) (with Z_f indicated in Table 16-1) in each case, to obtain the U_d values listed in Table 16-3.

We also obtained the corresponding values for U_d for the actinides from Brewer's term values using Eq. (16-22) and these are listed in Table 16-3. There was also sufficient data to give values of U_d based upon a starting configuration of $f^{Z_f+1}ds$ and this gave quite similar values.

The results would strongly suggest that it is no longer appropriate to treat the d-levels in these trivalent metals as in transition metals. Our treatment of the sd-hybridization has been as an expansion in r_d/r_0 , keeping the leading terms. For the first time here the two parameters are about the same size. Further we have treated U_d as a large energy so that the d-band shifted easily to match the Fermi energy of the remaining free-electron gas, but here it is only of the order of one electron volt. Finally, we may compare the d-band width with the occupied region of a free-electron band, which is listed as the final column in Table 16-3. Even placing all three electrons in the free band leaves it small compared to the d-band width. This supports the earlier treatment (Harrison (1983d)) of the f-shell metals using free-electron theory.

Since we will be treating these electrons as free, we will also be interested in predicted pseudopotential empty-core radii r_c , though in the end we adjust them to accord with the observed spacing, and it is these adjusted values which appear in Table 16-1 and 16-2. It will be interesting to compare these adjusted values with those obtained from the atom as a test of our representation of the electronic structure.

3. Electronic Structure of Rare Earths

We see from Table 16-1 that the W_f for the rare earths are all extremely small in comparison to U_f . Thus the $U_f - \sqrt{W_f^2 + U_f^2}$ from Eq. (16-5) is of the order or less than the 0.003 eV for cerium, or 0.03 eV if we use the LMTO f-band width. Cerium lies at the beginning of the series and where the effects are largest, and quite negligible. This is the strongly-correlated limit and we may neglect any contribution of the f-electrons to the bonding. The effects of the f-shells are presumably analogous to the effects of the d-shells in chromium shown in Fig. 15-14, but the deep f-shell contribution to the energy shows up only at much smaller spacings. We discuss first the behavior of the f-electrons and then the sd- or free-electron states.

A. The f-Electrons

In this strongly correlated limit, there will be some number of occupied f-electron states for each rare earth. As in Anderson's (1961) model, discussed in Section 16-1-D, this level ε_f lies below the Fermi energy ε_F since it would require energy to remove an f-electron and put it in the "metallic" band which we shall treat as a $Z = 3$ metal band. We neglect any effect of the hybridization, as suggested by Eq. (16-5) and the parameters of Table 16-1.

These f-electron states are therefore treated as the same as in the atom and the spins will align, according to Hund's rule, to the extent possible.

In Table 16-1 we listed the rare-earth metals, and the elements immediately preceding them, giving the number of valence electrons Z and the number of metallic electrons Z_m. We see that Cs, Ba, and La, are simple monovalent, divalent, and trivalent metals, but from then on, each additional valence electron is added to the f-shell. Since we neglect the coupling of these f-states to the metallic states, the electronic structure is that of a simple trivalent metal, with these f-electrons forming a local magnetic moment on each atom. That moment is one Bohr magneton (electronic moment) for Ce and increases by one at each step until europium, for which an additional electron is removed from the metallic band, making Z_m equal to two, and filling the last of the seven states of parallel spin. This "grabbing" of another electron is associated with the stability of the full shell, the exchange energy associated with seven electrons in the shell. A similar effect occurred in the 3d transition metals where we indicated in Fig. 15-12 that an additional electron in chromium was taken from the free band to complete the five levels of parallel spin. Such effects were not so systematic in the 4d- and 5d-metals.

After europium comes gadolinium, which also has seven f-states occupied and parallel spins. If we were to make an experimental estimate of U_f for gadolinium using Eq. (16-20), we obtain a very high value, near 12 eV, because the exchange energy makes the $Z_f = 7$ state so much lower in energy than the $Z_f = 6$ and $Z_f = 8$ states. The *effective* U_f for this system is anomalously high.

Beyond gadolinium as we proceed on in the series, the net spin is reduced by one as each f-electron is added with opposite spin. Again at ytterbium another electron is grabbed to fill the shell so that neither it nor lutetium have local moments. Lutetium begins the 5d transition-metal series.

The interaction between the moments on different atoms arises indirectly in exactly the way we obtained the V_{ddm} interaction in Chapter 15. It is called the Ruderman-Kittel (1954) interaction and considerably predates the theory discussed in Chapter 15. It arises from the terms $\Sigma_k \langle f|\delta V|k\rangle\langle k|\delta V|f'\rangle / (\varepsilon_f - \varepsilon_F)$ coupling states of the same spin on neighboring atoms, which shift the total energy depending upon the relative orientation of the spins. These interactions depend upon the distance between the coupled f-states and show the Friedel oscillations such as we displayed in Fig. 13-7 in Section 13-6. These are generally regarded to be responsible for the complex magnetic properties (see, for example, Kittel

(1963)). These magnetic properties, and their theory, have been reviewed in detail by Rocher (1962).

B. The Cohesive Energy

Our description of the electronic structure of the rare earths as trivalent (or in two cases divalent) simple metals, with the f-electrons retaining their atomic core-like character, would suggest that the cohesive energy might be quite simple. One might even suggest, as did Wills and Harrison (1984), that the trivalent rare earths might have cohesion close to that of trivalent yttrium (2.43 eV per atom), which we treated as a transition metal, and the divalent europium and ytterbium might have cohesion close to the smaller value for divalent barium (1.90 eV per atom). At the conclusion of this part we shall in fact apply the transition-metal form, Eq. (15-45), for the trivalent rare earths and see that it predicts values near two eV per atom for all of them. The actual cohesive energies, plotted in Fig. 16-3, show strong variation. Upon closer consideration we see that this structure is not the result of differences in the *metals* , but differences in the free-atom states.

A consideration of the atomic configuration of the atoms (Weast (1975), p. B3) indicates that except for lanthanum and cerium at the beginning of the series, gadolinium in the middle, and lutetium at the end, they are $f^{Z-2}s^2$ as atoms, divalent. These four are trivalent as atoms and form trivalent metals, as we have indicated, and indeed have high cohesive energies as seen in Fig. 16-3, two volts higher even than trivalent yttrium. Among the divalent atoms, we have indicated that only europium and ytterbium form divalent metals (by grabbing an additional f-electron). Thus only they should have the cohesion of divalent barium, qualitatively explaining the low value seen for them in Fig. 16-3. Between cerium and gadolinium, to form the metal we must first promote an electron from the f-shell to the d-shell, and can then form the trivalent metal from the resulting trivalent atom. At the first step in this series, the f^{Z-2} (praseodymium configuration) energy has just dropped below the df^{Z-3} (cerium configuration) energy so the promotion energy is small and cancels only a small part of the trivalent cohesion. The cohesive energy drops slightly. The cohesive energy continues to drop as we move through the series, the f^{Z-2} energy continues to drop, and the needed promotion energy increases. Indeed the term values for the f-level of the divalent free atoms Pr, Nd, Pm, Sm and Eu (Mann (1967)) drop by nearly a volt at each step, even more than the drop in cohesive energy. Both drops are on the scale of the exchange energy U_x which is causing the drop, of the

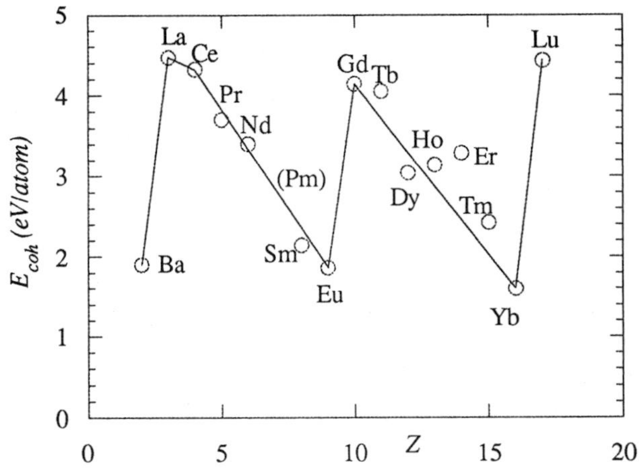

Fig. 16-3. The cohesive energy of the rare earths, from Kittel (1976). The line interpolates linearly the promotion required from divalent atoms to trivalent metals.

order of one tenth of U_f . To the extent that europium lies on the same straight line, the trivalent and divalent europium metal has about the same energy. The energy to promote an electron to form the trivalent atom just cancels the increase in cohesion of trivalent over divalent atoms. The same argument should apply moving to the right of gadolinium and successively filling the levels of opposite spin. Given the high experimental values for trivalent La, Ce, Gd and Lu, and low experimental values for divalent Ba, Eu, and Yb, the linear interpolation gives the line shown in Fig. 16-3.

We see that the variation of the cohesion is qualitatively understandable in terms of the electronic structure, but the main features are determined by variations in the properties of the atom rather than of the solid.

We return finally to the estimate of the cohesive energy treating the four metals La, Ce, Gd, and Lu as trivalent transition metals, though we have argued that the d-bands are so broad that this is questionable. We cannot use the simple-metal theory of Section 13-3-A since for trivalent metals the kinetic energy exceeds the lowering of the conduction-band minimum and in these metals energy is required to lift the extra electron to the s-level, rather than gained in dropping it from a p-state as in aluminum. We use Eq. (15-45) directly, taking $Z_d = 1.5$. Eq. (15-45) assumes two s-electrons in the free atom and the simple-metal energy is based upon those two electrons, even though the number of electrons after the atomic preparation is 1.5 metallic electrons. We also take the expansion $\sqrt{W_d^2 + U_d^2} - U_d \approx W_d - U_d$ and combine terms from the atomic preparation energy with the other terms in Eq. (15-45) to obtain

$$-E_{coh} = -0.634\frac{\hbar^2}{mr_0^2} - 0.463W_d + 0.513U_d + 1.785\frac{\hbar^2 r_d^6}{mr_0^8}. \quad (16\text{-}23)$$

Using the parameters from Table 16-1 (and W_d from Eq. (15.33)) this yields cohesive energies of 1.88, 1.95, 2.18, and 1.49 eV per atom for La, Ce, Gd, and Lu, respectively, rather constant and crudely in accord with the experimental values in Fig. 16-3, but far enough off to support our conjecture that treatment as a transition metal is not very appropriate.

C. The Volume-Dependent Energy

Our consideration of the d-shell parameters in Section 16-2-E suggested that the three valence electrons, outside the f-shell, be regarded as free, as we assumed earlier (Harrison (1983d)) for the f-shell metals. With that assumption, we use the free-electron energy of Eq. (13-18) (or Eq. (15-46)) (we are dropping the small second-order terms included in Section 14-2-C). Eq. (13-18) was

$$\underset{\text{total}}{\frac{E_{eg}}{ion}} = \underset{\text{kinetic}}{\frac{3Z\hbar^2 k_F^2}{10m}} + \underset{\text{pseudopotential}}{\frac{2Ze^2 r_c^2 k_F^3}{3\pi}} - \underset{\text{exchange}}{\frac{3Ze^2 k_F}{4\pi}} - \underset{\text{Madelung}}{\frac{Z^{5/3} e^2 k_F}{(18\pi)^{1/3}}\alpha}, \quad (6\text{-}24)$$

with each term labeled. If it is rewritten for $Z_m = 3$ or 2 and $\alpha = 1.8$ in terms of the atomic sphere radius r it becomes

$$\frac{E_{tot}}{ion} = 6.90\frac{\hbar^2}{mr^2} + 13.5\frac{e^2 r_c^2}{r^3} - 10.08\frac{e^2}{r}. \quad (trivalent)$$

$$\frac{E_{tot}}{ion} = 3.51\frac{\hbar^2}{mr^2} + 6\frac{e^2 r_c^2}{r^3} - 4.75\frac{e^2}{r}. \quad (divalent) \qquad (16\text{-}25)$$

The r_c were adjusted to give a minimum energy at the observed atomic radius in Harrison (1983d), and compared with values of r_c obtained earlier (Harrison (1980)) from adjustment of the pseudopotential core radii to obtain the experimental energy required to remove the last valence-electron from the atom. [This method used the energy of that state since it was an s-state in a simple potential, $-3e^2/r$ without appreciable affects from the potentials arising from other electrons.] Both sets of values are listed in Table 16-4 and are in remarkable agreement, considering that one came from atomic spectra and the other from the equilibrium spacing in the metal through the total energy. It provides a good test of our description. Radii for Cs and Ba (r_c seems large for Ba) were also included

The second derivatives of Eqs. (16-25) were also evaluated in order to obtain the bulk modulus. Those results (Harrison (1983d)) are also compared with experiment in Table 16-4.

D. Interatomic Interactions

Since we treated the sd-electrons as free, and neglected contributions of the f-levels to the energy of the metal, it is natural to suggest using the same interatomic interactions, $Z_m^2e^2\cosh^2\kappa r_c\, e^{-\kappa d}/d$, as for simple metals,

Table 16-4. Pseudopotential core radii for the rare-earth metals The first set was adjusted to give the correct atomic-sphere radius in the metal. The second set was adjusted to give the experimental ionization potential of the atom. The bulk modulus B was predicted from Eqs. (16-25) and (16-27) and compared with experiment from Gschneidner (1964).

	Z	Z_m	r_0 (Å)	r_c (Å) Fit to metal	r_c (Å) Fit to atom	$B(10^{12}$ ergs/cm$^3)$ Theory	$B(10^{12}$ ergs/cm$^3)$ Experiment
Cs	1	1	3.00	1.39	1.55	0.02	0.02
Ba	2	2	2.47	1.05	1.60	0.13	0.10
La	3	3	2.08	0.84		0.55	0.24
Ce	4	3	2.02	0.81	0.83	0.61	0.26
Pr	5	3	2.02	0.81	0.86	0.61	0.30
Nd	6	3	2.01	0.80	0.85	0.62	0.24
Pm	7	3					
Sm	8	3	1.99	0.79	0.83	0.64	0.29
Eu	9	2	2.27	0.95	0.81	0.18	0.15
Gd	10	3	1.99	0.79	0.71	0.64	0.39
Tb	11	3	1.95	0.77	0.75	0.69	0.40
Dy	12	3	1.96	0.78	0.61	0.68	0.39
Ho	13	3	1.95	0.77		0.69	0.40
Er	14	3	1.94	0.77	0.72	0.71	0.41
Yb	16	2	1.99	0.80	0.70	0.72	0.40
Lu	17	3	1.92	0.76		0.30	0.13

with $Z_m = 3$ and core radii from Table 16-1. That was done in Harrison (1983d), and calculating the bulk modulus entirely in terms of these two-center interactions as in Section 14-3-A yielded 0.39×10^{12} ergs/cm^3 for the bulk modulus of lanthanum, comparable to value 0.55×10^{12} ergs/cm^3 obtained from Eq. (16-25) and listed in Table 16-4. (It gave 0.015 and 0.098×10^{12} ergs/cm^3 for Cs, and Ba, respectively and would have given values similar to the 0.39×10^{12} ergs/cm^3 for the other rare earths because the r_0 's and r_c 's are so similar.) That approach again seems plausible and provides a method for estimating a wide variety of properties, but it has not been explored further.

E. Crystal Structure

The crystal structure varies between different close-packed structures through the series. The variation might seem surprising since the f-shell electrons are not appreciably taking part in the bonding. However, Duthie and Pettifor (1977) have noted that they change the potential seen by the other electrons, and in particular the added attractive potential near the nucleus to the right in the series lowers the s-state relative to the d-states, decreasing the role of the d-states in the binding (by about 1.5 eV across the series according to Mann's (1967) Hartree-Fock calculations). They were able to correlate the changes in structure with this trend.

4. Electronic Structure of the Actinides

From Table 16-2 we may see that in the actinides, in contrast to the rare earths, W_f and U_f are comparable and the bonding contribution of the f-levels given in Eq. (16-5) is not negligible as it was for the rare earths. Incorporating this term, the rest of the theory follows closely what we have given for the rare earths.

A. The Cohesive Energy

For the cohesive energy we may follow the procedure we used for the rare earths, having all of the same difficulties. In particular, there is a problem of promotion energy since while we think of each of the metals as having three free electrons, or an s^2d-configuration, the configuration of the atoms varies as indicated in Table 16-5. Thus a promotion is required from a d- to an f-state for thorium, and a promotion from an f- to a d-state for plutonium and americium. As with the rare earths this is consistent with the variation of the free-atom difference in d- and f-term values. In those three cases the promotion should decrease the cohesive energy, but with the limited experimental data, also in Table16-5, it would be difficult to detect that feature. Using the parameters from Table 16-3 we may evaluate the "transition-metal contribution" to the cohesion from Eq. (16-23), which is listed as "sd" in Table 16-5. As for the rare earths it is considerably smaller than the experimental cohesion. The contribution of the f-band formation is not negligible for these cases, evaluated using parameters from Table 16-1 and listed in Table 16-5, but it is not enough to account for the difference. It would be very slightly reduced by the inclusion of spin-orbit coupling, but not in an important way. Not so much is learned from the comparison.

Table 16-5. The configuration of the free atoms of the actinides, to be compared with $f^{Z-3}ds^2$ assume for the solid. The estimate (Eq. (16-23)) for the sd-contributions and for the f-contribution (Eqs. (16-5) and (16-15)) to the cohesion and the experimental value.

	Z	Atom Configuration*	Cohesion (eV/atom)		
			sd	f	Exper.**
Ac	3	ds^2	0.56	0.00	4.25
Th	4	d^2s^2	1.35	1.11	6.20
Pa	5	f^2ds^2	0.98	1.17	
U	6	f^3ds^2	0.76	1.35	5.55
Np	7	f^4ds^2	0.77	1.21	
Pu	8	f^6s^2	1.06	0.74	3.60
Am	9	f^7s^2	1.85	0.10	2.73
Cm	10	f^7ds^2	1.12	0.05	3.86

*Weast (1975), B4. **Kittel (1976)

B. Volume-Dependent Properties

We follow the treatment of the rare earths by replacing the simple-metal and atomic preparation energy by the free-electron energy given by Eq. (16-25) for the energy arising from the three s-like and d-like electrons. To this we add the bonding energy and repulsion for the f-shells from Eqs. (16-5) and (16-15). The resulting total energy per ion (or atom) in terms of the atomic-sphere radius r becomes

$$\frac{E_{tot}}{ion} = 6.90\frac{\hbar^2}{mr^2} + 13.5\frac{e^2 r_c^2}{r^3} - 10.08\frac{e^2}{r}$$

(16-26)

$$- \frac{Z_f(1 - Z_f/14)}{2}\left(\sqrt{\left(181\frac{\hbar^2 r_f^5}{mr^7}\right)^2 + U_f^2} - U_f\right) + 816\, Z_f\frac{\hbar^2 r_f^{10}}{mr^{12}} .$$

As for the rare earths we adjusted the pseudopotential core radius for each actinide so that this expression, using other parameters from Table 16-2, was minimum at the observed equilibrium spacing. The resulting r_c values were then listed in Table 16-2. These are rather close to the values obtained earlier from the third ionization potential of thorium and uranium, the only two for which data were found (Harrison (1980)). Those core radii were 0.59 Å for thorium and 0.72 Å for uranium, compared to the 0.88 and 0 .69 Å, respectively, obtained here. They are smaller than recent estimates from atomic states by Straub (1998), 1.93, 1.43, 1.39, 1.38, 1.36, 1.35, and 1.35 Å, respectively, for thorium through curium.

In order to understand the results, we have plotted the energy of Eq. (16-26) for Np, Pu, and Am in Fig. 16-4. (Problem 16-1 involves a similar plot for U.) We also plotted the f-shell and free-electron contributions separately; these will give the complete picture. We had fit r_c for plutonium to obtain a minimum at the observed spacing, but we see that it yields a deeper minimum at smaller spacing. Adding spin-orbit splitting raises the f-shell minimum about 0.3 eV, not enough to eliminate this minimum. This is an incorrect prediction, but a significant one. When two such minima occur, the lower is predicted to occur at vanishing pressure. If the plot is energy versus volume, then a tangent to the curve has a slope equal to the pressure which is required for a first-order phase transition (a discontinuous shift from one density to another) to the neighborhood of the second minimum, in this case a negative pressure would be required to bring plutonium to the observed spacing. Fig. 16-4 would indicate that a positive pressure would cause such a transition from the expanded americium to one at lower spacing. Having these two minima at similar energies is suggestive of the many phases which occur in plutonium, as is the inflection of the curves for americium, and the very flat energy as a function of spacing for neptunium and uranium (Problem 16-1). All four of these occur in more than one crystal structure. In Chapter 15 we discussed a similar double minimum occurring in our curves for chromium.

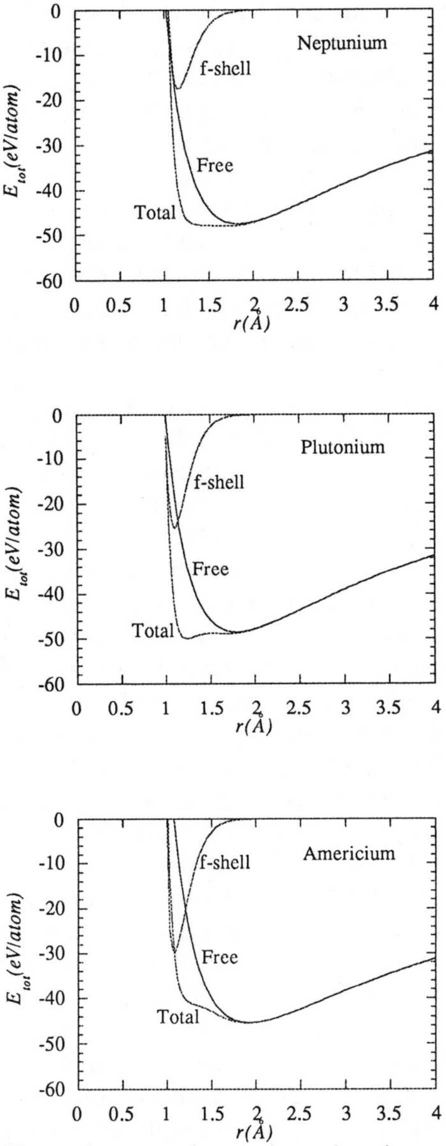

Fig. 16-4. The total energy of neptunium, plutonium, and americium as a function of atomic-sphere radius from Eq. (16-26) and Table 16-2. Also shown separately are the f-shell and free-electron contributions.

These irregularities in the curves, and variations in crystal structure, are arising because of an f-shell energy with a deep minimum and free-electron energy with a broad minimum at larger spacing. As r_f decreases with increasing atomic number through the actinide series, the minimum in the f-shell energy becomes deeper and moves to the left. Only when it is deep enough does if have an important effect and when it moves too far to the left it becomes lost in the rapidly rising free-electron repulsion. The effect is important only in the middle of the series and perhaps largest near plutonium.

We have not predicted the equilibrium spacings (since we adjusted r_c) but see the observed spacings, which reach a minimum near plutonium and then rise in americium and curium, are understandable in terms of this shifting role of the f-shell energy.

This rise in spacing at americium, as well as the phase transitions in plutonium, the rare-earth cerium, and the 3d-transition metals at the right end of the series, has traditionally been interpreted as arising from formation of correlated states for the electrons (or "electron localization"), and indeed that was our interpretation in Harrison (1983d) and (1984) . It now appears that these affects are not caused by such electron-electron interactions, though when the end result is a metal with spacings large enough that the f- or d-shell states become unimportant in the binding, it becomes likely that the effects of the electron-electron interaction will dominate their behavior. As in the rare earths, where this has clearly occurred, we saw that the f-electrons aligned their spins and became free moments in the metal. This same rare-earth-like behavior is expected in americium and the heavier actinides because of the large spacing, but was not the cause of that large spacing.

Earlier calculations by Brooks (1984) had suggested that the increased atomic volume in plutonium relative to neptunium was due to spin-orbit coupling, and the sensitivity which we noted for the double-minimum situation suggests that as small as spin-orbit effects are, they could be significant. Recent calculations by Kollar, Vitos, and Skriver (1997) included a *Gradient Expansion Method*, which mimics the effects of correlated states, to predict the equilibrium spacing of α–plutonium, the high-density phase represented in Fig. 16-7. They attributed the increased volume to an incipient localization associated with the lower coordination in this complex structure. This does not fit well with our view here, which seems quite insensitive to structure, and for which the correlations enter in a more continuous way through the appearance of U_f in Eq. (16-26).

We may of course also take the second derivative of Eq. (16-26) and predict the bulk modulus, and the third derivative to predict the Grüneisen constant, as in Eqs. (15-48) and (15-49) . For the bulk modulus it was

$$B = \Omega_0 \frac{\partial^2 E_{tot}}{\partial \Omega_0{}^2} = \frac{1}{9\Omega_0} r^2 \frac{\partial^2 E_{tot}}{\partial r^2} \ . \tag{16-27}$$

Values for the bulk modulus are similar to the earlier results (Harrison (1983d)) but differ slightly due to different values of r_f. The results with the new parameters are compared with experiment in Table 16-6. We see that in the center of the series, we predict values much too small. We see correspondingly in Fig. 16-4 that for neptunium the total energy is very flat near the equilibrium spacing; its curvature is near zero. Such flatness signals instabilities, and alternative structures. That of lowest energy may have a much higher curvature and bulk modulus. For the same reason, the spacing at this point is found to be very sensitive to small changes, such as elimination of the Coulomb energy, U_f, as seen in Problem 16-1.

A similar comparison of the Grüneisen constant, and the resulting thermal expansion coefficient, with experiment was made in Harrison (1983d) based upon the earlier r_f . The resulting values are listed in Table 16-7. The general magnitudes are correct, but not the trends through the series, perhaps for the reasons discussed for the small predicted bulk modulus.

Table 16-6. Bulk moduli (in 10^{12}ergs/cm^3) predicted from the second derivative of Eq. (16-26) with respect to r and Eq. (16-27). The value below is the experimental value from Kittel (1976).

	Ac	Th	Pa	U	Np	Pu	Am	Cm
Theory	0.53	0.39	0.30	0.11	0.12	0.36	0.65	0.55
Exper.	0.25	0.60		1.21	0.77	0.55	0.36	

Table 16-7. Volume derivative of the bulk modulus, $-(\Omega/2B)\partial B/\partial\Omega$, from Harrison (1983d), based on the theory described here with similar, but different, r_f . Also the resulting thermal expansion coefficient is compared with experiment from R. O. Elliot (private communication).

	Th	Pa	U	Np	Pu	Am
$-(\Omega/2B)\partial B/\partial\Omega$	1.46	1.15	0.78	0.30	0.37	1.48
α (10^{-6}/^{o}K)	12.2	11.4	12.1	6.4	6.5	7.1
$\alpha_{expt.}$	≈12	≈11	≈19	≈28	53.89	7.1

C. Interatomic Forces

We may obtain a representation of the energy as interatomic forces using Eq. (16-26) as we did for the transition metals in Section 15-5. The first three terms are simple-metal terms, which will be replaced by an interaction $Z_m{}^2e^2\cosh^2\kappa r_c e^{-\kappa d}/d$, based upon the electron gas of $Z_m = 3$ electrons per atom. The remaining terms are from the f-shells and can be divided by $X/2 = 6$, and the atomic sphere radius r written in terms of the internuclear distance d for fcc structures, $4\pi r^3/3 = d^3/\sqrt{2}$, to obtain the contribution to an effective nearest-neighbor interaction. The resulting interaction could be used in calculations such as for the vibration spectrum as for simple metals in Section 14-4, or for the structure of defects.

D. Crystal Structure

The crystal structures for the heavier actinides, americium and beyond, are close-packed and follow the scheme we have described for the transition metals and rare earths. However, the lighter actinides, thorium through plutonium, form in complicated structures unique to these systems, as we indicated in connection with Fig. 16-4. There have been efforts to rationalize these in terms of angular bonding terms from f-orbitals, or from the incipient localization of the f-states, but the analysis we have given would suggest that the real reason may be that with intermediate-width bands, irregular interactions arise, as for interatomic interactions deduced as we suggested from Fig. 16-4, and we may find a lower-energy structure (at given volume) by arranging different neighbors at different spacings. For an fcc structure, all nearest neighbors are at the same spacing - fixed by the volume per atom - and no advantage can be taken of the complexity of the interatomic interactions. The effect becomes largest at a specific point in each series, as for plutonium in Fig. 16-4. The idea would be that this occurs in the 3d-transition elements at manganese, not at all in the 4d nor 5d series, at cerium in the rare earths (where it produces phases at two different volumes) and for the light actinides from thorium through plutonium.

An alternative view of this, which may be closely related, has been given in an insightful study by Söderlind, Eriksson, Johannson, Wills, and Boring (1995). They note that the Madelung energy favors a simple close-packed (fcc, bcc or hcp) structure. For a free-electron metal, or a transition metal with a very broad band, this Coulomb energy dominates the determination of structure as we have indicated. For narrower bands there might be a tendency to favor a lower-symmetry structure with fewer atoms per primitive cell for just the reason it occurs in the Peierls distortion discussed in Section 18-4-B: by increasing the number of atoms in the cell additional Bragg planes are introduced, and if these occur at the Fermi surface the energy will be lowered by repopulating the states. Perhaps there is a range of band widths where this energy gain is large enough to overcome the increased Madelung energy. Söderlind, et al, tested this supposition in detail by comparing the energy of the complicated alpha-uranium structure with the face-centered-cubic (or body-centered cubic) structure as a function of atomic volume for uranium, for the transition metals niobium and iron, and for the simple metal aluminum. Plotting each energy as a function of band-width (f-band in the case of uranium, d-band in the case of the transition metals, and free-electron Fermi energy in the case of aluminum) they found that in all cases the alpha-uranium structure was stable at low enough band width, though of course only in the case of uranium was the band width in the observed structure sufficiently narrow. This could be the counterpart in wavenumber space of the real-space argument for complex structure based upon Fig. 16-4. In either case, it is suggestive that the complicated structures of the light actinides arise because the band widths are of the order of one to three eV for the f-electrons, and not because of the orbital symmetry of the f-states. This also makes the interesting prediction that these light-actinides would form in close-packed structure at sufficiently high pressure where the bands had broadened sufficiently.

5. Formation of Local Moments

In our treatment of the bonding properties of transition and f-shell metals we have used the Friedel model and included the Coulomb interactions using a generalization of the exact correlated energy of the Li_2 molecule. In the rare earths, the effects of the coupling V_{ffm} became negligible and the f-levels behaved as core-like localized atomic orbitals. This would also be true if one rare-earth atom were dissolved in a simple metal. In other cases where the Coulomb U is not so large, the isolated atom becomes simply a scattering center, as in the simple metal alloys described in Section 14-7. We address here the intermediate cases, with an isolated transition-metal or f-shell-metal atom in a simple-metal electron gas. The interatomic interactions which broadened the bands in our discussion of bonding properties are not present and the approach based upon the exact solution is not applicable. We also take the opportunity here to introduce phase shifts which have not been needed before, and the Friedel Sum Rule.

We described the physics of such an isolated state in Section 16-1-Da and we return to fill in the details here. This seems to require the use of the less-accurate Unrestricted Hartree-Fock description of the Coulomb interactions, but can provide additional insight into the correlated state and the possibility of treating properties of individual atoms.

As we discussed for the rare earths, when the correlated state is formed, the spins align - by Hund's rule - and there is a net moment associated with each atom, which is called a *local moment* . Thus it is customary to discuss the criterion for the formation of the correlated state as the criterion for formation of a local moment. The first treatment of such moments in metals was given by Anderson (1961). Essentially the same theory was applied to f-shell metals in Harrison (1984) and we follow that approach here. It will require introduction of the concept of phase shifts and of a resonant state, neither of which we have had occasion to use before. A more complete analysis has been given by Wiegmann and Tsvelick (1983). We shall make use of the results of the analysis in this section to define a "resonance width" in Eq. (16-36) which enters the criterion, Eq. (16-41), for the formation of correlated states , the counterpart of the condition $V_{ss\sigma} < U/2$ for the formation of a correlated state which we found for the diatomic molecule in Section 16-1. We shall also use it to discuss some special properties, in particular of americium, in Section 16-6.

A. Phase Shifts

In Eq. (15-9) we expanded free-electron plane waves in terms of spherical waves. These spherical waves, $j_\ell^m(kr)Y_\ell^m(\theta,\phi)$, are solutions of the Schroedinger Equation with a constant potential and are a good starting point if we wish to consider states associated with a single atom in a metal. We let the constant potential be the flat portion of the potential, the muffin-tin zero equal to the d-state energy in the case of transition metals. The spherical Bessel functions $j_\ell(kr)$ are solutions of the radial Schroedinger Equation (Eq. (15-7)) with energy $\hbar^2k^2/2m$ relative to this constant potential. We gave the form for these spherical Bessel functions at small r . At large r they are given by (Schiff (1968) p. 86)

$$j_\ell(kr) \rightarrow \frac{1}{kr}\cos(kr + \delta_\ell - \frac{1}{2}(\ell + 1)\pi) \qquad (16\text{-}28)$$

with δ_ℓ equal to zero. This is illustrated as the bold curve in Fig. 16-5. We can immediately see that this form is correct by considering the Eq. (15-7) at large r , where the second term, proportional to $1/r^2$, is negligible, and substituting $R_\ell(r) = \chi(r)/r$ .We find the general solution of the equation is $R_\ell(r) = [A\cos(kr) + B\sin(kr)]/(kr) = C\cos(kr + \delta)/(kr)$. A more complete analysis is needed to obtain the form of the constant inside, $-{}^1/_2(\ell + 1)\pi$.

This same argument for the form of the solution applies also if there is a spherically symmetric potential near the origin, within a muffin-tin sphere, but then the constant δ in $\cos(kr + \delta)$ will be different and can be written as Eq. (16-28) but now with $\delta_\ell \neq 0$. The result is that the effect of the potential at the origin on the state far from the origin is represented by the *phase shift* δ_ℓ which it causes, illustrated as the light curve in Fig. 16-5. This in fact represents most of the effect of the potential. We may construct states in a very large sphere, of radius R centered at this atom, with wavefunctions which go to zero at that spherical surface. Then the boundary condition is written in terms of this large-distance form, Eq. (16-28), as

$$kR + \delta_\ell - \frac{1}{2}(\ell + 1)\pi = n\pi \tag{16-29}$$

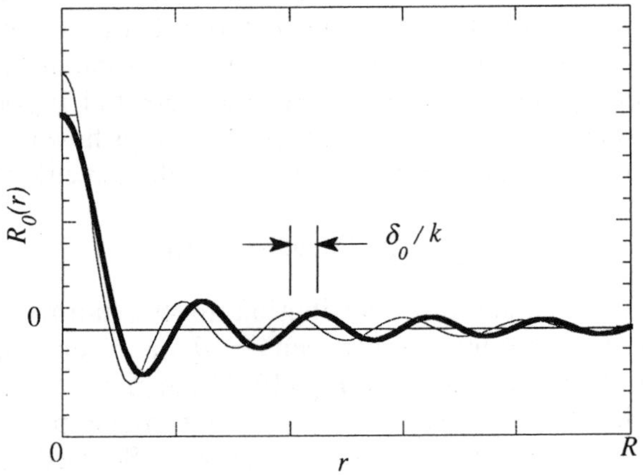

Fig. 16-5. A plot of $R_\ell(r)$ for $\ell = 0$. The bold curve is for no potential, so it is the spherical Bessel function. For the light curve an attractive potential has been introduced near $r = 0$, producing a phase shift δ_0 in the state. The wavenumber was then shifted so that the state again vanished at the surface, $r = R$.

with n any integer. Without the phase shifts δ_ℓ the states for each angular-momentum quantum numbers ℓ and m are closely spaced with wavenumber difference $\delta k = \pi/R$, and this remains true, to lowest order in $1/R$ with the δ_ℓ nonzero. If we normalize the state, the angular integral over the $Y_\ell^{m*}Y_\ell^m$ gives one, and the integral $\int r^2R_\ell(r)^2dr$ is dominated by large distances where the integrand is $(1/k^2)\, C^2\cos^2(\,kr\; + \delta_\ell - {}^1/_2(\ell + 1)\pi)$. The integral becomes $C^2R/(2k^2) = 1$ so the asymptotic form of the normalized state is $\sqrt{2k^2/R}\,\cos(\,kr\; + \delta_\ell\; - \;{}^1/_2(\ell + 1)\pi)/(kr)$ with the normalization constant independent of the phase shift as in Fig. 16-5, where the amplitude increased at small r but not at large distances. In order to satisfy the boundary condition at $r = R$ we must shift the wavenumber by Δk such that $(k + \Delta k)R + \delta_\ell = kR$, as was done in Fig. 16-5. This shift is small, so the energy of the state has been shifted by

$$\Delta\varepsilon = \frac{\hbar^2 k\Delta k}{m} = -\frac{\hbar^2 k\delta_\ell}{mR}\,. \tag{16-30}$$

These δ_ℓ are also called *scattering phase shifts* since the scattering of electrons by a center can be described entirely in terms of the states at large distance and the phase shifts characterize those states. From Eq. (16-30) we see that they also give us the shift in the energy of the metallic states [as in Fumi's method (Fumi (1955))] due to the potential introduced within a muffin-tin sphere. We next see that they also tell us how much extra charge is brought to the central region by the potential, the screening charge.

B. The Friedel Sum Rule

We may calculate the contribution to the normalization integral $\int r^2R_\ell(r)^2dr$ from the integral between $r = 0$ and some radius M , large enough that the asymptotic form, Eq. (16-28), applies and yet much smaller than R . Friedel (1952) evaluated this integral by the procedure which we copied in obtaining Eq. (15-26), writing the radial Schroedinger Equation for two different energies, subtracting, and integrating to this radius M . For small differences Eq. (15-26) becomes

$$-\frac{\hbar^2}{2m}M^2\left(R_\ell'\frac{\partial R_\ell(r)}{\partial r} - R_\ell\frac{\partial R_\ell'(r)}{\partial r}\right)_{r = M} = \Delta\varepsilon \int_{0,M} r^2R_\ell(r)^2dr\,. \tag{16-31}$$

The left side is evaluated using the asymptotic form of the wavefunction to find the excess normalization arising because the phase shift is nonzero, finding a contribution proportional to $\partial\delta_\ell/\partial\varepsilon$.

Friedel sought the total contribution for the states with energy less than the Fermi energy, so he summed over wavenumbers and over angular momentum, converting the sum over k to an integral and obtained the very simple result that the number of electrons accumulated locally is

$$\Delta N = \frac{2}{\pi}\sum_\ell (2\ell + 1)\, \delta_\ell(\varepsilon_F)\,, \tag{16-32}$$

the *Friedel Sum Rule* . This derivation did not require any assumption about the potential being small, only that it be restricted to lie near the origin. He also obtained and discarded the oscillatory terms which are the Friedel oscillations we have discussed. They do not affect the validity of Eq. (16-32) and it turns out do not disrupt the assumption of the potential being restricted to near the origin.

One aspect of Eq. (16-32) which we should note is that if we introduce an attractive potential strong enough to form a bound state, the state just above that bound state, of the same angular momentum, will be the lowest-energy metallic state. By following its evolution, looking at a figure such as Fig. 16-5, we see that it acquires a phase shift of π . This added shift in Eq. (16-32) counts the $2\ell + 1$ bound states of the same angular momentum, for both spins. One way of understanding the simple-metal pseudopotentials which we introduced in Section 15-1 is that they remove these extra π-phase shifts which count the core states, and give us the much smaller phase shifts which correspond to the same wavefunction (except maybe for sign) outside the core region, and therefore the same energies. Because of the smallness of the phase shifts, they can be treated in perturbation theory to obtain the small shifts corresponding to Eq. (16-30).

C. Resonant States

We can now describe in terms of phase shifts the nature of the d-like and f-like states in metals which we have been discussing. To be specific, the f-states in a metal correspond to phase shifts of the form (see Messiah (1958), p. 396ff or Harrison (1970), p. 197ff)

$$\tan\delta_f(\varepsilon) = \frac{\Gamma_f}{2(\varepsilon_f - \varepsilon)}\,, \tag{16-33}$$

with Γ_f a "resonance width" which is small and only weakly dependent upon energy. We may understand this form by noting that with Γ_f small, δ_f rises from near zero (with possibly some integral multiples of π) to near π over a narrow energy range of order Γ_f . From the Friedel sum rule, based on energy ε rather than specifically ε_F , we see that $2\ell + 1$ extra states of each spin and of angular momentum $\ell\hbar$ have been introduced at this center over a narrow energy range near ε_f, which is precisely what we did by adding atomic f-states to the free-electron, simple-metal basis. This has occurred by decreasing the spacing between the successive states allowed by the boundary condition at R to fit in the extra state for each Y_ℓ^m .

We may identify the width with our hybridization by noting that for ε well below ε_f , the energy shift from Eqs. (16-30) and (16-33) is

$$\Delta\varepsilon = -\frac{\hbar^2 k\delta_\ell}{mR} = \frac{\hbar^2 k\Gamma_f}{2mR}\frac{1}{\varepsilon - \varepsilon_f}$$
$$= \frac{\Sigma_f <kfm|H|f> <f|H|kfm>}{\varepsilon - \varepsilon_f} \qquad (16\text{-}34)$$

where in the final form we have written the normalized free-electron state $|kfm> = \sqrt{2k^2/R}\, j_3(kr)Y_3^m(\theta,\phi)$. This final expression is just the shift in that free-electron level, from second-order perturbation theory, due to the hybridization coupling between it and the local f-state of the same symmetry. We can use these two expressions directly to relate the resonance width to the square of the hybridization matrix element. We write out the matrix element between a plane wave, with wavenumber $\mathbf{k}$ along the z-axis, and the local f-state by expanding the plane wave in normalized states $|kf0>$ using Eq. (15-9). The only contribution comes from the local state of the $m = 0$ symmetry, and when it is squared we find

$$<<\mathbf{k}|H|f> <f|H|\mathbf{k}>> = \frac{4\pi(2\ell + 1)R}{2k^2\Omega} <kf0|H|f> <f|H|kf0> , \qquad (16\text{-}35)$$

with of course $\ell = 3$ and $|f>$ being the state with $m = 0$. We may substitute from Eq. (16-7) for the plane-wave matrix elements on the left and noting that $Y_\ell^0(0,\phi)^*Y_\ell^0(0,\phi) = (2\ell+1)/4\pi$ we may solve for $<kf0|H|f> <f|H|kf0>$, substitute it in Eq. (16-34) and solve for the resonance width as

$$\Gamma_f = \frac{1}{3\pi}\frac{\hbar^2 k^{2\ell+1} r_\ell^{2\ell-1}}{m} = \frac{1}{3\pi}\frac{\hbar^2 k^7 r_f^5}{m} . \tag{16-36}$$

k is to be taken as the wavenumber at resonance $k_f = \sqrt{2m\varepsilon_f/\hbar^2}$, where the free-electron band energy equals the resonant energy. This applies for all angular momentum, as written in the first form. The same form was derived for d-states in term of band widths in Harrison (1980), p. 511, with a leading coefficient 15% larger due to approximations not required here. This derivation is preferable in any case since it is based upon an isolated atom rather than a band.

We now have everything required to see if a correlated state, with a local moment, is formed.

D. The Criterion

We follow the analysis given in Harrison (1984), based upon Anderson's (1961) approach as applied to a degenerate resonance such as the f-shell. We use the Unrestricted Hartree-Fock idea in allowing resonances for some m- values to lie lower in energy, and therefore have greater occupation, than others, as we allowed the spin-up and spin-down occupation to be different on the two atoms in the Heitler-London problem of Section 16-1. We will in fact let Z_f resonances lie below the Fermi energy (called *occupied resonances*) and $14\text{-}Z_f$ lie above (called *empty resonances*). It is important that in the present case spin is not the distinction since Z_f is ordinarily not seven. For each set we may calculate the number of electrons in the resonance using the Friedel Sum Rule, Eq. (16-32). For the occupied resonance we write the phase shift at the Fermi energy $\delta(\varepsilon_F)$ as δ^- and for the empty resonance we write it δ^+ . Thus the total effective occupation of the resonance is written

$$Z_f{}^* = Z_f\frac{\delta^-}{\pi} + (14\text{-}Z_f)\frac{\delta^+}{\pi} , \tag{16-37}$$

which will be approximately Z_f . Now the energy for any electron occupying the resonance depends upon the occupation of the other thirteen resonances (but not its own resonance since no artificial self-interactions are introduced). Thus the energy of an occupied resonance contains $Z_f{}^* U_f$ minus its own contribution, or

$$\varepsilon_f{}^- = \varepsilon_f{}^0 + (Z_f^* - \delta^-/\pi)U_f\,,$$
$$\varepsilon_f{}^+ = \varepsilon_f{}^0 + (Z_f^* - \delta^+/\pi)U_f\,. \tag{16-38}$$

This allows different resonant energies for the two different sets of resonances. $\varepsilon_f{}^0$ should be calculated self-consistently. We note that if the resonances are well removed from the Fermi energy δ^- will be near π and δ^+ will be near zero, and the average of $\varepsilon_f{}^+$ and $\varepsilon_f{}^-$ from Eq. (16-42) is $\varepsilon_f{}^0 + (Z_f^* - {}^1/_2)\,U_f$. Harrison (1984) took the Fermi energy to lie midway between the two resonances in the absence of coupling so that

$$\varepsilon_f{}^0 \approx \varepsilon_F - (Z_f - {}^1/_2)\,U_f \tag{16-39}$$

and we do that also. For each actinide, with a particular $Z_f = Z - 3$, this sets the f-level relative to the Fermi energy as appropriate for that actinide. It is then shifted self-consistently according to Eq. (16-38) by the coupling to the metallic states.

The phase shifts themselves are given by Eq. (16-33), which becomes

$$\tan\delta^\pm = \frac{\Gamma_f}{2(\varepsilon_f{}^\pm - \varepsilon_F)}\,. \tag{16-40}$$

We may now substitute Eq. (16-37) (for Z^*) , Eq. (16-39) (for $\varepsilon_f{}^0$), and Eq. (16-40) (for $\varepsilon_f{}^\pm$) into the two equations, Eq. (16-38). We may then solve one for δ^- in terms of δ^+ , and the other for δ^+ in terms of δ^-, to obtain

$$\delta^- = \frac{\pi}{Z_f}\left(\frac{\Gamma_f}{2U_f\tan\delta^+} - (14 - Z_f)\frac{\delta^+}{\pi} + \frac{\delta^+}{\pi} + (Z_f - {}^1/_2)\right)$$
$$\delta^+ = \frac{\pi}{14 - Z_f}\left(\frac{\Gamma_f}{2U_f\tan\delta^-} - Z_f\,\frac{\delta^-}{\pi} + \frac{\delta^-}{\pi} + (Z_f - {}^1/_2)\right). \tag{16-41}$$

These are to be solved together for the $\delta^\pm$, and in fact the UHF solution for the Heitler-London problem of Section 16-1-B could also have been formulated as such a pair of equations, to obtain the minimum-energy solution.

In order to understand the criterion for forming a correlated state we follow Anderson (1961) and plot the two equations in Fig. 16-6 using parameters appropriate to an americium atom in a $Z_m = 3$ metal corresponding to the free electrons in americium. Γ_f was evaluated from Eq. (16-44) at the Fermi energy. For these parameters there are three solutions, which we shall discuss in the following subsection. That designated as "local" is the lowest-energy solution and corresponds to a correlated state. The solution with $\delta^+ = \delta^-$, designated "itinerant", is of course the ordinary Hartree-Fock, or band-like, solution. If we change parameters, this itinerant solution remains and if it is the *only* solution, there will be no correlated state. If there is more than one solution the correlated state will form, as when there is more than one solution of $\partial E(\gamma)/\partial\gamma = 0$ in Fig. 16-1a. The condition for a correlated state is immediately seen from Fig. 16-6 to be that the slope of the curve for the second of Eqs. (16-41) (for which δ^+ diverges as δ^- goes to zero) is greater than that of the curve for the first equation, at the point $\delta^+ = \delta^-$. Thus we evaluate the derivative $\partial\delta^+/\partial\delta^-$ of the two equations,

$$\frac{\partial\delta^+}{\partial\delta^-} = Z_f\left(-\frac{\pi\Gamma_f}{2U_f\sin^2\delta^+} - (14 - Z_f) + 1\right)^{-1},$$

$$\frac{\partial\delta^+}{\partial\delta^-} = \frac{1}{14-Z_f}\left(-\frac{\pi\Gamma_f}{2U_f\sin^2\delta^-} - Z_f + 1\right). \quad (16\text{-}42)$$

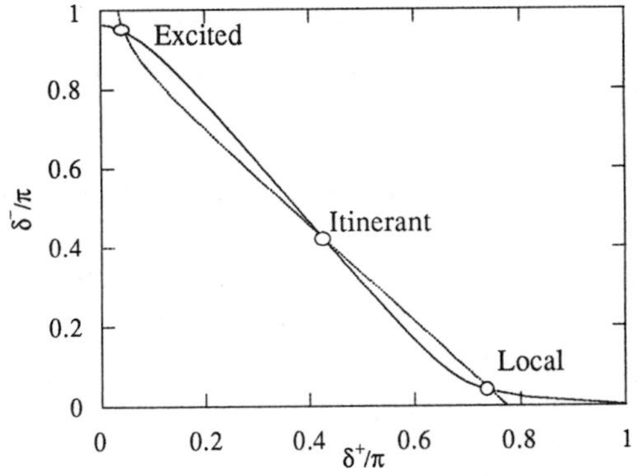

Fig. 16-6. Eqs. (16-41) are plotted for parameters appropriate to americium with $Z_f = 6$. The intersections of the two curves give solutions for the electronic structure. The solution "itinerant" corresponds to $\delta^- = \delta^+$, and is the ordinary band-like solution with unsplit resonances. The solution labeled "local" has lower energy, with six resonances well below the Fermi energy and eight above. The solution labeled "excited" has higher energy, with eight resonances occupied and six empty. All three have $Z_f^* \approx 6$.

We note that they are equal with $\delta+ = \delta- = \delta$ if $\pi\Gamma_f/(2U_f \sin^2\delta) = 1$ and the second becomes less negative if it is smaller than one. The condition for a correlated state becomes $\pi\Gamma_f/(2U_f \sin^2\delta) < 1$, or if we note that $Z_f^* \approx Z_f$ we may use Eq. (16-37) to write (Harrison (1984))

$$\Gamma_f \lessapprox \frac{2U_f}{\pi}\sin^2\left(\frac{\pi Z_f}{14}\right) \qquad (16\text{-}43)$$

as the condition for formation of a correlated state. For the single level treated by Anderson (1961) the 14 would be replaced by 2 and Z_f by one, so the factor $\sin^2(\pi Z_f/14) = 1$ and that factor did not appear. This factor drops to 0.05 at each end of the series, making it much more difficult to form a correlated state so it is an essential factor.

We should also emphasize that the derivation of this condition was based upon the properties of an isolated resonance, and did not depend upon there being an entire lattice of levels and the possibility of the formation of a band. However, in the case of such a lattice of levels, a band will be formed. Even in the case of correlated states, the coupling between adjacent occupied resonances, or between empty resonances, will broaden them into a band. If that band does not cross the Fermi energy, our treatment as resonances below and above the Fermi energy remains appropriate. If a band does cross the Fermi energy, the occupation of the resonance is modified and the calculation should be modified. In particular, if the band crosses the Fermi energy we may lower the band energy by populating lower band states and can more easily form the correlated state. This would increase the factor on the right in Eq. (16-43), particularly where one of the resonances is near the Fermi energy; that is, when Z_f is near one or fourteen.

We cannot directly relate the resonance width and the band width since they depend upon different parameters. However, they are both proportional to r_f^5 so we can provide an approximate relationship for the pure material. Γ_f depends upon the wavenumber k_f of the resonance and if the band width is small, this must be near the Fermi wavenumber k_F determining the free-electron density. If there are in fact about three free-electrons per atom, this k_F can be obtained in terms of r_0 using Eq. (13-2) as $k_F^3 = 27\pi/(4r_0^3)$. Then taking $k_f = k_F$ we may evaluate Eq. (16-36) as $\Gamma_f \approx 132.\hbar^2 r_f^5/(mr_0)^7$. Comparing with Eq. (16-12), $W_f = 181.\hbar^2 r_f^5/(mr_0)^7$ we obtain

$$\Gamma_f \approx 0.726\, W_f \qquad (16\text{-}44)$$

for the bulk close-packed metal. Then the criterion from Eq.(16-43) for formation of a correlated state becomes

$$W_f \lesssim 0.877\, U_f \sin^2\left(\frac{\pi Z_f}{14}\right). \qquad (16\text{-}45)$$

Similarly, for d-states we take the first form of Eq. (16-36) for Γ_d and Eq. (15-33) for W_d. We relate the Fermi wavenumber to the atomic-sphere radius for 1.5 free electrons per atom for transition metals as $k_F{}^3 = 27\pi/(8r_0{}^3)$ and take the wavenumber of the resonance at the Fermi wavenumber to obtain

$$\Gamma_d = 5.43\frac{\hbar^2 r_d{}^3}{m r_0{}^5} = 0.982\, W_d\,. \qquad (16\text{-}46)$$

The criterion for formation of correlated states from Eq. (16-43) becomes $\Gamma_d \lesssim (2U_d / \pi) \sin^2 (\pi Z_d / 10)$ for d-states, or

$$W_d \lesssim 0.648\, U_d \sin^2\left(\frac{\pi Z_d}{10}\right). \qquad (16\text{-}47)$$

It is interesting to compare these conditions, which ultimately came from the criterion for formation of an isolated local moment, with the conditions which we developed at the very end of Section 16-1 for formation of a correlated state from an entire band, $W < U/2$. There was no $\sin^2(\pi Z_f/14)$ factor in the condition, though an analogous factor of $Z_f(1\text{-}Z_f/14)$, going to zero at $Z_f = 0$ and 14, arose. Since it appeared both with the U_f and W_f it seems not to affect the relative importance of the two terms. The factor $\sin^2(\pi Z_f/14)$ does appear in the theory of photoemission of cerium by Gunnarsson and Schönhammer (1983). The condition for formation of the isolated local moment is a little less stringent, 0.648 and 0.877 being greater than 0.5, than for the band at half filling, when the $\sin^2$ is one. It becomes more stringent away from half filling as the $\sin^2$ factor becomes small. This finding would seem to be correct, and is consistent with the experimental finding that nickel atoms, with $Z_d \approx 8.5$, dissolved in copper do not form local moments until the nickel concentration reaches a critical value in the neighborhood of half nickel. We use the form based upon local moments, in its form based upon band widths as in Eqs. (16-45) and (16-47), for our discussion of f-shell metals.

In Harrison (1984) Γ_f was obtained a little more precisely from Eq. (16-36), using an ε_f which would lead to Z_f electrons per atom in the f-bands, and essentially the same r_f and U_f as in Tables 16-1 and 16-2, to see when the condition given in Eq. (16-43) was satisfied. This is very nearly the same as Eq. (16-45). It was found that all rare earths, except for cerium, satisfied the condition, and correlated states were predicted. Our neglect of any f-shell contributions to the bonding in Section 16-3 was consistent with such correlated states.

The case of cerium is interesting. The right side of Eq. (16-45) is 0.31 eV, based upon the U_f of Table 16-1, and exceeds the W_f for cerium from that table suggesting a local moment should form. However, we indicated that LMTO estimates give three times as large an estimate of W_f suggesting that no moment should form. As we indicated in Section 16-3-E cerium forms a high-density α-phase at low temperatures, in which no moments are believed to form. By room temperature it transforms to a low-density γ-phase, which apparently has a local moment just as the other rare earths do. Perhaps in this borderline case - borderline experimentally and theoretically - our parameters are inadequate. We would have hoped that a plot of the total energy for cerium, as for the actinides in Fig. 16-4, would have yielded two minima with Eq. (16-45) satisfied only for the minimum at larger r , but our parameters obtained independently did not do that. The r_0 we listed in Table 16-1 is the room temperature r_0 and we were therefore treating the room-temperature properties of cerium, with results consistent with that phase.

For the actinides, the condition Eq. (16-45) is clearly satisfied for americium, and heavier elements, marginally satisfied in plutonium, and not satisfied for the lighter actinides, completely in accord with the properties of these metals. However, as for the γ-phase of cerium, we no longer regard the Coulomb correlations as the cause of the low-density of these metals, but as a consequence of the low density. If we were to compress americium to the density for which its r_0 were equal to that of plutonium, the increased Γ_f would no longer satisfy the condition, Eq. (16-43), and we believe that at such a pressure americium would lose its local moment (a moment which we shall see has zero magnitude in any case).

The criterion has achieved something very useful. It tells us for each metal how we should think about its electronic structure and calculate its properties. The correlated state forms when the f-shells do not participate in an important way in the bonding, so when it has formed we can take it as a signal that ignoring the contribution to the energy $-\sqrt{W_f^2 + U_f^2} + U_f$ will probably not be a serious error, though it is of course always a real contribution to the energy.

6. Strongly Correlated Systems

We look finally at the electronic and magnetic properties of strongly correlated systems, systems with $U_\ell \approx W_\ell$.

A. The Electronic Structure of Americium

We shall give here the results of the analysis for americium from Harrison (1984), based upon slightly different parameters than given here. Eqs. (16-41) were solved, as we illustrated in Fig. 16-6 for newer parameters, to see that the local solution corresponded to a $\delta^- = 2.85$ for six localized states 1.44 eV below the Fermi energy, and eight states 1.81 eV above the Fermi energy. Even with the full band width of 1.26 eV for the itinerant solution, the bands are too narrow to overlap the Fermi energy. The itinerant solution gave $\delta^+ = \delta^- = 1.34$ with all fourteen resonances lying 0.10 eV above the Fermi energy, so the f-band would span it and be partially filled. However that solution is of higher energy.

The third solution, designated "excited", was not anticipated. It contains eight resonances 0.35 eV *below* the Fermi energy and six 2.26 eV above as indicated in the caption to Fig. 16-6. When we divided the resonances to six and eight, both solutions came out and this one is expected to have higher energy, and it does. It would be interesting to compare the energy of a solution assuming seven occupied resonances, but that has not been done, and which energy would be lowest would depend upon our choice of ε_f^0 made in Eq. (16-39). We regard that as a good choice and expect that it would yield a higher energy for seven occupied resonances. It is quite remarkable that for the three solutions illustrated in Fig. 16-6, almost the same Z_f^* is obtained for each, 6.04 for the local solution, 5.97 for the itinerant solution, and 6.14 for the excited solution. As more resonances are occupied, they move to higher energy, reducing their contribution. An assumed seven resonances, with the same ε_f^0, should give a Z_f^* midway between the $Z_f = 6$ and $Z_f = 8$, at $Z_f^* = 6.10$.

Having established that the correlated state with six resonances occupied is the correct description of the ground state, we may ask which of the fourteen resonances are occupied. Of course exchange energy favors occupying them with parallel spin, Hund's rule, as in the rare earths, giving a total spin of $S = 3$. We list the exchange energies for the actinides in Table 16-8. Based upon the value $U_x = 0.36\ eV$ for americium, flipping one spin would cost an energy $(Z_f - 1)U_x = 1.8$ eV. This should be very stable. Then with all but one m-value occupied for that spin, the total angular momentum is $L = 3$. Spin-orbit coupling favors the state of lowest J as seen in Section 16-2-C so we expect the ground state to have $J = 0$, with the energy increase in going to $J = 1$ being $2V_{SO} = 0.66$ eV from Table 16-2. This confirms the supposition by Skriver, Andersen, and Johansson (1978) of a localized, $J = 0$ ground state for metallic americium. With $J = 0$ it is isotropic and though the magnetic moments arising from spin would be estimated to be different from that arising from the orbital motion, the atom behaves as a system with zero moment.

Table 16-8. Exchange energy U_x for the f-levels in the actinides, obtained in Harrison (1984) (which included also values for the rare earths) from the work by Nugent, Baybarz, Burnett, and Ryan (1970).

	Pa	U	Np	Pu	Am	Cm
U_x (eV)	0.22	0.27	0.29	0.32	0.36	0.40

B. Antiferromagnetism

It would seem that in this Unrestricted Hartree-Fock context, the spins would always align on each atom if the resonances become split as we have described. Furthermore, we may expect that the moments on adjacent atoms will align antiparallel. We see this by forming Z_f occupied states of one spin, and $14 - Z_f$ empty orbitals of opposite of spin, higher in energy by U_f. Any occupied orbital will be coupled, by some combination of V_{ffm} 's, to every orbital of the same spin on the neighboring site, but only for the neighboring empty orbitals will there be an energy gain $(- V_{ffm}^2/U_f)$. Thus for an atom with orbitals of predominantly one spin, it will be energetically favorable to have the neighboring orbitals of that spin predominantly empty, for the moments to alternate from atom to atom. Thus antiferromagnetism is predicted for these localized systems.

This coupling, of order $- V_{ffm}^2/U_f$, can be very small, particularly in the rare earths. Then other contributions to the interaction, such as the Ruderman-Kittel interaction discussed in Section 16-3-A, can be larger and dominate the magnetic properties. Indeed the rare earths are far from simple antiferromagnets.

This predicted antiferromagnetism also reminds us of Eq. (15-66) which gave a criterion for alignment of the spins in the solid (ferromagnetism) on the assumption of band states each of a single spin. Such a state is favored when the bands are broad. On the other hand, Eqs. (16-45) and (16-47) are criteria for alignment of spins for isolated atoms, or when the bands are very narrow. As for many other problems we have treated, one can choose the approximate representation which is most appropriate to the system in question. For intermediate cases, where we do not know the magnetic state, we might try to compare the energy for ferromagnetic and antiferrromagnetic states, but because the approximations used for the two cases are so different, we could not be confident of the prediction.

Finally, we note that we used a very different description of antiferromagnetism for chromium in Section 15-2-D, where the spin fluctuations were not commensurate with the lattice. Had they come on alternate atoms, the description which we have given here could also have been used.

C. The Kondo Effect

We should mention briefly an important aspect of the local moments which arose initially from the observation at low temperatures of a minimum in the resistance as a function of temperature in systems which have local moments. Kondo (1966) correctly pointed out the origin of this minimum in the scattering of metallic electrons in which the orientation of the local moment is modified in an intermediate state of the scattering. This can lead to a scattering rate obtained in perturbation theory which diverges as the temperature goes to zero.

This divergence actually reflects an instability of the system with a local moment which can have various orientations, all of the same energy. As one would guess if the question arose, there is a lower-energy isotropic state which is a combination of the states of all orientations, referred to as the Kondo state. It is analogous to the ground state of the lithium molecule which we discussed in Section 16-1; that ground state was a mixture of the state with spin-up on the right and spin-down on the left and the state with spins reversed. We found it to have twice as deep an energy as the individual (Unrestricted Hartree-Fock) states with broken symmetry and a specific moment on each atom. Like superconductivity it has only a tiny effect on the total energy, but like superconductivity it is an intriguing phenomenon. It gives rise to a small peak in optical absorption taking electrons to the Fermi energy, though the electronic resonances are well separated from the Fermi energy as we have indicated.

D. Heavy-Fermion Systems

We have focused our attention on systems such as transition metals which are clearly describable by the energy-band picture or most rare earths which are clearly describable in terms of localized states. There are of course systems which lie at the boundary between those two extremes, though surprisingly few.

One such set of systems are called *heavy-Fermion systems* ; they are metallic alloy, or compound, systems containing rare-earth or actinides in which the Coulomb interactions between the electrons make major modifications in their properties without completely localizing the f-states. There really is no distinction between the two descriptions as long as one talks only of the ground state, and it is the excited states of the system which define the difference. The tendency is for heavy-Fermion systems to behave as local systems with respect to high-energy excitations and band-like with respect to very low-energy excitations. In this regard "high-energy" extends down to thermal energies at high temperatures. For the low-energy excitations, at low-temperature thermal energies, the velocities of the particles are greatly reduced even in comparison to band calculations which yield narrow bands and correspondingly low velocities for these systems. The term "heavy" refers to these low velocities. For a recent discussion of this renormalization of the bands which reduces the velocities, and for references, see Zwicknagl (1993).

Problem 16-1. Volume-dependence of the energy in uranium.

Evaluate and plot Eq. (16-26) using the parameters for uranium from Table 16-2. The author found a flat region similar to that in neptunium as shown in Fig. 16-4. The minimum may not come at the observed r_0 because of round-off errors in the various parameters.

To illustrate the sensitivity of this structure to the parameters, replot the curve with U_f equal to zero but other parameters the same as before.

It might be interesting to explore the energy for different structures in such a case, converting the f-electron part to interatomic interactions first, as described in Section 16-3-D. This has not yet been done.

CHAPTER 17

Transition-Metal (AB) Compounds

One path into transition-metal compounds is as a continuation of the isoelectronic compounds (compounds which can be obtained from each other simply by transferring protons between different nuclei) which we discussed in Chapter 9, Such a series is KF, CaO, ScN, and TiC. These are the octet compounds all with a total of eight valence electrons since the transfer of protons does not change the number of electrons. This is quite a narrow category, but informative. We shall discuss these octet compounds first after introducing the necessary parameters for all transition-metal compounds.

As with semiconductors and insulators, such AB compounds, with equal numbers of the two constituents, are particularly important and we go on to compounds with different total numbers of valence electrons. We shall focus in particular on the transition-metal carbides, nitrides, and oxides, in the rock-salt structure. We shall follow the tight-binding analysis of

Harrison and Straub (1987), but will make use of the results of more complete and accurate analyses. For the carbides and nitrides, the most complete studies have been given by Grimvall and coworkers, who studied successively the 3d- (Häglund, Grimvall, Jarborg, and Fernández Guillermet (1991)) 4d- (Guillermet, Häglund, and Grimvall (1992)), and 5d- (Guillermet, Häglund, and Grimvall (1993)) transition-metal compounds. They calculated the electronic structure using the Linear-Muffin-Tin-Orbital method (LMTO method, Andersen (1975)) and included even rock-salt structure compounds where that is not the stable phase. They also obtained experimental values from empirical extrapolations. They provided an interpretation (Häglund, Fernández Guillermet, Grimvall and Körling (1993)) along similar lines to that we use here, but referring to what we call the antibonding band as a nonbonding band and vice versa. Finally, Grimvall and coworkers made reference to an extensive set of computations on these systems, Ihara, Hirabayashi, and Nakagawa (1976), Williams, Kübler, and Gelatt (1979), Gelatt, Williams, and Moruzzi (1983), Papaconstantopoulos, Picket, Klein, and Boyer (1985), Schwarz (1987), Grimvall, Thiessen, and Fernández Guillermet (1987), Liu, Wentzcovitch, and Cohen (1988), Zhukov, Medvedeva, and Gubanov (1989), and Price and Cooper (1989).

For many of these compounds the question of correlated states, which we discussed in the preceding chapter, is also a central feature, particularly in the transition-metal monoxides, which form in the rock-salt structure. We discuss these in Section 17-4, and then the role of d-states in semiconductors in Section 17-5. In Chapter 18 we move on to f-shell-metal compounds and to dioxides, and then to trioxides which form in the perovskite structure and are among the most important transition-metal compounds. We end with the cuprates, related to the perovskites, which include the high-temperature superconductors.

17-1. Parameters for Compounds

The important electronic states in the transition-metal compounds are the transition-metal d-states and the p-states on the nonmetallic atoms. The d-states generally lie lower than the metal s-state, as we saw already in Fig. 1-2. Further we shall see that in compounds the d-bands generally begin at the d-state energy and spread upward, so the minimum of the conduction band is associated with d-states and in our treatment we shall ignore the s-states on the metallic atoms (as well as the s-states on the nonmetallic atoms, as with the simple compounds). Of course in occupying the bands we include all of the Z electrons, with Z given in Table 15-1 for each transition element, and all of the valence s- and p-electrons from the nonmetallic element.

Thus the important couplings are between p- and d-states on neighboring atoms. We gave in Eq. (16-16) the general form for coupling between states of any angular-momentum quantum number, taken from Wills and Harrison (1983). Taking $\ell' = 1$ for the p-states and $\ell = 2$ for the d-states, that equation becomes

$$V_{pdm} = \eta_{pdm} \frac{\hbar^2 \sqrt{r_p r_d^3}}{m d^4} \tag{17-1}$$

Furthermore, the coefficients for $m = 0$ and $m = 1$ are obtained from Eq. (16-17) (also from Harrison and Wills (1983)) as

$$\eta_{pd\sigma} = -\frac{3\sqrt{15}}{2\pi},$$

and (17-2)

$$\eta_{pd\pi} = \frac{3\sqrt{5}}{2\pi}.$$

This is a different form for the coupling than was used in *Electronic Structure* (Harrison (1980)) where it was assumed that the s- and p-states formed a free-electron band so that the coupling between them and the d-states could be obtained from the $\langle\langle k|H|d\rangle$ of Eq. (15-11). That analysis (Harrison (1980), p. 519; an error appears in the denominator of Eq. (20-51) which is corrected here) gives a form $V_{ldm} = \eta_{ldm} \hbar^2 r_d^{3/2}/(m d^{7/2})$ with a coefficient which differed by a factor of two from that obtained by fitting perovskite bands. The form, Eq. (17-1), used by Harrison and Straub (1987) seems definitely preferable.

We already have values for r_d given in Table 15-1 (we use the MTO values) for all transition-metal atoms. Straub (Harrison and Straub (1987)) obtained r_p values for the nonmetals using the Atomic Surface Method of Section 15-3-A. For oxygen, for example, they imagined oxygen atoms in a simple-cubic structure, noting that the nearest-neighbor tight-binding p-bands have both their minimum and maximum at $\mathbf{k} = [\pi/d, 0, 0]$, giving $\varepsilon_p \pm (2V_{pp\sigma} - 4V_{pp\pi})$. Using the expressions for $V_{pp\sigma} = (1/\pi)\, \hbar^2 r_p/md^3$ and $V_{pp\pi} = -(1/2\pi)\, \hbar^2 r_p/md^3$ from Eqs. (16-16) and (16-17) this gives a band width of $W_p = (8/\pi)\,\hbar^2 r_p/md^3$. It is interesting that the same band width is obtained for a body-centered-cubic structure (Harrison and Straub (1987)).

That band width is also given within the Atomic Surface Method by Eq. (15-29) which becomes

$$W_p \approx \frac{-\frac{\hbar^2}{m}4\pi r_0^2 \frac{\partial \rho_p}{\partial r}\Big|_{r=r_0}}{\int_{0,r_0} 4\pi r^2 \rho_p\, dr} . \tag{17-3}$$

If this value turned out to vary with atomic-sphere radius as $1/r_0^3$ the value of r_p obtained by equating this to $W_p = (8/\pi)\,\hbar^2 r_p/md^3$ would be independent of the choice of r_0 . Straub (Harrison and Straub (1987)) found that to be approximately true, and selected an r_0 where this *was* true. That was $r_0 = 1.19$ Å for oxygen, corresponding to a simple-cubic $d = 1.92$ Å . Then he could use Hartree-Fock atomic wavefunctions, and r_0 values chosen similarly for the other elements, to obtain the values of r_p listed in Table 17-1.

Harrison and Straub actually used the asymptotic form of the wavefunction, rather than the true wavefunction, which makes only a tiny difference for r_p . They then similarly used that form for the evaluation of the integrals in Eq. (17-3) to derive an approximate analytic form for r_p in terms of the free-atom term value. It was

$$r_p \approx \frac{\pi^2 5^5\, e^{-5}}{24}\sqrt{\frac{\hbar^2}{2m\varepsilon_p}} , \tag{17-4}$$

Table 17-1. r_p values, in Å, for atoms of nonmetallic elements, to be used to obtain matrix elements as in Eq. (17-1), or $V_{pp\sigma}$ and $V_{pp\pi}$ as described in the text. (After Harrison and Straub (1987)). ε_p values in eV from Table 1-1 are also listed for convenience.

	r_p	$-\varepsilon_p$		r_p	$-\varepsilon_p$		r_p	$-\varepsilon_p$
C	6.59	11.07	N	5.29	13.84	O	4.41	16.77
Si	13.7	7.59	P	11.4	9.54	S	10.1	11.60
Ge	14.4	7.33	As	13.2	8.98	Se	12.1	10.68
Sn	18.0	6.76	Sb	16.8	8.14	Te	15.9	9.54
Pb	19.8	6.53	Bi	18.9	7.79	Po	17.9	9.05

Table 17-2. Hartree-Fock d-state energies for the $d^{Z-2}s^2$ configuration obtained by Straub (Straub and Harrison (1985)) from Mann's (1967) calculations, correcting for the $d^{Z-1}s$ configuration which Mann treated in the case of those with an asterisk. The superscript zero indicates that they are to be corrected for configurations different from $d^{Z-2}s^2$. s-state energies, also from Mann, will be needed for estimating the cohesive energy of the compounds. Also listed for convenience are the Coulomb U_d and the d-state radius, r_d (from MTO theory), from Table 15-1.

Z	$-\varepsilon_d^0$ (eV)	U_d (eV)	r_d (Å)	$-\varepsilon_s$ (eV)
3 Sc	9.35	5.3	1.163	5.71
4 Ti	11.05	5.4	1.029	6.04
5 V	12.54	5.3	0.934	6.33
6 Cr	13.88*	5.1	0.939	6.59
7 Mn	15.27	5.6	0.799	6.84
8 Fe	16.54	5.9	0.744	7.08
9 Co	17.77	6.3	0.696	7.31
10 Ni	18.97	6.5	0.652	7.54
11 Cu	20.26*	6.9	0.688	7.72
3 Y	6.80	1.9	1.602	5.33
4 Zr	8.46	2.4	1.415	5.67
5 Nb	9.98*	2.7	1.328	5.95
6 Mo	11.49*	3.0	1.231	6.19
7 Tc	13.08	3.4	1.109	6.39
8 Ru	14.61*	3.7	1.083	6.58
9 Rd	16.14*	4.0	1.020	6.75
10 Pd	17.70*	4.3	1.008	6.91
11 Ag	19.23*	4.6	0.889	7.05
3 Lu	6.63	2.0**	1.603	5.42
4 Hf	8.14	2.8	1.455	5.71
5 Ta	9.57	3.0	1.346	5.97
6 W	10.97	3.1	1.268	6.19
7 Re	12.35	3.2	1.201	6.38
8 Os	13.74	3.4	1.142	6.56
9 Ir	15.12	3.5	1.085	6.71
10 Pt	16.47*	3.5	1.069	6.85
11 Au	17.78*	3.6	1.007	6.98

* Interpolated by Straub (Harrison and Straub (1987)) so that they are for the $d^{Z-2}s^2$ configuration.
** Corrected from what appears to have been a typographical error in Harrison and Straub (1987).

giving $r_p = 4.13$ Å for oxygen rather than the 4.41 Å obtained without that approximation. Such an approximation might be useful, but it will be preferable here to use the more accurate values from Table 17-1.

Harrison and Straub compared the resulting pd-matrix elements obtained using Eq. (17-1) and Table 17-1 with values obtained by fitting full band calculations (Mattheiss(1972)) for the perovskites, which we shall discuss in Section 18-3. For $SrTiO_3$ they obtained $V_{pd\sigma} = -2.30$, $V_{pd\pi} = 1.33$ eV in good accord with Mattheiss' values of -2.43 and 1.13, respectively.

We should also check consistency of the $V_{pp\sigma}$ and $V_{pp\pi}$ we use here and the universal parameters which we gave in Chapter 1 and have used throughout the book. Using $V_{pp\sigma} = (1/\pi)\, \hbar^2 r_p/md^3$ and Table 17-1 we obtain $V_{pp\sigma}$ for diamond silicon, germanium and tin, to be compared with the very similar universal $V_{pp\sigma} = 2.22\, \hbar^2/md^2$ (in parentheses), as 4.38 (7.13), 2.56 (3.06), 2.40 (2.84), and 1.99 (2.15) eV, respectively. The difference between the $1/d^3$ and $1/d^2$ dependence is not so great except for carbon. We see further that the r_p drop in approximately equal steps to the right of these column-four elements, as 14.4, 13.2, and 12.1 Å for Ge, As, and Se, as one might expect due to increasing nuclear charge. We would extrapolate a linear growth in values to the left so that for the semiconductor compounds the average (or geometric mean) of the values for the two constituents should be independent of polarity, as in our universal parameters. Finally, for the couplings introduced here $V_{pp\pi} = -\,^1/_2\, V_{pp\sigma}$ whereas the universal parameters gave $V_{pp\pi} = -\, 0.28 V_{pp\sigma}$ and a fit of free-electron bands to simple-cubic bands would have given $V_{pp\pi} = -\, ^1/_3 V_{pp\sigma}$. These are only moderately close, and just as the universal parameters are to be preferred for the semiconductors, values based upon r_p would seem appropriate for the transition-metal compounds.

We also need the d-state energies on the same scale with the p-state energies of the nonmetallic atoms. The Hartree-Fock values are listed in Table 17-2 all for the atomic configuration with two s-electrons. For the values indicated by an asterisk, the original calculations were performed for the configuration with a single s-electron, but for uniformity they are corrected here for the same s^2 configuration. The $\varepsilon_d\,(s^1)$ for a single s-electron is higher than the value listed by U_d obtained from same calculations, which is also listed in Table 17-2 as well as Table 15-1. In Section 17-3-D, after we have developed a Friedel-like model, we shall calculate the ε_d for the compounds and we use the resulting values here in the study of the bands.

Also listed for convenience are the d-state radii (the preferred values) from Table 15-1. These parameters, and those in Table 17-1 are all that we shall need in the analysis of transition-metal compounds in terms of p- and d-states. Here we shall focus on oxides, and in some cases nitrides, but gave parameters for the others where we had them. r_p values for the halogens could be estimated by extrapolation of Table 17-1.

17-2. The Monoxides, Nitrides and Carbides

A. The Octet Compounds

We may proceed from the electronic structure of KCl and CaO to that of ScN and TiC by continuing the transfer of protons illustrated in Fig. 9-1. However, in comparing the resulting ions with the neutral atom we see that in the case of ScN one of the electrons removed from Sc was a d-electron in the atom. Similarly in TiC, titanium carbide, two electrons are removed from d-states. We indicated in Section 9-1-D that we must take this into account in estimating the cohesion, and in Section 9-1-C that the lowest empty states in ScN, forming the conduction band, were d-like.

Harrison and Straub (1987) estimated the band gap for each of these, as we did for simple compounds, by subtracting Hartree-Fock term. This yields 15.85, 11.40, 4.06, and 0.02 eV, for KCl, CaO, ScN, and TiC, respectively (the second d-electron in TiC, at $\varepsilon_d^0 - U_d$, would be deep in the valence band). Experimentally they are 10.7, 7.7, unknown to us, and less than zero, respectively. It would appear that the agreement is comparable to that for the simple compounds.

They also made the simplest estimate of cohesion for these four systems by subtracting the term values for the final occupied states from those for the initially occupied states, as we did for alkali halides for the first entry in Table 9-5 and divalent compounds in Table 9-6. The second d-electron in TiC would remain in a d-state so that the cohesive energy for both ScN and TiC would be $2\varepsilon_s + \varepsilon_d^0 - 3\varepsilon_p$). This gives cohesive energies for the four compounds of 15.9, 22.9, 20.8, and 10.1 eV, respectively, compared to the experimental 7.6, 11.0, not known, and 14.6 eV. They also added Coulomb corrections essentially the same as those leading to the second entries in Tables 9-5 and 9-6 which were largest for CaO, which improved the agreement with experiment, but left it only semiquantitative. They added bonding terms, which they found to make small contributions to the cohesion except for TiC, where it increased the estimate to 24.0 eV, a considerable overestimate.

These bonding terms, along with an overlap repulsion of the type we introduced in Eq. (10-4), enabled them also to predict equilibrium spacings in these compounds. The most interesting, but hardly surprising, finding was that the principal drop in bond length as we move into the transition-metal series comes directly from the pd-coupling which we shall discuss for a wider variety of compounds. It even appeared that pd-coupling would be important in CaO, for which they had not included d-states, as we have not.

Octet compounds of transition metals are not a very great extension of the simple compounds, and there seem not to be very many beyond this one series, so we shall focus our discussion on a more general set of compounds.

B. Other Monoxides and Nitrides

The 3d-metals Ti, V, Mn, Fe, Co and Ni form simple oxides in the rock-salt structure. Chromium appears not to form such an oxide. We again think of these as ionic compounds, with each oxygen atom acquiring two additional electrons. That leaves in each case a partly filled d-shell on the metallic atom. There are a number of such series, where the oxygen is replaced by nitrogen or carbon. In SmS, SmSe, SmTe, oxygen is replaced by sulfur and the f-shell levels play the role played by the d-states in the 3d-monoxides (Batlogg, Kaldis, Schlegel, and Wachter (1976)).

There have been many studies of the electronic structure of transition metal compounds, mostly focusing upon the conducting properties and the electronic excitation spectrum, which was the focus of the studies by Mott (1974) and several recent studies by Fujimori and coworkers (Mizokawa and Fujimori (1996)). Our focus will be more on the bonding properties which have been much less studied, though there was substantial effort on the nitrides and carbides by Grimvall and coworkers, e. g., Häglund, Grimvall, Jarborg, and Guillermet (1991). This was based upon the same self-consistent-field approximation which we have used throughout this book, and addressed the cohesion but they did not seek to predict the equilibrium spacings as we shall attempt here. We shall also attempt to deal more completely with the electron-electron interaction.

The Coulomb interaction between electrons in the d-states plays two important roles in the electronic structure of the monoxides. The first is in shifting the value of the d-state energy ε_d , which we treat in detail in Section 17-3-D, and again in a different way for the total energy in Sections

17-3-E and 17-3-F . The second is in inducing a correlated state, as in the electronic structure associated with the f-shells in the rare-earth metals. We shall examine this sufficiently in Section 17-4 to see that it appears not to be important in the bonding and energetics, but can dominate the electronic properties.

Concerning the second aspect, in our earlier account of these systems (Harrison (1980)) we followed the analysis given by Koiller and Falicov (1974), in which the d-states on the transition-metal atom were regarded as localized and partly occupied, forming atomic spin-orbit-split states in the sense we discussed in Section 16-2-C. Then there were valence bands formed primarily from the oxygen p-states and conduction bands formed primarily from the transition-metal s-states - just as if local d-states were added to a calcium-oxide electronic band structure. However, Koiller and Falicov assumed that this s-band was broadened around the s-state energy, rather than being broadened entirely upward due to its dominant interaction with the oxygen p-states. We questioned the choice, but noted that it gave a good account of the conducting properties.

We now believe that their choice was indeed incorrect and that when these compounds are conducting the electrons reside in states derived from the transition-metal d-states. The metal-insulator transition then is between localized and itinerant properties of the d-electrons rather than due to an intra-atomic promotion of d-electrons to s-states. The same difference arose in our discussion of cerium in Chapter 16: an earlier view, that the transition from a low-density to a high-density fcc structure was due to promotion of f-electrons to the metallic band, has been supplanted by the view that it is associated with delocalization of the f-electrons. It would seem that in both cases either explanation could in principle have been correct. The question can be resolved when we have an understanding of the electronic structure and approximate values for the parameters. Then if fitting the numerical properties requires too large a change in the parameters, or an implausible postulate, alternative mechanisms should be explored. This now appears to be a place in *Electronic Structure and the Properties of Solids* (Harrison (1980)) where we failed to heed that message. We now provide a description which we believe to be based upon the correct physics, largely following the analysis of these systems by Harrison and Straub (1987), but incorporating Coulomb shifts which they did not include.

17-3. Quantitative Theory

A. The Essential Electronic Structure

We proceed with the rock-salt structure, with five d-states on each transition-metal compound, coupled to each of the three p-states on the neighboring oxygen (or other nonmetallic) atom. Then at each wavenumber in the calculation of energy bands there are two combinations of d-states which are uncoupled to the three oxygen p-states and form nonbonding bands, independent of wavenumber at the energy ε_d (the value of which will need to be determined self-consistently as we shall describe). This is exactly as in the alkali halides where two orientations of halogen p-states were uncoupled from the single alkali s-state and formed nonbonding bands. This is a mathematical feature of n states of one energy, coupled only to m states of different energy. If $n > m$ then there are $n - m$ combinations of the n states which can be chosen such that they are uncoupled to all of the m states. In the Hamiltonian matrix the corresponding $n - m$ by $n - m$ submatrix has equal diagonal elements and no off-diagonal elements, and so these are eigenstates. We saw it explicitly for wavenumbers in a [100] direction in rock salt in Section 9-2-B. When we include *second-neighbor* coupling in Part B, this will no longer be true because these nonbonding d-states will then be coupled to each other and to the other d-states.

The three remaining d-states are coupled to the three oxygen p-states to form three bonding and three antibonding bands, in close analogy to the single bonding and antibonding bands of the alkali halides. We shall outline the calculation of such bands for chromium nitride in the following subsection, and show them to the left in Fig. 17-1. This figure also includes more accurate bands and there is a very considerable difference, which we shall discuss in Part B.

Focusing on the simple bands to the left in Fig. 17-1, we note that the energy is measured from ε_d so the two nonbonding levels are the horizontal line at zero energy. The antibonding bands spread upward from this d-state energy and the antibonding bands spread down from the p-state energy, also similar to the alkali-halide bands. This is not an oxide, but the bands in all of the rock-salt structure transition-metal compounds are similar, while the occupation of the bands differs from one to another. These bands differ somewhat from those given by Harrison and Straub (1987) because the d-state energy is shifted upward by three eV from the value ε_d^0 in Table 17-2 by Coulomb effects, as we shall show in Part D, and they did not include this

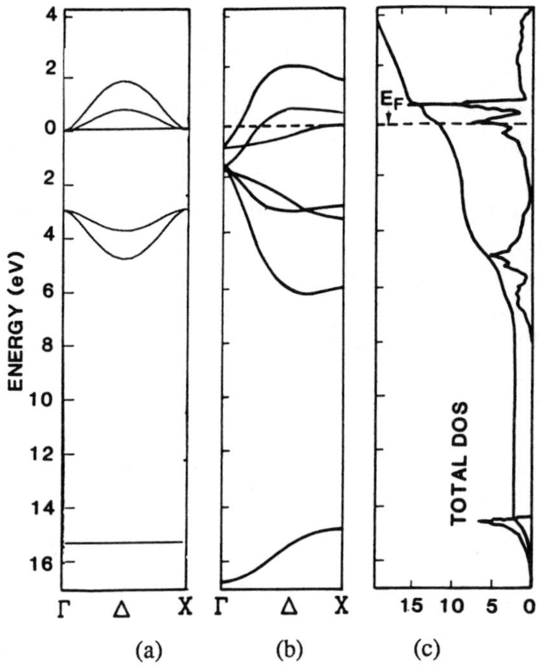

Fig. 17-1. The electronic energy bands of CrN, along a [100] direction. (a) Tight-binding bands for nearest-neighbor coupling only, (b) the self consistent augmented-plane-wave bands and (c) density of states (DOS in electrons per atom) given by Papaconstantopoulos, Pickett, Klein, and Boyer (1985). They also gave the integrated density of states, shown crossing the Fermi energy at eleven electrons per atom pair.

shift. However the essential features, and the resulting expected properties, are similar. We outline these briefly and then return to a more complete analysis.

The nitrogen atom in CrN has enough electrons to fill all but three of the p-like bonding states so each transition metal will give up its two s-electrons and one d-electron to fill those bonding bands. In the case of chromium, with a $Z = 6$, this leaves three electrons in the nonbonding d-bands, with the antibonding bands completely empty. This would be the case also for oxides of Sc, Ti, V, and Cr, and the 4d and 5d transition metals in the same columns of the Periodic Table. For nitrides (with one less electron per atom than oxygen) this would include Mn and the elements below it, Tc and Re. Harrison and Straub (1987) found that the only compounds in which there were more than the twelve electrons needed to fill the nonmetal s-states, the bonding states and the nonbonding states were the oxides of the 3d-transition metals. We shall see that such compounds only form when the d-states are unimportant in the bonding, so that occupation of nominally antibonding states is not a problem. In the case of 3d-transition-metal oxides this is true because of the small r_d as seen in Table 17-2, small r_p as seen in Table 17-1, and the large energy difference between d- and p-states. We shall nevertheless retain the antibonding states where they occur in order to see that this is true. In such cases the d-state spins will align as in the rare earths, forming a strongly correlated state. Similar considerations led Harrison and Straub to a very simple summary of transition-metal compounds in the rock-salt structure:

For compounds with a total of eight valence electrons (the octet compounds) the bonding bands are filled and the nonbonding and antibonding bands empty. They will be insulating unless ε_d lies below ε_p (as in TiC) so that the nonbonding band is part of the lower set of bands in Fig. 17-1. Results for the simplest theory of these were given in Part A of this section. No compounds are expected (e. g., ScC) with fewer than eight valence electrons.

For compounds with nine to twelve valence electrons (e. g., CrN), the bonding bands are full and the remaining electrons are in the nonbonding bands. Because the bands are partially occupied (or even if they are full, but the nonbonding bands are part of the upper set in Fig. 17-1) they are metallic, but because the nonbonding bands are narrow, they are poor metals.

Compounds in the rock-salt structure with thirteen to eighteen electrons (MnO, FeO, CoO, and NiO) are rare. They occur only when the contribution of the d-electrons to the bonding is not important, and they form strongly correlated states. Most usually neighboring spins align antiparallel so the crystals are antiferromagnetic insulators. As for f-shell metals we now regard the formation of the local moments as a consequence of the weak bonding, not a cause of it as assumed in Harrison and Straub (1987).

This is the essential electronic structure described by Harrison and Straub. We proceed now to make it more quantitative, largely following their approach.

B. Bands for the Compounds

We begin with the construction of the simple bands shown to the left in Fig. 17-1. The highest and lowest bands near the Fermi energy arise from σ-states, zero angular momentum around the wavenumber direction $\boldsymbol{k}$. Each d-state is coupled to a p-state a distance $d = a/2$ in the positive and negative direction along $\boldsymbol{k}$ by $\pm V_{pd\sigma}$. That d-state is not coupled to the other four neighboring p_σ-states in the lateral direction. The coupled levels differ in energy by $\varepsilon_d - \varepsilon_p$ (with ε_d to be determined) so the bands are given by

$$\varepsilon_k = \frac{\varepsilon_d + \varepsilon_p}{2} \pm \sqrt{\left(\frac{\varepsilon_d - \varepsilon_p}{2}\right)^2 + 4V_{pd\sigma}^2 sin^2(ka/2)}\,. \qquad (17\text{-}5)$$

The coupling is $V_{pd\sigma} = 1.49\ eV$ for CrN from Eqs. (17-1) and (17-2). The opposite sign of the coupling with the two neighbors led to the sine-like dependence. The bands which spread less from the atomic levels in Fig. 17-1 arise from π-states, and are doubly degenerate, two bonding bands of the same energy and two antibonding bands of the same energy, given by Eq. (17-5) with $V_{pd\sigma}$ replaced by $V_{pd\pi} = 0.86\ eV$ (again the matrix elements forward and back are of opposite sign). A Fermi energy in the nonbonding bands, as in this figure, is the usual situation as we indicated at the end of the last subsection. The nitrogen s-band was drawn as a constant since we include no coupling between these s-states and the chromium states.

We compare with the middle panel of Fig. 17-1 showing the bands obtained by Papaconstantopoulos, Pickett, Klein, and Boyer (1985) in a full band calculation. The similarity is only qualitative, as for the alkali halides, but by the time we sum over occupied states, the resulting total energies will be meaningful, as they were for the simple compounds. We may note in particular the total density of states ($n(\varepsilon)$ or DOS) calculated by Papaconstantopoulos, et al., for their bands, shown in the right panel in Fig. 17-1. It does not look very different from what we might expect from our own bands. There is of course a peak near -16 eV from the nitrogen s-like states, a peak near -5 eV from the p-like bands and two peaks near zero energy arising from the d-like bands. We see in the full bands however that the p-like and d-like bands are not separated by a gap as in the left panel. The crystal would be metallic, rather than insulating, even if there were only eight electrons to occupy these bands. For some properties our simplified bands would be inadequate but for the total energy they should do well as did our crude bands for the alkali halides.

In particular, we may focus on the integrated density of states $I(\varepsilon) = \int_{-\infty,\varepsilon} n(\varepsilon)\, d\varepsilon$ obtained by Papaconstantopoulos et al. (1985) and also shown in the right panel in Fig. 17-1. It rises from zero to two through the s-like band. It rises from two to eight over a narrow energy range from -6 to -3 eV, compared to the -4 to -3 eV suggested by our bands. It then remains rather flat till near zero where it rises by four through what we regard as nonbonding bands, another four from the π-like d-bands and then continues to rise through the antibonding bands. In some ways this density of states resembles our bands along Γ-X shown in the figure as much as the real bands from which the density-of-states was calculated. The difference of course comes from the bands in those parts of the Brillouin Zone not shown here. This gives us courage to proceed with our parameters and the simple bands.

We may, however, explore the bands a little more fully in the tight-binding context, introducing the second-neighbor coupling between d-states. This is straight-forward, but a little intricate. The second neighbors to each d-state are arranged as a face-centered cubic lattice. Then the coupling of the $3z^2$-r^2 d-state to each of the eight neighbors at [±1,0,±1]d is obtained from Table 15-2 , noting that the direction cosines are $l = \pm 1/\sqrt{2}$, $m = 0$, $n = \pm 1/\sqrt{2}$, as $V_8 = V_{dd\sigma}/16 + 3V_{dd\pi}/4 + 3V_{dd\delta}/16 = 5.82\, \hbar^2 r_d^3/(md_2^5)$ to be evaluated from Eq. (15-20) with second-neighbor spacing $d_2 = \sqrt{2}\, d$ in terms of the nearest-neighbor spacing d . The same matrix element applied to neighbors at [0,±1,±1] so the same V_8 applies to all eight neighbors giving a contribution $V_8 \cos kd$ to the d-state energy. Similarly the coupling to the four lateral neighbors at [±1,±1,0] is $V_4 = V_{dd\sigma}/4 + 3V_{dd\delta}/4 = -5.37\, \hbar^2 r_d^3/(md_2^5)$. Then the energy of this d-band state, including second-neighbor dd-coupling is

$$\varepsilon_d(k) = \varepsilon_d + 4V_4 + 8V_8 \cos kd, \tag{17-6}$$

to be substituted for ε_d in Eq. (17-5). Similarly we find for states of symmetry zx or yz $\varepsilon_d(k) = \varepsilon_d + 4\times 3.58\, \hbar^2 r_d^3/(md_2^5) + 4[3.58\, \hbar^2 r_d^3/(md_2^5) - 11.34\, \hbar^2 r_d^3/(md_2^5)]\cos kd$ to be substituted in Eq. (17-5) with $V_{pd\sigma}$ replaced by $V_{pd\pi}$. Finally, the nonbonding bands, which were independent of wavenumber before, now are $\varepsilon_d(k) = \varepsilon_d - 4\times 11.34\, \hbar^2 r_d^3/(md_2^5) + 8\times 3.58\, \hbar^2 r_d^3/(md_2^5) \cos kd$ for states of symmetry xy and are $\varepsilon_d(k) = \varepsilon_d + 4\times(30/\pi)\, \hbar^2 r_d^3/(md_2^5) - 8\times 1.64\, \hbar^2 r_d^3/(md_2^5) \cos kd$ for symmetry x^2-y^2.

These are all plotted in Fig. 17-2, where they are compared again with those from the full band calculation. Indeed the main part of the discrepancy between the first two panels in Fig. 17-1 has been eliminated by the inclusion of second-neighbor coupling. The broadening arising from the

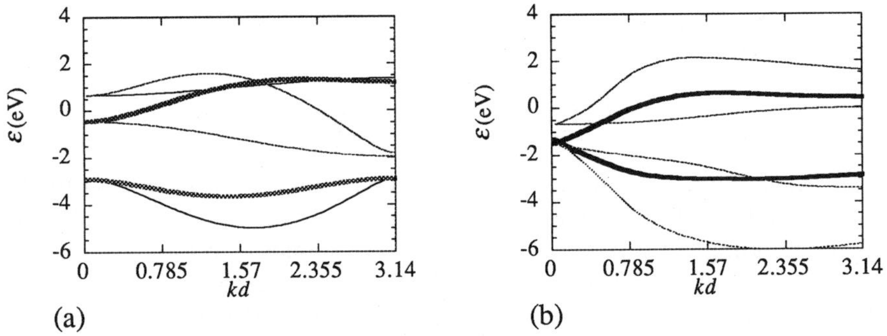

Fig. 17-2. The electronic energy bands of CrN near the Fermi energy (the zero of energy) as in Fig. 17-1. (a) Tight-binding bands with second-neighbor dd-coupling included, as well as nearest-neighbor pd-coupling. (b) The self consistent augmented-plane-wave bands given by Papaconstantopoulos, Pickett, Klein, and Boyer (1985). Doubly-degenerate (π-like) bands are shown as heavy lines.

second-neighbor coupling alone is similar to that from nearest-neighbor coupling to the p-states. This might suggest that it would be preferable to model all of the d-bands together with a Friedel model, but our discussion of the density of states from the right panel in Fig. 17-1 suggests continuing to treat the nonbonding bands separately, and very narrow, and that is what we shall do.

Much of the remainder of the discrepancy between Fig. 17-2a and 17-2b may arise from the second-neighbor coupling between p-states on the oxygen atoms. This would replace the ε_p in Eq. (17-5) by an $\varepsilon_p(k)$ analogous to Eq. (17-6). One effect of this second-neighbor p-state coupling would be to raise the energy of the lower set of bands at $\boldsymbol{k} = 0$, where they are purely p-like. It would be a considerable overestimate to use our universal parameters $V_{pp\sigma}$ and $V_{pp\pi}$ proportional to $1/d^2$ for this face-centered-cubic arrangement of second neighbors since the parameters were set initially to fit free-electron bands with nearest-neighbor coupling only. It would be preferable to use the form given in deriving Eq. (17-3), or one could adjust the second-neighbor coupling to bring the bands at $\boldsymbol{k} = 0$ into accord with the band calculation to obtain a simple analytical account of the band structure. It may be of more interest simply to note that the bands in Fig. 17-2a are derived independently of the band calculation and entirely in terms of our parameters. Similar accord could be expected for other transition-metal compounds.

C. Covalent and Polar Energies

For calculating the total energy, and in fact for computing the shift in the d-states due to charging, we need a sum over occupied states. We shall find that the bonding band is always completely occupied and the antibonding band almost always completely empty, as in the alkali halides. The coupling between second-neighbor d-states discussed in Part B would not in itself affect the total occupation of d-states. Thus it will be very convenient to introduce a covalent and a polar energy to represent the pd-coupling contributions, as we did for sp-coupling in simple ionic compounds (with no transition metals) in Section 10-2. Because there are three orbitals of each type coupled it will be better to use a moments method, as in the treatment of transition-metal bands in Section 15-4-A and as done for compounds by Harrison and Straub (1987), rather than the special-points method used for simple ionic compounds which here would lead to three bonding levels and three antibonding levels at each special point.

We calculate the second moment from all $8N_p$ (three p-states and five d-states for each of the N_p atom pairs) bands relative to the average between the ε_p and ε_d values, analogous to our calculation of the second moment for the metal d-bands in Section 15-4-A. Every diagonal element of the corresponding $8N_p$-by-$8N_p$ Hamiltonian matrix based upon the states of the atoms then is given by $\pm(\varepsilon_d - \varepsilon_p)/2$ and the contribution to the $\Sigma_{i,j} H_{ij}H_{ji}$ of these eight-N_p diagonal terms is $8N_pV_3{}^2$ with the polar energy

$$V_3{}^{pd} = \frac{\varepsilon_d - \varepsilon_p}{2} . \qquad (17\text{-}7)$$

The contribution to the second moment for each atom from the interatomic coupling with each of its $X = 6$ neighbors is given by $V_{pd\sigma}{}^2 + 2V_{pd\pi}{}^2$. Thus the second moment $M_2 = (1/(8N_p))\Sigma_{i,j} H_{ij}H_{ji}$ is $2 \times X(V_{pd\sigma}{}^2 + 2V_{pd\pi}{}^2)/8 + V_3{}^{pd\,2}$ with the leading factor of two for the two atoms per pair.

This is equal to the second moment of the eight bands. We may now drop the contribution of the two nonbonding bands, eliminating one quarter of the $V_3{}^{pd\,2}$ contributions to the sums, but then the total is divided by $6N_p$ rather than $8N_p$. This leaves a contribution of $V_3{}^{pd\,2}$ to the M_2 of the remaining six bands. However, the interatomic couplings did not contribute to those bands so the full contribution to $\Sigma_{i,j} H_{ij}H_{ji}$ remains, but is now divided by $6N_p$ rather than $8N_p$ for the moment of the remaining six bands. Thus the second moment of the bonding and antibonding bands is $M_2 = V_{2pd}^2 + V_{3pd}^2$ with the covalent energy given by

$$V_{2pd} = \sqrt{X\frac{V_{pd\sigma}^2 + 2V_{pd\pi}^2}{3}} = \left(\frac{75X}{4\pi^2}\right)^{1/2} \frac{\hbar^2\sqrt{r_p r_d^3}}{md^4} \qquad (17\text{-}8)$$

where in the last form we have used Eqs. (17-1) and (17-2). We shall give values of these for the nitrides in Table 17-3 and for the oxides in Table 17-4. These forms are analogous to Eq. (15-31) with now a polar contribution to M_2 and with $X = 6$ in the rock-salt structure. This is the result given by Harrison and Straub (1987), who then took the average energy of the bonding bands, relative to $(\varepsilon_d + \varepsilon_p)/2$, to be $-\sqrt{M_2} = -\sqrt{V_{2pd}^2 + V_{3pd}^2}$. They confirmed that if $V_{2pd} << V_{3pd}$ this result is consistent with second-order perturbation theory with V_{2pd} as the perturbation.

They also noted that an alternative approximation, analogous to the Friedel Model, may be more accurate. For this the density of states would be approximated by a constant density of states giving three bonding-band states per spin between the energy $-V_{3pd}$ and $-V_{3pd} - W_b$, relative to the average, and the symmetric constant density of states for the antibonding bands, with W_b adjusted to give the correct second moment. The density of states for the nonbonding states becomes a delta function. This Friedel model of the density of states is illustrated in Fig. 17-3a. They gave the corresponding equation for W_b , which can be written

$$W_b = [3W_b(V_{2pd}^2 + V_{3pd}^2) + V_{3pd}^3]^{1/3} - V_{3pd}, \qquad (17\text{-}9)$$

This can be solved numerically and the average energy of the bonding bands below the average $(\varepsilon_d + \varepsilon_p)/2$ would be $-\varepsilon_{bond} = -V_{3pd} - W_b/2$. They suggested that the artificiality of that density of states did not justify the added complexity and took the simpler average of $-\sqrt{V_{2pd}^2 + V_{3pd}^2}$. Results for the bond energy calculated using both approaches, shown in Fig. 17-3b , confirm that suggestion and we also make that approximation for the present. Then the sum over the one-electron energies for both spins of the three fully occupied bonding bands, relative to the average, is $-6\sqrt{V_{2pd}^2 + V_{3pd}^2}$ per atom pair. It will be desirable to use the more complete description when we calculate the effects of partly-filled bands.

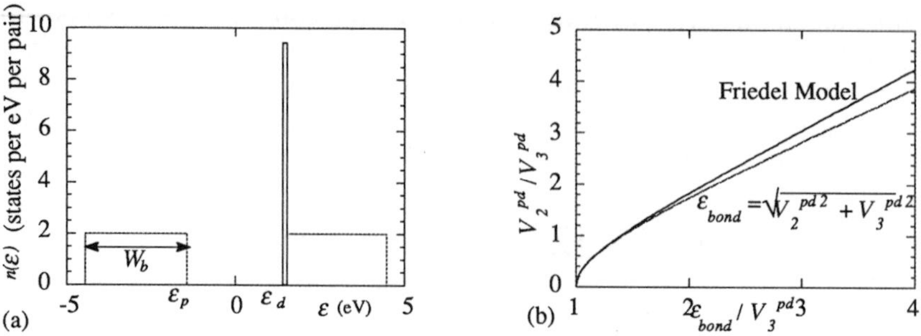

Fig. 17-3. a) The Friedel model for CrN. b) A comparison of the bond energy ε_{bond} , as obtained from the Friedel Model and Eq. (17-9), and from the simpler form we use here.

This description envisages a very narrow nonbonding band, though we saw in Part B that the nonbonding bands appeared only marginally narrower than the bonding, or antibonding, bands. We continue to think of it as narrow because of the density-of-states peak visible in the right panel of Fig. 17-1. However, it will be useful to quantify the width of that band arising from the second-neighbor coupling. We note again that in the rock-salt structure each metal atom has twelve neighboring metal atoms, arranged as a face-centered-cubic structure. We only have two nonbonding states, rather than five, but the averaged, squared matrix element is the same. Thus we may carry over Eq. (15-33) which related the band width to the second moment, $^1/_{12}W_d^2(nonbond) = M_2 = 12(V_{dd\sigma}^2 + 2V_{dd\pi}^2 + 2V_{dd\delta}^2)/5$, to obtain

$$W_d(nonbond) = 107.2\frac{\hbar^2 r_d^3}{m d_2^5} = 18.95\ \frac{\hbar^2 r_d^3}{m d^5}\ . \tag{17-10}$$

In the first form d_2 represents the second-neighbor distance, $\sqrt{2}d$. The coefficient in the second term is $45\ \sqrt{7}/(2\pi)$. For CrN, with $d = 2.11\ \text{Å}$ and $r_d = 0.939\ eV$ from Table 15-1, we obtain $W_d\ = 2.86\ eV$, almost as large as the width of the antibonding band W_b which we shall give as 3.01 eV in Table 17-3. It also appears roughly consistent with the middle panel of Fig. 17-1 showing the full bands for CrN. We shall see in Part E that this coupling makes only a small contribution to the energy of the entire set of pd-bands, largely because it shifts the states equally upward and downward, while the pd-coupling shifts all of the bands away from the average energy. The broadening of the nonbonding bands impacts the total energy very little, but we shall include its effects on the total energy.

D. The d-State Energy, ε_d

We see from Table 17-2 that U_d is large enough that there can be considerable shift in the energy ε_d if the net d-state charge on the transition-metal atom is significantly different from the $Z_d = Z - 2$ assumed for Table 17-2. Here we take the approach that if Z_d differs from $Z - 2$, ε_d will be shifted by $(Z_d - Z + 2)U_d$. When we treat the total energy in Part E we shall include the changes in the Coulomb energy, $Z_d^2 U_d/2$, due to changes in Z_d, which is an equivalent but more convenient way to include U_d in the total energy. In both cases this ignores correlation energies and self-interactions which we take up separately in Section 17-4.

The Z_d which enters is a weighted average, as in the transition metals themselves as discussed in Section 15-4-B. By introducing covalent and polar energies we have represented the electronic structure by four (including spin) nonbonding d-states, and six d-states, each coupled to one p-state by V_2^{pd}. For all compounds in the rock-salt structure there are six electrons in the bonding (predominantly p-like) states and we can set at the outset the number $Z_d(nonbond)$ of nonbonding electrons and the number of electrons, if any, Z_{ab} in antibonding states. (Z_{ab} is only nonzero in some 3d-oxides.) For a transition metal in column Z three electrons are transferred to each nitrogen (or two to each oxygen) leaving $Z - 3$ d-like electrons (or $Z - 2$ for oxides), up to four in nonbonding states and any additional in antibonding states. When we introduce the coupling V_2^{pd} we form bond orbitals, each of which can be written as $\cos\gamma$ (this corresponds to the u_- of Section 2-2-C) times a nonmetal p-state plus $\sin\gamma$ (this corresponds to the u_+ of Section 2-2-C) times a transition-metal d-state(writing them this way retains normalization as in Section 16-1-B). Eventually we shall minimize the energy with respect to γ. The antibonding orbitals are then given by $-\sin\gamma$ times the nonmetal p-state plus $\cos\gamma$ times the transition-metal d-state. Then we add the contributions to obtain the average number of electrons in d-states on the transition-metal atom as

$$Z_d = 6\sin^2\gamma + Z_d(nonbond) + Z_{ab}\cos^2\gamma \qquad (17\text{-}11)$$

and a d-state energy different from the value ε_d^0 listed in Table 17-2, based upon $Z_d = Z - 2$, as

$$\varepsilon_d = \varepsilon_d^0 + [Z_d - Z + 2\,]U_d\,. \tag{17-12}$$

As we indicated, this ignores the refinements in the energy levels associated with the correlated states as discussed in Sections 16-1 and 16-6 and for compounds in Section 17-4.

If we knew ε_d we could find the states, as in Section 2-2-C. From the $V_3^{pd} = (\varepsilon_d - \varepsilon_p)/2$ we would define a polarity $\alpha_p^{pd} = V_3^{pd}/\sqrt{V_2^{pd\,2} + V_3^{pd\,2}}$ and we have $cos\gamma = \sqrt{(1 + \alpha_p^{pd})/2}$ and $sin\gamma = \sqrt{(1 - \alpha_p^{pd})/2}$. These allow the evaluation of Z_d from Eq. (17-11), which then determines the ε_d from Eq. (17-12), so the two equations must be solved together.

We illustrate this for CrN with $Z = 6$, $Z_d(nonbond) = Z - 3 = 3$ and $Z_{ab} = 0$. Then $Z_d = 6sin^2\gamma + 3$ and we have $\varepsilon_d = \varepsilon_d^0 + (6sin^2\gamma - 1)U_d$. We may define a $V_3^{pd\,0} = (\varepsilon_d^0 - \varepsilon_p)/2$. Then $V_3^{pd} = V_3^{pd\,0} + 1/2U_d(2 - 3V_3^{pd}/\sqrt{V_2^{pd\,2} + V_3^{pd\,2}}\,)$, which may be iterated numerically. For CrN with $r_d = 0.939$ Å , $r_p = 5.29$ Å , and $d = 2.11$Å (and of course $X = 6$ for the rock-salt structure), we obtain $V_2^{pd} = 2.72$ eV. With ε_p for nitrogen from Table 1-1 and ε_d^0 for chromium from Table 17-2, $V_3^{pd\,0} = -0.02$ eV , and $U_d = 5.1$ eV from Table 17-2. This leads to $V_3^{pd} = 1.46$ eV . Thus the shifted d-state energy $\varepsilon_d = \varepsilon_p + 2V_3^{pd} = -10.92$ eV, three eV above its starting value of -13.88 eV. This is the value which was used in the bands shown in Figs. 17-1 and 17-2. There are 1.58 electrons in the chromium d-state from

Table 17-3. Parameters for the 3d-transition metal nitrides. Energies are in eV. V_3^{pd} was based upon a self-consistently determined ε_d , as described in the text and V_3^{pd0} is the value based directly on the ε_d^0 from Table 17-2. W_b is the subband (bonding or antibonding) width from Eq. (17-9), $W_d(nonbond)$ is from Eq. (17-10), and U_d the dd-repulsion.

	ScN	TiN	VN	CrN	MnN
Nonbonding electrons	0	1	2	3	4
d (Å)	2.22	2.11	2.11	2.11	(2.11)
V_2^{pd}	3.06	3.12	2.69	2.72	2.13
V_3^{pd0}	2.25	1.40	0.65	-0.02	-0.72
V_3^{pd}	2.51	2.17	1.70	1.46	1.08
W_b	2.74	3.05	2.76	3.01	2.41
$W_d(nonbond)$	4.21	3.76	2.81	2.86	1.76
U_d	5.3	5.4	5.3	5.1	5.6

this bonding band, in addition to the three nonbonding electrons. The extra 0.58 has raised the ε_d three eV. The corresponding calculation for the other nitrides yields ε_d values of -8.83 eV, -9.50 eV, -10.43 eV and -11.68 eV for ScN, TiN, VN, and MnN, respectively, leading to the $V_{3}pd$ listed in Table 17-3. The values obtained without these Coulomb shifts are listed as $V_{3}pd0$. The shift grows as Z increases across the series.

The corresponding parameters for the oxides in the rock-salt structure are given in Table 17-4. The values obtained for $V_{3}pd$ directly from the term values in Tables 17-1 and 17-2 without self-consistency are again listed as $V_{3}pd0$ and are typically 1.5 or 2 eV lower than the self-consistent $V_{3}pd$. Clearly the shifts are significant, as they were for the nitrides.

E. The pd-Bonding Energy and Cohesion

We proceed next to the total energy, including Coulomb effects, in order to describe the cohesion and volume-dependent properties. We follow generally the procedure of Harrison and Straub (1987) and our results would reduce to theirs if we were to take U_d equal to zero. However, the effects of U_d are considerable. In particular the sizable shifts in ε_d which we have just discussed make large changes in $V_{3}pd$ and subsequently in the predicted spacings and energy. We initially tried using the $V_{3}pd$ from Tables 17-3 and 17-4 but found that they shifted so greatly with spacing that we needed to evaluate the $V_{3}pd$ self-consistently at each spacing in order to predict the spacing for minimum energy. This is most directly done by writing the

Table 17-4. Parameters evaluated for the 3d-monoxides, as for the nitrides in Table 17-3.

	TiO	VO	CrO	MnO	FeO	CoO	NiO
d(Å)	2.12	2.05	2.13*	2.22	2.16	2.12	2.08
$V_{2}pd$	2.79	2.76	2.39*	1.59	1.59	1.55	1.52
$V_{3}pd0$	2.86	2.12	1.45	0.75	0.12	-0.50	-1.10
$V_{3}pd$	4.20	3.70	3.07	2.14	1.71	1.25	1.26
W_b	1.64	1.77	1.59	1.02	1.20	1.40	1.35
W_d(nonbond)	3.67	3.25	2.72	1.37	1.26	1.14	1.03
U_d	5.4	5.3	5.1	5.6	5.9	6.3	6.5

* The compound does not form and d was interpolated between VO and MnO.

electronic energy using V_3^{pd0} and then adding explicitly the Coulomb energy $Z_d^2U_d/2$, with the result minimized with respect to both charge distribution and spacing and that is how we proceeded.

Starting with free transition-metal atoms in the $d^{Z-2}s^2$ configuration, for the oxides we transfer the two s-electrons to oxygen p-states. For nitrides we also transfer a d-electron, leaving the others in what would be nonbonding states. This is preparation energy before the pd-coupling is introduced. If each of the six bond states is written $\cos\gamma\,|p\rangle + \sin\gamma\,|d\rangle$, before introducing the coupling, V_2^{pd} , we have $\cos\gamma = 1$ and $\sin\gamma = 0$. As we indicated, it is convenient to separate out the Coulomb energy using the Z_d , given in Eq. (17-11) in terms of $\cos^2\gamma$ and $\sin^2\gamma$ and the number of nonbonding electrons, as

$$E_{Coul} = \frac{Z_d^2U_d}{2} . \tag{17-13}$$

When we vary the bond states, letting $\sin^2\gamma$ differ from zero, to minimize the energy, there will be a contribution from this term equal to $Z_dU_d\,\delta Z_d$ in addition to the term $\varepsilon_d^0\,\delta Z_d$ which already included a term $(Z - 2)U_d\,\delta Z_d$. Therefore we should subtract that term so that the energy of the six occupied bonding electrons is given by

$$E_{bond} = 6[\cos^2\gamma\,\varepsilon_p - 2V_2^{pd}\cos\gamma\sin\gamma + \sin^2\gamma(\varepsilon_d^0 - (Z - 2)U_d] , \tag{17-14}$$

plus the Coulomb energy of Eq. (17-13). If there are Z_{ab} electrons in antibonding states, in addition to the bonding and nonbonding electrons, there is an additional $Z_{ab}[\sin^2\gamma\,\varepsilon_p + 2V_2^{pd}\cos\gamma\sin\gamma + \cos^2\gamma(\varepsilon_d^0 - (Z - 2)U_d]$ and Eq. (17-13), including those electrons in Z_d , remains appropriate and again $\cos\gamma = 1$ before coupling . We may add these together for general γ, and subtract the value of the same quantity with $\cos\gamma = 1$ to obtain the reduction in total due to the coupling. Including the Coulomb energy from Eq. (17-13), minus it's value for no coupling, this becomes

$$E_{pd\,bond} = (6 - Z_{ab})\,[-2V_2^{pd}\cos\gamma\sin\gamma + \sin^2\gamma(\varepsilon_d^0 - (Z - 2)U_d - \varepsilon_p)] + \frac{Z_d^2U_d}{2} - \frac{(Z-2)^2U_d}{2} \tag{17-15}$$

for oxides and the same form for nitrides with $Z - 2$ in the final term replaced by $Z - 3$. Straub and Harrison (1987) used this form with U_d taken equal to zero. If we minimize the full expression with respect to γ, for V_{2pd} evaluated at the equilibrium spacing, we obtain values of γ which may be substituted in Eq. (17-11) to obtain a Z_d which in Eq. (17-12) gives an ε_d leading to the V_{3pd} listed in Tables 17-3 and 17-4.

Harrison and Straub noted that with the pd-bonding energy there was an associated repulsion as there was for the dd-coupling in transition metals. It arises from the nonorthogonality, and is derived exactly as the dd-repulsion was derived in Eq. (15-42). Here the number of electrons per atom pair is the six electrons in the bonding band plus Z_{ab} in the antibonding band, rather than Z_d per atom. It is interesting that this repulsion term does not change as we add electrons to the nonbonding d-states, since the repulsion is proportional to the coupling V_{2pd}, and these states uncoupled by V_{2pd} do not contribute. The coefficients which enter are $(\sigma_{pd\sigma}\eta_{pd\sigma} + 2\sigma_{pd\pi}\eta_{pd\pi})/3$ rather than the corresponding expression for ddm-coefficients. Using the coefficients from Eqs. (17-2) and (16-19) they obtained a repulsive energy per atom pair to be added directly to Eq. (17-15) ,

$$E_{pd\ repuls.} = \frac{45\hbar^2 r_p r_d^3}{2\ \pi^2 m d^6}(6 + Z_{ab}) . \tag{17-16}$$

We treat separately the effects of second-neighbor coupling, which led to the bands in Fig. 17-2a. For such a partly-filled band, with $Z_{d(nonbond)}$ electrons in a band which can accommodate four electrons, we should use a Friedel Model, which gives an additional term in the energy

$$E_{nonbond} = -\frac{Z_{d(nonbond)}(1 - Z_{d(nonbond)}/4)}{2} W_{d(nonbond)} \tag{17-17}$$

with $W_{d(nonbond)}$ given by Eq. (17-10). We should in principle also include the corresponding repulsion, but since it drops as $1/d^8$ it would certainly be negligible at second-neighbor distances.

Following Harrison and Straub (1987) we start with the preparation energies mentioned just before Eq. (17-13). We transfer the transition metal s-electrons to p-states on the nonmetallic atom, $E_{sp\ trans} = 2(\varepsilon_p - \varepsilon_s)$. ε_p is obtained from Table 1-1 and ε_s from Table 17-2. Additional electrons may be transferred from the d-shell to complete the occupation of the p-shell, no electrons for oxides since the two s-electrons filled the shell, one electron for nitrides, $E_{pd\ trans} = \varepsilon_p - \varepsilon_d^0$, or two electrons for carbides, $E_{pd\ trans} = 2(\varepsilon_p - \varepsilon_d^0)$, and minus one electron for halides, $E_{pd\ trans} = -(\varepsilon_p - \varepsilon_d^0)$. At this stage we have proceeded as if both V_2^{pd} and U_d were zero but their effect is included in the pd-bonding contribution.

We may then add the sp-bonding term, $2V_3^{\ sp} - 2\sqrt{V_2^{sp\ 2} + V_3^{sp\ 2}}$ with $V_2^{sp} = \sqrt{6}V_{sp\sigma}$ and $V_3^{sp} = (\varepsilon_s - \varepsilon_p)/2$. When V_2^{sp} is very small compared to V_3^{sp} this can be expanded to second-order to equal the $E_{cov.}$ of Eq. (9-6) but we use the full form here. To this they added six times a repulsive energy plus the second-neighbor repulsion between the nonmetallic atoms using Eq. (10-10), with, in the case of the 3d nitrides and oxides, ε_{ig} the geometric mean of the neon and argon p-state energies,-19.29 eV. The sum of the sp-bonding term and the repulsions was called $E_{sp\ bond}$.

For the cohesion we shall add the Coulomb term associated with the p-states, the $\delta U(X)$ of Eq. (9-2) multiplied by the six electrons per nonmetallic atom. As for the simple ionic compounds, this was found to improve the agreement with experiment for the cohesion. However, the effective charge varies with spacing in such a way that this term has negligible effect on the dependence of the energy upon volume and we omit it for those calculations. We shall tabulate the resulting cohesive energy predictions for the nitrides in Table 17-5 after discussing the volume dependence.

F. The Volume-Dependent Properties

All of these terms are obtained explicitly in terms of our formulae, using numbers from Tables 1-1, 17-1, and 17-2, without the use of the equilibrium spacing for the compound in question. The repulsions were obtained with general coefficients using the spacing of the potassium halides in Eq. (10-1). Thus they can be used to predict the equilibrium spacing, the cohesion and the bulk modulus for any transition-metal compound in the rock-salt structure. To the extent that it succeeds this can be said to mean that we understand the electronic structure well enough to make all of the corrections starting with a potassium halide and ending with the any transition-metal compound. Of course the accuracy is not very high.

We begin with the prediction of equilibrium spacing. We add $E_{pd\ bond}$ from Eq. (17-15), $E_{pd\ repuls.}$ from Eq. (17-16), $E_{dd\ nonbond}$ from Eq. (17-17) and exactly the $E_{sp\ bond}$ which was used to predict the equilibrium spacings of the monovalent and divalent compounds in Tables 10-1 and 10-2. At each spacing d we minimize $E_{pd\ bond}$ with respect to γ (the only term depending upon γ) and then vary d to obtain the minimum energy.

For the nitrides this leads to the lower solid line in Fig. 17-4. Removing the nonbonding contribution leads to the dashed curve, the correction vanishing for empty and full nonbonding bands. It is indeed a small correction, but worth retaining. If we remove also the pd-bonding and repulsion terms (set r_d equal to zero) we obtain the upper solid curve near 2.55 Å, equivalent to the simple-compound predictions in Tables 10-1 and 10-2. We see that the pd-bonding describes qualitatively the smaller spacings and dependences upon element, though it still yields spacings 5% or 10% too large. We suspect that the discrepancy may come more from inadequacy in the effects of the sp-coupling than from the pd-coupling. The nitrides for transition metals to the right of manganese do not form but we continued the prediction to those compounds, including correlation energy as described in Section 17-4-A, finding that the spacings increase with increasing atomic number, much as we shall find for the oxides. It is interesting that the extrapolations made by Häglund, Grimvall, Jarborg, and Guillermet (1991) showed a decrease in spacing with further increase in atomic number.

We make the same plot for the oxides in Fig. 17-5. CrO and compounds to the left have electrons in the nonbonding bands, as for the nitrides in Fig. 17-4 and again show qualitative accord with what little experiment there is,

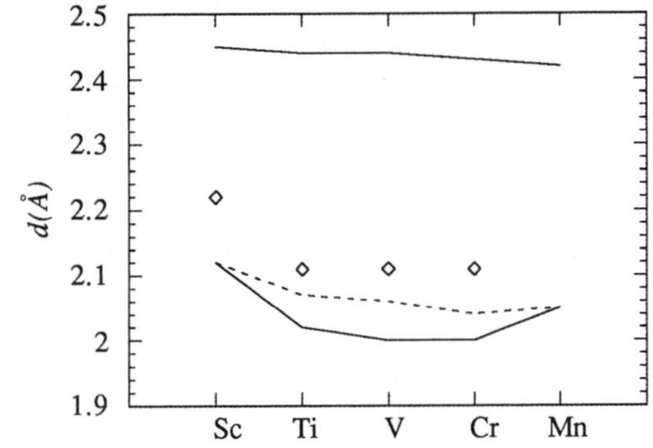

Fig. 17-4. Predicted equilibrium spacing for nitrides in the rock-salt structure, compared with experiment (points). The lower curve includes $E_{pd\ bond}$ from Eq. (17-15), $E_{pd\ repuls.}$ from Eq. (17-16) and $E_{d\ nonbond}$ from Eq. (17-17). For the dashed curve, $E_{d\ nonbond}$ has been omitted, and for the upper curve all three pd-terms have been omitted leaving only the sp-contributions.

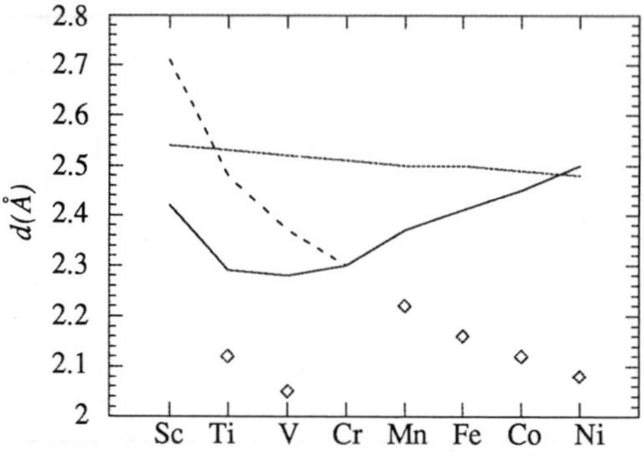

Fig. 17-5. Predicted equilibrium spacing for 3d-oxides in the rock-salt structure, compared with experiment (points). The curves and points are as in Fig. 17-4.

but are even less accurate than for the nitrides. To the right of CrO there would be electrons in the antibonding bands, which accounts for the rising predicted curve including pd-coupling. The fact that the experimental spacings are dropping with increasing atomic number is contrary to our predictions. One thinks immediately that this may have to do with electron correlations, which are expected to be strong in this region, and we discuss that in the following section. We shall find that their inclusion does not rectify this discrepancy. Further, the predictions based upon sp-bonding alone lie far above the experimental values, which suggests that it may be the sp-bonding terms which are most inadequate. We would look there if we sought to improve the predictions and if we were to model the electronic structure for accord with this feature, we would model the sp aspects of the electronic structure.

The predicted spacings for the nitrides are sufficiently close that it makes sense to extract predictions also for the cohesion and the bulk modulus from the electronic structure for the nitrides, as did Harrison and Straub (1987). Our predictions, which include Coulomb effects as we have described, are given in Table 17-5, separated by the different contributions. In the end the predicted cohesion is extraordinarily close to the experimental values, but we should note that the uncertain Coulomb term $6\delta U(X)$ is a very sizable contribution. The d-states are seen to contribute significantly to the bonding, but are not dominant. We saw that they produced a major reduction in

Table 17-5. The predicted cohesive energy, bond length (Å), and bulk modulus of the 3d-transition metal nitrides. Energies are in eV per atom pair and bulk modulus B is in Mbars (1Mbar = 10^{12} dynes/cm^2). Experimental values for cohesion were taken from Häglund, Grimvall, Jarborg, and Guillermet (1991). Calculations were carried out self-consistently. $E_{pd\ bond}$ includes the contributions from Eqs. (17-15) and (17-16). W_{pd} is the subband width from Eq. (17-13), and U_d the dd-repulsion.

	ScN	TiN	VN	CrN	MnN
Nonbonding electrons	0	1	2	3	4
$E_{sp\ trans}$	-16.26	-15.60	-15.02	-14.50	-14.0
$E_{sp\ bond}$	-2.20	-1.65	-1.37	-1.29	-2.01
$E_{pd\ trans}$	-4.50	-2.80	-0.13	0.04	1.44
$E_{pd\ bond}$	-0.14	-1.68	-2.92	-4.28	-4.16
$E_{nonbond}$	0	-1.20	-1.34	-1.07	0
$6\delta U(X)$	15.10	9.54	7.91	7.24	9.54
Total	- 8.00	-13.39	-14.04	-13.86	-9.19
$-E_{coh}(exper)$	(-13.4)	(-13.4)	(-12.5)	(-10.3)	(-8.2)
Predicted d	2.37	2.18	2.13	2.11	2.18
Experimental d	(2.22)	(2.11)	(2.11)	(2.11)	
Predicted B	0.53	1.27	1.63	1.91	1.47

predicted bond length, which was predicted to be near 2.5 Å in all cases if the pd-terms were dropped.

The bulk modulus equals the volume times the second derivative of the total energy with respect to volume, which is $1/(18d)$ times the second derivative of the energy per pair with respect to d. This has also been evaluated at the predicted spacing (dropping the Coulomb term $\delta U(X)$ of Eq. (9-2), as we did for the prediction of spacings) and is listed in Table 17-5, but no experimental values were available with which to compare. The predicted values are about half those obtained by Harrison and Straub (1987). However, they are sensitive to the way they are evaluated. Evaluating the derivative of the total energy at the observed 2.22Å for ScN, rather than the predicted 2.37Å yields 1.44 Mbar rather than 0.53 Mbar for the bulk modulus. Note that for the bulk modulus of the simple ionic compounds in Table 10-3 we readusted the sp-repulsion to fit the observed spacing. That would also be appropriate here, but it has not yet been done. Nor have we carried out the corresponding calculations for the oxides. Judging from Figs. 17-4 and 17-5, showing predictions of d, we would expect the accuracy for the cohesion to be less than that for the nitrides for the compounds up to CrO, for which there are no electrons in antibonding states, as there were none for the nitrides. For oxides of transition metals to the right of chromium it is clear that our predictions are very inaccurate. We of course suspected that the errors were from the omission of correlation energy, but including those effects in the following section indicates that they do not correct the discrepancy for the oxides.

Pettifor and Podloucky (1984) have also addressed the relative stability of different crystal structures which arise in these AB compounds. This involves other contributions and assumptions and we do not describe the theory here.

17-4. Correlated States

The discrepancies in the prediction of the equilibrium spacings, particularly for the oxides to the right of CrO, suggested that we look more carefully at the effects of Coulomb correlations, which we discussed for the f-shell metals in Chapter 16. In fact this can be done quite naturally in terms of the formulation of the energy in terms of covalent and polar energies in Section 17-3-C. We had seen that the band structure corresponded to three bonding bands and three antibonding bands, plus two nonbonding bands. Section 17-3-C reduces this to three d-levels on each transition-metal atom, each coupled to one p-state on its neighbors by V_{2pd}. In addition, each transition-metal atom may have electrons in nonbonding levels, uncoupled to the neighbors. There is no correlated motion associated with the nonbonding electrons; with no coupling the states are either occupied or not.

Rather than the single pair of electrons which we discussed for Li_2 in Section 16-1, we now have three pairs sharing the transition-metal atom. The other major difference is that we have a Coulomb energy U_d associated with the transition-metal atom, but regard any U_p associated with the nonmetallic neighbors as being canceled by a Madelung energy, so we have a one-sided Coulomb effect. This latter feature seems to have reduced the effects on the total energy enough to make them unimportant, so we only indicate the approach taken and the results.

A. The Correlation Energy

We included the correlated motion of the two electrons in each pd-pair, treating the Coulomb interaction between electrons in different pairs as a self-consistently determined field. This is the approximation which enabled us to treat what we regarded as the most essential aspect of the correlation exactly. For small U_d it corresponds to the band approximation. For large U_d the state contains the doubly-occupied p-states on the nonmetal, and some contribution with d-state occupation of one spin, but almost none with d-states of both spins occupied. In all compounds which occur there are enough electrons to put two in each of the three coupled pairs. In most cases there were enough electrons to fill the nonbonding levels, and to the right of CrO there were additional electrons which need to be added to the three coupled pairs. Such an additional electron can be thought of as producing a full pair with one hole, and no correlation energy, just as there would be no correlation energy for a single electron in the pair.

The various total numbers of electrons for different cases made the calculations intricate, but in each case we could solve the four simultaneous equations (for the ground state) for each pair as we did in Section 16-1 and obtain the total energy as a function of volume. We then predicted the equilibrium spacing for each case, giving curves analogous to those in Figs. 17-4 and 17-5. In fact, the curves for the oxides were almost the same and the predictions for the nitrides dropped 0.2 to 0.3 Å, values too small by a similar amount that they were too large before. The larger effect for the nitrides appears to arise from their smaller polarity. The inclusion of correlation did not remove the discrepancies for the oxides which we discussed in the last section.

Such a finding that the effects of electron correlations on the total energy are not important is not at all new. Although correlation energies are often on the order of an electron volt per electron, they are not so sensitive to the exact environment, and tend not to change greatly in going from the atom to the solid. Thus they tend to cancel in the energy *difference* , which is the cohesive energy. This may be related to the fact that the energy of a free-electron gas, obtained as an expansion is r_s (related to the electron density ρ by $4\pi r_s^3\rho/3 = 1$) contains kinetic energy ($\propto 1/r_s^2$), exchange ($\propto 1/r_s$) , and the correlation energy, given by Nosières and Pines (1958) as $(0.031 \ln(me^2r_s/\hbar^2) -0.115)me^4/(2\hbar^2)$, is much less sensitive to r_s than the leading terms in the energy. For this reason they tend to have small effects on the spacing.

A study of the states in these transition-metal compounds indicated that they are rather strongly correlated, and that the contribution to the energy is significant, though the difference may largely cancel in the cohesion. The more careful treatment of electron-electron interactions has little impact on the predicted spacing, and does not explain the decreasing experimental spacing from MnO to NiO seen in Fig. 17-5.

B. The Metal-Insulator Transition

A second aspect of the correlated state in the compounds is the possibility that it is insulating, whereas it will be metallic in the band description as we discussed in Section 17-2-B. We found in Section 16-1 that the f-resonance can be split into an occupied and an empty resonance, with the Fermi energy lying between. That left the system metallic because of the s-like, or free, electron states at the Fermi energy. However, these s-like levels are well-removed from the Fermi energy in the compounds, so such a separation can produce an insulating state where it would have been conducting without it. In this case the "correlated state" of Unrestricted Hartree-Fock theory has real physical meaning beyond an approximation to the real many-body state.

It would seem that the criteria for forming an insulating state are quite different for the different compounds. When there are just enough electrons to fill the oxygen (or nitrogen) states, and there is a gap between bands, the system will be insulating, whatever the degree of correlation of each pair. If there are one to four additional electrons, they will lie in nonbonding states, and the question is whether the bands they form are broader than the separation, U_d, of the empty bands from the occupied bands. The degree of correlation of the bonding-antibonding pairs we have discussed is not relevant since it would take considerably more energy to excite one of these correlated pairs to an excited correlated state, or to transfer an electron from the pair to a band state, than to transfer an electron between nonbonding bands. If there are more than enough electrons than needed to fill the correlated (principally oxygen or nitrogen) states *and* the nonbonding states, these extra electrons will lie in the antibonding bands. A set which would otherwise have been a correlated pair becomes a pair of electrons in a bonding band and a single electron per atom pair in an antibonding band since there is no correlated motion, as we indicated, for a single hole. The question then is whether the breadth of that antibonding band is greater or smaller than the U_d which raises the energy for three electrons per pair over that for two electrons per pair. There is indeed experimental evidence that bands can be separated by a Coulomb U but with the bands overlapping and a metallic state remaining (Tsujioka, Mizokawa, Okamoto, Fujimori, Nohara, Takagi, Yamaura and Takano (1997)).

We look first at compounds with partly-filled nonbonding bands, ScN to MnN and TiO to CrO. Noting the values of $W_d(nonbond)$ and U_d in Table 17-3 for the nitrides and Table 17-4 for the oxides, we see that in all of these cases $W_d(nonbond)$ is less than U_d so we should not be surprised if they were insulating. Experimentally the nitrides all appear to be metallic, even ScN where the nonbonding bands, which we expect to be empty, must cross the bonding bands. The conducting properties of the oxides experimentally are quite intricate (Adler (1968), Mott(1974)). TiO, in particular, is a reasonably usual metal (Adler (1968)), VO forms only with a high concentration (15%) of vacancies and CrO does not form. Our prediction of insulating properties fails, but not by a large margin since our $W_d(nonbond)$ values are more than half U_d in all cases and our choice of band width corresponds to a rather crude description. MnN and CrO, which would have just enough electrons to fill the nonbonding bands appear not to form. CrN is antiferromagnetic, which may be a consequence of a correlated state as we shall see explicitly for the cuprates in Section 18-4, though CrN is metallic, as we indicated.

We look next at the compounds to the right of these, with electrons in antibonding bands. No nitrides of these elements occur. Table 17-4 indicates that for the oxides the width of the antibonding band is less than a quarter of U_d and we correctly expect them all to be insulating. Note that these antibonding states lie predominantly on the transition-metal d-state so something close to the full U_d is appropriate in the comparison.

Our view of the electronic structure meets with limited success on the prediction of metallic or insulating properties. There are a number of approaches we have considered but not used here. One was that used by Harrison and Straub (1987) who included a factor $sin^2(\pi Z_d/10)$ in their criterion for formation of a localized state, similar to the $Usin^2(\pi Z_f/14)$ which we obtained in Section 16-5-D . We now think of that factor as arising from the proximity of the resonance to the metallic Fermi energy and not appropriate to the compounds we discuss here. We also considered a total band width for the pd-bands, lumping bonding, antibonding and nonbonding bands together, as might be suggested by the CrN bands in Fig. 17-2. This gave band widths dropping from 13 eV to 7.6 across the nitride series of Table 17-3, and from 17 to 6.4 across the oxide series of Table 17-4, and the comparison with U_d seemed less informative than what we obtained here. The study of metal-insulator transitions has been an active area of research for many years, and is discussed extensively in Mott's treatise (1974). We do not pursue it further here.

17-5 The Role of d-States in Semiconductors

We included only s- and p-states in our discussion of semiconductors, but of course d-states are present in these systems, particularly when the elements Cu, Zn, Ga and Ge, and the elements directly below them in the periodic table, are involved. Wei and Zunger (1988, 1998) have emphasized the possible importance of such d-states. They note in particular that the symmetry of the d-bands even at $\boldsymbol{k} = \boldsymbol{0}$ allows coupling with the p-bands at that point. This coupling shifts the valence-band maximum, upward since the d-states in the cases just mentioned lie lower, reducing the gap, and modifying band line-ups at heterojunctions. The coupling also affects the average energy of the bands, modifying the cohesive energy and the equilibrium lattice spacing, all effects considered by Wei and Zunger. It is also possible to estimate these effects using the methods described here, and in fact Wei and Zunger included such estimates also. Harrison and Straub (1998) redid Wei and Zunger's 1988 analysis, using more recent forms for the interactions and argued for a different choice of d-state energies. We describe that analysis here.

We have not to this point given d-state parameters for the elements such as zinc, gallium and germanium which enter semiconductors, and in fact some important questions arise in selecting such parameters. For the coupling we use the coupling $V_{pd\sigma}$ given in Eq. (17-1), and have the parameters r_p which enter them as listed in Table 17-1, along with the corresponding atomic p-levels. We have evaluated the r_d for the elements which enter, using the Atomic-Surface Method as described in Section 15-3-A. In particular, we combine the two equations for W_d , Eqs. (15-25) and (15-29), to obtain an expression for r_d

$$r_d^3 \approx \frac{-4\pi r_0{}^7 \left.\frac{\partial \rho_2}{\partial r}\right|_{r=r_0}}{6.83\int_{0 \text{ to } r_0} 4\pi r^2 \rho_2 \, dr} \tag{17-18}$$

We may evaluate this expression using the radial wavefunction $P_{n\ell}(r)$ tabulated by Mann (1967), in terms of which the electron density is $\rho_\ell(r) = P_{n\ell}{}^2(r)/(4\pi r^2)$. Then the numerator is obtained numerically from the tabulated $P_{n\ell}(r)$. The integral in the denominator is simply $\int_{0,r_0} P_{n\ell}^2(r)\, dr$, which is very near one since this is the normalization integral for $P_{n\ell}(r)$ if the upper limit is taken infinite. For the small correction, $\int_{r_0,\infty} P_{n\ell}^2(r)\,dr$ it is convenient to take a simple form for the tail of the wavefunction. We use the simple exponential form $Ae^{-\mu r}$ fit to $P_{n\ell}$ and its derivative. Then the integral is simply

$$\int_{0,r_0} P_{n\ell}^2(r)\,dr = 1 - \frac{P_{n\ell}^3(r_0)}{2\partial P_{n\ell}/\partial r} \tag{17-19}$$

and Eq. (17-18) becomes

$$r_d^3 \approx \frac{-r_0{}^7 \left.\dfrac{\partial(P_{n2}{}^2/r^2)}{\partial r}\right|_{r=r_0}}{6.83\left(1 - \dfrac{P_{n2}{}^3(r_0)}{2\partial P_{n2}/\partial r}\right)} \tag{17-20}$$

Actually Mann (1967) tabulated $P_{n\ell}(r)$ on a grid of x- values, $r = ux$ with $u = (9\pi^2/128Z)^{1/3}$. [Z here is the atomic number, 32 for Ge.] In terms of this $P(x)$ we have $P_{n\ell}(r) = P(x)$ and $\partial P_{n\ell}(r)/\partial r = (1/u)\partial P(x)/\partial x = (1/u)P'$ so that

$$\frac{r_d^3}{r_0^3} = \frac{2\,u\,(xP^2 - x^2PP')}{6.83(1 - u\,P^3/2P')}\ . \tag{17-21}$$

We need to choose a sphere radius for the evaluation and we chose for the 4d-states the radius of an atomic-sphere with volume equal to $a^3/16$ for germanium, obtained by dividing the cubic cell into a cell for each atom, and an equal number of cells of the same volume for the interstitial positions. This was 1.39 Å for the copper row, and 1.59 Å for the silver row (from the gray tin lattice distance), which we used also for the gold row. It was interesting that these radii were very nearly the same as the atomic sphere radii 1.41, 1.59, and 1.59 Å we listed in Table 15-1 for the pure elements copper, silver and gold. These noble metals are the only ones which appear in both tables and the values differ principally because we have used 6.83 for the band width in Eq. (17-20) rather than the 5.53 used by Straub and Harrison, for the reasons discussed following Eq. (15-25). The resulting values, along with the d-state energy from Mann (1967), are given in Table 17-6 .

Table 17-6 The Hartree-Fock d-state term values, and d-state radii from Eq. (17-21) , all based on the calculations of Mann (1967).

Element	ε_d (eV)	r_d (Å)
Cu	13.35	0.638
Zn	21.28	0.518
Ga	32.46	0.409
Ge	44.62	0.316
Ag	14.62	0.813
Cd	20.77	0.693
In	28.91	0.574
Sn	37.36	0.469
Au	14.17	0.914
Hg	19.42	0.811
Tl	26.34	0.705
Pb	33.31	0.607

Wei and Zunger (1988) used local-density theory, which yields d-state term values, and argued against the use of Hartree-Fock term values which have greater magnitude. We may test the two approaches by estimating the difference between the d-state term value and the valence-band maximum using Hartree-Fock term values (Table 1-1 for p-states and Table 17-6 for d-states) and Eq. (6-1) for the valence-band maximum. For the LDA calculation we take the difference ε_p - ε_d from Wei and Zunger (1988), using the same energy for the valence-band maximum relative to the p-state energy which we used for the Hartree-Fock case. We then compare directly with the experimental difference (from photo-emission) in energy tabulated by Wei and Zunger (1988). The comparison given in Table 17-7 would suggest that the Hartree-Fock values are much more appropriate and we proceed with them.

Our choice to list Hartree-Fock values for s- and p-states in Table 1-1, rather than the local-density values used in Harrison (1980), was determined by the fact that they were closer to the removal energy for electrons in the atom, as measured from the ionization potential listed in the CRC Handbook (Weast (1975)). This test of the Hartree-Fock values was made in Table 1-2. We now include also local-density values in the comparison for the 3s-3p-row of the periodic table in Table 17-8 . The preference for Hartree-Fock is not overwhelming, and it would be possible to scale up the exchange potential in the LDA calculation to improve the accord, but the comparison

Table 17-7 The d-state binding energy in eV, relative to the valence-band maximum, from semirelativistic local-density (LDA) calculations [atomic term value difference from Wei and Zunger (1988), corrected for small shifts (less than an eV) from p-p-coupling in the solid], from the corresponding Hartree-Fock (H-F) calculations, and from experiment.[a]

Compound	LDA	H-F	Experiment
GaP	12.83	22.25	18.76
GaAs	13.09	22.82	18.86
GaSb	13.47	23.70	18.96
InP	12.58	18.89	17.41
InAs	12.82	19.44	17.23
InSb	13.19	20.31	17.80
ZnS	2.89	9.29	9.03
ZnSe	3.37	10.24	9.20
ZnTe	3.96	11.42	9.84
CdS	4.47	8.88	9.64
CdSe	4.93	9.81	10.04
CdTe	5.50	10.97	10.49
HgS	2.61	7.54	
HgSe	3.07	8.47	8.05
HgTe	3.64	9.63	8.58

[a]Compiled by Wei and Zunger (1998) from data from Shevchik, Tejeda, Cardona, and Langer (1973), Ley, Pollak, Mcfeely, Kowalczyk, and Shirley (1974) and Shevchik, Tejeda, and Cardona (1974).

lends some support to our choice for Table 1-1 and for our use here of Hartree-Fock d-states.

We first consider the shift in the valence-band maximum due to the coupling with d-states, which we mentioned at the beginning of this Section. We consider the coupling of a p-state on a nonmetallic atom (anion) to a d-state on a neighboring metallic atom (cation). This is the p-state which is dominant at the valence-band maximum and the energy difference with that of the neighboring d-state is much smaller than for the other combination. A p-state oriented along the x-axis is coupled to the d-states on all of the four neighbors, but at $\boldsymbol{k} = 0$ we must add the same d-states from all four neighbors and only the d-state of symmetry yz survives. The coupling is evaluated using the Slater-Koster Tables (Table 15-2) giving a total coupling of

Table 17-8 . Estimates of the ionization potential of free atoms (in eV) taken as local-density-approximation values from Herman and Skillman (1963) calculations (LDA) and from Hartree-Fock calculations from Mann (1967), and experiment from Weast (1975).

Element	LDA	H-F	Experiment
Na	5.13	4.96	5.14
Mg	6.86	6.89	7.64
Al	4.86	5.71	5.98
Si	6.52	7.59	8.15
P	8.33	9.54	10.48
Cl	12.31	13.78	13.01
Ar	14.50	16.08	15.76

$$E_{pd} = \frac{4}{3}(V_{pd\sigma} - \frac{2}{\sqrt{3}}V_{pd\pi}) = -\frac{10\sqrt{15}}{3\pi}\frac{\hbar^2\sqrt{r_p r_d^3}}{d^4} \quad . \qquad (17\text{-}22)$$

In obtaining the final form we used Eqs. (17-19). In making the evaluation we shall of course use Table 17-1 for r_p and Table 17-6 for r_d.

The states at Γ of course contain p-states on the metallic (e. g., Ga) and on the nonmetallic (e. g., As), in addition to the d-state on the metallic atom. We neglect contributions from the d-states on the nonmetallic atom, which can be seen from Table 17-6, by extrapolation, to have smaller r_d and deeper energy, both reducing their effects. The p-states are coupled by $E_{pp} = -1.28\hbar^2/md^2$, as we found in Eq. (6-1)). Thus the band energies are found by diagonalizing the Hamiltonian matrix

$$H = \begin{pmatrix} \varepsilon_{p+} & E_{pp} & 0 \\ E_{pp} & \varepsilon_{p-} & E_{pd} \\ 0 & E_{pd} & \varepsilon_d \end{pmatrix}, \qquad (17\text{-}23)$$

or solving the secular equation from that matrix,

$$(\varepsilon_{p+} - \varepsilon)(\varepsilon_{p-} - \varepsilon)(\varepsilon_d - \varepsilon) - E_{pp}^2(\varepsilon_d - \varepsilon) - E_{pd}^2(\varepsilon_{p+} - \varepsilon) = 0 \ . \qquad (17\text{-}24)$$

This was evaluated numerically, with the intermediate eigenvalue representing the valence-band maximum. The parameters $\varepsilon_{p\pm}$ are obtained from Table 1-1, ε_d from Table 17-6 , and couplings from Eq. (17-22) and Eq. (6-1), using r_p from Table 17-1 and r_d from Eq. 17-6. d -values were listed in Table 2-1 (we used estimates 2.71 Å for HgS and 2.81 Å for HgSe).

The resulting shifts, listed in Table 17-9, are almost the same as the $E_{pd}{}^2/(\varepsilon_{p}$- - $\varepsilon_d)$ which would follow from lowest-order perturbation theory (perturbation theory gave shifts typically larger by 10%). The difference is small because perturbation theory overestimates the coupling by treating the valence band as purely from the $/p\text{->}$ state but also overestimates the energy denominator.

These shifts are much smaller than those suggested by Wei and Zunger (1998), principally because the ε_d which we used is considerably deeper, a choice which we supported by the comparison in Table 17-7. Nevertheless, it is a real contribution to the energy of the valence-band maximum and by raising the valence-band maximum reduces the gap. There is no compensating shift of the conduction band minimum because it is purely s-like and uncoupled to the d-bands at Γ. The largest shifts are for the mercury compounds where the gaps are smallest and the shifts are quite important in these cases. The shift also influences the values of band off-sets in heterojunctions (continuous crystal structure, changing from one compound to another across a plane) which we shall discuss in Section 19-4-B, and which were treated by Wei and Zunger (1998).

There is a second, but much smaller, contribution to these band off-sets arising from the d-states which we may note. We may imagine a heterojunction with a particular step in energy of the valence band (the valence-band offset Δ) in the absence of any coupling with the d-states. Adding that coupling in the bulk of the two materials will shift the valence band differently in the two materials as we have discussed here, modifying the band off-set. In addition, the coupling *across* the heterojunction will transfer different amounts of charge from the left than from the right, producing an interface dipole which gives an additional relative shift of the valence-band maxima. Such a shift is included in the full calculation by

Table 17-9 The shift upward of the valence-band maximum (eV) due to the coupling with d-states on the metallic atom. They were obtained from solution of Eq. (17-24), but are close to values obtained in simple perturbation theory.

GaP	0.033	InP	0.058	CdS	0.210	HgS	0.300
GaAs	0.026	InAs	0.053	CdSe	0.167	HgSe	0.230
GaSb	0.018	InSb	0.035	CdTe	0.116	HgTe	0.264

Wei and Zunger (1998) for cases in which they studied a multilayer system of alternating compounds, each separated by a heterojunction. When the calculation is performed self-consistently, it automatically includes any such shift, but it may not be easy to separate it from the other contributions.

One might at first imagine such a shift arising from the coupling between d-states of the metallic atom on one side and p-states on the nonmetallic atom on the other. However, since both sets are fully occupied the coupling produces no net charge transfer to lowest order. Transfer from occupied states on the right is canceled by transfer from occupied states on the left. Thus the only effects come more subtly from the mutual coupling of p-states on both atom types and d-states on the metallic atom, or from coupling with unoccupied s-states on either side. Such shifts could be calculated using the parameters given here, using the root-mean-square coupling $\sqrt{(V_{pd\sigma}^2 + 2V_{pd\pi}^2)/3}$ rather than the E_{pd} of Eq. (17-22). However, the effects appear to be even smaller than those from the shift of the valence-band maximum given above. Because the coupled states are both occupied we would expect the largest effect to come from the overlap repulsion arising from them, increasing the spacing and reducing the cohesion as suggested by Wei and Zunger (1988), but by a smaller amount due to our different choice of parameters.

Problem 17-1 Cohesion in the monoxides

Evaluate the contributions to the cohesion, and the total, for the oxides TiO, VO, and CrO, using the data from Table 17-4. This will give values such as the first seven listed for the nitrides in Table 17-5, but without the prediction of d and evaluated at the observed (or interpolated) d.

CHAPTER 18

Other Transition and f-Shell Compounds

The concepts and parameters introduced in Chapter 17 for the study of AB-compounds (monoxides, nitrides and carbides) are appropriate to a wide variety of compounds. In this chapter we extend the analysis to include f-electron parameters and discuss also more complex structures such as dioxides, trioxides, and multinary compounds.

18-1 Compounds of f-Shell Metals

In the f-shell metals the partly filled f-shell plays the role which the partly filled d-shells played in the transition-metal compounds. Harrison and Straub (1987) generalized the definition of the covalent energy of Eq. (17-8) to f-shell metals as

$$V_2{}^{pf} = \sqrt{X\frac{V_{pf\sigma}{}^2 + 2V_{pf\pi}{}^2}{3}} = \left(\frac{1225X}{\pi^2}\right)^{1/2}\frac{\hbar^2\sqrt{r_p r_f^5}}{md^5} . \qquad (18\text{-}1)$$

r_p values are again taken from Table 17-1 and r_f values were given in Tables 16-1 and 16-2. We see from the first expression that it equals the square root of the contribution of the coupling to the second-moment of the pf-bands.

A. Rare-Earth Compounds

Lanthanum, the last element before the rare-earth series, contains two valence s-electrons and one valence d-electron. It is thus electronically the same as scandium and can form a nitride, as discussed for ScN in Section 17-2, with no electrons in the lanthanum f, d, nor s-states. To the right of lanthanum we expect to gain one f-electron for each increase of one in atomic number. The energy $\varepsilon_f{}^0$ at which the last f-electron would be removed from the atom is lower by U_f than the energy at which an additional electron would be added, and the removal energy for the p-state on the nonmetallic atom lies always below this energy for adding, $\varepsilon_f{}^0 + U_f$. The coupling between the occupied p-states and any empty f-level will lower the energy of the p-level, contributing to the bonding of the compound. These six (including different spin states) p-levels become six bonding levels and the six f-levels to which they are coupled become antibonding levels. The remaining $14 - 6 = 8$ f-levels are the nonbonding levels. Any f-electrons for the atom will preferentially occupy one of the nonbonding levels since they are lower in energy than the antibonding levels. The electrons in these nonbonding states can then be expected to have their spins aligned, according to Hund's rule, due to their exchange interaction. Thus they form local moments on each atom equal to the electron moment times the number of f-electrons up to seven, then decreasing by one at each step.

The bonding contribution is quite small for the rare-earth compounds since the f-electrons are so core-like, as we indicated for the rare-earth metals. We may estimate it, as we did for transition-metal compounds in Section 17-3-E, by writing each of the six occupied pf-states as $\cos\gamma$ times a p-state plus $\sin\gamma$ times the f-state to which it is coupled. Then the energy for the six occupied pf-states, relative to the energy if all electrons were in the p-states, can be written as

$$E_{pf\,bond}(\gamma) = 6((\varepsilon_f{}^0 + U_f - \varepsilon_p)\sin^2\gamma - 2V_{2pf}\sin\gamma\cos\gamma) \qquad (18\text{-}2)$$
$$+ {}^{6\times5}/_2 U_f \sin^4\gamma .$$

This is the counterpart of Eq. (17-15) for the transition-metal compounds. The final Coulomb term omits the self-interaction. By using $\varepsilon_f{}^0$ from the free atom we are assuming a number of nonbonding f-electrons equal to the number in the free atom. We would need to modify this form if we considered rare earths beyond terbium so that electrons were placed in the antibonding states. We are not allowing the different electron states to have different coefficients, $\sin\gamma$ and $\cos\gamma$, and we are not attempting an exact calculation of the Coulomb correlations as we did for the two-level problem, but this simply means we may underestimate the additional bonding.

We may readily minimize Eq. (18-2) for any system, using the parameters we have tabulated. Probably the largest contribution arises for CeN since cerium as the largest r_f and therefore largest V_{2pf} , equal to 0.71 eV from Eq. (18-1) and with $d = 2.50$ Å, and $\varepsilon_f{}^0 + U_f - \varepsilon_p = 0.44\ eV$ is near zero, based upon the $\varepsilon_f{}^0 = -19.00\ eV$ (Mann (1967)), $U_f = 5.6\ eV$, and $\varepsilon_p = -13.84\ eV$. For that case we find the minimum at $\sin\gamma = 0.263$ with $E_{pf\,bond}(\gamma) = -1.57\ eV$. This is on the scale of the errors in our prediction of most transition-metal compounds (see Table 17-4), so it is not negligible but not large. Also it is smaller in other rare-earth compounds.

This solution for CeN shifts $6\sin^2\gamma = 0.415$ electrons into the f-levels on the cerium atom, raising the removal energy for the nonbonding level to -16.68 eV, just -2.84 eV below the original p-levels, and the energy at which electrons would be added to the nonbonding f-level to 2.76 eV above. In terms of this shifted $\varepsilon_f{}^{rem}$ we could estimate the bonding energy gain in perturbation theory as $-6\,V_{2pf}{}^2/((\varepsilon_f{}^{rem} - \varepsilon_p) = -6\times0.71^2/2.76 = -1.10\ eV$, comparable to the more accurate -1.57 eV obtained from Eq. (18-2). Recalculating for a variety of parameters suggested that this form generally gave a good estimate both of the contribution to the bonding and to the derivative with respect to spacing which helps determine the equilibrium lattice constant. Thus the contribution to the bonding properties of the pf-coupling can be simply estimated, but it requires a determination of the shifted f-level.

This view of the electronic structure differs from that of the earlier discussion by Harrison and Straub (1987) who sought a criterion for the formation of a "correlated state", for which the pd-coupling was neglected. They found such a correlated state for all of the rare-earth nitrides except for CeN, suggesting that this explained an anomalously low equilibrium distance for CeN. They indicated that there was experimental evidence for bands in this one case, which seems to have been just that the internuclear distance was smaller for that compound than for the surrounding ones. These nearest-neighbor distances (in Å, from Wyckoff (1963)) are LaN (2.63), CeN (2.50), PrN(2.58), NdN (2.57). We now have no explanation of that shorter spacing, but note that even if there were one electron in a nonbonding f-band, it should do little to the spacing. We do not know if CeN shows metallic conductivity, but we expect that it does not.

In all of these nitrides, the level at which electrons are added to the f-states is well above the nitrogen p-level and above the level for removal of an f-electron, so they should be simple insulators. We might form narrow bands for the occupied f-states and for the empty f-states, but they would not overlap each other. The effect of the f-levels on the bonding is small, and appreciable only for CeN. Again, we expect that those beginning with CeN should have local magnetic moments increasing by one electron to $3^1/_2$ Bohr magnetons at gadolinium, and decreasing after. This will also be true for the phosphides and arsenides, with even larger spacings, and thus even smaller covalent energies. The rare earths are behaving as simple trivalent metallic atoms and we might therefore not expect monoxides to form. Only in the special cases of EuO and YbO, for which the rare earth element tends to acquire a seventh, or fourteenth, f-electron to complete the shell (of the same spin) do the elements behave as divalent. These compounds would be expected to have the same electronic structure as CaO or BaO, except for the large moment in the case of EuO.

B. Actinide Compounds

In the actinides the 4f-states of the rare earths are replaced by the 5f-states of the next row. Brooks (1984) has made very extensive and accurate calculations on the electronic structure of a number of these compounds. There is also a wealth of information, theoretical and experimental, in Freeman and Darby (1974). Our effort here will be directed at a much simpler treatment based upon exactly the methods outlined in this book. We have no experimental information of the cohesive energies nor bulk modulus so we focus upon the bond-length determination.

The interesting question for these 5f-systems again is how large the role of pf-coupling is for the different compounds. We follow the analysis which we gave for the rare-earth compounds, again departing from the treatment by Harrison and Straub (1987) which focused on a criterion for omitting or including f-shell effects. However, we follow them in evaluating the

Table 18-1. Parameters (all in eV except for r_f) for the actinide metals from the Hartree-Fock term values from Mann (1967) who assumed configurations of $5f^{Z-3}6d7s^2$ except for $6d^27s^2$ for Th and $5f^{Z-2}7s^2$ for Pu and Am. For these three cases ε_f^0 values were interpolated from neighboring values to correspond to a $5f^{Z-3}$ configuration (the super zero indicates the $5f^{Z-3}$ configuration in all cases, even if a different configuration was assumed by Mann). U_f values and r_f values are from Table 16-2.

	ε_s	ε_d	$\varepsilon_f{}^0$	U_f	r_f (Å)
Ac	-4.38	-6.84	*	3.00	1.11
Th	-4.65	-8.05	-7.06	3.20	0.87
Pa	-4.49	-7.15	-8.67	3.35	0.70
U	-4.53	-7.25	-10.09	4.09	0.64
Np	-4.58	-7.32	-11.41	3.90	0.60
Pu	-4.36	*	-12.64	4.61	0.58
Am	-4.40	*	-13.81	4.96	0.58
Cm	-4.71	-7.42	-14.69	5.10	0.61

*No values available because they did not arise in Mann's configuration.

covalent energy $V_2{}^{pf}$ of Eq. (18-1), using $X = 6$ nearest neighbors, r_p values from Table 17- 1, and r_f values from Table 18-1. They also calculated the U_f values, and we list updated values in Table 18-1.

Harrison and Straub went on to predict equilibrium spacings, cohesive energies and bulk moduli for many of the compounds we shall include in Fig. 18-2, just as we did here for transition-metal compounds and rare-earth compounds. Again the only empirical parameters came from the sp-overlap repulsion, which was fit to the potassium halides as we did for transition-metal compounds in Section 17-3-F. We redo those calculations, including the self-consistent shift of the f-levels as in Eq. (18-2). These are important corrections.

The energy levels for an actinide compound, UN, are illustrated in Fig. 18-1 Cohesion arises in this case first from the transfer of electrons from the uranium atom to fill the nitrogen p-states. This E_{trans} , usually $3\varepsilon_p - 2\varepsilon_s - \varepsilon_d$, can be calculated directly from the energy levels ε_s and ε_d (and the configurations indicated) from Table 18-1 and ε_p from Table 1-1. In the case of thorium, one of the four electrons (of the $6d^27s^2$) is transferred to ε_f^0 , which is the correct value for adding the single electron there. In the case of Pu and Am, two electrons are transferred from ε_s for the actinide to the ε_p for the nitrogen, and a third from $\varepsilon_f^0 + U_f$ of the actinide (since it had configuration $5f^{Z-2}s^2$) to the ε_p of the nitrogen. We may think of the remaining f-electrons as in nonbonding states.

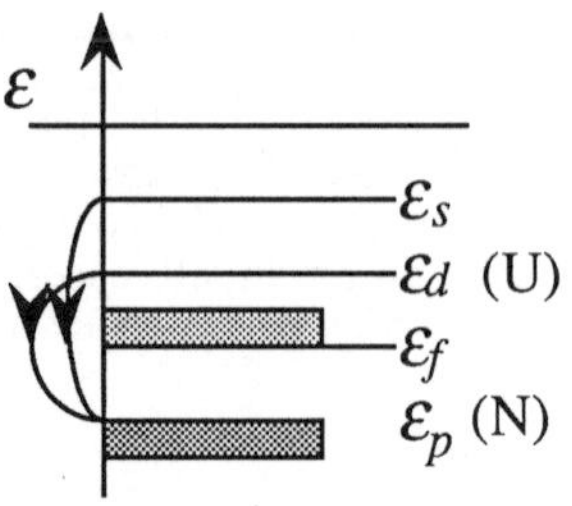

Fig. 18-1. A schematic energy-level diagram for uranium nitride. The upper three levels are atomic levels for a uranium atom, which contain two electrons in the s-state, one in the d-state and three in the f-states. Coulomb repulsions prevent greater occupations of the lower levels. In the compound the electrons from the s- and d-states are transferred to the nitrogen p-state. Coupling between the s- and p-states produces sp-bonding. Similarly, coupling between the f- and p-states broadens them into bands giving fp-bonding and shifting the average band energy as shown. In UN three electrons per atom pair remain in the nonbonding f-states.

We next add the sp-bonding energy per atom pair of

$$E_{sp\ bond} = 2V_3^{sp} - 2\sqrt{V_2^{sp\,2} + V_3^{sp\,2}} + E_{overlap}^{sp} \tag{18-3}$$

with $V_2^{sp} = \sqrt{6}V_{sp\sigma}$ and $V_3^{sp} = (\varepsilon_s - \varepsilon_p)/2$ from Section 10-2. This would be zero if there were no coupling (if$V_2^{sp} = 0$). The first two terms are the energy gained from the coupling between actinide s-states and p-states on the nonmetallic atom. The final term is the repulsive energy per atom pair,

$$E_{overlap}^{sp} = 6\eta_0 V_2^{sp\,4}\left(\frac{1}{|\varepsilon_{ig}|^3} + \frac{1}{16|\varepsilon_p|^3}\right) \tag{18-4}$$

from Eq. (10-10) with η_0 obtained for the row of the nonmetallic ion, obtained from the potassium halide and equal to 4.69 for KF, 7.93 for KCl, 9.02 for KBr, and 11.48 for KI.

We may illustrate the evaluation of Eqs. (18-4) and (18-3) for actinium nitride, which has no electrons in f-states, and for which these two are the dominant terms. We use $\eta_0 = 4.69$, and to obtain ε_{ig} we note that $\varepsilon_p = -23.14\ eV$ for neon in the nitrogen row and $\varepsilon_p = -11.65\ eV$ for radon in the actinium row, so $\varepsilon_{ig} = -17.40\ eV$ as the average between neon and radon. To obtain $V_3^{\ sp}$ we use $\varepsilon_p = -13.84\ eV$ for nitrogen and $\varepsilon_s = -4.38$ eV for actinium from Table 18-1, giving $V_3^{sp} = 4.73\ eV$. These are the only contributions to the energy which depend upon spacing for AcN. They yield $E_{bond}^{sp} = 2{\times}4.73 - 2\sqrt{V_2^{sp\,2} + 4.73^2} + V_2^{sp\,4}/166.3$ (in eV, with V_2^{sp} in eV). This is a minimum at $V_2^{sp} = 3.72\ eV$, or $d = 2.67\ Å$ for AcN. The corresponding calculation yields $d = 3.23\ Å$ for AcP and $d = 3.41\ Å$ for AcAs. We have neglected (as we did for the transition-metal compounds in the volume-dependent energy) the Coulomb corrections described in Section 10-5 and included in the second entry of Table 10-1, and we neglect the contribution from pd-coupling described in Section 17-2, which is small for these systems with $Z_d \approx 1$, and of course we have not included any effects of f-states.

For the other actinide nitrides, phosphides, and arsenides we must add the effects of pf-coupling from Eq. (18-2). We also add the pf-overlap repulsion, though it is not large, which was obtained by Harrison and Straub (1987) exactly as the pd-repulsion was obtained in Eq. (17-16). It is

$$E_{pf\,repuls.} = \frac{1050\hbar^2 r_p r_f^5}{\pi^2 m d^8}\,6\,. \tag{18-5}$$

The factor six was for the six electrons in the bonding band, since there are no electrons in antibonding bands, and the factor 1050 contains a factor six for the six nearest neighbors.

The sum of $E_{pf\ bond(\gamma)} + E_{pf\ repuls.} + E_{sp\ bond} + E_{overlap}^{sp}$ was minimized with respect to γ for a range of spacings d for the actinides using parameters essentially equal to those in Table 18-1. The spacing at minimum total energy gives the predictions for the equilibrium spacing shown in Fig. 18-2.

Without pf-bonding, the estimates of bond length would have been close to the corresponding AcN, AcP, or AcAs prediction, the d-value far to the left in Fig. 18-2, for which no pf-coupling was included in the estimate. It appears generally that we have obtained reduced spacings roughly in accord with experiment. This may be fortuitous, since the results are very sensitive to the value of r_f used, which is also the reason that the nitride predictions at the left end of the curve are irregular. The effect of pf coupling for thorium nitride was to increase the spacing, whereas in all other compounds it decreased it. That arose because r_f is larger for thorium, 0.87 Å from Table 18-1, somewhat higher than the next highest value, 0.70 Å for protactinium. In Fig. 18-3 we have plotted just the $E_{pf\,bond}(\gamma) + E_{pf\,repuls.}$ for ThN, but also with two reduced values for r_f. We can see that the spacing is increased from the pf-coupling, using our $r_f = 0.87\ Å$, but would have been contracted for the smaller values of r_f.

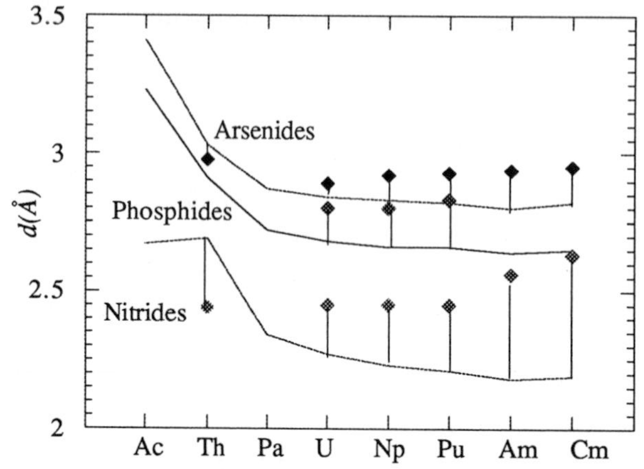

Fig. 18-2. Equilibrium spacings of actinide compounds calculated self-consistently, including sp-bonding and pf-bonding with the only experimental parameter the η_0 for the sp-repulsion, fitted to the potassium halides. Experimental values for the Th, U, Np, and Pu compounds came from Lam, Darby, and Nevitt (1974), the others were collected by Harrison and Straub (1987).

The qualitative message from Fig. 18-3 may be an important one. At the beginning of the series, when r_f is relatively large, the canceling effects of the bonding and the repulsive terms largely cancel for determining the spacing. Of course at small enough r_f both terms become small and the effects on the bond length are small. With r_f dropping significantly across the series this may be the reason for the smallest spacings at the center of the series, though our particular parameters and forms would predict the minimum further to the right.

This is quite a different explanation from the Harrison and Straub (1987) indication that Coulomb correlations are responsible for the rise in spacing at americium and curium. We did try to see if the effects of a correlated state should produce such an increase in spacing by replacing the bonding calculation which led to Eq. (18-2) by the counterpart of the exact calculation for the correlated state in Eq. (16-1). We could then minimize the total energy numerically with respect to the resulting two-electron states to obtain a total energy as a function of spacing. We applied it to PuN, PuP, and PuAs, and found the minimum very close to those obtained with Eq. (18-2). This argues strongly against our earlier interpretation of

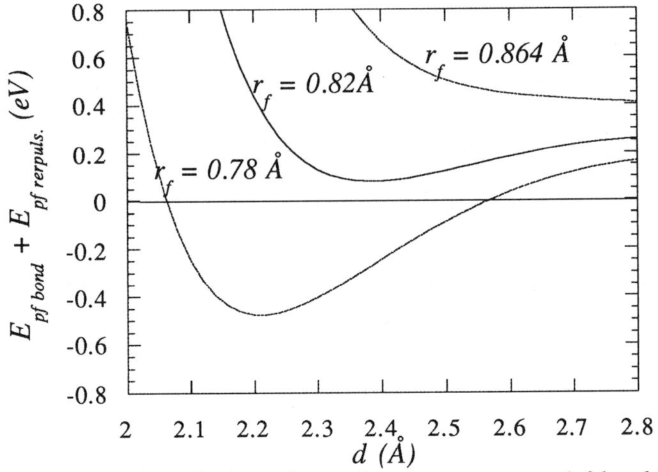

Fig. 18-3. The contribution of coupling between p- and f-levels to bonding in ThN, with r_f = 0.864 Å, as well as what it would have been with slightly reduced values of r_f.

the compound results, as well as the interpretation of the spacings of the pure actinide metals, in terms of electron correlations.

The same calculation which led to the predicted spacings in Fig. 18-2 gives also predictions of the cohesive energy, and if evaluated for neighboring spacings would give estimates of the bulk modulus. We do not tabulate these since we have no experimental information about either and the consideration of the spacing gives more insight into the electronic structure. However, we note in passing that our results from this self-consistent calculation were similar to those from Harrison and Straub (1987) which did not incorporate self-consistent calculations of ε_f. We found that the cohesion for the nitrides, phosphides, and arsenides of the actinides to the left was dominated by the transfer energy, $3\varepsilon_p - 2\varepsilon_s - \varepsilon_d$, near -10 eV in all cases. The other large term was the $E_{pf\ bond} + E_{pf\ repuls.}$ of Fig. 18-3. It was small at the left end of the series (actually positive for ThN, as seen in Fig. 18-3, but negative for other systems) but grew to -5.6 eV, -7.7 eV, and -7.6 eV for CmN, CmP, and CmAs, respectively, at the right end of the series. Harrison and Straub (1987), without self-consistency, found considerably larger values of cohesion.

2. Dioxides

An electronic structure similar to that of the AB compounds arises in the compounds with more general formulae, but it is somewhat more complicated. We treat only a few important cases, beginning with a dioxide.

A. TiO_2, Rutile

A familiar example is the mineral, rutile, TiO_2. Each titanium, with four valence electrons, formally gives its electrons to the two oxygen atoms, forming an insulator with empty titanium d-like conduction bands. The rutile structure may be thought of as a body-centered arrangement of TiO_2 molecules, or as parallel straight lines of TiO_2 across top- and bottom-face diagonals, but parallel to the other top-face diagonal midway between these two faces. The structure is not so essential in such ionic insulators and estimates of the electronic structure proceed much as for ionic crystals without transition metals and in simpler structures. However, it is appropriate to incorporate the shifts in the d-like states due to charging, as we did for the monoxides, and as we shall do for the one dioxide, RuO_2, which we shall discuss in detail.

B. RuO_2

Ruthenium dioxide, and many other dioxides, form in this rutile structure, but under pressure RuO_2 makes two structural phase transitions, finally forming in a fluorite structure (Haines and Legér (1993)). When the pressure is removed it remains in this metastable phase and was found to have a very large bulk modulus. Lundin, Fast, Nordström, Johansson, Wills, and Eriksson (1998) analyzed the electronic structure of this system with the purpose of understanding the large bulk modulus. We are not concerned about that aspect. In fact the bulk modulus for RuO_2 in the fluorite structure is similar to those we expected for the transition-metal nitrides in the rock-salt structure, discussed in the preceding section. These nitrides also had nearest-neighbor distances near the 2.06 Å of RuO_2 and large bulk moduli come naturally to such compact systems. However, we shall make use of the understanding of the electronic structure which they obtained.

Lundin, et al., performed full-potential local-density-approximation (LDA) calculations for several dioxides, including RuO_2, in the fluorite structure. The electronic structure is again dominated by the transition-metal d-states and oxygen p-states. There are eight valence electrons from the ruthenium and a total of eight p-electrons from the two oxygens. Lundin, et al., found that they were accommodated in three bonding bands, three non-bonding oxygen p-bands and two nonbonding d-bands. This is quite different from our description of the monoxides, since the five d-states, coupled to the six oxygen p-states, would be expected to yield only a single nonbonding p-band. However, we may understand the electronic structure by considering the energy bands at the center of the Brillouin Zone, $\mathbf{k} = 0$, where we shall find the corresponding four band energies, as they suggest. This enables us to understand and quantify the electronic structure using their view, but in terms of the parameters introduced in this book. Of course as we move away from $\mathbf{k} = 0$ new couplings arise and other couplings decrease between each of the d-states and p-states. Even in nearest-neighbor tight-binding theory these bands are not nonbonding at all wavenumbers as they were for the monoxides. The bands spread and could have overlapped each other. However, Lundin, et al., found that the four peaks remain distinct and we make use of that finding in interpreting the electronic structure in terms of the bands at $\mathbf{k} = 0$.

In the fluorite (CaF_2) structure (See Problem 9-1) the transition metal atoms form a face-centered-cubic lattice and one of the two oxygens per molecular unit completes a zincblende structure, displaced by $d = 2.06$ Å from each ruthenium in a [111] direction. The second oxygen lies the same distance in the - [111] direction, so each oxygen is tetrahedrally surrounded by ruthenium atoms and each ruthenium is surrounded by eight oxygens,

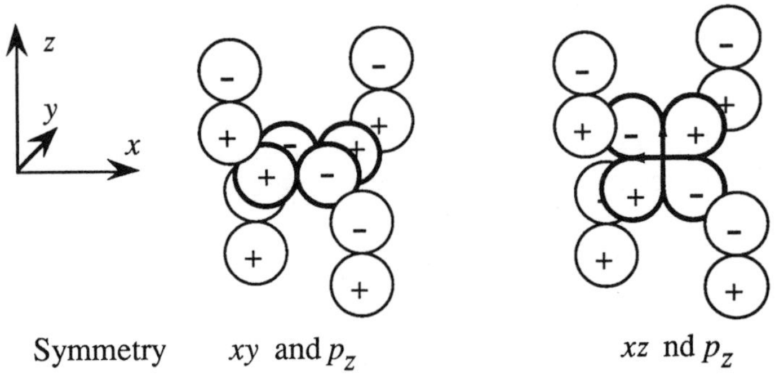

Fig. 18-4. A transition-metal d-state, the heavier circles, with symmetry written below, with oxygen p_z -states on one tetrahedron of neighboring oxygen states. For the xy - state the nearest overlapping lobes are of the same sign so all matrix elements add. For the xz- state, half the overlapping lobes are of the same sign and half of opposite sign so the net coupling is zero.

forming a cube of edge $a = 2d/\sqrt{3}$. We look specifically at $\mathbf{k} = 0$ and note that for each band the coefficient of the d-state is the same on every ruthenium atom and the coefficient of each p-state (p_x, p_y, or p_z) is the same for each tetrahedron of oxygen atoms (e. g., those in the - [111] direction from a ruthenium). This allows a coupling between a d-state of the form xy and p_z-states on an oxygen tetrahedron, as seen to the left in Fig. 18-4. The individual matrix elements are obtained from Table 15-2 as $E_{x,yz}$ (equivalent by rotating indices to $E_{z,xy}$) for $\ell = m = n = 1/\sqrt{3}$ as

$$E_{z,xy} = E_{x,yz} = \frac{V_{pd\sigma}}{3} - \frac{2V_{pd\pi}}{3\sqrt{3}} = -\frac{5\sqrt{15}}{6\pi}\frac{\hbar^2\sqrt{r_p r_d^3}}{md^4} . \tag{18-6}$$

The coupling between the two Bloch sums is four times the individual coupling and with $r_p = 4.41\ \text{Å}$ for oxygen, $r_d = 1.127\ \text{Å}$ for Ru from Table 15-1, and $d = 2.06\ \text{Å}$ from Lundin, et al. (1998), we obtain a coupling of $4E_{z,xy} = 4.37\ eV$.

The coupling between each tetrahedron of p-states and any other d-state is zero, as is seen for the xz and p_z coupling to the right in Fig. 18-4, and as can be seen for the other d-states of Fig. 15-2. However, this d-state has an equal coupling to the complementary oxygen tetrahedron (completing the surrounding cube of oxygen atoms). Thus the Hamiltonian matrix for the $\mathbf{k} = 0$ states contains a three-by-three submatrix of these three states,

$$H = \begin{pmatrix} \varepsilon_d & 4E_{z,xy} & 4E_{z,xy} \\ 4E_{z,xy} & \varepsilon_p & 0 \\ 4E_{z,xy} & 0 & \varepsilon_p \end{pmatrix} \tag{18-7}$$

uncoupled to all other states. It may be diagonalized immediately to obtain one state with energy ε_p and two others with energy

$$\varepsilon_\Gamma = \frac{\varepsilon_p + \varepsilon_d}{2} \pm \sqrt{\left(\frac{\varepsilon_d - \varepsilon_p}{2}\right)^2 + 2(4E_{z,xy})^2} . \tag{18-8}$$

There are two more sets of states of identical energy obtained from the zx , and yz d-states so that these three form the three bonding, antibonding, and non-bonding p-bands, indicated by Lundin, et al.. The $3z^2 - r^2$ and $x^2 - y^2$ d-states are uncoupled to the p-states at $\mathbf{k} = 0$ and form the non-bonding d-states also indicated by Lundin, et al.

To estimate the energies of these states we must evaluate ε_d for ruthenium in this compound, which we do as if the $\mathbf{k} = 0$ state were representative. We define a polar energy equal to $V_3 = (\varepsilon_d - \varepsilon_p)/2$ and a covalent energy equal to $V_2 = \sqrt{2}\,(4E_{z,xy}) = 6.18\ eV$; then the number of d-electrons per ruthenium in the compound is $4 + (1 - \alpha_p)\times 6/2 = 7 - 3\alpha_p$ with of course $\alpha_p = V_3/\sqrt{V_2^2 + V_3^2}$. The 4 is from the nonbonding d-states and the second term is from the d-state contribution to the six bonding states. Then the d-state energy can be written as the p-state energy of the oxygen (-16.77 eV) plus twice V_3 . It can also be written as the value from the Hartree-Fock calculation, for the d^6s^2 configuration (-14.59 eV from Table 17-1) plus the correction $(7 - 3\alpha_p - 6)U_d$ for the change in the number of d-electrons, with $U_d = 3.7\ eV$ from Table 15-1. These are solved together numerically to obtain $V_3 = 1.57\ eV$, or $\alpha_p = 0.246$ and $\varepsilon_d = -13.62\ eV$.

This gives a crude electronic structure which we may compare with the full calculation by Lundin, et al., shown in Fig. 18-5. We have found filled nonbonding d-bands (III) at -13.62 eV, with the Fermi energy at that point. The nonbonding oxygen p-bands (II) are below that Fermi energy by 0.23 Ryd.(we give values in Rydbergs for Fig. 18-5) at -16.77 eV. The bonding bands (I) are below the Fermi energy by 0.58 Ryd. at -21.58 eV and the antibonding bands (IV) above the Fermi energy by 0.35 Ryd. at -8.82 eV. These energies are in reasonable accord with the groups of bands obtained by Lundin, et al., as seen in the density of states in Fig. 18-5. Lundin, et al., of

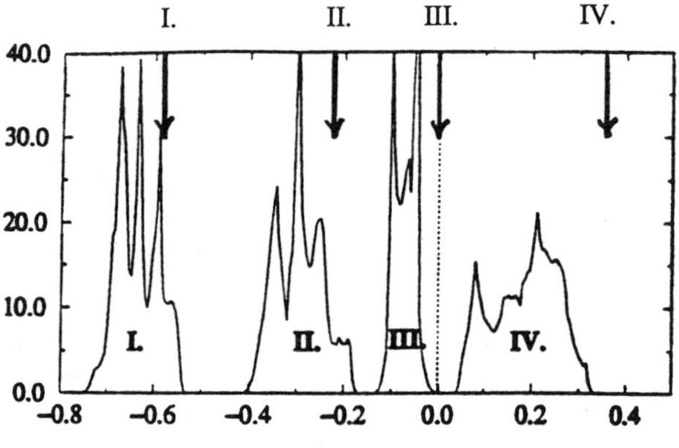

Fig. 18-5. The density of states of RuO_2 in the fluorite structure, calculated by Lundin, Fast, Nordström, Johansson, Wills, and Eriksson (1998), as a function of energy (in Rydbergs) measured from the Fermi energy. The arrows above indicate the energy of these four sets of bands obtained in the $\mathbf{k} = 0$ tight-binding calculation given in the text.

course set the Fermi energy at the top of the nonbonding d-band and if we were to adjust our non-bonding energy to the center of their nonbonding band, the agreement would be even better.

It is interesting to compare the electronic states in the nonbonding p-bands with those in the bonding bands. The basic difference is that in the bonding bands the oxygen p-lobes are of the same sign near the ruthenium, so that the coupling of the ruthenium d-states with all of the oxygen p-states adds up, while in the nonbonding states they are of opposite sign so that the coupling with the p-states canceled, the nonbonding condition. Lundin, et al., noted this fact in charge-density plots and associated the lobes of the same sign with oxygen-oxygen bonding. Our interpretation, based upon nearest-neighbor coupling only, is somewhat different and is consistent with lobes of the same sign also raising the energy of the antibonding states.

Lundin, et al. (1997) made similar calculations on other transition-metal dioxides in the same structure and on RuNF which is isoelectronic with RuO_2. All can be treated as we have done here. In the case of RuNF, the two oxygen tetrahedra are replaced by one of nitrogen and one of fluorine so that one of the ε_p's in Eq. (18-7) is the value for nitrogen and one is for fluorine. Again three eigenvalues are obtained for each coupled d-state.

Given the occupied bonding levels we found in Eq. (18-8) we may discuss the properties of these compounds as we did for the monoxides. Because of the gap between the nonbonding d-bands and the antibonding states (found by Lundin, et al., but not guaranteed by our calculation at $\boldsymbol{k} = 0$ only) RuO_2 should be insulating, or semiconducting. We expect dioxides with transition elements to the right of ruthenium in the periodic table to be unstable in this structure due to the occupation of antibonding states, but elements to the left of ruthenium might form in the fluorite structure and would be poor metals, with Fermi surface in the nonbonding d-bands.

3. Trioxides, the Perovskites

Some of the most important oxides are those in the perovskite structure. They are characterized ordinarily by three oxygen atoms and one transition-metal atom (called, for example, a titanate if the transition-metal atom is titanium) and frequently an additional simple-metal atom (such as strontium in $SrTiO_3$, strontium titanate). Such titanates and niobates are important electronic materials. The high-temperature superconductors of recent years are based upon cuprates of this type, with some number of oxygen atoms missing from the full perovsksite structure. It even appears that at high pressures aluminum silicate may take the perovskite structure and may be the most common crystal composing the earth.

Our analysis uses the same parameters and general approach which we have used for other transition-metal compounds, and is based upon the analysis give earlier in *Electronic Structure and the Properties of Solids* (Harrison (1980)). That analysis was begun before we had pd-matrix elements as described in Section 17-1. We attempted to interpolate matrix elements between those fit to the band structures of a number of perovskites calculated by Mattheiss (1972) with little success, but once theoretical forms were available interpolation was not really necessary, though in our study in Harrison (1980) we tuned to matrix elements to fit those calculation. Here we proceed directly with our parameter and compare results with Mattheiss (1972). We also shorten the discussion considerably.

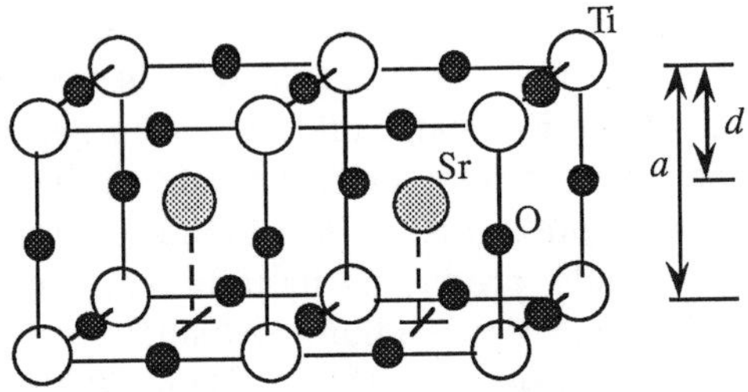

Fig. 18-6. The perovskite structure of $SrTiO_3$, a simple cubic arrangement of titanium atoms, with an oxygen on every cube edge of length a , and with a strontium atom at each cube center.

A. The Perovskite Structure

The perovskites typically have a formula ABO_3 , with B a transition-metal, and A an alkali metal or the divalent neighbor. (There are cases, such as $NaMgF_3$, which form in the same structure but do not contain transition metals.) The structure is illustrated in Fig. 18-6 for $SrTiO_3$, a prototype material, with transition-metal atoms forming a simple cubic lattice and oxygen atoms lying at the center of each cube edge. It was important conceptually to place the cube corners at the transition metal since the essential electronic structure will involve just the transition-metal d-states and oxygen p-states, so the grid lines join the essential states. The oxygen atoms were drawn as small (dark) circles to more easily see the geometry, though in fact they should be thought of as the large atoms, with the small transition-metal atoms wedged between. The A atom, strontium in Fig. 18-6, gives its electrons up to the BO_3 grid, and ordinarily has little other effect on the properties, though in some cases it is instrumental in determining some distortion of the structure.

The basic electronic structure is easily understandable. The oxygen p-states, at -16.77 eV from Table 1-1, can accommodate two electrons each, or a total of six electrons per formula unit in addition to those from the free-atom oxygens. In $SrTiO_3$, the strontium s-state, at -4.86 eV will provide two and the four valence electrons of titanium, at -6.04 eV for the s-state and -11.05 eV for the d-states, provides the rest. Thus viewing it as an ionic crystal we may think of a filled oxygen p-like valence band, with a titanium d-like conduction band, 5.72 eV above it. Of course, then, $SrTiO_3$ is a wide-gap insulator.

Not all perovskite compounds are insulating. The tungsten bronzes, Na_xWO_3, are an important metallic case. Tungsten oxide itself, WO_3, is insulating. It forms the structure in Fig. 18-6, but with no atoms at the Sr sites. It also distorts so that it is not cubic as shown, but retains the same connectivity of the grid. The tungsten has just the six valence electrons needed to fill the oxygen p-states and there is a gap below the tungsten d-states. However, when sodium atoms are added, they go to the cube-center position, that of strontium in Fig. 18-6, and the electrons go to the tungsten d-state nonbonding bands, producing metallic conductivity, increasing with increasing concentration x of sodium. At sufficiently large x the structure again becomes cubic as in Fig. 18-6. Rhenium trioxide, ReO_3, has one additional valence electron, in comparison to WO_3, and that electron must go into the d-bands. Similarly, $KTaO_3$ provides six valence electrons to fill the oxygen states and is insulating but $KMoO_3$ has an additional electron and is metallic. All of these compounds, except Na_xWO_3 for small x , are cubic and their bands were calculated by Mattheiss (1972).

In all cases the properties are consistent with the qualitative picture of the electronic structure as an ionic compound, with in some cases electrons in the nonbonding d-bands, just as in our description of the transition-metal monoxides. As a first approximation, the energy required to separate an insulating perovskite crystal into isolated atoms would be the sum of the energies gained in dropping the six electrons into the oxygen p-state, much as we estimated the cohesive energy of simple ionic compounds in Chapter 9. Such estimates for $SrTiO_3$ are probably of accuracy comparable to corresponding estimates for SrO and TiO_2. For Na_xWO_3 there is the additional contribution from dropping the s-electrons to the nonbonding d-bands. The heat of formation of these compounds is defined to be this energy of atomization minus the heat of atomization of the individual elements.

B. The Electronic Structure

We may go beyond the simplest ionic description by introducing interatomic coupling. The coupling in $SrTiO_3$ between the oxygen p-states and the strontium and titanium s-states will push them even further upward, contributing something to the cohesion by lowering the p-states, but of little interest otherwise. The coupling which is of more interest, as for the monoxides, is that between the transition-metal d-states and the oxygen p-states. The same matrix elements, term values, and U_d values enter which were given in Section 17-1. We may in fact define covalent and polar energies V_2^{pd} from Eq. (17-8) with coordination again $X = 6$. Values for the four perovskites we treat are given in Table 18-2. We may also obtain a $V_3^{pd\,0}$ from Eqs. (17-7) using the atomic d-state energy from Table 17-2.

Our first task is to recalculate ε_d self-consistently as we did for monoxides following Eq. (17-12). In this case $Z_d = 6\, sin^2\gamma + Z_d(nonbond)$, with $Z_d(nonbond)$ equal to zero for $SrTiO_3$ and $KTaO_3$, and one for $KMoO_3$ and ReO_3, and from Eq. (17-12) we obtain the $(\varepsilon_d - \varepsilon_d^0)/2$ which we add to $V_3^{pd\,0}$ to obtain V_3^{pd}. Using this V_3^{pd} we recalculate the polarity and the $sin\ \gamma$ to redetermine Z_d, repeating until we obtain self-consistency. The resulting values are also listed in Table 18-2. It is interesting to compare with the results of Mattheiss (1972). He performed a band calculation for the four compounds using atomic charge densities (corresponding in our case to using $V_3^{pd\,0}$) but then adjusted parameters in a tight-binding fit to the band

Table 18-2. Parameters for four cubic perovskites, obtained in the text. Below are given values derived from Mattheiss's fit to experiment.

	$SrTiO_3$	$KTaO_3$	$KMoO_3$	ReO_3
$d(Å)$	1.95	1.99	1.96	1.87
$V_3^{pd\,0}(eV)$	2.86	3.60	2.64	2.21
$V_2^{pd}(eV)$	3.90	5.38	5.00	5.81
$V_3^{pd}(eV)$	1.95	2.02	1.42	0.33
		From Mattheiss (1972)		
$V_2^{pd}(eV)$	3.91	5.15	4.66	5.96
$V_3^{pd}(eV)$	1.50	1.76	1.65	0.36

calculation in order to agree with various experiments on the compounds. Before adjustment his band energies corresponded to a $V_3^{pd\,0}$ of 3.17, 3.12, 2.33, and 1.72 eV for the four compounds (in the order of Table 18-2), rather close to our $V_3^{pd\,0}$ from Hartree-Fock atomic term values. We used his values after adjustment to obtain the parameters listed below in Table 18-2: $(\varepsilon_d - \varepsilon_p)/2$ for his V_3^{pd} and $\sqrt{6(V_{pd\sigma}^2 + 2V_{pd\pi}^2)/3}$ (as in Eq. (17-8)) for V_2^{pd}. (His ratios of $V_{pd\sigma}$ to $V_{pd\pi}$ were similar to the $-\sqrt{3}$ obtained in our treatment.) We would say that the agreement was remarkable for our completely independent elementary calculation. In particular, the rather large shifts in V_3^{pd} arising from our self-consistent calculation (compare $V_3^{pd\,0}$ and V_3^{pd} in Table 18-2) bring them into good accord with his adjustment to the experimental values.

We next look at the band structures themselves. At first we include only $V_{pd\sigma} = -(3\sqrt{15}/2\pi)\times V_2^{pd}/\sqrt{225/2\pi^2}$, $V_{pd\pi} = (3\sqrt{5}/2\pi)\times V_2^{pd}/\sqrt{225/2\pi^2}$ and the V_3^{pd} values from Table 18-2. For $SrTiO_3$ these are -2.14 eV, 1.23 eV, and 1.95 eV. With only these couplings, and wavenumbers parallel to a [100] direction, we obtain one band (symmetry Δ_1) with energy, relative to ε_p, of

$$\varepsilon_k = V_3^{pd} \pm \sqrt{V_3^{pd2} + 4V_{pd\sigma}^2 \sin^2 kd}\,, \tag{18-9}$$

a set of two degenerate bands (symmetry Δ_5) with energy

$$\varepsilon_k = V_3^{pd} \pm \sqrt{V_3^{pd2} + 4V_{pd\pi}^2 \sin^2 kd} \tag{18-10}$$

two bands with energy $2V_3^{pd}$ and five with energy equal to ε_p. These bands are plotted to the left in Fig. 18-7. The bands to the right are those obtained by Mattheiss (1972). The principal differences between the two sets of bands arise from the effects of couplings which we neglected in our analysis.

One such coupling is between the oxygen s-state and the titanium d-states, a coupling $V_{sd\sigma}$. We did not introduce this coupling when we introduced the $V_{pd\sigma}$ and $V_{pd\pi}$ in Section 17-1 because they are ordinarily of considerably less interest. It would be possible to introduce it, just as the pd-coupling was introduced in that section, leading to a coupling $V_{sd\sigma}$ equal to $-(\sqrt{5}/2\pi)\sqrt{r_d^3/r_s}/md^3$ with an s-state radius r_s fit to a tight-binding s-band just as r_p was obtained by Harrison and Straub (1987), but that has not been done. In Harrison (1980) we actually used a different formalism for all of

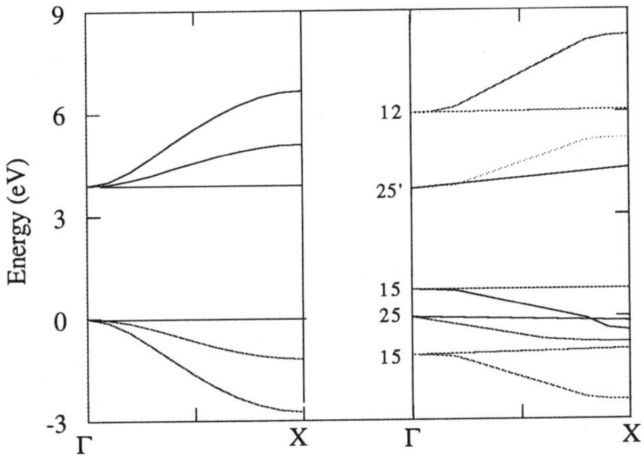

Fig. 18-7. The energy bands of $SrTiO_3$ for wavenumbers along a [100] direction. To the left are the bands obtained from Eq. (18-9) and (18-10) with our parameters. To the right are the bands, adjusted to fit experiment, given by Mattheiss (1972). The differences can be accounted for by adding couplings $V_{sd\sigma}$ and second-neighbor pp-coupling, as described in the text.

these couplings, $V_{sd\sigma}$, $V_{pd\sigma}$, and $V_{pd\pi}$, by beginning with the coupling $<<\boldsymbol{k}|H|d>$ of Eq. (15-11), and equating it to the coupling obtained with the plane wave replaced by a tight-binding sp-state. (This procedure is described in Harrison (1980), 519 ff.) It leads to all of the couplings, $V_{sd\sigma}$, $V_{pd\sigma}$, and $V_{pd\pi}$, being proportional to $\hbar^2 r_d^{3/2}/md^{7/2}$, with comparable coefficients. We do not believe that this procedure is as reliable and accurate as the approach used in Section 17-1 for the $V_{pd\sigma}$, and $V_{pd\pi}$, and since these are more important we have chosen to proceed that way. To obtain an approximate value for $V_{sd\sigma}$ we make use of the qualitative finding from the alternative approach that it should be comparable to $V_{pd\sigma}$

In contrast to the monoxides in the rock-salt structure, the coupling $V_{sd\sigma}$ shifts the d-band energy even at Γ . This is because with three oxygen atoms per primitive cell, there is a combination of oxygen s-states which has the symmetry Γ_{12} of two of the atomic d-states. (e. g., a d-state of the form x^2-y^2 is coupled to a sum of s-states in the x-direction and the negative of a sum of s-states in the y-direction (with a factor $1/\sqrt{2}$ so the pair is normalized in the cell.) That coupling at Γ is obtained using the Slater-Koster Tables (Table 15-2) as $\sqrt{6}\, V_{sd\sigma}$ so it shifts the Γ_{12} band upward by some

$$\delta\varepsilon_d = -\,\delta\varepsilon_s = \frac{6V_{sd\sigma}^2}{\varepsilon_d(Ti) - \varepsilon_s(O)}\ . \tag{18-11}$$

If $V_{sd\sigma}$ were equal to $V_{pd\sigma}$ at -2.14 eV, this shift would be 1.3 eV. The shift shown by Mattheiss' calculation in Fig. 18-7 is 2.2 eV, and correspondingly he used a larger coupling (-2.56 eV). This coupling in fact raises the entire set of bands beginning at the Γ_{12} state indicated in Fig. 18-7 by approximately this amount. The bands beginning at $\Gamma_{15'}$ are uncoupled (with d-states corresponding to yz, zx, and xy) and are not shifted. The coupling which shifts these three bands up by 2.2 eV, would of course lower the energy of three oxygen s-bands by the same amount, contributing $3/\delta\varepsilon_s/ = 6.6\ eV$ to the cohesion per formula unit. (A factor of two for spin was included, but the result divided by two to account for overlap repulsion.)

The other important coupling we have omitted is between oxygen p-states, which are second neighbors. These couplings were also necessary to obtain appropriate bands for the alkali halides as noted at the end of Section 9-2-B. p-states of given symmetry, say x , lying along one axis through a titanium are coupled to p-states of the same symmetry lying along another axis by $^1/_2(V_{pp\sigma} + V_{pp\pi})$. Such coupling shifts one Γ_{15} set up by $2\sqrt{2}\,(V_{pp\sigma} + V_{pp\pi})$ and the other down by the same amount (Harrison (1980), p. 446), leaving the Γ_{25} set unshifted, as seen in Mattheiss's bands. His splitting of 1.8 eV corresponds to a second-neighbor $V_{pp\sigma} + V_{pp\pi}$ of 0.16 eV, a factor of ten smaller than what would be obtained using the formula for nearest-neighbor matrix elements, but of course such formulae are quite inappropriate for second-neighbor coupling. These shifts do not affect the cohesion since both the upward-shifted and downward-shifted bands are occupied.

These additional couplings were included in Harrison (1980), producing bands which are in quite good accord with those of Mattheiss to the right in Fig. 18-7. To do that it was necessary to adjust the additional couplings, so there is little real test as there is in Table 18-2 where we compare with our own parameters. We find that our parameters are quite appropriate, and consistent with the best information on the electronic structure.

Note that two of the compounds listed in Table 18-2, potassium molybdate and rhenium trioxide have an additional electron per bonding unit.

The bands look very much like those given in Fig. 18-7, but the Fermi energy is above the $\Gamma_{25'}$ level and they are good metals. Although these bands appear in the figure to have energy almost constant, they are not nonbonding bands in the sense of the nonbonding d-bands in the monoxides. In the monoxides, with three valence p-states and five valence d-states, nearest-neighbor coupling leaves two d-bands completely independent of energy. Here with nine valence p-states and five valence d-states, there are four nonbonding p-bands, and all five d-bands are antibonding. This is also apparent in the more complete bands for these systems given by Mattheiss (1972), where we see that they rise parabolically in the [110] and [111] directions. Thus when they are occupied, they subtract from the bonding energy of the lower bands and affect the bonding properties.

An additional perovskite compound, $KNiF_3$, was treated by Mattheiss (1972) along with those we have discussed. An electron count puts eight electrons in the d-like bands. This would place so many electrons in antibonding bands that we would expect, as in the monoxides, that this could occur only because the coupling is weak (small r_d for Ni) and the coupled states well separated (ε_p deep for fluorine) so that the bonding energies are very small. Then we would also expect strongly correlated states for these d-electrons, for the spins to align, and to form an antiferromagnetic structure. That is indeed the case. The understanding of the monoxides again extends to the more complicated structure.

C. Bonding Properties

The perovskites are basically ionic compounds, but as in many other transition-metal compounds, the coupling is large enough that we cannot use the perturbation theory which sufficed for the alkali halides. Clearly one possibility for calculating the bonding properties is to construct a Friedel model for five bonding and five antibonding bands as in Section 17-3-C, with widths given by Eq. (17-9) in terms of our covalent and polar energies for the perovskites. We would then add a contribution for the coupling between the d-states and the oxygen s-states as $6\delta\varepsilon_s$ using Eq. (18-11). We would also include the same coupling between the transition-metal s-states and oxygen p-states which enters all the simple compounds, and an overlap repulsion associated with all of these couplings. Finally, one would anticipate including a coupling between the strontium, or other alkaline earth or alkali metal s-states and the oxygen p-states, taken from the theory of the simple ionic compounds. This has not yet been carried out.

An alternative approach for the pd-bonding was taken in Harrison (1980). It was noted that $V_{pd\sigma}$ and $V_{sd\sigma}$ were comparable and large, and that $V_{pd\pi}$ was smaller (by a factor $1/\sqrt{3}$). Thus one could construct sp-hybrids on the oxygen, each of which is strongly coupled to d-states on one neighboring atom, and almost uncoupled to those on the other. Then a bonding unit could be constructed consisting of a single transition-metal atom and six neighboring oxygen hybrids. In fact if the $V_{pd\pi}$ coupling was neglected, there are only two states of d-symmetry (symmetry Γ_{12}, or the orbitals of $x^2 - y^2$ and $3z^2 - r^2$ symmetry) which are coupled. The corresponding two antibonding states are empty in all cases, and the two bonding states occupied. All other p-states become nonbonding and occupied, and all other d-states are nonbonding and empty (or occupied by one electron in $KMoO_3$ and ReO_3). In this simplified representation of the electronic structure, the p-states on the oxygen atoms are fully occupied except for the reduction of oxygen charge from the two bonding states of the bonding unit. One could equivalently say that charge was extracted by the two "ghosts" of the d-states, the antibonding states of the bonding unit. Similarly, the only contribution to the bonding in this representation is from the lowering in energy of the two bonding states, or equivalently the rise in energy of the two spins for each of these two ghosts. Thus all of the bonding and dielectric properties are estimated in terms of a covalent and polar energy for these perovskite ghosts. Such an approach neglects a number of the couplings which we enumerated in the preceding paragraph, as well as the coupling $(\varepsilon_s(O) - \varepsilon_p(O))/2$ between two hybrids sharing the same oxygen atom. This approach was not carried much further and it is not clear how much accuracy is sacrificed in order to obtain the conceptual and computational simplicity of the model.

This simplification in terms of perovskite ghosts *was* used to evaluate the chemical grip described for the alkali halides in Section 10-4-B. These are the terms of fourth order in an expansion in the ratio of the coupling between neighboring orbitals to the energy difference. For empty states of angular-momentum quantum number ℓ , coupled by $-V_2$ to a set of occupied states on neighboring atoms, numbered by j , and lower in energy by $2V_3$, the angular term in the energy, of fourth order in the coupling, is

$$E_{grip} = \Sigma_{j>i} \frac{4V_2{}^4}{|2V_3|^3} P_\ell(\cos\theta_{ij})^2 . \tag{18-12}$$

Eq. (10-18) was a special case of this for $\ell = 1$. [Eq. (18-12), as well as Eq. (10-18) which is the evaluation for $\ell = 1$, appeared with a factor 2 rather than 4 in the first printing of the Freeman edition (hard-cover) of Harrison (1980), but was corrected in the Dover edition.]. The same formula applies if the central orbital is occupied and the surrounding orbitals are empty, being positive in both cases. For d-states $P_2(\cos\theta) = {}^3/_2\cos^2\theta - {}^1/_2$. There can be contributions for opposite neighbors (θ near 180°) and neighbors with θ near 90°. The latter are of more interest and for them $P_2(\theta)^2 = {}^1/_4 - {}^3/_2\delta\theta^2 + \ldots$

It is interesting that the angular term in E_{grip} , which is $-{}^3/_4(V_2{}^4/|V_3|^3)$ times a sum over squared angular deviations, favors distortion of the lattice away from right angles, as it did for the rock-salt structure. Thus it may be responsible for, or at least contribute to, the observed noncubic structure of WO_3. In the cubic perovskites, it will cause a reduction in the elastic constant c_{44} . In the corresponding distortion e_4 there are four angles at each transition metal changing by $\delta\theta = \pm e_4$, for a change in E_{grip} of $-3V_2{}^4/|V_3|^3)$ $e_4{}^2$ for each volume of $(2d)^3$ giving a contribution to c_{44} of ${}^3/_4V_2{}^4/(V_3{}^3d^3)$.

We must keep our parameters straight in making an evaluation of this term. Since the V_3 which enters is half the energy difference between the oxygen *hybrid* energy and the d-state energy, while the $V_3{}^{pd}$ was half the difference between the oxygen *p-state* energy and the d-state energy, we have

$$V_3 = V_3{}^{pd} + \frac{\varepsilon_p(O) - \varepsilon_s(O)}{4}, \qquad (18\text{-}13)$$

V_3 being larger than $V_3{}^{pd}$ by 4.31 eV. Then the contributions to c_{44} are comparable to those obtained earlier, -1.4×10^{11} ergs/cm^3 for $SrTiO_3$, and -5.04×10^{11} ergs/cm^3 for $KTaO_3$. (A correction would need to be made for the electrons in the d-bands for the other two compounds in Table 18-2.)

D. Madelung Energies

A relevant comparison is with the electrostatic contributions to c_{44} , which are positive. It was seen in Harrison (1980), 468ff, that the change in electrostatic energy under an e_4 distortion could be made by direct summation, without use of special techniques for calculations of Madelung constants for ideal structures (mentioned in Section 9-1-B and 10-5-B.) The contribution to the constant c_{44} was written

$$c_{44}{}^{electrostatic} = \alpha(Z,Q)\,\frac{e^2}{(2d)^4} \qquad (18\text{-}14)$$

Table 18-3. The electrostatic rigidity constant $\alpha(Q,Z)$ for the perovskite structure, with edge-center ion charge $-Q$, cube-center ion charge Z, and cube-corner ion charge $6 - Z$. The shear constant $c_{44} = \alpha(Q,Z)e^2/(2d)^4$ for nearest-neighbor distance d.

Z =	0	1	2	3
$Q = 0$	-0.593	1.075	5.29	6.68
$Q = 1$	10.95	14.00	14.24	11.68
$Q = 2$	29.39	28.48	24.76	18.24

for a perovskite structure of cube edge $2d$ in which the cube-edge-center ions have charge $-Qe$ (without charge transfer by coupling, $Q = 2$ for $SrTiO_4$), the cube-center ion has charge Ze ($Z = 2$ without transfer for $SrTiO_4$) and the cube-corner ion was given a charge $(6- Z)e$ (4 without transfer for $SrTiO_4$). A set of values are given in Table 18-3. Note that this choice of parameterization allows evaluation for formula units which are charged; only for $Q=2$ does this give charge neutrality for the unit cell so only the values for $Q = 2$ are relevant to the systems we are discussing. With a little algebra it allows any combination of the three charges, since scaling all charges (e. g., by a factor two), scales the total energy by that factor squared (e. g., four). Thus for $KNiF_3$ we want $Q = 1, Z = 1$, and a Ni charge of 2. We pick the $\alpha(2,2) = 24.76$, which is for a cube-corner charge of $6\text{-}2 = 4$, and divide all charges by two, or α by four, to obtain $\alpha = 6.19$.

For $SrTiO_3$, $Q = 2$ and $Z = 2$, and, using d from Table 18-2, Eq. (18-14) yields 24.7×10^{11} ergs/cm^3. For $KTaO_3$, for $Q = 2$ and $Z = 1$ we obtain 26.2×10^{11} ergs/cm^3. These are reduced by a factor of four if we allow charge transfer to reduce all charges by a factor of two. This reduced value is comparable to the contribution from the chemical grip for the case of $KTaO_3$, which was -5.0×10^{11}. When the $V_2{}^{pd}$ is so much larger than $V_3{}^{pd}$ as seen in Table 18-2, and when the parameters are entering with such high powers, as in Eq. (18-12), it would seem important to calculate the total energy much more carefully than in the approximation of a chemical grip which we used here. Until this is done, we cannot be certain if the destabilization of the cubic structure by the pd-coupling can dominate the electrostatic stabilization and be responsible for the observed distortions. It seems likely that our parameters are adequate, but our simplification of the electronic structure by omission of a number of terms may not be. This refinement would also be needed for prediction of the elastic constants for the cubic systems. The task remains to be done.

The electrostatic energies represented in Table 18-3 appear to be accurate and apply to a variety of systems. WO_3 and ReO_3 are both described by taking $Z = 0$ and $Q = 2$, with the rhenium charge reduced to six by the single electron in the d-state. Na_xWO_3 can be described with $Q = 2$ by letting $Z = x$. The values in the Table for $Q \neq 2$, so that the sum of the charges is not zero, can also be interesting. A uniform compensating background may be assumed, which does not change with distortion. In this way $Q = 0$, $Z = 3$, gives a cube-corner charge of $6 - Z = 3$, a body-centered cubic simple metal (of valence three). Dividing the α from Table 18-3 by nine, and noting that the cube edge is $2d$, gives the same value for α as that obtained (for $Z = 1$) by Fuchs (1936). Taking both Q and Z equal to zero leaves the cube corner with six charges. Dividing by 36 give α for a simple-cubic structure (edge $2d$), with the negative value in Table 18-3 indicating that it is unstable electrostatically.

E. The Electron-Phonon Interaction

In Harrison (1980), 471 ff, we calculated the electron-phonon interaction in the perovskites, following a scheme due to Barišić, Labbé, and Friedel (1970). This was coupling between states within the same band, assuming that the same single atomic orbital entered the two coupled states, and therefore the only change in tight-binding parameters due to a lattice distortion came from the $1/d^{2\ell+1}$ dependence of the interatomic coupling upon spacing d. For this reason its applicability was very limited, but results are sufficiently interesting that we quote them here, and refer to the derivation and discussion there.

With the assumption just given, the matrix element between a tight-binding state $|\boldsymbol{k}\rangle$ of wavenumber $\boldsymbol{k}$ and a tight-binding state $|\boldsymbol{k}'\rangle$ of wavenumber $\boldsymbol{k}'$, due to a lattice vibration for which the atomic displacements are $\delta\mathbf{r}_j = \boldsymbol{u}_Q\, e^{i\boldsymbol{Q}\cdot\boldsymbol{r}_j}/\sqrt{N}$ is given by

$$\langle\boldsymbol{k}'|H|\boldsymbol{k}\rangle = -\frac{(2\ell+1)\hbar}{d^2\sqrt{N}}\, i\boldsymbol{u}_Q\cdot(\boldsymbol{v}_k - \boldsymbol{v}_{k'})\,, \qquad (18\text{-}15)$$

if $\boldsymbol{Q} = \boldsymbol{k}' - \boldsymbol{k}$, or differs from this by a lattice wavenumber. Here $\boldsymbol{v}_k$ and $\boldsymbol{v}_{k'}$ are equal to the electron velocities of the states $|\boldsymbol{k}\rangle$ and $|\boldsymbol{k}'\rangle$, respectively, related to the interatomic matrix elements by

$$v_k = \frac{1}{\hbar}\frac{dE(k)}{dk} = \frac{i}{\hbar}\Sigma_j H_{ij}\, d_j\, e^{ik\cdot d_j}. \qquad (18\text{-}16)$$

This is essentially the form of the result later found by Varma and Weber (1977). They noted the very important feature that the matrix elements are large where the velocities are large, and therefore the densities of states are small. The matrix elements enter squared in scattering rates, and most other properties, while the density of states enters only linearly, so the important parts of the electronic structure are just the opposite of what one would guess if one judged the electronic structure on the basis of density of states only. This important aspect of the electron-phonon interaction presumably carries over to a more complete tight-binding analysis of the bands, including interband coupling. It is not specific to the model based upon a single orbital per band.

It is interesting that the essential form, Eq. (18-15), is consistent also with the electron-phonon interaction in simple metals which we found in Section 13-9. Writing Eq. (18-15) for simple metals, we replace v by $\hbar k/m$ and obtain

$$<k'|H|k> = \frac{(2\ell+1)\hbar^2}{md^2\sqrt{N}}\, iu_Q\cdot(k' - k\,). \qquad (18\text{-}17)$$

The corresponding matrix element from Eq. (13-50) was $-iu_Q\cdot(k' - k\,)\, w_q/\sqrt{N}$. For normal scattering (small Q) the pseudopotential approaches $-{}^2/_3\hbar^2 k_F{}^2/2m$ so the two expressions are the same except for a numerical factor of order one. It is interesting to note finally that this dependence upon $1/d^2$ in Eq. (18-17) carries over to d-states in Eq. (18-15), though the interatomic matrix elements are proportional to $1/d^5$.

4. The Cuprates

Copper-oxide-based compounds became suddenly important with the discovery of high-temperature superconductivity in $La_{2-x}Ba_xCuO_4$ by Bednorz and Müller (1986), and the subsequent discovery of even higher-temperature systems (Wu, et al. (1987)). Almost immediately there were band calculations on La_2CuO_4 by Mattheiss (1987) and by Yu, Freeman, and Xu (1987), and an elementary treatment of the electronic structure (Harrison

(1987a)). We outline the latter treatment, which was along the lines, and using essentially the parameters, which we have introduced here.

A. The Electronic Structure

All of the superconducting cuprates contain planes of CuO_2 which are just like the (001) planes containing the transition-metals in the perovskites, and shown in Fig. 18-6, but oxygen ions are missing from the regions between these parallel planes, and there are an assortment of other ions, such as barium, lanthanum, and yttrium, which contribute their electrons to the copper-oxide planes. The term values for these atoms between, as well as the copper s-state energy, are well above that of the copper d-states and the oxygen p-states, so the electronic structure is dominated by these two sets, as it was in the perovskites.

In the absence of coupling, in the parent compound La_2CuO_4, the six lanthanum electrons and one copper s-electron are sufficient to fill all but one of eight oxygen p-states, so the complex of copper-d, oxygen p-states is short one electron, and we expect one hole per copper atom in each CuO_2 plane.

The first task in obtaining the electronic structure is to calculate ε_d for the copper self-consistently, which could be done exactly as in Part 18-3-B, but the result is immediate. From Table 17-2 we see that if the copper d-shell were full, the d-state energy would be $\varepsilon_d^0 + U_d = -13.36$. If the hole were equally shared by the copper d-states and the oxygen p-states, it would lower the d-state energy by $U_d/2$ to -16.81 eV, almost identical to the oxygen p-state energy from Table 1-1 of -16.77 eV. But, when the d-state and p-state energies are equal, all band states *are* equally share by the two orbital types so this result is self-consistent, and does not even depend upon the Friedel model (Fig. 17-3) used to make the evaluation for the monoxides. The same conclusion was reached in Harrison (1987a) though it was based upon slightly different parameters. This is a significant simplification of the our view of the electronic structure, and certainly the magnitude of the tiny difference is not meaningful.

We proceed then with $\varepsilon_d = \varepsilon_p = -16.77\ eV$. The coupling $V_{pd\sigma}$ is given by Eqs. (17-1) and (17-2) as

$$V_{pd\sigma} = -\frac{3\sqrt{15}}{2\pi}\ \frac{\hbar^2\sqrt{r_p r_d^3}}{md^4} = -1.30\ eV \qquad (18\text{-}18)$$

based upon the r_p and r_d from Tables 17-1 and 17-2, and the equilibrium spacing of $d = 1.90$ Å . $V_{pd\pi}$ is positive and smaller by a factor $1/\sqrt{3}$ at 0.75 eV.

We may then construct energy bands for the CuO_2 planes, forming a square lattice of copper atoms, with an oxygen at the center of each edge. There *are* oxygen atoms out of the plane, above and below each copper atom, in La_2CuO_2 but they are more distant and we neglect their effect. We keep only the pd-coupling, as for the perovskite bands which we gave to the left in Fig. 18-7. Probably again the largest error is in neglecting the pp-coupling between second-neighbor oxygen states and the sd-coupling with the oxygen s-states. These could be carried over from the perovskite bands, but the principal conclusions are not sensitive to this.

The square structure with edge $2d$ gives a square Brillouin Zone, with edge $2\pi/2d$, as shown in Fig. 18-8. The 5 + 2×3 orbitals per copper atom broaden out into eleven bands. At each point in the Brillouin Zone only a single combination of the d-states of Fig. 15-2 enters each band, and those at the point X in the Brillouin Zone are used to label the bands. Along symmetry lines there is mixture of these orbitals, such as the $3z^2 - r^2$

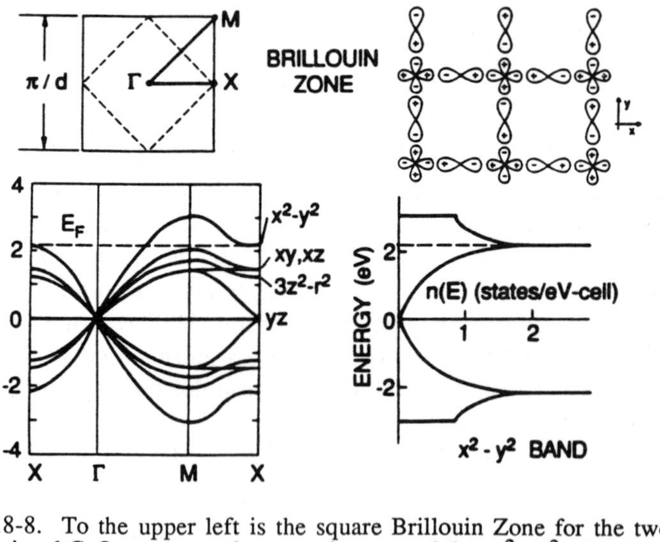

Fig. 18-8. To the upper left is the square Brillouin Zone for the two-dimensional CuO_2 structure shown to the upper right. $x^2 - y^2$ orbitals are drawn on the Cu sites and x- and y-orbitals are on the oxygen sites. Below to the left are approximate tight-binding bands (measured relative to $\varepsilon_d = \varepsilon_p$) including only pd-coupling. The Fermi energy is shown as E_F and the corresponding Fermi surface is the dashed line in the Brillouin Zone above. To the right of the bands is shown the density of states for the x^2-y^2 band. After Harrison (1987a).

orbital and the $x^2 - y^2$ orbital for wavenumbers along the x-axis. This mixing was neglected in Harrison (1987a), making the band calculations immediate and leading to the bands shown in Fig. 18-8. We note from the figure that the Fermi energy lies above all but the highest band, which in this approximation is given by

$$\varepsilon_k = \varepsilon_d - V_{pd\sigma}\sqrt{3(\sin^2 k_x d + \sin^2 k_y d)}\ . \tag{18-19}$$

This is the important band for understanding the properties.

In the same approximation the other bands have similar forms and this is often used as a model for the energy bands of the cuprates. If we had not taken ε_d exactly equal to ε_p there would have been an additional constant term under the square root in Eq. (18-19) and the bands would have been split at Γ rather than crossing the axis linearly, but again a small difference in ε_p and ε_d makes only a small difference in the bands where they are important.

We may note that if $k_y d = \pi/2 - k_x d$ then $\sin^2 k_y d$ becomes $\cos^2 k_x d$ and $\varepsilon_k - \varepsilon_d$ becomes constant at $-\sqrt{3}\, V_{pd\sigma}$. The dashed lines at the upper left in Fig. 18-8 are such lines, they separate half the Brillouin Zone from the other and are the Fermi surface for these simple bands when they are half occupied, one hole per formula unit, with all other bands fully occupied. This square Fermi surface is characteristic of half-filled bands in a square lattice, occurring also for simple s-bands, $\varepsilon_k = \varepsilon_s + 2V_{ss\sigma}(\cos k_x d + \cos k_y d)$. It is only approximately true in the case of the cuprates, but it provides a simple model Fermi surface and electronic structure. [The bands are also sometimes modeled by the form of the simple s-band just given.] By doping with further holes, increasing x from zero in $La_{2-x}Ba_xCuO_4$, the Fermi surface pulls away from this square. Experimental studies of the electronic excitations correspond to bands which are narrowed by a factor of order two, presumably due to strong electron-electron interactions which reduce the effects of the coupling.

We may note from Eq. (18-19) that the band reaches a maximum as a function of k_x along the line of $k_y = 0$. The maximum, at $k_x d = \pi/2$, is the point X in the Brillouin Zone. Thus from X the bands decrease in the x-direction and increase in the y-direction, so that there is a saddle point in the top band, Eq. (18-19), at this point. A saddle point in two dimensional bands gives a diverging density of states shown to the lower right in Fig. 18-8. For the simple bands and half filling, the Fermi energy comes just at that energy. Small corrections to the bands, or doping, can shift the Fermi energy relative to the divergence, but the divergence must be there because there must be a saddle point in that neighborhood.

If we were to construct a three-dimensional cuprate structure, composed of many parallel CuO_2 planes displaced by $2d$ in the z-direction, but neglect any coupling between them, we would have a cubic Brillouin Zone, with a square prism Fermi surface, with axis parallel to the z-axis. The properties would be the sum of the properties of the individual planes.

If we further introduced a weak coupling between planes, the energy would vary slightly along the z-direction so the cross-section would undulate along the z-direction. If we made alternate planes slightly different, the distance along z between equivalent points would double, the Brillouin Zone would halve. Then we would have two undulating cylinders, and in the basal plane of the zone, two similar, approximately square, Fermi lines. Such modifications occur in the many structures of the cuprates.

B. Peierls Instabilities and the Antiferromagnetic State

The band structure and metallic Fermi surface shown in Fig. 18-8 would not correspond to a stable physical system. For example, displacing alternate copper atoms by a small amount $\boldsymbol{u}$ in opposite directions would correspond to a vibrational distortion with wavenumber at M in the Brillouin Zone (See Fig. 18-8). This would certainly cost elastic energy proportional to u^2 , but this distortion would couple states on opposite sides of the square by an interaction proportional to u , would open a gap proportional to u , and if one integrates the energy shifts over the occupied part of the band one finds that it lowers the energy by an amount proportional to $u^2 ln/u/d/$. At small enough u this gain always wins and the system is unstable against the distortion. When the distortion occurs, a gap is opened along the Fermi line and the system becomes insulating. This is the familiar *Peierls* instability (see, for example, Peierls (1979)), or metal-insulator transition, of one-dimensional systems. It may be regarded as one-dimensional since a square Fermi line in a two-dimensional system corresponds to one-dimensional behavior, just as a collection of uncoupled planes was seen at the end of the last part to give flat Fermi surfaces in three dimensions. This particular insulating state is also sometimes called a *charge-density wave* since the filling of states up to the gap occurring at k_F in a one-dimensional system gives rise to a charge density varying with position x as $cos(2k_F x)$. It is analogous to the Jahn-Teller distortions discussed in Section 8-3-B, in which a system spontaneously lowers its symmetry. However, in this case it can be inhibited, either by small deviations from the straight Fermi line or by finite temperatures, both of which eliminate the logarithmic factor. It becomes a quantitative question whether it will occur.

Often there are competing instabilities (Zhong and Vanderbilt (1995)) and it is possible that some other instability is stronger. In the case of the cuprates it is generally believed that the strongest instability is against becoming an antiferromagnetic insulator. The result can be called a *spin-density wave* , as opposed to the charge-density wave described above. We make a quantitative analysis of such an instability here, following again Harrison (1987a).

In Part A we introduced the Coulomb repulsion U_d between two electrons occupying a d-state on the same copper atom, in this case two electrons of opposite spin occupying the d-state of symmetry $x^2 - y^2$. We may make the Unrestricted Hartree-Fock Approximation of Section 16-1-B by allowing unequal occupation of the two spins on each atom, with a net spin μ equal to the spin-up occupation minus the spin-down occupation on one atom, changing sign from atom to atom.

This problem is closely related to the determination of a correlated state in Section 16-1-B. In fact if we obtain the average of the energy over the bands using the special-points method (Section 9-2-C), it also becomes a two-level system. We note that if all of the bands in Fig. 18-8 were occupied, the average would be simply ε_d. We must subtract from that average the contribution of the *empty* portion in the highest band. For a square lattice, the special point for an average over an entire band lies at $\mathbf{k}^* = [1,1]\pi/2a$, equal to $[1,1]\pi/4d$ for the CuO_2 plane, but we seek an average over only the empty half of the band. By making alternate atoms different we actually change the primitive displacements which take the plane into itself from $[1,0]2d$ and $[0,1]2d$ to $[1,1]2d$ and $[1,-1]2d$. This rotates the Brillouin Zone by 45° and contracts it by a factor of $1/\sqrt{2}$ to coincide with the Fermi surface shown in Fig. 18-8. That is of course no accident since in forming the antiferromagnetic state we open a gap at the new Brillouin-Zone faces and form an insulator, just as in the Peierls transition. The reduction to the smaller Brillouin Zone is accomplished by translating the bands into the smaller Zone, where we can use a special point $\mathbf{k}^* = [0,1]\pi/4d$ for the smaller Zone, corresponding to the point $\mathbf{k}^* = [-2,-1]\pi/4d$ of the untranslated band of Eq. (18- 19). This corresponds to an energy $\varepsilon_d + 3\sqrt{2}\, V_{pd\sigma}/2$ to a be associated with the states which are empty. We thus take the covalent energy associated with the bonding in the structure as $3\sqrt{2}\, V_{pd\sigma}/2$.

For this system it seems easier to construct the energy of the state and subtract self-interactions counted twice, equivalent to the procedure (with the $cos^2\gamma$ occupation) used in Section 16-1-B. The net moment μ on each copper atom shifts the energy of the d-states of a given spin (which feel only the Coulomb repulsion of the states in the same orbital of opposite spin) up and down on alternate atoms by $\pm\mu U_d/2$. The energies at the special point are then $\varepsilon_d \pm \sqrt{9V_{pd\sigma}^2/2 + \mu^2 U_d^2/4}$, giving the answer above when $\mu = 0$, and the correct result when the coupling is zero. Thus the total energy per copper atom, relative to uncoupled d-electrons on the atoms, is

$$E_{tot} = -\sqrt{9V_{pd\sigma}^2/2 + \mu^2 U_d^2/4} + \mu^2 U_d/4 \ . \tag{18-20}$$

Minimizing this with respect to moment (equivalent to the procedure in Section 16-1-B) gives

$$\mu = \sqrt{1 - \frac{18V_{pd\sigma}^2}{U_d^2}} \ . \tag{18-21}$$

This also gives a criterion for the formation of the antiferromagnetic state,

$$U_d > \sqrt{18}\, V_{pd\sigma} \, , \tag{18-22}$$

satisfied with our parameters of$V_{pd\sigma} = 1.3\ eV$ and $U_d = 6.9\ eV$. In contrast to the analytic variation of energy with moment which we had in Section 16-1-B, there is a real phase transition here in going to the insulating state. We only approximate the energy as a two level system here, but it is a prediction of a real, observable transition.

With these parameters we predict a moment of $\mu = 0.60$ and an insulating gap of $^1/_2\,\mu U_d = 2.1\ eV$. [Note that the potential $\pm\mu U_d/2$ on alternate d-states couples the states at opposite sides of the square Fermi line by only $\pm\mu U_d/4$ because only half of the total probability density for each Fermi-line state lies on the d-states, the other half on oxygen p-states. This coupling shifts the states along the Fermi line by $\pm\mu U_d/4$ giving a gap along the entire line of $^1/_2\mu U_d$.] This splitting of 2.1 eV is larger than the 1.5 eV found in Harrison (1987a) because of use of slightly different parameters, but still of the order of the apparent insulating gap in La_2CuO_2 . We believe this to be an essentially correct description of the antiferromagnetic state in this system.

When the compound is doped p-type, by replacing enough of the lanthanum atoms by barium as $Ba_xLa_{2-x}CuO_4$, the antiferromagnetism disappears and the system becomes metallic. In terms of our analysis this is understood by retaining the alternate spin moments as barium is introduced, but a fraction x electrons per copper are now removed from the one electron per copper which produced the potential $^{1}/_{2}\,\mu U_d$, having the effect of replacing the U_d in Eq. (18-20) by $(1\text{-}x)U_d$. Using the criterion given above, we find that the antiferromagnetism should disappear for

$$x > 1 - \frac{\sqrt{18}\,V_{pd\sigma}}{U_d} \tag{18-23}$$

or x exceeding 0.20 for our parameters. The experimental number is closer to 0.15. It is in this metallic state that the high-temperature superconductivity occurs.

C. Phonons, Paramagnons, and Excitons

The formation of Cooper pairs in the superconducting state must derive from a net attractive interaction between electrons, an attractive term exceeding the Coulomb repulsion between electrons. One possibility of course is through the electron-phonon interaction, as in ordinary superconductors. Since the electrons responsible for the superconductivity are certainly those in the CuO_2 planes, we expect the important vibrational modes to be the modes of the plane, and the optical modes, with displacements along the CuO axes, to be the most strongly coupled as in other polar systems. The frequency of these modes was estimated in Harrison (1987a), and again in Problem 18-1 here, using the electronic structure described here. It is in reasonable accord with the experimental value (Stavola, Cava, and Rietman (1987)) $\omega = 1.14\times 10^{14}/sec.$ or $\omega/2\pi c = 603\ cm^{-1}$.

Optical modes for this frequency were assumed, along with an electron phonon interaction as described in Section 18-3-E, with electron-phonon matrix elements as in Eq. (18-15) of order $i\mathbf{q}\cdot \mathbf{u_Q}\, V_{pd\sigma}/\sqrt{N}$, to obtain (Harrison (1987a)) the dimensionless electron-phonon coupling constant of Eq. (13-51) . This λ_ϕ (the ϕ referring to phonons) was found to be equal to approximately 0.05 with parameters similar to what we use here. This is not a large value, and gives no indication of the extraordinary superconducting properties.

An alternative source of coupling associated with nearness to the antiferromagnetic transition was also considered in Harrison (1987a). This returned to the total energy of Eq. (18-20), when there was sufficient doping that no antiferromagnetic state was formed, according to the criterion in Eq. (18-23). Then the energy increases with the square of the moment μ , just as for a phonon the energy increases with the square of the amplitude u_Q . Such magnetic normal modes are called *paramagnons.* It was seen that the frequency ω_m of such magnetic modes is given by

$$\hbar^2 \omega_m{}^2 = 18 V_{pd\sigma}{}^2 \left(1 - \frac{(1-x)U_d}{\sqrt{18} V_{pd\sigma}} \right) . \qquad (18\text{-}24)$$

By comparison with Eq. (18-23) we see that the frequency goes to zero as we approach the antiferromagnetic transition. An electron-paramagnon interaction was defined in terms of the electron energy shift due to such fluctuating moments, and a dimensionless electron-paramagnon coupling constant evaluated in analogy to the dimensionless electron-phonon coupling constant. It was found to be $\lambda_m = 9V_{pd\sigma}{}^2(1 - x)U_d n(\varepsilon_F)/(2\hbar^2 \omega_m{}^2)$. This seemed to offer the possibility of high-temperature superconductivity, but it would also seem to apply to other systems which are not superconducting, and the suggestion has gone nowhere. It is quite distinct from a third mechanism proposed by Schrieffer, Wen, and Zhang (1988) and referred to as the *spin bag* , in which an evanescent antiferromagnetic state is assumed to exist , though it is not an antiferromagnetic insulator, and that electrons excited into the upper band reduce the gap (as we indicated doping does for the antiferromagnetic insulator), attracting other excited electrons to the place where the gap is reduced.

A fourth mechanism for attraction which might produce high-temperature superconductivity was suggested by Bardeen, Ginsberg, and Salamon (1987) who noted that the polarization of copper-oxygen bonds neighboring the CuO_2 planes, produced by one electron, would cause an attractive potential for another electron. Since the polarization of a bond is the admixture of antibonding state to the bonding state of a given bond site, this can be considered the virtual creation of an exciton, and it is physically the same as the excitonic mechanism suggested many years earlier by Little (1964). We explored this mechanism for attraction in detail (Harrison (1988)) in terms of the representation of the electronic structure which we have given here. It seemed a very possible explanation, well within the validity of the parameters of explaining the high superconducting transition temperature and related properties. One prediction it led to is contrary to contemporary experiment. The onsite repulsion U_d was avoided by the appearance of a superconducting order parameter $\Delta(\boldsymbol{k})$ which varied in phase over the square Fermi surface of Fig. 18-8 as $\Delta e^{i\boldsymbol{k}\cdot\boldsymbol{T}}$, where $\boldsymbol{T}$ is the vector distance between neighboring copper atoms. This had the effect of keeping the two electron forming a Cooper pair always on different rows of atoms, separated by this vector. It is now generally believed that this is accomplished by d-wave pairing, in which the order parameter would vary along a circular Fermi surface as $\Delta cos2\phi$, which could be written for the square Fermi surface as $\Delta cos\boldsymbol{k}\cdot\boldsymbol{T}$, with $\boldsymbol{T}$ equal to $2d$ along a [100] direction. The earlier analysis indicated that a lower-energy state could be obtained with the nodeless $\Delta e^{i\boldsymbol{k}\cdot\boldsymbol{T}}$, but the nodes seem now to be clearly observed by Shen and Dessau (1995).

On the question of the origin of the electron-electron attraction, which is more relevant to our discussion here, it should be noted that every real physical contribution to the electron-electron interaction *does* contribute to the formation of the superconducting state. The only real question is whether one mechanism is dominant, and that question has not yet been answered.

Problem 18-1. Optical Vibrational Modes in the Cuprates

Estimate the vibrational frequency for an oxygen atom vibrating between two copper atoms in a CuO_2 plane. Since the copper is so much heavier than the oxygen, this can be taken as an estimate of the optical mode vibration. You may assume that the dominant force is from a radial force constant, obtained from the volume dependence (or area dependence) of the total energy. A convenient path is through the total energy per copper, containing the effects of the $V_{sp\sigma}$ coupling between copper s-states and oxygen p-states (to second order in $V_{sp\sigma}$ as for simple ionic crystals) , the pd-coupling energy from Eq. (18-20) (with $\mu = 0$) and a $1/d^8$ overlap repulsion, with coefficient fit to the equilibrium spacing. The nearest neighbor radial force constant may be obtained from the second derivative of this energy with respect to d leading to a prediction of the frequency, to be compared with the value given in the text. [The estimate of the pd-coupling contribution differs from that given in Harrison (1987a) but the $V_{sp\sigma}$ term is dominant in any case.]

CHAPTER 19

Surfaces and Interfaces

Much of the analysis of surfaces and interfaces included in this chapter is based upon a review article by the author (Harrison (1994)), discussing the history of the development of the theory used in this text, and its application to surface properties. The studies of metals have mostly not been previously published. Since we cover surfaces for all classes of materials, we include it here as a separate chapter.

1. Surface Energies

The simplest view of surface energies is that they are the excess energy due to broken surface bonds. One takes the bond energy equal to the cohesive energy (heat of atomization) per nearest-neighbor bond and simply counts the number of broken bonds. Such a picture correctly indicates that the minimum-energy surface for silicon is the (111) surface, which is then the cleavage surface and the growth surface.

A. Elemental Semiconductors.

As appealingly simple as this view is, a slightly deeper look indicates that it will get very little else right. We look first at an elemental semiconductor such as silicon which we might expect to be the best case. It is helpful to review our discussion of the cohesive energy from Section 2-3.

We started with the free silicon atoms and calculated the change in energy as the solid is formed as the sum of the changes in the one-electron energies. The first step was the promotion of electrons in each atom to sp^3-hybrids, costing an energy $(\varepsilon_p - \varepsilon_s)/2$ per bond, equal to 3.61 eV for silicon. We then brought the atoms together to form bonds between each pair of hybrids, coupled by $-3.22\hbar^2/(md^2)$ equal to -4.44 eV for silicon. The gain in energy per bond was twice this, but approximately half of the total is canceled by the repulsion which prevents collapse of the crystal. Thus the contribution of these two terms in the cohesion, 3.61 eV and - 4.44 eV, nearly cancels. In fact, this cancellation is so great that the next-order corrections was required to obtain a meaningful cohesive energy. However, if we split the solid, forming two surfaces, we will still have expended the promotion energy for every broken bond at the surface since each dangling orbital must also be an sp^3-hybrid to remain orthogonal to the other three hybrids. Thus the cost in energy is 4.44 eV per bond for silicon, not the cohesive energy of 4.44-3.61 eV per bond. Use of the cohesive energy per bond would greatly underestimate the surface energy.

The presence of these surface energies much larger than the energy per bond has an important consequence. The formation of the ideal surface is so costly in energy that semiconductors find alternative arrangements to recover some of this lost energy. These new arrangements ordinarily reduce the translational symmetry which the ideal surface (the surface obtained by removing all atoms beyond the surface plane without displacing any of the remaining atoms), and are called *surface reconstruction* . They are clearly observed by low-energy electron diffraction (LEED). Reconstructions occur on essentially all surfaces of elemental semiconductors. We shall discuss a few important cases in Section 19-3. Their neglect in discussing the surface energy is another error in the simplest view described above. A very thorough overview of the theory and experiment concerning reconstructions on clean semiconductor surfaces has been given more recently by Duke (1996).

B. Ionic Crystals and Polar Semiconductors.

The simplest view of one cohesive-energy-per-bond lost for every broken bond fares as badly in other systems. In the tight-binding formulation of cohesion in ionic crystals such as rock salt in Section 9-1-D, the neutral atoms are brought together and at the equilibrium spacing an electron is transferred from the sodium to the chlorine for a gain in energy of $\varepsilon_s(Na) - \varepsilon_p(Cl)$ equal to 8.83 eV for NaCl, in comparison to the experimental cohesive energy of 8.04 eV per atom pair. (This was also the tight-binding estimate of the band gap, experimentally 8.5 eV for NaCl.) With the tight-binding view, there is no change in the energy due to the formation of the surface and this prediction of a negligible surface energy in comparison to the cohesion per bond is essentially true, and just the opposite of the case for elemental semiconductors. Because the expected surface energy is so small in ionic crystals, one does not expect reconstruction of the ideal surface, and it seems not to occur.

Born's approach to ionic-crystal cohesion, Section 9-1-E, based upon the cation ionization energies minus the anion electron affinity and the Madelung energy gives the same qualitative result. The Coulomb surface energy of $0.0422\ e^2/d$ per surface atom for a (100) surface, obtained by Madelung himself (Madelung (1918)), is tiny compared to the full Madelung energy of $1.75\ e^2/d$ per atom pair. The surface must be neutral, (the ideal (100) surface is, but a (111) surface would not be) or the Madelung energy would diverge for very large crystals (Harrison (1979)) We shall make this point more carefully in Section 19-3. The (100) surface has the lowest Madelung energy and is the growth and cleavage surface.

This same exclusion of charged surfaces applies also to polar semiconductors such as gallium arsenide (Harrison (1979)). Thus the (111) surface which we found for silicon is not allowed for gallium arsenide, the lowest energy surface of which is a (110) surface, neutral but with a slightly higher broken-bond density than the (111) surface.

C. Metal Surfaces.

Tight-binding theory also allows a formulation of surface energy for metals since it is quite straightforward to terminate a tight-binding solid at a surface. It is much less clear how to terminate a free-electron model at the surface. The analysis given here was motivated by contemporary active research in Giant Magnetoresistance, to be discussed in Part D of this Section, and the need for a clearer understanding of metallic surfaces and interfaces. The only parts of this study of metals which were published were given in Harrison and Kozlov (1992) and Kozlov and Harrison (1993). Closely related studies were carried out at that time by Edwards, and coworkers (Edwards and Mathon (1991), Edwards, Mathon, Muniz, and Phan (1991)) whose principal interest was in quantum-well states formed by finite slabs of metals.

To see how we do treat finite slabs, it may be helpful to return to the one-dimensional chain illustrated in Fig. 1-4. In that case we constructed tight-binding s-states of the form $\Sigma_{j=1,N}\, u_j|j\rangle$ based upon atomic s-states $|j\rangle$ on each of the atoms, separated by d. The variational solution for the states gave conditions, Eq. (1-12), on the coefficients, u_j,

$$\varepsilon_s u_j + V_{ss\sigma}(u_{j+1} + u_{j-1}) = \varepsilon u_j \; . \tag{19-1}$$

For periodic boundary conditions, the solution could be written immediately as coefficients which varied along the chain as $u_j = e^{ikdj}/\sqrt{N}$ with the periodic boundary conditions allowing only k such that kNd is an integral multiple of 2π. It is also possible to add states of $\pm k$, of the same energy, to obtain eigenstates with coefficients $u_j = \sqrt{2/N}\, sinkdj$ or states with $u_j = \sqrt{2/N}\, coskdj$, which also satisfy periodic boundary conditions if the same k are used. However, of more interest, we can construct states for a *finite* chain of N atoms, numbered $j = 1$ to N by picking $kd(N+1)$ equal to an integer times π with $u_0 = u_{N+1} = 0$ since then all N of the Eqs. (19-1) are satisfied with the energy again given by $\varepsilon_k = \varepsilon_s + 2V_{ss\sigma}\, coskd$, Eq. (1-13). Since the coefficients on atoms $j = 0$ and $j = N+1$ are zero, removing those atoms and those beyond them does not affect the solution. This finding may seem remarkable, that the energy band description for this finite chain is complete and exact, no matter how short the chain. It is only the allowed wavenumbers which are affected by the chain length. In that sense the band description is valid right up to the surface of a solid, and no matter how thin is the system. Even for one atom, $k = \pi/(2d)$ is the only state and $e_k = \varepsilon_s$ is its energy.

In any range of wavenumber we now have twice as many solutions, since an odd number of half-wavelengths in the length $((N+1)d)$ is allowed, as well as the even number required for periodic boundary conditions, but now we do not distinguish positive and negative k so for large N the density of states is approximately the same. In Problem 19-1 we see how such a truncation of the chain modifies the energy, which can be regarded as the energy required to break the chain of N atoms bent into a circle. It is this procedure which we now generalize to three-dimensional slabs.

We consider a tight-binding s-band with a single atom per primitive cell, and for simplicity let this initially be a simple-cubic crystal. From an infinite crystal we select a large number N_p of atomic planes parallel to a (001) surface we wish to construct. We consider wavefunctions which for motion parallel to the surface planes correspond to periodic boundary conditions, fixing k_x and k_y, on the lateral surfaces of the slab. k_z, perpendicular to the slab surface, may be chosen such that the sine is zero for $n = 0$ and for $n = N_p + 1$. Then removal of the atoms for $n < 1$ and $n > N_p$ from the infinite crystal leaves these as exact eigenstates of the now finite slab.

The wavenumbers k_z are restricted by $k_z(N_p + 1)s$ being equal to a positive integer times π, with s the spacing between planes ($s = d$ for this simple-cubic case). For each such k_z there is a two-dimensional band of states with energy varying in terms of the k_x and k_y allowed by periodic boundary conditions over lengths L_x and L_y in those directions. If that lateral motion were free-electron-like, as we shall assume in the end, the number of such states with transverse wavenumber $k_t = \sqrt{k_x^2 + k_y^2}$ in the range dk_t would be given by $2(L_xL_y/(2\pi)^2)2\pi k_t dk_t$, for two spins. This corresponds to a constant density of states $n(E)$ in each subband of $m/\pi\hbar^2$ per unit energy, per unit area of cross-section (in the xy-plane). We have obtained the essential structure of the subbands and allowed wavenumbers using tight-binding theory, but it could be applied to bands of any form and we choose free-electron bands for simplicity and their relevance to metals.

This is illustrated for $N_p = 4$ in Fig. 19-1. The lowest subband begins at the energy of the band at which $k_z \times 5s = \pi$ and provides a constant density of states at higher energies, as shown. The second subband begins where $k \times 5s = 2\pi$ and provides an equal contribution for energies above that, etc. States are occupied in the first three subbands. For $N_p = 8$ there would be twice as many subbands, each again contributing $m/(\pi\hbar^2)$ to the density of states above its energy minimum. As we increase the number of planes, the number of partially occupied subbands increases in proportion so that the number of occupied states per unit energy per plane stays approximately constant, but fluctuates up and down around the average as the number of occupied subbands at the Fermi energy increases. Further, with for example one electron per atom, the total energy per electron fluctuates up and down as the total number of planes increases, as emphasized by Edwards, Mathon, Muniz, and Phan (1991).

We have an exact solution for fixed N_p but we will be interested in large N_p for which there will be a term in the energy *per atom* independent of N_p which determines the cohesive energy per atom. There will be a smaller term in the energy per atom proportional to $1/N_p$, corresponding an energy

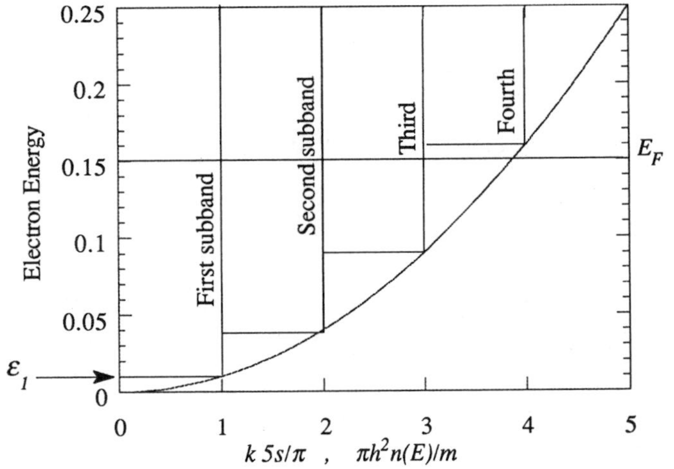

Fig. 19-1. The curve is the energy band (in units of $\hbar^2(2\pi/s)^2/2m$) for wavenumbers k perpendicular to a (100) slab. For four planes in the slab four subbands arise, with the lowest energy of each occurring at the corresponding horizontal line. Each subband contributes a constant density of states (per unit area), $n(E)$ represented by the rectangle labeled "First subband", etc. It is plotted as the abscissa in units of $m/\pi\hbar^2$ as a function of energy as the ordinate. States are occupied in the first three subbands for the Fermi energy E_F shown. The energy ε_1 of the bottom of the first subband is indicated.

per surface atom independent of N_p which determines the surface energy. There will be an additional term in the energy proportional to $1/N_p^2$ which determines the interaction energy between the two surfaces of the slab, to which we shall return.

The first task in obtaining the total energy is to sum the energy of occupied states as a function of N_p , keeping the number of electrons per atom fixed. We note that the state at the bottom of the first band has one half-wavelength equal to the thickness $(N_p + 1)s$ and has energy $\varepsilon_1 = \hbar^2\pi^2/(2m(N_p + 1)^2 s^2)$, indicated in Fig. 19-1. Similarly the n 'th band has energy $\varepsilon_n = \varepsilon_1 n^2$. All subbands are filled to the same Fermi energy E_F , as indicated in the figure. We define n_F , which is to be determined and which is not necessarily an integer, so that

$$E_F = \varepsilon_1 n_F^2. \qquad (19\text{-}2)$$

Then the n 'th band is seen to be filled over an energy range

$$R_n = E_F - \varepsilon_n = (n_F^2 - n^2)\,\varepsilon_1 \ . \tag{19-3}$$

Then the total number of electrons for a slab of area A is

$$N = \frac{mA}{\pi\hbar^2}\sum_n R_n = \frac{m\varepsilon_1 A}{\pi\hbar^2}\sum_n (n_F^2 - n^2) \ . \tag{19-4}$$

This expression is used, for given N, to fix the Fermi energy and therefore n_F. The total energy of the slab, relative to all electrons having the Fermi energy, is the number in each subband, $R_n m/(\pi\hbar^2)$, times the average energy below the Fermi energy, $-R_n/2$, summed over subbands. Thus relative to all electrons at rest it is

$$E_{tot}^{slab} = N\varepsilon_F - \frac{mA}{2\pi\hbar^2}\sum_n R_n^2 = N\varepsilon_F - \frac{mA\,\varepsilon_1^{\,2}}{2\pi\hbar^2}\sum_n (n_F^2 - n^2)^2 \ . \tag{19-5}$$

The sums in Eqs. (19-4) and (19-5) are over n for which the expression in the brackets is positive. We evaluate both by expanding $(n_F^2 - n^2)^2$ and performing the appropriate sums over n, n^2, n^4, or a constant. Writing n_B for the highest integer less than n_F (i. e., $n_B = Int(n_F)$) we have

$$N = \frac{mA\,\varepsilon_1}{\pi\hbar^2}\left(n_B n_F^2 - \frac{2n_B^3 + 3n_B^2 + n_B}{6}\right) \tag{19-6}$$

and

$$E_{tot}^{slab} = \frac{mA\,\varepsilon_1^2}{2\pi\hbar^2}\left(n_B n_F^4 - \frac{6n_B^5 + 15n_B^4 + 10n_B^3 - n_B}{30}\right) \tag{19-7}$$

In Eq. (19-7) we used Eq. (19-6) to evaluate the term $N\varepsilon_F$. A more useful expression may be the energy per electron,

$$\frac{E_{tot}^{slab}}{N} = \frac{\varepsilon_1}{10}\left(\frac{30n_B n_F^4 - 6n_B^5 - 15n_B^4 - 10n_B^3 + n_B}{6n_B n_F^2 - 2n_B^3 - 3n_B^2 - n_B}\right) \tag{19-8}$$

An important point is that these results are exact for the model. We can take N_p very large and extract bulk and surface properties, and we will also be able to obtain the smaller oscillating terms. At large N_p the Fermi energy E_F approaches the bulk Fermi energy, but ε_1 becomes very small, so n_F , and $n_B = Int(n_F)$, become large. To lowest order in N_p we may drop all but the first two terms in the numerator and the denominator in Eq. (19-8) to obtain an energy per electron of $^3/_5\varepsilon_1 n_F = ^3/_5\varepsilon_F$, exactly the bulk free-electron energy obtained at the beginning of Section 13-3-A. Thus Eq. (19-8) could be used to give exactly the theory of the cohesive energy and bulk modulus given in Sections 13-3-A and B.

To obtain the next order term, which will give us the surface energy, we sketch n_B as a function of n_F in Fig. 19-2 to see that n_B - n_F fluctuates, but averages to $-^1/_2$. Thus we may replace n_B by $n_F - ^1/_2$ in Eqs. (19-6) and (19-7). Then we expand Eq. (19-6) to obtain n_F , keeping terms to first and zero-order in N_p . This is substituted into Eq. (19-7) to obtain the slab energy to first and zero-order in N_p , the zero-order term, when divided by the area A and by two, since two surfaces are created, is the contribution of this energy to the surface energy. It can be written in terms of the bulk Fermi wavenumber and energy as

$$E_{surf.} = \frac{k_F{}^2 E_F}{8\pi}\left(1 - \frac{8k_F s}{15\pi}\right) . \tag{19-9}$$

The first, positive, term is larger by a factor two than given in Harrison

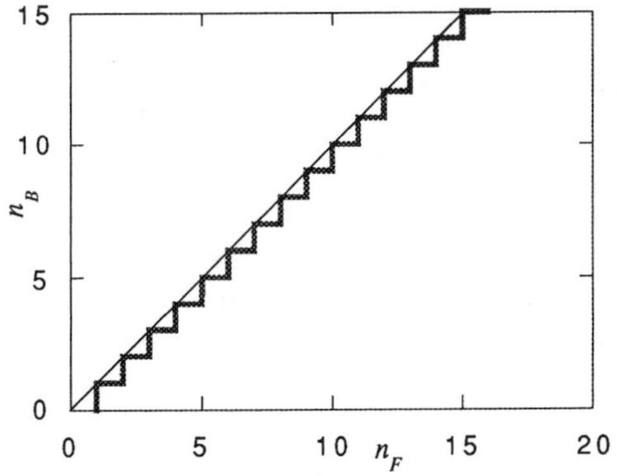

Fig. 19-2. The heavy line is a plot of n_B as a function of n_F with the number of planes in the slab N_p increasing to the right. The light line is n_F.

(1994) but appears to be correct. [It is possible that the difference arose from holding k_F fixed, rather than the number of electrons per atom, as the number of atom planes was varied. That will be appropriate in the following section.] It is the extra kinetic energy arising from cutting off the free-electron gas, The second term is smaller and arises from some decompression of the electron gas when part of the crystal is removed; we would have missed it in a pure free-electron picture using a thickness of $N_{p}s$. This term indicates that the surface energy is lowered when the spacing between planes is increased, corresponding to more closely-packed planes. A (111) surface is favored for an fcc lattice.

The derivation of Eq. (19-9) using this combined tight-binding-free-electron approach is only a first step in a possible study of metal surfaces and interfaces. We should add the overlap repulsion between nearest-neighbor atoms, and its elimination at the surface would reduce the surface energy of Eq. (19-9). It would also be interesting to minimize the energy with respect to displacement of the atomic planes to learn about the surface structure and possible distortions extending into the metal. Such studies seem not to have been carried out.

It is interesting again to compare with the "broken-bond" picture of surface energies. For simple metals we found in Eq. (13-7) the cohesive energy per atom was $A(Z)E_F$, with $A(Z) = 0.314$ for monovalent metals. For a close-packed structure there are six broken bonds per atom, and on a (100) surface four bonds are broken for every surface atom and there is one surface atom per area d^2 . For a monovalent metal $k_F{}^3d^3 = 3\sqrt{2}\pi^2$ so broken bonds predict a surface energy of $^1/_2\ ^1/_6\ A(1)\ E_F\ ^4/_{d^2} = 0.0087$ $E_Fk_F{}^2$. Had we ignored the overlap repulsion, this would have been doubled to $0.017\ E_Fk_F{}^2$. For this same system, the bracket in Eq. (19-9) is 0.583 and the contribution to the surface energy is predicted to be 0.023 $E_Fk_F{}^2$. Remarkably in this one case the two views of surface energy seem to give similar results, though it might have been thought to be the least likely case.

Perhaps the most important message applies to all types of surfaces. It is that the simplest view - in this case broken bonds - can be very far off the mark, but the almost-as-simple tight-binding view, based upon free-electron fits to the parameters, is generally qualitatively correct, and frequently semiquantitative It does of course not compete for accuracy with complete local-density calculations.

D. Interaction Between Surfaces and Giant Magnetoresistance

If we carry the expansion of the energy in $1/N_p$ to higher order we would have obtained energy depending upon the spacing between planes, corresponding to an interaction between opposite surfaces. Interestingly enough this interaction has become of considerable interest recently. This interest arose from experimental observations of oscillatory exchange coupling through simple-metal layers between magnetic metal layers (Bennett, Schwarzacher, and Egelhoff (1990), Parkin, Bhadra, and Roche (1991), Mosca, Petroff, Fert, Schroeder, Pratt, and Loloee (1991), Johnson, Coehoorn, de Vries, McGee, van de Stegge, and Bloemen (1992)), with periods most precisely measured by Ungaris (1991) using a wedge of material between the magnetic layers. The origin of this interaction was interpreted by Edwards and Mathon (1991) and Edwards, Mathon, Muniz, and Phan (1991) as arising from the formation of quantum-well states of just the kind we described in the treatment of simple-metal slabs in the preceding subsection. These can arise, for example, in a multilayer structure with a copper slab sandwiched between two (100) iron layers. The iron layers are ferromagnetic, with bands for the minority spin shifted up in energy and those for majority spin shifted downward, as described in Section 15-6-B. The minority-spin bands of the iron are believed to have a gap at the Fermi energy for wavenumbers normal to the (100) interface, so that the electrons of that same spin in the copper can be confined to the copper layer if the magnetization is parallel in the two iron layers. Then the slab energy of Eq. (19-7) applies to these copper electrons. If N_p is chosen such that the slab energy is low, the parallel orientation of the two iron spins will be favored, while if it is high, the antiparallel orientation is favored.

This interaction has been correctly and elegantly calculated by Edwards and Mathon (1991) by Edwards, Mathon, Muniz, and Phan (1991) and by Jones and Hanna (1993); here we carry out a simpler, but essentially equivalent, formulation. We imagine N_p planes of copper, between two thick iron layers, each with N_p' planes. For parallel alignment, the minority-spin electrons are confined to N_p layers, but the majority-spin electrons to N_p+2N_p' layers. For antiparallel alignment, electrons of each spin are confined to N_p+N_p' layers. We require $E_{tot}^{slab}(N_p) + E_{tot}^{slab}(N_p+2N'_p) - 2E_{tot}^{slab}(N_p+N'_p)$ as N'_p becomes very large, as twice (because E_{tot}^{slab} included a factor of two for spin) the excess energy for parallel alignment. The terms in $E_{tot}^{slab}(N)$ which are linear in N, or constant, cancel in this expression leaving only the interaction. In fact, when N is very large, only these terms survive so we may simply subtract the linear and constant terms from $E_{tot}{}^{slab}(N_p)$ to obtain the energy for parallel alignment minus that for antiparallel arrangement.

It would be appropriate to proceed as we did for the slabs in Part C, using Eq. (19- 6) to fit n_F to the original number of electrons per atom in the slab. We did that, but found that though it was necessary for surface energies, the interaction energy was distinguishable only at very small N_p from the result obtained by fitting n_F to the bulk electron density. The latter approach is very much simpler and we use it here. Then from Eq. (19-2) we have

$$n_F = \frac{(N_p + 1)k_F\, s}{\pi} . \tag{19-10}$$

and Eq. (19-7) can be rewritten as

$$\frac{E_{tot}{}^{slab}}{A} = \frac{k_F{}^2 E_F}{4\pi}\left(n_B - \frac{6n_B{}^5 + 15n_B{}^4 + 10n_B{}^3 - n_B}{30\, n_F{}^4}\right) \tag{19-11}$$

For the bulk expression, obtained at very large N_p , the expression in brackets approaches $n_B - 6n_B{}^5/30n_f{}^4 + 15n_B{}^4/30n_F{}^4 \approx 4n_F/5 - 1/2$ which we subtract within the bracket in Eq. (19-11) to obtain the energy for parallel magnetization minus that for antiparallel magnetization of the iron slabs. (We also now divide the result by two since $E_{tot}{}^{slab}$ added energy for two spins per state.) This interaction was plotted in Fig. 19-3 for $k_F s = 2.544$ appropriate to copper (100) slabs. In using Eq. (19-11) the result may be obtained as a continuous function of N_p as shown in the figure, and we also indicated by points the meaningful values for which N_p is an integer.

The result is quite interesting, as were the earlier findings by Edwards, et al. (1991). Indeed the parallel and antiparallel magnetizations are alternately favored as we increase the number of planes. The period of the function is small, determined by the wavelength of an electron at the Fermi surface. This is seen from Eq. (19-10), since n_F increases by an integer whenever $N_p k_F s$ increases by π, or when $N_p s$ increases by a half wavelength. When the result is evaluated at the discrete spacings of the real lattice, a much longer period shown by the points appears. The period is said to be "aliased". Indeed it was the longer periods which were observed in the experiments described at the beginning of this subsection.

There has been an alternative interpretation of the experiments as arising from a superposition of oscillatory interactions as for example in the theory by Bruno and Chappert (1991). As we saw in Section 13-6-D, the screening

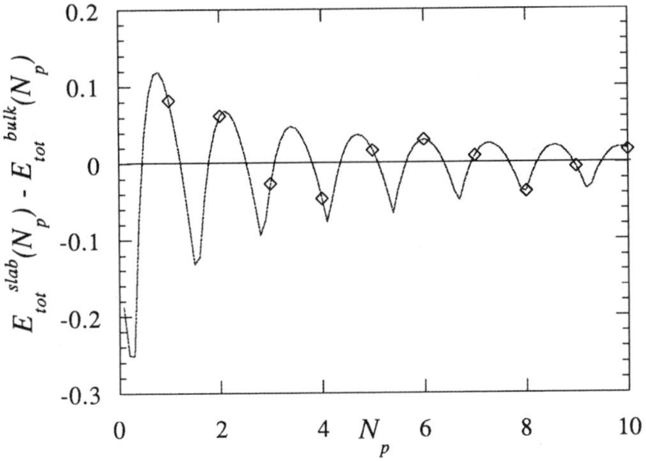

Fig. 19-3. A plot of the energy per unit area (in units of $E_F k_F^2/8\pi$) of a copper slab of N_p layers (each of (100) orientation) between two iron slabs with parallel magnetization, minus that for antiparallel magnetization, obtained From Eq. (19-11).

field obtained from a full quantum-mechanical calculation shows Friedel oscillations, and these same oscillations appear in the interatomic interactions. Now, if one iron atom in a free-electron gas has its d-electron spins oriented in the positive z-direction, it provides an exchange potential $-U_x$ seen by any free electron of the same spin. (U_x here is between s- and d-electrons rather than that between two d-electrons as in Section 15-6-A.) The probability density of electrons of this spin will show Friedel oscillations around that atom. Then a second iron atom of the same spin orientation will have its energy shifted by this same $-U_x$ times the local probability density for that spin. This is just as in the effective interactions between atoms which we discussed in Chapter 14 and we obtain a term in the energy proportional to $U_x^2 \cos(2k_F r)/(k_F r)^3$. When this is positive, antiparallel moments are favored; when it is negative parallel moments are favored. These may be summed over planes of parallel moments on each side of a copper slab to simulate the ferromagnetic iron and an interaction analogous to that in Fig. 19-3 obtained. It of course shows the same oscillation periods of a half electron wavelength. We in fact made such an evaluation, corresponding to Fig. 19-3 and found very similar, though more nearly sinusoidal, curves. Interestingly enough the phase was reversed for the two models, one being maximum where the other was minimum. They could have been brought into phase by taking the spacing between iron layers as N_ps, but the real spacing between iron layers is $(N_p + 1)s$.

In spite of this discrepancy, it is probably best to regard both interpretations of the interaction between magnetization across the slab as valid, but as applying in different limits. It was possible to make a determination as to which limit was most appropriate in real systems using first-principles, local-density calculations of the interaction by van Schilfgaarde (van Schilfgaarde and Harrison (1993)). The idea was that if the exchange coupling between iron d-states and the s-like band states was sufficiently weak, the perturbation theory assumed in summing the oscillatory interactions must be valid and the theory must apply. In addition the strength of the coupling would grow as the square of the exchange interaction. If the coupling is sufficiently strong, the absence of bands at the Fermi energy which leads to the quantum-well picture must apply and in addition, the interaction will become independent of the magnitude of the exchange interaction. Van Schilfgaarde's calculations gave a good quantitative account of the observed coupling between iron moments through an intermediate chromium (rather than copper) layer and therefore corresponded to a realistic treatment of exchange. In his calculation he could simply vary the strength of the exchange away from the realistic value to obtain the dependence of the difference in energy of the parallel and antiparallel magnetizations upon exchange interaction. He indeed found it to be quadratic at small couplings, but to have very nearly saturated at the realistic exchange strength, suggesting that at least in this case, the interpretation in terms of the quantum-well states which led to Fig. 19-3 is the more realistic.

The reason that this coupling between layers is important is that with parallel spins in the iron, and electrons of minority spin confined to the copper, we may have very high conductivity of the system parallel to the planes. The highly conducting copper and highly resistive iron are electrically in parallel leading to a high total conductivity. With antiparallel minority spins all electrons (except those moving exactly parallel to the plane) will move to and from the iron. The resistances average and the conductivity is very low. Thus, a system may be designed to have the magnetization antiparallel in the iron layers, and a low conductivity. An applied magnetic field can align the magnetizations and dramatically increase the conductivity. This would be called a giant negative magnetoresistance. Such a property can be used, for example, in a device to read magnetic data stored on disks.

2. Photothresholds, Workfunctions, and Electron Affinities

The tight-binding electron states are based upon free-atom term values, which approximate the removal energy of an electron from the atom, as we stated in Section 1-1. We might expect that the eigenvalues , as modified for the solid, might give good estimates of the removal energies for the solid. This removal energy is called the *photothreshold* , the energy of the photon required to remove an electron from the highest occupied state. (Note that this term differs from the optical threshold, or absorption threshold, discussed in Section 6-2, which pertains to electron transitions within the solid.) For variation from one material to another this is qualitatively the case, but there is an important correction which is easy to understand, though it is technically a many-body effect.

A. Corrections to the Tight-Binding Photothreshold

We may see that the removal energy for an electron from an atom is considerably reduced if that atom is near a metal or dielectric medium. This is done in steps for an atom near, but not touching, a metal. We first carry the atom far from the metal, at no cost in energy, remove an electron, requiring an energy approximately equal to the magnitude of the free-atom term value, and then return the charged atom to the vicinity of the metal. It is this last step which gives the correction. The atom feels an attractive image force, given by $e^2/(2z)^2$ if the atom is a distance z from the metal. Integrating the effect as we bring the atom to the surface, a distance d away, we find an energy gain $e^2/(4d)$. Of course, removing the electron directly from the atom near the metal, without removing the atom first, must give the same result since these are two reversible paths between the same initial and final states. Therefore, this is the reduction in the removal energy. It seems clear that a similar reduction is appropriate for the removal of an electron from an atom imbedded in the surface or a removal of an electron from the surface.

Had the same calculation been performed for an atom near a dielectric, it would have been the same except that the image force is reduced by a factor $(\varepsilon-1)/(\varepsilon+1)$, which is very close to one for a semiconductor with a large dielectric constant ε , so the effect is essentially the same. The value of this correction is about 1.5 eV for silicon, with a d of 2.35 Å .

B. Photothresholds in Semiconductors.

We obtained in Eq. (6-1) the valence-band maximum E_v in terms of the term values ε_{p+} for cation (Ga) and ε_{p-} for the anion (As) as

$$E_v = \frac{\varepsilon_{p+} + \varepsilon_{p-}}{2} - \sqrt{\left(\frac{\varepsilon_{p+} - \varepsilon_{p-}}{2}\right)^2 + \left(1.28\,\frac{\hbar^2}{md^2}\right)^2} \quad . \tag{19-12}$$

This did not include the shift in the energy due to nonorthogonality which we saw in Section 2-2-E shifted both the bonding and antibonding levels upward by SV_2. This did not influence properties such as the band gaps, but would reduce the photothreshold. If we take that as an estimate of the reduction, with the nonorthogonality $S \approx 0.5$ as we saw there, this would reduce the photothreshold by another 2.22 eV for silicon. Thus our estimate for silicon would be reduced from the 9.35 eV of Eq. (19-12) to 5.63 eV, in better agreement with the experimental threshold of 5.10 eV than we might expect given the crudeness of the corrections we have made and the fact that photothresholds depend significantly on the crystallographic orientation of the surface and upon surface reconstructions which we discuss in the following Section.

A plot of the experimental photothreshold against similar values was made by Ciraci (Harrison and Ciraci (1974)) and we redo that plot in Fig. 19-4 using the values for Eq. (19-12) given in Table 6-1. The plot is remarkably linear, and in fact the difference is an approximate shift of 4.2 eV as indicated by the line drawn in the figure. That shift is larger than our estimate for silicon of 1.5 + 2.2 = 3.7 eV, but not greatly. Both corrections decrease with increasing spacing, which might account for the slope in Fig. 19-4 being slightly less than one, but we have not explored the question further.

There are other quantities related to the photothreshold. One is the *electron affinity*, the lowest energy at which an electron can be added to the semiconductor (or insulator) from the outside. It is of course smaller by the band gap, and in making that estimate we must include the Coulomb enhancement of the gap which we discussed in Section 5-5-A. There is also the *work function*, the energy required to remove an electron at the Fermi energy in the semiconductor. It is approximately equal to the photothreshold

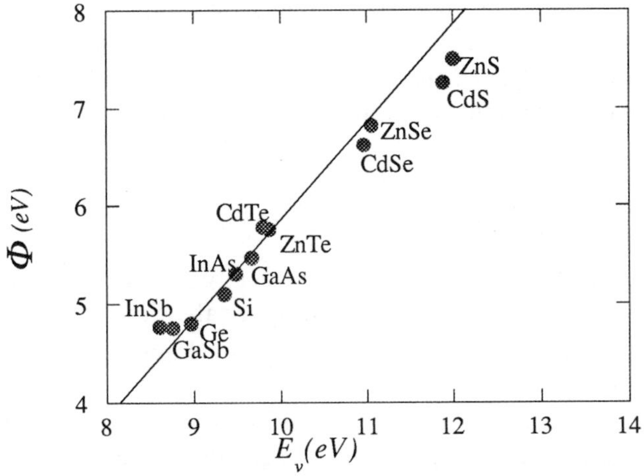

Fig. 19-4. A plot of the experimental photothreshold (from Gobeli and Allen (1965) and Swank (1967)) against the energy of the valence-band maximum from Eq. (19-12). The line is $\Phi = E_v - 4.2\ eV$. An experimental value of 7.82 eV for ZnO is not shown since the E_v value is 17.2 eV, out of the figure and off the line.

for heavily doped p-type materials and approximately equal to the electron affinity for heavily doped n-type materials. It is not so suitable for characterizing semiconductors because it depends upon doping and upon temperature, but it is the quantity most often discussed for metals.

C. Metallic Work Functions

In metals the minimum energy required to remove an electron is for an electron at the Fermi energy of the crystal, so the photothreshold is equal to the work function for the metal. Similarly, an electron can be added at this energy so the electron affinity is also equal to the work function.

The same kind of tight-binding approach which we used for semiconductors should be equally applicable to metals. We in fact have an immediate estimate of the workfunctions of the simple metals from the tight-binding theory given in Section 13-3. It is the energy to remove an s-electron from the free atom $|\varepsilon_s|$ plus the energy difference between the s-state energy and the bottom of the band, $3\pi^2\hbar^2/(4md^2)$ as in Eq. (13-6), minus the Fermi energy $E_F = \hbar^2 k_F{}^2/2m$. We should again correct for the upward shift of the levels due to the nonorthogonality which gave rise to the overlap repulsion, $^1/_2(3\pi^2\hbar^2/(4md^2 - {}^3/_5E_F)$, corresponding to the correction applied in Eq. (13-7), and the Coulomb correction which may be estimated as $e^2/(4d)$ as for the semiconductors. The net result is

$$W = -\varepsilon_s - \frac{e^2}{4d} + \frac{3\pi^2\hbar^2}{8md^2} - \frac{7\hbar^2k_F{}^2}{20m} . \tag{19-13}$$

This can readily be evaluated using the ε_s values from Table 1-1 and the r_0 values from Table 13-1; we did all evaluations for a close-packed structure so $d = (4\sqrt{2}\pi/3)^{1/3}r_0$. The results are compared with experimental values in Fig. 19-5.

It is best to consider the comparison by groups. We see that the alkali metals, where one might expect the best agreement, are in reasonable accord, both as to trend and magnitude, particularly in view of the size of the contributing terms. The four terms in Eq. (19-13) for lithium, for example, are 5.34 - 1.16 + 2.91 - 3.32 = 3.77 eV. The alkaline earths (Ca, Sr, Ba) do even better. The noble metals and divalent metals of type B (Mg, Cd, Hg) are in reasonable accord quantitatively, but experimentally show much greater variation from one to another than do the predictions. Experimental values for Be varied from 3.17 eV to 3.92 eV, and we selected the "preferred value" from Weast (1975) for the plot. The predictions are considerably above experiment for the trivalent and tetravalent metals. For silicon and germanium the predictions are also much larger than the predictions made treating them as semiconductors in the preceding subsection. The experimental values were in fact for the semiconductors but they are not far from the experimental values for tin and lead, which were for the metallic state.

We would regard the predictions as semiquantitatively correct, and perhaps as good as could be anticipated for such a simple theory. The real work functions on single crystals vary as much as an electron volt from one crystallographic face to another (Weast (1975)), due to surface effects which we have completely ignored, and the experiments were generally on polycrystalline surfaces. We provide a very elementary theory but, as for the cohesion, it appears to contain the essential features of the problem for the simple metals as well as for semiconductors. We have not compared the corresponding picture for insulators with experiment, and the theory has not been extended to transition metals.

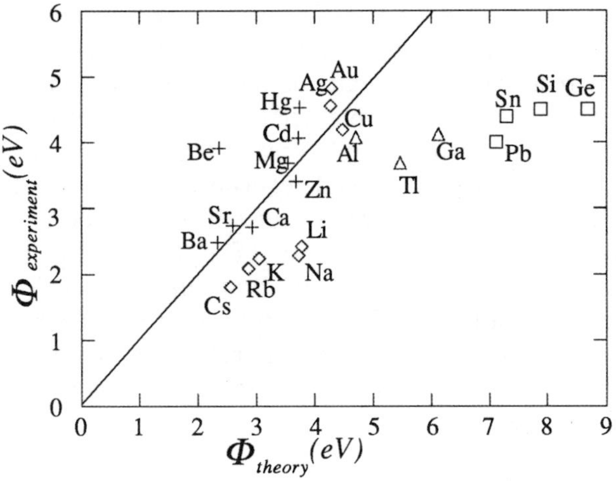

Fig. 19-5. Experimental values of the work function of the simple metals [taken from the CRC Handbook (Weast (1975))] plotted against the predictions of Eq. (19-13).

3. Reconstruction at Solid Surfaces

Our discussion of surface energies proceeded as if the arrangements of the atoms at the surface was the same is if all atoms beyond the surface plane of the bulk crystal were removed, and those remaining retained their original positions. In ionic crystal where the surface energy is small this can be very nearly true, and it is often approximately true for metals. We noted in Part 19-1-C that a more complete calculation could be carried out, allowing the last layer of atoms at a surface, and other layers, to move away from their original positions. In some cases it even appears that the energy may be lower if alternate rows of atoms are removed at the surface, or other rearrangements which lower the symmetry of the surface. However, in semiconductors, where we have seen that the surface energy is very large compared to the energy per bond broken, it is almost universal that rebonding occurs and the symmetry of the surface is changed. This rearrangement is called *reconstruction* of the surface.

There has been intense activity in studying reconstruction for many years. One major impact was the observation by Binnig, Rohrer, Gerber, and Weibel (1983) of the seven-by-seven reconstruction with the scanning tunneling microscope, which demonstrated convincingly for the first time that this microscope had truly atomic resolution. The fact that this showed the adatoms which had been predicted for this reconstruction using tight-binding theory (Harrison (1976b)) was of much less consequence. This same systematic study of general reconstructions by tight-binding theory (Harrison (1976b), (1980)) turned out not to have been so successful on other systems. However, the use of tight-binding has taught some interesting lessons. We recount that history briefly, centering on reconstruction of silicon surfaces noting what new came from each case. We make no attempt to cover generally this very large field. An authoritative account of semiconductor reconstruction in general is given by LaFemina (1992).

A. The Silicon (111) 2×1 and General (110) Surfaces

The freshly cleaved (111) silicon surface has been known for forty years to show a 2×1 reconstruction (Schlier and Farnsworth (1959), Haneman (1961), Lander, Gobeli and Morrison (1963)). Tight-binding theory suggests an immediate explanation (Harrison (1976b)). On the (111) surface sp^3-hybrids from each surface atom form bonds with the three neighboring atoms below the surface, leaving a single dangling sp^3-hybrid and one electron to occupy it. The same is true of the (110) surface. This is a prime candidate for a "negative-U center", as we discussed in Section 8-3 : If we were to allow alternate atoms to move slightly in and slightly out (from their relaxed equilibrium position), the energy will increase in proportion to the square of the displacement. This follows from the fact that these are displacements from the minimum-energy positions. It is also consistent with an elasticity view. However, the hybrid energies shift alternately up and down *in proportion* to the displacement. This results as the hybrid energy varies continuously from the s-state energy for back-bond angles equal to 90° to the p-state energy for back-bond angles equal to 120°. The reason that the total energy still varies in proportion to the square of the displacement is that a single electron occupies each and the linear shifts cancel. However, if we shift an electron from each hybrid rising in energy to each hybrid lowering in energy, the net energy decreases linearly with displacement and we are guaranteed a gain in energy, as in the Jahn-Teller Effect discussed in Section 8-3. As long as the maximum gain in energy as a function of displacement exceeds the cost in Coulomb energy, a screened U, from doubling up the electrons in one hybrid, this is the expected surface, reconstructed in a two-by-one pattern. When this occurs it is as if the repulsive energy U were instead a negative energy, attracting the electrons to occupy the same hybrid.

Such a distortion had early been suggested by Haneman (1968) for somewhat analogous reasons. A similar argument could be made in terms of a Peierls-like distortion opening a gap in the half-filled surface band. There seemed little doubt that this was the explanation, although calculations in M. L. Cohen's group (private communication) indicated that the repulsive U exceeded any gain which could be obtained. He proposed an alternative, antiferromagnetic rearrangement which did not involve this U.

At this point, Pandey (1981) suggested a totally different *π-bonded chain model* with alternate five- and seven-member silicon rings. This suggestion came purely from a theoretical effort, interestingly enough mostly carried out using the simple tight-binding model which Pandey had used earlier (Pandey and Phillips (1974)). This enabled him to explore a large range of geometries, and when he found the most promising one he carried out careful and complete local-density calculations to confirm this startlingly different suggestion. There is no doubt now that he was correct, and the π-bonding energy gained from this structure overwhelms the distortion energy it requires, just as the π-bonding energy in graphite more than makes up for the broken σ-bonds. For tight-binding theory there was the lesson that the Coulomb U's and Madelung shifts do not necessarily cancel at the surface, as we found in Section 9-1 that they do for bulk crystals.

The corresponding in-out relaxation can occur if the outmoving atom has extra nuclear charge, as for a column-V atom on the (110) surface of a polar semiconductor. This is not considered a reconstruction since the translational symmetry is not modified, but the corresponding distortion does occur at a gallium arsenide (110) surface with the arsenic moving out, and gallium moving in, as expected. It presumably does not occur on the (110) surfaces of silicon for the same reason it did not on the (111).

B. The Silicon (100)

On the (100) surfaces of silicon each surface atom has two neighbors below the surface so there are a pair of electrons to occupy the surface bands. This is quite a different situation. Nonetheless a 2×1 reconstruction is observed and was attributed by Lander and Morrison (1962) to alternate vacancies. This seemed likely to us also, and we studied a range of vacancy-based reconstructions using tight-binding theory (Harrison (1976b), (1980)). The central point was that making a vacancy on a (100) surface did not break extra bonds: the surface atom with two broken bonds was removed, leaving two broken bonds behind. The modified geometry allowed for other paths to lower energy.

There had been an earlier proposal by Schlier and Farnsworth (1959) which we regarded as a misuse of tight-binding theory, though it turned out to be correct. They thought of the two broken bonds for the surface atom as two dangling hybrids, each with a single electron. Then two neighboring surface atoms might lower their energy by moving together, like hydrogen atoms, forming two-electron bonds and leaving a single dangling bond on each atom. In fact the situation is more analogous to two helium atoms. The remaining two orbitals on the silicon atom at a (100) surface , after the formation of the two bonds below the surface, are one even in reflection about the surface normal, an sp-hybrid, and one odd in reflection, a p-state. The sp-orbital is several volts lower in energy than the p-state and is doubly occupied. Bringing the two together will initially split the lower state, but with both occupied the shifts cancel and one expects a net upward shift (arising partly from elastic energy and partly from the nonorthogonality which produced the overlap repulsion). The upper state is also split, but unoccupied so it does not affect the energy. This is as for a pair of helium atoms. Only if the distortion proceeds to the point that the bonding combination of upper states drops below the antibonding combination of lower states, can the energy drop. This is the circumstance discussed by Woodward and Hoffmann (1971) in *Conservation of Orbital Symmetry* in which absorption of light is required to take an electron into an antibonding state before the bonding can occur, and in cases such as helium it does not go in any case. It hardly seemed a likely explanation but it could not be ruled out and it is now established theoretically and experimentally (See LaFemina (1992) for details.). It was learned that with competing plausible alternatives, one can not be confident in predicting the outcome without quantitative comparisons.

C. Polar Surfaces

The (100) surface of a compound semiconductor has another problem since it tends to lead to an unacceptable charged surface as indicated in Section 19-1-B. For gallium arsenide the last plane would be entirely arsenic or entirely gallium. We may see why this is a problem by constructing the surface using the theoretical alchemy which we introduced in Section 1-4-E. We may begin with an ideal germanium (100) surface,

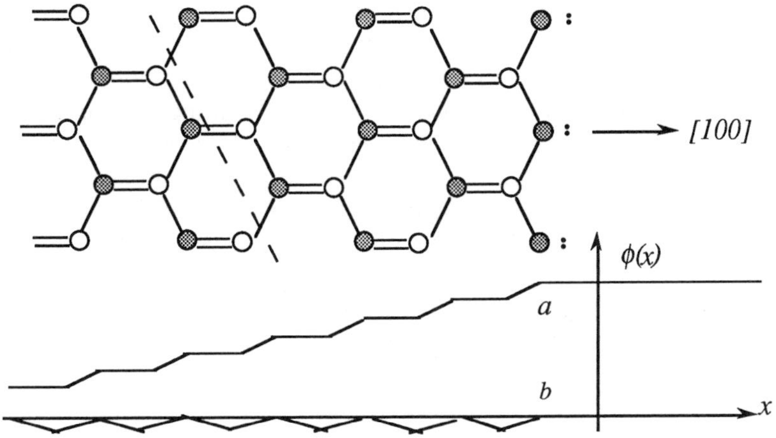

Fig. 19-6. A germanium crystal, with a (100) surface on the right, and two electrons in the dangling hybrid. Since every atom is neutral the electrostatic potential would be flat outside, and on average flat inside the crystal. We then freeze the electrons and transfer one proton from each open circle (the nucleus becomes a Ga nucleus) to each shaded circle (the nucleus becomes an As nucleus). The potential, averaged over planes of constant x, then takes the form labeled a below. If half the arsenic atoms are removed, it takes the form labeled b below. The dashed line is a (110) plane.

with each surface atom placing a single electron in each of the two back bonds, and leaving two in the dangling sp-hybrid, as shown in Fig. 19-6. Certainly every atom is neutral and we take the potential to be constant everywhere. We now freeze the electronic states and transfer protons between neighboring atoms to produce gallium arsenide, but the surface layer is inevitably of one atom type as seen in Fig. 19-6 where it is chosen to be As. To obtain the potential averaged over planes parallel to the surface, we average the extra charge over such planes and each sheet of charge causes a discontinuity in the field, $-\partial\phi/\partial x$, as shown in curve a . We have taken the field to be zero outside the surface and we inevitably find the average field within the crystal to be nonzero. The extra field energy is a contribution to the surface energy which diverges as the crystal becomes large to the left. It will not be a possible state of the bulk system since the energy can always be lowered by neutralizing that charge, even if it costs a band gap of energy for each carrier required. Even that represents such a large surface energy that the surface shown does not form. The crystal will grow with (110) surfaces, indicated by a dashed line in the figure, in which the planes parallel to that surface are all neutral, or else a reconstruction will occur at the surface which restores neutrality.

One such reconstruction is obtained by removing one half of the arsenic atoms in the surface layer (Harrison (1976b)). Then the change in slope at that surface is half as large as shown in curve b , while all other changes in slope remain the same. That is seen in curve b to leave an average field of zero within the crystal, and is therefore an allowed geometry. To be sure there is an alternating field between planes, but that is a contribution to the bulk energy, unaffected by the surface. It is seen that this reconstruction actually produces a dipole layer, by shifting the average potential inside the crystal relative to that outside and this shift would be of opposite sign for a gallium surface. Such shifts are not seen experimentally, suggesting that a more complicated reconstruction occurs, such as removing three quarters of the surface arsenic layer and one quarter of the gallium layer below it. (Harrison (1979)).

This same analysis of course applies to the surface of an ionic crystal. In rock salt a (110) surface will be entirely sodium ions or entirely chlorine ions and is not allowed. The crystal grows with (100) surfaces and cleaves in (100) planes.

We might at first think that there is a way around this compelling argument in gallium arsenide by charge redistribution within the bonds, but that is not the case. We may see this by allowing the bonds in Fig. 19-6 to relax toward the arsenic, typically shifting the effective charge on the arsenic to $Z^* \approx -1$ on arsenic as we saw in Section 4-1, rather than the +1 which comes from the proton transfer. This does not change the sign of the surface charge, as we might expect, but leaves it unchanged. Looking atom by atom in Fig. 19-6 we see that it adds $+2e$ to every gallium atom and $-2e$ to every arsenic atom except in the surface layer, where it adds only $-e$. The added charges are just like proton charges in the reconstructed GaAs structure mentioned above, with half the arsenic atoms removed; it produced no average field within the bulk and correspondingly here the redistribution does not modify the average field within the bulk.

For the geometry shown in Fig. 19-6 we placed two electrons in each dangling sp-hybrid, since that hybrid energy is well below the tight-binding valence-band maximum. Then the internal fields corresponded to $+{}^{1}/{}_{2}e$ per surface atom. To add half an electron per surface atom, which would be required to neutralize the surface, we would need to add electrons to the empty π-states at the energy $\varepsilon_p(As) = -8.98\ eV$. From Table 6-1 we see that the valence-band maximum in GaAs is at -9.64 eV, so that it is possible that neutrality could be accomplished electronically, particularly in n-type GaAs with a Fermi energy near the conduction-band minimum at -7.58 eV. We are not aware of experimental studies of such doping-dependent surface structures, and we have not pursued it theoretically. We note from Fig. 19-6 that it breaks no extra bonds to make a vacancy on this surface; the surface arsenic had two broken bonds and removing it leaves a single broken bond on the two gallium atoms below. For that reason we earlier expected (Harrison (1980)) that the equilibrium structure on polar (100) surfaces would have vacancies at half the surface sites, leading to a neutral surface with doubly occupied sp-hybrids on the arsenic atoms and empty sp-hybrids on the gallium. This seems to have been the case. Ideal (111) surfaces of polar semiconductors are also systems with net surface charge, and require reconstruction, but just what reconstruction is much less clear.

D. The Silicon (111) 7×7

The real challenge was the exotic 7×7 reconstruction observed on (111) surfaces after annealing at high temperature, observed first by Schlier and Farnsworth (1959). Lander and Morrison (1963) early proposed a pattern of thirteen vacancies in the 7×7 surface unit cell, but this required the breaking of three bonds for each vacancy and seemed unlikely. We took a clue from the findings of Lander and Morrison (1963) that added aluminum and phosphorous atoms on silicon (111) surfaces appeared to form on bridging sites, over three dangling hybrids, to replace the three broken bonds by three bent bonds and a single broken one, forming a $\sqrt{3}\times\sqrt{3}$ pattern. We suggested (Harrison (1976b)) that silicon adatoms might similarly be energetically favorable, with a concentration determined by a compromise between the energy gained by each and a repulsion between them arising from the distortion of the substrate. The pattern could not be predicted, but the same pattern as that of the thirteen vacancies proposed by Lander and Morrison seemed plausible.

This aspect of the 7×7 pattern was strikingly confirmed by the scanning tunneling microscope images by Binnig, Rohrer, Gerber, and Weibel(1983), but with one of the adatoms missing. Subsequent studies by Takayanagi, Tanishiro,. Takahashi, and Takahashi (1985) established that there was in addition a stacking fault buried beneath one half of the 7×7 cell. This stacking fault does not occur on the germanium (111) 2×8 structure which similarly is a pattern of adatoms (Takeuchi, Selloni, and Tosatti (1992)). In fact, M. B. Webb, private communication, had on the basis of LEED studies suggested this stacking fault for the silicon 7×7 structure earlier to the author. It seemed to us difficult to believe that if such a stacking fault were there, the crystal could successfully grow as an unflawed crystal, but that appears to be the case.

4. Interfaces

Interfaces between two solids are intrinsically much simpler than interfaces with the vacuum because the environment on the two sides is more similar. The important message at the outset for interfaces is the same as for surfaces: as we noted following Eq. (19-1), the energy bands remain valid up to the interface; it is just the allowed wavenumbers, the normalization and the phase of the states that is determined by the geometry. At interfaces the questions concern the line-up in energy of the bands on the two sides relative to each other, and the matching conditions on the wavefunctions. We address both here.

One of the simplest interfaces between crystals is that for which the crystal structure is continuous across an interface, but the constituent atoms are different. Such an interface is called a *heterojunction* and we consider it first. Most other interfaces involve defects and are more complicated. For a planar heterojunction the system retains translational symmetry parallel to the plane so that it is still possible to construct Bloch functions, or tight-binding states, with transverse wavenumbers $\boldsymbol{k}_t$ parallel to the plane of the junction.

For the variation of the wavefunctions perpendicular to the plane of the heterojunction it is again helpful to think in terms of the one-dimensional chain, with tight-binding. Now, however, the chain is represented by different tight-binding parameters ε_s and $V_{ss\sigma}$ on the two sides of the interface. Eq. (19-1) is again the correct set of equations to solve for the states and energies but now for an interface between $j = 0$ and $j = 1$, there will be an ε_{s-} for $j \leq 0$ and a different ε_{s+} for $j > 0$, and the $V_{ss\sigma}$ will depend upon which states are coupled. Again with may think of a completely defined band structure on each side, given by Eq. (1-13),

$$\varepsilon_k = \varepsilon_{s\pm} + 2V_{ss\sigma\pm} \cos kd \ . \tag{19-14}$$

The parameters ε_{s+} and ε_{s-} are different for the two sides, as are the $V_{ss\sigma\pm}$. On each side the coefficients are given by $u_j = A \sin(kdj + \phi)$ and again the A, k, and ϕ are different on the two sides. They may be determined by the Eqs. (19-1) near the interface and by the boundary conditions at the outside surfaces.

A. Matching Wavefunctions at Heterojunctions

It is frequently the case in semiconductors that the states of interest are sufficiently near to band edges that an effective-mass approximation is appropriate, as discussed in Chapter 6. Then we may also think of the determination of the parameters A and ϕ as a matching of the functions of position, $\psi(z) = A\, sin(kz+ \phi)$ at the interface, but it is not immediately obvious what matching conditions are to be used. We would like conditions, analogous to the matching of the wavefunction and its gradient in ordinary quantum theory, for these effective-mass wavefunctions. It has long been known (Harrison (1961)) that if the mass changes across the interface, it cannot be correct to simply match value and slope of the effective-mass wavefunction. The reason is that then the current density j, obtained from

$$j = -\frac{e\hbar}{2m^*\, i}(\psi^* \nabla\psi - \psi\nabla\psi^*) \qquad (19\text{–}15)$$

(we have not derived this form, but note it is real and for a free electron is $-ep/m^*$, with p the momentum) would not be continuous if ψ and $\partial\psi/\partial z$ were continuous but m^* was not. Requiring that charge not disappear as time proceeds *could* be used as one matching condition (requiring the normal component of $\psi^* \nabla\psi/m^*$ to be continuous across an interface), but a second unknown condition is needed.

One way of learning what matching conditions are appropriate is to construct tight-binding states across an interface using Eq. (19-1), since we know how to do that. The same system could also be treated as an effective mass problem near the band edge. Then one could determine what conditions would have given the correct matching. Such a calculation was carried out, using the model represented by Eq. (19-1), by Harrison and Koslov (1992). An alternative treatment was given by Balian, Bessis and Mezincescu (1995). The result was more complicated than might have been anticipated. In general ψ on one side of the interface was a constant times that on the other, plus another constant times the slope on the other side. Similarly the gradient on one side depended upon both the slope and value on the other.

$$\psi_+(0) = C_1\psi_-(0) + C_2\, d\frac{\partial\psi_-(0)}{\partial z},$$

$$d\frac{\partial\psi_+(0)}{\partial z} = C_3\psi_-(0) + C_4\, d\frac{\partial\psi_-(0)}{\partial z}. \qquad (19\text{-}16)$$

These constants C_j depended upon the effective masses on the two sides through the tight-binding parameters. They also depended upon the coupling matrix element $V_{ss\sigma}$ *across* the interface as well as upon those in the bulk on the two sides. The latter are chosen to give the desired effective masses but the matrix element across the interface could be chosen independently. The only circumstance in which the cross terms between slope and wavefunction disappeared, $C_2 = C_3 = 0$, and therefore the only circumstance for a simple result, is if the coupling across the interface is the geometric mean of the bulk values on the two sides, $V_{ss\sigma} = \sqrt{V_{ss\sigma+}V_{ss\sigma-}}$, a perfectly plausible choice. In this case, $C_1 = C_4 = \sqrt{m^*_+/m^*_-}$, the condition for a matching on the two sides of $\psi/\sqrt{m^*}$ and of $\partial\psi/\partial z / \sqrt{m^*}$. For the three dimensional case it is a matching of $\psi/\sqrt{m^*}$ and of the normal component of $\nabla\psi / \sqrt{m^*}$. This obviously satisfies current conservation and presumably is the best choice if we wish to use effective-mass wavefunctions.

The matching condition for the simple chain becomes significantly more complicated if we allow second-neighbor coupling between neighbors. This in fact comes up automatically in three dimensions even with nearest-neighbor interatomic coupling. For example, in a face-centered-cubic crystal with wavenumbers in a [110] direction, there are four nearest neighbors with phase factors $e^{ikd/2}$ and one with phase factor e^{ikd} , as well as four with phase factors $e^{-ikd/2}$ and one with phase factor e^{-ikd} and two with phase factor unity, so that in the corresponding one-dimensional equations (Eq. (19-1)) there are second-neighbor couplings. If we return to the surface problem with second-neighbor couplings, truncating a chain at $j = 1$ and $j = N$ does not leave solutions of the form $sinkdj$ for Eq. (19-14) since it eliminates terms from $j = -1$ and $j = N + 2$ which were present in the infinite chain. The consequences of these couplings have been explored by Kozlov and Harrison (1993). More distant couplings give more complicated forms for the bands (e. g., $\varepsilon_k = \varepsilon_s + 2V_{ss\sigma}(1 + 4cos(kd/2) + coskd)$ for $\boldsymbol{k}$ parallel to [110] in fcc crystals) as well as more matching conditions, but then multiple solutions decaying exponentially away from the heterojunction also arise so that these extra conditions can be satisfied. We do not go into these complications here. If one is satisfied with the simplification of effective-mass wavefunctions, one may also be satisfied with the less complete matching conditions.

B. Semiconductor Band Line-Ups

We have seen that the bulk band structure remains meaningful on the two sides of the interface, and a central question is how these two band-structures line up. Since in tight-binding theory all of our band structures were referenced to the atomic term values, it would be natural - though we shall see incorrect - to assume that these term values provided the appropriate scale, and that was in fact the approach used in Harrison (1980). The correct approach with a line-up arising from interface dipoles was given by Tejedor and Flores (1978) and by Tersoff (1984).

The situation is perhaps most clearly understood by imagining a single semiconductor, say silicon, and following an argument given by Harrison and Tersoff (1986). At a plane through this crystal we produce an interface by shifting all term values to the right upward in energy by Δ. Obviously the bands to the right are simply raised by Δ producing this same discontinuity in the conduction bands and in the valence bands. However, this shift in term values could also be accomplished by *applying* an electrostatic potential Δ across that plane and in that case we know that the bonds will polarize, reducing the net discontinuity by a factor of the reciprocal of the dielectric constant to Δ/ε, much smaller than Δ. In fact, it was the bonds which cross the plane which are polarized. The different energy of the two hybrids entering the bond makes them polar and the resulting dipole is responsible for the shift. Even if the interface were spread out over a number of atomic layers, the argument with the dielectric constant tells that the net offset of the hybrid energies is reduced by a factor $1/\varepsilon$. To the extent that the dielectric constant is very large compared to one, the hybrid energies on the two sides are brought into coincidence by this dipole.

Thus the correct view of the band line-ups is that the hybrid energies on the two sides are brought nearly into coincidence, and the bands shift with those hybrid energies. In the case of compound semiconductors we may imagine both kinds of bonds crossing the interface and take the average hybrid energy as the reference level. Indeed, the average hybrid energy in a compound is the average of the valence s- and p-state (with three-times the weighting for p-states) energies for the constituents.

The scheme for predicting band line-ups becomes straightforward. We list in Table 19-1 the valence-band maximum (Eq. (6-1) and Table 6-1) minus the average hybrid energy (from Table 1-1). This difference for system A minus that for system B gives the valence band maximum for A relative to that of B assuming that the "neutral points", defined as the average hybrid energy, align. Small corrections could be made for the finite

Table 19-1. The energy of the valence band maximum, E_v (in eV), measured from the average hybrid energy ε_h obtained from Table 1-1. These may be subtracted for two semiconductors to predict the relative positions of the valence-band maximum at a heterojunction. We also list the internuclear distances, d , which are relevant to heterojunction formation. Band gaps are listed in Table 6-1.

	d (Å)	$E_v - \varepsilon_h$		d (Å)	$E_v - \varepsilon_h$
C	1.54	-2.03	BeO	1.65	-4.02
Si	2.35	+0.04	BeS	2.10	-1.77
Ge	2.44	+0.32	BeSe	2.20	-1.31
Sn	2.80-	+0.33	BeTe	2.40	-0.98
SiC	1.88	-1.32	MgTe	2.76	-1.54
			ZnO	1.98	-4.16
BN	1.57	-2.62	ZnS	2.34	-2.16
BP	1.97	-0.74	ZnSe	2.45	-1.71
BAs	2.07	-0.42	ZnTe	2.64	-1.40
AlN	1.89	-2.72	CdS	2.53	-2.18
AlP	2.36	-0.76	CdSe	2.63	-1.75
AlAs	2.43	-0.46	CdTe	2.81	-1.46
AlSb	2.66	-0.23			
GaN	1.94	-2.55	CuF	1.84	-5.47
GaP	2.36	-0.66	CuCl	2.34	-3.18
GaAs	2.45	-0.34	CuBr	2.49	-2.58
GaSb	2.65	-0.14	CuI	2.62	-2.24
InN	2.15	-2.59	AgI	2.80	-2.26
InP	2.54	-0.77			
InAs	2.61	-0.47			
InSb	2.81	-0.28			

dielectric constant, so that the matching is not complete, but the accuracy of the prediction is sufficiently limited that this small correction may not be significant.

In proceeding this way we are of course assuming that there are no extra charges at the interface, ruled out for the same reasons which ruled out charged surfaces in Section 19-3-C. We are also assuming that there are no surface dipoles of the sort generated by structural surface dipoles, as were illustrated in the curve b in Fig. 19-6. Such added dipoles are of course conceivable in real interfaces, as seen for example, by Spicer, Chye, Skeath,

Su, and Lindau (1979), by Waldrop, Sullivan, Grant, Kraut, and Harrison (1992), and by Grant, Waldrup and Kraut (1979).

Experimental values for the band line-ups are collected in many of the references given in the following subsection on "Other Views". In Harrison (1980), for example, experimental values for the germanium valence-band maximum, relative to that for silicon, run from 0.24 eV to 0.17 eV, while we predict -0.01 eV. For germanium relative to GaAs, experimental values run from 0.36 to 0.76, while Table 19-1 predicts 0.66 eV. An experimental value for GaAs, relative to $Ga_{0.8}Al_{0.2}As$, was 0.03 eV, while Table 19-1 would predict 0.02 eV. InP relative to CdS was found experimentally to be 1.63 eV, while Table 19-1 gives 1.41 eV. Some of the inconsistencies from one measurement to another may arise because the line-ups are shifted by different extrinsic effects in different laboratories. Values from Table 19-1 may prove as useful a guide as any, but one must expect variations in any real laboratory system.

We might also ask for effects of band shifts due to nonorthogonality, which we included for the photothresholds and workfunctions in Sections 19-2-B and 19-2-C. These are shifts in the average energy of the hybrids, so they shift both hybrids equally in any one bond and do not introduce a bond dipole. Thus these shifts do not in themselves affect the band line-ups. However, when the strain is different in different parts of the system, the hybrids may be shifted differently in the two regions and yet there is no interface dipole produced to screen the effect. It was for these reason that there were nonorthogonality contributions to each deformation potential we have calculated. We use these deformation potentials to discuss such strained systems in Section 19-4-D.

C. Other Views

This picture also allows us to understand the other views which have been taken of the band line-ups. The earliest treatment was by Anderson (1960), who used experimental electron affinities to define the line-up. We saw in Section 19-2 that there are corrections to work functions, photothresholds, and electron affinities arising from the surfaces, and from surface reconstructions, which do not enter here so Anderson's is probably the least reliable method. The "natural" band line-ups mentioned above took the tight-binding bands directly (Harrison (1977)) using the valence-band maximum of Eq. (6-1). This is more reliable than matching experimental electron affinities and in fact the average hybrid energies for different semiconductors are frequently very close to each other so this approach often gives values close to those obtained by matching hybrid energies. However, when it differs from matching average hybrids, the latter is the correct way. Another early view is based upon Heine's concept of "metal-induced gap states" (Heine (1965)). The idea is that metallic - or semiconductor - states on the right side of an interface, which lie in the gap of a semiconductor on the left side, will have exponentially decaying tails extending into the semiconductor on the left. If these states lie near the conduction-band on the left they will introduce negative charge in the semiconductor, producing a dipole which raises the semiconductor bands on the left. Similarly if they lie near the valence band they produce positive charge shifting the bands downward. Thus these gap states tend to lead towards a matching of some "neutrality level" lying in the gap in each semiconductor. This was in fact the basis of Tersoff's 1984 approach to determining band line-ups. We now believe that one should distinguish between these long-range tails and the polarization of interface bonds. If one does, the effects of the long-range tails are very small (Harrison (1985b)). In addition any intuition based upon these metal-induced gap states will tend to make one guess effects incorrectly. Shifts of bands due to these gap states are dominated by any bands of low effective mass while the real shifts are dominated by high-mass bands since the center of gravity of the bands is the relevant measure of the neutrality point which in fact determines the band line-ups. Further, this view would seem to rule out a system such as InAs-GaSb where the conduction band in the former is found to lie below the valence-band maximum in the latter - a "Type-II line-up" (e. g., Sai-Halasz, Esaki, and Harrison (1978)). In fact the neutrality level does not need to lie in the gap, so this is possible. The fact that InAs-GaSb is the one combination with this Type-II line-up follows from the neutrality-level approach, and the natural-band-line-up approach (Harrison (1985b)).

At any real interface there may be defects and it is certainly true that they may control the band line-up, as suggested by Spicer, Chye, Skeath, Su, and Lindau (1979). The effects of such defects can readily be treated in tight-binding theory, as we have done in this chapter, but it remains to the experimentalist to determine what defects are present. Intentional defects, or interlayers, can also be used to control band line-ups, as discussed by Franciosi and Van de Walle (1996), Saito, Hashimoto, Ikoma (1994), Nicolini, Vanzetti, Mula, Bratina, Sorba, Franciosi, Peressi, Baroni, Resta, Baldereschi, Angelo, and Gerberich (1994),and Saito and Ikoma (1995).

Perhaps because of unintentional defects there has been some spread in experimental line-ups produced by different workers. There have been a wide range of semiempirical and theoretical (e. g., Cardona and Christensen

(1987)) models for determining heterojunction line-ups. From the tight-binding point of view, it is the average hybrid energy which aligns and any theory consistent with that can be viewed as a correct theory. One of particular interest seemed at first not to be consistent: it was found that transition-metal levels could serve as reference levels for band lineups [Zunger (1985), Caldas, Fazzio, and Zunger (1984), Langer and Heinrich(1985), and Tersoff(1986b)] as if these levels provided some absolute reference for the energy bands. Upon careful examination with a self-consistent tight-binding approach (Tersoff and Harrison (1987)) it was found that in contrast, these transition-metal levels were fixed relative to the same dangling-hybrid energies by a charge-neutrality condition arising from the large Coulomb U of the transition metal atom. Thus these levels *should* provide a suitable reference, but only because they were consistent with the same hybrid-matching view.

C. Strain-Layer Superlattices

When semiconductors are grown epitaxially on substrates of different lattice spacing, the added material takes on the transverse spacing of the substrate, unless the added layer becomes so thick that it is energetically favorable to create misfit dislocations which allow the added material to take its native spacing. Without dislocations the strain will shift the band structure according to the deformation potentials which we described in Chapter 7. We illustrate it for a special case, GaP on GaAs (100). It is quite simple to estimate the changes for any system.

The gallium arsenide bond length is 2.44Å, corresponding to a cube edge a of 5.63 Å. Gallium phosphide has a bond length of 2.36 Å, near that of silicon, and a cube edge of 5.45 Å. We take a (001) surface of a gallium arsenide substrate with the z-axis perpendicular to it. Then a thin layer of GaP will acquire a strain $e_1 = e_2 = 2.44/2.36 - 1 = 0.034$. We may determine the strain e_3 which will lead to a stress-free surface (or a surface with no normal stress at a subsequent gallium arsenide layer). That stress, $\sigma_3 = c_{11}e_3 + c_{12}(e_1 + e_2) = 0$. This yields immediately $e_3 = (2c_{12}/c_{11})e_1 = -0.030$, where we have used the experimental elastic constants for GaP from Table 3-1.

These strains provide a net dilatation of $\Delta = 2\times 0.034 - 0.030 = 0.038$. Using the deformation potential constants from Table 7-1 we obtain a shift of the conduction band at Γ downward by 0.26 eV and a downward shift of the valence band by 0.07 eV. This is an indirect band gap semiconductor with a conduction-band minimum near X ; again using Table 7-1 we find that the dilatation lowers those minima by 0.16 eV. This decrease of the net gap by 0.09 eV is significant, even on the scale of the experimental gap of 2.38 eV.

Of more interest are the effects due to shear. We may add to the effects of the dilatation $\Delta = 0.038$, the effects of a shear with $e_1 = - e_3 = 0.021$ and an additional shear $e_2 = -e_3 = 0.021$ which add up to the strains calculated above. These shears do not affect the conduction band at Γ to first order in the shear but they do shift the minima near X according to the deformation-potential constant $D_X(shear) = 5.34$ eV given following Eq. (7-9). These shears cause a raising of the conduction band minima lying in the plane of the interface by 5.34×0.021 = 0.11 eV and a lowering of the conduction-band minima in the direction normal to the interface by 5.34×0.043 = 0.23 eV These are large shifts and at room temperature, $k_BT = 0.025$ eV, the ratio of the occupation of electrons in the two orientations of minima differs by a factor $e^{-0.341/0.025} = 10^{-6}$. Almost all of the electrons will be in the normal valleys (near the two X-points normal to the surface) with very small effective masses for motion along the layer. Such an effect could be important in enhancing the mobility of the conduction electrons. It is interesting that with our parameters an epitaxial layer of silicon on germanium, with very nearly the same lattice-parameter mismatch and a similar $D_{X1}(shear)$ should provide a very similar enhancement of the mobility.

The shear distortion is also important for the valence band. We saw following Eq. (7-11) that with $e_2 = - e_1$ and e_1 positive the band based upon p_x-states shifts by $D_v(shear)\, e_1$, and that based upon p_y-states shifts by the negative of this, with $D_v(shear) = -3.27$ eV for GaP from Table 7-1. (This neglected spin-orbit coupling, to which we return.) Correspondingly superimposing the two shears $e_1 = -e_3 = 0.0213$ and $e_2 = -e_3 = 0.0213$ for GaP on GaAs, the band-edge states based upon p_x- and p_y-states are lowered by $3.27 \times 0.0213 = 0.070$ eV and those based upon p_z- states are raised by twice this. As expected, the lowering of the symmetry of the lattice has lifted the three-fold degeneracy of the top of the valence band.

If we consider states near the top of the band, but with wavenumber in the xy-plane, the states which are raised in energy (p_z-states) are oriented perpendicular to their wavenumber and are heavy-hole states for propagation in the plane of the layer. Holes will concentrate in this highest band and because of the high mass will have low mobility. The alternative case, with reversed strain, can be much more interesting in providing holes of high mobility.

In this case of a thin layer of GaP on GaAs, the splitting due to the strain is considerably larger than the spin-orbit splitting in the GaP. (From Table 5-2 the spin-orbit parameter λ is 0.058 eV for Ga and 0.022 eV for P and the weighting for the valence-band maximum, Eq. (6-28), favors the phosphorus leading to a λ for the compound of 0.027 eV.) Thus it is a good approximation to estimate the spitting due to strain without spin-orbit splitting. We can neglect the effect of spin-orbit splitting or could include its effect in perturbation theory.

If the spin-orbit splitting had been large compared to the strain splitting, we should proceed as in Section 7-2-D and obtain the term linear in strain. For the bands at energy $E_v + \lambda$ we write the state again as $\cos\theta|1\rangle + \sin\theta|2\rangle$, and calculate the first order term in strain which in this case ($e_3 = -e_1 = -e_2$) equals $D_v(shear)e_1(\cos^2\theta - \sin^2\theta)$. The maximum and minimum are at $\theta = 0, \pi/2$ and are $\pm D_v(shear)e_1$. The sum of all three shifts must be zero for this splitting so there is no linear shift of the unsplit band. We would not have obtained the correct result by superimposing the two strains because of this necessity of selecting the correct combination of degenerate states.

The correct combination was $|1\rangle$ (corresponding to $\cos\theta = 0$) for the state of energy $E_v + D_v(shear)e_1$ and contains no $|p_z\rangle$ so it has the lighter mass. The combination at $E_v - D_v(shear)e_1$ is the state $|2\rangle$ which will have a heavier mass. As in the analysis without spin-orbit splitting, with the negative $D_v(shear)$ for all materials listed in Table 7-1 except for CuBr, holes would accumulate in the high-mass band.

If the strain had been reversed, raising the light-hole band above the heavy-hole band, there would have been an additional complication. Simply shifting the bands of Fig. 6-4 b in this way would lead to bands crossing at some wavenumber near the top of the band. It is possible that such a crossing could occur along a symmetry line, as do the lower conduction bands for tin along the line Γ-X in Fig. 5-6. In more general directions the crossing will be avoided, as are the lower conduction-band crossings along Γ-X in the InSb bands of Fig. 5-6. For many properties, we will be interested only in the states quite near the band maximum in any case.

D. Metal-Semiconductor Interfaces

We use the same basic approach for aligning the bands at a metal-semiconductor interface. We may in fact imagine two semiconductors in contact and simply let the gap in the second semiconductor go to zero, with its neutrality level remaining near the gap and becoming the Fermi level.

Thus the Fermi level in the metal plays precisely the role of the dangling-hybrid energy and at the interface we expect the Fermi level of the metal to come at the neutral point of the semiconductor. Indeed this is essentially the idea used in the formulation of a neutral point for a semiconductor, Tersoff (1986b). Thus at the interface with a metal, the numbers E_v - $\bar{\varepsilon}_h$ from Table 19-1 become the energy of the valence band relative to the Fermi energy of the metal. We obtain the conduction-band edge relative to the Fermi energy in the metal by adding the gap (from Table 6-1) to the value in Table 19-1.

Deep in the semiconductor, if no voltage is applied, the Fermi level must equal to the Fermi level in the metal. For n-type semiconductors that Fermi level is near the conduction-band edge. Then the band diagram must look as illustrated in Fig. 19-7. An applied voltage raising the energy on the left appears as a field in the central region where there are few carriers (because the Fermi energy is far from the band edges) and more and more electrons in the bulk semiconductor have sufficient energy to overcome the barrier, called a *Schottky barrier*, appearing at the interface. On the other hand a voltage lowering the Fermi energy on the left leaves the barrier for flow of electrons from the metal unchanged, and little current flows. This provides the rectifying character of metal-semiconductor contacts.

It has in fact been long know that the Schottky-barrier heights for different semiconductors correlated with the band line-ups between those semiconductors, and this feature follows immediately from the point of view described here.

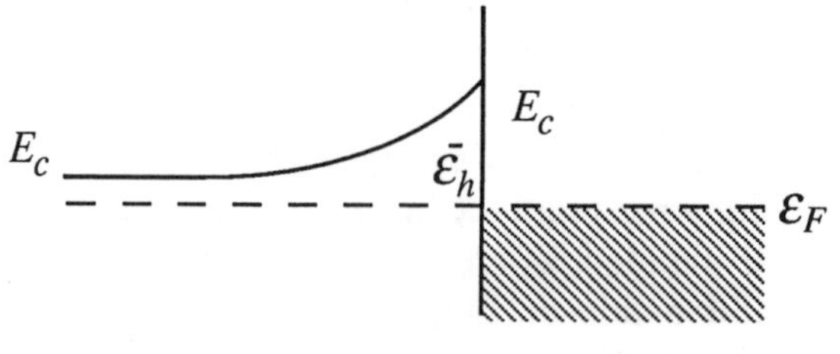

Fig. 19-7. At a semiconductor metal interface the neutrality level of the semiconductor, $\bar{\varepsilon}_h$, in tight-binding theory, lines up with the Fermi energy ε_F of the metal at the interface. In thermal equilibrium the Fermi energy must be a constant of the system so charge redistribution shifts the conduction band edge E_c deep within the semiconductor to its bulk position, the transition region typically extending over hundreds of Ångstroms. In n-type material that produces a *Schottky barrier* , equal to E_c - $\bar{\varepsilon}_h$, to electron flow into, but not out of, the semiconductor.

F. Adsorbed Atoms on Semiconductors

Also of interest is the question of individual atom adsorption on a semiconductor surface. This does not concern interfaces, but perhaps fits best here. There has been a great amount of study of such systems, both experimentally and theoretically. On the (100) surface of silicon alone there have been studies of acceptors such as Al, Ga, and In, which adsorb onto the silicon surface (Nogami, Baski, and Quate (1991), Northrup, Schabel, Karlsson, and Uhrberg (1991)) and donors such as P, As, and V, which incorporate into the surface (Yu and Oshiyama (1993), Wang, Chen, and Hamers (1994)). There are also studies of antisite defects near the surface by Van Schilfgaarde, Weber ,and Newman (1994).

A careful tight-binding view, rather than general bonding considerations, can be essential. If we ask for example whether an oxygen atom should adsorb on a gallium or on an arsenic site at a gallium arsenide surface, a general consideration might have us compare the gallium-oxygen bond energy with the arsenic-oxygen bond energy and guess that the oxygen would go to the gallium. However, if we start with the tight-binding view of this surface as a doubly occupied arsenic dangling hybrid and an empty gallium dangling hybrid, we are led to expect the oxygen to go instead to the arsenic. For this case it is the only place to acquire electrons, irrespective of the question of reactions of oxygen with elemental gallium and arsenic.

In another instance a tight-binding consideration is needed even to classify different surface-bond types. We can imagine a covalent bond to a surface forming as an atom approaches the surface and the coupling produces occupied bonding states and empty antibonding states. If the main contribution comes from the transfer of an electron across to a state of lower energy we would classify it as an ionic bond. However, a third possibility arises if the coupling between occupied levels on both sides splits them, but for example the occupied level on one side rises and the electron is transferred to a lower level on the same side. Such a *rearrangement bond* occurs because of a redistribution of electrons in the substrate induced by the presence of the surface atom, rather than by a direct bonding between the substrate and surface. The bonding due to the image charge in a metal may be thought of as such a rearrangement bond. It may also be useful to think of bonding for example of inert gas-atoms to transition metals as arising from redistribution between s-like and d-like states in the substrate.

Another interesting case is the bonding of various atoms to simple metals, treated by Lang and Williams (1976) and discussed in Harrison (1980). One may look at the energies of the various states obtained from their calculation and find, for example, that for nitrogen on aluminum, the nitrogen s-state drops in energy while the p-state remains near its atomic value. It would however be a mistake to associate the bonding with the nitrogen s-state, since if it were removed from the problem - by artificially making ε_s very deep in a tight-binding calculation - one would find the bonding energy little changed. It would seem better to say that the real bonding came from coupling with the p-state, but that its asymmetric environment coupled the fully occupied s-state and fully occupied p-states, pushing them apart. The net effect on the energy is small because both are occupied, but the rising of the p-state cancels against the lowering of the same p-state due to bonding .

Such arguments can be made quantitative with universal parameters, and care with the Coulomb shifts. Not so much has been done using this approach (Klepeis and Harrison (1988)), but in one study (Klepeis and Harrison (1989)) found that for Cs and Au on gallium arsenide (110), the arsenic site is favored at very low coverages but at coverages of the order of one-tenth monolayer the gallium site is favored. The latter was consistent with tunneling-microscope studies by Feenstra (1989), but the former has not been tested. Such a prediction requires at least a systematic tight-binding study, if not a full local-density calculation.

Problem 19-1. Return to the lithium chain

In preparation for treating truncated systems we return to the lithium chain discussed in Section 1-3, and the lithium molecule discussed in Section 1-2. For the molecule we wrote the energy $2V_{ss\sigma} + C/d^4$ and chose C such that the minimum energy with respect to d came at $d = d_0 = 2.67$ Å .

For the lithium chain we obtained a total energy per atom, relative to isolated atoms, of $-(5\pi^2/24)\,\hbar^2/(md^2)$ plus a repulsive term which canceled half of that energy. Assume that the same C is applicable for the infinite chain as for the molecule and predict the spacing in the chain which will have minimum energy. It is somewhat larger than the spacing in solid lithium.

[This same set of assumptions could allow us to find the change in spacing for short chains, or departures from uniform spacings at the end of a chain.]

Problem 19-2. Surface energies and surface interactions

Now let the chain from Problem 19-1 be a finite chain of N atoms, as discussed in Section 19-1-C, with a fixed spacing d appropriate to the infinite chain. You need only work out the problem for N an even integer. Then the sum of free-electron energies made for the finite chain will become a constant times a sum $1^2 + 2^2 + \ldots (N/2)^2$, for which there is a formula in most sets of mathematical tables.

Use this to evaluate the total energy as a function of N exactly, including the C/d^4 terms for each set of nearest neighbors. Expand the result in $1/N$, which should give a leading term (in the total energy for N atoms) proportional to N, appropriate to the result obtained earlier for the long chain with periodic boundary conditions. There is also a constant term equal to the "bond-breaking" energy, which can be compared with the energy per bond for the long chain.

The remaining terms may be regarded as interaction between the ends, decreasing in powers of $1/N$. [Redoing the problem for odd N would require an additional singly occupied state added to the sum above. This would lead to an oscillatory interaction as we found in three dimensions in Section 19-1-D.]

CHAPTER 20

Other Systems

The variety of solids found in nature, or in the laboratory, is extraordinary. We may have given the appearance of covering all in treating metals, semiconductors, and insulators, but even then transition-metals and f-shell metals entered as special cases. We turn now to a few systems for which new features arise, so they did not fit well in the categories already discussed. The same fundamental concepts, the tight-binding basis in particular, remain appropriate, but the outlooks which make them simple are often different.

1. Nontetrahedral Covalent Solids

A. Two Nearest Neighbors, S, Se, and Te

We begin with covalent solids with other than four nearest-neighbors. When there are less than four neighbors, we may still construct sp^X-hybrids

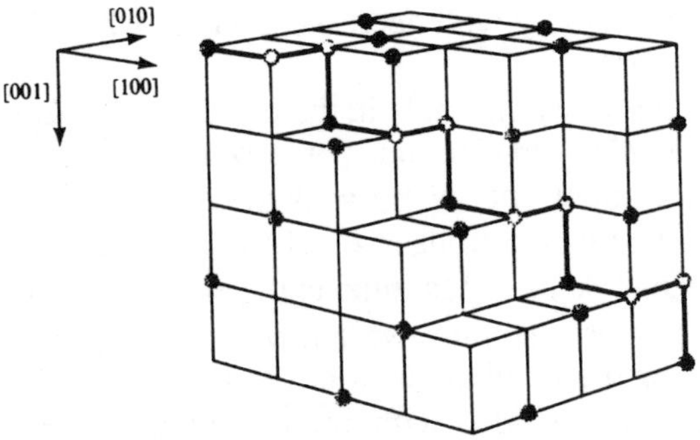

Fig. 20-1. The crystal structure of Se and Te can be obtained beginning with a simple-cubic structure, as shown. One chain is indicated by heavy lines, with atoms forming a spiral around a [111] axis. In the selenium and tellurium structure the chains are stretched along their axis so that the angles between bonds are greater than 90° and the spacing between atoms in different chains becomes larger than that within each chain. (After Harrison (1980), p. 93)

to the X nearest neighbors, but will have combinations left over. This was the case at semiconductor surfaces, for the oxygen atoms in SiO_2, and for graphite, to which we return in Section 20-3.

This is also the case for Se and Te, and apparently for one structure of S, for which each atom has two nearest neighbor at approximately 90° from each other. The resulting chains spiral around a [111] direction in an approximately cubic grid, as shown in Fig. 20-1. In the real structure of Se, (these structures were obtained from Wyckoff (1963)) the chains are stretched along their axis so that the angles become 105° and the distance between the nearest atoms in neighboring chains (3.46 Å) is considerably larger than the distance within chains (2.32 Å). For Te the angle is stretched only to 102° and the Te-Te distance within a chain is 2.86 Å.

We may understand the electronic structure by constructing sp-hybrids from any atom to its two neighbors in the chain. If the angle between these neighbors is 105°, each must be a combination $\cos\gamma|s\rangle + \sin\gamma p\rangle$ with $\cos^2\gamma + \sin^2\gamma \cos 105^o = 0$ for orthogonality. The solution is $\sin\gamma = 0.89$, $\cos\gamma = 0.45$. Such hybrids are 80% p-state. The coplanar dangling hybrid must be 60% s-state in order to be orthogonal to the others. With six electrons from each atom, one is contributed to the bond with each of the two neighbors, two electrons go into this dangling hybrid, and two into the π-state perpendicular to the plane of the three hybrids. It is the doubly occupied dangling hybrid, with energy dropping rapidly with decreasing angle, which has brought that angle down towards 90°.

We might in fact predict the angle in the Bond-Orbital Approximation by writing the energy of the two electrons per hybrid as $2\cos^2\gamma\,\varepsilon_s + 2\sin^2\gamma\,\varepsilon_p - 2V_2$, with the covalent energy $V_2 = -V_{ss\sigma}\cos^2\gamma + 2V_{sp\sigma}\cos\gamma\,\sin\gamma + V_{pp\sigma}\sin2\gamma$. Then the two electrons in the dangling hybrid have energy $2[(1 - 2\cos^2\gamma)\varepsilon_s + (2 - 2\sin^2\gamma)\varepsilon_p]$. Minimizing the energy with respect to angle, using the parameters for selenium, we obtain a minimum too close to 90° (91.5°) as we did for the bond angle at oxygen in SiO_2, again because the hybrid energy rises so rapidly with increasing angle θ (through the dependence of γ on θ) while V_2 varies little.

In spite of this inaccuracy in the prediction of angle we can, as in SiO_2, use hybrids, at the observed angle, to directly address other bonding properties, such as the radial force constants and the angular rigidity and hope to have comparable success to that we had for SiO_2. We could also estimate dielectric properties and incorporate interbond coupling for an elementary description of the energy bands, or could incorporate also the coupling of the π-states with the second-neighbor bonding and antibonding states on nearest neighbor atoms, etc. All aspects are accessible using the parameters and matrix elements from Chapter 1, and approximations such as we used for SiO_2 in Chapter 12. We have not carried out such studies, though analogous studies have been made by Joannopoulos, Schlüter, and Cohen (1974).

B. Multicenter Bonds

The electronic structure cannot so directly be generalized to crystals with more than four nearest neighbors since, with only four orbitals, we cannot construct orthogonal hybrids to all of the equivalent neighbors, and the two-center Bond-Orbital Approximation is not possible. Nor would adding an excited state (s or d), which is in fact not important in the electronic structure, allow a meaningful approximation. When this situation arose for ionic crystals, an alternative approximation was available based upon the fact that the states on one atom type were far removed in energy from those on the other atom type, and perturbation theory could be used. When there are more than four neighbors for covalent solids that option is not available.

A study of such systems was undertaken by van Schilfgaarde (van Schilfgaarde and Harrison (1986)) and we follow his approach. Perhaps the most important finding in van Schilfgaarde's study was that, not only are two-center bonds ruled out with coordination greater than four, but that it is no longer possible to separate out the nonorthogonality of the orbitals as an effective two-body repulsion. Such a separation for four-fold coordination actually *follows* from a Bond-Orbital Approximation since each pair of neighbors is treated independently. It also follows for ionic crystals to the extent that second-order perturbation theory applies. It cannot be justified either way here and in fact van Schilfgaarde found that a separate overlap repulsion did not give good predictions. We made that separation in Section 2-5-B and found it useful for discussion of structural energy differences, but do not make it here. We simply keep the nonorthogonality in the calculation as in Section 1-2-A. A very careful reformulation of the entire tight-binding scheme keeping the nonorthogonality explicitly has been made more recently by Menon (1997) and by Menon and Subbaswamy (1997). It was found to enhance the transferability of the matrix elements to different situations, in particular as applied silicon. Menon and Subbaswamy used it as a basis for molecular-dynamics calculations.

We again expand the states for the solid in terms of atomic orbitals $|\psi\rangle = \Sigma_{j=1,N}\, u_j\, |j\rangle$ as in Section 1-3, but no longer regard the atomic orbitals $|j\rangle$ as orthogonal. Then when we write the total energy as in Eq. (1-18) we keep the nonorthogonality terms

$$\varepsilon = \frac{\langle\psi|H|\psi\rangle}{\langle\psi|\psi\rangle} = \frac{\Sigma_{ij}\, u_i{}^*u_j\, H_{ij}}{\Sigma_{ij}u_i{}^*u_jS_{ij}}\ , \tag{20-1}$$

where we have written the matrix elements of the Hamiltonian $H_{ij} = \langle i|H|j\rangle$. As before we take H_{ii} as the free-atom term value and H_{ij} are nearest-neighbor matrix elements to be determined. Similarly, $S_{ij} = \langle i|j\rangle$ and equals one for $i = j$ and is the nonorthogonality for nearest-neighbor orbitals. Minimizing the energy, as in Eq. (1-19), gives the set of equations (numbered by i)

$$\Sigma_j\, (H_{ij} - \varepsilon S_{ij}) = 0\ . \tag{20-2}$$

Extended Hückel-Theory (Hoffmann (1963)) assumes the H_{ij} are proportional to $S_{ij}(H_{ii} + H_{jj})$, obtains the S_{ij} by direct integration of tabulated atomic wavefunctions, and solves Eqs. (20-2). For the parameters, we instead follow van Schilfgaarde and Harrison (1986) in working back from our universal parameters using that Extended Hückel Theory to obtain H_{ij} and S_{ij} We then assume that the H_{ij} and S_{ij} associated with atomic states transfer to the other structures, though the universal matrix elements ($V_{ss\sigma}$, etc.) may not, and again solve Eq. (20-2)

For the elemental tetrahedral solid we used Extended Hückel Theory to say that the coupling between two hybrids is

$$-V_2 = K\varepsilon_h S_2 \tag{20-3}$$

so that both V_2 and S_2 had the same $1/d^2$ dependence and the overlap repulsion from two bond electrons becomes $V_0 = -2S_2V_2 = -2V_2^2/(K\varepsilon_h)$. Van Schilfgaarde and Harrison (1986) also did that and minimized the total energy per bond, $E_{tot}/bond = 2V_1 - 2V_2 - 3V_1^2/(2V_2) - 2V_2^2/(K\varepsilon_h)$ from Eq. (2-30) at the equilibrium spacing to obtain

$$K = 1.63 \text{ for diamond,} \qquad K = 1.08 \text{ for silicon,}$$
$$K = 1.07 \text{ for germanium, and} \qquad K = 0.93 \text{ for tin,} \tag{20-4}$$

to be used for each entire row. These are remarkably close to the values 1.75, 1.06, 0.99, and 0.83 which we gave after Eq. (1-16), obtained by fitting universal parameters to Eq. (20-3) with $S = 0.45$. Then the universal matrix elements are taken to be from the generalization of Eq. (20-3) as

$$V_{\ell\ell'm} = -K S_{\ell\ell'm}\frac{\varepsilon_\ell + \varepsilon_{\ell'}}{2}, \tag{20-5}$$

which gives us values for $S_{\ell\ell'm}$ in terms of our universal parameters and the term values from Table 1-1.

We had originally defined our universal matrix elements for orthogonalized orbitals from the splitting in Eq. (1-13) as

$$V_{ss\sigma} = \frac{\langle s|H|s'\rangle - S_{ss\sigma}\varepsilon_s}{1 - S_{ss\sigma}^2}. \tag{20-6}$$

It would be natural to generalize this as $V_{\ell\ell'm} = [\langle \ell,m|H|\ell',m\rangle - (\varepsilon_\ell + \varepsilon_{\ell'})S_{\ell\ell'm}/2]/[1 - S_{\ell\ell'm}^2]$ and use it to evaluate the matrix elements $\langle \ell,m|H|\ell',m\rangle$ which we seek. However, it would not then be true that $-V_2 = [\langle h|H|h'\rangle - \varepsilon_h S_2]/[1 - S_2^2]$. Therefore, van Schilfgaarde and Harrison kept the S_2 in the denominator and substituted for $S_{\ell\ell'm}$ from Eq. (20-5) to obtain

$$\langle \ell\, m|H|\ell'\, m\rangle = V_{\ell\ell'm}\,(1 + {}^1/_K - S_2^2). \tag{20-7}$$

from which the $<\ell\, m|H|\ell'\, m>$ can be obtained, and which satisfies $-V_2 = [<h|H|h'> - \varepsilon_h S_2]/[1 - S_2^2]$.

Having the $S_{\ell\ell' m}$ and the $H_{\ell\ell' m} = <\ell\, m|H|\ell'\, m>$ we may solve Eq. (20-2) for the electronic energy eigenvalues, and total energies, as we do for one case in Problem 20-1. It is the $H_{\ell\ell' m}$ and $S_{\ell\ell' m}$ which we might reasonably expect to be transferable to systems of higher coordination, allowing improved theoretical treatment of a very large number of systems. This expectation seemed to be confirmed by van Schilfgaarde's calculations on Zn_3P_2 and boron, which we discuss next, and by Problem 20-1.

C. Zn_3P_2

A good case for testing the multicenter-bonding theory given in the preceding Part is Zn_3P_2 , in which the zinc atoms are tetrahedrally coordinated, but the phosphorus atoms have six zinc neighbors and independent bonds cannot be formed. We may nevertheless form a bonding unit, as we did for impurities in Section 8-2 and SiO_2 in Section 11-2. If sp^3-hybrids are formed on the six zinc atoms surrounding a given phosphorus, the bonding unit consisting of the six inward-pointing Zn hybrids and phosphorus s- and p-states (a $Zn_{6/4}P$ unit) bonding units will be only weakly coupled to neighboring bonding units. One such bonding unit is illustrated in Fig. 20-2. The structure itself is rather complicated, 40 atoms per primitive cell, but can be described as face-centered-cubic stacking of such units in various orientations. Three other II-V compounds, Cd_3P_2, Zn_3As_2, and Cd_3As_2 have similar structures.

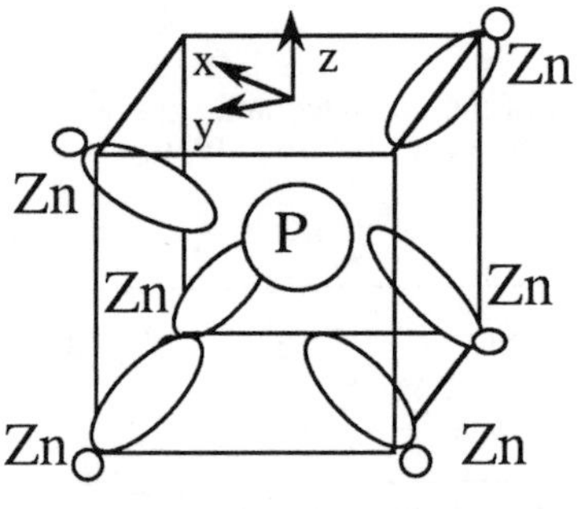

Fig. 20-2. A single bonding unit of Zn_3P_2, with a central phosphorus atom and inward-pointing sp^3-hybrids from the six surrounding zinc atoms. Eliminating two zincs could produce a tetrahedral structure; adding two could produce a fluorite structure.

The solution for the states of this bonding unit proceeds much as that for tetrahedral impurities in Section 8-2. Combinations of the six zinc hybrids are made which have s- p_x, p_y, and p_z symmetry (with the x-axis parallel to the line between the two vacancies on the upper plane of Fig. 20-2, and y-axis along the line between the two atoms on that plane), and two remaining nonbonding combinations. Then bonding and antibonding combinations of each are formed, by solving quadratic equations as with two-center bonds. Finally, a small residual coupling between the states of s- and p_z-symmetry was included by perturbation theory. There are just enough electrons to fill the bonding levels, leaving the nonbonding and antibonding levels empty. These then broaden into bands, if coupling between bonding units is included, forming a semiconductor.

Van Schilfgaarde evaluated these energies from Eq. (20-2), neglecting the coupling between bonding units and using the $H_{\ell\ell'm}$ and $S_{\ell\ell'm}$, obtained essentially as in Part B, to obtain the energy as a function of the phosphorus-zinc spacing d and thus a prediction of the spacing, the cohesion, the bulk modulus, and the Grüneisen constant. His predictions are compared with experiment in Table 20-1. (Theoretical estimates, except of the spacing, were evaluated at the predicted equilibrium spacing.) Note that the prediction of spacing is made without any experimental input for this compound, as is customary in Extended Hückel Theory where nonorthogonality provides the needed overlap repulsion explicitly. The agreement in all cases is very good.

Van Schilfgaarde also evaluated the energy as a function of spacing using the universal $V_{\ell\ell'm}$ (as for orthogonal orbitals) directly but adding a repulsion of $6V_0(d) = 12\ S_2V_2$ and found that it did not even produce a minimum in energy at any spacing. If he scaled this V_0 down by a factor of $3/4$ to obtain a minimum at the observed spacing, he obtained quite good predictions for

Table 20-1. Comparison of the predicted and experimental parameters for Zn_3P_2 (after van Schilfgaarde and Harrison (1986)).

	d(Å)	Cohesion (eV)	B/B(Ge)[a]	γ
Multicenter Bond	2.44	9.5	4.4	1.8
Experiment	2.48	10.3	5.2	1.4

[a] The bulk modulus is shown relative to that of Ge (for the same theory, and for experiment) since the theory seemed to underestimate the bulk modulus by a factor of about three for both cases.

the other three quantities, 10.6 , 5.1, and 1.9, respectively, suggesting that the error in using V_0 is principally in the scale.

He went on to include metallization in the estimate, which made only small contributions. This is as expected because such corrections decrease with the size of the bonding unit. He also generated second-neighbor matrix elements by fitting free-electron bands to a tight-binding model with second neighbors, much as we fit free-electron bands to a tight-binding model with only nearest neighbors. This seemed necessary to obtain good energy bands, but did not greatly affect the other predictions.

D. Boron

Elemental boron contains regular icosahedra of twelve boron atoms, with each boron having five nearest neighbors in this icosahedral unit. The stable structure of the element is essentially a face-centered-cubic arrangement of such icosahedra. The presence of the five nearest neighbors to each boron indicates multicenter bonding as described in Part B and its treatment was undertaken by van Schilfgaarde (van Schilfgaarde and Harrison (1985), (1986)).

The electronic structure was explored by treating first the isolated icosahedra, and then the coupling between neighboring icosahedra. Because one cannot construct independent bonds, the calculation of the states needed to be done for the entire icosahedron, making the solution for the eigenvalues more difficult than for the smaller unit shown in Fig. 20-2 for Zn_3P_2. However, it was possible to simplify the tight-binding electronic structure considerably using an approach suggested by Longuet-Higgens and de V. Roberts (1955) in which inward and outward directed hybrids at each atom were formed using the s-state and a radial (with respect to the icosahedron) p-state. They then sought orbitals based upon the twelve inward-directed hybrids and the twenty four (two for each boron atom) tangential hybrids, finding that this gave 13 deep bonding levels and 23 high antibonding levels. Thus 26 of the total of 36 valence electrons are accommodated in these deep bonding levels. The remaining ten electrons were used in two kinds of bonds between icosahedra, constructed from the outward-directed hybrids. Six of these were simple two-center bonds with neighboring icosahedra. In addition six outward-directed hybrids took part in bonds between three icosahedra (called delta bonds), with each icosahedron contributing two-thirds of an electron to each. Thus the analysis made by van Schilfgaarde utilized twelve-atom icosahedral bonding units based upon the inward and tangential states, plus three simple bonds per icosahedron, plus two delta bonding units per icosahedron.

It is possible to simplify the calculation of the icosahedral bonding unit using group theory (which we shall not do, but describe for the benefit of those conversant with it) and the high symmetry of the icosahedron, but the irreducible representations for this group include a four-dimensional and a five-dimensional representation. It was interesting that van Schilfgaarde (van Schilfgaarde and Harrison (1985)) found that the calculation was easier using only five-fold rotations around a single axis and the inversion through the center of the icosahedron. Then combinations of orbitals of the form $\Sigma e^{-i\phi n}$, with ϕ being the angle to each atom measured around this single axis, could be made with $\phi = 0, \pm\pi/5$, and $\pm 2\pi/5$, as in the atomic chain in Section 1-3. Each of these states could be made either to be odd or even with respect to an inversion through the center of the octahedron. Since combinations of different ϕ do not couple to each other, it was possible to reduce the calculation to the solution of quadratic equations, derived from Eq. (20-2), using values for the $H_{\ell\ell' m}$ and $S_{\ell\ell' m}$ obtained as in Part B. It may frequently be true that discarding some of the symmetry can simplify the calculation in this way; it is also be easier to treat a simple six-member ring (or the π-states in benzene) as in Section 1-3, with such propagating solutions, rather than working with the two-dimensional irreducible representations of the hexagon.

Having obtained formulae for the energies of these 13 occupied states of the octahedron, the total energy was written as a function of the boron-boron distance d , which allows prediction of the bond length, cohesion, spring constant and Grüneisen constant. These are compared with experiment in Table 20-2, in accord comparable to other systems, a factor of two in the cohesion as for other first-row elements.

Table 20-2. Comparison of the predicted and experimental bond length, cohesion, nearest-neighbor spring constant and Grüneisen constant for boron (after van Schilfgaarde and Harrison (1986)).

	d(Å)	Cohesion (eV)	k (eV/Å^2)	γ
Multicenter Bond	1.85	11.9	274	1.8
Experiment	1.78	5.77[a]	263	-

[a] The experimental value includes the other two-center and three-center bonding units in the total energy, presumably considerably smaller than the internal bonding of the icosahedron.

Van Schilfgaarde also made these evaluations using an overlap repulsion of $2\,S_2V_2$ rather than including the nonorthogonality $S_{\ell\ell' m}$ explicitly. This gave values for the four parameters of Table 20-2 of 1.54 Å, 27.0 eV, 1266 eV/Å^2 and 2.0. These very poor predictions confirm the finding for Zn_3P_2 that the use of the tetrahedral overlap repulsion is quite inadequate for multicenter bonds of this type. He also tried scaling the V_0 to obtain the correct spacing and found values for the cohesion (19.6 eV) and spring constant (714 eV/Å^2) still in poor accord with experiment.

More extensive studies of boron by van Schilfgaarde and Harrison (1985) included prediction of the two-center and three-center bond lengths, the properties of the compound B_4C, and vibrational modes of the icosahedral molecular ions $B_{12}H_{12}^{=}$, $B_{12}D_{12}^{=}$ and $B_{12}Cl_{12}^{=}$.

2. Inert-Gas Solids and Molecular Crystals

Since inert-gas atoms have full valence shells, there is no ordinary bonding between them and since they are not charged there is no Coulomb attraction between them. In this section we do not discuss the inert-gas helium, which has a full s-shell, rather than a full p-shell, and it is a very interesting special case of its own. For the other inert gases there are covalent contributions to the bonding from the coupling between the occupied valence p-states and the empty excited states on neighboring atoms, but these seem not sufficient to bring the atoms close together and the dominant contribution to cohesion seems to be the remote van-der-Waals interaction which we shall estimate in Part D.

In the same way, molecules may have all bonding orbitals occupied (and perhaps some nonbonding orbitals) so that direct bonding between molecules is not important, and they are held together in a solid by van-der-Waals forces. We discuss the central features of both of these systems here, but have not made very extensive applications to molecular crystals.

A. Electronic Structure of Inert-Gas Solids

We have listed the valence term values for the inert-gas atoms in Table 1-1, $\varepsilon_p = -16.08\ eV$, $\varepsilon_s = -34.76\ eV$ for Ar. We imagine bringing such atoms together as a face-centered cubic solid, (the observed structure for all but helium) but not close enough that the coupling between states on adjacent atoms is important. Then to excite an electron so that it would conduct electricity we must free it from the atom, requiring 16.08 eV in argon. That would not free the electron from the crystal because it would be bound by image forces to the crystal as discussed in Section 19-2-A. In Table 20-3 we list the nearest-neighbor distance (fcc), the magnitude of the p-state energies from Table 1-1, and the experimental ionization energy of the atom to which it approximately corresponds. These last two are a purely theoretical and an experimental estimate of the band gap.

Pantelides (1975b) noted that this band gap for the solid varies almost exactly from one element to the other as $1/d^2$. This is seen in the essentially constant values of $-md^2\varepsilon_p/\hbar^2$ in the final column of Table 20-3, in spite of a variation in ε_p by a factor of two. This is not surprising in the context of other systems we have discussed, but is quite startling as a correlation for the free-atom ionization energy. This can be made plausible, as he suggested, by thinking of the valence p-state as resembling a free-electron state within a sphere, and that sphere radius being related to the internuclear distance in the atom as γd with γ a constant of order one. It is of some interest to carry that further since it gives approximate wavefunctions which could be used for other properties. Free-electron states in a sphere can be taken as a constant times $j_\ell(kr)\,Y_\ell^{\,m}(\theta,\phi)$, as in the expansion of plane waves in terms of these spherical Bessel functions and spherical harmonics in Eq. (15-9). For a p-like state $Y_\ell^{\,0}(\theta,\phi)$ is a constant times $\cos\theta$, and

$$j_1(kr) = \frac{\sin kr}{(kr)^2} - \frac{\cos kr}{kr}\ . \qquad (20\text{-}8)$$

The lowest energy p-state confined to a sphere of radius γd would have its first node at γd, corresponding to $\gamma kd = 4.49$ obtained from Eq. (20-8). We might think of this as a pseudowavefunction for the free-atom p-state as discussed in Section 1-6-A. Then we would estimate the kinetic energy for

Table 20-3. Properties of inert-gas solids. d is the internuclear distance, all in the face-centered-cubic structure. ε_p is the free-atom term value from Table 1-1, E_i is the experimental ionization energy of the free atom, and E_{ex} the experimental exciton energy for the solid.

	d(Å)	$\|\varepsilon_p\|$ (eV)	E_i (eV)	E_{ex} (eV)	$-md^2\varepsilon_p/\hbar^2$
Ne	3.13	23.14	21.56	-	29.8
Ar	3.76	16.08	15.76	12.1	29.8
Kr	4.01	14.26	14.00	10.2	30.1
Xe	4.35	12.44	12.13	8.4	30.9

Sources: d from Kittel (1976); E_i from Moore (1949, 1952); E_{ex} from Baldini (1962)

this state as $\hbar^2 k^2/2m$ (the extra kinetic energy associated with the orthogonality to the core states is canceled by the extra potential arising from the core) and use the virial theorem (Section 2-2-E) to suggest a potential energy minus twice this, or an energy $\varepsilon_p = -\hbar^2(4.49/\gamma d)^2/2m$. Using the final column in Table 20-3 gives a value of $\gamma = 0.58$ for all four elements. This provides the correlation of ε_p with $1/d^2$, and perhaps a reasonable sphere radius at 58% of the internuclear distance.

It is interesting also to construct an s-state for this sphere, and we do that for a 2s-state, in the same sphere as for the 2p-state from Eq. (20-8). This does not correspond to the 1s-state envisaged in pseudopotential theory, but one might easily guess that the kinetic energy for the pseudowavefunction $\phi(r)$ shown in Fig. 1-12 was better given by a free-electron state with one node than with none. With this choice $\gamma kd = 2\pi$ and the s-state energy is found to be $(2\pi/4.49)^2 = 1.96$ times the ε_p just obtained. This match to the general $\varepsilon_s = 2\varepsilon_p$, which holds also for the term values for the inert-gas atoms in Table 1-1, suggests that such states could be useful for estimates of other properties.

B. Excitons

If we were instead to excite an electron from the valence 3p-state to the excited 4s-state, it would take less energy. Such a state in a solid is called an *exciton*. We could take a Bloch sum of such states, $\Sigma_j |\Psi^*{}_j\rangle e^{i\mathbf{k}\cdot\mathbf{r}_j}$, with $|\Psi_j{}^*\rangle$ representing a crystal with all electrons in the ground state except for one such exciton on the atom at position $\mathbf{r}_j$. If we included some coupling between exciton states on neighboring atoms such an exciton would propagate through the crystal, carrying energy but no charge. Excitons are more usually described in terms of an electron in a conduction band bound to a hole in the valence band in a hydrogen-like state which can also move through the crystal; the physics is essentially the same.

We could estimate the energy of such an exciton with the atomic calculation for that excited state. [We hesitate to associate it with the 1s-state in the free-electron estimate above, which would give -7.7 eV below the vacuum level.] It is better to estimate it as the energy for the transition from the $3s^23p^6$ configuration to the $3s3p^54s$ configuration from Moore's *Atomic Energy Levels* (Moore (1949), (1952)). Those experimental tables give an excitation energy of 11.55 eV, 4.2 eV below the experimental ionization potential of 15.76 eV. We have included the experimental energy of the exciton E_{ex} in the solid in Table 20-3, and see that the value of 12.1 eV for solid argon is in reasonable accord with the 11.55 eV we obtained from Moore for the atom.

Thus our picture of the electronic structure of argon is a filled narrow p-band with a free-electron-like conduction band some 16.08 eV above it, and a bound exciton state some 4 eV below the gap. The optical properties should show an absorption peak near 12 eV and a continuum absorption beginning some 4 eV above that. In this case, it is perhaps better to say that the internuclear distance in the solid scales as the inverse square root of the atomic term values, rather than that the band gap scales as the inverse square of the internuclear distance since we think of the atomic size as determined by the valence states rather than the other way around.

C. Dielectric Polarizability

We may think of the dielectric constant as the response of polarizable atoms at the nuclear positions. This polarizability of argon should be dominated by the coupling, for a field along the z-axis, between the valence 3p-electrons with zero angular momentum along the z-axis and the excited 4s-state. There are additional contributions from the coupling to higher s-states and d-states. Our earlier calculations [based upon the Orbital Correction Method (Harrison (1973b)), discussed in Harrison (1980), 546, ff] have suggested sizable contributions from coupling to the continuum of free-electron states. We are motivated to make some study of this partly because of the role the polarizability will play in the van-der-Waals interaction.

The polarizability α of an atom could be calculated by writing the shift in energy in the presence of an electric field $\boldsymbol{E}$ as $-{}^1/_2\alpha E^2$ (analogous to Eq. (4-18) for semiconductors) and equating this to the shift in energy due to coupling with the excited states $|s^*\rangle$ calculated in perturbation theory and including a factor of two for spin,

$$-\frac{1}{2}\alpha E^2 = 2\Sigma_{s^*} \frac{\langle s^*|eEz|p\rangle^2}{\varepsilon_p - \varepsilon_{s^*}} . \tag{20-9}$$

We have suggested a form for the p-state given by Eq. (20-8) with $k = 4.49/\gamma d$ and the integration continuing only to $r = \gamma d$. A reasonable approximate model for the excited states might be the free-electron form $\sin k'r\ /k'r$, summed over all k'. In any case, the matrix elements are expected to be of the order of γeEd and the denominator to be of order of $\varepsilon_p = \hbar^2k^2/2m$ so that the polarizability should be of the form

$$\alpha = \frac{me^2}{4.49^2\hbar^2}(\gamma d)^4 \tag{20-10}$$

with γ the same for all inert gas elements. (The numerical factors could have been chosen different ways, and we have chosen them so that the approximate magnitudes are correct.) The values obtained with Eq. (20-10) are compared with experiment in Table 20-4. The experimental values vary more nearly as d^7 than the predicted d^4. With that discrepancy, and our rather arbitrary choice of scale factor (taking the matrix elements equal to γeEd) we cannot regard the theory as very successful, but perhaps the physics is essentially correct, and the experimental values will be useful. There were similar discrepancies for the dielectric constants of semiconductors discussed in Section 4-2. It could be interesting to explore the same theory for the polarizability of other closed-shell ions, such as Na^+ and Cl^-, many of which have experimental values tabulated by Kittel (1976), p. 411. It might clarify the discrepancy in the dependence upon d, but that has not yet been done.

Table 20-4. Polarizability of inert-gas atoms, α in Å^3, from Eq. (20-10) and from experiment

	d(Å)[a]	Eq.(20-10)	Experiment[a]
Ne	3.15	1.04	0.39
Ar	3.76	2.12	1.62
Kr	3.99	2.69	2.46
Xe	4.33	3.72	3.99

[a] From Kittel (1976).

D. Van-der-Waals Interaction

When two systems do not overlap each other, so that there is no covalent bonding, there remains a coupling from the quantum-mechanical fluctuations of each system, producing dipoles which induce dipoles on the neighboring system, and an attraction arises from the interaction of these dipoles. A very clear and clean treatment of such interactions between dipole oscillators has been given by Kittel (1976). p. 78. We provide the corresponding treatment here as a calculation of the reduction in zero-point vibrational energy, relate it to the more direct quantum calculation, and therefore to the interaction between atoms.

We imagine two oscillators, separated by a distance r , with displacements along this axis of u_1 and u_2 . Each has spring constant κ and mass M , so that they have frequencies $\omega_0 = \sqrt{\kappa/M}$. If each also has charge e , and therefore dipole eu_i , there will be an electrostatic energy of interaction between them of $-2e^2u_1u_2/r^3$ (from the formula for the interaction of dipoles, or the u_1u_2 term in the change in electrostatic energy calculated directly). These coupled oscillators have one normal mode in which $u_1 = -u_2$ with frequency $\omega_+ = \sqrt{(\kappa + 2e^2/r^3)/M}$. A second normal mode, with $u_1 = u_2$, has frequency $\omega = \sqrt{(\kappa - 2e^2/r^3)/M}$. We may write the zero-point energy $^1/_2\hbar(\omega_+ + \omega_-)$, and expand to second-order in $2e^2/r^3$ to obtain a lowering in that zero-point energy of

$$\delta E = -\frac{1}{2}\hbar\omega_0\frac{e^4}{\kappa^2 r^6} = -\frac{1}{2}\frac{\hbar\omega_0\,\alpha^2}{r^6}. \tag{20-11}$$

In the final form we wrote the result in terms of the polarizability α of each of these oscillators, the dipole $eu_i = e^2E/\kappa$ divided by a classical field E which induces it. We see from Eq. (20-11) that because the zero-point motion of the two oscillators is correlated by the coupling, the energy is lower than without it. This is the same physical mechanism as the induced dipoles discussed at the beginning of this Part.

An alternative, and more complicated, way of deriving this formula would have been to consider the coupling between the quantum-mechanical ground state of each of the two oscillators, $\phi_0(u_1)\phi_0(u_2)$, and an excited state, $\phi_1(u_1)\phi_1(u_2)$, at energy higher by $2\hbar\omega$, due to the coupling $-2e^2u_1u_2/r^3$. Calculated to second order it must give the same result. Thus we could calculate the coupling between two inert-gas atoms in the ground state due to the electron-electron interaction in just the same way. The energy difference between the two states would have the electronic excitation energy replacing $\hbar\omega_0$, but the result can be written in terms of the polarizability in the same way. Thus Eq. (20-11) can be directly generalized to the van-der-Waals interaction.

We note first that had we looked at oscillators moving perpendicular to the axis between them, the interaction would have been $e^2u_1u_2/r^3$ and the lowering would have been one quarter as large as in Eq. (20-11). Thus the interaction for an oscillator in three dimensions is three halves that in Eq. (20-11). Similarly, there are excitations of all three p-states on the two atoms increasing the result by a factor of $^3/_2$. We have included a factor of two for the contribution of the two spins to the polarizability α as in Eq. (20-9), and we should add contributions to the van-der-Waals energy from four combinations of spin on the two atoms (since there are terms in the energy for example from a spin-up electron on the right and a spin-down electron on the left). Since the van-der-Waals energy is proportional to α^2 the same formula in terms of polarizability remains appropriate. Finally we replace $\hbar\omega$ by the excitation energy, which we suggested in Part C could be taken as- ε_p, and include the factor $^3/_2$, and the van-der-Waals interaction between two inert-gas atoms from Eq. (20-11) becomes

$$V_{vdW}(r) = \frac{3\varepsilon_p \alpha^2}{4r^6}. \tag{20-12}$$

The ε_p is of course negative and α is the polarizability, as in Table 20-4.

This same form is appropriate for the coupling between molecules, which ordinarily have filled bonding shells and then do not have important covalent interactions between separated molecules. In that case, each mode of polarization should be considered separately, as in Eq. (20-11), and the energy ε_p for each contribution is replaced by the excitation energy which dominates that particular mode of polarization. In the case of molecules one must also consider whether there are intrinsic electrostatic dipoles (as in CO) or electrostatic quadrapoles (as in CO_2) or other contributions to an interaction.

Before comparing this van-der-Waals attraction with experiment, it is useful to add the repulsive interaction between inert-gas atoms.

E. Lennard-Jones Interaction

Already in the 1920's this $1/r^6$ attraction between atoms was known, and Lennard-Jones (1924), (1925) introduced a repulsion, which he took to vary as $1/r^{12}$, fit both to experimental properties of the gas phase, and estimated a range of properties of the inert-gas solid. This use of twice the exponent for the repulsion is convenient, as it has been in our other analyses, but most properties are not so sensitive to the exact choice. We return to our usual use of d to represent internuclear distances and write the interaction in a conventional form,

$$V(d) = 4\varepsilon\,[(\sigma/d)^{12} - (\sigma/d)^6\,], \tag{20-13}$$

with the second term of course being the van-der-Waals interaction. In terms of the two parameters, σ and ε, the minimum occurs at $d = 2^{1/2}\sigma$

and the minimum energy is $-\varepsilon$. Bernardes (1958) made a more recent fit of these parameters to the observed deviations of the properties of the gas from those of an ideal gas, and his values are listed in Table 20-5. Also listed are the values these predict for the equilibrium spacing ($2^{1/6}d$) and cohesion (6ε for a face-centered-cubic structure). This relation between the two experimental quantities is very well satisfied for Eq. (20-13).

It is also of interest to see how well our prediction of the van-der-Waal interaction, Eq. (20-12), did. From Eq. (20-13) we see that it is $4\varepsilon\sigma^6$ which is to be compared with our prediction from Eq. (20-12) of $^3/_4\ \varepsilon_p\alpha^2$. Values of $^3/_4\ \varepsilon_p\alpha^2$ and $4\varepsilon\sigma^6$ in parentheses (both in eV-Å^6) for the four elements are obtained from Tables 20-3, 4, and 5 (using the experimental α). They are 2.63 (5.24), 31.7 (64.3), 64.7(132.4), and 148.5 (318.0). In view of the extraordinary range of values, one would take the theory as appropriate, if not very accurate. Using the experimental polarizability and ε_p has underestimated the attraction by a factor of two. The fact that it is so close to a factor of two suggest that we may have made an error in one of the many factors of two which enter, such as those from spin, but if so we have not found the error. There are other possible sources of error, such as our choice of $-\varepsilon_p$ as the characteristic energy denominator.

There is an alternative formulation of the interatomic interactions by Gordon and Kim (1972), (1976), which we introduced for closed-shell ions in Section 10-1-A. This was based upon superimposing charge densities for the two neighboring ions and calculating the energy locally as for a free-electron gas. We saw in Section 10-1-A that the kinetic-energy contribution gave a good description of the repulsion in ionic crystals, varying approximately as $1/d^8$ near the equilibrium spacing. When applied to inert-gas atoms, at their much larger spacing, it yields a repulsion varying approximately as $1/d^{12}$ as we assumed here. Further, Gordon and Kim

Table 20-5. Lennard-Jones parameters for the inert-gas atoms, obtained from gas data by Bernardes (1958), and the equilibrium spacing and cohesive energy predicted by them. The final columns are the experimental values for the solid.

	σ(Å)	ε(eV)	d(Å) Theory	d(Å) Exp.	Cohesion (eV/atom) Theory	Cohesion (eV/atom) Exp.
Ne	2.74	0.0031	3.08	3.15	0.019	0.02
Ar	3.40	0.0104	3.82	3.76	0.062	0.080
Kr	3.65	0.0140	4.10	3.99	0.084	0.116
Xe	3.98	0.0200	4.47	4.33	0.120	0.17

(1972) included a correlation energy from the free-electron gas which gave an attraction at large distances, and in fact produced a net interaction very close to the empirical Lennard-Jones interaction near the observed spacing. Indeed, the van-der-Waals attraction *is* a correlation energy, viewed in a somewhat different way so both pictures can make sense. The Gordon-Kim approach certainly becomes inappropriate at larger spacing where its exponential decay does not accord with the true $1/d^6$ behavior. However, near the equilibrium spacing it gave quite a good description as was described in Harrison (1980) 292ff.

For most numerical purposes the Lennard-Jones form in Eq. (20-13) with the parameters from Table 20-5 is much easier to use than tabulated interactions so we have not emphasized the Gordon-Kim approach here.

F. Molecular Crystals

The essential feature from this analysis of the van-der-Waals interaction is the understanding of this attraction which is present when other couplings between systems are absent. We evaluated it for harmonic oscillators, and extended it to inert-gas atoms. It is also present as an interaction between two bulk metals or dielectrics, where Kittel (1976) , p. 319, formulates the interaction as the shift in surface plasmon energies.

In molecular crystals we must estimate other interactions which might be present, as we indicated in Part D for CO and CO_2. Experimentally we should ordinarily be able to tell from the spacings whether other interactions are important. For the inert-gas solids in Table 20-5 we saw that the spacings are very large in comparison to the spacings of ionic crystals from the same row; e. g., a spacing of 3.15 eV for neon in comparison to the isoelectronic NaF with a spacing of 2.32 Å, suggesting that only van-der-Waals forces are important. It may also be the case that the spacings are determined by van der Waals interactions, but the orientation of the molecule is fixed by electrostatic asymmetries, or even by asymmetries in the van der Waals interaction due to anisotropic polarizabilities.

One of the compounds we discussed as a covalent insulator in Section 12-12-C, thallium-arsenic-selenide, contained a group , $AsSe_3$, which clearly formed a separate molecular unit, but with a charge of three extra electrons. Thus this molecular ion forms an ionic crystal with three thallium single charged positive ions. This can also be the case for groups such as phosphate groups, though we described $AlPO_4$ in Section 12-12 as a polar version of the SiO_2 covalent crystal. Presumably the P-O distance is less than the Al-O distance so treating it as consisting of Al^{3+} and $PO_4{}^{3-}$ ions might be defensible.

The principal message here with respect to molecular crystals is that there can be covalent, or ionic, or electrostatic multipole interactions between molecules, but there will always be the small van-der-Waals interaction. It becomes essential when other interactions are negligible. In the following section we will see that the coupling between carbon planes in graphite is such a case.

3. π-Bonded Systems

We introduced π-bonding states for the nitrogen molecule already in Section 1-4-B. The energy of these states was $\varepsilon_p \pm V_{pp\pi}$ .The most familiar solid-state π-bonded system is graphite, which we have discussed at many points in this book. In particular in Problems 1-1 and 1-2 we considered those aspects of its electronic structure which are describable as direct generalization of the sp^3-hybrid bonds for the semiconductors to sp^2-hybrid bonds in graphite. Then if we consider a graphite plane lying perpendicular to the y-axis, as in Problem 2-1, we construct hybrid states from the states $|s\rangle$, $|p_x\rangle$. and $|p_z\rangle$ and three bonds around each atom based upon even combinations of the hybrid states directed into that bond. This utilizes three of the four electrons from each atom, and the remaining electron contributes, much as in a metal, to bands based upon the orbitals $|p_y\rangle$. In partly occupying these bands, we gain energy as we did in the formation of bonds for the lithium chain in Section 1-3. It is systems of this kind which we discuss in this section.

A. Graphite

a. Energy Bands. We begin with a construction of the bands based upon the $|p_y\rangle$, or π- states. The structure is shown in Fig. 20-3, redrawn from Problem 2-1. There is one π-orbital per atom so the states will be a combination of a Bloch sum $\sqrt{1/N_c}\, e^{i\mathbf{k}\cdot\mathbf{r}_j}|p_y(\mathbf{r}-\mathbf{r}_j)\rangle$ for the shaded atoms, and another Bloch sum of the same form for the open-circle atoms, just as the bonding bands for a semiconductor band were obtained in Section 5-2-A as a combination of the four bond Bloch sums for that case. The coupling between them is obtained as in Eq. (5-7) as a sum over couplings to each of the three shaded atoms neighboring an open-circle atom as

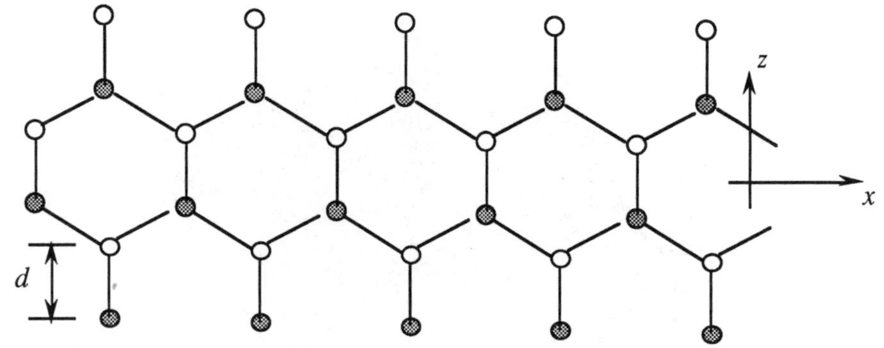

Fig. 20-3. The structure of a graphite plane. All bond angles are 120°. The differently shaded atoms are not translationally equivalent.

$$V_{12} = V_{pp\pi}\Sigma_{j=1,2,3}e^{i\mathbf{k}\cdot\mathbf{r}_j} = V_{pp\pi}\,[2e^{ik_zd/2}(\cos(\sqrt{3}k_xd/2) + e^{-ik_zd}] \quad ,(20\text{-}14)$$

and the energy bands are given by

$$\varepsilon_k = \varepsilon_p \pm \sqrt{V_{12}{}^*V_{12}} \qquad (20\text{-}15).$$

These are plotted in Fig. 20-4. This band has cone-shaped peaks at each corner of the hexagonal Brillouin Zone. There are just enough electrons to fill the band so the Fermi energy comes just at the peak, and just below the upper band which is the reflection through a horizontal plane containing the peaks.

The density of states at the Fermi energy is zero, but it is nonzero just away from the peaks. Furthermore, if we were to include additional graphite planes to form the three-dimensional crystal, and the very small coupling between these distant planes, the bands would vary slightly with k_y . The energy of this peak (and actually its exact position in the zone) would vary with k_y , rising above the Fermi energy at some places and dropping below it at others. This produces tiny needle-shaped Fermi surfaces lying along the Brillouin-Zone edges parallel to the y-direction with something like 3×10^{-5} electrons, and an equal number of holes, per atom. Such a system is called a *semimetal* . Bismuth and antimony are also semimetals, treated by Liu and Allen (1995) and by Xu, Wang, Ting, and Su (1994), but we shall not discuss them here.

A more accurate band calculation for graphite has been given by Painter and Ellis (1970), with references to other work on the electronic structure.

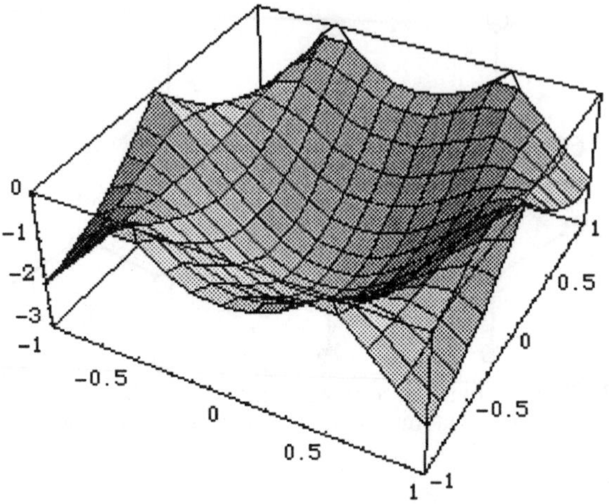

Fig. 20-4. The lower of the two energy bands for the π-states in graphite obtained from Eq. (20-15). Peaks occur at the corners of the hexagonal Brillouin Zone, inscribed in this box. An identical band (empty) is the reflection of this band through the top plane of the box.

There are σ-bonding and σ-antibonding bands, similar to those for semiconductors (except only three rather than four per atom pair), which lie well above and below the Fermi energy, as well as these π-bands which overlap the bonding bands below and the antibonding bands above.

b. The π-Bonding Energy. With such a set of bands, half filled, it is natural to estimate the bonding energy using the second moment, as we did for semiconductors in Section 2-5 and as we did with the Friedel Model for transition-metal bands. Our bands were calculated in Part a. from a Hamiltonian matrix for which all the diagonal elements were ε_p and for each atom there were couplings of $V_{pp\pi}$ to three neighbors, giving a second moment for the bands (measured from ε_p) of $\sqrt{3V_{pp\pi}{}^2}$. If we take this to be the average of the occupied band, measured from ε_p, we gain an energy of $\sqrt{3}V_{pp\pi}$ per atom from the π-bonding. The use of tight-binding with universal parameters has probably introduced more error than this approximation to the energy, so we may regard this approximation as adequate.

We note that this is an example of the resonant bonding which we discussed in Section 10-2 for ionic crystals. If we choose to think in terms of the usual two-electron bonds, which in the case of π-bonds in nitrogen gain energy $2Vpp\pi$, we would now say there is one π-bond for every two atoms,

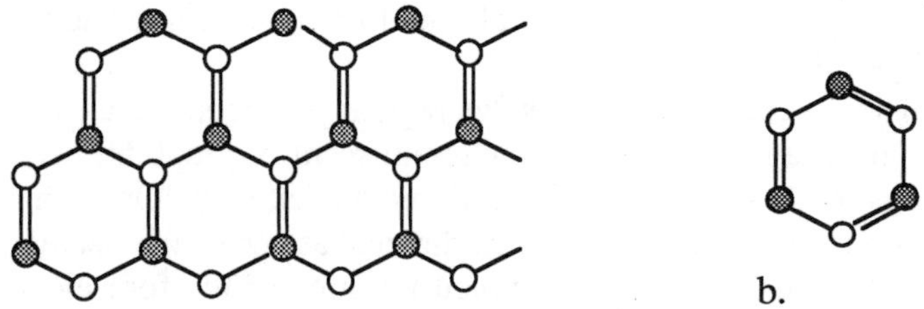

Fig. 20-5. Resonant π-bonds are shown as a second line between atoms. In graphite, Part a., each resonant two-electron bond resonates between three bond sites, the other two corresponding to this figure rotated ±120°. In benzene, Part b., each two-electron bond resonates between two bond sites. The resonance enhances the bonding by a factor of $\sqrt{3}$ in Part a., and by $\sqrt{2}$ in Part b.

and that each bond resonates among three sites, as illustrated in Fig. 20-5. Then the enhancement of the bond due to its resonance is by a factor of the square root of the number of sites between which it resonates, or $2\sqrt{3}V_{pp\pi}$ per atom pair. There is no real difference between the different bonds in one figure so it is not clear how useful the concept is.

In any case we may treat graphite, using the same approximations we used for semiconductors, treating the σ-bonding terms based upon sp^2-hybrids, and then add the π-bonding term to the energy. In the graphite structure there is a stacking of such planes, at the very large spacing of some $2.4d$. This is so large that we expect the bonding between planes to be dominated by the van-der-Waals interaction discussed in Section 20-2-D. We would need to recalculate that in terms of a polarizability perpendicular to the planes. Certainly that is not the only contribution to the bonding between planes since, as we noted, the coupling between π-orbitals on adjacent planes does deform the bands slightly, and the lowering in energy as the electrons readjust further lowers the energy of the system. However, that covalent contribution - or metallic contribution - is certainly very small.

B. Fullerenes

In 1985, Kroto, Heath, O'Brian, and Smalley (1985) found clusters of 60 carbon atoms in soot, which they surmised were spheres of carbon in exactly the shape of a soccer ball, with a carbon atom at each juncture of the seams. The structure is reminiscent of designs by the architect Buckminster-Fuller, and the system is sometimes referred to as a fullerene, and the spheres called bucky-balls.

The sphere consists of twelve regular pentagons, each surrounded by five regular hexagons. This is very reminiscent of graphite, bent into a ball and the electronic structure can be interpreted that way. On each carbon atom we may construct three sp^2-hybrids, directed at one of the nearest neighbors, and form σ-bonds which are occupied by three of the four electrons from each atom. The remaining electrons are in π-states, oriented radially, and almost parallel to the neighboring π-states, very much like the radial states in boron discussed in Section 20-1-D.

The π-states inevitably provide the highest occupied and lowest unoccupied levels as in graphite and are the interesting states. Certainly the lowest π-state has the full symmetry of the ball, with energy about the $\varepsilon_p + 3V_{pp\pi}$ of graphite. The other states might be expected to have approximately the symmetry of the spherical harmonics, $Y_\ell^m(\theta,\phi)$, and indeed that is the case. They increase in energy with increasing ℓ, and with sixty electrons they will fill the states with $\ell = 0$, 1, 2, 3, and 4, accommodating $2\ell + 1$ in each, for fifty electrons. The remaining ten go into states from $\ell = 5$. In fact with icosahedral symmetry of this system, the highest degeneracy which occurs is five levels (as in atomic d-states) and the eleven $\ell = 5$ π-states separate into a set of five degenerate levels, lowest in energy, and two sets of three-fold degenerate levels higher in energy, as was seen in an Extended-Hückel-Theory calculation by Haddon, Brus, and Raghavachari (1986) They found the separation between the occupied and empty levels to be $0.757\ V_{pp\pi}$ (or 0.757β in usual Extended-Hückel-Theory notation). There are two bond types (between a hexagon and a pentagon or between two hexagons) with slightly different lengths, 1.405 Å and 1.426 Å, for an average $V_{pp\pi}$ of 2.4 eV so the gap should be approximately 1.6 eV.

These C_{60} molecules can condense into a solid, forming a face-centered-cubic structure, with each molecule fixed at low temperature, but free to rotate in its site at higher temperature. In contrast to the boron icosahedra, there are no covalent bonds between adjacent spheres; like graphite sheets we think of the bucky balls as held together largely by van-der-Waals forces. Also as in graphite there *is* some coupling between spheres and a consequent broadening of the sphere orbitals into bands. The system is insulating because of the gap between empty and full states, but can be doped by, for example, the addition of sodium to become a conductor, and in fact a superconductor at low temperatures.

More recently it has been found that other analogous structures form, for example as tubes of graphite, as we might imagine the chicken-wire-like sheets of Fig. 20-3 rolled and rejoined to form tubes. These seem to be axially shifted before rejoining so that the structure has a spiral nature, which complicates the band structure. It often comes as concentric tubes, with a spacing between tubes comparable to the spacing between graphite planes. There has also been suggestion that the ends of the tubes can be closed by bucky-ball-like structures.

C. The Heme Group

We discuss briefly the heme complex, illustrated in Fig. 20-6, because it is such an important system and because it illustrates how we can proceed analytically with such a complicated system. It is the functional group in hemoglobin. (Weissbluth (1974) gives a much more extensive discussion of the electronic structure of hemoglobin). Oxygen attaches to an iron atom at the center and is carried to the cells in the body. Essentially the same complex, with iron replaced by magnesium and with different groups attached at the edges, is the functional group in chlorophyll. We shall carry the analysis far enough to see just which states of the complex without the iron are coupled to the iron d-states. These are the steps which would precede an attempt to understand the working of the complex, but we have not gone that next step.

There are three p-states and one s-state on each of the four nitrogens and twenty carbons shown, 96 orbitals, in addition to the iron states and the states

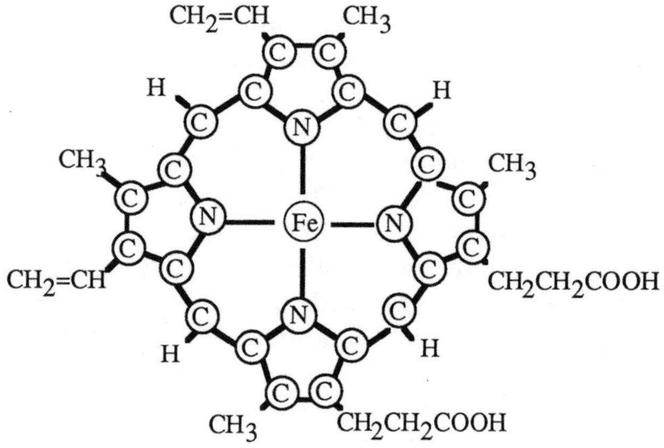

Fig. 20-6 The heme complex, after Weissbluth (1972).

on the outer groups shown. This is a very large number of orbitals, but as in graphite these will divide into σ-bonding states well below the Fermi energy, and antibonding σ-states well above and the interesting states will arise from π-states on the central carbon and nitrogen atoms. We need to look at the σ-bonding states first to see to see how many electrons are available to the π-bands, and to see the extent to which these π-states are isolated from the rest of the system. We carry this far enough to see which orbitals appear to be the ones important in the coupling to the iron, and therefore the ones of importance in the functioning of the heme group.

We again construct sp^2-hybrids on each of the carbons and nitrogens shown as circles in the figure, directed at the three neighbors to each atom, form a bonding and antibonding combination with each. We place two electrons in the bond, one from each of the two atoms forming that bond. The bond between each carbon and a hydrogen, or other attached group, is also such a σ-bond in the sense we discuss in the next section, with one electron contributed to it from the carbon atom. For the line between the nitrogen and the central iron, the nitrogen sp^2-hybrid has energy $(\varepsilon_s + 2\varepsilon_p)/3 = -17.97\ eV$, so deep that it will be doubly occupied, taking two electrons from the nitrogen, and leaving the iron with its original number of electrons. Thus each nitrogen contributes two electrons to such a hybrid, two electrons to σ-bonds and one electron to the π-states, with each carbon also contributing one electron to the π-states. These twenty-four π-states with their twenty-four electrons provide the highest occupied and lowest unoccupied molecular states and are the important states in the problem. We discuss the iron atom afterward.

Concerning the isolation of this set, we note that for an attached CH_3 group, we would construct sp^3-hybrids on the carbon and all its orbitals would become part of σ-bonds far from the Fermi energy, leaving the π-states at the center of the group unaffected. The same is true for the attached CH_2CH_2COOH groups, but not for the $CHCH_2$ groups for which both carbons have three neighboring atoms. They will be coplanar with the central group, forming σ-bonds with their neighbors, but each also containing a π-electron, coupled to the central group. We really should be dealing with twenty-eight π-orbitals, and no longer with the square symmetry of the central group, but we stay with twenty-four at first. We shall discuss briefly the effect of these added orbitals later.

As in our discussion of boron in Section 20-1-D, we could proceed with the central twenty-four atoms using the full symmetry of the square, reducing the problem in this case down the solution of a five-by-five matrix at most, or we can focus on the consequences of subgroups of the symmetry operations. We make the latter choice.

In particular we determine states on a subgroup, the *pyrrole* which consists of a nitrogen atom and the four carbons with which it makes a pentagon. The energy of the nitrogen π-state lies 2.77 eV below that of the four carbon π-states. Each nitrogen lies at a distance of 2 Å from the iron position, which will be of interest later. The two nitrogen-carbon bond lengths are 1.38 Å, the next carbon-carbon bond lengths are 1.46 Å, and that across the base of the pentagon is 1.32 Å. (Weissbluth (1974)) The coupling between these neighboring states, given by $V_{pp\pi} = -0.63\hbar^2/md^2$, is -2.52 eV, -2.25 eV, and -2.76 eV for these three sets.

We can easily solve for the five eigenstates, but the important feature can already be seen from these parameters. The coupling between the two carbon atoms at the base of the pentagon is almost identical to the difference in energy between the nitrogen and carbon atomic states, so the even combination of those carbon states has the same energy as the nitrogen π-state. If the coupling between the base carbons and the central carbons (-2.25 eV) were equal to the coupling between the central ones and the nitrogen (-2.52eV), which they almost are, one eigenstate for the five-atom pyrrole has equal coefficients on the two carbon π-states of the base, an equal coefficient of opposite sign on the nitrogen π-state and no contribution from the intermediate carbon atoms; we call it a "floating orbital" since, with zero coefficient on the other two carbon atoms it would be completely uncoupled to rest of the system of π-states. Its energy is equal to that of the nitrogen p-state. We proceed with that approximate equality and *thus the four floating orbitals are eigenstates of the entire heme group of 24 orbitals* . (We have ignored for the moment the effect of the two $CHCH_2$ attachments, which would still however leave two of the orbitals floating.) Two other states for the pyrrole have all three coefficients on the base atoms and nitrogen equal, and in either a bonding or antibonding relation to an even combination of the other carbon π-states. There are two more states which are odd under a reflection through the axis of the pentagon containing the nitrogen, and contain no contribution from the nitrogen π-state.

We may now proceed to the solution for the complete 24-orbital system. We have in fact the four floating-orbital solutions, of energy $\varepsilon_p(N)$. We may note further that every other eigenstate must be orthogonal to each floating state, and therefore must have a coefficient on the nitrogen orbital equal to the sum of the coefficients on the two carbon atoms of the base. Thus the state can be specified by giving the coefficients u_j for the twenty carbon atoms represented by circles in Fig. 20-6. These form a ring, so it is possible to use a set of equations such as Eq. (1-19), but with the coupling to the nitrogen state included in terms of the carbon coefficients, to determine u_{j+1} from u_j and u_{j-1} . It is easy to write a program which obtains the solutions. The program begins with a low energy ε , and first for a solution even with respect to one pentagon base, so $u_1 = u_2$, and work around the ring. For an eigenstate we must come back to the same value u_1 after completing the round. We raise ε until this occurs to find the first eigenstate, and continue to raise ε for successive even states. The same procedure gives the odd states. We ran such a program, actually taking the other carbon-carbon and carbon-nitrogen matrix elements the same as the base matrix element, which makes little difference. The eigenvalues resulting are listed in Table 20-6, designated by the symmetry of the states. The states of symmetry A are nondegenerate and even under reflection through a diagonal in Fig. 20-6. The states of symmetry B are nondegenerate and odd under reflection through the diagonal. States with a sub-one are even under reflection through a vertical or horizontal axis, and with a sub-two are odd under those reflections. States E are doubly degenerate and can be chosen one even under horizontal reflection and odd under vertical reflection and the second the other way around.

We see that if we place two electrons (spin-up and spin-down) in the lowest energy states until we have 24 electrons, corresponding to a neutral complex, we occupy all the states in the left-hand column, and none in the right column.

We now reinsert the iron atom at the center. The $\varepsilon_d{}^0$ for iron is -16.54 eV from Table 17-2 (for 6 electrons in the d-shell, 2 in the s-shell), with a shift in d-state energy of $U_d = 5.9\ eV$ from Table 15-1 if one electron is

Table 20-6. The energies of the π-states, relative to the nitrogen p-state energy, in eV, for the heme complex (with the iron removed and the $CHCH_2$ groups omitted).

A_1	-4.49	A_1	1.72
E	-4.16	E	3.38
B_1	-3.60	B_2	4.49
B_2	-1.69	B_1	6.37
E	-0.61	E	6.90
A_1, B_1, E (floating)	0.00	A_2	7.45
A_2	1.05	A_1, B_2, E	8.31

transferred from the s-shell to the d-shell. Three of the five d-states are even under reflection in the plane of Fig. 20-6, and are not coupled to the π-states. We imagine them to be doubly occupied, as are the nitrogen hybrids directed toward them, and coupled to them. The other two d-states have symmetry E (they are of the form xz, and yz if the z-axis is taken perpendicular to the plane of Fig. 20-6) and are coupled only the levels in Table 20-6 which are of E symmetry. That coupling is through the interatomic iron-nitrogen matrix element $V_{pd\pi} = (3\sqrt{5}/2\pi)\hbar^2 \sqrt{r_p r_d^3}/(md^4)$ from Eq. (17-2), equal to 0.94 eV with the parameters from Tables 15-1 and 17-1 and the 2.00 Å spacing (Weissbluth (1974)).

This is as far as we have carried the analysis, and it is far from done. The d-state energy with six electrons is some 3 eV below the nitrogen level, and the floating levels. The coupling will transfer electrons into the d-states, causing them to rise to the neighborhood of the Fermi energy, and the floating levels are the closest π-levels to the Fermi energy. Our feeling has been that these floating levels, which may be a unique feature of the heme group, and their coupling to the important d-states are the features which may prove essential to the operation of the complex, but that is far from established. In particular, it may be important to introduce the effects of the $CHCH_2$ groups, which we see from Fig. 20-6 break the symmetry of the pyrroles to the left and to the top. We have so far done little more than organize the electronic structure of this complicated system, reducing it down to a small enough set of orbitals than it might be possible to make progress in understanding it.

4. Systems With Hydrogen

We found already that the tight-binding theory with universal parameters was not as accurate for systems from the carbon row of the periodic table as it was for heavier materials. We might expect that extension of the same procedure to hydrogen would be even less accurate. We shall see that that is the case and in addition, in hydrogen (or helium) there is no valence p-state. Further, in the case of hydrogen, the repulsion between the nucleus and any neighbors arises entirely from the Coulomb repulsion between the proton and the other nuclei, in contrast to all other systems where the electronic kinetic energy was responsible for the dominant repulsion. This will suggest an alternative view of systems containing hydrogen. We begin with the H_2 molecule and first try direct application of our tight-binding theory. It is of course also possible to include hydrogen in any full local-density calculation of electronic structure to predict the properties with some confidence. (See, for example, Van de Walle (1998) and references therein.)

A. H_2

If we were to think of the hydrogen molecular states as a bonding combination of atomic s-states, the principle parameter would be the coupling $V_{ss\sigma}$. The formula $V_{ss\sigma} = -1.42\hbar^2/md^2$ yields 20. eV at the observed spacing of 0.74 Å. This would suggest an ionization potential for the molecule to be $|\varepsilon_s + V_{ss\sigma}| = 13.6\ eV + 20.0\ eV = 33.6\ eV$, in comparison to the observed 15.4 eV. If we proceed as for Li_2 and assume that nonorthogonality of the s-states produces a repulsion canceling half of the bonding energy, it would also raise the bond energy by $|V_{ss\sigma}|/2$ raising the estimate to 23.6 eV. This would also lead to a cohesive energy of $V_{ss\sigma} = 20\ eV$ per molecule, much larger than the experimental value of 4.5 eV. Perhaps these are not far from the errors we might extrapolate from the corresponding errors (a factor of two in cohesion) for the carbon-row atoms. However, we should explore an alternative view, that of the "united atom".

In the united atom approach for this case we proceed much as in the theoretical alchemy which we introduced in Section 1-4 for constructing covalent and ionic solids. We begin with helium, which is isoelectronic with H_2. We then imagine freezing the electrons, splitting the nucleus into two protons, and let them approach their positions of electrostatic equilibrium. This is quite simple and straightforward if we take the helium s-orbitals to be given by $\sqrt{\mu_s^3/\pi}\, e^{-\mu_s r}$. Then we may integrate to obtain the number of electrons Q_s contained within a sphere of radius r as

$$Q_s(r) = 2[1 - (1 + 2\mu_s r + 2\mu_s{}^2 r^2)e^{-2\mu_s r}]. \tag{20-16}$$

Thus if the two protons are at positions $x = \pm r$, the force on a proton due to the electronic charge and the second proton is $-Q_s e^2/r^2 + e^2/(2r)^2$ or at equilibrium $Q_s(r) = 1/4$. From Eq. (20-16) we find that this gives $\mu_s r = 0.610$.

Taking μ_s by setting the helium s-state energy of -24.98 eV (Table 1-1, which is close to the negative of the experimental ionization of helium of 24.5 eV) equal to $-\hbar\mu_s{}^2/23m$, we obtain $\mu_s = 2.56\ Å^{-1}$ and a proton-proton distance of $2r = 0.48\ Å$. . If we instead used the experimental ionization energy for H_2, 15.4 eV given above, we obtain an equilibrium spacing of 0.61 Å, a little closer to the experimental spacing of 0.74 Å. In either case this is quite a crude representation of the electronic structure, but would seem to be a much better starting point than the outlook used for the heavier elements and the universal matrix elements. It also provides the starting point for a path to the other properties such as cohesion and the vibrational frequency of H_2, but we have not pursued it.

B. Central Hydrides

We may similarly construct the central hydrides, hydrogen fluoride, water, ammonia and methane from neon using this theoretical alchemy. We freeze the neon charge density, remove one proton from the nucleus and let it fall to its equilibrium position corresponding to HF. We remove a second and let the two go to their equilibrium distributions, on opposite sides of the remaining oxygen nucleus, forming water, but without the appropriate angle of 104.4° between protons. We shall return to the angular forces shortly. A third proton has minimum electrostatic energy with the three forming an equilateral triangle with the remaining nitrogen nucleus at its center, a flattened ammonia. A fourth proton removed from the nucleus produces methane, with the appropriate tetrahedral geometry.

We may make quantitative estimates of the geometry with a direct generalization of the approach we used for helium. We again take s-states as simple exponentials and obtain a charge within the radius r given by Eq. (20-16), with μ_s evaluated for the s-state energy of neon, -52.53 eV. For the p-states we let the electron density be proportional to $r^2 e^{-2\mu_p r}$ with μ_p fit to the neon p-state energy of -24.14 eV. We obtain a charge for the p-states, analogous to Eq. (20-16) given by

$$Q_p(r) = 6[1 - (1 + 2\mu_p r + 2\mu_p^2 r^2 + {}^4/_3\mu_p^3 r^3 + {}^2/_3\mu_p^4 r^4)e^{-2\mu_p r}] \quad (20\text{-}17)$$

For the case of HF, the proton, at a distance r from the fluorine nucleus feels a force $(7 - Q_s(r) - Q_p(r))e^2/r^2$, so we predict an equilibrium spacing for $Q_s(r) + Q_p(r) = 7$. Using Eqs. (20-16) and (20-17) this leads to the $d_1 = 1.44$ Å for HF listed in Table 20-7. For H_2O, the prediction is $Q_s(r) + Q_p(r) = 6 + {}^1/_{2^2} = 6.25$. For NH_3 it is $5 + {}^1/_{\sqrt{3}} = 5.58$, and for CH_4 it is $4 + {}^3/_4\sqrt{{}^3/_2} = 4.92$, with the results listed as d_1 in Table 20-7. If we scale the s- and p-state energies with the experimental ionization energy, listed in Table 20-7, we obtain the predictions listed as d_2 in that table. Neither predictions is very impressive, and not as good as the corresponding predictions for helium, but again we regard it as a good starting point for a path toward understanding the properties of these simple systems.

Table 20-7. Ionization energy ε_I for the central hydrides, isoelectronic with neon. Bond lengths d_1 estimated using the charge density for neon, and d_2 estimated using the charge density for the molecule, and the experimental bond length d_{exp} .

	ε_I (εV) [a]	d_1 (Å)	d_2(Å)	$d_{exp.}$(Å) [b]
Ne	21.55			
HF	15.77	1.44	1.68	0.92
H_2O	11.53	1.21	1.65	0.96
NH_3	10.2	1.06	1.54	1.01
CH_4	12.89	0.94	1.22	1.1

[a]Weast (1975)

[b]Pauling (1960), p.226

One evidence for the shortcomings of this simplest model is that it predicts the wrong geometry for water and ammonia, which experimentally have asymmetric structures with large electric dipoles. It is interesting that we are guaranteed to predict the symmetric geometries if our description of the electronic states is based solely upon atomic s- and p-states, since these produce a spherical charge density. Thus rationalizations of the geometry of water as placing protons near the "lobes" of p-states at 90° from each other are not tenable. On the other hand, expansion of the states in terms of atomic states on the central atom *and* s-states on the hydrogen will rationalize the bond angles near 90° for water and ammonia. So also would the Orbital Correction Method (Harrison (1973b) also discussed in Harrison (1980) Appendix E) in which corrections to the orbitals are expanded in orthogonalized plane waves, in much the same way we included the coupling between d-states and OPW's in Section 15-1-C. An application of this theory to HF and CH_4 by Meserve (1975) suggested that it would have given a good prediction for the bond angle in water, but that case was not carried out. Incorporation of such terms also reduces the predicted equilibrium spacings of Table 20-7, as found by Meserve and as expected since they lower the energy by an amount which rises from zero when d is increased from zero.

It is interesting to note in passing that the model without orbital corrections gives a prediction of the molecular dipole for HF, H_2O, and NH_3 if the observed geometry is used. The prediction is of the right sign, but slightly too large since the orbital corrections will shift the electronic charge slightly in the direction of the dipole. This is quite a different view from a chemical view sometimes expressed in which the dipole arises from "lone-pair" electrons which are not compensated by a proton.

Again, the simple united-atom, with freely moving protons and orbital corrections, seems a very reasonable approach for studying these central hydrides, but may not have been carried far to date.

C. Hydrocarbons

The general view taken for the central hydrides would seem appropriate for more complicated molecules with hydrogen. In ethane for example, C_2H_6, the three protons near one carbon atom see a potential of the type represented in Eqs. (20-16) and (20-17), plus the deviation of the potential from the other CH_3 which would be a simple $- e^2/d$ if it were spherically symmetric at a distance d. Thus the dominant terms are very much the same as for CH_4, and the geometry is very nearly the same, with the three protons at one end rotated 60° with respect those at the other, to reduce the Coulomb energy. A system such as ethylene, C_2H_4, does not have enough electrons to fill the sp^3 shell of the two carbon atoms, and with two electrons missing there are two p-states empty, in fact from the same π-orbital with opposite spins with the protons lying in the same plane as that of the occupied π-orbitals, again to reduce the electrostatic energy. One can think of extending the same approach to other systems such as HNO_3, but we have not followed the approach very far. A recent related study was made by Porezag, Frauenheim, Köhler, Seifert, and Kaschner (1995)

In the study of the heme group in the preceding part we did note that there were "σ-bonds" between the carbon atoms of the central group and attached hydrogen atoms. In the context of the united atom we would describe these bonds differently, but the effect is the similar. We construct sp^3-hybrids on the carbon atoms and bonds with its neighbors, with a dangling hybrid directed at the hydrogen neighbor. Then an approximate potential is constructed for the hydrogen, as in the C_2H_3 discussed above. Again the π-bonds of the central group are isolated from the rest of the system since there is no coupling except through hybrids directed into other bonds. Thus the approximation we made there was tenable.

D. Shared-Hydrogen Bonding

There is a final role of hydrogen in bonding in which it forms a bridge between other systems. (See for example Pauling (1960), 449ff, for a discussion of the hydrogen bond.) One important such case is in the bonding between base pairs in DNA. Another is in the bonding of water molecules in ice. The idea at first seems very plausible in terms of the united-atom picture. Two atoms are brought together and a proton seeks a low potential between the two where the two negative charge densities overlap. At closer inspection, it is not so obvious.

Imagine a proton bound in one neutral molecule, and we bring up a second molecule, neutral already without the proton in question. If this second molecule were a neon atom, there would be no attraction of the proton to that atom. The electrostatic field from a neon atom arises from the charge of $8e$ from the core, minus the field from a charge of $[2Q_s(r) + 6Q_p(r)]e$, as in Eqs. (20-16) and (20-17). The field, and the force on the external proton, is always outward, because the negative charge at any r is always less than the core charge, and the proton is repelled from the overlap region. Of course the *electronic* charge from the first atom is always attracted to the neon (aside from the important effects of kinetic energy from nonorthogonality) so there is electrostatic attraction between the atoms, but the proton in question does not contribute to it, but moves away from the second atom.

If on the other hand, the second molecule were an HF, then the hydrogen from; the HF could move to the opposite side of the fluorine, and the proton on the first molecule *would* be attracted toward the HF, would occupy an intermediate position, lowering its energy and contributing to the bonding. This is illustrated in Fig. 20-7 where we show the energy of a proton in the electrostatic field of a neon atom, calculated using Eqs. (20-16) and (20-17). It is consistently repulsive as expected. Also shown is the electrostatic potential for a united-atom HF, with the hydrogen (second proton) at the opposite side of the fluorine, at the distance d_1 from Table 20-7. Holding all radial distances between protons and the nucleus of the united atom representing its molecule fixed, electrostatic energy favors the proton on the first atom moving to the intermediate region. For the total energy it would of course be necessary to include the interactions involving the two overlapping electron clouds, but the exercise shows that the hydrogen bond is a possibility in this context.

The effect is even larger with two or three protons on the second atom as also shown in Fig. 20-7 for a water molecule and an ammonia molecule, but calculated as for the HF, with the proton distances taken as the d_1 of Table 20-7 and the bond angle to the hydrogens taken as 90°. This could then be taken as a starting approximation for the bonding between water molecules in ice. Of course the hydrogen bonding will be very similar if the second molecule is replaced by a hydrocarbon or other material where the ligands are other than hydrogen atoms. The hydrogen bond is quite weak, experimentally some 0.2 to 0.4 eV, as would be suggested by Fig. 20-6, but that makes it important in biological systems where the hydrogen bonds can

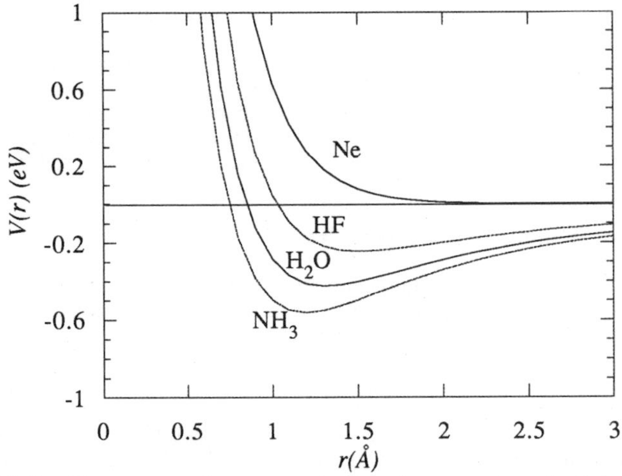

Fig. 20-7. The energy of a proton in the electrostatic potential of the charge density of a neutral neon atom, as a function of the distance r to the neon nucleus. The curve HF is the corresponding curve for a united-atom hydrogen fluoride, with the hydrogen at the opposite side of the fluorine. The H_2O and NH_3 curves are the corresponding interactions of the proton with those neutral molecules.

readily be made and broken with available energies. We have only introduced this very important subject, and have not ourselves followed it further.

Problem 20-1. Multicenter bonds

We have taken $V_{ss\sigma} = -1.32\hbar^2/md^2$ for systems with four or less neighbors. In a simple cubic structure, with six nearest neighbors, we could proceed as for multicenter bonds. We do that here for silicon in a simple-cubic structure and construct the s-like bands as we did for a chain in Section 1-3-A. It is still possible to write the states as $\Sigma_j e^{i\boldsymbol{k}\cdot\boldsymbol{r}_j}|j\rangle/\sqrt{N}$ but with nonorthogonal orbitals we must use Eq. (20-1) which gives

$$\varepsilon_k = \varepsilon_s + \frac{4 + 2coskd}{1 + (4 + 2coskd)S_{ss\sigma}}H_{ss\sigma} \tag{20-1pr}$$

for electrons moving in a [100] direction in a simple-cubic structure (with four neighbors in the plane perpendicular to $\boldsymbol{k}$).

We first need parameters from Section 20-1-B. It probably is best to take a larger spacing due to the higher coordination than the 2.35 Å for tetrahedral silicon. Take $d = (6/4)^{1/4} \times 2.35$ Å, from Eq. (2-58). We need S_2 for this spacing from Eq. (20-4) and (20-

5), using of course $V_2 = 3.22\hbar^2/md^2$. We also need numerical values for the $S_{ss\sigma}$ which can be obtained from Eq. (20-5).

The goal now is to see if this leads to a free-electron band width, which should be appropriate for simple-cubic silicon. $H_{ss\sigma}$ can be obtained. [Units of energy of $\hbar^2/md^2$ can be used from now on if desired.] Evaluate the band width from $k = 0$ to $k = \pi/d$ using Eq. (20-1pr). By what percentage does it differ from the free-electron difference? What would be the percentage difference is you took $S_{ss\sigma} = 0$ and used $V_{ss\sigma}$ rather than $H_{ss\sigma}$? (In this case both are quite close.)

Problem 20-2. Van-der-Waals interaction

We consider the van-der-Waals interaction between two Li_2 molecules, which was discussed in Section 1-2. We can obtain the polarizability from the formula for bond polarization, Eq. (4-9), but with fields parallel to the bond, eliminating two factors of $1/\sqrt{3}$. It becomes $\delta p = e^2 d^2 E/(2V_2) \equiv \alpha E$, with $d = 2.67Å$ for Li_2 The coupling V_2 was $/V_{ss\sigma}/$ for that case and the excitation energy was $2V_2$. In this two-orbital approximation the polarizability is zero for electric fields perpendicular to the axis.

a)Write the van-der-Waals interaction energy, for two molecules oriented parallel to their internuclear separation, by generalizing Eq. (20-11), in analogy with our generalization to inert-gas atoms. Substitute for all expressions, except d and r to obtain a result in terms of fundamental constants.

b) Write the same for two molecules, parallel to each other, but perpendicular to the internuclear axis.

c) Estimate the difference in direct static Coulomb interaction between these two orientations of the molecules, by modeling each molecule as charges ofZ^*e, located at $\pm d/2$ from the center of gravity of each molecule, and a charge $-2Z^*e$ at the center of gravity of each molecule, with those centers of gravity for the two molecules separated by r. This produces a quadrapole field. Expand and obtain the leading term in powers of d/r. For the real molecules, the r-dependence is presumably correct, but the Z^* would be quite small and representing deviations of the charge density from that of superimposed atoms. It nonetheless probably dominates the anisotropy of the total interaction.

Author Index

Abarenkov, Igor See Heine, V.

Abd-Elmiguid, Mohsen M., Hugo Pattyn, and Shmuel Bukshpan, 1994, *Microscopic Observation of the s→d transition in metallic cesium under high pressure*, Phys. Rev. Letters **72**, 502. *576*

Adler, D., 1968, *Insulating and Metallic States in Transition-Metal Oxides*, Solid State Physics **21**, ed. by F. Seitz, D. Turnbull, and H. Ehrenreich, p. 1. *671*

Ahuja, Rajeev, Olle Eriksson, J. M. Wills, and Börje Johansson, 1995, *Theoretical confirmation of the high -pressure simple cubic phase in calcium.* Phys. Rev. Letters **75**, 3473. *578*

Aigran, P., and M. Balkansky, 1961, *Selected Constants Relative to Photoconductors*, Pergamon, Press, New York. *167*

Allen, F. G. See Gobeli, G. W.

Allen, P. B., See Chadi, D. J., Chakraborty, B.

Allen, Roland E. See Liu, Yi

Almqvist, L. See Stedman, R.

Althoff, J. D. See Moriarty, John A.,

Anastassakis, E., and M. Cardona, 1985, Phys. Stat. Solidi (b) **129**, 101. *163*

Andersen, O. K., 1973, Solid State Commun. **13**, 133. *69, 83, 531, 536, 542, 543, 556*

Andersen, O. K., 1975, Phys. Rev. **B12**, 3060. *565, 642*

Andersen, O. K., and O. Jepsen, 1977, Physica (Utrecht) **91B**, 317. *538, 539, 548, 560, 563*

Andersen, O. K., W. Klose, and H. Nohl, 1978, Phys Rev. **B17**, 1209. *531, 543, 544*

Anderson, O. K. See Jepsen, O., Mackintosh, A. R., Savrasov, S. Y., Skriver, H. L.

Anderson, P. W., 1961, Phys. Rev. **124**, 41. *598ff, 612, 626ff*

Anderson, R. L., 1960, *Proceedings of the International Conference on Semiconductors, Prague, 1960*, Czechoslovakian Academy of Sciences, Prague, p. 563. *743*

Andreoni, W. See Blöchl, Peter E.

Boyce, J. B. See Mikkelsen, J. C.
Boyer, J. L. See Papaconstantopoulos, D. A.
Boykin, T. B., 1997, *A more complete treatment of spin-orbit effects in tight-binding models,* Phys. Rev. **B56**, 9613. See also Boykin, et al., 1997. *240*
Boykin, Timothy B. , Gerhard Klimeck, R. Chris Bowen, and Roger Lake, 1997, Phys. Rev. **B56**, 4102. *240*
Brajczewska, Marta See Fiolhais, Carlos
Bratina, G. See Nicolini, R.
Bratkovsky, A. M. See Horsfield, A. P.
Braunstein, R. and E. O. Kane, 1962, J. Phys. Chem. Solids **23**, 1423. *201*
Brebrick, R. F. See Su, Ching-Hua
Brewer, Leo, 1971, *Energies of the Electronic Configurations of the Lanthanide and Actinide Neutral Atoms,* J. Opt. Soc. Am., **61**, 1101. *605, 607, 608, 611*
Brock, J., 1981, Private communication. *355*
Brooks, M. S. S., 1984, J. Phys. F. **14**, 639; **14**, 653, and **14**, 857. *622, 682*
Brudervoll, T., D. S. Citrin, N. E. Christensen, and M. Cardona,1993, *Calculated band structure of zinc-blende-type SnGe,* Phys. Rev. **B48**, 17128. *240*
Bruno, P., and C. Chappert, 1991, Phys. Rev. Letters **67**, 1602. *724*
Brus, L. E. See Haddon, R. C.
Bryant, G. W., and G. D. Mahan, 1978, Phys. Rev. **B17**, 1744. *598*
Bukshpan, Shmuel, See Abd-Elmiguid, Mohsen M.
Burnett, J. L. See Nugent, L. J.
Burstein, E. See Lucovsky, G.
Callaway, Joseph, 1976, *Quantum Theory of the Solid State*, Academic Press, New York. *212*
Callaway, Joseph See Rath, J.
Caldas, M. J., A. Fazzio, and A. Zunger, 1984, J. Appl. Phys. **45**, 671. *745*
Car, R. See Blöchl, Peter E., Sarnthein, Johannes
Cardona, M., 1982, in *Topics in Applied Physics* **50** , ed. by M. Cardona and G. Güntherodt, Springer Verlag, Heidelberg, p. 19. *269*
Cardona, M., and N. E. Christensen, 1987, Phys. Rev. **B35**, 6182. *255, 258ff, 744*
Cardona, M., and G. Güntherodt, 1982, eds. *Topics in Applied Physics* **50** , Springer Verlag, Heidelberg *269*
Cardona, M. See Anastassakis, E. , Brudervoll, T., d'Amour, H., Laude, L. D. , Lew Yan Voon, L. C., Pollak, F. H., Shevchik, N. J., Shields, A. J.
Cargill, G. S. III See Erbil, A.
Carlsson, A. E., 1985, *Simplified electrostatic model for band-gap underestimates in the local density approximation*, Phys. Rev. **B31**, 5178.. *207, 208, 209, 342, 343*
Carlsson, A. E., 1996, *Quantum-mechanical derivation of angular and torsional forces in well-bonded systems,* Phys. Rev. **B54**, 13656. *109, 130*
Cava, R. J. See Stavola, M.
Celotta, R. J. See Unguris, J.
Chadi, D. J., 1977, *Spin-orbit splitting in crystalline and compositionally disordered semiconductors*, Phys. Rev. **B16**, 790. *201*
Chadi, D. J., 1992, *Antisite-Interstitial Complex Model for EL2,* in *The Proceedings of the 21st International Conference on the Physics of Semiconductors*, Beijing, China, 1992, ed. by Ping Jiang and Hou-Zhi Zheng, World Scientific, Singapore, 1513. *312*
Chadi, D. J., and K. J. Chang, 1988, Phys. Rev. Letters **60**, 2187. *312*
Chadi, D. J., and M. L. Cohen, 1973, *Electronic structure of $Hg_{1-x}Cd_xTe$ alloys and charge-density calculations using representative k points*, Phys. Rev. **B7**, 692. *128, 348*
Chadi, D. J., and M. L. Cohen, 1975, Phys. Stat. Solidi (b) **68**, 405. *190, 196*
Chadi, D. J., and W. A. Harrison, 1985, eds. *Proceedings of the International Conference on the Physics of Semiconductors*,(Springer, New York). *207, 208, 209*
Chadi, D. J., and R. M. Martin, 1976, *Calculation of Lattice Dynamical Properties from Electronic Energies: Application to C, Si, and Ge*. Solid State Commun. **19**, 643. *128*
Chadi, D. J., R. M. White and W. A. Harrison, 1975, *Theory of the Magnetic Susceptibility of Tetrahedral Semiconductors,* Phys. Rev. Letters **35**, 1372. See also Sukhatme and Wolff (1975) *158*
Chakraborty, B., W. E. Pickett, and P. B. Allen, 1976, Phys Rev. **B14**, 3227. *481*
Chan, C. T. See Tang, M. S.
Chang, K. J., See Chadi, D. J.
Chappert, C. See Bruno, P.

Chelikowsky, J. R., 1981, Phys. Rev. Letters **47**, 387. *452*

Chelikowsky, James R., and Marvin L. Cohen, 1976, *Nonlocal Pseudopotential Calculations for the Electronic Structure of Eleven Diamond and Zinc-Blende Semiconductors,* Phys. Rev. **B14,** 556. *152, 190, 194, 197, 203-205, 510*

Chelikowsky, James R., and Marvin L. Cohen, 1992, *Ab Initio Pseudopotentials and Structural Properties of Semiconductors,* in *Handbook on Semiconductors,* ed. by P. T. Landsberg, Elsevier, Amsterdam, Vol. **1**, p. 59. *203*

Chelikowsky, James R., and M. A. Schlüter, 1977, Phys. Rev. **B15,** 4020. *406, 422ff*

Chelikowsky, James R., N. Troullier, and Y. Saad, 1994, *The finite difference-pseudopotential method: electronic structure calculations without a basis,* Phys. Rev. Letters **72**, 1240. *203*

Chen, A.-B., Arden Sher and W. T. Yost, 1992, *Elastic Constants and Related Properties of Semiconductor Compounds and Their Alloys,* in *Semiconductors and Semicmetals, Vol. 37,* Academic Press, New York. *304*

Chen, C. T. See Wang, C. Z., Xu, C. H.

Chen, X. See Wang, Y.

Ching, W. Y., Fanqi Gan, and Ming-Zhu Huang, 1995, *Band theory of linear and nonlinear susceptibilities of some binary ionic insulators,* Phys. Rev. **B52**, 1596. *382*

Ching, W. Y. See Xu, Yong-Nian

Christensen, N. E., S. Satpathy, and Z. Pawlowska, 1987, *Bonding and ionicity in semiconductors,* Phys. Rev. **B36**, 1032. *143*

Christensen, N. E., See Brudervoll, T., D. , Cardona, M., Lew Yan Voon, L. C.

Chu, C. W. See Wu, M. K.

Chye, P. W. See Spicer, W. E.

Ciraci, S. See Harrison, W. A.

Citrin, S. See Brudervoll, T.

Chou , Chi-Lung, 1994, private communication. *318*

Coehoom, R. See Johnson, M. T.

Coffa, S., C. Rafferty, T. de la Rubia, and P. Stolk, 1997, eds. Materials Research Society Symposium Series **469**.

Cohen, M. H., 1962, in *Metallic Solid Solutions* (J. Friedel and A. Guinier, eds.), Benajmin, New York, p. XI-1. *489*

Cohen, M. H., and L. M. Falicov, 1961, Phys. Rev. Letters **7,** 231. *236*

Cohen, M. L., private communication, *733*

Cohen, M. L., See Chadi, D. J., Chelikowsky, James R., Corkill, Jennifer L., Froyen, Sverre, Joannopoulos, J. D., Liu, A. M., Rubio, A.

Colombo, L. See Molteni, C.

Cook, E. L. See Strehlow, W. H.

Cooper, B. R. See Price, D. L.

Corkill, Jennifer L., and Marvin L. Cohen, 1993, *Structural, bonding, and electronic properties of IIA-IV antifluorite compounds.* Phys. Rev. **B48**, 17138. *389*

Corckill, Jennifer L. See Rubio, A.

Cousins, C. S. G., L. Gerward, L. J. Staun, B. Selsmark, and B. J. Sheldon, 1987, J. Phys. C **20**, 29.

Cowan, R. D., 1967, Phys. Rev. **163**, 54. *598*

Curl, R. F. See Kroto, H. W.

Cyrot-Lackmann, F., 1970, J. Physique **C1**, 67. *578*

Dabrowski, J., and M. Scheffler, 1988, Phys. Rev. Letters **60**, 2183. *312*

Dabrowski, J., and M. Scheffler, 1989, Phys. Rev. **B40**, 10391, Materials Science Forum **38-41**, 51. *312*

d'Amour, H., W. Denner, H. Schulz, and M. Cardona, 1982, Acta Crystallogr. **B.** *116*

Darby, J., B. See Freeman, A. J., Lam, D. J.

Das Sarma, S. See Khor, K. E.

Davenport, J. W. See Watson, R. E., Weinert, M.

de Boer, F. R. See Miedema, A. R.

Decarpigny, J. N. See Lannoo, M.

de Chatel, P. F. See Miedema, A. R.

De la Rubia, T. See Coffa, S.

Del Sole, R., and Raffaello Girlanda, 1993, Phys. Rev. **B48**, 11789. *210*

Denner, W. See d'Amour, H.

Dessau, D. S. See Shen, Z.-X.

de Vries, J. J. See Johnson, M. T.

Dick, B. G., 1965, *Shell and Exchange Charge Models of Interatomic Interactions in Solids*, in *Lattice Dynamics*, edited by R. F. Wallis, Pergamon Press, Oxford. *132*

Dolling, G., 1962, *Inelastic Scattering of Neutrons in Solids and Liquids,* Vol **II**, p. 37 International Atomic Energy Agency, Vienna. *124*

Dorantes-Dávila, J., and G. M. Pastor, 1995, *Alternative local approach to nonorthogonal tight-binding theory: Environment dependence of the interaction parameters in an orthogonal basis,* Phys. Rev. **B51**, 16627. *69*

Gelatt, C. D. , Jr., A. R. Williams, and V. L. Moruzzi, 1983, Phys. Rev. **B27**, 2005. *642*
Gelatt, C. D. See Williams, A. R.
Gerber, Ch. See Binnig, G.
Gerberich, W. W. See Nicolini, R.
Gerward, L. See Cousins, C. S. G.
Gianino, P. D. See Bendow, B.
Gillespie, R. J., *Molecular Geometry* ,1972, Van Nostrand Reinhold, London. *428, 429*
Ginsberg. D. M. See Bardeen, J.
Girifalco, L. A., 1976, Acta Metall. **24**, 523. *524, 525*
Girlanda, Raffaello See Del Sole, R.
Glötzel, D., and L. Fritsche, 1977, Phys. Status Solidi **B79**, 85. *603*
Gobeli, G. W., and F. G. Allen, 1965, Phys. Rev. **137** A245. *729*
Gobeli, G. W. See Lander, J. J.
Goedecker, S., and M. Teter, 1995, *Tight-binding electronic structure calculations and tight-binding molecular dynamics with localized orbitals,* Phys. Rev. **B51**, 9455. *12*
Gold, A. V., 1958, Phil. Trans. Roy. Soc. (London), **A251**, 85. *468, 470*
Gordon, R. G., and Y. S. Kim, 1972, J. Chem. Phys. **56**, 3122. *355, 768, 769*
Gordon, R. G., and Y. S. Kim, 1976, J. Chem. Phys. **65**, 379. *355, 768, 769*
Grant, R. W., J. R. Waldrup, and E. A. Kraut, 1979, *XPS Measurements of Abrupt Ge-GaAs Heterojunction Interfaces*, J. Vac. Sci. Technol. **15**, 1451 . *743*
Grant, R. W. See Waldrop, J. R.
Grimvall, G., M. Thiessen, and A. Fernández Guillermet, 1987, Phys. Rev. **B36,** 7816 . *642*
Grimvall, G. See Guillermet, A. Fernández , Häglund, J.
Grosso, G., and C. Piermarocchi, 1995, *Tight-binding model and interactions scaling laws for silicon and germanium,* Phys. Rev. **B51**, 16772. *20*
Gschneider, K. A., Jr., 1964, in *Solid State Physics,* edited by F. Seitz and D. Turnbull, Academic, New York, **16**, 275. *617*
Gubanov, V. A. See Zhukov, V. P.
Guillermet, A. Fernández , J. Häglund, and G. Grimvall, 1992, Phys. Rev. **B45**, 11557. *642*
Guillermet, A Fernández, J. Häglund, and G. Grimvall,1993, *Cohesive properties and electronic structure of 5d-transition-metal carbides and nitrides in the NaCl Structure,* Phys. Rev. **B48**, 11673. *642*
Guillermet, A. Fernández See Grimvall, G., Häglund, J.,
Guinier, A., 1962, ed. *Metallic Solid Solutions,* Benajmin, New York, *489*
Gunnarsson, O. and K. Schönhammer, 1983, Phys. Rev. Letters **50**, 604. *635*
Gunnarsson, O. See Svane, A.
Güntherodt, G. See Cardona, M.
Haddon, R. C. , L. E. Brus, and K. Raghavachari, 1986, *Electronic Structure and Bonding in Icosahedral* C_{60}, Chem. Phys. Letters **125**, 459. *774*
Hafner, J., 1987, *From Hamiltonians to Phase Diagrams,* Solid State Sciences **70**, Springer-Verlag, Heidelberg. *438, 453, 488, 513, 521ff, 542, 578*
Häglund, J., G. Grimvall, T. Jarborg, and A. F. Guillermet, 1991, Phys. Rev. **B43**, 14400. *642, 648, 667, 667*
Häglund, J., A. Fernández Guillermet, G. Grimvall and M. Körling, 1993, Phys. Rev. **B48,** 116. *642*
Häglund, J. See Guillermet, A. Fernández
Haines, J., and J. M. Legér, 1993, Phys. Rev. **B48**, 13344 (1993). *688*
Hamann, D. R. See Biswas, R.
Hammers, R. J. See Wang, Y.
Haneman, D., 1961, Phys. Rev. **121**, 1093. *732*
Haneman, D., 1968, Phys. Rev. **170**, 705. *733*
Hanna, C. B. See Jones, Barbara
Harrison, W. A., 1956a, *Mobility in Zinc Blende and Indium Antimonide,* Phys. Rev. **101**, 903. *279*
Harrison, W. A., 1956b, *Scattering of electrons by lattice vibrations in nonpolar crystals,* Phys. Rev. **104**, 1281. *270, 280, 283*
Harrison, 1960, *Electronic Structure of Polyvalent Metals*, Phys. Rev. **118**, 1190. *470*
Harrison, W. A., 1961, Tunneling *from an Independent-Particle Point of View*, Phys. Rev. **123**, 85. *739*
Harrison, W. A., 1963, *Electronic Structure and the Properties of Metals. I. Formulation,* Phys. Rev. **129**, 2503 *II. Application to Zinc,* Phys. Rev. **129**, 2512. *435, 445, 484, 507*
Harrison, W. A., 1964, *Theory of Sodium, Magnesium and Aluminum*, Phys. Rev. **136**, A1107. *484, 488*
Harrison, W. A., 1966, *Pseudopotentials in the Theory of Metals,* W. A. Benjamin, Inc.,

Popovíc, Z, V. See Shields, A. J.
Porbansky, E. M. See Trumbore, G. A.
Porezag, D., Th. Frauenheim, Th. Köhler, G. Seifert, and R. Kaschner, 1995, *Construction of tight-binding-like potentials on the basis of density-functional theory: Application to carbon*, Phys. Rev. **B51** , 12947. *783*
Pratt, W. P., Jr. See Mosca, D. H.
Price, D. L., and B. R. Cooper, 1989, Phys. Rev. **B39,** 4645. *642*
Quate, C. F. See Nogami, J.
Queisser, H. J., 1990, *The Conquest of the Microchip*, Harvard University Press, Cambridge. *430*
Rafferty, C. See Coffa, S.
Raghavachari, K. See Haddon, R. C.
Ram-Mohan, L. R. See Lew Yan Voon, L. C.
Ransil, Bernard J., 1960, *Studies in Molecular Structure. II. LCAO-MO-SCF Wave Functions for Selected First-Row Diatomic Molecules*, Rev. Mod. Phys., **32,** 245. *26*
Rath, J. and J. Callaway, 1973, Phys. Rev. **B8,** 5398. *550ff*
Rebane, Y. T., 1993, Phys. Rev. **B48** , 11772. *245*
Ren, S.-Y., and W. A. Harrison, 1981, *Semiconductor Properties Based upon Universal Tight-Binding Parameters*, Phys. Rev. **B23,** 762. *151, 171, 173, 174, 220, 221, 225*
Resta, R., 1991, Phys. Rev. **B44,** 11035. *270, 283*
Resta, R. See Nicolini, R.
Rhyner, J. See Monnier, R.
Rice, T. M. See Monnier, R.
Richtmyer, F. K., and E. H. Kennard, 1947, *Introduction to Modern Physics* , McGraw-Hill, New York. *252*
Rietman, E. A. See Stavola, M.
Roberts, M. de V. See Longuet-Higgens, H. C.
Roche, K. P. See Parkin, S. S. P.
Rocher, Y. A., 1962, Adv. in Phys. (GB) **11,** 233. *614*
Rohrer, H. See Binnig, G.
Rose, J. H., and H. B. Shore, 1991, Phys. Rev. **B43,** 11605. *561*
Rose, J. H., and H. B. Shore, 1993, *Uniform electron gas for transition metals: Input parameters,* Phys. Rev. **B48,** 18254. *561*
Rubio, A., Jennifer L. Corkill, Marvin L. Cohen, Eric L. Shirley, and Steven G. Louie, 1993, Phys. Rev. **B48,** 11810. *210*
Ruderman, M. A., and C. Kittel, 1954, Phys. Rev. **96,** 99. *613*
Ryan, J. L. See Nugent, L. J.
Saad, Y. See Chelikowsky, James R.
Sai-Halasz, G. A., L. Esaki, and W. A. Harrison, 1978: *InAs-GaSb superlattice energy structure and its semiconductor-semimetal transition*, Phys. Rev. **B 18,** 2812. *744*
Saito, T., Y. Hashimoto and T. Ikoma, 1994, *Band discontinuity at the (311) A GaAs/AlAs interface and possibility of its control by Si insertion layers* , Phys. Rev. **B50,** 17242. *744*
Saito, T. Y., and T. Ikoma, 1995, *Effect of P-insertion layer on band discontinuities at (100) GaAs/AlAs Interfaces* , Proceedings of the 14th Electronic Materials Symposium (Izu-Nagoka, Japan, July 5-7, 1995), p. 85. *744*
Salamon, M. See Bardeen, J.
Samara, G. A., 1976, Phys. Rev. **B13,** 4529. *395*
Sanchez-Dehesa, J., C. Tejedor, and J. A. Verges, 1982, *Self-Consistent Calculation of the Internal Strain Parameter of Silicon*, Phys. Rev. **B26,** 5960. *117, 118*
Sarnthein, Johannes, Alfredo Pasquarello, and Roberto Car, 1994, *Structure and Electronic Properties of Liquid and Amorphous SiO_2: An Ab Initio Molecular Dynamics Study,* Phys. Rev. Letters **74,** 4628. *400*
Satpathy, S. See Christensen, N. E.
Savrasov, D. Y. See Savrasov, S. Y.
Savrasov, S. Y., D. Y. Savrasov, and O. K. Andersen, 1994, *Linear-response calculations of electron-phonon interactions,* Phys. Rev. Letters, **72,** 372 . *541*
Schabel, M. C. See Northrup, J. E.
Scheffler, M., 1987, (private communication), based on the method described by Scheffler, Vigneron, and Bachelet (1982). *297*
Scheffler, M., J. P. Vigneron, and G. B. Bachelet, 1982, Phys. Rev. Letters **49,** 1785. *297*
Scheffler, M. and R. Zimmermann, 1996, ed. *Proceedings of the 23rd International Conference on the Physics of Semiconductors, Berlin, July 22-26, 1996,* World Scientific, Singapore. *319ff*
Scheffler, M. See Dabrowski, J.

Schiff, L. I, 1968, *Quantum Mechanics*, McGraw-Hill, New York. *66, 199, 200, 220, 231, 534, 536, 537, 626*
Schlegel, A. See Batlogg, B.
Schlier, R. E., and H. E. Farnsworth, 1959, J. Chem. Phys. **30**, 917. *732, 734, 737*
Schlüter, M., and C. M. Varma, 1981, Phys. Rev. **B23**, 1633. *526*
Schlüter, M. A., See Chelikowsky, James R., Joannopoulos, J. D., Sham, L. J.
Schönberger, U., and F. Aryasetiawan, 1995, *Bulk and surface electronic structures of MgO*, Phys. Rev. **B52,** 8788. *384*
Schönhammer, K. See Gunnarsson, O.
Schrieffer, J. R., X.-G. Wen, and S.-C. Zhang, 1988, Phys. Rev. Letters **60**, 944. *712*
Schroeder, P. A. See Mosca, D. H.
Schule, D. R. See Bartels, R. A.
Schulz, H. See d'Amour, H.
Schwarz, K. , 1987, CRC Crit. Rev. Solid State Mater. Sci. **13**, 211. *642*
Schwarzacher, W. See Bennett, W. R.
Segall, Benjamin See Kim, Kwiseon
Segmuller, A. and H. R. Neyer, 1965, *Internal Strain in Elastically Strained Germanium and Silicon. II. General Relations, Transverse and Longitudinal Case*, Physik Kondensierten Materei **4**, 63. *116*
Seifert, G. See Porezag, D.
Seitz, F., 1940, *Modern Theory of Solids* , McGraw-Hill, New York. *108*
Seitz, F., 1948, Phys. Rev. **73**, 549. *283*
Seitz, F., and D. Turnbull, 1964, eds. *Solid State Physics* , Academic, New York. *498, 617*
Seitz, F., D. Turnbull, and H. Ehernreich, 1969, eds. *Solid State Physics* , Academic, New York. *210, 451*
Seitz, F. See Wigner, E. P.
Selloni, A. See Takeuchi, N.
Seong, Hyangsuk , and Laurent J. Lewis, 1995, *Tight-binding molecular-dynamics study of point defects in GaAs,* Phys. Rev. **B52**, 5675. *316*
Selsmark, B. See Cousins, C. S. G.
Selwood, P., 1956, *Magnetochemistry*, Interscience, New York. *160*
Sham, L. J., and M. Schlüter, 1983, Phys. Rev. Letters **51**, 1888. *207*
Sham, L. J. See Kohn, W.
Sharma, R. R., 1979, Phys. Rev. **B19**, 2813. *546*
Sheldon, B. J. See Cousins, C. S. G.
Shen, Z.-X., and D. S. Dessau, 1995, Phys. Rep. **253**, 1 . *713*
Sher, Arden See Chen, A.-B., Shih, C. K., van Schilfgaarde, M.
Shevchik, N. J., J. Tejeda, and M. Cardona, 1974, Phys. Rev. **B9**, 2627. *675*
Shevchik, N. J, J. Tejeda, M. Cardona, and D. W. Langer, 1973, Phys. Status Solidi B **59**, 87. *675*
Shields, A. J., Z. V. Popovíc, M. Cardona, J. Spitzer, R. Nötzel, and K. Ploog, 1994, *Resonant interference effects in the phonon Raman spectra of (311) GaAs/AlAs superlattices*, Phys. Rev. **B49** ,7564. *269*
Shih, C. K., W. E. Spicer, W. A. Harrison, and A. Sher, 1985, *Bond-length relaxation in pseudobinary alloys,* Phys. Rev. **B31**, 1139. *297*
Shirley, D. A. See Ley, L.
Shirley, Eric L. See Rubio, A.
Shockley, W. See Bardeen, J.
Shore, H. B. See Rose, J. H.
Silcox, J. See Muller, D. A.
Simmons, G., and H. Wang, 1971, *Single-Crystal Elastic Constants and Calculated Aggregate Properties*, 2nd ed., MIT Press, Cambridge. *365, 585*
Singh, D. J. See Muller, D. A.
Sirota, N. N., 1972, ed. *Chemical Bonds in Solids*, Plenum, New York. *160*
Skeath, P. R. See Spicer, W. E.
Skillman, S. See Herman, F.
Skriver, H. L., 1983, private communication. *604*
Skriver, H. L., 1984, *The LMTO Method, Muffin-Tin Orbitals and Electronic Structure*, Springer-Verlag, Berlin. *603, 605, 611*
Skriver, H. L., 1985, Phys. Rev. **B31**, 1909. *548*
Skriver, H. L., O. K. Andersen, and B. Johansson, 1978, Phys. Rev. Letters **41**, 42. *638*
Skriver, H. L. See Kollar, J.
Slater, J. C., and G. F. Koster, 1954, Phys. Rev. **94**, 1498. *544, 545, 546*
Smalley, R. E. See Kroto, H. W.
Smargiassi, Enrico See Blöchl, Peter E.
Smith, E. D. See Perdew, J. P.
Söderlind, P., Olle Eriksson, Börje Johannson, J. M. Wills, and A. M. Boring, 1995, *A unified picture of the crystal structure of metals*, Nature **374**, 524. *624, 625*
Sokel, R. C., 1978, Thesis, Stanford University, discussed also in Harrison (1980). *113, 114, 116, 118, 120, 127, 170*
Sommerfeld, A., 1928, Z. Physik **47**, 1. *434*

Zheng, Hou-Zhi See Jiang, Ping
Zhong, W., and D. Vanderbilt, 1995, *Competing Structural Instabilities in Cubic Perovskites,* Phys. Rev. Letters **74**, 2587. *709*
Zhukov, V. P. , N. I. Medvedeva, and V. A. Gubanov ,1989, Phys. Status Solidi **B151**, 407. *642*
Ziman, J. M., 1961, Phil. Mag. **6**, 1013. *507*
Ziman, J. M., 1969, ed. *The Physics of Metals*, Cambridge University Press, New York. *559ff, 590*
Zimmermann, R. See Scheffler, M.
Zunger, A., 1985, Annual Rev. Mater. Sci. **15,** 411. *745*
Zunger, A., J. P. Perdew, and G. L. Oliver, 1980, Solid State Commun. **34**, 933. *598*
Zunger, A. See Caldas, M. J., Perdew, J. P., Wei, S.-H.
Zwicknagl, G., 1993, Physica Scripta **T49**, 34. *640*

Subject Index

Simple Compounds

$$V_{ss\sigma} = -1.32\,\frac{\hbar^2}{md^2}$$

$$V_{sp\sigma} = 1.42\,\frac{\hbar^2}{md^2}$$

$$V_{pp\sigma} = 2.22\,\frac{\hbar^2}{md^2}$$

$$V_{pp\pi} = -\,0.63\,\frac{\hbar^2}{md^2}$$

Semiconductors

$$V_1 = \frac{\varepsilon_p - \varepsilon_s}{4}$$

$$V_2 = 3.22\,\frac{\hbar^2}{md^2}$$

$$V_3 = \frac{\varepsilon_h^+ - \varepsilon_h}{2}$$

Transition Metals

$$V_{dd\sigma} = \frac{-45}{\pi}\,\frac{\hbar^2 r_d^3}{md^5}$$

$$V_{dd\pi} = \frac{30}{\pi}\,\frac{\hbar^2 r_d^3}{md^5}$$

$$V_{dd\delta} = \frac{-15}{2\pi}\,\frac{\hbar^2 r_d^3}{md^5}$$

Transition-Metal Compounds

$$V_{pd\sigma} = -\,\frac{3\sqrt{15}}{2\pi}\,\frac{\hbar^2\sqrt{r_p r_d^3}}{md^4}$$

$$V_{pd\pi} = \frac{3\sqrt{5}}{2\pi}\,\frac{\hbar^2\sqrt{r_p r_d^3}}{md^4}$$

Simple Metals:

$$k_F^3 = \frac{9\pi}{4r_0^3} \qquad \kappa^2 = \frac{4e^2 k_F m}{\pi\hbar^2} \qquad w_q = -\,\frac{4\pi Z e^2 \cos q r_c}{\Omega_0(\kappa^2 + q^2)} \qquad V(d) = \frac{Z^2 e^2 \cosh^2 \kappa r_c\, e^{-\kappa d}}{d}$$

Note: $\frac{\hbar^2}{m} = 7.62$ eV-Å^2 $\quad e^2 = 14.4$ eV-Å

Transition Metals

		D3	D4	D5	D6	D7	D8	D9	D10	D11	
	Z	Sc 21	Ti 22	V 23	Cr 24	Mn 25	Fe 26	Co 27	Ni 28	Cu 29	
	ε_s eV	-5.71	-6.04	-6.33	-6.59	-6.84	-7.08	-7.31	-7.54	-7.72*	
	ε_d eV	-9.35	-11.05	-12.54	-13.88	-15.27	-16.54	-17.77	-18.97	-20.26	
	U_d eV	5.3	5.4	5.3	5.1	5.6	5.9	6.3	6.5	6.9	
	r_o Å	1.81	1.61	1.49	1.42	1.43	1.41	1.39	1.38	1.41	
	r_d Å	1.163	1.029	0.934	0.939	0.799	0.744	0.696	0.652	0.688	
	At.Wt.	44.96	47.90	50.94	52.00	54.94	55.85	58.93	58.71	63.54	
		Y 39	Zr 40	Nb 41	Mo 42	Tc 43	Ru 44	Rh 45	Pd 46	Ag 47	simple metals
		-5.33	-5.67	-5.95	-6.19	-6.39	-6.58	-6.75	-6.91	-7.05*	
		-6.80	-8.46	-9.98	-11.49	-13.08	-14.61	-16.16	-17.70	-19.23	
		1.9	2.4	2.7	3.0	3.4	3.7	4.0	4.3	4.6	
		1.99	1.77	1.62	1.55	1.50	1.48	1.49	1.52	1.59	
		1.602	1.415	1.328	1.231	1.109	1.083	1.020	1.008	0.889	
		88.91	91.22	92.91	95.94	(99)	101.1	102.9	106.4	107.9	
4f metals		Lu 71	Hf 72	Ta 73	W 74	Re 75	Os 76	Ir 77	Pt 78	Au 79	
		-5.42	-5.71	-5.97	-6.19	-6.38	-6.56	-6.71	-6.85	-6.98*	
		-6.63	-8.14	-9.57	-10.97	-12.35	-13.74	-15.12	-16.47	-17.78	
		2.0	2.8	3.0	3.1	3.2	3.4	3.5	3.5	3.6	
		1..92	1.75	1.62	1.56	1.52	1.49	1.50	1.53	1.59	
		1.603	1.455	1.346	1.268	1.201	1.142	1.085	1.069	1.007	
		175.0	178.5	180.9	183.8	186.0	190.2	192.2	195.0	197.0	

*For $d^{9.5}s^{1.5}$

The Solid-State Table

	IB	IIB	IIIB	IV	V	VI	VII	VIII	IA	IIA
								He 2	Li 3	
								-24.98	-5.34	
								-	-	
								-	8.17	
								-	1.72	
								-	0.92	
								4.	6.94	
		Be 4	B 5	C 6	N 7	O 8	F 9	Ne 10	Na 11	
		-8.42	-13.46	-19.38	-26.22	-34.02	-42.79	-52.53	-4.96	
		-5.81	-8.43	-11.07	-13.84	-16.77	-19.87	-23.14	-	
		10.25	10.26	11.76	13.15	14.47	15.75	15.00	6.17	
		1.25	-	1.54*d*	-	-	-	-	2.08	
		0.58	-	6.59&	5.29&	4.41&	-	-	0.96	
		9.01	10.81	12.01	14.01	16.00	19.00	20.18	22.99	
	Z	Mg 12	Al 13	Si 14	P 15	S 16	Cl 17	A 18	K 19	Ca 20
	ε_s eV	-6.89	-10.71	-14.79	-19.22	-24.02	-29.20	-34.76	-4.01	-5.32
	ε_p eV	-3.79	-5.71	-7.59	-9.54	-11.60	-13.78	-16.08	-	-
	U eV	7.28	6.63	7.64	8.57	9.45	10.30	11.12	5.56	6.40
	r_o Å	1.77	1.58	2.35*d*	-	-	-	-	2.57	2.18
	r_c Å	0.74	0.61	13.7&	11.4&	10.1&	-	-	1.20	0.90
	At.Wt.	24.31	26.98	28.09	30.97	32.06	35.45	39.95	39.10	40.08
	Cu 29	Zn 30	Ga 31	Ge 32	As 33	Se 34	Br 35	Kr 36	Rb 37	Sr 38
	-6.49	-7.96	-11.55	-15.16	-18.92	-22.86	-27.01	-31.37	-3.75	-4.86
	-3.31	-3.98	-5.67	-7.33	-8.98	-10.68	-12.44	-14.26	-	-
	7.07	7.83	6.61	7.51	8.31	9.07	9.78	10.48	5.02	5.71
	1.41	1.53	1.67	2.44*d*	-	-	-	-	2.75	2.25
	-	0.59	0.59	14.4&	13.2&	12.1&	-	-	1.38	1.14
	63.54	65.37	69.72	72.59	74.92	78.96	79.91	83.80	85.47	87.62
	Ag 47	Cd 48	In 49	Sn 50	Sb 51	Te 52	I 53	Xe 54	Cs 55	Ba 56
	-5.99	-7.21	-10.14	-13.04	-16.03	-19.12	-22.34	-25.70	-3.37	-4.29
	-3.29	-3.89	-5.37	-6.76	-8.14	-9.54	-10.97	-12.44	-	-
	6.34	6.95	6.00	6.73	7.39	8.00	8.58	9.13	5.05	5.70
	1.60	1.73	1.84	2.80*d*	-	-	-	-	2.97	2.47
	-	0.65	0.63	0.59	16.8&	15.9&	-	-	1.55	1.60
	107.9	112.4	114.8	118.7	121.8	127.6	126.9	131.3	132.9	137.3
	Au 79	Hg 80	Tl 81	Pb 82	Bi 83	Po 84	At 85	Rn 86	Fr 87	Ra 88
	-6.01	-7.10	-9.83	-12.49	-15.19	-17.97	-20.83	-23.78	-3.21	-4.05
	-3.31	-3.83	-5.24	-6.53	-7.79	-9.05	-10.34	-11.65	-	-
	6.75	7.33	6.30	7.03	7.68	8.28	8.85	9.39	4.93	5.54
	1.59	1.77	1.89	1.93	-	-	-	-	-	-
	-	0.66	0.60	0.57	18.9&	17.9	-	-	-	-
	197.0	200.6	204.4	207.2	209.0	210	210	222	223	226

Transition metals

f-shell metals, Chapter 16

&: p-state radius. *d:* nearest-neighbor distance in tetrahedral structure.